C0-ALU-731

Springer
Tokyo
Berlin
Heidelberg
New York
Hong Kong
London
Milan
Paris

T. Okuda, N. Manokaran,
Y. Matsumoto, K. Niiyama,
S.C. Thomas, P.S. Ashton (Eds.)

Pasoh

Ecology of a Lowland Rain Forest in Southeast Asia

With 179 Figures

Springer

Toshinori Okuda, Ph.D.
National Institute for Environmental Studies (NIES)
16-2 Onogawa, Tsukuba, Ibaraki 305-8506, Japan

N. Manokaran, Ph.D.
Forest Research Institute Malaysia (FRIM)
Present address: No.4, Jalan SS 22/29, Damansara Jaya
47400 Petaling Jaya Selangor, Malaysia

Yoosuke Matsumoto, Ph.D.
Forestry and Forest Products Research Institute (FFPRI)
1 Matsunosato,Tsukuba, Ibaraki 305-8687, Japan

Kaoru Niiyama, Ph.D.
Forestry and Forest Products Research Institute (FFPRI)
1 Matsunosato,Tsukuba, Ibaraki 305-8687, Japan

Sean C. Thomas, Ph.D.
Faculty of Forestry, University of Toronto
33 Willcocks St., Toronto, Ontario M5S 3B3, Canada

Peter S. Ashton, Ph.D.
Organismic and Evolutionary Biology, Harvard University
Cambridge, Massachusetts 02138, U.S.A.

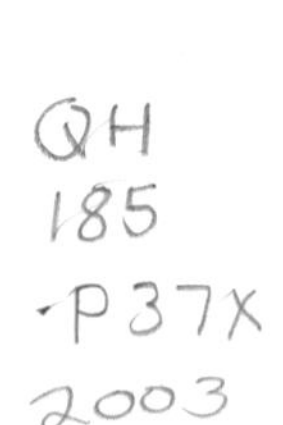

Cover: Young fruits of *Dipterocarpus sublamellatus*

Photo credit: Shinya Numata

ISBN 4-431-00660-5 Springer-Verlag Tokyo Berlin Heidelberg New York

Library of Congress Cataloging-in-Publication Data applied for.

Printed on acid-free paper

Typesetting: Camera-ready by the editors and authors
Printing and binding: Nikkei Printing, Japan
SPIN: 10917305

Editorial Board

Foreword

The Pasoh Forest Reserve (Pasoh FR) has been a leading center for international field research in the Asian tropical forest since the 1970s, when a joint research project was carried out by Japanese, British and Malaysian research teams with the cooperation of the University of Malaya (UM) and the Forest Research Institute (FRI, now the Forest Research Institute Malaysia, FRIM) under the International Biological Program (IBP). The main objective of the project was to provide basic information on the primary productivity of the tropical rain forest, which was thought to be the most productive of the world's ecosystems.

After the IBP project, a collaborative program between the University of Malaya and the University of Aberdeen, Scotland, UK, for post-graduate training was carried out at Pasoh. Reproductive biology of some dipterocarp trees featured in many of the findings arrived at through the program, contributing greatly to progress in the population genetics of rain forest trees. Since those research programs, a part of the Pasoh forest and its field research station have been managed by FRIM. In 1984, FRIM started a long-term ecological research program in Pasoh FR with the Smithsonian Tropical Research Institute (STRI) and Harvard University, establishing a 50-ha plot and enumerating and mapping all trees 1 cm or more in diameter at breast height. A recensus has been conducted every 5 years. During the original census, 800 tree species and more than 300,000 individuals were recorded in the plot. In 1991, the National Institute for Environmental Studies (NIES), Japan; FRIM and UPM (Universiti Pertanian Malaysia, now Universiti Putra Malaysia) initiated another joint research project. The program (NIES-FRIM-UPM joint research project) includes the broad fields of ecology, forestry, meteorology and hydrology with a focus on biodiversity and sustainable management of tropical rain forests. On the Japanese side, besides NIES, many other research organizations have been involved in this project: Forestry and Forest Products Research Institute (FFPRI), Japan; Japan International Research Center for Agricultural Sciences (JIRCAS); Japan Science and Technology Corporation (JST); and Japan International Cooperation Agency (JICA). Also included are a number of universities in Malaysia (e.g., Universiti Malaya, Universiti Kabangsan Malaysia, Universiti Teknologi Malaysia) and in Japan. Besides these programs, a number of diverse research projects have also been carried out on topics such as mechanisms for the maintenance of tropical biodiversity, genetic variation in rain forest trees, plant-animal interactions, and the role of tropical rain forests as a carbon sink. Malaysian, European, and American universities and organizations have conducted the programs.

The primary objective of the present publication is to make this large body of scientific information available to a wider audience worldwide. I hope that the information provided in this book will contribute greatly to the protection of biodiversity and will lead to progress in sustainable management of the tropical lowland rain forest.

I would like to congratulate Drs. T. Okuda, N. Manokaran, Y. Matsumoto, K. Niiyama, S. C. Thomas and P. S. Ashton for their painstaking efforts in bringing this book to publish, and I thank all scientists and support staff involved in the research projects in Pasoh FR for their commitment, dedication, and effort.

Dato' Dr. Abdul Razak Mohd. Ali

Director General
Forest Research Institute Malaysia

January 2003

Preface

Tropical rain forests, the most species-rich ecological systems on our planet, have decreased at an alarming rate over the last few decades. Pasoh Forest Reserve (Pasoh FR), located in Peninsular Malaysia, is easily the best-studied remnant of lowland tropical forest in Southeast Asia. More than 1,000 tree species have been recorded in Pasoh FR. The forest is also world-famous for its bird fauna and is home to nearly 500 species of ants. The main objective of this book is to provide an overview of the ecology and natural history of Pasoh FR as well as the ecology of lowland rain forests in Southeast Asia. Secondly, the book aims to provide a better foundation for sustainable use and management of tropical forest, considering not only timber resources, but also values relating to the protection of biodiversity, recreation, socioeconomics and the spirit.

Since the 1970s, a large number of biological research projects have been carried out at Pasoh FR. The International Biological Program (IBP) began under the collaboration among Japanese, British and Malaysian research institutes and universities, and focused mainly on the estimation of primary and secondary production of the forest. The results of studies conducted under the IBP program were published in the *Malayan Nature Journal* in 1978. Since that time a very large number of additional studies have built on the foundation of the IBP work, focusing on such diverse topics as mechanisms for the maintenance of tropical biodiversity, genetic variation in rain forest trees, plant-animal interactions and the role of tropical rain forests as a carbon sink. While researchers from more than a dozen countries have worked at Pasoh FR, a large proportion of the research and publications have been by Malaysian and Japanese scientists. Also many of the results of the research at Pasoh FR have not been published in a consolidated form and hence are not widely known outside of Southeast Asia. The mission of this book is to make this scientific information available to a wider audience worldwide.

This book comprises six parts: Physical settings and environment (Part I); Vegetation structure, diversity and dynamics (Part II); Plant population and functional biology (Part III); Animal ecology and biodiversity (Part IV); Plant-animal interactions (Part V); and Anthropogenic impacts and forest management (Part VI). Part I includes introductory information on Pasoh FR, and the climatological and physical background of the forest, such as topography and soils. Part II comprises chapters on adaptation and phenological and morphological patterns of some species groups that are predominant in Pasoh FR. Chapters in Part III focus on the reproduction and regeneration strategies, and the physiological characteristics of some tree species. Part IV includes studies on the ecology of animals such as birds, mammals, reptiles, insects and spiders. In Part V are chapters on insect herbivory on plants, ant-plant interactions and impacts of wild pigs on vegetation. Part VI covers the topics of regeneration of the tropical lowland rain forest, logging impacts on wildlife animals, and implications and suggestions for future studies of sustainable forest management that could be undertaken. The final section is a list of the references to research undertaken in the Pasoh FR.

We would like to express our sincerest appreciation to Dr. Y. Oshima, Professor Emeritus of Waseda University; Dr. E. Soepadomo and Dr. Abd. Rahim Nik at the Forest Research Institute Malaysia for their valuable comments on the contents of this book; and to Messrs J. Kuraoka, K. Iwamoto, Y. Sakurai, R. Abe and to Drs. K. Yoshida, S. Numata, Y. Tang and Y. Niiyama for technical assistance that helped to facilitate the editorial process. Our sincerest gratitude also goes to Dr. A. Furukawa (Nara Women's University) and Dr. Y. Morikawa (Waseda University) for their great efforts for the initial establishment of the NIES-FRIM-UPM joint research project that contributes greatly to the research output of Pasoh FR studies, many of which are now presented as chapters in this book. Special thanks should go to Dr. Naoki Adachi (formerly of the Japan Science and Technology Corporation), Dr. Y. Maruyama (Forestry and Forest Products Research Institute), Dr. N. Osawa (Kyoto University), Dr. S. Ichikawa (Japan Wildlife Research Center), Mr. T. Sugimoto (CTI Engineering Co. Ltd), Mr. S. M. Wong (FRIM) and Mr. E. S. Quah (FRIM) for their technical support that greatly facilitated the research conducted in Pasoh FR. We also would like to thank the National Institute for Environmental Studies (NIES) and the Forestry and Forest Products Research Institute (FFPRI) and the Ministry of Environment, Japan, for the financial support to publish this book.

Editors:
Toshinori Okuda
N. Manokaran
Yoosuke Matsumoto
Kaoru Niiyama
Sean C. Thomas
Peter S. Ashton

Contents

Part IV: Animal Ecology and Biodiversity

Color Plates

Processed by T. OKUDA, K. YOSHIDA, N. ADACHI, and VISONTECH INC.

Color plate 1
Land-use map in the Pasoh Forest Region based upon Landsat TM data acquired in 1996.

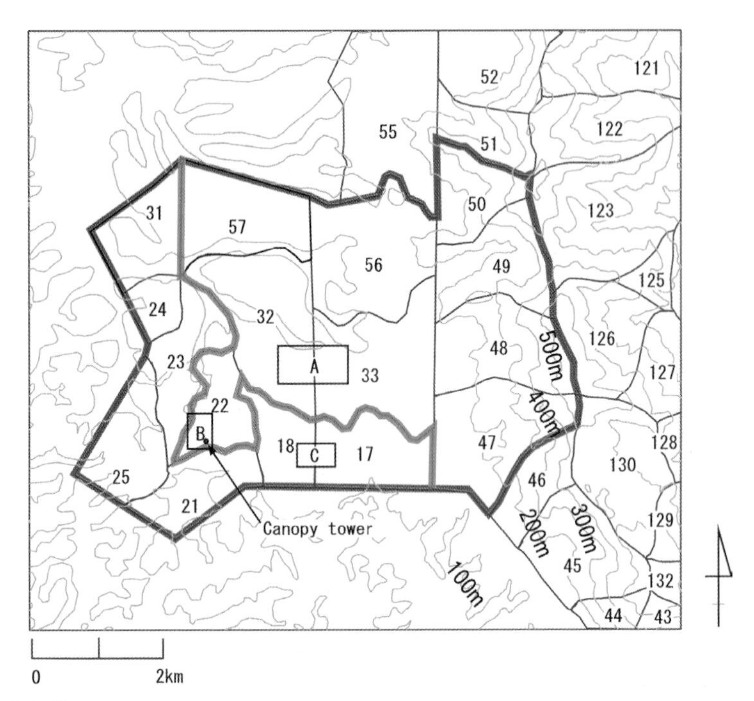

Color plate 2
Location and definition of the Pasoh Forest Reserve (Pasoh FR). The numbers indicate the Forest Compartments. The area described here as the "Pasoh Forest Reserve" varies with the literature; it has variously been called or cited as "Pasoh", "Pasoh Forest", or "Pasoh Forest Research Site". In this book, however, the Pasoh FR is expediently defined as the area bounded by the solid green line, which is about 2400 ha altogether. The boundary between the primary (unlogged) and regenerating forest (logged in the 1950s) is shown by the solid red line (see Chapter 2 for details). The northern section of the reserve (a part of Compartments 56 and 57) had been logged in 1980s in order to construct logging road to access other forest compartments. Letters indicate locations of major study plots, A: 50-ha plot, B: IBP plot (in some chapters it is described as Plot-1, e.g. in Chapter 39), and C: regenerating forest plot. The canopy walkway towers described in some chapters are located within the area of the IBP plot (see Chapters 1, 2 and 39 for details).

Color plate 3
A view of the Pasoh Forest Reserve from the canopy tower with meteorological observation equipment (left) and canopy walkway (right) located in the Pasoh FR. The three canopy towers (two of them are 32 m tall, the other is 52 m) are interconnected by the canopy walkway at a height of 30 m.

Part I

Physical Settings and Environment

1 Pasoh Research, Past and Present

Peter S. Ashton[1], Toshinori Okuda[2] & N. Manokaran[3]

Abstract: The Pasoh forest reserve (Pasoh FR), albeit in an atypically dry region of Peninsular Malaysia, is now one of the most extensive remaining examples of the mixed dipterocarp forests which once covered the lowlands west of the main range. Its gentle topography provides the most extensive relatively homogeneous terrain, and thereby opportunity to examine large scale community patterns. The Pasoh FR has become the paradigm for lowland dipterocarp forest through publications emanating from a succession of independent research projects there, and is now the leading center for international field research in the Asian tropical forest. We here describe the history of this work and its present status.

Key words: Peninsular Malaysia, dipterocarp forest, Asian tropical forest, research history, plant geography, International Biological Program (IBP), University of Malaya, University of Aberdeen, Forest Research Institute Malaysia (FRIM), Smithonian Tropical Research Institute (STRI), Arnold Arboretum of Harvard University, National Institute for Environmental Studies (NIES), Universiti Putra Malaysia (UPM).

1. INTRODUCTION

Pasoh Forest Reserve (Pasoh FR) is set in the center of southern Peninsular Malaysia, towards the southern end of the main mountain range (See Color plate 1). Here, the range is divided. To the west, the southwest monsoon must first cross Sumatra's high Barisan range, then the Malacca Straits at their narrowest before being forced up over the wall of the Ulu Langat mountains, which crest at Gunung Nuang, 1,496 m and Gunung Besar Hantu, 1,462 m to the west of Pasoh FR. Although the eastern ribbon of the Main range, which marks the eastern border of the Pasoh FR, is here only 600–640 m tall the northeast monsoon must pass over the two highest mountains in the peninsula, Gunung Tahan, 2,190 m away north of the Pahang River drainage and Gunung Benom, 2,108 m, and the mass of the Terengganu hills before reaching Pasoh FR. The intermontane valley in which Pasoh FR is set is therefore a rainshadow with mean annual rainfall hovering below two meters through much of the area, exceptionally low for the humid tropics and the lowest in all Malaysia. Nevertheless, mean monthly rainfall exceeds expected evapotranspiration in ten months of the year; the minimum recorded annual rainfall was > 1,500 mm (Chap. 4, 6) and the region has not been known this century to experience the prolonged, El Niño related killing droughts which have increasingly afflicted regions to the east and south. Blocks of weathered laterite in the soil surface of the ridges within the Pasoh FR bespeak a more seasonal climate in the past, possibly as recently as the last glacial period when pine savanna flourished near Kuala Lumpur (Morley & Flenley 1987). It is unlikely, then, that the Pasoh FR

[1] Harvard University, Cambridge, Massachusetts, 002138, USA.
E-mail : Peter_Ashton/FS/KSG@Ksg.harvard.edu
[2] National Institute for Environmental Studies (NIES), Japan.
[3] Forest Research Institute Malaysia (FRIM), Malaysia.

as currently composed is older than the Holocene although there must have been refugia in the Peninsula from which its flora immigrated.

The rivers of the flatlands of Negeri Sembilan east of the western ridge of the main range drain eastward into the Pahang river. Perhaps because the low rainfall makes annual harvesting and thus permanent cultivation of cereals feasible, this region is one of the few in the Peninsula to have traditionally supported irrigated rice, brought in centuries ago by Minangkabau settlers who had left behind a similar climate in east Sumatra. Minangkabau Malays did not settle in the northern and eastern part of the valley, where the alluvial soils are sandier and less fertile. Here, proto-Malay Semelai and, to the south, Mon-Khmer speaking Jakun aborigines were until recently the principal inhabitants. Most of the 27,000 ha of uncultivated forest land in the region immediately east of the Kuala Pilah-Durian Tipus road was designated Forest Reserve early this century. All but 16,000 ha were degazetted for smallholder oil palm cultivation during the seventies. The area described as the "Pasoh Forest Reserve" varies with the literature; it has variously been called or cited as "Pasoh", "Pasoh Forest", or "Pasoh Forest Research Site". In this book, however, the Pasoh FR is expediently defined as the area bounded by the solid green line, which is about 1,700 ha altogether representing Compartments 17–18, 21–25, 31–33, and 56–57 (see Color plate 2). When the unlogged compartments 46–50 in the hill forest are included with this, the total area is about 2,400 ha, as previously described many times in the literature .

The Pasoh FR comprises low hills and alluvium rising to a granite ridge along its eastern border. It consists of an east-west oriented oblong block. A buffer zone of 700 ha within the western and southern margins was logged in the fifties according to the Malayan Uniform System (Chap. 2) which, in principle, mandates removal of all canopy trees. There is evidence that felling was in practice more selective. Most of the lowland overlies Triassic sediments yielding sandy clay ultisol soils. Westward flowing streams, which later flow northeast into the Pertang (Pahang) valley, drain these lowlands on the north and southern sides. The hills are low, generally rising less than 30 m above the plain. They bear yellow-red clay loam ultisols. Eroded laterite fragments patchily occur, mainly on crests. Bordering the streams is recent clay alluvium bearing more or less swampy hydromorphic gley soils; swamp becomes extensive along the northern border. Most of the alluvium is raised c.a 2 m above the recent alluvium and represents an earlier deposition surface. It bears sandy and sandy clay entisol soils, which structurally resemble the adjacent ultisols.

2. PLANT GEOGRAPHY

Although the granite hills that surround Pasoh FR, and to the east are included within it, are clothed with hill dipterocarp forest with the pale grey-green crowns of *Shorea curtisii* dominating the ridges, all research has focused on the Lowland Dipterocarp forest of the undulating land and well drained alluvium. In the old Forest Reserve and the area of the research forest the predominant type is Red Meranti-Keruing, which was the leading and from a timber standpoint most valuable lowland dipterocarp forest community throughout the Peninsula (Wyatt-Smith 1954, 1963, 1987 ed.2 1995). In north-east Negeri Sembilan, including Pasoh FR, this type has been recognized in forest inventories for the abundance of the shade tolerant heavy hardwood emergent dipterocarp *Neobalanocarpus heimii* and for species of *Dipterocarpus* notably *D. costulatus, D. cornutus, D. sublamellatus, D. kunstleri* and *D. crinitus* (Wyatt-Smith 1987 ed. 2 1995). Some parts of the forest, notably on ridges and on the sandy raised alluvium in the research forest are particularly rich in another dipterocarp heavy hardwood, *Shorea maxwelliana.* These species are well represented in the research forest, the latter being locally abundant on the raised alluvium. A further, main canopy, dipterocarp *Hopea dyeri* is also particularly common in the Pasoh area. Otherwise, the forest is not known to be exceptional for the southern Peninsula, either in composition or diversity.

Only 40 km to the east is Tasek Bera, a shallow lake at the base of whose organic sediments mangrove pollen dominates (Morley 1981). Here the westward flowing Muar headwaters meet those of the Teriang, a tributary of the eastward flowing Pahang river. It is clear that there were marine transgressions, which separated the southern Peninsula as an island during maximum interglacial eustatic sea level, as late as eight thousand years ago. Astonishingly, normally coastal tree species in the peninsula still survived until recently in the forests formerly surrounding the lake, including *Dipterocarpus kerrii* and *Planchonella obovata,* as well as *Terminalia bellirica*, which is most abundant in somewhat seasonal areas. Another coastal species and apparent relict, *Shorea glauca,* has been found at Pasoh itself.

3. HISTORY OF RESEARCH

3.1 Early silvicultural work

The early scientific activity at Pasoh FR was confined to silvicultural operations; these have been referred to by Wyatt-Smith (1987 ed.2 1995). He records a first cleaning of the understorey, and poison girdling of trees of lesser economic importance that are competing for space with commercial species, known as CG 1, in part of the Reserve. This was followed by a second, similar manipulation, known as CG 2 in 1932, and a final CG 3 in 1937. A felling, in which all exploitable timber was removed, continued between 1932–1938, but never entered the area of the current research forest. In 1956, the post-felling stocking was assessed by half chain square sampling, LS1/2. In this, continuous transects of half chain square plots are censused for individuals of commercial species > 6 inches girth[1]. Commercial trees are identified by species. Wyatt-Smith mentions that this was carried out in Compartments 1, 2, 5, 6, 7 and 11. A milleacre survey of seedling regeneration was also carried out. Species composition generally proved excellent and size class stocking good, but percentage stocking was variable. Post-harvest treatments were carried out on the basis of that survey. In 1957–1958, a 10% total enumeration of trees > 4 foot girth (38.8 cm diameter) was carried out. Shortly after this, logging of compartments adjacent and in the buffer zone of the Pasoh FR commenced. Records of these operations are available at the Negeri Sembilan State Forest Department Headquarters, in Seremban.

Direct involvement of researchers of Forest Resarch Insitute (FRI), presently known as FRIM at Pasoh FR began in 1961, when silviculturist Wong Yew Kwan censused ten, north-south oriented, 10 × 1 chain (1 acre, 0.2 ha) plots within the area of the present primary research forest. Trees > 1 foot diameter (9.7 cm diameter)

[1] A chain is 22 yards or 100 links in the English system of measurement. A mile is 80 chains. An acre is 10 square chains. A 'milleacre plot' (see next page) is 10 links square, therefore one thousandth of an acre. An inch is 2.59 cm; it is serendipitous that 12 inches (1 foot) girth equals 9.7 (or approximately 10) cm diameter.

at 4. 5 feet (1.5 m) above the ground ('breast height') were included. From an analysis of these data, Wong & Whitmore (1970) concluded that the limited distances that tropical tree seeds are generally dispersed causes a gradual change in species composition with distance that obscures any habitat-related floristic variation that might be present.

3.2 The IBP era

In 1970, FRI negotiated, on behalf of the University of Malaya, the Japanese Society for the Promotion of Science and the Royal Society of London with the Conservator of Forests, Negeri Sembilan to establish a site for research at Pasoh FR under the International Biological Program (IBP). This collaboration, under the auspices of the national IBP committees of Malaysia, Japan and the United Kingdom aimed at documenting the primary productivity of tropical rain forest, long thought to be the most productive of the world's ecosystems. Over seventy researchers participated. The program established the Pasoh FR as the site for leading international research which it remains today, now as formerly thanks to the foresight and good will of the State Forest Department of Negeri Sembilan, for which it technically remains part of their production forest estate. Pasoh FR has become the most intensively researched, and therefore the best understood rain forest in all Asia.

The IBP research by T. Kira, H. Ogawa (Kira 1978a,b; Kira & Ogawa 1971) and their colleagues at Pasoh FR (Kato et al. 1978; Koyama 1978; Yoda 1978a; Yabuki & Aoki 1978) demonstrated, by the most painstaking and complete assays ever undertaken, and using three methodologies, that tropical forest approaching dynamic equilibrium differs little in net primary productivity from its temperate deciduous counterparts. The difference is in gross production rates; tropical forests achieve high gross assimilation rates, which are balanced by very high respiration rates.

The IBP program at Pasoh FR served as umbrella for a wealth of other research, the environmental and productivity research of which was summarized in 26 papers presented at a meeting in 1974 at Kuala Lumpur (Soepadmo 1978), notable among which was Manokaran's (1979, 1980) work on nutrient cycling, Yoda's (1978b) on variation in space and time in the light climate beneath the canopy, and Abe and Matsumoto's (Abe 1978a, b, 1979, 1980; Abe & Matsumoto 1978, 1979; Matsumoto 1976, 1978a,b) on termite ecology. Ashton (1976), who expanded five of Wong's plots to two hectares and recorded spatial positions of the trees on behalf of IBP researchers, concluded that habitat does indeed influence forest composition, though the relative importance of such deterministic factors and stochastic influences as dispersal limitation remains unresolved. Manokaran and his colleagues have continued to census these plots at regular five-year intervals, so that they have continued to yield new insights from what is now one of the longest and most consistent data sets on the dynamics of species rich tropical forests (Manokaran 1998; Manokaran & Kochummen 1987; Manokaran & Swaine 1994). The continuous monitoring of growth and mortality within Wong's plots since 1961 serve as but one example of why the value of Pasoh FR for rain forest research will continue to grow, exponentially, into the future.

3.3 UM-Aberdeen Project

Between 1973–1978 a collaborative program of research through graduate training was carried out at Pasoh FR, between the University of Malaya and the University of Aberdeen, Scotland, under the leadership of P. S. Ashton and E. Soepadmo. It's theme was the reproductive biology of rain forest trees, about which little was known at that time. The first evidence was found for the preponderance of self-incompatibility (Appanah 1981; Chan & Appanah 1980), yet apomixis by means of adventive embryony was revealed to be widespread (Ha et al. 1988; Kaur et al. 1978, 1986). Gan (née Yap) (Gan et al. 1977, 1981) performed the first research into the population genetics of rain forest trees, finding evidence of both high variability and panmixis. The reproductive biology of dipterocarps was decyphered (Appanah 1981, 1985; Appanah & Chan 1981; Ashton et al. 1988; Chan 1980, 1981, 1982; Chan & Appanah 1980; Gan et al. 1977; Kaur et al. 1978; Yap 1982).

3.4 FRIM-STRI Project

In 1984, the Forest Research Institute Malaysia (FRIM) began a long term collaborative research program with the Smithsonian Tropical Research Institute (STRI) and Harvard University, based on the Pasoh FR. Initially led by Dr. Wan Razalli of FRIM with Dr. J. E. Klahn on behalf of Drs. P. S. Ashton and S. P. Hubbell at Harvard and STRI respectively, soon Drs. N. Manokaran and J. V. LaFrankie took over. Since 1998 the National Institute of Environmental Studies of Japan has become a partner, under the leadership of Dr. T. Okuda. The overall aim of the program is to understand the mechanisms by which tree species populations are maintained in mixture in biodiverse tropical forest, the more precisely to manage them. The core of the approach is comparative study of statistically comparable forest samples established along the environmental gradients along which species composition and richness most vary, and replicated across the three tropical continents. Samples are large enough to include at least one hundred individuals of at least half the species for demographic analysis. To achieve this, samples include all trees $\geq$ 1 cm, which are identified and recensused for diameter at 5-year intervals, in samples generally of 16–50 ha. In order to analyze between-tree and between-population physical interactions, data are spatially explicit, and single large rectangular plots are used. The plot is of 50 hectares, 500 × 1000 m, and contained 320,903 individuals and 822 species in the second recensus conducted in 1995 (Manokaran et al. 1999).

Pasoh FR represented the second such plot, the first having been started in 1981 on Barro Colorado Island, the research forest of STRI, under Dr. Hubbell. In 1992, the Center for Tropical Forest Science (CTFS) at STRI, a worldwide partnership between scientists and institutions and STRI, willing to use the same methodology and gain by the sharing of a common database, was established to assist in coordinating a rapidly growing movement (Ashton et al. 1999). Pasoh FR is now the flagship site of a regional network of collaborating scientists and institutions with ten sites, representing forests along gradients of rainfall seasonality, canopy disturbance, soil nutrient levels and insularity. Pasoh FR is unique among these in its uniformity of habitat, permitting analysis of a very large sample of a single forest community. Protocols for sampling, and models for publication of basic data on composition and structure at the Asian sites have been developed here (Appanah et al. 1993; Kochummen 1997; Kochummen et al. 1991, 1992; Manokaran & LaFrankie 1991; Manokaran et al. 1990, 1991, 1993). Pasoh FR, as a member of CTFS is partner in a group of sixteen sites worldwide.

Publications, based in part or whole on the 50-ha plot of Pasoh FR now include studies of forest dynamics (Appanah & Weinland 1993; Manokaran & Swaine 1994; Manokaran et al. 1993), spatially explicit simulation models of biodiverse tropical forest (Liu & Ashton 1998) including consequences of timber harvesting (Liu & Ashton 1999) and impact of forest edges (Liu & Ashton 1999), optimization of timber harvesting and carbon sequestration (Boscolo in press; Boscolo & Buongiorno 1997; Boscolo & Vincent in press; Boscolo et al. 1997), non-timber forest products (LaFrankie 1994; Lim & Ismail 1994; Saw et al. 1991); also fundamental studies of community structure (Bellehumeur et al. 1997; Condit et al. 1996; He & LaFrankie in press; He & Legendre 1996; He et al. 1994, 1996, 1997 and in press; Okuda et al. in press; Plofkim et al. 2000a,b), density dependent sapling mortality and recruitment (Condit et al. in press; Okuda 1995, 1997; Wills in press; Wills & Condit in press a,b), and comparative ecology of congeneric tree species (Rogstad 1990; Thomas 1993, 1996a-c; Thomas & Ickes 1995; Thomas & LaFrankie 1993). There have also been studies of sampling design (Condit et al. 1998; Gimaret-Carpentier et al. 1998; Hall et al. 1998; Hill-Rowley et al. 1996).

3.5 NIES-FRIM-UPM Joint Research Project

Since 1991, the National Institute for Environmental Studies (NIES), Tsukuba, Japan has developed a program of rain forest research in collaboration with Forest Research Institute Malaysia (FRIM) and Universiti Putra Malaysia (UPM) as part of the global environment research sponsored by the Japan Environment Agency in 1991, one of whose field sites is Pasoh FR. The project included not only biological studies, which focus on an inventory-like study of flora and fauna, but also meteorological and hydrological studies such as measuring CO_2 flux. Besides the organizing institution above, the project has been supported by the research staffs of other Japanese institutions including the Forestry and Forest Product Research Institute (FFPRI), Tsukuba; Japan International Research Center for Agricultural Sciences (JIRCAS), Tsukuba; Hokkaido University, Tohoku Univeristy, Akita Prefectural University, Fukushima University, Tsukuba University, University of Tokyo, Tokyo Metropolitan University, Tsuru University, Waseda University, Chiba University, Nagoya University, Gifu University, Nihon Fukushi University, Kyoto University, Kobe University, Osaka City University, Shinshu University, Nara Woman's University, Hiroshima University, Shimane University, Kyushu University, Kumamoto Prefectual University, Japan Wildlife Research Center (JWRC), Japan Science and Technology Corporation (JST), Japan International Cooperation Agency (JICA), Nikkei Research Co. Ltd and CTI Engineering Co., Ltd. Phases 1 and 2, from the beginning until 1996, consisted of a diverse array of research in many fields, such as the structure and dynamics of tropical forests, micrometeorology, hydrology, the ecology of small and medium-sized mammals, birds, and insects, plant and animal interactions, plant physiology and ecology, plant genetics, and the process of decomposition of fallen trees, leaves, and branches by soil organisms and fungi.

Three canopy towers, two of which are 32 m and one is 42 m in height were established in 1992 by NIES and FFPRI (Color plate 3). The towers are interconnected by a canopy walkway and have been used for the study of vertical variation of insect composition, bird and small mammal trapping and observation, physiological response of canopy tree leaves to light intensity and other environmental factors, and vertical variation of CO_2 intensity from forest floor to above the canopy. The tallest tower was extended to 52 m in height in 1995 in order to conduct more precise measurement of CO_2 flux. A permanent tree demography plot 6 ha in size was established near the canopy tower in 1994. All trees > 5 cm in diameter within the plot were identified and mapped, and a recensus was conducted every 2 years.

The plot and its demography data have been used for studies of small mammal behavior and their habitat use, seedling and sapling establishment in relation to the forest floor environment, spatial heterogeneity of microenvironment, gap formation and closure processes and genetic diversity of adult and juvenile trees. Since Pasoh FR is mostly covered by lowland rain forest, a study site in Semangkok forest 60 km north of Kuala Lumpur (hill dipterocarp forest) was chosen for the comparative study of forest structure and dynamics between the two forest types. Research at both sites was until now concerned with natural forests, but Phase 3, beginning in 1996 posed the problems of finding something to serve as an index of tropical forest degradation. New test areas were established in selectively logged forests in the 1950's, adjoining the natural forests within the two existing test areas (see Color plate 2). The project then initiated comparisons and research on the natural and the regenerating forests in each of the various research fields, thereby providing for project convergence on the common objectives of assessing the degradation of forest structure and function following selective logging. Along these lines, canopy structure and tree species composition, insect, bird and small mammal communities were studied and compared between the primary and regenerating forests. In Phase 4 of the project, starting in 1999, the project shifted to applied science focusing on the ecological service value of the forests CO_2 sequestration, soil and watershed conservation, recreational amenity etc., which is expected to give an insight for sustainable forest management.

3.6 Malaysia-UK Project

A memorandum of understanding signed between the Government of Malaysia and the Government of the United Kngdom in May 1992 led to the initiation of a project on Conservation, Management and Development of Forest Resources that was largely based in Pasoh FR. The joint research program, which involved several institutions in both Malaysia and UK, was designed to focus on widely debated issues related to the conservation of biological diversity and sustainable forest management, a theme that featured prominently in Agenda 21, the Convention on Biological Diversity and Forest Principles that was endorsed at the Earth Summit in Rio de Janeiro on June 1992. Research carried out in Pasoh FR under this program has included studies of the impact of logging on populations and diversity of tree species (Manokaran 1998), epiphytic cryptogams (Wolseley et al. 1998), ants (Bolton 1996), ectomycorrhizal fungi (Watling et al. 1998), palms (Nur Supardi et al. 1998), woody climbers (Gardette 1998) and herpetofauna (Kiew et al. 1998).

3.7 Graduate researches

The Pasoh FR and its plots are a valuable resource for graduate research. Theses which have availed of it include the six arising from the universities of Aberdeen and Malaya rain forest project (Appanah 1979; Chan 1977; Ha 1978; Singh (née Kaur) 1977; Yap, S. K. 1976; Yap, Y. Y. 1976), those of H. M. S. Amir (University of Malaya, 1989); S. H. Rogstad 1986; A. S. Moad 1992 and S. C. Thomas 1993 (Harvard University), and currently J. Ahmad (Aberdeen University 2001), K. Ickes (Louisiana State University in progress), M. Potts (Harvard University 2001), A. M. Abdullah (Universiti Putra Malaysia 1998), Miyamoto (Hiroshima University 1998), Numata, S. (Tokyo Metropolitan University 1998, 2001); Yasuda, M. (University of Tokyo 1993, 1998); Osada, N. (Kyoto University 1999), Yasuda, Y. (Chiba University 1999); Obayashi, K. (Tsukuba University 2000); Adachi, M. (Gifu University 2002); Naito, Y. (Kyoto Univesity 2002).

REFERENCES

Abdullah, A. M. (1998) Radiation interception by a tropical lowland evergreen forest and productivity of its tree species. Ph.D. diss., Univ. Putra (Pertanian).

Abe, T. (1978a) Studies on the distribution and ecological role of termites in a lowland rain forest of West Malaysia. (1) Faunal composition, size, colouration and nest of termites in Pasoh Forest Reserve. Kontyû 46: 273-290.

Abe, T. (1978b) The role of termites in the breakdown of dead wood in the forest floor of Pasoh Forest Reserve. Malay. Nat. J. 30: 391-404.

Abe, T. (1979) Studies on the distribution and ecological role of termites in a lowland rain forest of West Malaysia. (2) Food and feeding habits of termites in Pasoh Forest Reserve. Jpn J. Ecol. 29: 121-135.

Abe, T. (1980) Studies on the distribution and ecological role of termites in a lowland rain forest of West Malaysia. (4) The role of termites in the process of wood decomposition in Pasoh Forest Reserve. Revue d'Ecologie et de Biologie du Sol 17: 23-40.

Abe, T. & Matsumoto, T. (1978) Distribution of termites in Pasoh Forest Reserve. Malay. Nat. J. 30: 325-334.

Abe, T. & Matsumoto, T. (1979) Studies on the distribution and ecological roles of termites in a lowland rain forest of West Malaysia. (3) Distribution and abundance of termites in Pasoh Forest Reserve. Jpn. J. Ecol. 29: 337-351.

Ahmad, W. J. W. (2001) Habitat specialization of tree species in a Malaysian tropical forest. Ph.D. diss., Univ. Aberdeen.

Adachi, M. (2002) CO_2 efflux from soil in tropical rainforest, Malaysia, M. thesis, Gifu Univ.

Amir, H. M. S. (1989) Site fertility and carrying capacity in two Malaysian tropical forest reserves. Ph.D. diss., Univ. Aberdeen.

Appanah, S. (1979). The Ecology of insect pollination of some tropical rain forest trees. Ph.D. diss., Univ. Malaya.

Appanah, S. (1981) Reproductive biology of some Malaysian dipterocarps. II. Flowering biology. Malay. For. 44: 37-42.

Appanah, S. (1985) General flowering in the climax rain forests of southeast Asia. J. Trop. Ecol. 1: 225-304.

Appanah, S. & Chan, H. T. (1981) Thrips: The pollinators of some dipterocarps. Malay. For. 44: 234-252.

Appanah, S. & Weinland, G. (1993) A preliminary analysis of the Pasoh 50 hectare demography Plot: I. Dipterocarpaceae. Research Pamphlet 112. Forest Research Institute Malaysia, Kepong.

Appanah, S., Gentry, A. H. & LaFrankie, J. V. (1993) Liana diversity and species richness of Malayan rain forests. J. Trop. For. Sci. 6: 116-123.

Ashton, P. S. (1976) Mixed dipterocarp forest and its variation with habitat in the Malayan lowlands: A re-evaluation at Pasoh. Malay. For. 39: 56-72.

Ashton, P. S., Boscolo, M., Liu, J. & LaFrankie, J. V. (1999) A global programme in interdisciplinary forest research: the CTFS perspective. J. Trop. For. Sci. 11:180-204.

Ashton, P. S., Givnish, T. J. & Appanah, S. (1988) Staggered flowering in the Dipterocarpaceae: New insights into floral induction and the evolution of mast fruiting in the aseasonal tropics. Am. Nat. 132: 44-66.

Bellehumeur, C., Legendre, P. & Marcotte, D. (1997) Variance and spatial scales in a tropical rain forest: Changing the size of sampling units. Plant Ecol. 130: 89-98.

Bolton, B. (1998) A preliminary analysis of the ants (Formicidae) of Pasoh Forest Reserve. In Lee, S. S., Dan, Y. M., Gauld, I. D. & Bishop, J. (eds). Conservation, management and development of forest resources. Proc. of the Malaysia-United Kingdom Programme Workshop, 21-24 October 1996, Kuala Lumpur, Malaysia, pp.84-95

Boscolo, M. Non-convexities and multiple use management in tropical forests. In Losos, E. C. & Leigh, E. G. Jr. (eds). Forest diversity and dynamism: Findings from a network of large-scale tropical forest plots. University of Chicago Press (in press).

Boscolo, M. & Buongiorno, J. (1997) Managing a tropical rain forest for timber, carbon storage and tree diversity. Commonwealth For. Rev. 76: 246-254.

Boscolo, M., Buongiorno, J. & Panayotou, T. (1997) Simulating options for carbon sequestration through improved management of a lowland tropical rainforest. Environ. Dev. Econ. 2: 241-263.

Boscolo, M. & Vincent, J. R. Promoting better logging practises in tropical forests: A simulation analysis of alternative regulations. Land Economics (in press).

Chan, H. T. (1977) Reproductive biology of some Malaysian dipterocarps. Ph.D. diss., Univ. Aberdeen.

Chan, H. T. (1980) Reproductive biology of some Malaysian dipterocarps. II. Fruiting biology and seedling studies. Malay. For. 43: 438-451.

Chan, H. T. (1981) Reproductive biology of some Malaysian dipterocarps. III. Breeding systems. Malay. For. 44: 28-36.

Chan, H. T. (1982) Reproductive biology of some Malaysian dipterocarps. IV. An assessment of gene flow within a natural population of *Shorea leprosula* using leaf morphological characters. Malay. For. 45: 354-360.

Chan, H. T. & Appanah, S. (1980) Reproductive biology of some Malaysian dipterocarps. I. Flowering biology. Malay. For. 43: 132-143.

Condit, R., Loo de Lao, S., Leigh, E. G., Foster, R. B., Sukumar, R., Manokaran, N. & Hubbell, S. P. (1998) Assessing forest diversity from small plots: Calibration using species-individual curves from 50 ha plots. In Dallmeier, F. & Comisky, J. A. (eds). Forest biodiversity research, monitoring and modeling: Conceptual background and old world case studies. Man and the Biosphere Series 20. Parthenon and UNESCO, Paris, pp.247-268.

Condit, R., Hubbell, S. P., LaFrankie, J. V., Sukumar, R., Manokaran, N., Foster, R. B. & Ashton, P. S. (1996) Species-area and species-individual relationships for tropical trees: A comparison of three 50 ha plots. J. Ecol. 84: 549-562.

Condit, R., Hubbell, S. P., Foster, R. B., Manokaran, N., Ashton, P. S. & LaFrankie, J. V. A direct test for density dependent population change in two rainforest plots. In Losos, E. C. & Leigh, E. G. Jr. (eds). Forest diversity and dynamism: Findings from a network of large-scale tropical forest plots. University of Chicago Press, Chicago (in press).

Gimaret-Carpentier, C., Pélissier, R., Pascal, J. P. & Houllier, F. (1998) Sampling strategies for the assessment of tree species diversity. J. Veg. Sci. 9: 161-172.

Gan, Y. Y., Robertson, F. W., Ashton, P. S., Soepadmo, E. & Lee, D. W. (1977) Genetic variation in wild populations of rain forest trees. Nature 269: 323-325.

Gan, Y. Y., Robertson, F. W. & Soepadmo, E. (1981) Isozyme variation in some rain forest trees. Biotropica 13: 20-28.

Gardette, E. (1998) The effect of selective timber logging on the diversity of woody climbers at Pasoh. In Lee, S. S., Dan, Y. M., Gauld, I. D. & Bishop, J. (eds). Conservation, management and development of forest resources. Proc. of the Malaysia-United Kingdom Programme Workshop, 21-24 October 1996, Kuala Lumpur, Malaysia, pp.115-125.

Ha, C. O. (1978) Embryological and cytological aspects of the reproductive biology of some understorey rain forest trees. Ph.D. diss., Univ. Malaya.

Ha, C. O., Sands, V. E., Soepadmo, E. & Jong, K. (1988) Reproductive patterns of selected understorey trees in the Malaysian rain forest: The sexual species. Bot. J. Linn. Soc. 97: 295-316.

Hall, P., Ashton, P. S., Condit, R., Manokaran, N. & Hubbell, S. P. (1998) Signal and noise in sampling tropical forest structure and dynamics. In Dallmeier, F. & Comisky, J. A. (eds). Forest biodiversity research, monitoring and modeling: conceptual background and old world case studies. Man and the Biosphere Series 20. Parthenon/UNESCO, Paris, pp.63-78.

He, F-L., & LaFrankie, J. V. Scale dependence of tree abundance and tree species richness in a tropical rain forest. In Losos, E. C. & Leigh, E. G. Jr. (eds). Forest diversity and dynamism: Findings from a network of large-scale tropical forest plots. University of Chicago Press, Chicago (in press).

He, F-L. & Legendre, P. (1996) On species-area relations. Am. Nat. 148: 719-737.

He, F-L., Legendre, P., Bellehumeur, C. & LaFrankie, J. V. (1994) Diversity pattern and spatial scale: A study of a tropical rain forest in Malaysia. Environ. Ecol. Sci. 1: 265-286

He, F-L., Legendre, P. & LaFrankie, J. V. (1996) Spatial pattern of diversity in a tropical rain forest in Malay. J. Biogeogr. 23: 57-74.

He, F-L., Legendre, P. & LaFrankie, J. V. (1997) Distribution patterns of tree species in a Malaysian tropical rain forest. J. Veg. Sci. 8:105-114.

He, F-L., Legendre, P. & LaFrankie, J. V. Stand structure analysis of a tropical rain forest with respect to abundance and diameter size. J. Trop. Ecol. (in press).

Hill-Rowley, R. L., Kirton, L., Ratnam, L. & Appanah, S. (1996) The use of a grid-based geographic information system to examine ecological relationships within the Pasoh 50 ha research plot. J. Trop. For. Sci. 8: 570-572.

Kaur, A., Ha. C. O., Jong, K., Sands, V. E., Chan, H. T., Soepadmo, E. & Ashton, P. S. (1978) Apomixis may be widespread among trees of the climax rain forest. Nature 271: 440-441.

Kaur, A., Jong, K., Sands, V. E. & Soepadmo, E. (1986) Cytoembryology of some Malaysian dipterocarps, with some evidence of apomixis. Bot. J. Linn. Soc. 92: 75-88.

Kato, R., Tadaki, Y. & Ogawa, H. (1978). Plant biomass and growth increment studies in Pasoh forest. Malay. Nat. J. 30: 211-224.

Kiew, B. H., Lim, B. L. & Lambert, M. R. K. (1998) To determine the effects of logging and conversion of primary forest to tree crop plantation, on Herpetofaunal diversity in Peninsular Malaysia. In Lee, S. S., Dan, Y. M., Gauld, I. D. & Bishop, J. (eds). Conservation, management and development of forest resources. Proc. of the Malaysia-United Kingdom Programme Workshop, 21-24 October 1996, Kuala Lumpur, Malaysia, pp.126-140.

Kira, T. (1978a) Primary productivity of Pasoh forest–A synthesis. Malay. Nat. J. 30: 291-298.

Kira, T. (1978b) Community architecture and organic matter dynamics in tropical lowland rain forests of southeast Asia with special reference to Pasoh forest, west Malaysia. In Tomlinson, P. B. & Zimmerman, M. H. (eds). Tropical trees as living systems. Cambridge University Press. pp.561-590.

Kira, T. & Ogawa, H. (1971) Assessment of primary production in tropical and equatorial forests. In Duvigneaud, P. (ed). Proc. Brussels Symp., 1969. Productivity of forest ecosystems. UNESCO, Paris. pp.309-321.

Kochummen, K. M. (1997) Tree Flora of Pasoh Forest. Forest Research Institute Malaysia. Kepong. 461pp.

Kochummen, K. M., LaFrankie, J. V. & Manokaran, M. (1991) Floristic composition of Pasoh Forest Reserve a lowland rain forest in Peninsular Malaysia. J. Trop. For. Sci. 3: 1-13.

Kochummen, K. M., LaFrankie, J. V. & Manokaran, M. (1992) Representation of Malayan trees and shrubs at Pasoh Forest Reserve. In Yap, S. K. & Lee, S. W. (ed). In harmony with nature. Proc. of an International symposium on the conservation of biodiversity. The Malayan Nature Society. pp.545-554.

Koyama, H. (1978) Photosynthesis studies in Pasoh forest. Malay. Nat. J. 30: 253-258.

LaFrankie, J. V. (1994) Population dynamics of some tropical trees that yield non-timber forest products. Econ. Bot. 48: 301-309.

Lim, H-F. & Ismail, J. (1994) The uses of non-timber forest products in Pasoh Forest Reserve, Malaysia. Research Pamphlet 113. Forest Research Institute Malaysia, Kepong.

Liu, J. & Ashton, P. S. (1998) FORMOSAIC: An individual-based spatially explicit model for simulating forest dynamics in landscape mosaics. Ecol. Model. 106: 177-200.

Liu, J. & Ashton, P. S. (1999) Simulating effects of landscape contexts and timber harvest on tree species diversity. Ecol. Appl. 9: 186-201.

Manokaran, N. (1979) Stemflow, throughfall and rainfall interception in a lowland tropical rain forest in Peninsular Malaysia. Malay. For. 42: 174-201.

Manokaran, N. (1980) The nutrient contents of precipitation, throughfall and stemflow in a lowland tropical rain forest in Peninsular Malaysia. Malay. For. 43: 266-289.

Manokaran, N. (1998). Effect, 34 years later, of selective logging in the lowland dipterocarp forest at Pasoh, Peninsular Malaysia, and implications on present-day logging in the hill forests. In Lee, S. S., Dan, Y. M., Gauld, I. D. & Bishop, J. (eds). Conservation, management and development of forest resources. Proc. of the Malaysia-United Kingdom Programme Workshop, 21-24 October 1996, Kuala Lumpur, Malaysia, pp.41-60.

Manokaran, N. & LaFrankie, J. V. (1991) Stand structure at Pasoh Forest Reserve, a lowland rain forest in Peninsular Malaysia. J. Trop. For Sci. 3: 14-24.

Manokaran, N. & Kochummen, K. M. (1987) Recruitment, growth and mortality of tree species in a lowland dipterocarp forest in Peninsular Malaysia. J. Trop. Ecol. 3: 315-330.

Manokaran, N. & Swaine, M. D. (1994) Population dynamics of trees in dipterocarp forests of Peninsular Malaysia. Malayan Forest Records 40. Forest Research Institute Malaysia, Kepong.

Manokaran, N., LaFrankie, J. V., Kochummen, K. M., Quah, E. S., Klahn, J. E., Ashton, P. S. & Hubbell, H. P. (1990) Methodology for the fifty-hectare research plot at Pasoh Forest Reserve. Research Pamphlet 104. Forest Research Institute, Kepong.

Manokaran, N., LaFrankie, J. V. & Ismail, R. (1991) Structure and composition of dipterocarpaceae in a lowland rain forest in Peninsular Malaysia. In Tjitrisoma, S. S., Umay, R. C. & Umboh, I. M. (eds). Proc. of the fourth round-table conference on Dipterocarps. Biotrop Special Publications 41. SEAMEO-BIOTROP, Bogor, Indonesia. pp.317-331.

Manokaran, N., Raman Kassim, A., Hassan, A., Quah, E. S. & Chong, P. F. (1993) Short-term population dynamics of dipterocarp trees in a lowland rain forest in Peninsular Malaysia. J. Trop. For. Sci. 5: 97-112.

Manokaran, N., LaFrankie, J. V., Kochummen, K. M., Quah, E. S., Klahn, J. E., Ashton, P. S. & Hubbell, S. P. (1999) The Pasoh 50-ha forest dynamic plot: 1999 CD-ROM version.

Matsumoto, T. (1976) The role of termites in an equatorial rain forest ecosystem of west Malaysia. I. Population density, biomass, carbon, nitrogen and calorific content and respiration rate. Oecologia 22: 153-178.

Matsumoto, T. (1978a) Population density, biomass, nitrogen and carbon content, energy value and respiration rate of four species of termites in Pasoh Forest Reserve. Malay. Nat. J. 30: 335-351.

Matsumoto, T. (1978b) The role of termites in the decomposition of leaf litter on the forest floor of Pasoh study area. Malay. Nat. J. 30: 405-413.

Miyamoto, K. (1998) Natural regeneration of dipterocarps in a tropical lowland forest of Malaysia. M. thesis, Hiroshima Univ.

Moad, A. (1992) Dipterocarp juvenile growth and understory light in Malaysian tropical forest. Ph.D. diss., Harvard Univ.

Morley, R. J. (1981) Palaeoecology of Tasek Bera, a lowland swamp in Pahang, west Malaysia. Singapore J. Trop. Geogr. 2: 50-56.

Morley, R. J. & Flenley, J. R. (1987) Late cainozoic vegetational and environmental changes in the Malaya Archipelago. In Whitmore, T. C. (ed). Biogeographical Evolution of the Malay Archipelago. Oxford Monographs on Biogeography 4. Oxford Scientific publications. pp.50-59.

Naitou, Y. (2002) Reproductive ecology of *Shorea acuminata* (Dipterocarpaceae), M. thesis, Kyoto Univ.

Numata, S. (1998) Herbivory and defenses at early stages of regeneration of dipterocarps in Malaysia. M. thesis, Tokyo Metropolitan Univ.

Numata, S. (2001) Comparative ecology of regeneration process of dipterocarp trees in a lowland rain forest, Southeast Asia. Ph.D. diss., Tokyo Metropolitan Univ.

Nur Supardi, M. N., Dransfield, J. & Pickersgill, B. (1998) Preliminary observations on the species diversity of palms in Pasoh Forest Reserve, Negeri Sembilan. In Lee, S. S., Dan, Y. M., Gauld, I. D. & Bishop, J. (eds). Conservation, management and development of forest resources. Proc. of the Malaysia-United Kingdom Programme Workshop, 21-24 October 1996, Kuala Lumpur, Malaysia, pp.105-114.

Obayashi, K. (2000) Genetic diversity and outcrossing rate between undisturbed and selective disturbed forests of *Shorea curtisii* (Dipterocarpaceae) in Peninsular Malaysia. M.

thesis, Tsukuba Univ.
Okuda, T., Kachi, N., Yap, S. K. & Manokaran, N. (1995) Spatial pattern of adult tree and seedling survivorship in *Pentaspadon motleyi* in a lowland rain forest in Peninsular Malaysia. J. Trop. For. Sci. 7: 475-489.
Okuda, T., Kachi, N., Yap, S. K. & Manokaran, N. (1997) Tree distribution pattern and fate of juveniles in a lowland tropical rain forest - implications for regeneration and maintenance of species diversity. Plant Ecol. 131: 155-171.
Okuda, T., Hussain, N. A., Manokaran, N., Saw, L-G., Amir, H. M. S. & Ashton, P. S. Local variation of canopy structure in relation to soils and topography and the implications for species diversity in a rainforest in Peninsular Malaysia. In Losos, E. C. & Leigh, E. G. Jr. (eds). Forest diversity and dynamism: Findings from a network of large-scale tropical forest plots. University of Chicago Press, Chicago (in press).
Osada, N. (1999) Stand level leaf phenology and leaf dynamics in a tropical rain forest in Peninsular Malaysia. Ph.D. diss., Kyoto Univ.
Plotkin, J. B., Potts, M. D., Yu, D. W., Bunyavejchewin, S., Condit, R., Foster, R., Hubbell, S., LaFrankie, J., Manokaran, N., Lee, H. S., Sukumar, R., Nowak, M. A. & Ashton, P. S. (2000) Predicting speceis diversity in tropical forests. Proc. Am. Nat. Acad. Sci. 97: 10850-10854.
Plotkim, J. B., Potts, M. D., Leslie, N., Manokaran, N., LaFrankie, J. V. & Ashton, P. S. (2000) Species-area curves spetial aggregation, and habitat specialization in tropical forests. J. theor. Biol. 207: 81-99.
Potts, M. D. (2001) Species spatial patterning in tropical rainforests. Ph.D. diss., Harvard Univ.
Rogstad, S. H. (1986) A biosystematic investigation of the *Polyalthia hypoleuca* complex (Annonaceae) of Malesia. Ph.D. diss., Harvard Univ.
Rogstad, S. H. (1990) The biosystematics and evolution of the *Polyalthia hypoleuca* complex (Annonaceae) of Malesia. II. Comparative distributional ecology. J. Trop. Ecol. 6: 387-408.
Saw, L. G, LaFrankie, J. V., Kochummen, K. M. & Yap, S. K. (1991) Fruit trees in a Malaysian rain forest. Econ. Bot. 45: 120-136.
Singh, A. (1977). Embryological and cytological studies on some members of the Dipterocarpaceae. Ph.D. diss., Univ. Aberdeen.
Soepadmo, E. (1978) The Malaysian international biological program synthesis meetings: Selected papers from the symposium, Kuala Lumpur, 1974, IBP-PT. Malay. Nat. J. 30: 119-450.
Thomas, S. C. (1993) Interspecific allometry in Malaysian rain forest trees. Ph.D. diss., Harvard Univ.
Thomas, S. C. (1996a) Asymptotic height as a predictor of growth and allometric characteristics in Malaysian forest trees. Am. J. Bot. 83: 556-566.
Thomas, S. C. (1996b) Relative size of onset of maturity in rain forest trees: a comparative analysis of 37 Malaysian species. Oikos 76: 145-154.
Thomas, S. C. (1996c). Reproductive allometry in Malaysian rain forest trees: biomechanics versus optimal allocation. Evol. Ecol. 10: 517-530.
Thomas, S. C. & Ickes, K. (1995). Ontogenetic changes in leaf size in Malaysian rain forest trees. Biotropica 27: 427-434.
Thomas, S. C. & LaFrankie, J. V. (1993) Sex, size and inter-year variation in flowering among dioecious trees of the Malayan rain forest understory. Ecology 74: 1529-1537.
Watling, R., Lee, S. S. & Turnbull, E. (1988) Putative ectomycorrhizal fungi of Pasoh Forest Reserve, Negeri Sembilan. In Lee, S. S., May, D. Y., Gauld, I. D. & Bishop, J. (eds). Conservation, management and development of forest resources. Proc. of the Malaysia-United Kingdom Programme Workshop 21-24 October 1996, Kuala Lumpur, Malaysia. pp.96-104.
Wills, C. Comparable non-random forces act to maintain diversity in both a new and an old world rainforest plot. In Losos, E. C. & Leigh, E. G. Jr. (eds). Forest diversity and dynamism: Findings from a network of large-scale tropical forest plots. University of Chicago Press, Chicago (in press).

Wills, C. & Condit, R. Frequency and density dependent interactions within tree species in a moist lowland tropical rainforest lead to the maintenance of diversity, and are far stronger than between-species interactions. Ecology. (in press[a])

Wills, C. & Condit, R. A balance between non-random processes and chance in the maintenance of diversity in two tropical rainforests. Proc. Natl. Acad. Sci. USA (in press[b])

Wolseley, P., Ellis, L., Harrington, A. & Moncrieff, C. (1998) Epiphytic cryptogams at Pasoh Forest Reserve, Negeri Sembilan, Malaysia-quantitative and qualitative sampling in logged and unlogged plots. In Lee, S. S., Dan, Y. M., Gauld, I. D. & Bishop, J. (eds). Conservation, Management and Development of Forest Resources. Proc. of the Malaysia-United Kingdom Programme Workshop, 21-24 October 1996, Kuala Lumpur, Malaysia, pp.61-83

Wong, Y. K. & Whitmore, T. C. (1970) On the influence of soil properties on species distribution in a Malayan lowland dipterocarp forest. Malay. For. 33: 42-54.

Wyatt-Smith, J. (1954) Forest types in the Federation of Malaya. Malayan Forester 17: 83-84.

Wyatt-Smith, J. (1963) Manual of Malayan silviculture for inland forest. 2 vols. Malayan Forest Records 23. Forest Research Institute Malaysia, Kepong. 2nd ed. 1995.

Wyatt-Smith, J. (1987) Red Meranti-Keruing forest. Forest Research Institute Research Pamphlet 101. Kepong, Malaysia.

Yabuki, K. & Aoki, M. (1978) Micrometeorological assessment of primary production rate of Pasoh forest. Malay. Nat. J. 30: 281-290.

Yap, S. K. (1976) The reproductive biology of some understorey fruit tree species in the lowland dipterocarp forest of west Malaysia. Ph.D. diss., Univ. Malaya.

Yap, S. K. (1982) The phenology of some tree species in a lowland dipterocarp forest. Malay. For. 45: 21-35.

Yap, Y. Y. (1976) Population and phylogenetic studies on species of Malaysian rainforest trees. Ph.D. diss., Univ. Aberdeen.

Yasuda, M. (1993) Ecology of small mammals in a lowland tropical rain forest in SE Asia-seed consumption and seed dispersal on the forest floor. M. thesis, Univ. Tokyo.

Yasuda, M. (1998) Community ecology of small mammals in a tropical rain forest of Malaysia with special reference to habitat preference, frugivory and population dynamics. Ph.D. diss., Tokyo Univ.

Yasuda, Y. (1999) Observational study on short-term and long-term variations in CO_2 flux above forests. Ph.D. diss., Chiba Univ.

Yoda, K. (1978a) Respiration studies in Pasoh forest plants. Malay. Nat. J. 30: 259-280.

Yoda, K. (1978b) Three-dimensional distribution of light intensity in a tropical rain forest in West Malaysia. Malay. Nat. J. 30: 161-177.

2 Logging History and Its Impact on Forest Structure and Species Composition in the Pasoh Forest Reserve — Implications for the Sustainable Management of Natural Resources and Landscapes

Toshinori Okuda[1], Mariko Suzuki[1], Naoki Adachi[1], Keiichiro Yoshida[1], Kaoru Niiyama[2], Nur Supardi Md. Noor[3], Nor Azman Hussein[3], N. Manokaran[3] & Mazlan Hashim[4]

Abstract: A part of the Pasoh Forest Reserve (Pasoh FR) was once logged under a logging regime called the Malayan Uniform System (MUS) in the 1950s. The core area of the reserve is a residual unlogged (primary) forest that shows the typical structure and species composition of lowland dipterocarp forest; the logged area of the reserve is also a relict area of regenerating lowland forest. In this chapter, we review and summarize previous studies of logging impacts on the forest structure and total aboveground biomass by comparing the primary and regenerating forests of this reserve. We also studied landscape changes in the Pasoh Forest Region in order to discuss the relationship between logging history in this region and its impacts on the forest. From a chronological analysis of the changes in land use in this region, we found that ca. 50% of the forested area had been converted to either oil palm or rubber plantations from 1971 to 1996. Almost all of the lowland dipterocarp forest that had developed in the flat and alluvial topography had vanished from this region, except in the Pasoh FR. Thus, very little area was left to be managed by the MUS approach, which was originally designed for extracting timber with a longer logging cycle (> 70 years) in this type of forest. By examining the canopy and stand structure and the species composition of these forests, we found a greater density of semi-medium (6–10 cm in diameter) and medium trees (10–30 cm), a higher density of canopy-forming trees with relatively smaller crowns, and a higher density of non-commercial canopy-forming trees in the regenerating forest. These findings suggest that the MUS was incompletely implemented, since this system originally aimed to encourage the development of a uniform forest structure with a large number of sound commercial timber trees by removing non-commercial trees. Owing to the high density of canopy-forming trees, which probably resulted from incomplete post-logging thinning and vegetation-control operations, structural development was delayed in the regenerating forest. In addition, the species composition and the distribution of wildlife in the regenerating forest differed from those in the primary forest. We also found that the total aboveground biomass in the regenerating forest had not fully recovered to the level in the primary forest even 40 years after logging. We suggest that "follow-up operations" should be undertaken, with a special concern for encouraging the structural development

[1] National Institute for Environmental Studies (NIES), Tsukuba 305-8506, Japan.
E-mail: okuda@nies.go.jp
[2] Forestry and Forest Product Research Institute (FFPRI), Japan.
[3] Forest Research Institute Malaysia (FRIM), Malaysia.
[4] Universiti Teknologi Malaysia (UTM), Malaysia.

of the stand, which we consider to be crucial for ecologically sustainable management.

Key words: aboveground biomass, canopy structure, land-use changes, Malayan Uniform System (MUS), selective logging, stand structure.

1. INTRODUCTION

Logging has a major impact on the structural and compositional development of forests (Cannon et al. 1994; Johns 1997; Kasenene 1984; Pinard & Putz 1996). Although the selective logging that has been widely employed in the Asian tropics removes fewer trees per unit area than clearcutting, the extraction of large trees usually damages neighboring trees and influences the development of the understory vegetation. The removal of vegetation and the construction of logging roads, extraction trails, and landings causes soil erosion and subsequently affects the watershed ecosystem (Abdul Rahim & Harding 1992; Bornhan et al. 1987; Brooks & Spencer 1997; Oyebande 1988). Under such selective logging, the forest is repeatedly logged after the initial operation. The logging cycle has sometimes been based on early estimates of the typical growth rate of tropical trees and on average levels of felling damage to residual vegetation (e.g. Dawkins 1959). However, we have little scientific evidence about whether the forest could eventually recover its original timber productivity under such a logging cycle. Moreover, "sustainable forest management" usually ignores the recovery of the structure and composition of a forest, which relate strongly to the sustainability of biodiversity (Chapman & Chapman 1997; Chapman & Onderdonk 1998). Few studies have demonstrated the extent to which species richness and wildlife habitat can recover after logging, or how these characteristics change from their original state (population sizes, distribution patterns, etc.). For example, the removal of large trees during selective logging creates a homogeneous canopy structure of even-aged cohorts of canopy-forming trees, leading to the low frequency of gap formation in the canopy of younger regenerating forests (Brokaw 1982; Chapman & Chapman 1997; Knight 1975; Lang & Knight 1983). This in turn would eventually affect the spatial heterogeneity of the forest environment, species richness and composition, and wildlife habitat. To provide proper management protocols for use in future harvesting, promoting both sustainable timber resource use and ecologically sustainable management, we require detailed research on the recovery of the structure and composition of regenerating forests.

Part of the Pasoh Forest Reserve (Pasoh FR) was once selectively logged from 1955 to 1959 under the logging regime called the Malayan Uniform System (MUS, Fig. 1). Most ecological studies have been conducted in the core area (about 600 ha) of the reserve, where no trace of anthropogenic activity has so far been found (Lee et al. 1995), but a few studies have been undertaken in logged areas to obtain data on the differences from virgin forest. A tree census conducted in 1989 in the reserve and its vicinity (Manokaran 1998) indicated the presence of considerable regeneration of commercial timber species (e.g. Dipterocarpaceae) in the regenerating secondary forest even 34 years after logging. Okuda et al. (2000, in press[a]) studied the canopy and stand structure, species composition, and total aboveground biomass in the Pasoh FR and reported distinct differences in these characteristics between the primary and regenerating forests. In this chapter, we have reviewed and summarized these studies, with a focus on how the logging history and landscape changes in the Reserve relate to the logging impacts and

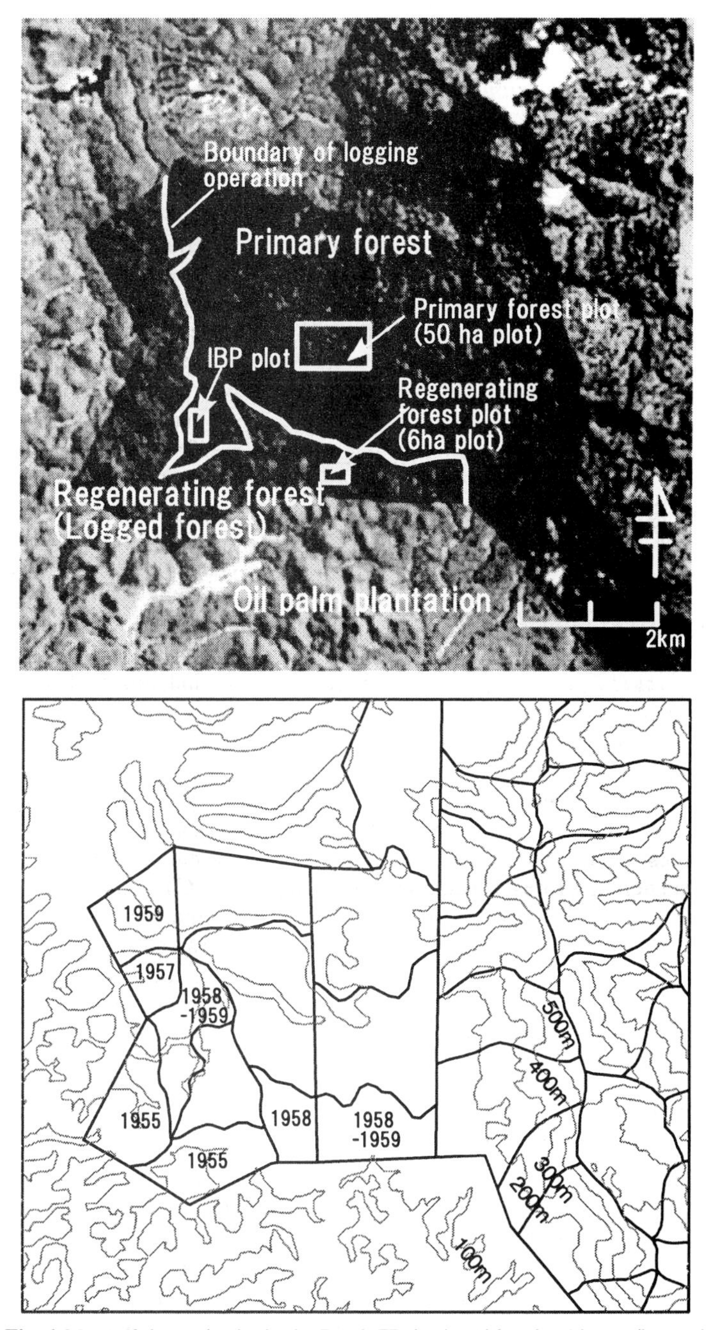

Fig. 1 Maps of the study site in the Pasoh FR (top) and logging history (bottom). The tree census plots were located in the primary forest (50 ha in size) and the regenerating forest (6 ha in size). The International Biological Program (IBP) plot (6 ha in size) was at the southwestern end of the primary forest in the reserve, and tree census data obtained in this plot were used for the comparative analysis of total aboveground biomass (see the text and Fig. 6 for details).

subsequent regeneration and growth of the forest. In addition to these previous studies, we studied the landscape changes in the Pasoh FR and its vicinity (hereafter called as Pasoh Forest Region) in order to investigate the Region's logging history.

2. LOGGING HISTORY AND REGIME IN THE PASOH FOREST REGION

The MUS was commonly used from the 1950s through the 1970s in Peninsular Malaysia. This logging regime involved removing the mature crop in a single operation that harvested trees of all species ≧ 45 cm in DBH (diameter at breast height) and releasing selected natural regeneration of various ages, most of which were light-demanding (shade-intolerant), hardwood species with medium-light density wood (Thang 1987, 1997). Wyatt-Smith (1963) described the MUS as felling of the upper canopy (which consists of the economic crop), followed immediately by girdling the remaining large unmerchantable canopy trees with herbicides. This treatment was extended to all smaller trees and saplings with a DBH ≦ 15 cm, except for economically valuable species of sound form. According to the MUS formulations, such thinning was to be done at 20, 35 and 55 years after logging (Wyatt-Smith 1963). Thus, the MUS was a system for converting the virgin tropical lowland rain forest (a rich, complex, multi-species and multi-aged forest) into a more-or-less even-aged forest that would contain a greater proportion of commercially valuable species (Wyatt-Smith 1963).

However, the MUS has not been used in Peninsular Malaysia since the 1970s, because the lowland dipterocarp forests, which were all easily accessible, had been cleared and converted into oil palm and other plantations. Since little forest was left that could be managed by the MUS approach, the approach was virtually abandoned in the lowland forests. In the remaining virgin forests, which were found on steeper slopes in more remote areas, the MUS proved unsuccessful because of uneven stocking and the uncertainty of natural regeneration. The damage

Table 1 Comparison of two logging regimes that were once operated in the Pasoh FR and its vicinity and that are still in use in some areas.

Logging system	Malayan Uniform System (MUS)	Selective Management System (SMS)
Period of operation	1948 to late 1970s	1978 onward
Logging operation	Monocyclic	Polycyclic
Area of suitability	Lowland dipterocarp forest	Hill dipterocarp forest
Final goals of the logging system	Conversion of virgin lowland forest into a more or less even-aged forest containing a greater proportion of commercial timber species.	Optimization of forest management goals supported by economical tree felling operations, the sustainability of timber production and the minimum cost for forest development.
Logging regime; post- and pre-logging operations	Removing all commercial timber trees (≥ 45 cm in DBH). Extraction of unwanted non-commercial timber species and defective relic trees.	Pre-felling inventories to determine minimum diameter cutting limits and number of trees to be felled. Cutting limits in practice: > 45–50 cm in DBH.
Logging cycle	60–80 years	25–35 years

caused by harvesting on these steep slopes also created a high erosion risk, and subsequently led to heavy seedling mortality (Burgess 1968, 1971; Whitmore 1975). Moreover, in such hill forests, various other factors also made the MUS unsuitable; for example, the secondary growth of various pioneer species was remarkable after the canopy had been opened, and it was not practical to delay harvesting until an adequate stock of the commercially valuable trees prescribed originally by the MUS approach had regenerated (Thang 1997). That is, the monocyclic approach advocated by the MUS was not sufficiently economically attractive to encourage large-scale investment by the forestry sector, because the investment was rendered uneconomical by the environmental difficulties encountered in trying to apply the approach in the hill forests (Thang 1997).

The requirement for economical, sustained-yield forestry could only be achieved by shortening the logging cycle or increasing the volume harvested per rotation (Thang 1997). Given this problem, a new logging regime called the Selective Management System (SMS) was introduced, and logging subsequently shifted to the hill dipterocarp forests. In the SMS, the smallest trees that can be harvested and the number of trees to be felled are determined by pre-felling inventories that estimate the stand's population parameters (e.g. DBH classes, species composition, stem density, soil type, slope, elevation and hydrological network). Based on this pre-logging information and other relevant market and socioeconomic information, a logging regime is formulated that meets the goals of the SMS. In practice, the regime targets trees > 45–50 cm in DBH for both dipterocarp and non-dipterocarp species, with a logging cycle of 25–35 years (Table 1).

3. LANDSCAPE CHANGES IN THE PASOH FOREST REGION

Fig. 2 shows the landscape changes between 1971 and 1996 in the Pasoh Forest Region. Each map covers a square area 60 km on each side (3°15' N to 2°42' N, 102°03' E to 102°35' E). The 1971 landscape map was made from topographical maps based on the geographical surveys conducted from the late 1950s until 1965 and revised in 1971. The 1985 map was compiled from aerial photographs taken from 1983 to 1986 (most of the study area was photographed in 1985). The other two maps were created from Landsat MSS and TM images acquired in 1979 and 1996, respectively. The total forested area, including regenerating logged forest, decreased from 2,362 km^2 (65.6% of the total area) to 1,057 km^2 (29.4% of the total) (Fig. 3). The deforestation rate during this period averaged about 2.2% yr^{-1}, which was higher than the annual deforestation rate for rain forest in insular and continental southeast Asia (1.2 to 1.5% yr^{-1}) from 1981 to 1990 (FAO 1993). In contrast, the area covered by oil palm plantations increased from 176 to 741 km^2 during this period, and the area of rubber plantations increased from 889 to 1,215 km^2 (Fig. 3). These landscape changes followed the general trends in land use throughout Malaysia, where oil palm plantations increased in area from 5,690 km^2 in 1975 to 21,980 km^2 in 1992 (Iwasa 1997). In fact, many of the lowland forests were converted into oil palm plantations. Most of the large denuded area in the middle-right parts of the 1979 map in Fig. 2 was either oil palm or rubber plantations by the time of the surveys that led to the 1985 map. The converted areas consisted mainly of flat alluvial topography and had originally been covered by lowland dipterocarp forest, which today only exists in the Pasoh FR. In contrast, most of the forest that remained in 1996 was distributed in the hilly parts of the region, where the development and management of oil palm plantations were considered to be overly difficult. That is, areas where oil palm plantations could be easily developed were

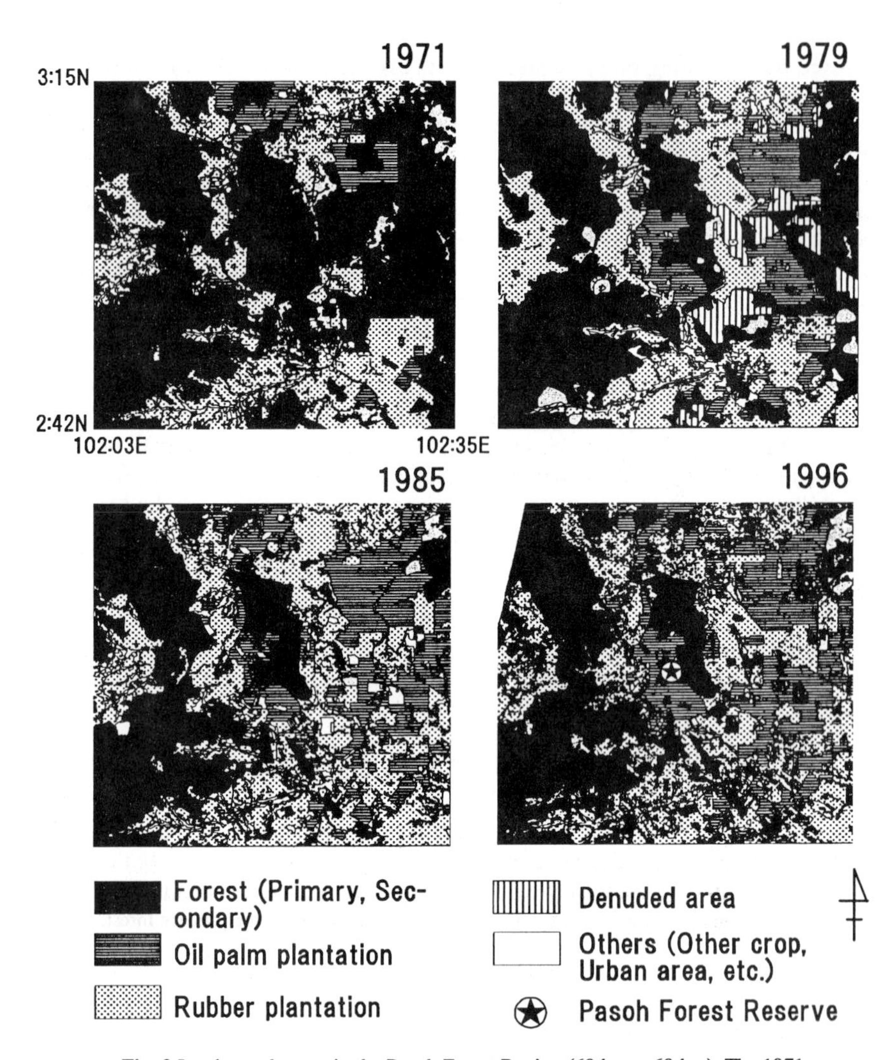

Fig. 2 Land-use changes in the Pasoh Forest Region (60 km × 60 km). The 1971 and 1985 maps were based on a topographic map, whereas the 1979 and 1996 maps were created from image analyses of Landsat MSS and TM data (respectively). The upper left corner of the 1996 map was trimmed off because of a lack of Landsat image data.

already mostly deforested by the early 1980s, and similar landscape changes affected almost all of the peninsular part of the country as a result of a policy to develop rural areas by increasing the production of agricultural crops.

As mentioned earlier, these landscape changes from the 1970s to the late 1980s meant that the MUS was hardly used in this region, since few suitable areas remained in which this logging practice could be applied. Thus, the Pasoh FR is a unique ecosystem because it preserves both a pristine, almost extinct virgin lowland forest and a regenerating forest logged by using the MUS.

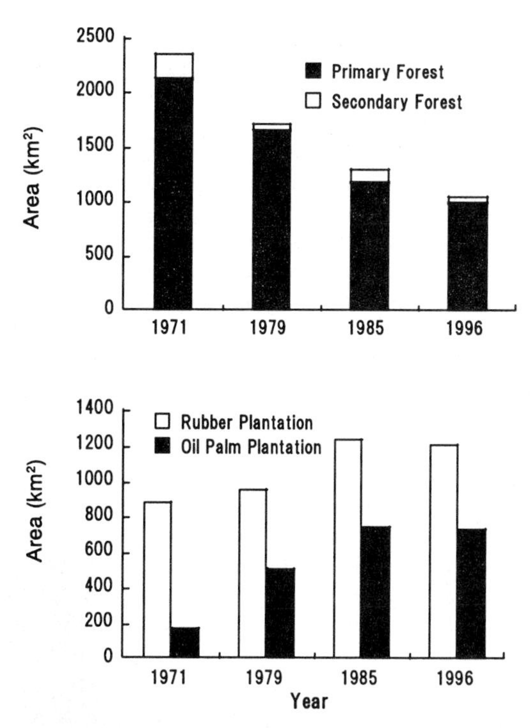

Fig. 3 Changes in the forest (top) and plantation areas (bottom) from 1971 to 1996 in the Pasoh Forest Region shown in Fig. 2. Secondary forest, which includes logged and regenerating forests, was identified mainly from aerial photographs or satellite images by the presence of logging roads or evidence of related activities. Thus, regenerating forests in which recovering vegetation masks the denuded area and logging roads may not have been included in the total area of secondary forest, and may instead have been allocated to the primary forest.

4. LOGGING IMPACTS OF THE MUS IN THE PASOH FOREST

4.1 Tree census, development of a canopy-surface model and estimation of aboveground tree biomass

The major and detailed studies described in this section will be published or have already been published elsewhere; thus, we will only summarize their methodologies and the results (Manokaran et al. 1998; Okuda et al. in press[a]) of comparisons of species composition, stand structure, canopy surface structure, and total aboveground biomass in primary and regenerating forests.

To evaluate the impact of logging in lowland dipterocarp forests, we compared the canopy and stand structures, the species composition, and the aboveground biomass in primary and regenerating forests that had been selectively logged from 1955 to 1959. Two study plots were established within the Reserve: one 50-ha plot (1,000 m × 500 m) lay in primary forest toward the center of the reserve, and a second 6-ha plot (300 m × 200 m) lay in part of a regenerating forest that had been logged under the MUS regime. The tree census in the regenerating forest plot was conducted between October 1997 and February 1999 by following the methods of Manokaran et al. (1990). The same approach was previously used by the Forest Research Institute of Malaysia and the Smithsonian Institution in 1985 to establish and census the 50-ha plot in the primary forest. In both plots, all woody plants ≧ 1 cm in DBH were identified and tagged. The DBH of each plant was measured and its position was also mapped to the nearest 10 cm. The census data of the primary forest was based on the 10-year recensus of the forest, between November 1995 and November 1997 (Manokaran et al. 1999).

For the comparative analyses of the canopy structures of the primary and regenerating forests, a three-dimensional (3-D) canopy surface model was developed by using aerial triangulation based on the aerial photographs taken at the center of the Pasoh FR in February 1997. These photographs covered both primary and regenerating forest plots at a 1:6,000 scale. The 3-D canopy-surface models were developed by using an analytical stereo-plotter that permitted stereoscopic analysis of these photographs. The height of the canopy surface was measured at 2.5 m intervals throughout the plots.

For the estimation of total aboveground biomass (TAGB), tree diameter (DBH) and tree height (H), allometric relationships were established individually for the two forest types (eqs. 1 and 2). Heights of individual canopy trees visible in the aerial photographs were measured by using aerial triangulation, then H was regressed against the DBH of the corresponding trees measured through the ground surveys.

$$1/H = 1/(1.5 \times DBH) + 1/H_{max} \quad \text{Primary forest} \quad \text{(eq. 1)}$$
$$1/H = 1/(2.2 \times DBH) + 1/H_{max} \quad \text{Regenerating forest} \quad \text{(eq. 2)}$$

where H_{max} represents the maximum canopy height within a forest stand; H_{max} was 61 m in the primary forest and 47 m in the regenerating forest.These equations differ slightly from eq. 3, which was established by Kato et al. (1978) by using destructive sampling in the primary forest of the Pasoh Forest Reserve.

$$1/H = 1/(2.0 \times DBH) + 1/61 \quad \text{(eq. 3)}$$

In the present study, we employed new equations (eqs. 1 and 2) rather than using eq. 3 because the growth process of the trees differed between the two forests (Okuda et al. 2000) and the value of TAGB is markedly influenced by H_{max}. Thus,we considered that the DBH versus H allometric equations would need to be applied differently to the two forests. The aboveground biomass of all trees (≧ 5 cm in DBH) was obtained by the summation of the dry weights of the main stems (Ws), branches (Wb) and leaves (Wl), which were all derived from previously obtained relationships among DBH, H, Ws, Wb and Wl (Kato et al. 1978). TAGB was then calculated for each forest plot.

Tree size classes were defined according to DBH as follows, unless otherwise specified: ≦ 3 cm = saplings; 3–6 cm = small trees; 6–10 cm = semi-medium-sized trees; 10–30 cm = medium-sized trees; 30–60 cm = semi-large trees; and ≧ 60 cm = large trees.

The data discussed in the following four sections of this chapter (Species composition, Stand structure, Canopy surface structure and Total aboveground biomass) were based on the studies done by Okuda et al. (2000, in press[a]), and all statistical tests were done by ANOVA, unless otherwise specified.

4.2 Species composition

Tree species richness in tropical rain forest is exceptionally high, and species composition and tree growth vary greatly in relation to microenvironmental conditions such as gap size, soil nutrient content and topography (e.g. Ashton & Hall 1992; Brown & Whitmore 1992; Denslow & Hartshorn 1994; Denslow 1980, 1987; Denslow et al. 1990). As a result, simple comparisons of species composition between local patches may not provide sufficient information for discussions of the effects of ecological attributes on differences in species composition. Nevertheless, the large trees of commercial timber species were selectively removed with the goal of encouraging the juveniles of these species to grow in the regenerating forests. Therefore, some differences should still be detectable, at least in terms of the stem density of the commercial timber species, if the logging had fully followed the MUS regime. According to Manokaran (1998), the density of dipterocarps in the 30–50 cm DBH class, most of which consisted of commercial timber species, was distinctly higher in the regenerating forest plots than in the primary forest plots. Similarly, the stem density of this family in the medium-sized class (10–30 cm) was significantly higher ($F_{1,222} = 41.25, P < 0.0001$) in the regenerating forest than in the primary forest.

However, the basal area of dipterocarp trees ≧1 cm in DBH accounted for 27.3% of the total basal area in the primary forest and was not significantly ($P >$ 0.05) different from the values in the regenerating forest (30.8%). Thus, it may not be appropriate to say that the MUS logging increased the relative dominance (basal area) of this family across all size classes. Such an indistinct difference between the two forests in dipterocarp basal area resulted from the greater number of large trees in this family in the primary forest; this offset the lower stem density there. Although the tree density and the basal area of medium-sized to semi-large trees (10–60 cm in DBH) of this family were always significantly greater ($P < 0.0005$) in the regenerating forest, the stem density of larger trees (80–100 cm in DBH) in this family was significantly lower ($F_{1,222} = 5.5, P < 0.05$) in the regenerating forest. Above this size class (≦ 190 cm), the stem density was always higher in the primary forest, but the differences were not significant. Moreover, non-dipterocarp families (e.g. the Euphorbiaceae and the Fagaceae) contributed markedly to the total basal area and the stem density in the largest size classes in the regenerating forest (100–120 cm). Dipterocarp species did not completely dominate this size class in the regenerating forest (< 50% of both stem density and basal area), but accounted for the majority (almost 100% of the total basal area, and 80%–100% of total stem density) of this size class in the primary forest. These facts indicate that the logging operation did not completely follow the MUS approach, in which non-commercial trees of canopy species were supposed to have been eliminated in order to provide more open space for the commercial timber species, which were mostly dipterocarps during the era of the logging operations in the Pasoh FR.

The five most abundant families in terms of stem density for trees ≧1 cm in DBH did not differ between the primary and regenerating forests; the most common family was the Euphorbiaceae, followed by the Dipterocarpaceae, in both forests (Table 2). The ranking of the three next-most-abundant families (the Annonaceae, Rubiaceae and Burseraceae) differed slightly between the two forests, but the proportion of the total stem density accounted for by each family was 5%–7%. The top five families accounted for more than 40% of the total stem density in both forests, and this proportion was not markedly different between the two forests.

Table 2 Comparison of tree densities and basal areas for major families in the primary (50 ha) and regenerating forest (6 ha) plots. Values in parentheses indicate that the family was not ranked within the most five abundant families in either stem density or basal area.

	Stem density (ha^{-1})				Basal area (m^2 ha^{-1})			
	Primary forest		Regenerating forest		Primary forest		Regenerating forest	
Family name	Density	% of total	Density	% of total	Basal area	% of total	Basal area	% of total
Euphorbiaceae	906.8	14.1	1013.7	16.7	2.4	7.4	3.4	10.0
Dipterocarpaceae	587.4	9.2	638.3	10.5	9.1	27.3	10.5	30.8
Annonaceae	470.8	7.3	287.8	4.7	1.1	3.4	(1.4)	(4.1)
Rubiaceae	381.8	5.9	283.2	4.7	(0.8)	(2.3)	(1.1)	(3.2)
Burseraceae	341.0	5.3	454.5	7.5	2.0	6.1	2.3	6.7
Leguminosae	(216.0)	(3.4)	(194.5)	(3.2)	2.8	8.5	1.5	4.4
Myrtaceae	(197.0)	(3.1)	(190.2)	(3.1)	1.1	3.4	(0.7)	(2.0)
Fagaceae	(94.9)	(1.5)	(69.5)	(1.2)	(1.1)	(3.3)	1.8	5.4
Total (all families)	6418.1	-	6066.7	-	33.1	-	34.0	-

These trends for the stem density of the top five families agree with the trends reported by Manokaran (1998). Thus, dominance of stands by these major families appears to be common throughout the lowland dipterocarp forests and might not have been greatly influenced by the MUS logging operations.

4.3 Stand structure

The density of stems for all tree size classes in the primary forest was 6,418 trees ha^{-1}, which did not differ significantly from the value in the regenerating forest (6,067 trees ha^{-1}) ($F_{1,222} = 2.26, P > 0.05$). However, when we compared each DBH class, there were some notable differences in stem density between the two forests. For example, the density of saplings (< 3 cm in DBH) was significantly ($F_{1,222} = 18.09, P < 0.0001$) higher in the primary forest. The stem densities of the larger trees (90–110 cm in DBH) were also significantly higher ($F_{1,222} = 8.98, P < 0.005$) in the primary forest. In contrast, the stem densities of semi-medium (6–10 cm in DBH) and medium-sized trees (10–30 cm) were significantly lower in the primary forest ($F_{1,222} = 64.91$ for semi-medium and $F_{1,222} = 98.93$ for medium-sized trees, $P < 0.001$ for both sizes). In contrast, for trees larger than this size class, the densities were always higher in the primary forest, but the differences were not significant ($P > 0.05$).

Basal area followed similar trends. The total for all DBH classes combined was around 33.1 m^2 ha^{-1} in the primary forest, versus 34.0 m^2 ha^{-1} in the regenerating forest. This difference was not significant ($F_{1,222} = 0.62, P > 0.05$). The basal area for all size classes < 30 cm in DBH, which accounted for just above 50% of the total basal area, was always significantly ($P < 0.01$) higher in the regenerating forests than in the primary forest. However, the basal area for trees < 3 cm in DBH was significantly ($F_{1,222} = 9.68, P = 0.0002$) higher in the primary forest. In the size classes ≧ 70 cm in DBH, basal area was always higher in the primary forest, but, as was the case for stem density, these differences were significant for only two size classes ($F_{1,222} = 9.02, P = 0.003$ for the size class of 90–100 cm in DBH, and $F_{1,222} = 4.45, P = 0.036$ for the size class of 100–110 cm in DBH).

These findings suggest that most of the regeneration or growth that arose

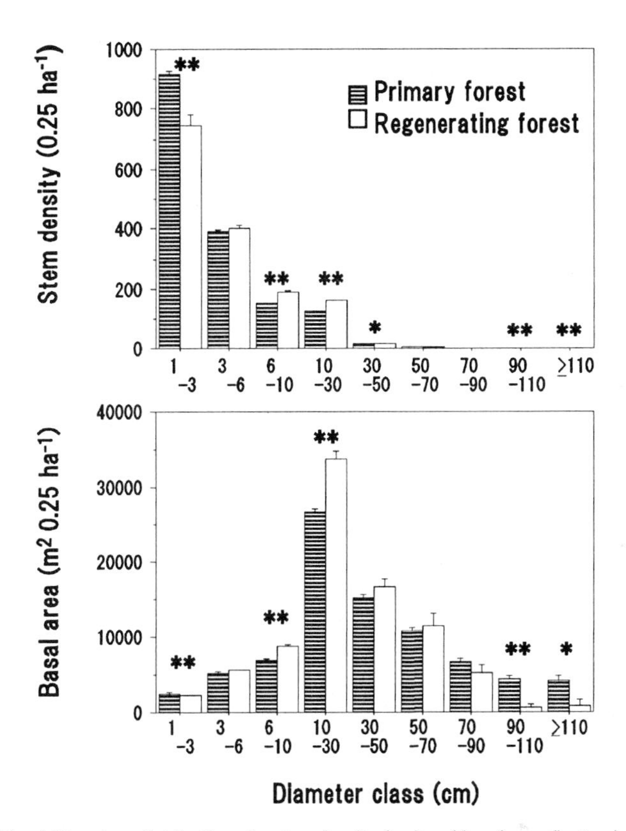

Fig. 4 Size-class distributions for stem density (top) and basal area (bottom). Both values were obtained in each sub-plot (50 m × 50 m) and were averaged over both forest plots. Statistical differences are shown by asterisks: ** = $P < 0.001$, * = $P < 0.05$. The stem density in the 30–50 cm size class was significantly higher ($P < 0.05$) in the regenerating forest, although the difference appears indistinct in the bar graph. Similarly, the stem densities in the 90–110 cm and > 110 cm size classes were significantly higher ($P < 0.001$) in the primary forest, although the difference is also not visible on the bar graphs.

after logging was clustered in the semi-medium and medium-sized classes rather than in the largest or smallest (sapling) classes (Fig. 4). These clustered semi-medium and medium-sized trees consisted mainly of canopy or emergent applicants that had originally been remnant cohorts.

4.4 Canopy surface structure

The canopy surface structure differed distinctly between the two forest plots. The mean canopy height and mean height of the tallest canopy in the primary forest were significantly higher than those in the regenerating forest (Table 3). In addition, the variance of canopy height in the primary forest was more than double the value in the regenerating forest (data was not shown in the table), and the canopy surface area in the primary forest was ca. 1.5 times the value in the regenerating forest (Table 3). These findings imply that the primary forest had higher heterogeneity and exhibited more complexity in terms of canopy height.

Table 3 Comparison of structural aspects of the canopy surface between the primary and regenerating forests in the Pasoh FR.

	Primary forest	Regenerating forest	*df*	*F*	*P*
Mean canopy height (m)	27.4	24.8	190,400	651.9	< 0.0001
Mean height (m) of tallest canopy within a subplot (50 m × 50 m)	46.5	41.1	1, 222	29.13	< 0.0001
Mean crown size of canopy trees (m^2)	94.5	42.9	14,805	371.85	< 0.0001
Mean canopy surface area (m^2 per 2.5 m × 2.5 m subplot)*	17.4	12.1	189,598	1723.13	< 0.0001
Mean Canopy Surface area (m^2/ha)	27844.0	19288.0			

* The canopy surface area was obtained in each perpendicularly mapped subplot (2.5 m × 2.5 m) and averaged among these subplots in each forest; these values were then expressed on a per-hectare basis.

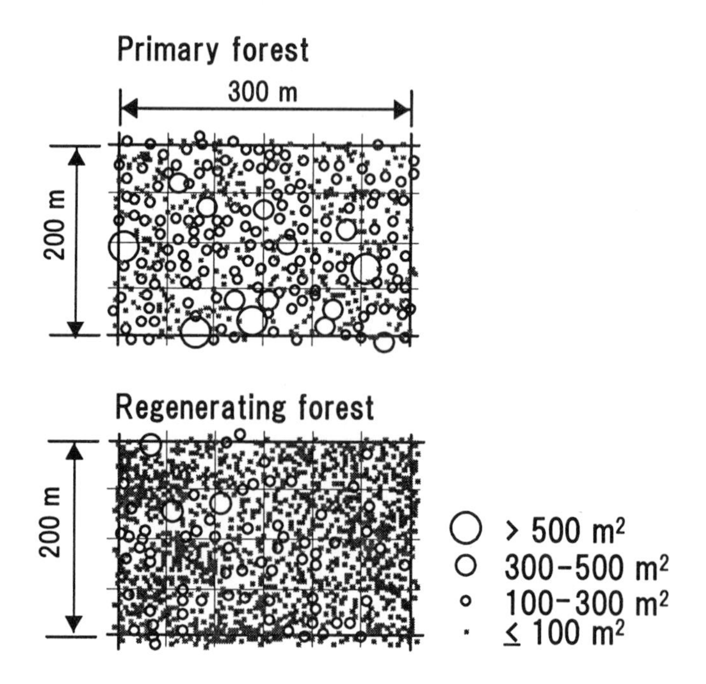

Fig. 5 Crown distribution for visible canopy trees identified from aerial photographs of the primary and regenerating forest plots. The size of each circle represents the average diameter of individual crowns.The upper graph shows crown distribution in a part of primary forest plot.

The crown sizes of individual canopy trees were also significantly larger in the primary forest ($F_{1,4805} = 371.85, P < 0.0001$) (Table 3). As was seen for canopy height, the variance of crown size in the primary forest was much greater than in the regenerating forest (data was not shown in the table). The number of trees with a crown size ≧ 300 m^2 (about 20 m in diameter) averaged 2.6 ha^{-1} in the primary forest versus 0.5 ha^{-1} in the regenerating forest. In contrast, the number of trees with smaller crowns (≦ 100 m^2, about 11 m in diameter) averaged 49.5 ha^{-1} in the primary forest versus 177.5 ha^{-1} in the regenerating forest. Overviews of the crown-size distribution in a southwestern part of the 50-ha primary forest plot and in the regenerating forest plot are shown in Fig. 5. As mentioned above, the canopy layer in the regenerating forest was characterized by a high density of trees with relatively small crowns, whereas the primary forest's canopy included more trees with relatively large crowns (mostly emergent tree species). Some short trees with small crowns were not visible in the primary forest, and their outlines could not be identified from the aerial photographs because taller canopy trees with large crowns concealed them. These "invisible" trees were not considered in the crown mapping. Thus, the density of canopy trees identified in the aerial photography did not necessarily represent the actual density of stems in the emergent, canopy, and sub-canopy layers; this number could instead be derived from tree census data. Nevertheless, the canopy surface layer, which represents the uppermost layer of the forest, was densely packed with many trees with relatively small crowns in the regenerating forest, whereas the canopy of the primary forest was characterized by unevenness in crown size and high convexity (i.e. many emergent crowns).

4.5 Total aboveground biomass (TAGB)

TAGB in the primary forest was 310 Mg ha^{-1}, and was significantly higher ($F_{1,1398} = 5.05, P = 0.025$) than the value in the regenerating forest (276 Mg ha^{-1}) when the two plots were subdivided on a 20 m × 20 m grid and the average values in these subplots were compared. These biomass figures were about 10% smaller than those estimated by using eq. 3, which was derived by using destructive sampling (Kato et al. 1978).

Since TAGB depended strongly on the density of large trees (e.g. those with DBH > 100 cm) in a given area, and since this density varied even within the primary forest, the presence of even one such large tree greatly increased the value of TAGB per hectare. In fact, the density of such trees was higher in the hilly parts of the primary forest plot than in the parts with flat, alluvial topography (Okuda et al. in press[b]). For this reason, we chose a 6-ha area at the center of the 50-ha primary forest plot (the "core area") in which the topographic features strongly resembled those in the regenerating forest plot. In this core area, the TAGB value was 365 Mg ha^{-1}. When we broke both plots down into 150 subplots (20 m × 20 m in size) and compared the mean TAGB per subplot, TAGB was significantly ($F_{1,1398} = 20.25, P < 0.0001$) greater in the core area than in the regenerating forest. In addition, we estimated the TAGB in another plot that had been established in the primary forest, where mapping and DBH measurements have been performed since the International Biological Program (IBP) began in the early 1970s (IBP plot, see Fig. 1). Within this plot (hereafter called the IBP plot), Kato et al. (1978) and Kira (1978) conducted destructive sampling to measure TAGB. In accordance with our DBH-H allometric relationship (eq. 1), TAGB in the IBP plot was 383 Mg ha^{-1}, and this was also significantly higher ($F_{1,1398} = 21.26, P < 0.0001$) than in the regenerating forest plot, but not significantly different from TAGB in the selected area of the primary forest plot (the 50 ha plot) ($F_{1,1398} = 0.44$, $P = 0.51$). These results suggest that TAGB was higher in the primary forest than in regenerating forest under similar topographic and edaphic conditions, and that TAGB in the regenerating forest had not yet fully recovered to the level seen in the primary forest.

In addition to the impacts of logging on TAGB, we observed that the distribution pattern for TAGB per unit area (e.g. ha^{-1}) was highly heterogeneous.

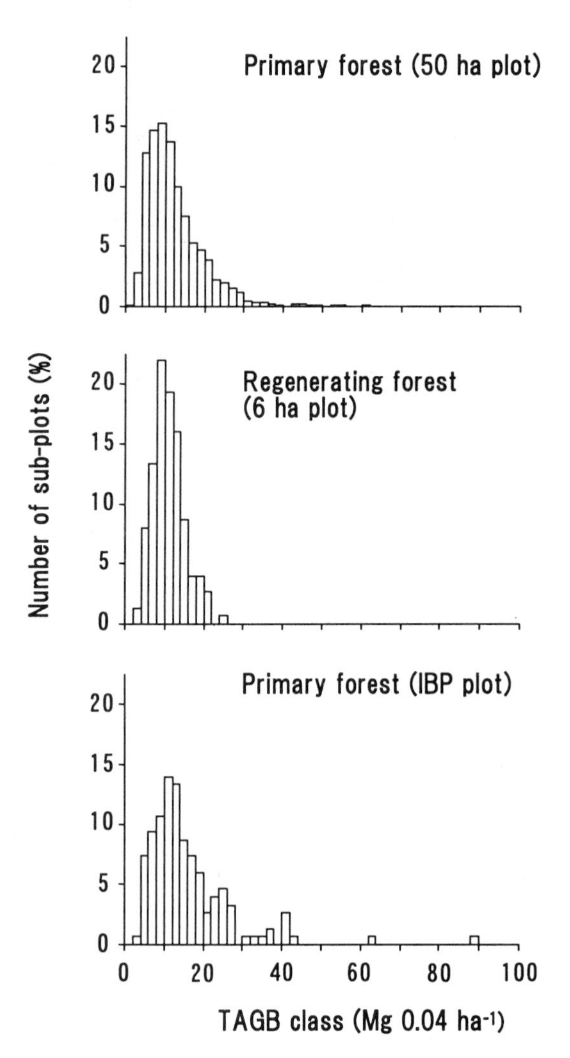

Fig. 6 Frequency distribution maps for total aboveground biomass (TAGB) in the primary and regenerating forests. The TAGB values were obtained in each subplot (20 m × 20 m in size), and the results were then classified into various TAGB classes. The Y-axis indicates the proportion of the subplots identified in each TAGB class in the primary forest (N = 1,250) and in the regenerating forest (N = 150).

The TAGB measured by destructive sampling in part of the IBP plot was 430 Mg ha^{-1}, but this value was derived from a very homogeneous stand with a high density of large trees in part of the primary forest. The TAGB ha^{-1} in the primary forest (the 50 ha plot) plot ranged from 216 to 402 Mg ha^{-1} (Fig. 6), whereas the value in the IBP plot (estimated from the tree census data in 1998 and the DBH-H allometry relationship) ranged from 341 to 390 Mg ha^{-1}. Yoneda et al. (1994) reported an almost 200% difference in TAGB between plots with the lowest and highest values in the Sumatran rain forest. In addition, TAGB in the Sarawak dipterocarp forest averaged 650 Mg ha^{-1} (Anderson et al. 1983; Proctor et al. 1983). These values demonstrate that carbon stocks (biomass) vary greatly among tropical rain forests, even though an ostensibly representative average value has generally been used for the evaluation of the carbon stock and sequestration potential in tropical forest ecosystems and of emission controls or carbon taxes. Nevertheless, the data suggest that the spatial heterogeneity of TAGB in the primary forest might have been reduced by the MUS logging regime as a result of the reduced structural heterogeneity of the forest after the selective removal of large trees (Fig. 6).

5. IMPLICATIONS OF THE OBSERVED LOGGING IMPACTS FROM APPLICATION OF THE MUS

The basal area and stem density of medium-sized (10–30 cm in DBH) dipterocarp trees was higher in the regenerating forest than in the primary forest. Although not all dipterocarp species followed this trend, some common dipterocarp species (e.g. *Shorea acuminata*, *S. dasyphylla*, *S. leprosula*, *S. lepidota*, *S. maxwelliana* and *S. parvifolia*) maintained or even increased their stem densities in the regenerating forest. In this sense, the original aim of the MUS approach was partially achieved (Manokaran 1998). However, this is not the case if maturity of the forest and structural aspects of the canopy are considered. In the MUS approach, the seedlings and saplings of non-commercial tree species should be repeatedly removed to provide open space for the seedlings and saplings of commercial timber species. If these operations had been properly implemented, the stem density of saplings of many dipterocarp species would have been higher in the regenerating forest than in the primary forest. However, this was not the case; the density of saplings and small trees (< 6 cm in DBH) of *Dipterocarpus cornutus*, *Neobalanocarpus heimii*, *Shorea acuminata*, *S. bracteolata*, *S. guiso*, *S. leprosula*, *S. lepidota*, *S. ovalis*, *S. parvifolia* and *S. pauciflora*—most of which are categorized as commercial species—were higher in the primary forest than in the regenerating forest. The indistinct increases and distinct decreases in the abundance of saplings in the regenerating forest imply that the MUS approach was not performed adequately.

Additional evidence suggests that the MUS strategy failed to accomplish some of its goals in the Pasoh FR. As mentioned earlier, regular thinning should have been conducted repeatedly after logging (Wyatt-Smith 1963), but such post-logging operations were never practiced (Manokaran 1998). Consequently, the failure of this practice resulted in higher densities of medium-sized trees (10–30 cm in DBH) in the regenerating forest than in the primary forest, and the large trees that form the emergent layers were apparently absent in the regenerating forest. We presume that the high density of such medium-sized trees and the high density of trees with relatively small crowns in the canopy layer resulted from a failure to regularly thin the forest, which in turn prevented the normal development of TAGB that was observed in the primary forest.

6. CONCLUSIONS AND FURTHER REMARKS CONCERNING SUSTAINABLE FOREST MANAGEMENT

The results of our study and others demonstrate that the MUS was incompletely practiced in the Pasoh FR and its vicinity. This outcome arose mainly from changes in the landscape management policy throughout the country. For example, the lowland dipterocarp forests in which timber production could have been adequately undertaken under this logging regime had been mostly converted into oil or rubber plantations. As a result, the MUS could no longer be performed, because no remaining forest was suitable for management by this approach, and selective logging shifted from lowland forests to hill forests, where the SMS was introduced. Consequently, the regenerating part of the Pasoh FR is a relict patch that was once logged under the MUS regime but that did not subsequently undergo the prescribed form of management (thinning to promote the development of commercial timber species). The regenerating forest exhibited a somewhat different species composition from the primary forest (e.g. based on the proportion of total stem density accounted for by the Dipterocarpaceae); this in turn implies that the forest showed only some of the characteristics that the MUS had originally aimed to produce. However, it was also apparent that the structural development of the regenerating forest had been retarded because of the failure to perform subsequent management after the initial logging.

This retarded development of the canopy and forest structure appeared to influence the heterogeneity of the forest microenvironment, of wildlife distribution, and of wildlife habitat. As shown in the other chapters of this book (Chaps. 31,36, 37, 38), the species compositions of insects (understory butterflies and soil microarthropods), of medium-sized and small mammals, and of birds were found to differ dramatically between the primary and regenerating forests (Fukuyama et al. 1998; Miura & Ratnam 1998; Nagata et al. 1998; Yasuda et al. 1998). The wild boar (*Sus scrofa*) and pig-tail monkeys (*Macaca nemestrina*) were observed more often in the regenerating forest than in the remaining primary forest, whereas dusky leaf monkeys (*Presbytis obscura*), which spend most of their time in the canopy layer, were less frequently observed in the regenerating forest. Although some differences in the animal species compositions may have arisen from certain animals being attracted by the persistent food resources of nutrient-rich seeds in the oil palm plantations adjoining the regenerating forest (see Chap. 35), we cannot deny the possibly that the observed changes in canopy structure and the lack of an emergent layer were responsible for these changes. In addition to these effects on the forest's fauna, the microclimate (i.e. light intensity) was reported to differ distinctly between the regenerating and primary forests (S. Numata unpubl. data, NIES, Japan). Changes in the structure of the canopy layer after logging will change the light spectrum on the forest floor, and this, in turn, will eventually affect plant growth and the competitive relationships between saplings and seedlings (Lee et al. 1997).

If ecologically sustainable management (Whitman et al. 1997) aims for better use of timber and other natural resources, special care is needed to maintain the functional aspects of the forest, and particularly the forest's structure. It may be necessary to thin the forest by artificially removing canopy trees stuck in a state that shows no signs of evolving towards maturity, while minimizing damage to the residual trees, in order to recreate the complexity of the canopy structure and the heterogeneity of the forest-floor light environment that exist in the primary forest. These conditions would, in turn, promote highly diverse regeneration, including regeneration by gap (pioneer) species (e.g. Denslow 1980; Denslow et al. 1990).

Nevertheless, the logging impacts seen in the regenerating forest after implementation of the MUS regime appear to have been much lower than those seen under the current logging system (the SMS regime). Large amounts of tree regeneration and minimal damage to the forest floor (undergrowth and soils) were observed in the regenerating forest in the Pasoh FR. The properties of the residual trees at the Pasoh FR are, however, unlikely to be preserved after the damage caused by intensive logging, excessive timber extraction, and soil disturbance from logging operations, which can affect as much as 80% of the timber resources and other ecological values of the forest (Manokaran 1998). In fact, the current logging regime (SMS) being used in the hill forests near the study area extracted more than 20 timber trees (excluding non-target trees) per hectare, which was much greater than the estimated number of trees extracted based on the number of stumps seen in the regenerating forest in the Pasoh FR. As for the logging cycle, the 70-year felling rotation suggested by the MUS approach proved to be insufficiently long to maximize timber production (Kurpick et al. 1997). Needless to say, the present logging cycle (25 to 30 years) used in the SMS is too short to permit a healthy stocking of regeneration (Manokaran, 1998). Furthermore, many of the timber and canopy-forming species reproduce only at intervals of 2 to 10 years in mass fruiting ("masting") events (e.g. Appanah 1985, 1993; Ashton et al. 1988). These phenological characteristics must be considered not only for maintaining the quantity and quality of the next crop, but also for maintaining the genetic diversity that maintains the adaptive potential of the offspring of these trees (Bawa 1991; also see Chap. 21).

ACKNOWLEDGEMENTS

We thank Messrs. Eng Seng Quah, Abd. Wahab bin Nali, Ahmad bin Awang, Abd. Samad bin Latif, Chan Yee Chong, Mohd. Yunus bin Hitam, Sahrie bin Mohd. Som, and Zamri bin Ahmad, Forest Reserch Institute Malaysia (FRIM), for their assistance with the field surveys and collection of tree census data at the Pasoh FR. The present study was a part of a joint research project between the NIES, FRIM and UPM (Grant No. E-1 from the Global Environment Research Program, Ministry of the Environment, Japan). The 50-ha forest plot in the Pasoh FR is an ongoing project initiated by the FRIM. Supplementary funding was provided by the National Science Foundation (USA); the Conservation, Food, and Health Foundation, Inc. (USA); the United Nations, through its Man and the Biosphere (MAB) program; UNESCO-MAB grants; UNESCO-ROSTSEA; and the continuing support of the Smithsonian Tropical Research Institute (Barro Colorado Island, Panama) and the Center for Global Environmental Research (CGER) at Japan's National Institute for Environmental Studies (NIES).

REFERENCES

Abdul Rahim, N. & Harding D. (1992) Effects of selective logging methods on water yield and stream flow parameters in Peninsular Malaysia. J. Trop. For. Sci. 5: 130-154.

Anderson, J. M., Proctor, J. & Vallack, H. W. (1983) Ecological studies in four contrasting lowland rain forests in Gunung Mulu National Park, Sarawak. III. Decomposition processes and nutrient losses from leaf litter. J. Ecol. 71: 503-527.

Appanah, S. (1985) General flowering in the climax forest of Southeast Asia. J. Trop. Ecol. 1: 225-240.

Appanah, S. (1993) Mass flowering of dipterocarp forests in the aseasonal tropics. J. Biosci. 18: 457-474.

Ashton, P. S., Givnish, T. J. & Appanah, S. (1988) Staggered flowering in the Dipterocarpaceae:

new insights into floral induction and the evolution of mast fruiting in the seasonal tropics. Am. Nat. 132: 44-66.

Ashton, P. S. & Hall, P. (1992) Comparisons of structure among mixed dipterocarp forests of north-western Borneo. J. Ecol. 80: 459-481.

Bawa, K. S. & Krugman, S. L. (1991) Reproductive biology and genetics of tropical trees in relation to conservation and management. In Gomez-Pompa, A., Whitmore T. C. & Hadley, M. (eds). Rain Forest Regeneration and Management. Parthenon Publ. Group, Park Ridge, NJ, pp.119-136.

Bornhan, M., Johari, B. & Quah, E. S. (1987) Studies on logging damage due to different methods and intensities of forest harvesting in a hill dipterocarp forest of Peninsular Malaysia. Malay. For. 50: 135-147.

Brokaw, N. V. L. (1982) Tree falls: frequency, timing and consequences. In Leigh, E. G Jr., Rand, A. S. & Windsor, D. W. (eds). The ecology of a tropical forest: seasonal rhythms and long term changes. Smithsonian Institution Press, Washington, DC., pp.101-108.

Brooks, S. M. & Spencer, T. (1997) Changing soil hydrology due to rain forest logging: An example from Sabah Malaysia. J. Environ. Manage. 3: 297-310.

Brown, N. D. & Whitmore, T. C. (1992) Do dipterocarp seedlings really partition tropical rain forest gaps? Philos. Trans. R. Soc. London Ser. B 335: 369-1992.

Burgess, P. F. (1968) An ecological study of the hill forests of the Malay Peninsula. Malay. For. 31: 314-325.

Burgess, P. F. (1971) The effect of logging in hill dipterocarp forests. Malay. Nat. J. 24: 231-237.

Cannon, C. H., Peart, D. R., Leighton, M. & Krtawinata, K. (1994) The structure of lowland rainforest after selective logging in West Kalimantan, Indonesia. J. For. Ecol. Manage. 67: 49-68.

Chapman, C. A. & Chapman L. J. (1997) Forest regeneration in logged and unlogged forests of Kibale National Park, Uganda. Biotropica 29: 296-412.

Chapman, C. A. & Onderdonk, D. A. (1998) Forests without primates: Primate/plant codependency. Am. J. Primatol. 45: 127-141.

Dawkins, H. C. (1959) The volume increment of natural tropical high-forest and limitations on its improvement. Empire For. Rev. 38: 175-180.

Denslow, J. S. (1980) Gap partitioning among tropical rain forest trees. Biotropica 12 (Suppl.): 47-55.

Denslow, J. S. (1987) Tropical rainforest gaps and tree species diversity. Annu. Rev. Ecol. 18: 431-451.

Denslow, J. S., Schultz, J. C., Vitousek, P. M. & Strain, B. (1990) Growth responses of tropical shrubs to tree fall gap environments. Ecology 71: 165-179.

Denslow, J. S. & Hartshorn, G. S. (1994) Tree-fall gap environments and forest dynamic processes. In McDade, L. A., Bawa, K. S., Hespenheide, H. A. & Hartshorn, G. S. (eds). La Selva: Ecology and natural history of a neotropical rain forest. University Chicago Press, Chicago, pp.120-127.

FAO (1993) Forest Resource Assessment 1990: Tropical Countries. U.N. Food and Agric. Org., Rome.

Fukuyama, K., Maeto, K., Kirton, L. G. & Sajap, A. S. (1998) Understory butterflies and soil micro-arthropoda as an indicator group. In Okuda, T. (ed). Research Report of the NIES/FRIM/UPM Joint Research Project 1998. National Research Institute for Environmental Studies, Tsukuba, Japan, pp.128-135.

Iwasa, K. (1997) The Rural Development and Small Farmers in Malaysia-An Examination of the FELDA Project. J. Agrarian Hist. 157: 1-16.

Johns, A. G. (1997) Timber production and biodiversity conservation in tropical rain forests, Cambridge University Press, Cambridge, 225pp.

Kasenene, J.M. (1984) The influence of selective logging on rodent populations and the regeneration of selected tree species in the Kibale forest, Uganda. Trop. Ecol. 25: 179-195.

Kato, R., Tadaki, Y. & Ogawa, H. (1978). Plant biomass and growth increment studies in Pasoh Forest. Malay. Nat. J. 30: 211-224.

Kira, T. (1978) Community architecture and organic matter dynamics in tropical lowland rain forests of Southeast Asia with special reference to Pasoh Forest, West Malaysia. In Tomlinson, P. B. & Zimmermann, M. H. (eds). Tropical trees as living systems. Cambridge University Press, New York, pp.561-590.

Knight, D. H. (1975) A phytosociological analysis of species-rich tropical forest on Barro Colorado Island, Panama. Ecol. Monogr. 45: 259-284.

Kurpick, P., Kurpick, U. & Huth, A. (1997) The influence of logging on a Malaysian dipterocarp rain forest: a study using a forest gap model. J. Theor. Biol. 185: 47-54.

Lang, G. E. & Knight, D. H. (1983) Tree growth, mortality, recruitment, and canopy gap formation during a 10-year period in a tropical moist forest. Ecology 64: 1075-1080.

Lee, S. S., Chan, H. T., Kirton, L. G, Lim, B. L., Ratnam, L., Saw, L. G & Francis, C. (1995) A guide book to Pasoh. FRIM Technical Information Handbook, No. 3. Kepong, Malaysia, 73pp.

Lee, D. W., Oberbauer, S. F., Krishnapilay, B., Mansor, M., Mohamad, H. & Yap, S. K. (1997) Effects of irradiance and spectral quality on seedling development of two Southeast Asian *Hopea* species. Oecologia 100: 1-9.

Manokaran, N. (1998) Effects, 34 years later, of selective logging in the lowland dipterocarp forest at Pasoh, Peninsular Malaysia, and implications on present day logging in the hill forests. In Lee, S. S., May, D. Y., Gauld, I. D. & Bishop, J. (eds.), Conservation, Management and Development of Forest Resources. Forest Research Institute Malaysia, Kepong, Malaysia, pp.41-60.

Manokaran, N., LaFrankie, J. V., Kochummen, K. M., Quah, E. S., Klahn, J. E., Ashton, P. S. & Hubbell, S. P. (1990) In Chan H. T. (ed). Methodology for the fifty hectare research plot at Pasoh Forest Reserve. Research Pamphlet (No. 104). Forest Research Institute Malaysia, Kepong, pp.1-69

Manokaran, N., LaFrankie, J. V., Kochummen, K. M., Quah, E. S., Klahn, J. E., Ashton, P. S. & Hubbell, S. P. (1999) The Pasoh 50-ha Forest Dynamic Plot: 1999 CD-ROM version. Forest Research Institute Malaysia, Kepong, Malaysia.

Miura, S. & Ratnam, L. C. (1998) Effect of deforestation, disturbance and fragmentation on the mammalian community in a tropical forest. In Okuda, T. (ed). Research Report of the NIES/FRIM/UPM Joint Research Project 1998. National Institute for Environmental Studies, Tsukuba, Japan, pp.98-99.

Nagata, H., Zubaid, A. M. A. & Azarae, H. I. (1998) Edge effects on the nest predation and avian community in Pasoh Nature Reserve. In Okuda, T. (ed). Research Report of the NIES/FRIM/UPM Joint Research Project 1998. National Institute for Environmental Studies, Tsukuba, Japan, pp.100-104.

Okuda, T., Suzuki, M., Adachi, N., Manokaran, N., Supardi, N. N. & Awang, M. (2000) Estimation of tree above-ground biomass in a lowland dipterocarp rainforest, by 3-d photogrammetric analysis. In Okuda, T. (ed). Research Report of the NIES/FRIM/ UPM Joint Research Project 2000. National Research Institute for Environmental Studies, Tsukuba, Japan, pp.17-28.

Okuda, T., Nor Azman, H., Manokaran, N., Saw, L. Q., Amir, H. M. S. & Ashton, P. S. Local variation of canopy structure in relation to soils and topography and the implications for species diversity in a rain forest of Peninsular Malaysia. In Losos, E. C. & Leigh, E. G. Jr. (eds). Forest diversity and dynamism: Findings from a network of large-scale tropical forest plots. University of Chicago Press, Chicago (in press[b]).

Okuda, T., Suzuki, M., Adachi, N., Quah, E. S., Nor Azman, H., Manokaran, N. Effect of selective logging on canopy and stand structure and tree species composition in a lowland dipterocarp forest in Peninsular Malaysia. For. Ecol. Manage. (in press[a]).

Oyebande, C. (1988) Effects of tropical forest on water yield. In: Reynolds, E. C. & Thompson, F. B. (eds). Forests, Climate, and Hydrology Regional Impacts. U.N. University, Tokyo. pp.16-50.

Pinard, M. A. & Putz, F. E. (1996) Retaining forest biomass by reducing logging damage. Biotropica 28: 278-295.

Proctor, J., Anderson, J. M., Fogden, S. C. & Vallack, H. W. (1983) Ecological studies in four

contrasting lowland rain forests in Gunung Mulu National Park, Sarawak. II. Litter fall, litter standing crop and preliminary observations on herbivory. J. Ecol. 71: 261-283.
Thang, H. C. (1987) Forest management systems for tropical high forest, with special reference to Peninsular Malaysia. For. Ecol. Manage. 21: 3-20.
Thang, H. C. (1997) Concept and basis of selective management system in Peninsular Malaysia. Proceedings of the Workshop on "Selective Management Systems and Enrichment Planting", Malaysia Forestry Department, Kuala Lumpur.
Whitman, A. A., Brokaw, N. V. L. & Hagan, J. M. (1997) Forest damage caused by selection logging of mahogany (*Swietenia macrophylla*) in northern Belize. For. Ecol. Manage. 92: 87-96.
Whitmore, T. C. (1975) Tropical Rain Forests of the Far East. Clarendon Press, Oxford, 498pp.
Wyatt-Smith, J. (1963) Malayan Uniform System. In Barnard, R. C. (ed). Manual of Malayan Silviculture for Inland Lowland Forest, vol. I, Part III-4, FRIM, Kepong, Malaysia, pp.1-14.
Yasuda, M., Ishii, N. & Nor Azman, H. (1998) Distribution of small mammals and their habitat preference in the Pasoh Forest Reserve. In Okuda, T. (ed). Research Report of the NIES/FRIM/UPM Joint Research Project 1998. National Institute for Environmental Studies, Tsukuba, Japan, pp.105-127.
Yoneda, T., Ogino, K., Kohyama, T., Tamin, R., Syahbuddin & Rahman, M. (1994) Horizontal variance of stand structure and productivity in a tropical foothill rain forest, West Sumatra, Indonesia. Tropics 4: 17-33.

3 The Trees of Pasoh Forest: Stand Structure and Floristic Composition of the 50-ha Forest Research Plot

Stuart J. Davies[1], Nur Supardi Md. Noor[2], James V. LaFrankie[3] & Peter S. Ashton[1]

Abstract: Stand structure and floristic composition of the 50-ha Forest Research Plot in Pasoh Forest Reserve (Pasoh FR) are described. Pasoh FR was found to have extremely high tree species diversity. In the 50-ha plot there were 338,924 trees with a total basal area of 1,659 m^2, comprising 818 tree species in 295 genera and 81 families. The Euphorbiaceae (85 species) was the most species-rich family. The Dipterocarpaceae dominated the forest with 27% of the basal area, and eight of the top 10 basal-area contributing species. The Fabaceae, Euphorbiaceae and Burseraceae were the next most important large trees in the plot. *Shorea* had the highest basal area and stem number. *Syzygium* was the richest genus with 45 species. As with other Asian tropical forests there were many speciose genera in the plot; 11 genera had ≧ 12 species. Floristic composition and stand structure varied across the 50-ha plot in relation to edaphic and topographic variation. Multivariate analyses revealed three main community types: a swamp community in the lowest areas of the plot, a hill community, and an alluvial forest community in mid to lower elevations in the plot. In addition, the alluvial community appears to be divisible into three types based on differences in soil properties. In addition to describing species characteristic of each of these community types, we also describe some of the distinctive life-history and evolutionary characteristics of the Pasoh FR. This chapter provides a basis for future work at Pasoh FR.

Key words: ballistic dispersal, Dipterocarpaceae, forest structure, pioneer trees, species richness, tree density, tree species diversity.

1. INTRODUCTION

The 50-ha forest research plot was established in Pasoh Forest Reserve (Pasoh FR), Negeri Sembilan (2°58' N, 102°18' E) between 1985 and 1988. Based on these data, the first detailed descriptions of the forest botany and stand structure were published in 1990 (Kochummen et al. 1990; Manokaran & LaFrankie 1990). In this chapter we provide a overview of the floristic composition of the Pasoh FR based on the 1995 recensus. Although the overall structure and composition of the forest have not changed much since the original census, this paper provides an updated description of Pasoh FR, and is designed to provide a background for many of the other ecological papers presented in this compilation. In addition, we present some new analyses on the floristic and structural variation in the 50-ha plot. We also

[1] Center for Tropical Forest Science-Arnold Arboretum Asia Program, Harvard University, Cambridge, MA 02138, USA.
[2] Forest Research Institute Malaysia (FRIM), Malaysia.
[3] Center for Tropical Forest Science-Arnold Arboretum Asia Program, National Institute for Education, Singapore.

highlight some of the interesting floristic and structural trends in the 50-ha plot and how they are related to edaphic and topographic variation in the plot. This description is designed to serve as an impetus for more detailed studies of variation in floristic and structural characteristics of the Pasoh FR forest in relation to environmental variation.

2. METHODS AND SITE DESCRIPTION

A detailed description of Pasoh FR was given in Chaps. 1, 2 and 3. Following a similar project in Barro Colorado Island (BCI), Panama (Hubbell & Foster 1983), the Pasoh study encompasses a large forest area (50-ha) to facilitate demographic analyses of a large proportion of the tree species, and to enable analyses of the importance of spatial heterogeneity of resource availability for tree species and forest community distribution patterns (Ashton et al. 1999; Condit 1995).

The 50 ha plot is located at c. 100 m above sea level on a relatively flat plain of Pleistocene alluvium, interspersed with a series of low hills of Triassic sediments and granite. The plot sits between two small meandering streams. Elevation within the plot has a maximum range of only 25.5 m, with the lowest areas in the swampy ground in the northwest of the plot, and the high area on the hill in the central east of the plot (Fig. 1). The soils are derived shale or granite and vary from to clay-rich to quite sandy across the plot (Kochummen et al. 1990; Manokaran et al. in press; Okuda et al. in press; Wong & Whitmore 1970).

The methods for the establishment of the 50-ha plot followed that for BCI (see Condit 1998; Manokaran et al. 1990). For detail description of tree census, please refer to Chap. 2 and other references (Manokaran et al. 1999). In 1990 and 1995, recensuses of the 50-ha plot were undertaken in which all extant trees were remeasured and their identifications confirmed. Newly recruited trees ($\geqq$ 1 cm DBH, Diameter at Breast Height) were mapped, tagged and identified as for all trees in 1985. All tallies presented here for the forest are as of the third census of the Pasoh FR plot in 1995. The data set used for the present analysis of stem abundance includes coppices which emerged after the main stem was damaged. DBH of these coppices was not measured during the tree census so they are not included in estimates of basal area. The results of analyses shown in this chapter were therefore slightly different from those shown in Chap. 2 in which the same data set was used but the coppices were all excluded from the analysis. A census of the plot was recently completed in 2001, but these data are not yet ready for analysis.

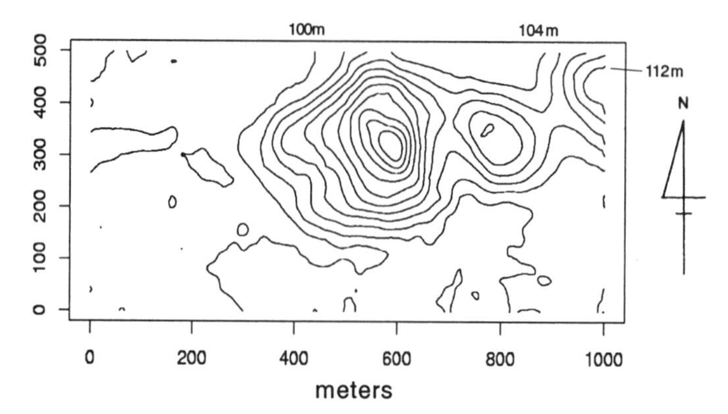

Fig. 1 Topographic map of the Pasoh 50-ha plot. Contour intervals are 2 m apart.

3. FLORISTIC COMPOSITION AND STAND STRUCTURE

The 50-ha plot included 338,360 trees ≧ 1 cm DBH (mean = 6767 trees/ha), comprising 81 families, 295 genera, and 818 species. Total basal area for all trees in the 50-ha plot was 1658.73 m^2 (mean = 33.17 m^2 ha^{-1}). Almost 80% of the trees were < 5 cm DBH, and only 721 trees (~26 trees/ha) were > 60 cm DBH (Table 1). For further details of stand structure in the plot, see Chap. 2.

Table 1 Size class distribution for all trees ≧ 1 cm DBH in the 50-ha plot at Pasoh. Number and percent of individual trees in increasing size classes given for all trees, and mean density per hectare given for individual hectares. Diameter size classes given as ≧ 1–< 2 cm DBH, ≧ 2–< 5 cm DBH, etc. Trees < 1 cm DBH were previously ≧ 1 cm but have since been damaged in some way.

Size class (cm DBH)	Tree number	%	Mean tree number/ha	(±*SD*)
<1	564	0.17		
1-2	139,784	41.24	2,807	(635.3)
2-5	126,884	37.44	2,537.7	(347.7)
5-10	43,298	12.78	866	(97.4)
10-20	19,514	5.76	390.3	(36.4)
20-30	4,933	1.46	98.7	(11.6)
30-60	3,226	0.95	64.5	(8.1)
> 60	721	0.21	14.4	(4.3)
All trees	338,924			

Table 2 Family composition of the 50-ha plot after the 1995 recensus. The 20 most important families are listed for total basal area (m^2), stem number and species number. Percentage contributions are in parentheses. Some values in stem density and basal area in this table are different from those shown in Chap. 2 (Table 2), since the data set used for the statistics in this chapter included coppices (new sprouts) emerged from broken stems which were all excluded from the analyses in the Chap. 2. See for detail in the text.

Family	Basal area (m^2)	Family	Stem No.	Family	Species No.
Dipterocarpaceae	453.21 (27.3)	Euphorbiaceae	45,436 (13.4)	Euphorbiaceae	85 (10.4)
Fabaceae	141.47 (8.5)	Dipterocarpaceae	31,178 (9.2)	Lauraceae	49 (6.0)
Euphorbiaceae	120.46 (7.3)	Annonaceae	24,752 (7.3)	Myrtaceae	48 (5.9)
Burseraceae	100.91 (6.1)	Rubiaceae	20,506 (6.1)	Rubiaceae	47 (5.8)
Myrtaceae	56.96 (3.4)	Burseraceae	17,701 (5.2)	Annonaceae	42 (5.1)
Annonaceae	55.98 (3.4)	Sapindaceae	16,997 (5.0)	Meliaceae	42 (5.1)
Fagaceae	54.01 (3.3)	Myristicaceae	14,391 (4.3)	Clusiaceae	35 (4.3)
Anacardiaceae	46.97 (2.8)	Ebenaceae	14,377 (4.2)	Anacardiaceae	32 (3.9)
Myristicaceae	46.24 (2.8)	Myrsinaceae	11,767 (3.5)	Myristicaceae	31 (3.8)
Sapindaceae	44.42 (2.7)	Fabaceae *	11,335 (3.3)	Dipterocarpaceae	30 (3.7)
Rubiaceae	37.54 (2.3)	Clusiaceae	10,874 (3.2)	Fabaceae	28 (3.4)
Clusiaceae	37.44 (2.3)	Myrtaceae	10,389 (3.1)	Burseraceae	25 (3.1)
Sterculiaceae	33.26 (2.0)	Meliaceae	9,020 (2.7)	Moraceae	25 (3.1)
Ixonanthaceae	29.91 (1.8)	Violaceae	8,542 (2.5)	Ebenaceae	23 (2.8)
Moraceae	27.96 (1.7)	Anacardiaceae	7,518 (2.2)	Sapindaceae	20 (2.4)
Sapotaceae	27.23 (1.6)	Lauraceae	7,114 (2.1)	Fagaceae	15 (1.8)
Ebenaceae	24.05 (1.5)	Melastomataceae	6,875 (2.0)	Sapotaceae	15 (1.8)
Oxalidaceae	24.01 (1.5)	Flacourtiaceae	6,364 (1.9)	Melastomataceae	15 (1.8)
Polygalaceae	23.98 (1.5)	Ulmaceae	5,358 (1.6)	Sterculiaceae	14 (1.7)
Meliaceae	22.15 (1.3)	Fagaceae	5,137 (1.5)	Flacourtiaceae	13 (1.6)

*: Leguminosae

Table 3 Generic composition of the 50-ha plot after the 1995 recensus. The 20 most important genera are listed for total basal area (m^2), stem number and species number. Percentage contributions are in parentheses.

Genus	Family	Basal area (m^2)	Genus	Family	Stem No.	Genus	Family	Species No.
Shorea	Dipterocarpaceae	306.5 (18.5)	*Shorea*	Dipterocarpaceae	20,960 (6.2)	*Syzygium*	Myrtaceae	45 (5.5)
Dipterocarpus	Dipterocarpaceae	71.17 (4.3)	*Aporosa*	Euphorbiaceae	17,604 (5.2)	*Diospyros*	Ebenaceae	23 (2.8)
Syzygium	Myrtaceae	54.99 (3.3)	*Diospyros*	Ebenaceae	14,377 (4.2)	*Aglaia*	Meliaceae	22 (2.7)
Neobalanocarpus	Dipterocarpaceae	46.83 (2.8)	*Ardisia*	Myristicaceae	11,465 (3.4)	*Garcinia*	Clusiaceae	17 (2.1)
Dacryodes	Burseraceae	37.69 (2.3)	*Knema*	Myristicaceae	10,643 (3.1)	*Shorea*	Dipterocarpaceae	14 (1.7)
Koompassia	Fabaceae	35.41 (2.1)	*Dacryodes*	Burseraceae	10,496 (3.1)	*Litsea*	Lauraceae	14 (1.7)
Ixonanthes	Ixonanthaceae	29.91 (1.8)	*Syzygium*	Myrtaceae	9,911 (2.9)	*Mangifera*	Anacardiaceae	13 (1.6)
Quercus	Fagaceae	27.06 (1.6)	*Xerospermum*	Sapindaceae	9,318 (2.7)	*Aporosa*	Euphorbiaceae	13 (1.6)
Santiria	Burseraceae	26.92 (1.6)	*Rinorea*	Violaceae	8,542 (2.5)	*Knema*	Myristicaceae	13 (1.6)
Artocarpus	Moraceae	24.34 (1.5)	*Anaxagorea*	Annonaceae	7,919 (2.3)	*Memecylon*	Melastomataceae	12 (1.5)
Canarium	Burseraceae	24.25 (1.5)	*Memecylon*	Melastomataceae	6,075 (1.8)	*Ficus*	Moraceae	12 (1.5)
Diospyros	Ebenaceae	24.05 (1.4)	*Baccaurea*	Euphorbiaceae	5,575 (1.6)	*Polyalthia*	Annonaceae	11 (1.3)
Sarcotheca	Oxalidaceae	24.01 (1.4)	*Gironniera*	Ulmaceae	5,355 (1.6)	*Baccaurea*	Euphorbiaceae	11 (1.3)
Xanthophyllum	Polygalaceae	23.98 (1.4)	*Aglaia*	Meliaceae	4,879 (1.4)	*Horsfieldia*	Myristicaceae	11 (1.3)
Cynometra	Fabaceae	23.22 (1.4)	*Xanthophyllum*	Polygalaceae	4,738 (1.4)	*Artocarpus*	Moraceae	10 (1.2)
Aporosa	Euphorbiaceae	20.70 (1.2)	*Lepisanthes*	Sapindaceae	4,693 (1.4)	*Xanthophyllum*	Polygalaceae	10 (1.2)
Knema	Myristicaceae	20.46 (1.2)	*Garcinia*	Clusiaceae	4,171 (1.2)	*Dacryodes*	Burseraceae	9 (1.1)
Xerospermum	Sapindaceae	19.75 (1.2)	*Barringtonia*	Lecythidaceae	4,120 (1.2)	*Canarium*	Burseraceae	8 (1.0)
Gironniera	Ulmaceae	19.13 (1.2)	*Canarium*	Burseraceae	4,086 (1.2)	*Calophyllum*	Clusiaceae	8 (1.0)
Scaphium	Sterculiaceae	18.70 (1.1)	*Neobalanocarpus*	Dipterocarpaceae	3,523 (1.0)	*Drypetes*	Euphorbiaceae	8 (1.0)

The 30 species of Dipterocarpaceae dominated the stand volume, accounting for 27.3% of the total basal area (Table 2). The Euphorbiaceae with 85 species was the richest family in the plot, had the highest number of stems in the plot with 13.4% of total stem numbers, and contributed third most to total basal area (7.3%). The Lauraceae, Myrtaceae and Rubiaceae were also exceptionally species-rich with > 45 species, and 15 families had ≧ 20 species in the plot. The Fabaceae and Burseraceae were also among the most important families in basal area contribution (6–9%), although the Fabaceae were far less abundant than the Burseraceae.

Shorea was the most important genus in the 50-ha plot in terms of tree number (20,960 trees, 6.2% of all trees), and basal area (306.5 m^2, 19% of total basal area) (Table 3). *Shorea* was the fifth most diverse genus in the plot with 14 species (1.7% of all species). *Syzygium* was by far the richest genus with 45 species. *Diospyros*, *Aglaia* and *Garcinia* were also exceptionally species-rich with > 20 species, and 11 genera had ≧ 12 species. For basal area contribution, two other dipterocarp genera, *Dipterocarpus* (4.3%) and *Neobalanocarpus* (2.8%), were the second and fourth most important genera. Several genera of small to subcanopy trees had substantial numbers of trees in the forest (e.g. *Aporosa*, *Diospyros,* and *Ardisia*).

The 20 most abundant species and those contributing most to total basal area are listed in Table 4. The seven species with the largest contribution to basal area were all dipterocarps, as were eight of the top 10 basal area contributors. The light-red meranti, *Shorea leprosula*, and the balau, *Shorea maxwelliana*, were the largest contributors to basal area with 3.5% and 3.3% of total basal area respectively. In tree numbers, individual dipterocarp species were less dominant (Table 4), with several very common canopy and understory shade tolerant, non-dipterocarps, the more important of which were *Xerospermum noronhianum*, *Rinorea anguifera*, *Anaxagorea javanica* and *Ardisia crassa*.

Table 4 Total basal area and tree numbers for the 20 most important species (≧ 1 cm DBH) in the Pasoh 50-ha plot. For basal area comparisons the timber grouping and taxonomic section (for species of *Shorea*) are listed. Percentage contributions are in parentheses.

Basal Area

Species	Family	Basal area (m^2)	Timber Group	*Shorea* section
Shorea leprosula Miq.	Dipterocarpaceae	57.24 (3.5)	Light Red Meranti	Mutica
Shorea maxwelliana King	Dipterocarpaceae	54.44 (3.3)	Balau	Barbata
Neobalanocarpus heimii (King) Ashton	Dipterocarpaceae	46.83 (2.8)	Balau	-
Shorea pauciflora King	Dipterocarpaceae	42.14 (2.5)	Dark Red Meranti	Brachypterae
Shorea acuminata Dyer	Dipterocarpaceae	38.50 (2.3)	Light Red Meranti	Mutica
Dipterocarpus cornutus Dyer	Dipterocarpaceae	38.49 (2.3)	Keruing	-
Shorea lepidota (Korth.) Bl.	Dipterocarpaceae	37.59 (2.3)	Light Red Meranti	Mutica
Koompassia malaccensis Maing. ex Benth.	Fabaceae	35.41 (2.1)		
Ixonanthes icosandra Jack	Ixonanthaceae	29.90 (1.8)		
Shorea parvifolia Dyer	Dipterocarpaceae	25.03 (1.5)	Light Red Meranti	Mutica
Quercus argentata Korth.	Fagaceae	23.96 (1.4)		
Cynometra malaccensis Meeuwen	Fabaceae	23.22 (1.4)		
Xerospermum noronhianum Bl.	Sapindaceae	19.35 (1.2)		
Dacryodes rugosa (Bl.) Lam	Burseraceae	17.57 (1.1)		
Dipterocarpus costulatus V. Sl.	Dipterocarpaceae	17.39 (1.0)	Keruing	-
Sarcotheca griffithii (Planch. ex Hk. f.) Hall f.	Oxalidaceae	17.38 (1.0)		
Millettia atropurpurea (Wall.) Benth.	Fabaceae	17.32 (1.0)		
Scaphium macropodum (Miq.) Beumee ex Heyne	Sterculiaceae	15.87 (1.0)		
Shorea macroptera Dyer	Dipterocarpaceae	15.82 (1.0)	Light Red Meranti	Mutica
Pimelodendron griffithianum (M.A.) Benth.	Euphorbiaceae	14.95 (0.9)		

Stem Number

Species	Family	Stem No.
Xerospermum noronhianum Bl.	Sapindaceae	9,282 (2.7)
Rinorea anguifera (Lour.) O.K.	Violaceae	8,260 (2.4)
Anaxagorea javanica Hk. f. & Thoms.	Annonaceae	7,919 (2.3)
Ardisia crassa C.B. Clarke	Myrsinaceae	7,362 (2.2)
Aporosa microstachya (Tul.) M.A.	Euphorbiaceae	6,810 (2.0)
Shorea maxwelliana King	Dipterocarpaceae	6,556 (1.9)
Dacryodes rugosa (Bl.) Lam.	Burseraceae	5,579 (1.6)
Knema laurina (Bl.) Warb.	Myristicaceae	4,480 (1.3)
Gironniera parvifolia Planch.	Ulmaceae	3,933 (1.2)
Barringtonia macrostachya (Jack) Kurz	Lecythidaceae	3,873 (1.1)
Neobalanocarpus heimii (King) Ashton	Dipterocarpaceae	3,523 (1.0)
Baccaurea parviflora (M.A.) M.A.	Euphorbiaceae	3,502 (1.0)
Scaphocalyx spathacea Ridl.	Flacourtiaceae	3,395 (1.0)
Diospyros scortechinii Bakh.	Ebenaceae	3,297 (1.0)
Ixonanthes icosandra Jack	Ixonanthaceae	3,160 (0.9)
Alangium ebenaceum (Clarke) Harms	Alangiaceae	2,927 (0.9)
Pimelodendron griffithianum (M.A.) Benth.	Euphorbiaceae	2,638 (0.8)
Aidia wallichiana Tirveng.	Rubiaceae	2,636 (0.8)
Lepisanthes tetraphylla Radlk.	Sapindaceae	2,622 (0.8)
Phaeanthus ophthalmicus (Roxb. Ex Don) Sinclair	Annonaceae	2,592 (0.8)

4. COMMUNITY DIVERSITY IN THE 50-HA PLOT

Multivariate analyses found a strong gradient in floristic composition across the plot (Fig. 2). For this analysis the plot was divided into 200 × 0.25-ha subplots (50 m × 50 m) and a range of multivariate statistical techniques were used to investigate spatial patterns of floristic variation. Clustering and ordination analyses found the principal division of community types to be related to the topographic gradient in the 50-ha plot. The results presented here were based on hierarchical clustering of

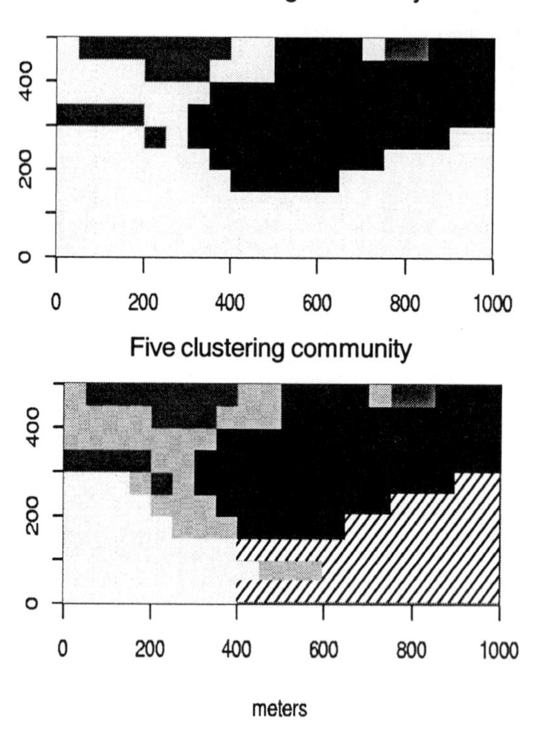

Fig. 2 Cluster analysis of species composition for 200 × 0.25-ha sub samples of the Pasoh 50-ha plot. Analyses used Ward's similarity index with three or five clusters. For the three cluster result (Three clustering community), the hill community is in black, the swamp in dark gray, and the alluvial community in light gray. For the five cluster result (Five clustering community), hill and swamp communities are the same, and the alluvial community is divided into the sandy alluvium in very light gray, the mixed alluvium in mid-gray, and the clay alluvium in hatched.

200 subplots (50 m × 50 m) using Ward's method following data standardization (Legendre & Legendre 1998). Analyses were performed using JMP version 4.0.4 (SAS Institute, Cary, NC, USA). Analyses based on other clustering algorithms produced qualitatively similar results. The three main community types found in the first three clusters were (shown as "Three clustering community" in Fig. 2): the swamp community in the lowest areas in the north and north-western corner of the plot, the hill community concentrated around the low ridge in the center and east of the plot, and the alluvial forest community which includes the remaining forest in mid to lower elevations in the plot.

The second main floristic division appears to be related to differences in soil properties (shown as "Five clustering community" in Fig. 2). The second division in the multivariate analyses divided the alluvial community into three types. Field observations suggest that the alluvium community varies distinctly between a sandy alluvium in the west of the plot, a clay alluvium in the east of the

Table 5 Species composition of the five community types as defined by the cluster analysis in Fig. 2. Relative density (percent of stems adjusted for community area) of the most important species in each community type is given. '%C1' is the relative species density of each species in community type 1- swamp, etc. '*N*' is the stem number in the 50-ha plot. In addition, the most important species in the combined Alluvium community of Types 3, 4 and 5 is presented.

SPECIES	*N*	%C1	%C2	%C3	%C4	%C5	%C1,4,5
COMMUNITY 1 - Swamp							
Chionanthus ramiflorus	67	97.5	0.0	0.0	2.5	0.0	
Diospyros cauliflora	315	93.5	1.0	0.2	3.8	1.5	
Diospyros andamanica	402	92.5	1.4	0.6	4.2	1.2	
Drypetes rhakodiskos	123	91.8	1.1	0.4	6.4	0.3	
Polyalthia lateriflora	128	91.4	1.1	0.0	6.6	1.0	
Saraca thaipingensis	243	90.9	0.9	0.0	7.7	0.5	
Glycosmis sapindoides	141	90.2	2.6	0.4	5.0	1.8	
Drypetes microphylla	60	89.8	0.5	2.6	5.1	2.1	
Iguanura wallichiana	395	87.1	1.0	0.4	6.5	4.9	
Mussaendopsis beccariana	55	84.8	0.6	1.1	0.0	13.6	
COMMUNITY 2 - Hill							
Cleistanthus myrianthus	98	0.0	100.0	0.0	0.0	0.0	
Pentace strychnoidea	599	0.0	98.8	0.3	0.4	0.5	
Chisocheton patens	99	0.0	92.7	0.0	4.4	2.9	
Elateriospermum tapos	223	0.0	89.3	0.0	9.4	1.3	
Anisophyllea corneri	714	0.5	85.3	4.1	6.9	3.1	
Cleistanthus malaccensis	59	0.0	84.0	3.0	3.5	9.5	
Shorea hopeifolia	52	0.0	81.0	3.3	7.8	7.9	
Teijsmanniodendron coriaceum	231	0.0	80.0	0.7	7.9	11.3	
Alseodaphne species 1	208	0.0	79.4	1.6	7.8	11.2	
Kayea kunstleri	340	7.3	76.3	0.0	6.3	10.1	
COMMUNITY 3- Sandy Alluvium							
Dipterocarpus crinitus	250	0.9	0.4	95.0	1.4	2.3	
Pavetta graciliflora	366	6.3	1.0	79.6	11.9	1.3	
Dracaena tetrastachys	539	6.0	3.7	76.4	8.8	5.1	
Pavetta species 1	681	3.9	5.6	73.6	14.7	2.3	
Syzygium duthieana	177	3.9	9.2	71.9	10.6	4.3	
Alchornea rugosa	93	4.5	2.2	68.2	23.4	1.7	
Syzygium species 10	24	0.0	18.3	68.1	5.8	7.8	
Harmandia mekongensis	47	0.0	2.5	66.9	2.8	27.8	
Vatica maingayi	99	8.9	9.9	66.5	14.7	0.0	
Diospyros rufa	132	5.5	14.4	65.4	6.0	8.8	
COMMUNITY 4 - Mixed Alluvium							
Neolamarckia cadamba	10	17.1	0.0	7.9	75.0	0.0	
Koilodepas longifolium	222	10.4	12.1	13.1	59.2	5.3	
Stelechocarpus cauliflorus	12	17.7	17.5	0.0	58.3	6.6	
Osmelia maingayi	10	34.6	8.5	0.0	56.9	0.0	
Ficus variegata	12	16.5	8.1	15.1	54.2	6.1	
Homalium caryophyllaceum	16	24.3	9.0	0.0	53.3	13.5	
Pouteria malaccensis	55	0.0	41.6	2.4	50.4	5.7	
COMMUNITY 5 - Clay Alluvium							
Euonymus javanicus	25	0.0	0.0	4.9	0.0	95.1	
Gordonia singaporiana	24	10.5	0.0	0.0	0.0	89.5	
Cleistanthus sumatranus	482	0.0	24.3	0.6	0.7	74.5	
Calophyllum depressinervosum	46	6.1	21.0	5.6	6.7	60.7	
Archidendron microcarpum	271	9.9	6.0	6.6	17.7	59.9	
Hopea dryobalanoides	971	13.2	3.8	0.9	22.7	59.4	
Rhodamnia cinerea	245	9.1	10.3	10.3	16.7	53.7	
Actinodaphne macrophylla	77	12.8	11.9	4.4	17.6	53.3	
Horsfieldia polyspherula	128	22.0	9.5	9.3	7.0	52.2	
Kokoona reflexa	233	3.4	14.4	15.5	15.5	51.2	
COMBINED COMMUNITIES 3, 4 and 5 - Alluvium							
Dipterocarpus crinitus	250	0.9	0.4	95.0	1.4	2.3	98.7
Rinorea sclerocarpa	244	1.7	0.0	62.2	35.3	0.9	98.3
Anisoptera megistocarpa	64	0.0	4.8	53.6	0.0	41.7	95.2
Alchornea rugosa	93	4.5	2.2	68.2	23.4	1.7	93.3
Pavetta graciliflora	366	6.3	1.0	79.6	11.9	1.3	92.8
Shorea maxwelliana	6556	5.4	2.7	49.3	24.8	17.8	91.9
Shorea ochrophloia	507	3.6	5.1	49.7	19.9	21.7	91.3
Pavetta species 1	681	3.9	5.6	73.6	14.7	2.3	90.6
Dracaena tetrastachys	539	6.0	3.7	76.4	8.8	5.1	90.3
Croton argyratus	1272	7.3	3.0	0.9	40.6	48.2	89.8
Syzygium duthieana	177	3.9	9.2	71.9	10.6	4.3	86.9
Hopea mengarawan	438	4.9	8.2	39.1	14.1	33.6	86.8
Drypetes kikir	232	7.6	5.9	51.1	20.3	15.1	86.5
Heritiera simplicifolia	453	3.1	11.3	39.5	33.2	13.0	85.7
Dipterocarpus costulatus	672	3.8	11.4	30.2	10.8	43.9	84.9

plot, and a intermediate or slightly wetter community in the northwest of the plot which we term here mixed alluvium. The mixed alluvium includes parts of the small temporary stream that runs through the plot. Since our sample plot sizes for analysis were 0.25 ha the analysis did not define this community very clearly.

Although it is clear that there is considerable overlap in species distributions among these five community types, especially for the alluvial communities, there are species with strong associations to each of the community types. Species characteristic of the five communities are shown in Table 5. The swamp and hill communities are characterized by a considerable number of habitat specialists. For these species the vast majority of their individuals are restricted to these habitats. The swamp community has a characteristic group of species, including *Saraca thaipingensis*, a species well known to associate with stream sides (Corner 1940; Wyatt-Smith 1964), *Diospyros andamanica*, and the palm *Iguanura wallichiana* whose distribution tightly fringes the stream sides. The hill community includes *Cleistanthus myrianthus*, *Pentace strychnoidea*, *Anisophyllea corneri*, and *Elateriospermum tapos* a species widely reported from well-drained ridges throughout Peninsular Malaysia (Ho et al. 1987).

Taking the alluvial community as a single community (combining communities 3, 4 and 5; Table 5) shows that many species are strongly associated with this community. Some characteristic species include *Shorea maxwelliana* which has previously been described for alluvial communities in Pasoh FR (Ashton 1976), *Dipterocarpus crinitus* and *Hopea mengarawan* which were mentioned for poorly drained land by Wyatt-Smith (1963), and *Pavetta graciliflora*. The three alluvial forest groups are not so strongly characterized by individual species, however each sub-community has a group of species disproportionately represented there (Table 5).

The extent to which these community types correspond to differences in soil properties as distinct from topography requires further study. In addition, it is important that studies verifying these descriptive findings incorporate the effects

Table 6 Comparison of stand structure and diversity for the five community types defined by the cluster analysis presented in Fig. 2. *N* is the number of 0.25 ha subplots. Basal area, stem number and species number are mean values (±1 *SE*) based on *N* 0.25-ha subplots in each community type. Different letters indicate significantly different means.

ALL TREES ≧1 cm DBH

Cluster	*N*	Basal Area (m^2/0.25 ha)	Stem No.	Species No.	Fisher's α
1 Swamp	17	7.68 (0.25) c	1689 (49.7) bc	337 (4.3) a	127.2 (2.52) a
2 Hill	69	8.89 (0.13) a	1588 (20.1) c	331 (2.5) a	127.6 (1.11) a
3 Sandy Alluvium	37	8.64 (0.23) ab	1913 (54.3) a	308 (2.9) b	105.1 (1.08) b
4 Mixed Alluvium	31	7.95 (0.19) bc	1751 (46.6) ab	335 (3.0) a	124.8 (2.28) a
5 Clay Alluvium	46	7.58 (0.13) c	1643 (48.4) bc	323 (2.6) a	122.4 (1.71) a

TREES >10 cm DBH

Cluster	*N*	Basal Area (m^2/0.25 ha)	Stem No.	Species No.	Fisher's α
1 Swamp	17	6.12 (0.26) bc	158.8 (4.0) a	96.9 (2.9) a	110.1 (7.43) a
2 Hill	69	7.48 (0.13) a	139.4 (1.5) bc	91.7 (0.9) ab	116.2 (2.31) a
3 Sandy Alluvium	37	6.98 (0.24) ab	144.7 (2.2) b	86.5 (1.3) c	93.6 (3.24) b
4 Mixed Alluvium	31	6.41 (0.20) bc	143.7 (2.4) bc	92.4 (1.7) ab	111.8 (4.00) a
5 Clay Alluvium	46	6.05 (0.14) c	136.3 (2.0) c	88 (1.3) bc	108 (3.34) a

of limited dispersal in assessing whether these species are actually specialists of their respective habitats.

5. SPATIAL VARIATION IN FOREST STRUCTURE AND SPECIES DIVERSITY

Stand structure and species diversity varied significantly across the 50-ha plot (Table 6, Fig. 3). Mean tree density per hectare was 6778 (*SD* = 1001, range: 4978 to 9353 trees), and mean basal area was 33.17 m^2/ha (*SD* = 2.9; range: 27.2–39.1 m^2/ha). For all stems, the mean number of species per hectare was 501 (*SD* = 23.7, range: 446–539 species/ha), and mean Fisher's α (Fisher et al. 1943) was 125.8 (range: 98–141). For trees ≧ 10 cm DBH, the mean number of species per hectare was 217.8 (range: 182–261), and the mean Fisher's α was 127.1 (range: 87–187). The single hectares with the highest species richness and Fisher's α values were in areas around the base of the hill community where it meets the alluvial community. Further analysis is required to assess whether this is due to the mixing of two sets of species, or whether there are a large number of species specialized to this particular microenvironment.

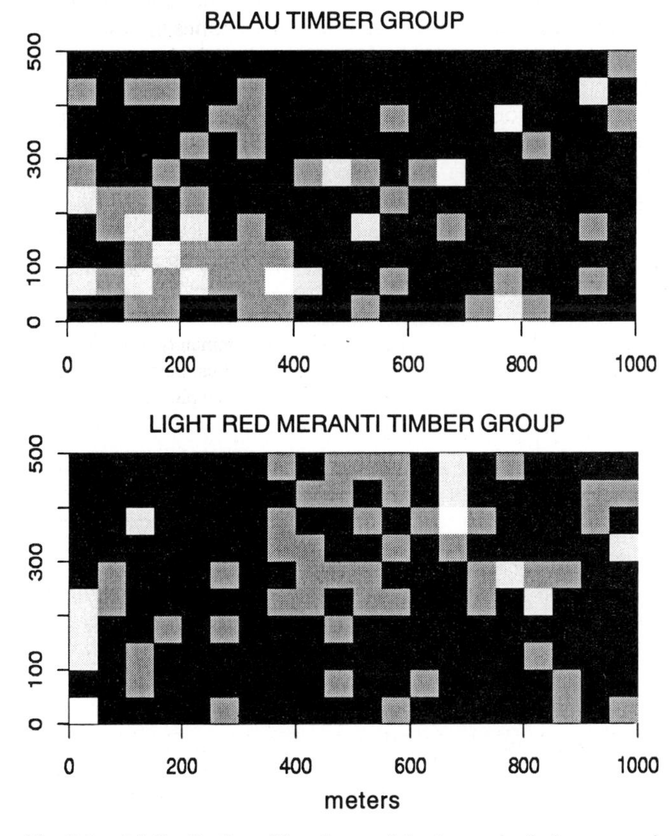

Fig. 3 Spatial distribution of basal area of the two main timber groups in the Pasoh 50-ha plot. Lighter colors indicate higher basal areas per 0.25 ha subplots.

To examine spatial variation in structure and diversity across the plot, we mapped density and diversity (Fisher's α) of 0.25-ha samples of the 50-ha plot and found variation associated with the five floristic communities described in the previous section. Total basal area and stem density differed significantly among the five identified community types in the 50-ha plot (Table 6). The hill community had among the highest basal areas and the lowest stem densities among the five communities. The swamp and sandy alluvium communities had relatively high stem densities, particularly of stems in the smaller size classes. This may reflect greater rates of recent disturbance in these two areas of the plot, where large numbers of trees have recruited over recent decades. It has previously been pointed out that disturbance rates by windfall seem to have increased in the Pasoh FR in recent decades (Manokaran & LaFrankie 1990). Whether the rates of disturbance are higher in the swampy areas of the plot is unknown, but current tree size distributions suggest that this may have been the case.

Species diversity also varied significantly among the five identified community types in the 50-ha plot. The sandy alluvium community had significantly fewer species and significantly lower Fisher's a than the swamp and hill communities for all trees and for trees >10 cm DBH (Table 6). The values of Fisher's α shown here are different from those of other studies conducted in the same plot (Manokaran et al. in press; Okuda et al. in press) due to the different sample areas used in this study. Nevertheless, the general trends of variation of species diversity in relation to edaphic factors shown in the results of the present study correspond well with recent analyses done by Okuda et al. (in press) on the tree species diversity in relation to topography and soils at Pasoh FR. The mixed alluvium and clay alluvium communities had diversity levels more similar to the swamp and hill communities.

There were also substantial differences in the floristic composition of the standing basal area across the plot. The light-red meranti timber group of *Shorea* had substantially higher basal areas on the higher topographic positions within the plot, whereas the heavy hardwood balau timber group, comprising principally *Shorea maxwelliana* and *Neobalanocarpus heimii*, had higher basal areas in the lower alluvial areas in the plot (Fig. 3). This result confirms the earlier finding of Ashton (1976) who found a similar pattern for smaller plots scattered through the landscape at Pasoh FR. The result could reflect significant differences in forest dynamics and rates of primary productivity between hill and alluvial forests (see also Ashton 1976). We expect that the hill forest dominated by the light-red meranti species of *Shorea* should have considerably higher rates of growth and productivity than the alluvial forest dominated by the heavy-hardwood balau species.

6. DISTINCTIVE LIFE HISTORIES OF PASOH TREE SPECIES

In this section we describe some of the more distinctive life-history and evolutionary characteristics of the forest that could form the basis of interesting future studies and for which the Pasoh FR 50-ha plot provides an ideal study site.

6.1 Pioneers

We estimate that there are 69 pioneer species in the Pasoh FR 50-ha plot. For this estimate we have taken a fairly broad view of pioneer species, and quite a few of the species are small-gap species or more typical of building-phase communities. These 69 species accounted for approximately 12,217 stems in the 1995 census which accounts for c. 3.6% of the trees in the plot. Table 7 provides a list of the pioneer species with more than 54 trees in the plot as of 1995. It is at once noticeable that

Table 7 List of the species typical of early successional to building-phase communities in the Pasoh 50-ha plot. Only species represented by at least 55 trees in the plot are listed.

Species	Family	Number of trees	Percentage of all trees
Croton argyratus	Euphorbiaceae	1,272	0.38
Epiprinus malayanus	Euphorbiaceae	1,057	0.31
Buchanania sessifolia	Anacardiaceae	973	0.29
Nauclea officinalis	Rubiaceae	733	0.22
Paropsia vareciformis	Flacourtiaceae	727	0.21
Parkia speciosa	Fabaceae	624	0.18
Artocarpus nitidus	Moraceae	537	0.16
Croton laevifolius	Euphorbiaceae	498	0.15
Pternandra coerulescens	Melastomataceae	458	0.14
Prunus grisea	Rosaceae	406	0.12
Dillenia sumatrana	Dilleniaceae	377	0.11
Carallia brachiata	Rhizophoraceae	340	0.10
Pternandra echinata	Melastomataceae	328	0.10
Artocarpus integer	Moraceae	320	0.09
Dyera costulata	Apocynaceae	287	0.08
Artocarpus rigidus	Moraceae	276	0.08
Rhodamnia cinerea	Myrtaceae	245	0.07
Artocarpus lowii	Moraceae	216	0.06
Glochidion wallichiana	Euphorbiaceae	197	0.06
Adenanthera bicolor	Fabaceae	196	0.06
Bhesa paniculata	Celastraceae	196	0.06
Sarcotheca monophylla	Oxalidaceae	183	0.05
Artocarpus dadah	Moraceae	171	0.05
Macaranga hypoleuca	Euphorbiaceae	169	0.05
Melicope glabra	Rutaceae	156	0.05
Vernonia arborea	Asteraceae	147	0.04
Alstonia angustiloba	Apocynaceae	105	0.03
Fagraea volubilis	Gentianaceae	103	0.03
Cratoxylum formosum	Clusiaceae	82	0.02
Macaranga conifera	Euphorbiaceae	78	0.02
Macaranga recurvata	Euphorbiaceae	76	0.02
Callicarpa maingayi	Verbenaceae	66	0.02
Decaspermum fruticosum	Myrtaceae	60	0.02
Mussaendopsis beccariana	Rubiaceae	55	0.02

the typical pioneer trees common throughout Western Malesia along roadsides and in secondary successional forests are relatively poorly represented within the 50-ha plot. Quite a number of these species are present in and around Pasoh FR, but in relatively small numbers. Some of these species include: *Endospermum diadenum*, *Albizia pedicellata*, *Sapium* (*Balakata*) *baccatum*, *Macaranga hosei*, *Macaranga gigantea*, *Neolamarckia cadamba*, *Campnosperma auriculata*, *Vitex pinnata*, *Ficus grossularioides* and *Trema tomentosa*. Further studies on what constrains the abundance of these species in the primary forest at Pasoh FR would be extremely interesting.

6.2 Ballistically Dispersed Species

A large number of the trees in the understory of Pasoh FR are represented by species with a highly characteristic dispersal syndrome in which the seed is forcibly or ballistically dispersed as the fruit matures and dries. These species are often

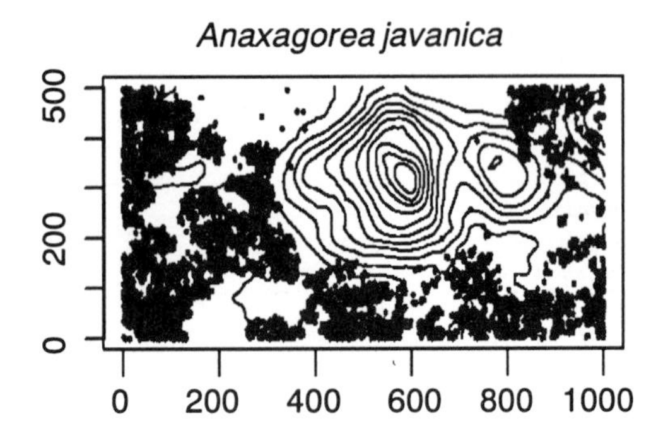

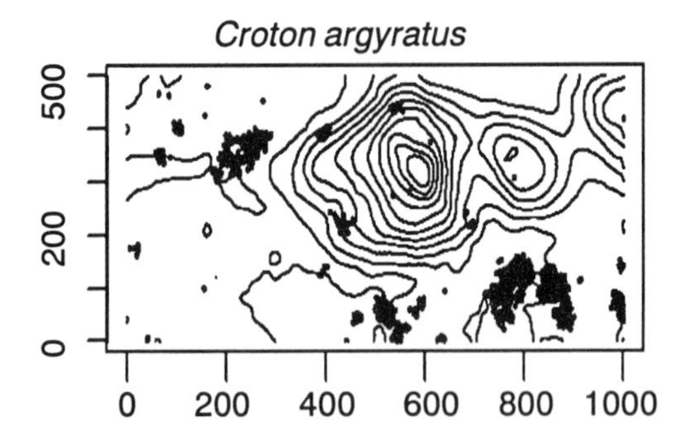

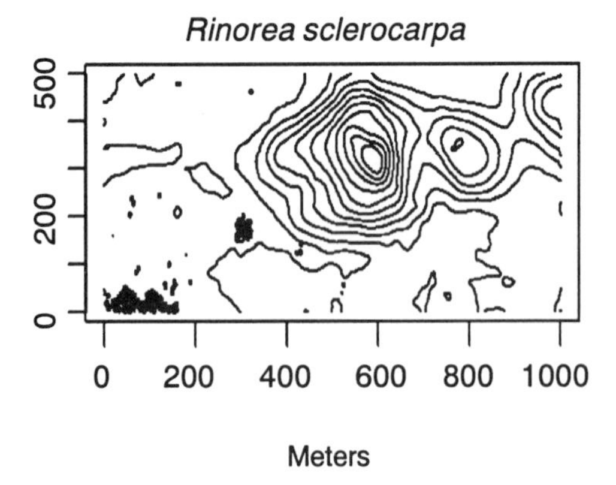

Fig. 4 Spatial distributions of three ballistically dispersed tree species in the 50-ha plot in Pasoh FR. The three species represent three independent origins of ballistically dispersed seeds in three separate families: Annonaceae, Euphorbiaceae and Violaceae respectively.

Table 8 List of species with ballistically dispersed seeds in 50-ha plot in Pasoh FR. The number of stems and percent of total stem number in the plot listed for each species.

Species	Family	Stem No. (%)
Anaxagorea javanica Bl.	Annonaceae	7,919 (2.34)
Cleistanthus maingayi Hk. f.	Euphorbiaceae	1,087 (0.32)
Cleistanthus myrianthus (Hassk.) Kurz	Euphorbiaceae	98 (0.03)
Cleistanthus sumatranus (Miq.) M.A.	Euphorbiaceae	482 (0.14)
Croton argyratus Bl.	Euphorbiaceae	1,272 (0.38)
Croton laevifolius Bl.	Euphorbiaceae	498 (0.15)
Elateriospermum tapos Bl.	Euphorbiaceae	223 (0.07)
Epiprinus malayanus Griff.	Euphorbiaceae	1,057 (0.31)
Koilodepas longifolium Hk. f.	Euphorbiaceae	222 (0.07)
Macaranga lowii King ex Hk. f.	Euphorbiaceae	2,178 (0.64)
Mallotus leucodermis Hk. f.	Euphorbiaceae	480 (0.14)
Mallotus penangensis M.A.	Euphorbiaceae	1,420 (0.42)
Trigonostemon laevigatus M.A.	Euphorbiaceae	142 (0.04)
Trigonostemon malaccanus M.A.	Euphorbiaceae	1,250 (0.37)
Rinorea anguifera (Lour.) O.K.	Violaceae	8,260 (2.44)
Rinorea sclerocarpa (Burgersd.) Jacobs	Violaceae	244 (0.07)

Table 9 List of monotypic genera in 50-ha plot in Pasoh FR. The number of stems and percent of total stem number in the plot listed for each species.

Species	Family	Stem No. (%)
Drimycarpus luridus (Hk. f.) Ding Hou	Anacardiaceae	245 (0.07)
Dendrokingstonia nervosa Hk. f. & Thoms.	Annonaceae	345 (0.10)
Monocarpia marginalis (Scheff.) Sinclair	Annonaceae	1,058 (0.31)
Aralidium pinnatifidum Miq.	Araliaceae	3 (0.00)
Pteleocarpa lamponga (Miq.) Bakh. ex Heyne	Boraginaceae	31 (0.01)
Triomma malaccensis Hk. f.	Burseraceae	503 (0.15)
Ctenolophon parvifolius Oliv.	Ctenolophonaceae	181 (0.05)
Neobalanocarpus heimii (King) Ashton	Dipterocarpaceae	3,523 (1.04)
Cheilosa malayana (Hk. f.) Corner ex Airy Shaw	Euphorbiaceae	38 (0.01)
Elateriospermum tapos Bl.	Euphorbiaceae	223 (0.07)
Scaphocalyx spathacea Ridl.	Flacourtiaceae	3,395 (1.00)
Pseudoclausena chrysogyne (Miq.) Mabb.	Meliaceae	259 (0.07)
Harmandia mekongensis Baill.	Olacaceae	47 (0.01)
Ochanostachys amentacea Mast.	Olacaceae	1,296 (0.38)
Champereia manillana (Bl.) Merr.	Opiliaceae	1,217 (0.36)
Gynotroches axillaris Bl.	Rhizophoraceae	254 (0.07)
Gardeniopsis longifolia Miq.	Rubiaceae	2 (0.00)
Jackiopsis ornata (Wall.) Ridsd.	Rubiaceae	1 (0.00)
Metadina trichotoma (Zoll. & Mor.) Bakh. f.	Rubiaceae	2 (0.00)
Trigoniastrum hypoleucum Miq.	Trigoniaceae	367 (0.11)

also characterized by highly clumped spatial distributions of the mature trees (Fig. 4, Table 8). An interesting aspect of this life-history type is that it appears to have evolved independently in several different families, and within some of these families, explosive fruits may have independently evolved several times (e.g. Euphorbiaceae). Further work on what constrains the spatial distribution of the conspicuous clumps of these species is required.

6.3 Monotypic Genera

Above we described the group of genera that are characterized by having an enormous diversity of sympatric species within the 50-ha plot (e.g. *Syzygium*). The diverse genera are often thought to be characteristic of the hyper-diverse forests of the lowlands in Southeast Asia. In addition to this group of highly diverse genera within the 50-ha plot at Pasoh FR, there are also a well represented group of genera that are globally monotypic (Table 9). They come from a wide range of families, some of which are otherwise highly diverse. Some of these species of monotypic genera naturally fit within other genera already represented in the forest (e.g. *Neobalanocarpus heimii*) and might therefore be interpreted as taxonomic artifacts. Others, however, appear to represent highly distinct evolutionary units, that in some cases have extremely broad geographic and ecological distributions (e.g. *Gynotroches axillaris*). These examples suggest more interesting ecological and evolutionary mechanisms. Further work with these taxa may reveal an interesting basis for evolutionary stasis in an otherwise highly diverse system.

7. CONCLUSIONS

As Kochummen et al. (1990) pointed out the Pasoh FR is representative of what Wyatt-Smith (1963) termed the Red Meranti-Keruing forest type. The high representation of *Shorea* species from section *Mutica* and species of the genus *Dipterocarpus* illustrate this clearly. Similar levels of diversity and standing basal area have been reported in several other studies of red meranti-keruing forest (Ho et al. 1987; Poore 1968). Pasoh FR has a relatively low standing basal area compared to some other forests in the region.

Floristic composition and stand structure in the 50-ha plot varied significantly with respect to topography and probably with respect to soils. This result supports the findings of Ashton (1976) and suggests that the sample sizes of Wong & Whitmore (1970) were too small to detect the environment-related variation in Pasoh FR. Further work is clearly required to test whether topography and soils have independent effects on forest structure and composition at Pasoh FR. These studies will also need to account for the limited dispersal ability of many of the species in this forest.

ACKNOWLEDGEMENTS

The 50-ha forest plot at Pasoh FR is an ongoing project of the Malaysian Government, directed by the Forest Research Institute Malaysia through its Director-General, Dato' Abdul Razak Mohd. Ali. The project was initiated under the leadership of Drs. N. Manokaran, P. S. Ashton and S. P. Hubbell. The project is now a collaboration of the Forest Research Institute Malaysia, under the leadership of Dr. A. Rahim Nik, the Center for Tropical Forest Science-Arnold Arboretum Asia Program of the Smithsonian Tropical Research Institute and Harvard University, USA, and the National Institute for Environmental Studies, Japan. The late Dr. K. M. Kochummen, while on a fellowship at STRI, supervised the species identification and personally examined all trees over 10 cm DBH. Funds for the project are gratefully acknowledged from: the National Science Foundation, USA (BSR Grant No. INT-84-12201 to Harvard University through Drs. P. S. Ashton and S. Hubbell); Conservation, Food and Health Foundation, Inc., USA; the United Nations, through the Man and the Biosphere program (UNESCO-MAB grant Nos. 217.651.5, 217.652.5, 243.027.6, 213.164.4, and also UNESCO-ROSTSEA grant No. 243.170.6).

REFERENCES

Ashton, P. S. (1976) Mixed dipterocarp forest and its variation with habitat in Malayan lowlands: a re-evaluation of Pasoh. Malay. For. 39: 56-72.

Ashton, P. S., Boscolo, M., Liu, J. & LaFrankie, J. V. (1999) A global programme in interdisciplinary forest research: The CTFS perspective. J. Trop. For. Sci. 11: 180-204.

Condit, R. (1995) Research in large, long-term tropical forest plots. Trends Ecol. Evol. 10: 18-22.

Condit, R. (1998) Tropical forest census plots: Methods and results from Barro Colorado Island, Panama and a comparison with other plots. Springer-Verlag, Berlin.

Corner, E. J. H. (1940) Wayside trees of Malaya. Government Printer, Singapore. 861pp.

Fisher, R. A., Corbet, A. S. & Williams, C. B. (1943) The relation between the number of species and the number of individuals in a random sample of an animal population. J. Anim. Ecol. 12: 42-58.

Ho, C. C., Newbery, D. M. & Poore, M. E. D. (1987) Forest composition and inferred dynamics in Jengka Forest Reserve, Malaysia. J. Trop. Ecol. 3: 25-56.

Hubbell, S. P. & Foster, R. B. (1983) Diversity of canopy trees in a neotropical forest and implications for conservation. In Sutton, S. L., Whitmore, T. C. & Chadwick, C. (eds.). Tropical rain forest: ecology and management. Serial Publication No. 2 of the British Ecological Society, UK, pp.25-41.

Kochummen, K. M., LaFrankie, J. V. & Manokaran, N. (1990) Floristic composition of Pasoh Forest Reserve, a lowland rain forest in Peninsular Malaysia. J. Trop. For. Sci. 3: 1-13.

Legendre, P. & Legendre, L. (1998) Numerical ecology (2nd ed.). Elsevier, Amsterdam, The Netherlands. 853pp.

Manokaran, N., LaFrankie, J. V., Kochummen, K. M., Quah, E. S., Klahn, J. E., Ashton, P. S. & Hubbell, S. P. (1990) Methodology for the fifty hectare research plot at Pasoh Forest Reserve. Research Pamphlet No. 104. Forest Research Institute Malaysia, Kepong, Malaysia.

Manokaran, N. & LaFrankie, J. V. (1990) Stand structure of Pasoh Forest Reserve, a lowland rain forest in Peninsular Malaysia. J. Trop. For. Sci. 3: 14-24.

Manokaran, N., LaFrankie, J. V., Kochummen, K. M., Quah, E. S., Klahn, J. E., Ashton, P. S. & Hubbell, S. P. (1999). The Pasoh 50-ha forest dynamic plot: 1999 CD-ROM version. Forest Research Institute Malaysia, Kepong, Malaysia.

Manokaran, N., Quah, E. S., Ashton, P. S. & LaFrankie, J. V. Pasoh forest dynamics plot, Malaysia. Pasoh forest dynamics plot, Malaysia. In Losos, E. C. & Leigh, E. G. Jr. (eds.). Forest diversity and dynamism: Findings from a network of large-scale tropical forest plots. University of Chicago Press, Chicago (in press).

Okuda, T., Nor Azman, H., Manokaran, N., Saw, L. Q., Amir, H. M. S. & Ashton, P. S. Local variation of canopy structure in relation to soils and topography and the implications for species diversity in a rain forest of Peninsular Malaysia. In Losos, E. C. & Leigh, E. G. Jr. (eds.). Forest diversity and dynamism: Findings from a network of large-scale tropical forest plots. University of Chicago Press, Chicago (in press).

Poore, M. D. (1968) Studies in Malayan rain forest. I. The forest on Triassic sediments in Jengka Forest Reserve. J. Ecol. 56: 143-196.

Wong, Y. K. & Whitmore, T. C. (1970) On the influence of soil properties on species distribution in a Malayan dipterocarp rain forest. Malay. For. 32: 42-54.

Wyatt-Smith, J. (1963) Manual of Malayan silviculture for inland forests. Malayan Forest Records No. 23. Forest Research Institute Malaysia, Kepong.

Wyatt-Smith, J. (1964) A preliminary vegetation map of Malaya with descriptions of the vegetation types. J. Trop. Geogr. 18: 200-213.

4 Rainfall Characteristics of Tropical Rainforest at Pasoh Forest Reserve, Negeri Sembilan, Peninsular Malaysia

Shoji Noguchi[1], Abdul Rahim Nik[2] & Makoto Tani[3]

Abstract: We investigated the rainfall at the Pasoh Forest Reserve (Pasoh FR), Peninsular Malaysia. Pasoh FR is located in the Southwest rainfall regime, in which the average annual rainfall (1,500–2,000 mm) is less than that in other regions of Peninsular Malaysia. Monthly rainfall in 1996 and 1997 ranged from 2.2 to 206.7 mm with a mean of 115.6 mm. The rainfall in 1997 was much smaller due to the El Niño Southern-Oscillation (ENSO) event. The longest period of dry days was 49 days. Dry periods as well as fluctuation in rainfall are major factors affecting the growth of vegetation. A distinct diurnal cycle in rainfall, in which 52% of the rainfall occurred between 13:00 and 19:00 h, was apparent. The frequency of the amount of rainfall in each event was an inverse J-shaped type distribution. The amount of rainfall in one event ranged from 1.2 and 93.1 mm with a mean of 11.4 mm and a median of 5.6 mm. The rainfall was characterized by a short duration (range:1.0–22.0 h, mean: 3.8 h) and high intensity. The maximum hourly rainfall intensity during a rain event ranged from 0.6 to 63.8 mm h^{-1} with a mean of 7.8 mm h^{-1} and a median of 3.8 mm h^{-1}.

Key words: amount, diurnal cycle, duration, rainfall intensity, seasonal variation, tropical rain forest.

1. INTRODUCTION

Rainfall data can be applied immediately to water resource planning and consumption. The Drainage and Irrigation Department and Malaysian Meteorological Service collect rainfall at more than 650 stations, which cover most of Peninsular Malaysia. However, there are a few stations in the hilly forested regions (Abdul Rahim 1983).

A tropical rain forest contributes to the prevention of sediment disaster as well as to the conservation of water resources. Understanding hydrological characteristics in a tropical rain forest is very important from a scientific point of view. Hydrological characteristics such as streamflow, soil moisture, suspended sediment yield, and nutrient cycling depend on the rainfall characteristics in tropical rain forests (Baharuddin & Abdul Rahim 1994; Noguchi et al. 1997; Zulkifli 1996). Therefore, it is important to understand the rainfall characteristics in order to elucidate the hydrological processes in a tropical rain forest.

The Pasoh Forest Reserve (Pasoh FR) has been a site of ongoing research on lowland rain forest ecology since the early 1970s. Rainfall data obtained from this site provides important information on ecology as well as hydrology. Hydrological observations for elucidation of rainfall-runoff responses, soil water

[1] Forestry and Forest Products Research Institute (FFPRI), Ibaraki 305-8687, Japan.
E-mail: noguchi@affrc.go.jp
[2] Forest Research Institute Malaysia (FRIM), Malaysia.
[3] Kyoto University, Japan.

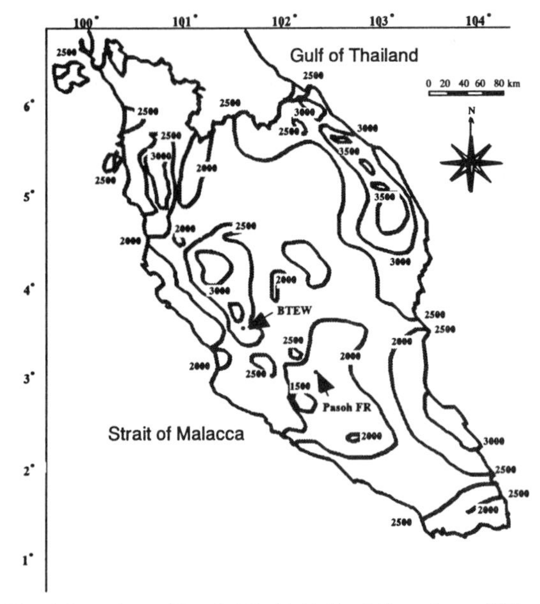

Fig. 1 Locations of Pasoh FR, Bukit Tarek Experimental Watershed (BTEW), and mean annual rainfall (mm) in Peninsular Malaysia (modified from Robiah et al. 1988).

storage, and nutrient cycling have been conducted at Bukit Tarek Experimental Watershed (BTEW) in Peninsular Malaysia (Noguchi et al. 1997, 2000; Zulkifli 1996). A comparison of the rainfall characteristics at Pasoh FR and BTEW would be helpful for understanding the hydrological characteristics at Pasoh FR.

The primary objective of this study was to analyze the temporal distribution of rainfall at Pasoh FR. A comparison of rainfall characteristics (duration, amount and intensity) at Pasoh FR and BTEW based on 3-year records was also carried out.

2. METHODS AND MATERIALS

The Pasoh FR is located in Negeri Sembilan of Peninsular Malaysia (2°59' N, 102°19' E: Fig. 1). The core area (600 ha) of the reserve (2,450 ha) is covered with a primary lowland mixed dipterocarp forest, which consists of various species of *Shorea* and *Dipterocarpus* (Manokaran et al. 1992). The emergent layer averages 46 m and the main canopy is 20–30 m in height (Manokaran & Swaine 1994). A 52-m high tower was established in the core area for meteorological observation (Color plate 3). Rainfall data was collected at 30-min intervals at a height of 52.6 m using a tipping-bucket rain gauge. The amount of rainfall collected at the tower was calibrated by a tatlizing rain gauge at a climate station in Pasoh FR. Details of the locations and the tower in Pasoh FR are shown in the Color Plates and Chaps. 1 and 2.

Bukit Tarek Experimental Watershed (BTEW) is located in Selangor Darul Ehsan, Peninsular Malaysia (3°31' N, 101°35' E: Fig. 1). The forest was logged in the early 1960s by MUS (Malayan Uniform System) and has now been fully regenerated. The vegetation is dominated by *Koompassia malaccensis*, *Eugenia* spp. and *Canarium* spp. Rainfall data were obtained at 10-min intervals using a weighting-type recording rain gauge and a tipping bucket rain gauge near weir C1. Details of the locations BTEW presented by Noguchi et al. (1996).

A rain event is defined as rainfall of more than 1 mm with an interval of more than six hours from the last recorded rainfall using hourly data.

3. RESULTS AND DISCUSSION

3.1 Diurnal and seasonal variations in rainfall

Figuar 2 shows monthly rainfall based on 13-year records (1983–1995) at Federal Land Depelopment Authority (FELDA) Pasoh Dua, which is located 3 km to the south of the tower in Pasoh FR. The annual rainfall at this site ranged from 1,468.6 to 2,349.5 mm with a mean of 1,810.7 mm. The monthly rainfall ranged from 3.4 to 430 mm with a mean of 150.9 mm. The monthly rainfall had a two-peak distribution (Mar.–May and Sep.–Dec.), suggesting that the climate in this site is influenced by both southwest and northeast monsoons. Typically, 42.6% of the annual rainfall occurred during the northeast monsoon (November to March), 39.1% during the southwest monsoon (May to September), and the remaining 18.4% during the transitional months (April and October). The monthly variation in rainfall is similar to that found for the southwest regime in Peninsular Malaysia as described by Dale (1959).

Diurnal and seasonal variations in rainfall in 1996 and 1997 at Pasoh FR are shown in Fig. 3. Annual rainfall in 1996 and 1997 were 1,610 and 1,182 mm, respectively. The variation in monthly rainfall in 1996 was similar to that in average monthly rainfall at Pasoh Dua. Annual rainfall at Pasoh FR was normal in 1996 but much less than normal in 1997. The much smaller rainfall in 1997 was caused by the El Niño Southern-Oscillation (ENSO) event (Toma et al. 2000; Chap. 6). Monthly rainfall in 1996 and 1997 ranged from 2.2 to 206.7 mm with a mean of 115.6 mm. The variation in soil water storage in a tropical rain forest corresponds to the fluctuation in rainfall (Noguchi et al. 1997, 2000). The variation in monthly rainfall at Pasoh FR might also reflect the variation in soil water storage.

Toma et al. (2000) divided various 30-day rainfall totals into three categories: wet (more than 100 mm), moist (60–100 mm), and dry (less than 60 mm). The percentages of wet and dry days in 1996 were 65.6% and 10.9%, respectively. In

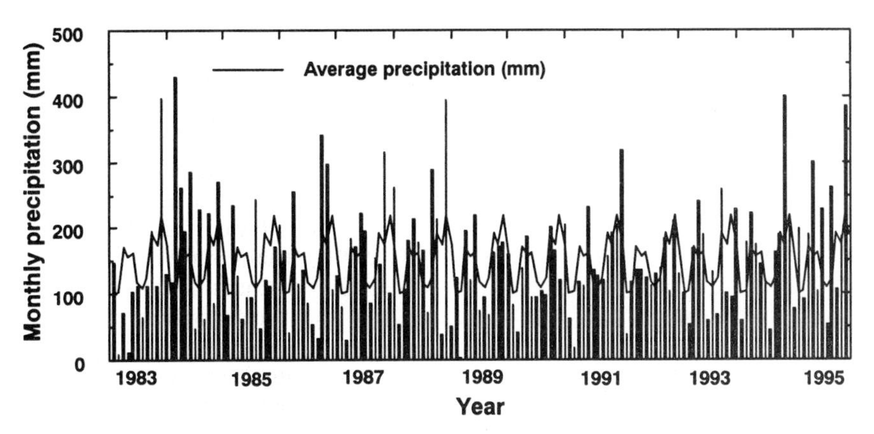

Fig. 2 Variation in monthly rainfall at Pasoh Dua based on 13-year records (1993-1995).

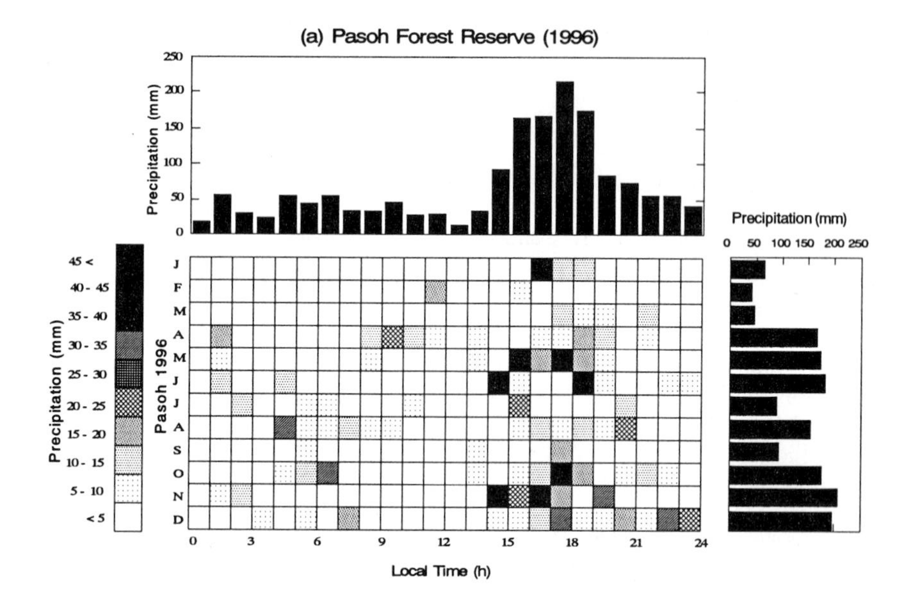

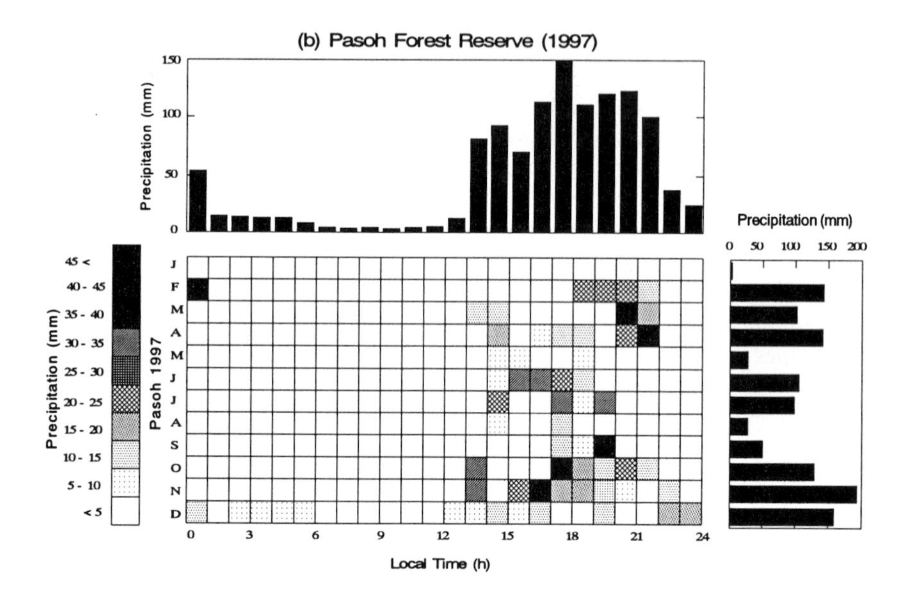

Fig. 3 Total rainfall at each time of day in each month at Pasoh FR in 1996 and 1997.

1997, the percentage of wet days (47.1%) decreased but the percentage of dry days (33.4%) increased because of the ENSO event. The longest period of dry days was 49 days, lasting from August 21 to October 8, 1997. On the other hand, annual rainfall at Bukit Soeharto Education Forest in East Kalimantan was more than 2000 mm, but the recorded period of dry days was longer (e.g. more than three months) than that at Pasoh FR. Toma et al. (2000) pointed out that high mortality rate of *Macaranga* spp. trees was observed during droughts. Thus, it is thought that dry periods have had a great effect on growth of vegetation.

Knowledge of the diurnal cycle of rainfall is important for evaluation of daily evapotranspiration because the rainfall time of a day and successive sunshine duration affect evapotranspiration (Oki & Musiake 1994). Fig. 3 shows the total rainfall at each time of day in each month at Pasoh FR. In 1996 and 1997, 63.6% and 55.4% of the total rainfall occurred during the daytime (07:00–19:00 h), respectively. A distinct diurnal cycle in rainfall, in which 52.7% and 52.1% of the rainfall occurred between 13:00 h and 19:00 h, is apparent. In 1996, 13.2% of the total rainfall occurred between 19:00 h and 22:00 h, whereas 29.1% of the total rainfall occurred between the same hours in 1997. Convectional storms are caused by differential solar heating of the ground and lower air layers, which typically occur during afternoons when warm moist air covers an area (Hewlett 1969). In this regard, most afternoon rainstorms at Pasoh Dua can be classified as convectional storms.

3.2 Characteristics of rainfall events

Noguchi et al. (1996) have already reported the rainfall characteristics in Bukit Tarek Experimental Watershed (BTEW). In the present study, we compared rainfall characteristics at Pasoh FR and BTEW over a period of 36 months.

The total numbers of rain events at Pasoh FR and BTEW were 366 and 555, respectively. The frequency of the amount of rainfall in each event at both Pasoh FR and BTEW showed an inverse J-shaped type distribution (Fig. 4), and the ranges of values were almost the same. However, the mean and median amounts of rain events at Pasoh FR were smaller than those at BTEW (Table 1).

Fig. 5 shows the percentages of rainfall in size classes at Pasoh FR and BTEW. Events of more than 25 and 50 mm of rainfall comprised 43.2% and 15.4% of the total rainfall at Pasoh FR and 57.9% and 17.8% of the total rainfall at BTEW,

Table 1 Statistical properties of amount, duration and intensity of rainfall.

	Pasoh FR			BTEW		
	Amount (mm)	Duration (h)	Intensity (mm h^{-1})	Amount (mm)	Duration (h)	Intensity (mm h^{-1})
Minimum	1.2	1.0	0.6	1.0	1.0	0.5
Maximum	93.1	22.0	63.8	96.0	18.0	76.5
Points	366	366	366	555	555	555
Mean	11.4	3.8	7.8	14.3	2.7	11.6
Median	5.6	2.0	3.8	8.0	1.0	6.5
RMS	17.7	5.20	12.1	21.3	3.87	17.5
Std. deviation	13.5	3.56	9.30	15.8	2.78	13.1
Std. Error	0.707	0.186	0.486	0.672	0.118	0.556
Skewness	2.36	2.34	2.36	1.79	2.30	1.82
Kurtosis	6.90	6.66	6.99	3.77	6.37	3.69

Pasoh FR: Pasoh Forest Reserve; BTEW: Bukit Tarek Experimental Watershed; Intensity: Maximum hourly rainfall intensity during a rain event.

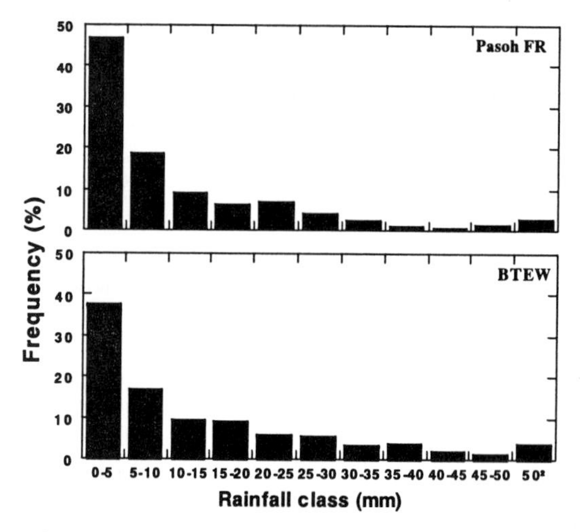

Fig. 4 Frequency distribution of the amounts of rainfall in each rain event at Pasoh FR and Bukit Tarek Experimmental Watershed (BTEW).

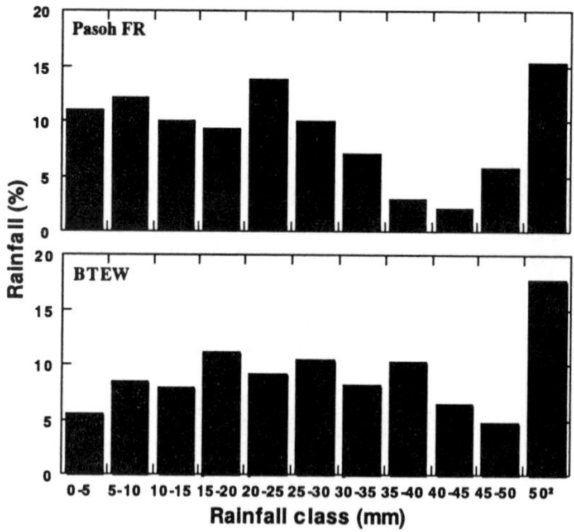

Fig. 5 Percentages of rainfall in certain-sized rain events in each rain event at Pasoh FR and BTEW.

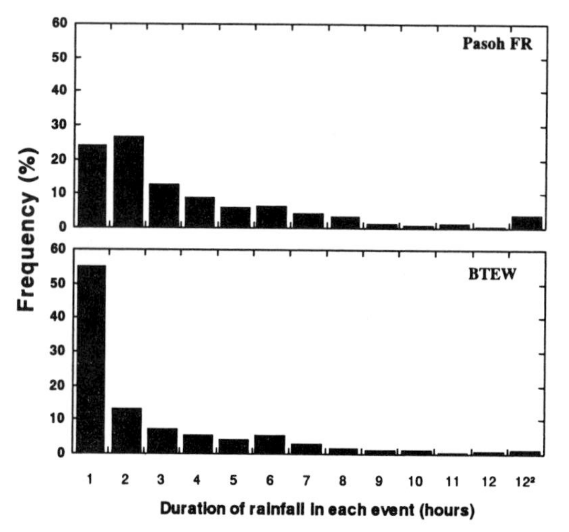

Fig. 6 Frequency distribution of duration of rainfall in each rain event at Pasoh FR and BTEW.

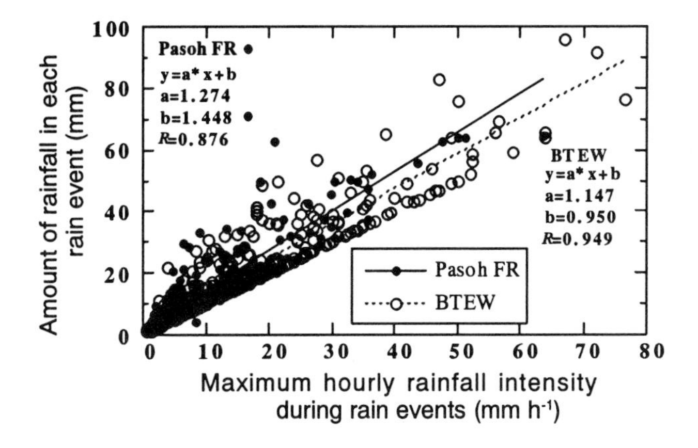

Fig. 7 Relationship between amount of rain and maximum rainfall per hour at Pasoh FR and BTEW.

respectively. Rain events of more than 25 and 50 mm constituted 12.3% and 2.7% of the total number of events at Pasoh FR and 20.7% and 4.0% of the total number of events at BTEW, respectively (Fig. 4). The smallest rain events (0–5 mm) constituted 46.7% and 37.8% of the total number of rain events at Pasoh FR and BTEW, respectively (Fig. 4). These small rain events (0–5 mm) produced 11% of the total rainfall at Pasoh FR, about two-times greater than that at BTEW (Fig. 5).

The frequency distributions of duration of rainfall in each rain event at Pasoh FR and BTEW are shown in Fig. 6. At Pasoh FR and BTEW, 50.8% and 68.2% of the total number of events occurred within 2-h periods. The duration of rainfall at Pasoh FR showed a wider range and a 1-h longer average than those at BTEW (Table 1). Figure 7 shows the relationship between total amount of rainfall in each rain event and maximum hourly rainfall intensity during the rain event. The amount of rainfall in each rain event was found to be proportional to the maximum hourly rainfall intensity at both sites. There was a higher positive correlation at BTEW (R = 0.949) than at Pasoh FR (R = 0.876). The mean and median values of maximum hourly rainfall intensity at Pasoh FR were smaller than those at BTEW (Table 1). However, the intensity of rainfall at Pasoh FR was also relatively high. The high intensity of rainfall has caused serious soil erosion on bare ground in this region.

4. CONCLUSION

Rainfall characteristics of tropical rainforest were investigated at the Pasoh FR in Negeri Sembilan, Peninsular Malaysia over a 3-year period. The rainfall was characterized by its short duration (mean: 3.8 h) and high intensity (mean: 7.8 mm h^{-1}). There was a distinct diurnal cycle in rainfall, in which 52% of the rainfall occurred between 13:00 and 19:00 h. Pasoh FR is located in the Southwest rainfall regime, in which the average annual rainfall (1,500–2,000 mm) is less than that of other regions of Peninsular Malaysia. The rainfall in 1997 (1,182 mm) was much lower than the norm due to El Niño Southern-Oscillation event. The longest period of dry days was 49 days. Soil moisture corresponds to the fluctuation of rainfall. Therefore, such dry periods as well as fluctuation in rainfall are major factors affecting the growth of vegetation.

ACKNOWLEDGEMENTS

We thank the Forestry Department of Negeri Sembilan and Director General of FRIM for giving us permission to work in the Pasoh FR. We also thank the staff of the Hydrology Unit of FRIM for collecting rainfall data. This study was carried out as a joint research project between NIES, FRIM and UPM (Global Environmental Research Program granted by the Japanese Environment Agency, Grant No. E-3).

REFERENCES

Abdul Rahim, N. (1983) Rainfall characteristics in forested catchments of Peninsular Malaysia. Malay. For. 46: 233-243.

Baharuddin, K. & Abdul Rahim, N. (1994) Suspended sediment yield resulting from selective logging practices in a small watershed in Peninsular Malaysia. J. Trop. For. Sci. 7: 286-295.

Dale, W. L. (1959) The rainfall of Malaysia, Part I. J. Trop. Geogr. 13: 23-37.

Hewlett, J. D. (1969) Principle of Forest Hydrology. The University of Georgia Press, 183pp.

Manokaran, N. & Swaine, M. D. (1994) Population dynamics of trees in dipterocarp forests of Peninsular Malaysia. Malayan Forest Records No. 40. Forest Research Institute Malaysia, Kuala Lumpur, 173pp.

Manokaran, N., LaFrankie, J. V., Kochummen, K. M., Quah, E. S., Klahn, J. E., Ashton, P. S. & Hubbell, S. P. (1992) Stand table distribution of species in the 50ha research plot at Pasoh Forest Reserve. FRIM Research Data No. 1. Forest Research Institute Malaysia, Kuala Lumpur.

Noguchi, S., Abdul Rahim, N., Sammori, T., Tani, M. & Tsuboyama, Y. (1996) Rainfall characteristics of tropical rain forest and temperate forest : Comparison between Bukit Tarek in Peninsular Malaysia and Hitachi Ohta in Japan. J. Trop. For. Sci. 9: 206-220.

Noguchi, S., Abdul Rahim, N., Zulkifli, Y., Tani, M. & Sammori, T. (1997) Rainfall-runoff responses and roles of soil moisture variations to the response in tropical rain forest, Bukit Tarek, Peninsular Malaysia. J. For. Res. 2: 125-132.

Noguchi, S., Zulkifli, Y., Baharuddin, K., Tani, M., Tsuboyama, Y. & Sammori, T. (2000) Seasonal soil water storage changes in a tropical rain forest in Peninsular Malaysia. J. Jpn. Soc. Hydrol. Water Resour. 13: 206-215.

Oki, T. & Musiake, K. (1994) Seasonal change of the diurnal cycle of rainfall over Japan and Malaysia. J. Appl. Meteorol. 33: 1445-1463.

Robiah bt. Bani, Leong, T. M., Yip, H. W. & Kelsom bt. Alias (1988) Mean monthly, mean seasonal and mean annual rainfall map for Peninsular Malaysia. Water Resour. Pub. 19: 16.

Toma, T., Marjenah & Hastaniah (2000) Climate in Bukit Soeharto, East Kalimantan. In Guhardja, E., Fatawi, M., Sutisna, M., Mori, T. & Ohta, S. (eds). Rainforest Ecosystems of East Kalimantan: El Niño, Drought, Fire and Human Impacts. Ecological Studies vol. 140. Springer-Verlag, Tokyo, pp.13-27.

Zulkifli, Y. (1996) Nutrient cycling in secondary rain forest catchments of Peninsular Malaysia. Ph.D. diss., Univ. Manchester, 380pp.

5 Soil Nutrient Flux in Relation to Trenching Effects under Two Dipterocarp Forest Sites

Tamon Yamashita[1] & Hiroshi Takeda[2]

Abstract: The standing stock and annual flux of soil macro-nutrients were studied under two dipterocarp forest sites one in a dipterocarp plantation and other in a lowland dipterocarp forest of Peninsular Malaysia. We measured the chemical properties of soils and the standing stock of the soil macro-nutrients such as available P, exchangeable (ex-) K, ex-Na, ex-Mg, ex-Ca, total C and total N of soil profiles taken from a depth of 50 cm. The annual flux of inorganic ions at a soil depth of 20 cm was estimated for ammonium-N, nitrate-N, phosphate-P, Na, K, Mg and Ca using the ion exchange resin (IER) core method. Cores of IER was buried under the trenched and untrenched sites of two forest sites for a year in order to clarify the soil-plant interactions in soil nutrient dynamics. The C/N ratio was higher in the plantation forest (11 to 14) than in the lowland forest (8 to 11). Total exchangeable cations of the plantation and lowland forest soils were 4.5 kmol (+) ha^{-1} and 5.9 kmol (+) ha^{-1}, respectively. The annual flux rate of cations and anions in the plantation forest was 2.97 kmol (+) ha^{-1} yr^{-1} and 1.12 kmol (+) ha^{-1} yr^{-1} for the untrenched sites and 2.79 kmol (+) ha^{-1} yr^{-1} and 2.16 kmol (+) ha^{-1} yr^{-1} for the trenched sites. In the lowland forest, those figures were 5.06 kmol(+) ha^{-1} yr^{-1} and 0.08 kmol (+) ha^{-1} yr^{-1} for the untrenched sites and 3.27 kmol (+) ha^{-1} yr^{-1} and 0.30 kmol (+) ha^{-1} yr^{-1} for the trenched sites. Trenching caused decreases in cation flux both in the plantation and lowland forests. But Mg flux increased in the plantation forest and ammonium-N flux showed no drastic changes at any of the sites. On the other hand, trenching increased the anion flux in both forests. We will discuss the effects of trenching treatment on nutrient dynamics in soil system of tropical rain forest.

Key words: dipterocarp plantation, ion exchange resin (IER), lowland dipterocarp forest, soil macro-nutrient, trenching experiment.

1. INTRODUCTION

During recent decades, many tropical rain forests have disappeared as a result of human activities. Virgin natural forests have become plantations or arable land. The changes in land use affect not only the above-ground system but also the below-ground system. Early work of Bormann and his colleagues showed that clear-cutting of forest land increased losses of soil nutrients in the USA (Bormann & Likens 1979). Since then, nutrient cycling has been an important topic, even in the study of tropical forest ecosystems (Jordan 1985). Bruijnzeel (1991) reviewed the nutrient balance in tropical forest ecosystems and stated the need for more and careful studies of nutrient budgets in tropical forests.

In any type of forest, the recycling system is a critical pathway to conservation of nutrients within the ecosystem. The interaction between plants and soil plays a central role in nutrient recycling, because plant uptake actively

[1] Education and Research Centre for Biological Resources, Shimane University, Matsue 690-8504, Japan. E-mail: tamonyam@life.shimane-u.ac.jp

[2] Kyoto University, Japan.

regulates nutrient cycling and plant litter exerts indirect effects. In this sense, it is essential that we try to learn what is happening in the rooting zone of a forest, including the organic layer and top soil.

The effects of plant roots on nutrient flux through soil profiles or the availability of nutrients in soil, have commonly been studied using soil trenching (Vitousek et al. 1982). Comparing nutrient dynamics in a root-excluded area with the dynamics in an untrenched area gives us useful information on the effects of plant roots. The ion exchange resin (IER) core method is one way to measure soil nutrient flux (e.g. Binkley & Matson 1983). IER consists of cation exchange resin (CER) and anion exchange resin (AER). Normally, a mixed-bed resin (CER + AER) is used for the IER core method.

In this study, we set up the IER cores in soil at a depth of 20 cm, with and without root zone trenching. We selected two different dipterocarp forests as study sites, to determine the effects of differences in climatic conditions and site quality on the nutrient dynamics. Finally, we examined the effects of plant uptake on the soil nutrient flux in two kinds of dipterocarp forests.

2. STUDY SITE

This study was carried out both in a lowland dipterocarp forest in the Pasoh Forest Reserve (Pasoh FR), about 70 km southeast of Kuala Lumpur (2°58–59' N and 102° 16–20' E) and in a plantation of *Dipterocarpus baudii* Korth (Dipterocarpaceae). The plantation of *D. baudii* was established at the campus of the Forest Research Institute Malaysia (FRIM site), located in Kepong (3°14' N and 101°38' E) near Kuala Lumpur, Peninsular Malaysia. One quadrat of 20 m × 20 m in size was marked out at each site in areas showing similar topography.

The annual precipitation during the 12 month period August 1992 through July 1993 was 1,720 mm in Pasoh FR (Sulaiman et al. 1994) and 2,471 mm in FRIM site (data provided by Dr. Abdul Rahim Nik, FRIM). The mean annual temperature was 25.8°C at Pasoh FR (Sulaiman et al. 1994) and 27.4°C at FRIM site (data provided by Dr Abdul Rahim Nik, FRIM). Figure 1 shows the climatic conditions at the study sites during that period. There are two peaks in the precipitation distribution for both Pasoh FR and FRIM site and 3 months in which the Pasoh FR had less than 100 mm rainfall. Precipitation in each of the peak periods during November–December 1992 and May–June 1993 in FRIM site was more than 600 mm, while Pasoh FR received at most 400 mm. Pasoh FR is drier and marginally cooler than FRIM site.

3. MATERIALS AND METHODS

3.1 Soil sampling and chemical analysis

We collected duplicate soil cores from each plot in August, 1992, using a soil core sampler with a 25 cm^2 cross-section The soil cores were divided into 4 according to soil depth: 0 to 10 cm, 10 to 20 cm, 20 to 30 cm and 30 to 50 cm. Each individual sample was sealed in a plastic bag and immediately returned to the FRIM laboratory where they were air-dried in the laboratory before being sent to Kyoto University, Japan.

Using the air-dried soil samples, we analyzed total carbon (T-C), total nitrogen (T-N), available phosphorus (av-P), exchangeable potassium (ex-K), exchangeable sodium (ex-Na), exchangeable calcium (ex-Ca) and exchangeable magnesium (ex-Mg). T-C and T-N were analyzed by CN coder (MT600, Yanaco Co., Kyoto). The av-P was determined by the Bray II method (Olsen & Sommers 1982). To analyze exchangeable cations, air-dried soil samples were subjected to 1M-ammonium acetate extraction. The extracts were analyzed by atomic absorption or flame emission spectrometry (Baker & Suhr 1982). The pH of the water and 1M-KCl suspensions was measured by pH meter equipped with a glass electrode (F-13, Horiba Co., Kyoto).

3.2 Trenching experiment with ion exchange resin core

The soil nutrient fluxes at 20 cm depth were measured by the ion exchange resin (IER) core method (Binkley & Matson 1983; Haibara et al. 1990), because the topsoil is usually an important layer for biological activity including nutrient flux.

We used a mixed bed of equal quantities of Amberlite IRA-400 (Organo Co.,

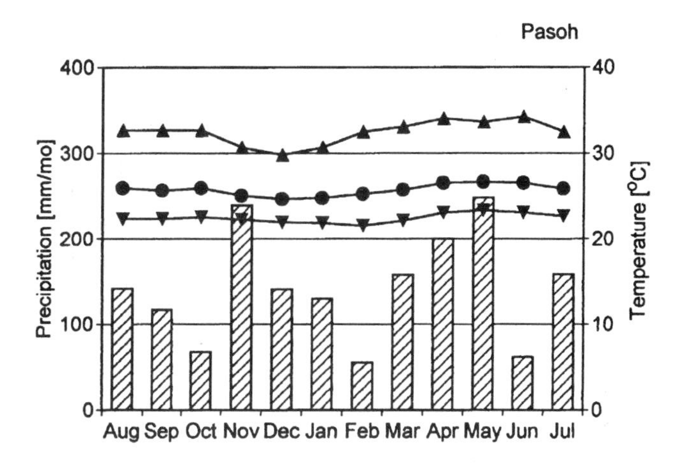

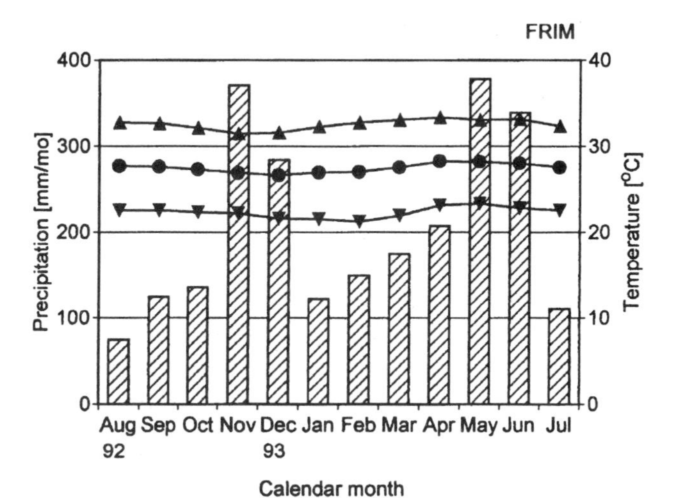

Fig. 1 Climatic condition of the lowland forest in Pasoh FR (upper) and the plantation forest in FRIM site (lower) from August 1992 to July 1993. Bar: the monthly precipitation; ▲: mean maximum temperature; ▼: mean minimum temperature; ●: mean temperature.

Tokyo) for the anions and Amberlite IR-120B (Organo Co., Tokyo) for the cations. The IER were mixed thoroughly and placed in 3 cm long PVC tubes with a 25 cm^2 internal cross-section. These had previously been washed thoroughly using deionized water. Both ends of the PVC tube were sealed with fine-mesh polyethylene. To examine the effects of plant uptake on the soil nutrient flux through the soil profile, we prepared 40 IER cores and set them in the field at 20-cm soil depth with and without root zone trenching. To prevent plant uptake, we adopted a "plastic tube method" (Fig. 2). A washed PVC tube 20 cm long was driven into the soil and pulled out, providing a soil-filled PVC tube. The IER core, with the same diameter, was attached to the soil-filled tube and fastened tightly with adhesive tape. The soil-filled tube with the IER core was then reburied. We also had to prepare the untrenched control site. Ideally, a control site should remain entirely intact, but this was difficult to achieve when setting up the IER core. In our experiment, the control site was prepared as follows. Washed PVC tubes were driven into the soil and pulled out, leaving a hole in the soil surface. An IER core was placed in each hole which was then refilled with the soil, retrieved from the PVC tube. Ten prepared IER cores were buried at the trenched site, and the other 10 cores were buried at the untrenched site in each plot. Following the recommendations of Binkley & Hart (1989) that the duration of the IER experiment should be 1 to 12 months, the IER cores were buried in August 1992 and retrieved a year later.

The retrieved IER cores were washed in deionized water, extracted with 1M-HCl for adsorbed ions and analyzed by the ICP-AES method using an ICP-AE spectrometer (SPS1500VR, Seiko Instrument Inc., Japan). Adsorbed inorganic N was extracted with 2M-KCl and analyzed colorimetrically. Ammonium (NH_4^+-N)

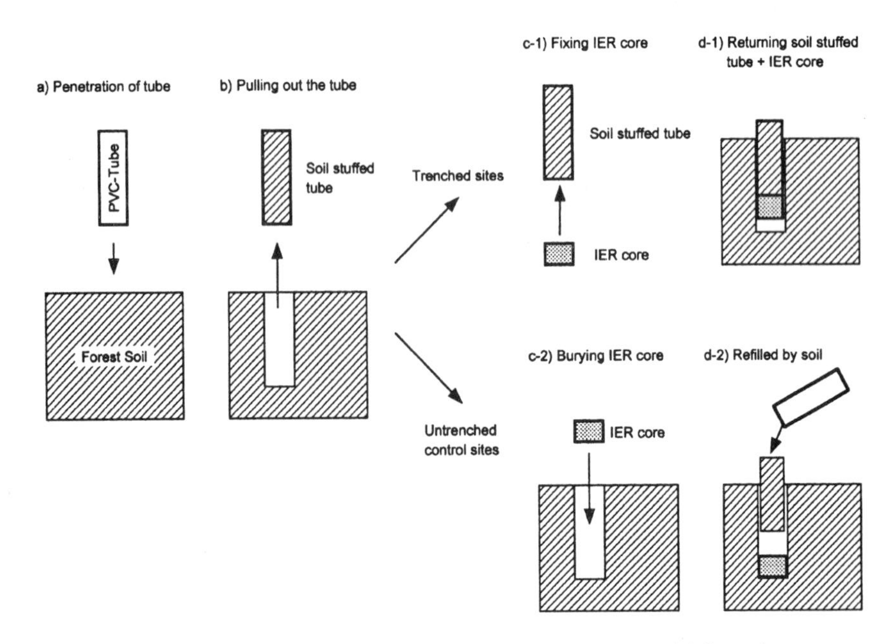

Fig. 2 Schematic diagram of the installation of IER core. Protocol follows the order from a) to b). To make the trenched site goes to c-1) to d-1), and to make the untrenched site goes to c-2) to d-2).

was measured by the indophenol blue method, and nitrate (NO_3^--N) was reduced to nitrite using zinc powder and determined by the modified Griess-Ilosvay method (Keeney & Nelson 1982). The extracts were highly acidic due to the properties of resin. As the colorimetric method was affected by the acidity of the matrix, the extracts were neutralized in a portion of alkaline solution prior to analysis.

3.3 Statistical analysis

Although we set only one comparable plot at each site, the effects of trenching and site quality on nutrient flux were tested by two-way ANOVA, and mean comparison was performed using Tukey's HSD test using SYSTAT for Macintosh (SYSTAT 1992, Evanston).

4. RESULTS

4.1 Standing stock of soil nutrients

Soil macro-nutrient levels and some soil chemical properties are shown in Table 1. For all but Ca, the cation concentrations in the mineral soils were lower at FRIM site than at Pasoh FR. In terms of profile distribution of cations in mineral soils, Na, Mg and Ca tended to concentrate in the 0 – 10 cm layer, especially Na and Mg for Pasoh FR and Ca for FRIM site. Standing stock of T-C in mineral soil 0 – 50 cm deep was similar between Pasoh FR and FRIM site; 42 Mg C ha^{-1} and 44 Mg C ha^{-1} respectively. The standing stock of T-N was higher and that of P was lower at Pasoh FR, especially in the top 20 cm of the soil. FRIM site soil contained 13.8 kg P ha^{-1}, twice as much as at Pasoh FR. The standing stock of total cations in mineral soil was lower at FRIM site than at Pasoh FR. The C/N ratio ranged from 8.1 to 11.3 at Pasoh FR and from 11.3 to 14.0 at FRIM site. This ratio was lower at Pasoh FR in all profiles, reflecting a higher N content than FRIM site. The pH (H_2O) was similar (4.07 – 4.34), but the pH (KCl) was lower at Pasoh FR throughout the profile. This led to larger Δ pH at Pasoh FR.

Although we recognized a significant A_0 layer at the Pasoh FR site, unfortunately we only have chemical data for the mineral soils. At the FRIM site, we determined the contribution of the A_0 layer to the standing stock of nutrients. The A_0 layer is an important source for soil macro-nutrients. It temporarily stores materials that are provided as litterfall both from above-ground and below-ground. It releases nutrients gradually and decomposes organic matter. T-C, T-N and Na were accumulated mainly in the mineral soil. P, K and Mg had accumulated partly in the A_0 layer. More than half of the Ca was accumulated in the A_0 layer. Assuming that all the mineral elements in the A_0 layer will be mineralized in the future, one-third of the total cations are stored in the A_0 layer.

4.2 Nutrient flux

The flux of the mineral ions, including inorganic N (NH_4^+-N and NO_3^--N) and inorganic P (PO_4^{3-}-P), is shown in Table 2. The flux of inorganic N ranged from 6.29 kg N ha^{-1} yr^{-1} at the untrenched site of Pasoh FR to 27.3 kg N ha^{-1} yr^{-1} at the trenched site of FRIM site. The flux of NH_4^+-N exceeded NO_3^--N at Pasoh FR and vice versa at FRIM site. Trenching caused an increase in flux of inorganic N. The NO_3^--N flux at the trenched site was two times higher than the untrenched site at both Pasoh FR and FRIM site. Though no significant trenching effect was observed ($P > 0.05$), the inorganic N flux was significantly larger at FRIM site than at Pasoh FR ($P < 0.05$). The flux of PO_4^{3-}-P ranged from 0.05 kg P ha^{-1} yr^{-1} at the untrenched Pasoh FR

Table 1 Concentration (Conc.) and standing stock (SS) of soil chemical elements and soil properties of a lowland dipterocarp forest in Pasoh and a *D. baudii* plantation in FRIM site. The values for P, K, Na, Mg, Ca, and cations (K+Na+Mg+Ca) are expressed based on a total value for A_0 layer and an exchangeable or available value for mineral soil. The " Δ pH" shows the difference between pH (H_20) and pH (KCl). The "A_0" is an organic layer. The "0–10", "10–20", "20–30" and "30–50" indicates the mineral soil layer at each depth (cm). "Sum A" is total amounts of elements in the A_0 and mineral soil. "Sum B" is total amounts of elements in the mineral soil only. "NA" means "data not available".

(a) Pasoh FR Conc.

	% dry mass		$mg\ kg^{-1}$	$cmol(+)\ kg^{-1}$								
	T-C	T-N	P	K	Na	Mg	Ca	cations	C/N	$pH(H_20)$	pH(KCl)	Δ pH
A_0	NA	NA	NA	NA	NA	NA	NA	NA	NA	NA	NA	NA
0-10	1.8	0.16	2.6	0.02	0.22	0.148	0.045	0.433	11.3	4.16	3.41	-0.75
10-20	0.9	0.1	0.6	0.017	0.106	0.066	0.034	0.222	9.1	4.21	3.48	-0.73
20-30	0.8	0.09	2.4	0.013	0.079	0.04	0.022	0.154	8.5	4.24	3.54	-0.7
30-50	0.8	0.08	1.6	0.012	0.081	0.054	0.023	0.171	8.1	4.30	3.61	-0.7

(b) Pasoh FR SS

	$Mg\ ha^{-1}$		$kg\ ha^{-1}$					$kmol(+)\ ha^{-1}$
	T-C	T-N	P	K	Na	Mg	Ca	cations
A_0	NA	NA	NA	NA	NA	NA	NA	NA
0-10	15.4	1.36	2.22	6.7	42.7	15.7	8.1	2.88
10-20	8.9	0.99	0.64	6.6	24.6	8.0	6.7	1.74
20-30	7.8	0.91	2.27	5.4	18.6	5.0	4.4	1.26
30-50	9.6	1.18	2.56	7.4	28.8	10.1	7.1	2.03
Sum B	41.7	4.44	7.70	26.1	114.7	38.8	26.3	7.91

(c) FRIM Conc.

	% dry mass		$mg\ kg^{-1}$	$cmol(+)\ kg^{-1}$								
	T-C	T-N	P	K	Na	Mg	Ca	cations	C/N	$pH(H_20)$	pH(KCl)	Δ pH
A_0	45.1	1.08	650	5.37	0.3	10.6	49.9	66.2	41.9	NA	NA	NA
0-10	1.6	0.11	5.8	0.011	0.127	0.078	0.089	0.305	14	4.07	3.47	-0.6
10-20	0.9	0.08	2.7	0.008	0.062	0.036	0.028	0.135	12.2	4.23	3.91	-0.33
20-30	0.7	0.06	1.8	0.008	0.048	0.028	0.032	0.116	11.3	4.32	4.06	-0.26
30-50	0.7	0.06	2.2	0.008	0.055	0.031	0.035	0.128	11.8	4.34	4.02	-0.32

(d) FRIM SS

	$Mg\ ha^{-1}$		$kg\ ha^{-1}$					$kmol(+)\ ha^{-1}$
	T-C	T-N	P	K	Na	Mg	Ca	cations
A_0	2.2	0.05	3.19	10.3	0.3	6.32	49.0	3.24
0-10	13.8	0.99	5.1	3.9	25.4	8.35	15.7	2.67
10-20	12.4	1.02	3.62	4.5	19.2	5.99	7.6	1.82
20-30	8.1	0.71	2.21	3.9	13.2	4.03	7.6	1.38
30-50	9.3	0.79	2.86	4.1	16.7	4.94	9.3	1.7
Sum A	45.8	3.56	17	26.7	74.8	29.6	89.2	10.8
Sum B	43.6	3.51	13.8	16.4	74.5	23.3	40.2	7.56

Table 2 Nutrient ions flux at 20 cm deep soil of the lowland forest at Pasoh FR and the plantation at FRIM site. The "sum of cations" indicates the sum of K, Na, Mg, and Ca. Different superscripts within lower columns indicate significant difference between means at $P < 0.05$.

(a) Pasoh FR (unit = $kg\ ha^{-1}\ yr^{-1}$)

	NH_4-N	NO_3-N	PO_4-P	K	Na	Mg	Ca
Untrenched	5.18	1.11	0.05	93.1	17.4	13.6	8.72
Trenched	7.45	3.23	0.69	51.4	7.6	10.3	4.93

(b) Pasoh FR (unit = $kmol\ (+/-)\ ha^{-1}\ yr^{-1}$)

	NH_4-N	NO_3-N	PO_4-P	K	Na	Mg	Ca	Sum of cations
Untrenched	0.37^{a}	0.08^{a}	0.005^{a}	2.38^{a}	0.75^{a}	1.12^{a}	0.44^{a}	4.69^{a}
Trenched	0.53^{a}	0.23^{a}	0.067^{b}	1.31^{b}	0.33^{b}	0.85^{a}	0.25^{a}	2.74^{b}

(c) FRIM (unit = $kg\ ha^{-1}\ yr^{-1}$)

	NH_4-N	NO_3-N	PO_4-P	K	Na	Mg	Ca
Untrenched	8.22	14.7	0.67	31.9	7.42	7.06	13.4
Trenched	8.34	29	0.91	0.64	5.97	18.6	7.8

(d) FRIM (unit = $kmol\ (+/-)\ ha^{-1}\ yr^{-1}$)

	NH_4-N	NO_3-N	PO_4-P	K	Na	Mg	Ca	Sum of cations
Untrenched	0.59^{a}	1.05^{a}	0.064^{a}	0.82^{a}	0.32^{a}	0.58^{a}	0.67^{a}	2.39^{a}
Trenched	0.60^{a}	2.08^{a}	0.088^{a}	0.02^{b}	0.26^{a}	1.53^{b}	0.39^{b}	2.20^{a}

site to 0.91 kg P ha^{-1} yr^{-1} at the trenched FRIM site. PO_4^{3-}-P flux also increased at the trenched sites of both plots. There was a significant difference between Pasoh FR and FRIM site, and between the untrenched and trenched sites ($P < 0.01$).

The total mineral cation (K, Na, Mg and Ca) flux ranged from 2.20 kmol (+) ha^{-1} yr^{-1} in the trenched site of FRIM site to 4.69 kmol (+) ha^{-1} yr^{-1} in the untrenched site of Pasoh FR. Values were significantly higher at Pasoh FR than at FRIM site ($P < 0.01$) and also significantly higher in the untrenched samples at both sites ($P < 0.05$). Individually, the flux of K and Na at Pasoh FR was significantly higher than at FRIM site ($P < 0.05$). The flux of Mg and Ca showed no significant differences between Pasoh FR and FRIM site ($P > 0.05$). The flux of K, Na and Ca was significantly reduced by trenching in comparison with the untrenched site ($P < 0.05$), while no clear difference was observed in the flux of Mg, especially at Pasoh FR.

In sum, trenching caused an increase in the flux of anions (NO_3^--N and PO_4^{3-}-P) and the decrease in the flux of some cations (K, Na and Ca).

The composition of cations, expressed as the proportion of a cation to total cations, % kmol (+) / kmol (+), was also affected by the trenching experiment (Fig. 3). At FRIM site, the proportions of K and Ca decreased and that of Mg increased, in response to trenching. In the untrenched treatment, the cation proportions ranged from 13% for Na to 35% for K. In contrast, there were rather larger variations in the proportions of the cations in the trenched site: 1% for K to 69% for Mg. The proportion of Mg in the trenched site was three times that in the untrenched site and K decreased from 35% in the untrenched site to 1% in the trenched site. The differences in Na and Ca between the untrenched and trenched sites were relatively

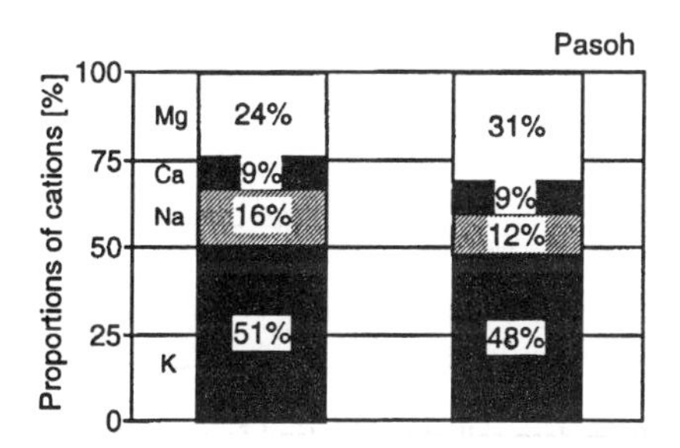

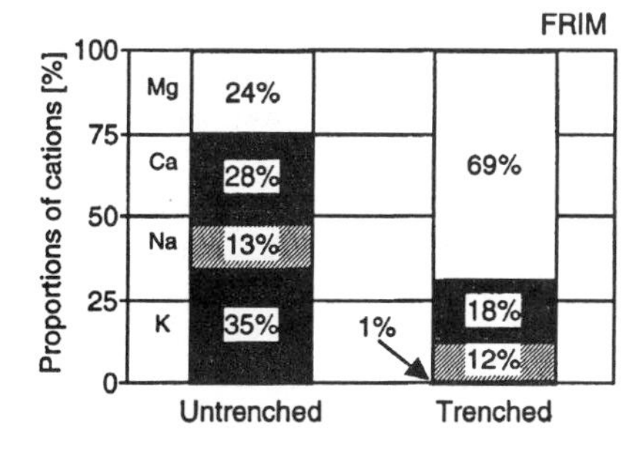

Fig. 3 Composition of cations in soil solution, which adsorbed on IER, at Pasoh FR (upper) and at FRIM site (lower). The proportion is the ratio of each cation (kmol (+) ha^{-1} yr^{-1}) to the sum of K, Na, Ca and Mg (kmol (+) ha^{-1} yr^{-1}).

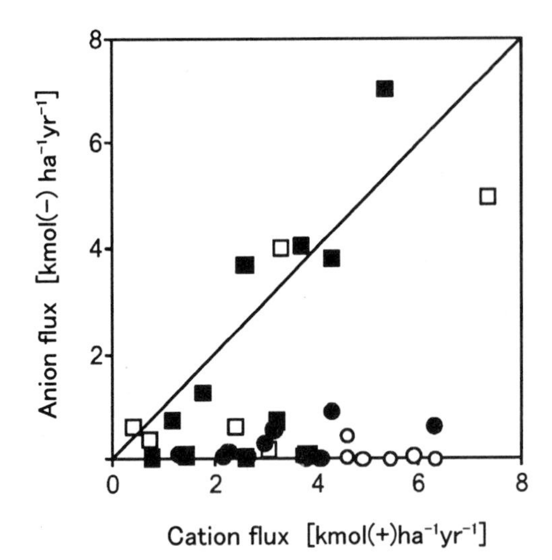

Fig. 4 Anion flux (kmol(-) ha^{-1} yr^{-1}) in relation to the cation flux (kmol(+) ha^{-1} yr^{-1}) in Pasoh FR and FRIM site. The line represents "the cation flux = the anion flux". ○ : untrenched site at Pasoh FR; ● : trenched site at Pasoh FR; □: untrenched site at FRIM; ■ : trenched site at FRIM.

small. At Pasoh FR, the proportion of K became almost 50% of the total cations in the untrenched site and in the trenched site. There was no drastic change in the proportion of each cation. The proportion of Mg slightly increased and that of the other elements slightly decreased in the trenched site.

The balance of cations (NH_4^+-N, K, Na, Mg and Ca) and anions (NO_3^--N and PO_4^{3-}-P) is shown in Fig. 4. The flux of cations and anions was almost balanced at FRIM site, but was not so at Pasoh FR. The flux of cations greatly exceeded the flux of NO_3^--N and PO_4^{3-}-P at Pasoh FR. Trenching reduced the cation flux but caused no changes in the balance of cations and anions at Pasoh FR.

5. DISCUSSION

5.1 Soil chemical properties and soil leachate

Comparative values of soil nutrient flux and stock are available for a number of tropical rain forest sites. Ohta & Syarif (1996) showed that, in terms of concentration, the range of T-C was 0.1% to 7% dry mass, T-N was 0.02% to 0.3% dry mass and the mineral cation was N.D. to 1.0 cmol (+) kg^{-1} at various dipterocarp forest soils in Kalimantan, Indonesia. Sakurai et al. (1998) and Yamashita et al. (1999) reported the soil nutrient status of tropical forests in Thailand. The sum of mineral cations (K + Na + Mg + Ca) recorded ranged from 0.2 cmol (+) kg^{-1} to 2.1 cmol (+) kg^{-1}. In our case, deeper soil showed a lower concentration than the value reported in Thailand (Table 1). However, as reported by Ohta & Syarif (1996), the soils of lowland forests in Kalimantan are low in mineral concentration, similar to those in our study. Medina & Cuevas (1989) summarized the nutrient inventory in Amazonian forests. Although soil depths were not consistent among study sites, the top soil contained 0.7 to 2.5 Mg N ha^{-1}, 21 to 434 kg P ha^{-1}, 19 to 69 kg K ha^{-1}, 3 to 133 kg Ca ha^{-1} and 7 to 24 kg

Mg ha^{-1}. Our data are within the range of humid tropics (Table 1). This is attributable to the higher precipitation in humid tropics compared with the seasonally dry tropical forests in Thailand. The exchangeable cations are generally thought to be leached out in humid tropics.

Nutrient ion flux in soil solution, or leachate, is estimated in Table 2. Under the untrenched site, soil solution carries about 6 kg N ha^{-1} yr^{-1}, 93.1 kg K ha^{-1} yr^{-1}, 17.4 kg Na ha^{-1} yr^{-1}, 13.6 kg Mg ha^{-1} yr^{-1} and 8.72 kg Ca ha^{-1} yr^{-1} at Pasoh FR and about 20 kg N ha^{-1} yr^{-1}, 31.9 kg K ha^{-1} yr^{-1}, 7.42 kg Na ha^{-1} yr^{-1}, 7.06 kg Mg ha^{-1} yr^{-1} and 13.4 kg Ca ha^{-1} yr^{-1} at FRIM site. The leachate at 120 cm deep is reported to carry 4.0 kg N ha^{-1} yr^{-1}, 4.6 kg K ha^{-1} yr^{-1}, 24.0 kg Na ha^{-1} yr^{-1}, 1.2 kg Mg ha^{-1} yr^{-1} and 2.2 kg Ca ha^{-1} yr^{-1} in a Guyanan forest (recalculated from Brouwewr & Riezebos 1998). In the case of a less-weathered temperate forest sol, the flux of the soil nutrients at 20 cm deep was 4 to 40 kg N ha^{-1} yr^{-1}, 40 kg K ha^{-1} yr^{-1}, 8 to 10 kg Na ha^{-1} yr^{-1}, 10 kg Mg ha^{-1} yr^{-1} and 40 kg Ca ha^{-1} yr^{-1} (recalculated from Haibara et al. 1990).

As for inorganic N, our results are higher than the Guyanan forests and similar to temperate forest. Since the inorganic N content of the soil solution decreased from upper soil to deeper soil due to plant uptake, microbial immobilization and fixation by soil particles, it is possible that a deeper position of a leachate collection in the Guyanan forests could cause a lower value in inorganic N. The flux of mineral cations is dependent on the nature of the bedrock, the extent of weathering and the characteristics of precipitation. Probably the Guyanan forests, the temperate forests, Pasoh FR and FRIM site have different conditions. But the values of these forests show no critical differences. Under the intact forest conditions, the flux of mineral cations seems to be rather stable.

5.2 Trenching effects

Trenching caused significant changes in the flux of P, K, Na and Ca, but not N and Mg. There was no significant difference in the inorganic N flux between the untrenched and trenched sites, as the flux of NH_4^+-N was small and the variation in that of NO_3^--N was high within each site. The flux of Mg decreased at Pasoh FR and significantly increased at FRIM site in response to trenching treatment.

The mobility of positively charged NH_4^+-N in soil profiles is lower than that of negatively charged NO_3^--N (Barber 1962). The cation exchange site dominates the surface of a soil particle and adsorbs NH_4^+-N. Produced NH_4^+-N is fixed on the surface of a soil particle, but NO_3^--N is easily leached. If plant uptake ceases in the rooting zone where the NO_3^--N is produced, NO_3^--N leaches down to a deeper soil profile (Bormann & Likens 1979). Vitousek & Matson (1985) showed low NO_3^--N losses in some forests after disturbance. They suggested that a plantation with enough organic matter can control nitrogen losses by means of microbial immobilization processes.

Soils of Pasoh FR and FRIM site do not have so much organic matter (Table 1). Microbial immobilization seems not to be able to completely regulate the NO_3^--N loss even in untrenched sites at both Pasoh FR and FRIM site. Although the deficiency of organic matter as a driving force for the immobilization process causes a large loss (29 kg N ha^{-1} yr^{-1}) of NO_3^--N, especially at FRIM site, this loss is smaller than expected value mentioned later. The extent at which immobilization and nitrification occurs may be variable among sites. Disturbance could cause more variability in immobilization and nitrification potential, which are governed by biological processes, in trenched sites. In addition, an anaerobic denitrification could occur in the soil where the reductive condition is occasionally created after heavy rain. Thus, the variation in the flux of NO_3^--N was high in trenched sites and no significant difference in the flux of NO_3^--N between untrenched and trenched site was observed at both Pasoh FR and FRIM site.

The logged forest in Guyana loses more than 100 kg N ha^{-1} yr^{-1} (recalculated from Brouwer & Riezebos 1998). N mineralization potential of tropical forest soil in Costa Rica is 217 to 1140 kg N ha^{-1} yr^{-1} (Matson et al. 1987, Vitousek & Denslow 1986). This value is much higher than the values given for temperate forests (Pastor et al. 1984). A tropical forest that has high N mineralization potential, like the Costa Rican forest, might have a potential to lose much inorganic N after a disturbance, due to its low immobilization potential.

The flux of PO_4^{3-}-P is larger in trenched sites than in untrenched sites. Trenching treatment significantly increases the PO_4^{3-}-P flux. As with NO_3^--N, PO_4^{3-}-P is negatively charged but its mobility should be low in soil profiles. Even if there is inadequate organic matter in the soil, PO_4^{3-}-P is likely to stay at the original place after the disturbance. However, the flux of PO_4^{3-}-P actually increases in trenched sites. Fresh organic matter like detached roots is supposed to act as P source in trenched site as reported in Jordan (1985). Both N and P are mineralized by biological processes. There might be a competition for inorganic N and P among heterotrophs, plants and autotrophs under certain conditions. When the competition is modified by removing plant roots, heterotrophs can utilize excess N or P using ambient organic matter. At both Pasoh FR and FRIM site, immobilization potential was low due to the low availability of organic matter. Thus, the flux of PO_4^{3-}-P also increases in response to trenching treatment despite its low mobility.

Logging has been found to cause an increase in cation loss in a northern hardwood forest (Bormann & Likens 1979) and a Guyanan tropical forest (Brouwer & Riezebos 1998). Cation concentrations in the leachate from closed forest in Guyana are smaller than values for Pasoh FR or FRIM site, except for Na. Cation concentrations in the leachate from logged forest increased in comparison with closed forest in Guyana, but were similar to the untrenched site of Pasoh FR or FRIM site. Clear trenching effects were detected in the loss of sum of cations in our study. However, trenching effects on the loss of Mg was insignificant. The loss of Mg from trenched site was greater than from untrenched site at FRIM site. Although the loss of Na and Ca was under clear trenching effect, significance for the loss of Ca was lower than for that of alkaline metals due to lower mobility of alkaline earth metals than alkaline metal. Unlike the other cations, the flux of Mg increased in response to trenching as observed in other disturbed site. When plant uptake stops, the excess Mg which is to be taken up by plant goes down to deeper soil profile. Exclusion of the plant roots results in the retention of the mineral cations, especially of K, within the soil system either at Pasoh FR and FRIM site. In other words, the existence of living plant roots accelerates the release of K from the soil surface, or living plant roots excrete K from the plant to the soil. When the forest soil is disturbed, the extent of the increment or decrement in the cation flux depends upon the condition of each forest ecosystem.

The excess nutrients in trenched sites are supposed to leach out from rooting zone if soil has no conservation system. Assuming that nutrients returned by litterfall reflect the amount of nutrients taken up by plants, the nutrients returned by litterfall approximate the excess nutrient flux in trenched sites over the flux in untrenched sites. Two years average of the return of nutrients in litterfall was 130 kg ha^{-1} yr^{-1} for N, 5.5 kg ha^{-1} yr^{-1} for P, 25 kg ha^{-1} yr^{-1} for K, 1.9 kg ha^{-1} yr^{-1} for Na, 14 kg ha^{-1} yr^{-1} for Mg and 87 kg ha^{-1} yr^{-1} for Ca at FRIM site (Yamashita et al. 1995).

Pasoh FR produces 8.6 tonnes ha^{-1} yr^{-1} of fine litter (Yamashita & Takeda 1998). If the litter material of Pasoh FR contains the same level of nutrients as the litter of *Dipterocarpus baudii* at FRIM site, the forest floor would receive 105 kg ha^{-1} yr^{-1} for N, 4.3 kg ha^{-1} yr^{-1} for P, 19 kg ha^{-1} yr^{-1} for K, 1.0 kg ha^{-1} yr^{-1} for Na, 11 kg ha^{-1} yr^{-1} for Mg and 61 kg ha^{-1} yr^{-1} for Ca. If all the nutrients that are to be taken up by plant leach out in response to trenching treatment, the nutrient flux in trenched sites would reach at the similar level as the amount returned by litterfall. However, the actual flux of N, P and Ca in trenched site is much lower than expected value. The flux of these three nutrients is maintained at low level even after disturbance. The flux of K is greater than expected value in both sites at Pasoh FR. The magnitude of the flux of Mg and Na is the similar order as expected value. As results, the flux of K, Na and Mg compares to the plant requirement, but the flux of N, P and Ca is maintained at similar or low level even after disturbance.

5.3 Site effects

Differences between Pasoh FR and FRIM site are recognized in vegetation type and climatic condition. These sites also differ in soil chemical properties (Table 1). Pasoh FR and FRIM site probably show different soil type.

The flux of inorganic N is dominated by NH_4^+-N at Pasoh FR and by NO_3^--N at FRIM site, and is larger at FRIM site than at Pasoh FR (Table 2). While the flux of NH_4^+-N showed similar values between Pasoh FR and FRIM, the flux of NO_3^--N showed a significant difference between sites. The production of NO_3^--N is a secondary process of N mineralization. That is, organic N is mineralized to NH_4^+-N and then nitrified to NO_3^--N. In addition, apparent nitrification does not always occur in all kinds of forest soils (Hirobe et al. 1998). Differences in climatic condition could be noted as one of factors to cause higher NO_3^--N loss at FRIM. The FRIM site has more precipitation and slightly higher temperatures than Pasoh FR (Fig. 1). Continuously moist conditions and higher temperatures may make the environment suitable for N mineralization and nitrification at FRIM site, while the drier condition in Pasoh FR leads to less microbial activity. The nitrification potential at Pasoh FR is supposed to be lower than at FRIM site. Although it was not only the quantity (i.e. low T-C and T-N concentration) but the quality (i.e. low C-to-N ratio) of the organic matter, at both Pasoh FR and FRIM site, that is insufficient to immobilize inorganic N in a microbial pool, the low potential for nitrification causes lower NO_3^--N flux at Pasoh FR than at FRIM site.

The flux of PO_4^{3-}-P is larger at FRIM site than at Pasoh FR. The standing stock of available P in soil is also larger at FRIM site than at Pasoh FR (Table 1). There is a positive relationship between the standing stock and the flux of P. This suggests that P availability is higher at FRIM site than at Pasoh FR. High P availability should promote N mineralization and subsequent nitrification at FRIM site.

The flux of mineral cations is greater at Pasoh FR than at FRIM site (Table 2). The standing stock of the exchangeable mineral cation in soil is larger at Pasoh FR than at FRIM site, except for Ca (Table 1). There is no general trend in the relationship between the standing stock and annual flux of soil nutrients, but the more the individual exchangeable cation accumulates, the greater the annual flux with exception of Mg (Fig. 5). The larger annual flux of cations at Pasoh FR is due to larger K and Na fluxes. The flux of K and Na is dependent on the standing stock in soil. The standing stock of K and Na at Pasoh FR was larger than that at FRIM site, and the flux of K and Na at Pasoh FR was larger than that at FRIM site. Despite the similar size of the standing stock in comparison with other elements at both Pasoh FR and FRIM site, the K flux is relatively large. This is due to the higher mobility of K in soil profiles and the existence of external sources of K like precipitation and throughfall.

The flux of cations and anions was balanced at FRIM site. But anions (NO_3^--N and PO_4^{3-}-P) were at low concentrations relative to cations at Pasoh FR (Fig. 4). This imbalance may be attributable to the lack of acids like bicarbonate or oxalic acid from the anion value. When we consider the ion balance of leachate after disturbance, it is suggested that we had better clarify the contribution of dissolved organic ions in leachate.

6. CONCLUDING REMARKS

The exclusion of plant root causes increase in the flux of anions and decrease in the flux of cations. However, the extent of increase was too small to account for the excess amount of nutrient that is created by ceased uptake. Inorganic N could be maintained at low level by means of immobilization probably and denitrification in trenched sites. In case of inorganic P, microbial immobilization and fixation by soil materials might keep the flux of inorganic P low. Although some are more and some are less than estimates for plant requirement, the flux of cations decrease in response to trenching treatment, except for Mg. Since the exclusion of plant roots causes decrease in the flux of cations, it is proposed that living root accelerates the nutrient release from exchange site or act as cation source in soil at least for a short period

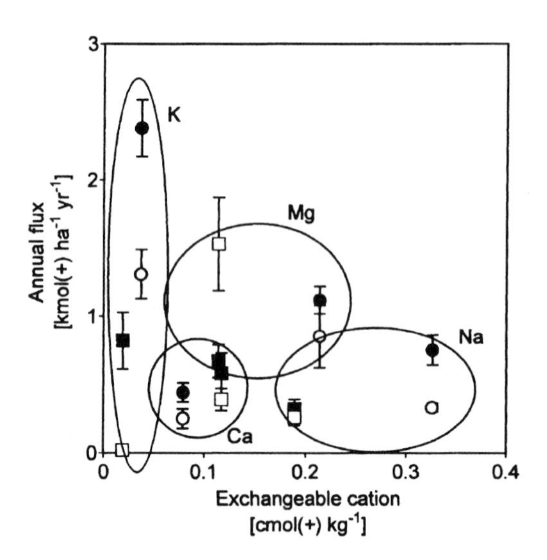

Fig. 5 Annual flux of cations in relation to the standing stock in soil as an exchangeable form. Vertical bars show one standard error. ●: Untrenched at Pasoh FR; ○: Trenched at Pasoh FR; ■: Untrenched at FRIM site; □: Trenched at FRIM site.

after disturbance.

ACKNOWLEDGEMENTS

The authors are grateful to Drs. L. G. Kirton (FRIM), J. Intachat (FRIM), A. Furukawa (NIES), Y. Tsubaki (NIES) and K. Takamura (NIES) for arranging this study in the project. Messrs. Noraffandy Othman, Suffian Mohammad and Abdul Rani Hussain rendered invaluable field and laboratory assistance. Drs. N. Tokuchi (Kyoto Univ.) and K. Nishimura (Kyoto Univ.) gave us technical advice on IER experiments. Dr. T. Masunaga (Shimane Univ.) made helpful comments on an earlier draft of the manuscript. Mr. I. Yamamoto (Organo Co.) kindly provided the regenerated IER for us. The present study is a part of joint research project between the NIES, FRIM and UPM (Global Environment Research Program granted by Japan Environment Agency, Grant No. E-2).

REFERENCES

Baker, D. E. & Suhr, N. H. (1982) Atomic adsorption and flame emission spectrometry. In Page, A. L., Miller, R. H. & Keeney, D. R. (eds). Methods of Soil Analysis, Part 2, Chemical and Microbiological Properties (2nd ed.), ASA-SSSA, Madison, USA. pp.13-27.

Barber, S. A. (1962) A diffusion and mass-flow concept of soil nutrient availability. Soil Sci. 93: 39-49.

Binkley, D. & Hart, S. C. (1989) The components of nitrogen availability assessments in forest soils. Adv. Soil Sci. 10: 57-112.

Binkley, D. & Matson, P. (1983) Ion exchange resin bag method for assessing forest soil nitrogen availability. Soil Sci. Soci. Am. J. 47: 1050-1052.

Bormann, F. H. & Likens, G. E. (1979) Pattern and Process in a Forested Ecosystem. Springer-Verlag, Berlin, 253pp.

Brouwer, L. C. & Riezebos, H. T. (1998) Nutrient dynamics in intact and logged tropical rain forest in Guyana. In Schulte, A. & Ruhiyat, D. (eds). Soils of Tropical Forest Ecosystems - Characteristics, Ecology and Management. Springer-Verlag, Berlin, pp.73-86.

Bruijnzeel, L. A. (1991) Nutrient input-output budgets of tropical forest ecosystems: a review. J. Trop. Ecol. 7: 1-24.

Haibara, K., Kawashima, Y. & Aiba, Y. (1990) Use of ion exchange resin (IER) to study the movement of elements in forest soil. Jpn. J. Ecol. 40: 19-25 (in Japanese with English summary).

Hirobe, M., Tokuchi, N. & Iwatsubo, G. (1998) Spatial variability of soil nitrogen transformation patterns along a forest slope in *Cryptomeria japonica* D.Don plantation. Eur. J. Soil Biol. 34: 123-131.

Jordan, C. F. (1985) Nutrient Cycling in Tropical Forest Ecosystems. John Wiley & Sons, Chichester, 190pp.

Keeney, D. R. & Nelson, D. W. (1982) Nitrogen-inorganic forms. In Page, A. L., Miller, R. H. & Keeney, D. R. (eds). Methods of Soil Analysis, Part 2, Chemical and Microbiological Properties, (2nd ed.), ASA-SSSA, Madison, USA, pp.643-698.

Matson, P. A., Vitousek, P. M., Ewel, J. J., Mazzarino, M. J. & Robertson, G. P. (1987) Nitrogen transformations following tropical forest felling and burning on a volcanic soil. Ecology 68: 491-502.

Medina, E. & Cuevas, E. (1989) Patterns of nutrient accumulation and release in Amazonian forests of the upper Rio Negro basin. In Proctor, J. (ed). Mineral Nutrients in Tropical Forest and Savanna Ecosystems. Blackwell Scientific Publications, Oxford, pp.217-240.

Ohta, S. & Syarif, E. (1996) Soils under lowland dipterocarp forests-characteristics and classification. In Schulte, A. & Schöne, D. (eds). Dipterocarp Forest Ecosystems—Towards Sustainable Management. World Scientific, Singapore, pp.29-51.

Olsen, S. R. & Sommers, L. E. (1982) Phosphorus. In Page, A. L., Miller, R. H. & Keeney, D.

R. (eds). Methods of Soil Analysis, Part 2, Chemical and Microbiological Properties, (2nd ed.), ASA-SSSA, Madison, USA. pp.403-430.

Pastor, J., Aber, J. D., McClaugherty, C. A. & Melillo, J. M. (1984) Aboveground production and N and P cycling along a nitrogen mineralization gradient on Blackhawk Island, Wisconsin. Ecology 65: 256-268.

Sakurai, K., Tanaka, S., Ishizuka, S. & Kanzaki, M. (1998) Differences in soil properties of dry evergreen and dry deciduous forests in the Sakaerat Environmental Research Station. Tropics 8: 61-80.

Sulaiman, S., Abdul Rahim, N & LaFrankie, J. V. (1994) Pasoh Climatic Summary (1991-1993). FRIM-Research Data 3, 31pp.

SYSTAT (1992) SYSTAT: Statistics, Version 5.2 Edition, Evanston, IL: SYSTAT Inc., 724pp.

Vitousek, P. M. & Denslow, J. S. (1986) Nitrogen and phosphorus availability in treefall gaps of a lowland tropical rainforest. J. Ecol. 74: 1167-1178.

Vitousek, P. M. & Matson, P. A. (1985) Disturbance, nitrogen availability, and nitrogen losses in an intensively managed loblolly pine plantation. Ecology 66: 1360-1376.

Vitousek, P. M., Gosz, J. R., Grier, C. C., Melillo, J. M. & Reiners, W. A. (1982) A comparative analysis of potential nitrification and nitrate mobility in forest ecosystems. Ecol. Monogr. 52: 155-177.

Yamashita, T. & Takeda, H. (1998) Decomposition and nutrient dynamics of leaf litter in litter bags of two mesh sizes set in two dipterocarp forest sites in Peninsular Malaysia. Pedobiologia 42: 11-21.

Yamashita, T., Takeda, H. & Kirton, L. G. (1995) Litter production and phenological pattern of *Dipterocarpus baudii* in a plantation forest. Tropics 5: 57-68.

Yamashita, T., Nakanishi, A., Tokuchi, N. & Takeda, H. (1999) Chemical properties and nutrient accumulation in two Siamese forest soils. Appli. For. Sci., Kansai 8: 89-94 (in Japanese with English summary).

6 Characteristics of Energy Exchange and Surface Conductance of a Tropical Rain Forest in Peninsular Malaysia

Makoto Tani[1], Abdul Rahim Nik[2], Yoshikazu Ohtani[3], YukioYasuda[3], Mohd Md. Sahat[2], Baharuddin Kasran[2], Satoru Takanashi[1], Shoji Noguchi[4], Zulkifli Yusop[5] & Tsutomu Watanabe[3]

Abstract: Energy exchange above tropical rain forest was studied using micro-meteorological monitoring from a 52 m tower established in the Pasoh Forest Reserve (Pasoh FR) in Peninsular Malaysia. The meteorological conditions were comparatively drier during the first half of the year and wetter toward the end of the year due to the seasonal variation of rainfall. The five-year observational period from 1995 to 1999 included a low rainfall duration due to the El Niño from 1997 to 1998. The latent heat flux estimated by the Bowen ratio method occupied a dominant portion of the energy exchange even in the driest condition in early 1998. Although evapotranspiration from the dry canopy tended to be smaller in this period than in a wet period during the end of 1998, the surface conductance estimated using the Penman Monteith Equation was consistently controlled by the same function of solar radiation and specific humidity deficit. This suggests the evaporation did not suffer from severe stress of soil water even in the driest condition.

Key words: energy flux, evapotranspiration, meteorology, surface conductance, tropical rainforest.

1. INTRODUCTION

Understanding the effects of tropical rain forest on climate and water resources at both local and global scales is one of the most important current environmental issues (Nobre et al. 1991). Intensive observational studies for estimating energy exchange including heat and vapor fluxes between tropical forest and the atmosphere are critical to develop such an understanding. Although forest hydrology has provided annual evapotranspiration estimates from annual water budgets in small experimental catchments (Bruijnzeel 1990), it is difficult to understand flux responses to many variables such as meteorology, physiology and soil moisture. Micro-meteorological observations are necessary to understand such flux responses. However, such data have been obtained from limited regions such as the Amazonian rain forest in South America (Shuttleworth 1988). This study demonstrated that a large percentage of incoming net radiation was used as latent heat by evapotranspiration and that actual evapotranspiration in each month was nearly equal to its potential value.

In South East Asia, only a few findings have been obtained from

---

[1] Kyoto University, Kyoto 606-8502, Japan. E-mail: tani@kais.kyoto-u.ac.jp
[2] Forest Research Institute Malaysia (FRIM), Malaysia.
[3] Forestry and Forest Products Research Institute (FFPRI), Japan.
[4] Japan International Research Center for Agricultural Sciences.
[5] Universiti Teknologi Malaysia (UTM), Malaysia.

observational studies estimating fluxes from tropical forests. For a tropical savannah with a long dry season (November–March), Pinker et al. (1980) presented a large difference in the radiative energy into sensible and latent heats between dry and wet seasons. For a tropical rain forest in Java, Indonesia, Calder et al. (1986) reported a slightly larger annual evapotranspiration amount than that of the Amazonian forest. However, evapotranspiration from dry canopy was estimated indirectly from soil moisture measurements over a short duration because no meteorological variables were monitored above the canopy.

A study estimating energy exchange above forest has been conducted under NIES-FRIM-UPM Joint Research Project using an observation tower established at Pasoh Forest Reserve (Pasoh FR) in Peninsular Malaysia. This paper reports meteorological observation results during the recent five years and analyzes the characteristics of energy exchange above the forest and the characteristics of surface conductance of the forest using the meteorological data sets.

2. SITE DESCRIPTION

This study was conducted in the Pasoh FR, which is located near Simpang Pertang in Negeri Sembilan about 140 km south east of Kuala Lumpur in Peninsular Malaysia (2°59' N, 102°18' E). An intensive research project by the International Biological Programme (IBP) was concentrated in this forest reserve from 1970 to 1974 (e.g. Soepadmo 1978). Micro-meteorological measurements for a short duration were reported as one of the findings from the project (Aoki et al. 1975) though the energy exchange was not estimated.

The core area (600 ha) of the reserve is covered with a primary lowland mixed dipterocarp forest, which consists of various species of *Shorea* and *Dipterocarpus*. The continuous canopy height is about 35 m, although some emergent trees exceed 45 m. Based on the empirical equations obtained for the Pasoh FR by Kato et al. (1978), the leaf area index (LAI) estimated from tree diameter observations (Niiyama, unpublished) was 6.52. The core area is surrounded by a buffer zone (650 ha) of regenerating logged over forest and primary hill dipterocarp forest (1,000 ha) (Manokaran & Kochummen 1994; Soepadmo 1978). The area has a gently undulating topographical feature. The altitude of the core area ranges from 75 to 150 m a.s.l. Soils belong to Durian Series characterized by the presence of a band of laterite and compact structure derived from shales within the area. The A horizon is thin (0–2 cm) and deeper soils are bright yellowish or reddish brown and light to heavy clay. Lateritic boulders are abundant below 30 cm depth and increasing with depth (Yoda 1978).

3. OBSERVATION DESIGN

A 52 m tower (Chaps. 1, 2) established near the IBP Plot 1 in the core area was used for our observation. The tower is located at the top of gentle hill. Meteorological factors were monitored by sensors installed at the 52 m height. They consisted of downward and upward solar radiations (albedometer EKO MR-22), net radiation (EKO MF-40), air temperature and humidity (Visala HMP-35C), wind direction and wind velocity (Campbell 03001, the threshold of wind speed is 0.5 m s^{-1}) and rainfall (Yokogawa Weathac B-011-00). Except for rainfall, the above factors were measured every 15-second and their 30-minute averages were recorded in a data logger (Campbell CR10X), which also recorded the maximum wind velocity and 30-minute rainfall totals. Accurate vertical profiles of air temperature, vapor pressure and wind velocity necessary for an energy flux estimation were measured by two ventilated psychrometers (EKO MH-020S) with platinum resistance temperature devices at 43.6 and 52.6 m, and four three-cup anemometers (Ikeda WM-30P, the threshold is 0.3 m s^{-1}) at 43.6, 46.6, 49.1 and 52.6 m. The 30-minute averages of the ventilated psychrometers were recorded using the same CR10 data logger. The 30-minute cumulated pulses from the four anemometers (1 pulse for 1 m run-of-wind) were counted by two additional loggers (Kona Sapporo DS-64K).

The net radiometer was calibrated in Tsukuba, Japan by comparing its values with the accurate values authorized by Japan Meteorological Agency. The values obtained from the two ventilated psychrometers and the four anemometers were sometimes compared to each other by installing them at the same height.

The vertical profiles of air temperature, vapor pressure and wind velocity at 6 heights (1.0, 17.0, 33.0, 41.5, 49.1 and 52.6 m) measured on the 4th and 5th of March 1995 was also used to understand the heat storage within canopy (Ohtani et al. 1997). The heat flux into the soil was neglected in the routine of energy flux estimation since its magnitude was smaller than these three storage terms.

4. DATA ANALYSIS

4.1 Energy budget

The Bowen ratio method was used for estimating sensible and latent heat fluxes from the forest canopy. The energy budget above the canopy is described as:

$$R_n = H + lE + Q_s + A \tag{1}$$

where R_n(kW m^{-2}) is the net radiation, H and lE are the sensible and latent heat fluxes, Q_s is the stored energy increase within a forest canopy, and A is the net rate of energy absorption by photosynthesis and other biochemical processes. Q_s consists of:

$$Q_s = Q_a + Q_w + Q_b + Q_g \tag{2}$$

where Q_a and Q_w are the sensible and latent heat storage increases in the canopy air (kW m^{-2}), and Q_b and Q_g are the heat storage increases within the biomass of the canopy and the soil, respectively.

Since A is assumed to be around 3% of R_n (Jarvis et al. 1976), this term was neglected. R_n and Q_g were obtained from the routine monitoring. Q_a, Q_w and Q_b were estimated from the air temperature and vapor pressure monitored at the reference height of 52.6 m (Ohtani et al. 1997). The empirical equations used here were developed based on the measurement of vertical profiles of air temperature and vapor pressure within the canopy carried out in March 1995 as described earlier:

$$Q_a = 0.0172(dT/dt) \tag{3}$$

$$Q_w = 0.0253(dV/dt) \tag{4}$$

$$Q_b = 0.0180(dT/dt) \tag{5}$$

where T and V are the air temperature (°C) and vapor pressure (hPa), and dT/dt and dV/dt are their increases in an hour, respectively. Thus, the available energy Q distributed to sensible and latent was calculated as:

$$Q = R_n - Q_s \tag{6}$$

4.2 Bowen ratio method

Sensible and latent heat fluxes are estimated from the available energy using Bowen ratio method as:

$$H = \frac{B}{1+B}Q \tag{7}$$

$$lE = \frac{1}{1+B}Q \tag{8}$$

Bowen ratio, B, is calculated from differences in potential temperature and vapor pressure between the two heights:

$$B = \frac{c_p}{l}\frac{\theta_1 - \theta_2}{q_1 - q_2} \tag{9}$$

where θ is the potential temperature (°C), q is the specific humidity (kg kg^{-1}), c_p is the specific heat of air (kJ kg^{-1} K^{-1}) and l is the latent heat of vaporization (kJ kg^{-1}). The two heights indicated by suffixes 1 and 2 were assigned to 43.6 and 52.6 m in our site.

4.3 Estimation of surface conductance

Effects of environmental variables on transpiration are represented by the control of the surface conductance, which means the integrated behavior of stomatal conductance at an individual leaf scale. The surface conductance (g_c m s^{-1}) is calculated by Penman Monteith Equation (Monteith 1965) as:

$$g_c = \{[(\Delta l / c_p)B - 1]r_a + [(\rho l / lE)(q_{SAT}(T) - q)]\}^{-1} \tag{10}$$

where Δ is the slope of saturated specific humidity curve against temperature (kg kg^{-1} K^{-1}), ρ is the density of air (kg m^{-3}), r_a is the aerodynamic resistance (m^{-1} s), and q and $q_{SAT}(T)$ are the specific humidity at the reference height (52.6 m) and the saturated specific humidity (kg kg^{-1}) for the air temperature (T) at the reference height. The value of g_c can be calculated from the r_a value since Bowen ratio (B) and lE are calculated in Eqs. 8 and 9. Considering that sensitivity of the aerodynamic resistance (r_a) to the evapotranspiration was estimated small for Amazonian rain forest (Shuttleworth 1988), r_a was assumed to be equal to the aerodynamic resistance for the momentum transfer (r_m) as:

$$r_a = r_m = \frac{u_r}{u_*^2} \tag{11}$$

where u_r is the wind velocity at the reference height (m s^{-1}) and u_* is the friction velocity (m s^{-1}).

The friction velocity was calculated from the profile of wind velocity measured at the four heights above the canopy in terms of the correction of atmospheric stability.

$$u_* = ku_r\left[\int_{\varsigma_0}^{\varsigma_r}\frac{\phi_m}{\varsigma}d\varsigma\right]^{-1} \tag{12}$$

where

$$\varsigma = (z - d) / L \qquad (13)$$

$$L = \frac{-u_*^3 \rho}{k g \left(\frac{H}{T_k c_p} + 0.61E \right)} \qquad (14)$$

ϕ_m is the non-dimensional universal function for momentum, k is Karman's constant (= 0.4), $\zeta_0 = z_0/L$, z_0 is the roughness length, $\zeta_r = (z_r-d)/L$, z_r is the reference height (52.6 m) , d is the zero plane displacement, L is the Obukov's stability length, T_k is the air temperature at the reference height in absolute temperature (K) and E is the vapor flux (kg m^{-2} s^{-1}). In the calculation process, a constant value of the zero plane displacement (d) was given first as 33.0 m considering the canopy height of 43.0 m.

Estimating u_* through the above equations requires an iterative calculation process using universal functions. Although the universal functions have not been established for tall vegetation such as forest (Viswanadham et al. 1987), simple non-dimensional universal functions for short vegetation were employed here to produce our estimation of g_c. The functions for unstable conditions (Dyer & Hicks 1970) are:

$$\phi_m = (1 - 16\varsigma)^{-1/4} \qquad (15)$$

For stable conditions, the following equation is used here (Kondo et al. 1978).

$$\phi_m = 1 + \frac{7\varsigma + 70\varsigma^3}{1 + 3\varsigma + 10\varsigma^2} \qquad (16)$$

4.4 Periods for flux analyses

Analyses of energy fluxes were applied to the records in the daytime from DOY (Day of the year) 65 (March 6) to 95 of 1998 for the dry period and from DOY 338 (December 4) to 356 of 1998 for the wet period. Days with rainfall and days with rainfall on their previous days were both removed from the analyses to avoid influences of evaporation from the wet canopy. Twenty and eleven days were screened for the analyses in the dry and wet periods, respectively.

5. RESULTS AND DISCUSSION

5.1 Meteorological characteristics

Monthly mean values of meteorological variables monitored at the tower from March 1995 to December 1999 are shown in Fig. 1. Table 1 summarizes the annual mean values. In addition to the tower data, rainfall records monitored at the nearest meteorological observatory, Federal Land Development Authorities (FELDA) Pasoh Dua, located at the 3 km south of our tower site were employed in Fig. 1 to know the seasonal variations of rainfall in the normal year. The monthly rainfall values averaged from 1983 to 1997 were illustrated with those monitored at the tower. Fluctuations of Southern Oscillation Index (SOI) are also plotted to survey El Niño and La Niña conditions (Data were obtained from the web page of Long Paddock Historical SOI Data, http:// www.dnr.qld.gov.au/longpdk).

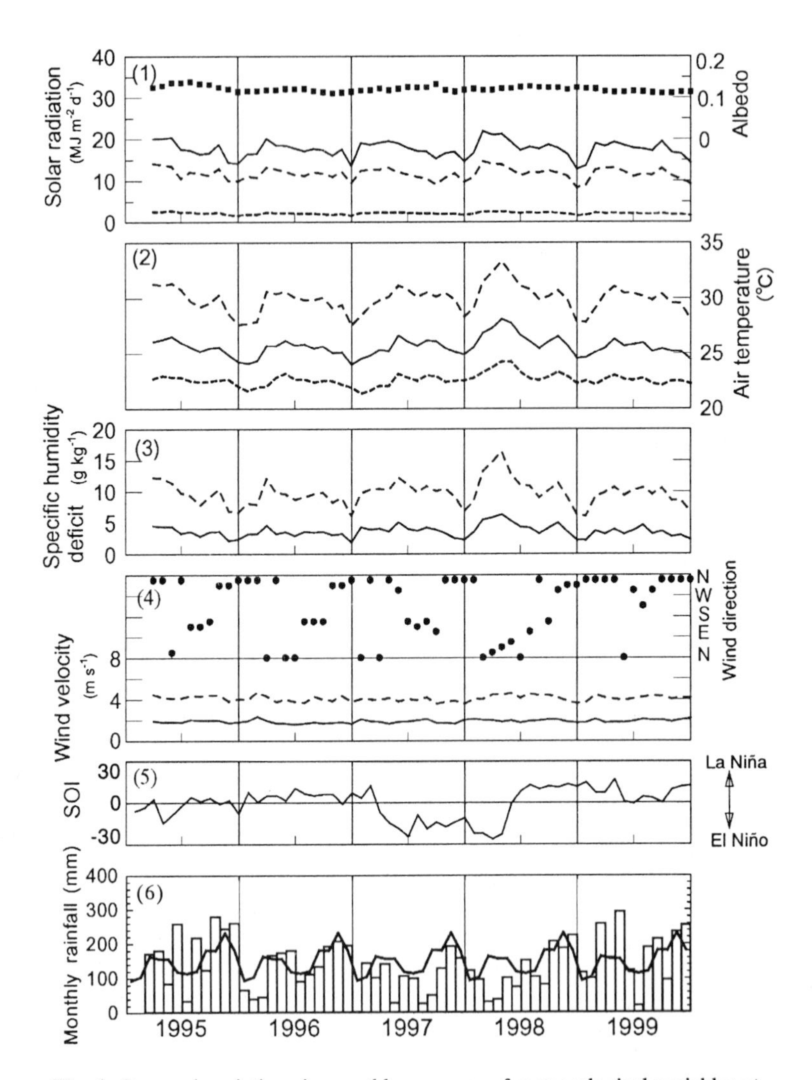

Fig. 1 Seasonal variations in monthly averages of meteorological variables at Pasoh Tower.

(1) ——— Daily total of downward solar radiation; -------- Daily total of upward solar radiation; ··············· Daily total of net radiation; ■ Daily average of albedo

(2) ——— Daily average of air temperature; -------- Daily maximum of air temperature; ··············· Daily minimum of air temperature

(3) ——— Daily average of specific humidity deficit; -------- Daily maximum of specific humidity deficit

(4) ● Daily vector average of wind direction; ——— Daily average of wind velocity; -------- Daily maximum of wind velocity

(5) ——— Southern Oscillation Index (SOI)

(6) Bar: Monthly rainfall; ——— Monthly rainfall in the normal year at FELDA Pasoh Dua

Table 1 Annual mean or total values of meteorological factors.

Year	Rainfall	Solar rad.	Albedo	Net rad.	Air temperature			Spec. humid. deficit		Wind velocity	
		MJ $m^{-2}d^{-1}$		MJ $m^{-2}d^{-1}$	Mean	Max	Min	Mean	Max	Mean	Max
	($mm\ y^{-1}$)	($mm\ y^{-1}$)		($mm\ y^{-1}$)	(deg C)			($g\ kg^{-1}$)		($m\ s^{-1}$)	
1995(Mar.-)	1855	17.26 (2585)	0.130	11.77 (1764)	25.5	29.8	22.6	3.35	9.26	1.90	4.22
1996	1610	17.30 (2590)	0.117	11.61 (1738)	25.2	29.4	22.4	3.29	9.03	1.81	4.04
1997	1182	17.59 (2634)	0.121	11.43 (1712)	25.5	29.9	23.2	3.74	10.10	1.87	3.90
1998	1426	18.21 (2729)	0.115	12.07 (1809)	26.4	30.7	22.5	4.40	11.09	1.93	4.22
1999	2065	17.36 (2600)	0.119	11.48 (1719)	25.4	29.7	22.6	3.26	9.13	1.93	4.07
1996-99	1571	17.62 (2638)	0.118	11.65 (1744)	25.6	29.9	22.6	3.67	9.84	1.89	4.07

'Mean', 'Max'and 'Min' indicate annual averages of daily mean, maximum and minimum values, respectively. Max' in wind velocity indicates annual averages of daily maximum values in 30-minute average wind velocity. Number in parentheses indicate energy flux value converted to mm y^{-1} of water equivalent using latent heat of vaporization.

Although seasonal variations of meteorological variables in Pasoh FR were small under the climate of tropical rain forest, radiative energy was comparatively high at the first half of the year and decreased toward the end of the year (Fig. 1). A similar trend was detected in the seasonal fluctuations of air temperature and specific humidity deficit. These characteristics may be due to the seasonal variations of rainfall in this region: a major rainy season is produced by the northeast monsoon from October to December while the generally weak southwest monsoon yields only a small peak of rainfall from March to May. Thus, the maximum and minimum monthly rainfall values in the normal year were recorded in November and in January (Fig. 1), and the annual amount was only 1,804 mm, lower than that in other regions of Peninsular Malaysia. It can be summarized that climate in Pasoh FR is characterized by low rainfall due to the inland location and that it was dry at the first half of a year, getting wetter toward the end of the year.

Four-year averages of meteorological variables in our tower observation period are also listed in Table 1. The annual rainfall average (1,571 mm) was much smaller than that of the normal year in FELDA Pasoh Dua (1,804 mm). The average net radiation 11.65 MJ m^{-2} d^{-1} coincides with the annual water vaporization energy of 1,744 mm y^{-1}, which is larger than the annual rainfall. This indicates the Budyko's radiation dryness was larger than unity, suggesting too dry condition for tropical rain forest. The rainfall was larger in 1995 and 1999 than in the normal year, similar to the normal year in 1996 but much smaller in 1997 and 1998. A tendency of high radiation, high air temperature and high specific humidity deficit was recorded in 1998 and was particularly remarkable from March to May in 1998 (Fig. 1). This was caused by unusual small rainfall in this duration, which clearly coincides to the El Niño from 1997 to early 1998 as shown in Fig. 1.

Details of radiative energy are described next. The averaged albedo was about 0.120, which is comparable to values observed in other tropical rain forests such as 0.1225 in Ducke, Amazon (Shuttleworth et al. 1984) and 0.12 – 0.13 in Nigeria (Oguntoyinbo 1970). Relationships of net radiation to solar radiation calculated from their hourly values in the daytime were expressed in the following regression equation ($R^2 = 0.996$):

$$R_n = 0.820 S_d - 0.0342 \tag{17}$$

where S_d is the downward solar radiation (kW m^{-2}). The relationship for the Amazonian forest was expressed by Shuttleworth et al. (1984) as:

$$R_n = 0.858 S_d - 0.035 \tag{18}$$

Both of the forests had very similar relationships. However, the annual total of net radiation 4,255 MJ m^{-2} in our site was larger than that in other tropical rain forests, that is about 3,600 MJ m^{-2} in Ducke, Amazon (Shuttleworth 1988) or 3,730 MJ m^{-2} in Java (Calder et al. 1986), probably because the annual rainfall amount during the observation period at Pasoh FR was very small compared to that in the normal year and was much smaller than the other forests (2,650 mm in Amazon and 2,850 mm in Java).

The wind velocity (Fig. 1) did not have a remarkable seasonal variation, and the monthly mean values of the daily average and the maximum were about 2 m s^{-1} and 4 m s^{-1}, generally weak under the tropical climate (Table 1). The predominant wind direction was north in general although the southern wind was dominant only from June to September.

5.2 Typical diurnal variations in fluxes

Figure 2 shows typical diurnal variations in heat fluxes with meteorological variables on rainless days, from DOY of 92 (April 2) to 94 in 1998, while Fig. 3 shows those on DOY of 345 (December 11), 355 and 356 in 1998. The atmospheric condition was very dry in the former period and wet in the latter. In the dry condition as shown in Fig. 2, the specific humidity deficit increased up to over 16 g kg^{-1} in the afternoon especially on fine days of DOY 92 and 94 caused by the increase of air temperature and the decrease of specific humidity. In spite of a dry condition, the latent heat flux was dominant throughout the three days (Fig. 2). The averaged fractions of latent heat to the net radiation in the daytime on DOY of 92, 93 and 94 were 0.65, 0.68 and 0.65, respectively. The value averaged over 20 days without influences of rainfall from DOY 65 to 95 was 0.68. In the wet period, the fractions on DOY of 345, 355 and 356 were 0.71, 0.73 and 0.75, and the value averaged over 11 days from DOY 338 to 356 was 0.73, which means higher fractions of the radiative energy were used by evapotranspiration in this wet period than in the dry period. However, it is important that the fraction did not drastically decrease due to the dry condition. This fraction value was similar to the value of 0.698 obtained from an Amazonian forest in Ducke (Shuttleworth et al. 1984). These data demonstrates that most of radiative energy is used by evapotranspiration in tropical rain forests even in dry seasons.

Daily variations of the surface conductance are also plotted in Figs. 2 and 3, showing that it reached a peak after a sharp increase in the morning and gradually decreased in the afternoon. This diurnal pattern was the same as that of stomatal conductance of an individual leaf measured with a porometer in this forest (Furukawa et al. 1994, 2001). A similar pattern was widely reported from temperate and tropical forests (Shuttleworth 1989). For example, an averaged daily variation obtained from Ducke was expressed as:

$$g_c = 12.17 - 0.531\,(t_s - 12) - 0.233\,(t_s - 12)^2 \qquad (19)$$

where t_s is the local time of day in hours. Curves calculated in Eq. 19 were also illustrated on each day in Fig. 3. Because the local time in Malaysia, t_m is one-hour later than the solar time there, $t_s = t_m - 1$ is substituted into Eq. 19 for the curves in Figs. 2 and 3. The patterns for our site and for Ducke are similar to each other, but it can be found that g_c for our site is larger than that for Ducke in the morning. This is caused by most of radiative energy is used for the latent heat flux in the rising limb of radiation in the morning (Figs. 2 and 3). One of the causes for the large g_c in the morning may be attributed to evaporation from free surface water due to the formation of dew in the previous night time (Kelliher et al. 1995) though we have no direct evidence for this phenomenon at Pasoh FR.

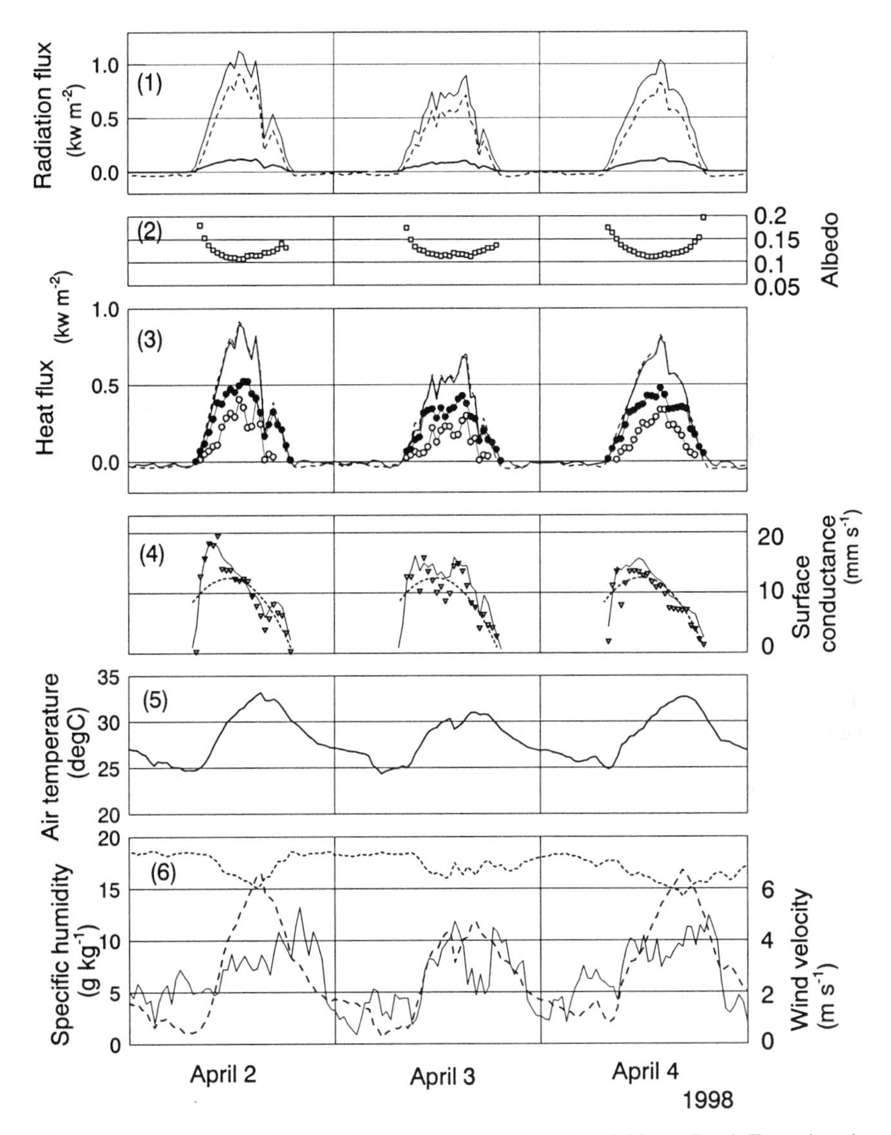

Fig. 2 Diurnal variations of energy fluxes with meteorological variables at Pasoh Tower in a dry period.
(1) ———— Downward solar radiation; ———— Upward solar radiation; ·············· Net radiation
(2) □ Albedo
(3) ·············· Net radiation; ———— Available energy; ○ Sensible heat flux; ● Latent heat flux
(4) ▽ Surface conductance estimated from the observation by Eq. 10; ———— Surface conductance calculated from environmental variables by Eq. 20; ·············· Averaged daily variation of surface conductance for Amazonian forest calculated by Eq. 19
(5) ———— Air temperature
(6) ·············· Specific humidity; – – – – Specific humidity deficit; ———— Wind velocity

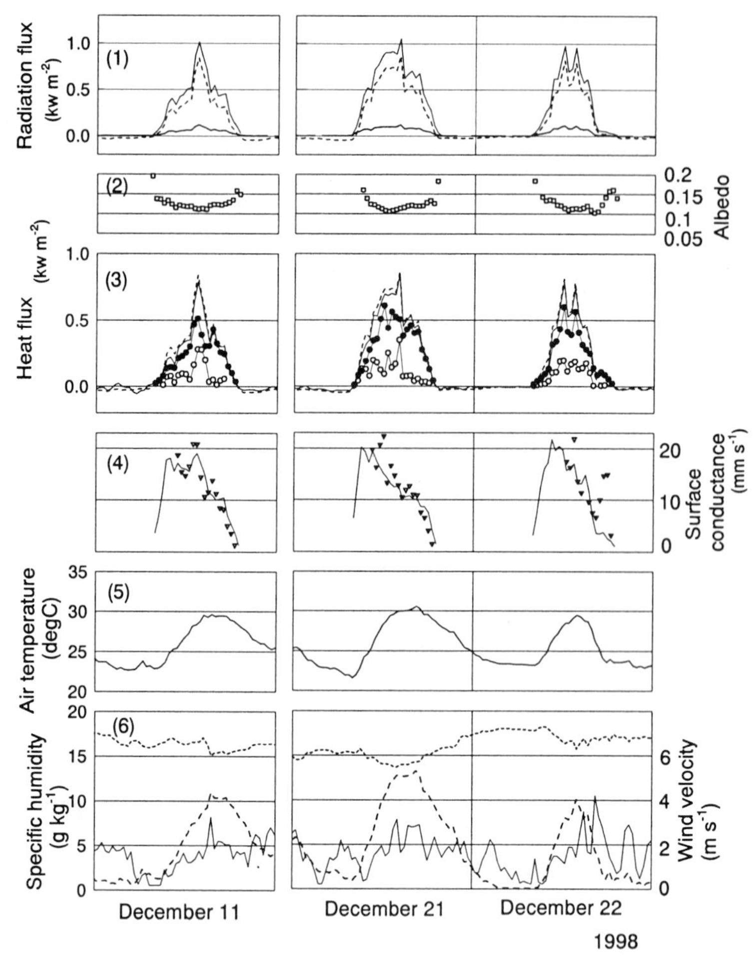

Fig. 3 Diurnal variations of energy fluxes with meteorological variables at Pasoh Tower in a wet period. Symbols are the same as Fig. 2.

5.3 Dependence of surface conductance on the environment

This section attempts to evaluate responses of surface conductance to the environmental variables in Pasoh FR. The surface conductance which represents the integrated behavior of the stomatal conductance at the canopy scale depends upon the environmental variables such as solar radiation, specific humidity deficit, temperature and soil moisture deficit (Jarvis 1976; Stewart 1988). For tropical forests, however, effects of some variables may be less critical for the responses. Previous results from Amazonian rain forests indicated that no substantial soil moisture deficits occurred (Dolman et al. 1991; Shuttleworth 1989). The annual variation range of air temperature was small only from 20 to 35℃ in our forest (Fig. 1). Hence, we omitted controls of soil moisture deficit and air temperature, and attempted to simply examine effects of solar radiation and specific humidity deficit on surface conductance.

Figs. 4 and 5 show responses of surface conductance to specific humidity deficit classified with the solar radiation range in the dry and wet periods, respectively. Plots of the surface conductance in Fig. 4 are distributed in a high range of specific humidity deficit due to a continuous dry condition since 1997 (see Fig. 1), while those in Figure 5 are in its low range. Curves in these figures indicate functional relationships of the responses (Jarvis 1976) expressed as:

$$g_c = g_0 g_r g_d \tag{20}$$

$$g_r = \frac{S_d / (S_d + a_1)}{1/(1 + a_1)} \tag{21}$$

$$g_d = \exp(a_2 \delta q) \tag{22}$$

where g_r is the radiation stress function, g_d is the specific humidity deficit stress function, g_0 is a constant representing the maximum surface conductance, and a_1 and a_2 are empirical constants. The constants, g_0, a_1, and a_2, fitted to the observational results throughout both the dry and wet periods are 35 mm s^{-1}, 0.6 kW, and -0.08 g^{-1} kg , respectively. The curves agree consistently with the responses of surface conductance in Figs. 4 and 5. The daily variations of g_c calculated in Eq. 20 also show good agreements with those observed in Figs. 2 and 3. These results mean the surface conductance was sufficiently controlled by solar radiation and humidity deficit through a simple functional relationship though the driest conditions in 1998 in our 5-year observational period was included.

Functional relationships in Eq. 20 obtained from our forest and from an Amazonian rain forest in Ducke (Dolman et al. 1991) are illustrated in Fig. 6. The constants, g_0, a_1 and a_2, for Ducke were 20.8 mm s^{-1}, 0.25 kW, and -0.064 g kg^{-1}, respectively. Figs. 4 and 5 indicate the differences in surface conductance between our site and Ducke is large particularly at small specific humidity deficits. The large differences in g_c occurred in the morning may be attributed to evaporation from dew as mentioned before (Figs. 2 and 3). In addition to this, however, some differences in canopy structure and/or physiological processes between these forests may influence surface conductance. Gash et al. (1989) compared the surface conductance of temperate pine forests between two sites, and found large differences in g_0 values (e.g. 33 mm s^{-1} for one site and 23 mm s^{-1} for the other) although the responses to environmental variables were similar at each site. They suggested the differences might originate from physiological structure or behavior and the contribution of transpiration from the bracken understorey. More detailed physiological data as well as flux data may be necessary to determine mechanisms for differences in surface conductance characteristics between different tropical rain forests.

5.4 Discussion on high evapotranspiration in a dry period

Our analyses demonstrate that responses of the surface conductance to specific humidity deficit in a dry period were consistently expressed with the same simple functional relationship in Eq. 20 as those in a wet period (Figs. 4 and 5). It may be surprising that high evapotranspiration rates or large fractions of the latent heat flux to the radiative energy could be maintained (Fig. 2) considering very small amounts of rainfall in 1997 and 1998 in Pasoh FR. Trees there may tolerate droughts because of the climatic environment characterized by low rainfall (Fig. 1). Discussion will be made on a background of high evapotranspiration in the dry period.

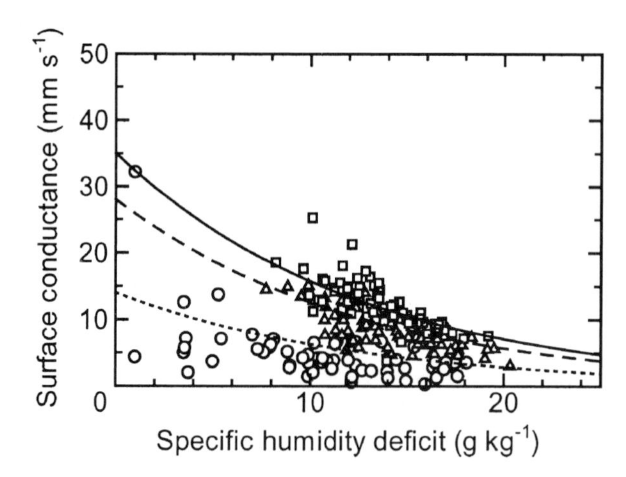

Fig. 4 Relationship of surface conductance to specific humidity deficit as a function of downward solar radiation S_d in a dry period.Symbols are estimated from the observation. □ $S_d \geqq 0.8$ kW s^{-1}; △ $0.8 > S_d \geqq 0.4$; ○ $0.4 > S_d$. Curves are calculated by from the environmental variables by Eq. 20.
——— $g_0 = 35$ mm s^{-1}, $a_1 = 0.6$ kW and $a_2 = -0.08$ g^{-1} kg and $S_d = 1$ kW s^{-1};
– – – – the same except for $S_d = 0.6$; ············ the same except for $S_d = 0.2$.

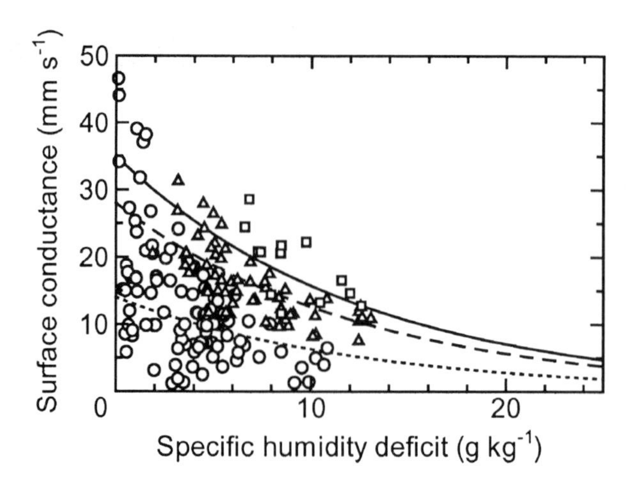

Fig. 5 Relationship of surface conductance to specific humidity deficit as a function of downward solar radiation S_d in a dry period. Symbols and curves are the same as Fig. 4.

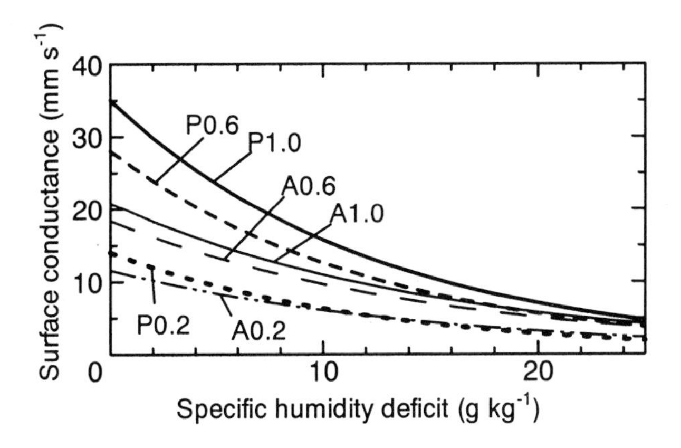

Fig. 6 Comparisons of functional relationship in Eq. 20 between Pasoh and Amazonian forests. P and A indicate Pasoh and Amazon respectively. Numbers are values of solar radiation S_d in kW s^{-1}. The other parameter values in Eq. 20 are the same as Fig. 4 for Pasoh FR, and those for Amazon are g_0 = 20.8 mm s^{-1}, a_1=0.25 kW and a_2 = -0.064 g^{-1} kg.

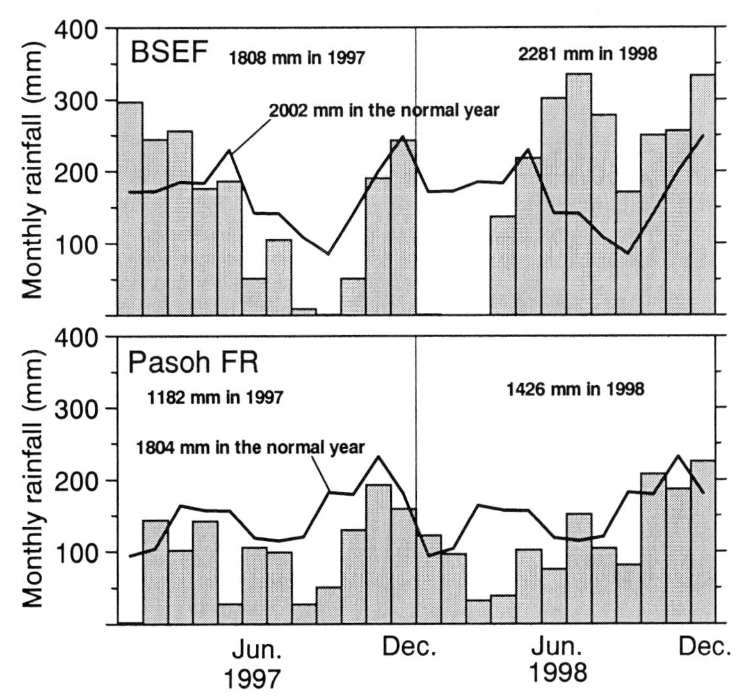

Fig. 7 Variations of monthly rainfall in Pasoh FR and Bukit Soeharto Education Forest (BSEF, East Kalimantan). Lines indicate monthly rainfall in the normal year. Numbers are the annual total. The values in normal year for Pasoh are calculated from the records in FELDA Pasoh Dua. Data of BSEF are after Toma et al. (2000a).

Some studies from Amazon suggested that maintenance of high transpiration from a tropical rain forest was supported by deep root system (Hodnet et al. 1996; Nepstad et al. 1994), which may partly cause our result. In addition to this, we should take it into consideration that rainless period was not long enough for a clear decreasing of evapotranspiration in Pasoh FR. From results on a physiological study in Bukit Soeharto Education Forest (BSEF, East Kalimantan), Indonesia, Ishida et al. (2000) reported that a severe drought occurred in 1997 and 1998 there due to El Niño and that stomatal conductance of the top canopy leaves of an evergreen pioneer tree, *Macaranga conifera,* did not decrease only with increasing specific humidity deficit but also with decreasing soil water potential in the dry period in early 1998. Although this result was not derived from a natural forest, many trees including dipterocarp trees besides *Macaranga* species partly shed their leaves (Toma et al. 2000b). Fig. 7 shows monthly rainfall amounts in Pasoh FR and BSEF (Toma et al. 2000a) in 1997 and 1998 with those in the normal year. Although the annual amounts of rainfall in the two years as well as those in the normal year were larger in BSEF than in Pasoh FR, rainfall condition during the El Niño in BSEF was characterized by two long rainless durations from August to October in 1997 and from January to March in 1998. The severe damage of trees in BSEF may strongly depend on these long rainless conditions. On the other hand, distributions of monthly rainfall were less fluctuant and rainless durations were shorter in Pasoh FR though the annual amounts in 1997 and 1998 there were much smaller than those in BSEF (Fig. 7). Therefore, the continuous supply of small rainfall amount during the driest periods may barely maintain the high evapotranspiration rate in Pasoh FR even though trees there tend to tolerate droughts due to the normal low rainfall climate. We should note that trees might suffer from severe water stress if a longer rainless duration visits Pasoh FR in the future. It will be particularly important for our coming research in Pasoh FR to detect effects of droughts on evapotranspiration using continuous long-term records on meteorology.

6. CONCLUSION

Findings obtained from meteorological observations in Pasoh FR are summarized below.

(1) Climate in Pasoh FR is characterized by low rainfall due to the inland location. Rainfall amount during observational period from 1995 to 1999 was generally small due to the El Niño event of 1997 and 1998.

(2) Seasonal variations in meteorological variables were not remarkable, but it was dry at the first half of a year, getting wetter toward the end of the year.

(3) Forest albedo and the relationship of net radiation and solar radiation were similar to those previously obtained from other tropical rain forests.

(4) Latent heat flux was dominant in the energy exchange even under dry conditions.

(5) The surface conductance was controlled mainly by solar radiation and specific humidity deficit and was consistently expressed with a simple functional relationship throughout dry and wet periods.

(6) A comparison of rainfall characteristics with another tropical rain forest in East Kalimantan suffering severe damage by a drought suggested that high evapotranspiration under dry conditions in Pasoh FR may barely be maintained by continuous supply of small rainfall amount.

ACKNOWLEDGEMENTS

We thank the Forestry Department of Negeri Sembilan and Director General of FRIM for giving us permission to work in the Pasoh FR, the Japanese Meteorological Agency for calibration of net radiometer and Dr. Kaoru Niiyama, FFPRI, for providing the tree diameter data. The staffs of Hydrology Unit and Pasoh Station of FRIM are greatly acknowledged for collecting the data and maintaining the meteorological observation system. They are Rajendran Pavadai, Ahmad Shahar Mohamad Yusof, Ibharim Hashim, Hashim Mohd, Abu Husin Harun, Hamidon Ahmad (Hydrology) and Ahmad Awang (Pasoh). This study is conducted as a part of joint research project between the NIES, FRIM and UPM (Global Environmental Research Programme granted by the Japanese Environment Agency, Grant No. E-3).

REFERENCES

Aoki, M., Yabuki, K. & Koyama, H. (1975) Micrometeorology and assessment of primary production of a tropical rain forest in West Malaysia. J. Agrc. Meteorol. 31: 115-124.

Bruijnzeel, L. A. (1990) Hydrology of moist tropical forest and effects of conversion: a state-of-knowledge review, UNESCO, Paris, 24pp.

Calder, I. R., Wright, I. R. & Murdiyarso, D. (1986) A study of evaporation from tropical rain forest -West Java. J. Hydrol. 89: 13-31.

Dolman, A. J., Gash, J. H. C., Roberts, J. & James, W. (1991) Stomatal and surface conductances of tropical rainforest. Agric. For. Meteorol. 54: 303-318.

Dyer, A. J. & Hicks, B. B. (1970) Flux-gradient relationships in the constant flux layer. Quart. J. R. Met. Soc. 96: 5715-5721.

Furukawa, A., Awang, M., Abdular, A. M., Yap, S. K., Toma, T., Matsumoto, Y., Maruyama, Y. & Uemura, A. (1994) Photosynthetic and stomatal characteristics of tropical tree species. In Research Report of the NIES/FRIM/UPM Joint Research Project 1993: National Reserch Institution Enviromental Studies, Tsukuba, pp.47-57.

Furukawa, A., Toma, T., Maruyama, Y., Matsumoto, Y., Uemura, A., Ahmad Makmam, A., Muhamad, Awang,M. (2001) Photosynthetic rates of four tree species in the upper canopy of a tropical rain forest at the Pasoh Forest Reserve in Peninsular Malaysia. Tropics. 10: 519-527.

Gash, J. H. C., Shuttleworth, W. J., Lloyd, C. R., Andre, J. C. & Goutorbe, J. P. (1989) Micrometeorological measurements in Les Landes Forest during HAPEX-MOBILITY, Agric. For. Meteorol. 46: 131-147.

Hodnet, M. G, Tomasella, J., Marques Filho, A. de O. & Oyama, M. D. (1996) Deep soil water uptake by forest and pasture in central Amazonia: predictions from long-term daily rainfall data using a simple water balance model. In Gash, J. H. C., Nobre, J. M., Roberts, J. M. & Victoria, R. L. (eds). Amazonian Deforestation and Climate, Institute of Wiley, Chichester, pp.79-99.

Ishida, A., Toma, T. & Marjenah (2000) Leaf gas exchange and canopy structure in wet and drought years in Macaranga conifera, a tropical pioneer tree. In Guhardja, E., Fatawi, M., Sutisna, M., Mori, T. & Ohta, S. (eds). Rainforest Ecosystems of East Kalimantan: El Niño, Drought, Fire and Human Impacts. Ecological Studies vol. 140. Springer-Verlag, Tokyo, pp.129-142.

Jarvis., P. G. (1976) The interpretation of variations in leaf water potential and stomatal conductances found in canopies in the field. Philos. Trans. R. Soc. London Ser. B 273: 593-610.

Jarvis, P. G., James, G. B. & Landsberg, J. J. (1976) Coniferous forest. In Monteith, J. L. (ed). Vegetation and the atmosphere, vol. 2. Academic Press, New York, pp.171-236.

Kato, R., Tadaki, Y. & Ogawa, H. (1978) Plant biomass and growth increment studies in Pasoh Forest. Malay. Nat. J. 30: 211-245.

Kellihar, F. M., Leuning, R., Raupach, M. R. & Schulze, E. D. (1995) Maximum conductances for evaporation from global vegetation types. Agric. For. Meteorol. 73: 1-16.

Kondo, J., Kanechika, O. & Yasuda, N. (1978) Heat and momentum transfers under strong stability in the atmospheric surface layer. J. Atmos. Sci. 35: 1012-1021.

Manokaran, N. & Kochummen, K. M. (1994) Tree growth in primary lowland and hill dipterocarp forests, J. Tropic. For. Sci. 6: 332-345.

Monteith, J. L. (1965) Evaporation and environment. Symp. Soc. Exp. Biol. 19:206-234.

Nepstad, D. C., de Carvalho, C. R. Davidson, E. A., Jipp, P. H., Lefebvre, P. A., Negreiros, D. G H., da Silva, E. D. & Stone, T. A. (1994) The role of deep roots in the hydrological and carbon cycles of Amazonian forests and pastures. Nature 372: 666-669.

Nobre, C. A., Sellers, P. J. & Shukla, J. (1991) Amazonian deforestation and regional climate change. J. Clim. 4: 957-988.

Oguntoyinbo, J. S. (1970) reflection coefficient of natural vegetation, crops and urban surfaces in Nigeria. Quar. J. R. Met. Soc. 96: 430-441.

Ohtani, Y., Okano, M., Tani, M., Yamanoi, K., Watanabe, T., Yasuda, Y. & Abdul Rahim, N. (1997) Energy and CO_2 fluxes above a tropical rain forest in Peninsular Malaysia. - Under estimation of eddy correlation fluxes during low wind speed conditions. J. Agric. Meteorol. 52: 453-456.

Pinker, R. T., Thompson, O. E. & Eck, T. F. (1980) The energy balance of a tropical evergreen forest. J. Appl. Meteorol. 19: 1341-1350.

Shuttleworth, W. J. (1988) Evaporation from Amazonian rainforest. Proc. R. Soc. London Ser. B 233: 321-346.

Shuttleworth, W. J. (1989) Micrometeorology of temperate and tropical forest. Phils. Trans. R. Soc. London Ser. B 324: 299-334.

Shuttleworth, W. J. Gash, J. H. C., Lloyd, C. R., Moore, C. J., Roberts, J., Marques, A. de O., Fisch, G, Silva, V. de P., Ribeiro, M. N. G, Molion, L. C. B., de Sa, L. D. A., Nobre, J. C., Cabral, O. M. R., Patel, S. R. & Moraes, J. C. (1984) Observations of radiation exchange above and below Amazonian forest. Quart. J. R. Met. Soc. 110: 1163-1169.

Soepadmo, E. (1978) Introduction to the Malaysian I. B. P. Synthesis Meetings. Malay. Nat. J. 30: 119-124.

Stewart, J. B. (1988) Modelling surface conductance of pine forest. Agric. For. Meteorol. 43: 19-35.

Toma, T., Marjenah & Hastaniah (2000a) Climate in Bukit Soeharto, East Kalimantan, In Guhardja, E., Fatawi, M., Sutisna, M., Mori, T. & Ohta, S. (eds). Rainforest Ecosystems of East Kalimantan: El Niño, Drought, Fire and Human Impacts. Ecological Studies vol. 140. Springer-Verlag, Tokyo, pp.13-27.

Toma, T., Matius, P., Hastaniah, Kiyono, Y., Watanabe, R. & Okimori, Y. (2000b) Dynamics of burned lowland dipterocarp forest stands in Bukit Soeharto, East Kalimantan, In Guhardja, E., Fatawi, M., Sutisna, M., Mori, T. & Ohta, S. (eds). Rainforest Ecosystems of East Kalimantan: El Niño, Drought, Fire and Human Impacts. Ecological Studies vol. 140. Springer-Verlag, Tokyo, pp.107-119.

Viswanadham, Y., Sa, L. D. de A., Silva, V. de P. & Manzi, A. O. (1987) Ratios of eddy transfer coefficients over the Amazon forest. In Forest Hydrology and Watershed Management, IAHS Publ. No. 167. IAHS Press, Wallingford, pp.365-373.

Yoda, K. (1978) Organic carbon, nitrogen and mineral nutrient stock in the soils of Pasoh Forest. Malay. Nat. J. 30: 229-251.

7 Soil and Belowground Characteristics of Pasoh Forest Reserve

Tamon Yamashita[1], Nobuhiko Kasuya[2], Wan Rasidah Kadir[3], Suhaimi Wan Chik[3], Quah Eng Seng[3] & Toshinori Okuda[4]

Abstract: We describe the soil and belowground characteristics of the Pasoh Forest Reserve (Pasoh FR), Peninsular Malaysia. Soil survey was conducted using the Malaysian classification system in primary and regenerating forests of Pasoh FR. The physical and chemical properties of various soil horizons were measured at the selected soil pits. Soil N dynamics as a soil biological process was also studied in a range of forest environments, including gap and closed forest. The fine root biomass in the topsoil was also quantified in primary forest. Pasoh FR has at least 18 soil types. The soils of Pasoh FR are whitish to yellowish in color rather than reddish. When compared to the other Southeast Asian tropical forest soils, the particle size distribution is characterized by lower sand and higher silt contents. Chemically, the Pasoh FR soil accumulates greater amounts of Al. Most CEC (cation exchange capacity) are occupied by Al. The high Al content leads to lower P availability. The pool of inorganic N at 0–10 cm soil depth ranges from 14.8 to 23.9 $\mu g\ N\ g^{-1}$. Net N mineralization rate in topsoil in the primary forest is estimated to be 100 $kg\ N\ ha^{-1}\ yr^{-1}$. Nitrification is pronounced at uppermost layer. The fine root biomass (FRB) less than 2, 3 and 5 mm in diameter (d) are 624, 751, 970 $g\ m^{-2}$, respectively. Within the top 20 cm, the FRB (< 1 mm in d) constitutes 73% of the total FRB (< 5 mm in d) from 0–4 cm and about 40% in subsequent layers. The Pasoh FR soil is supposed to be infertile. Net N mineralization is observed mainly in the topsoil. These facts suggest that the FRB (< 2 mm in d) constitutes a major part of the total, especially in the top soil to effectively absorb mineral nutrients released from decomposing organic matter.

Key words: chemical properties, fine root biomass, inorganic N pool, nitrification, N mineralization, physical properties.

1. INTRODUCTION

Soil is the fundamental resource for terrestrial ecosystems. Agricultural and forestry production are dependent upon soil fertility. Hence, soil supports not only plants and animals, but also human life. During recent decades, especially in the tropics, clearance and conversion of large areas of forest to more productive uses to support human life has proceeded at unprecedented rates (FAO 1995). Proper assessment and evaluation of soil are vital to rational and sustained use of forest resources.

Tropical soils are supposedly red, old, deeply weathered, leached, acidic, infertile, lateritic and unable to support intensive cultivation (e.g. Sibirtzev 1914 cited in Richter & Babbar 1991). In reality, the soils observed in humid tropics are

[1] Education and Research Centre for Biological Resources, Shimane University, Matsue 690-8504, Japan. Email: tamonyam@life.shimane-u.ac.jp
[2] Kyoto Prefectural University, Japan.
[3] Forest Research Institute Malaysia (FRIM), Malaysia.
[4] National Institute for Environmental Studies (NIES), Japan.

very diverse (Richter & Babbar 1991). Soils show diversity in both type and chemical, physical and biological properties and thus range from highly productive to almost totally infertile. Although diverse, most soils in humid tropics can be classified as highly weathered Oxisols, Ultisols or Alfisols. These groups account for about 80% of all soils in these regions (Sanchez 1976). Ultisols are the dominant soil group in the humid regions of Southeast Asia (Lal 1987) and are comparatively younger and have higher cation exchange capacity (CEC) than the inherently low fertility Oxisols. Characteristics of dominant soils in humid tropics were recently reviewed by Kauffman et al. (1998). They observed that a large portion of the humid tropics region was covered by soils of low fertility. They also referred to the importance of organic matter in such soils, as a nutrient source for plants. Nutrients in organic matter, especially N and P, are released through the processes of decomposition and mineralization.

Following the early work of Allbrook (1972), soil survey and soil analyses have been conducted intermittently in the Pasoh Forest Reserve (Pasoh FR). After permanent plots, such as the 50-ha Plot, were established, detailed soil maps have become available. In this chapter we review the soil characteristics of the Pasoh FR. We first use the distribution of soils in some permanent plots to demonstrate the diversity of Pasoh FR soils. Secondly, we describe the physico-chemical properties of several soil types and their interrelationships. Thirdly, we discuss biological properties, including N dynamics that are driven by microbial processes, and the distribution of fine roots, which are a hidden but active part of plants.

2. STUDY SITE

The Pasoh FR is 2,450 ha in area and is surrounded by oil palm plantations. The vegetation is classified as a lowland dipterocarp forest of Keruing (*Dipterocarpus*) -Meranti (*Shorea*) type. The Pasoh FR is located 70 km southeast of Kuala Lumpur, Peninsular Malaysia (2°59' N and 102°18' E).

Climatic conditions in the Pasoh FR are shown in Fig. 1. As no long-term climatic records are available, a recently published short-term record (Sulaiman et al. 1994) is used. The annual precipitation is about 1,600 mm and the mean

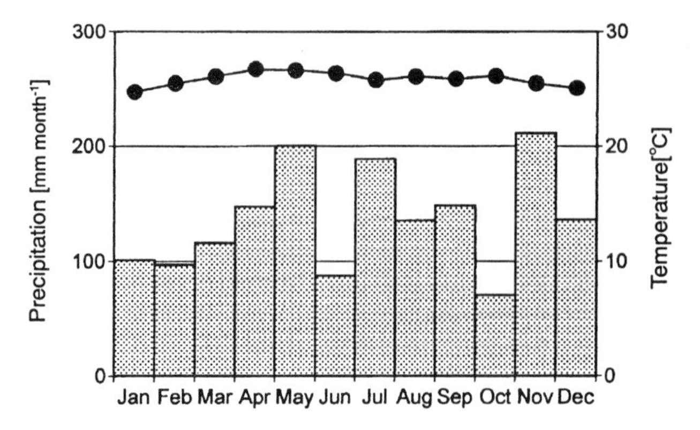

Fig. 1 Climatic condition of the Pasoh FR. Dots with line show the mean temperature and bars show the monthly precipitation during 1991 to 1993.

temperature is 26℃. During the dry season the precipitation is less than 100 mm per month and so the Pasoh FR seems to be drier than other tropical rain forests. The fluctuation of air temperature is limited within the narrow range and the deviation from the mean temperature is relatively small.

In this chapter we focus on three study sites, the 50-ha Plot, Plot 1 and Plot 2. These plots are shown as "primary forest plot", IBP plot and regenerating forest plot respectively in Chap. 2. The 50-ha Plot is covered by primary forest and is located in the center of the forest. Plot 1 is also covered by primary forest. By contrast, Plot 2 is characterized by regenerating forest and is located in the buffer zone of the Pasoh FR. In addition to these permanent plots, we used three more sites for physico-chemical analyses and root biomass measurement. Samples for physico-chemical analyses were taken also from a primary forest site other than the three permanent plots in the Pasoh FR that were mentioned above (the Plot 1, Plot 2 and 50-ha Plot). Mineral nitrogen pool was measured at the Plot 1, Plot 2 and a logged forest which was located in Serting area near Pasoh FR. In the logged forest in Serting, we also collected materials of termite mound for the measurement of the mineral nitrogen pool. Net nitrogen mineralization rate and fine root biomass were measured at another site in a primary forest which was established at the end of access-road to Pasoh FR and used in Yamashita & Takeda (1998).

3. DISTRIBUTION OF SOIL TYPES

En route to the 50-ha Plot, one of the permanent plots in the forest, one notices that the soil color is somewhat whiter than that commonly perceived to be associated with tropical soils. The "typical" soil color in tropical regions has often been considered, incorrectly, to be lateritic red (Richter & Babbar 1991). Of course, the Pasoh FR does have some areas where the soils are reddish in color. However, the range of reddish soils is restricted to a specific area.

There are several world wide soil classifications. The FAO/UNESCO system and Soil Taxonomy system are commonly used. Additionally, countries have their own local classification systems. In this section, we present the results of soil

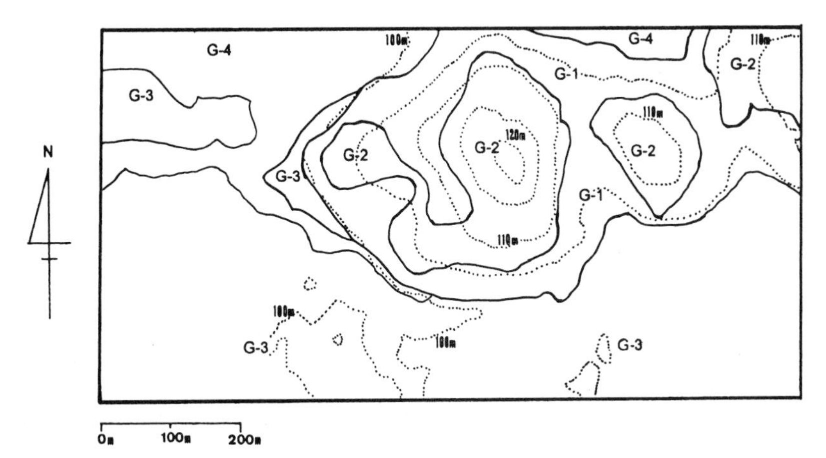

Fig. 2 Distribution of soil types within the 50-ha plot. The symbols 'G-1', 'G-2', 'G-3' and 'G-4' indicate group 1, group 2, group 3 and group 4, respectively. Dotted lines indicate the contour lines at 5 m intervals.

Table 1 Proportional areas of each soil type in the 50-ha Plot of Pasoh FR (The "na" indicates "not available").

Group	Soil type		Proportion to total area
	Malaysian method	Soil Taxonomy	
Group 1	Bungor	Typic Paleudults	20.8%
Group 2	Terap	na	10.8%
	Gajah Mati	Orthoxic Tropudults	5.7%
Group 3	Tebok	Typic Paleudults	8.6%
	Tebok (ms)	na	2.3%
	Tawar	Typic Kanhapludults	18.9%
	Tawar (p)	na	19.4%
Group 4	Awang	Aquic Paleudults	1.9%
	Alma	na	1.3%
	Kampong Pusu	Aeric Plinthic Kandiaquults	2.2%
	Kampong Pusu (cs)	na	8.1%

Table 2 Vertical distribution of soil color of each soil type in the 50-ha Plot. Soil color is described in the way of Munsell's soil color charts. Data of KPU (cs) is not available, but soil color is similar to KPU.

Group	Soil type	Top soil	Semi top soil	Deep soil
Group 1	BGR	10YR5/3	10YR5/8	10TR6/8
Group 2	TRP	7.5YR3/2	10YR5/4	10YR6/6
	GMI	10YR5/8	—	7.5YR6/8
Group 3	TBK	7.5YR4/3	7.5YR5/6	7.5YR6/8
	TBK(ms)	10YR5/4	10YR5/6	10YR6/8
	TWR	10YR3/3	10YR5/4	10YR6/8
	TWR(p)	10YR4/2	10YR5/3	10YR7/3
Group 4	AWG	10YR5/3	10YR6/4	10YR7/4
	AMA	10YR5/2	2.5Y7/2	2.5Y8/2
	KPU	5YR4/1	5YR7/1	5YR7/2
	KPU(cs)	—	—	—

surveys conducted at the 50-ha Plot in primary forest, Plot 1 in another area of primary forest and Plot 2 in regenerating forest, using the Malaysian classification method (Paramananthan 1983).

3.1 The 50-ha Plot

The 50-ha Plot is in a primary forest. A detailed soil map shows that four groups with 11 types of soil are contained within that plot (Fig. 2). Group 1 has only one soil type, Bungor (BGR), a well-drained soil developed from shale. Group 2 consists of Terap (TRP) and Gajah Mati (GMI) series, on lateritic materials. Group 3 which has four soil types, Tebok (TBK), Tebok medium sand variants (TBK (ms)), Tawar (TWR) and Tawar pale variants (TWR(p)) series, is a series of a moderately well or well drained soils. Group 4 has developed in alluvial or riverine areas and is divided into Awang (AWG), Alma (AMA), Kampong Pusu (KPU) and Kampong Pusu coarse sand variants (KPU (cs)) series.

Table 3 Description of soil profiles of Plot 1 and 2. Si, C and L in Texture indicates silty, clay and loam, respectively. Med, Cor, Gr, SA, A and Bl in Structure indicates medium, coarse, granular, subangular, angular and blocky, respectively. Root is shown by [size] / [frequency] and F, M and C in size indicate fine, medium and coarse, respectively.

Horizon	Depth [cm]	Texture	Soil Color	Mottle	Structure	Hardness	Root	Boundary
PLOT 1								
Malacca (Petroplinthic Haplorthox)								
A	0-5	C	10YR 4/2	-	Fine Gr	Friable	FM / few	clear
B11t	5-20	C	10YR 6/6	-	Med SA Bl	Friable	MC / few	gradual
B12	20-60	Si L	7.5YR 6/8	-	Med SA Bl	Friable	FM / common	diffuse
B13	60-80+	Si L	7.5YR 6/8	-	Med SA Bl	Friable	M / few	-
Bungor lateritic phase								
A	0-5	Si C L	10YR 5/3	-	Fine Gr	Friable	FM / frequent	clear
B11	5-25	Si L	2.5YR 7/6	-	Med SA Bl	Friable	MC / common	diffuse
B12	25-55	Si L	2.5YR 7/6	-	Med SA Bl	Friable	F / few	diffuse
B21	55-80	Si L	2.5YR 7/6	10YR 7/2	Med SA Bl	Slightly Firm	F / few	-
B22	80-90	Si L	2.5YR 7/6	10YR 7/2	Med SA Bl	Slightly Firm	F / few	gradual
B23	90-120+	Si L	2.5YR 7/6	10YR 7/2	-	Firm	F / few	-
PLOT 2								
Kuah (Typic Paleudults)								
A	0-10	L	10YR 4/3	-	Med SA Bl	Friable	FC / frequent	gradual
B	10-20	L	7.5YR 6/8	-	Med SA Bl	Friable	M / common	diffuse
BC1	20-70	Si L	7.5YR 6/8	-	Med SA Bl	Firm	M / few	diffuse
BC2	70-90+	Si L	7.5YR 6/8	-	Med SA Bl	Firm	M / very few	-
Kuala Brang								
A	0-10	C L	10YR 4/3	-	Med Gr	Friable	FC / frequent	gradual
B11	10-70	Si L	7.5YR 6/8	-	Med SA Bl	Friable	FM / common	diffuse
B12	70-90	Si L	7.5YR 6/8	-	Med SA Bl	Slightly Firm	M / few	diffuse
B21	90-110+	Si L	7.5YR 6/8	-	Med SA Bl	Firm	-	-
Bungor (Typic Paleudults)								
A	0-15	Si L	10YR 5/6	-	Med SA Bl	Friable	M / common	clear
B11	15-70	Si L	10YR 6/8	-	Med SA Bl	Friable	M / few	diffuse
B12	70++	Si L	10YR 6/8	-	Med SA Bl	Slightly Firm	M / few	-
Kampong Pusu (Aeric Plinthic Kandiaquults)								
Ap	0-10	Si L	10YR 5/3	-	Med SA Bl	Friable	M / common	clear
B11	10-60	Si L	10YR 7/1	10YR 6/8	Med SA Bl	Firm	MC / few	diffuse
B12	60-80	Si L	2.5Y 7/0	-	Cor A Bl	Firm	M / few	gradual
B13	80-100	Si C L	2.5Y 3/0	-	Cor A Bl	Firm	M / few	gradual
B14	100-120+	Si L	5Y 7/4	-	Cor A Bl	Firm	-	-

Unfortunately, not all soils classified under the Malaysian system are necessarily coordinated to the Soil Taxonomy (Soil Survey Staff 1998). Nevertheless, most soils of the 50-ha Plot are Ultisols (Table 1). Among the Ultisols, Udults are the dominant types in Groups 1, 2 and 3. The soils of Group 4 are influenced by water movement and KPU and AWG can be classified as Aquults and Aquic Paleudults, respectively (Jabatan Pertanian 1993).

Based on the detailed soil map, we estimated the areas of each soil type (Table 1) and described representative soil profiles and the vertical distribution of soil color (Table 2). The dominant soil group is Group 3, which occupies half of the plot. The soil color of this group is yellowish brown with the exception of TBK, which is bright brown or orange at deeper horizons. Soils from Group 4, which occur only along a narrow band, are periodically waterlogged during the rainy season. Soil color is gray except for AWG.

3.2 Plot 1

Plot 1, also in primary forest, is 6 ha in area and consists of only two soil types, the Malacca (MAL), and Bungor lateritic phase (BGR/l) series, belonging to Group 1 (Fig. 3). The plot is located on a gentle slope and MAL is distributed on the upper slopes and the BGR/l series on the lower slopes. MAL is the dominant soil type. The results of the soil survey are shown in Table 3. MAL is orange at deeper horizons and yellowish brown with clay texture in the upper horizons. BGR/l is a red colored soil with gray mottle at depth. Both soils have topsoil with a fine granular structure, and exhibit sub-angular blocky structures in deeper horizons.

3.3 Plot 2

Plot 2, in regenerating forest, is also 6 ha in area and consists of four soil types belonging to two groups (Fig. 4). Kuah (KUA), Kuala Brang (KBG), BGR and KPU series are recognized at this site. KUA is distributed on a broad ridge. The KBG series is found on upper slopes and BGR on lower slopes. KPU occurs in the poorly drained swampy area. KBG occupies the largest area of Plot 2, and KUA has the most limited distribution. The results of soil survey are shown in Table 3. As in the 50-ha Plot, KPU exhibits white or gray coloration reflecting poor drainage. BGR has silty loam texture at all horizons. Other soil series also show silty loam textures at deeper horizons. Other than the topsoil of KBG, which is granular, all horizons in this plot show subangular or angular structure.

4. SOIL PHYSICO-CHEMICAL PROPERTIES

Local variation in soil properties is at least as great in the tropics as elsewhere in the world (Richter & Babbar 1991). The physico-chemical properties of soils in the Pasoh FR can also be expected to exhibit diverse characteristics. Under similar climatic conditions, the study plots in the Pasoh FR have nevertheless developed not only various soil types but also differing soil properties depending upon differences in geology, biota, hydrology, and microclimate.

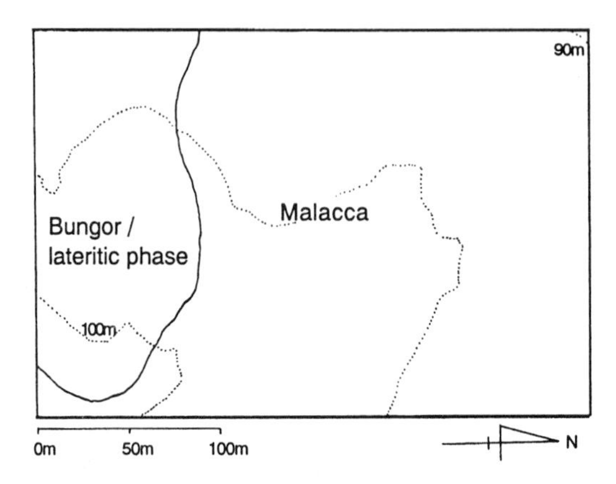

Fig. 3 Distribution of soil types within the Plot 1. Dotted lines indicate the contour lines at 5 m intervals.

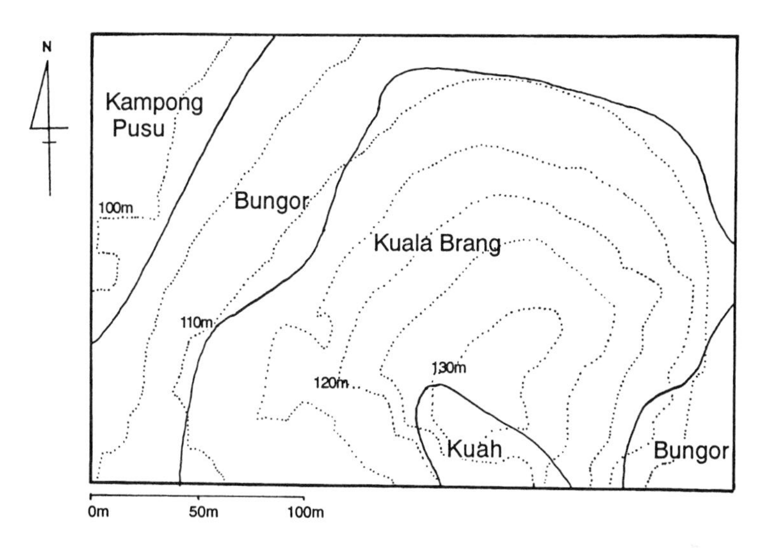

Fig. 4 Distribution of soil types within the Plot 2. Dotted lines indicate the contour lines at 5 m intervals.

4.1 Physical properties

An understanding of the physical properties of tropical soils is important both for sustaining high levels of production and for preserving the stability of ecological environments (Lal 1987). Physical properties can be divided into structural and functional properties. Particle size distribution, bulk density and other factors contribute to structural properties. In general, the higher potential for exchange sites on clay surfaces renders clay-rich soils more fertile than those dominated by coarser particles. The functional properties include water holding capacity and consistency.

We determined the vertical particle size distribution at Plots 1 and 2 and for several other soil types in the Pasoh FR (Fig. 5). MAL and BGR/l from Plot 1 and the deeper horizon of KPU from Plot 2 are low in sand (coarse + fine sand < 20%). The upper horizons of AWG series soils exhibit higher sand contents (> 60%). The other soil types show intermediate sand content. The clay content of the profiles ranges widely from 0 to 70%. Apart from MAL, the mean clay contents through the soil profiles in Plots 1 and 2 were around 20%. Silt contents of Padang Besar (PBR), Berserah (BER) and AWG are less than 20%.

Lal (1987) noted that 85% of the Ultisols and Alfisols in Northeast Brazil contained less than 20% silt and that 20% of soils had clay contents of between 0 to 20%. A recently developed Inceptisol in Venezuela contains greater amounts of silt (Pla Sentis 1977 cited in Lal 1987). The sedentary soils of West Africa are characterized by low silt content (Mbagwu et al. 1983), whereas loess soils from the sahel zone contain high amounts of silt and fine sand and are easily compacted. Soils from the Lambir Hills Forest in Sarawak have 62.4 to 87.1% sand, 5.2 to 15.9% silt and 7.5 to 21.6% clay (Ishizuka et al. 1998). Proportions of the three soil particle sizes in the Sakaerat Forest in Thailand range from 38.9 to 75.1% for sand, 5.5 to 16.6% for silt and 10.6 to 50.1% for clay (Sakurai et al. 1998). Ohta & Effendi (1992a) reported higher silt content in finer textured soils from East Kalimantan than from the Lambir or Sakaerat Forests. However, the East Kalimantan soil silt content is

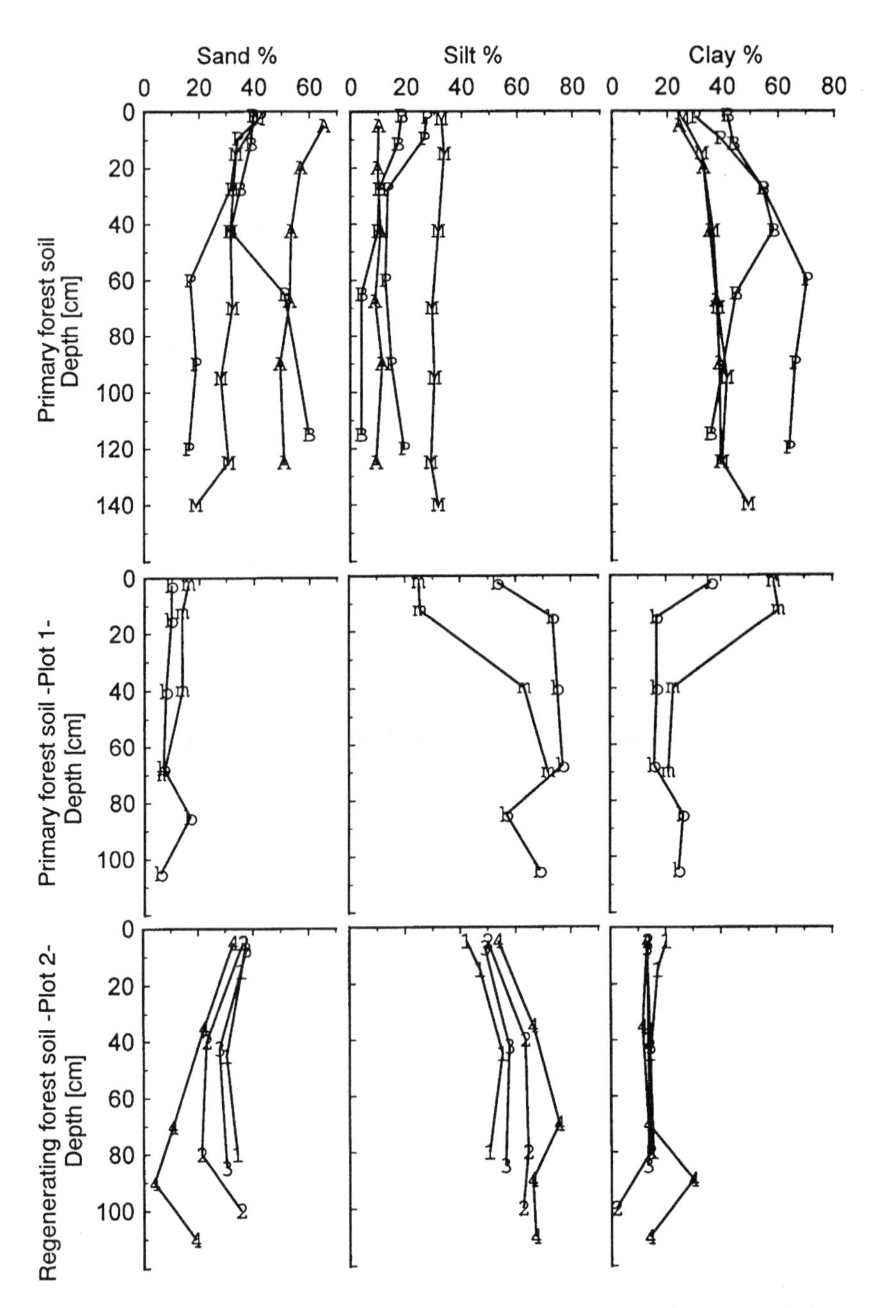

Fig. 5 Vertical distribution of particle size in four soil types in the primary forest (other than the 50-ha Plot or Plot 1), Plot 1, and Plot 2.The symbols 'P', 'B', 'M' and 'A' in the primary forest indicate Padang Besar, Berserah, Musang and Awang. The symbols 'b' and 'm' in the Plot 1 indicate Bungor lateritic phase and Malacca. The symbols '1', '2', '3' and '4' in the Plot 2 indicate Kuah, Kuala Brang, Bungor and Kampong Pusu.

still lower than that of the soils in Plots 1 and 2 of the Pasoh FR. Although the sand and silt content of BER and AWG are within the same range as those of soils from the Sakaerat Forest, the other soil series overall are characterized by lower sand content and higher silt content than those reported from Sakaerat or Lambir Hills Forests.

4.2 Chemical properties

The major cations are derived from weathering of parent materials. Weathering processes are physically, chemically and biologically driven. The exchange sites on the surface of clay minerals or organic materials adsorb cations, forming cation pools and these supply nutrients for most forest plants. The sizes of exchangeable cation pools in the soil are good indices for cation availability in forest ecosystems. In some types of soil, the capacity for soil nutrient retention originates primarily from organic matter (Kauffman et al. 1998). Organic matter is also an important

Table 4 Chemical properties of four soils in primary forest stands. 'av-P' indicates available P by Bray (II) method. BS indicates base saturation rate and is defined as (K+Na+Mg+Ca) × 100 × CEC^{-1}.

	Total [%]			mg kg^{-1}	cmol (+) kg^{-1}						[%]	
Horizon	C	N	C/N	av-P	K	Na	Mg	Ca	Al	CEC	BS	pH
Padang Besar (Orthoxic Tropudults)												
Ai	2.29	0.16	14.3	3.9	0.31	0.13	0.61	0.49	2.37	5.17	29.8	4.5
AB	0.83	0.07	11.9	1.4	0.08	0.09	0.23	0.05	3.21	7.67	5.9	4.4
B21t	0.42	0.06	7.0	0.9	0.14	0.18	0.10	0.08	3.58	5.34	9.4	4.7
B22t	0.42	0.04	10.5	0.5	0.10	0.12	0.07	0.05	3.59	5.94	5.7	4.7
B23t	0.31	0.04	7.8	0.3	0.09	0.09	0.06	0.03	3.28	6.45	4.2	4.6
BC	0.23	0.03	7.7	0.4	0.09	0.10	0.07	0.08	3.62	11.56	2.9	4.6
Berserah (Typic Kandiults)												
Ai	3.37	0.26	13.0	3.4	0.20	0.26	0.56	0.06	2.28	6.35	17.0	4.0
AB	1.56	0.14	11.1	1.5	0.12	0.14	0.39	0.06	1.84	3.77	18.8	4.2
B21t	0.60	0.06	10.0	0.5	0.05	0.06	0.10	0.08	1.62	3.50	8.3	4.4
B22t	0.66	0.06	11.0	0.3	0.04	0.04	0.05	0.05	2.10	3.78	4.8	4.3
B23(Qq)	0.47	0.04	11.8	0.9	0.06	0.07	0.03	0.04	1.77	2.39	8.4	4.5
BC	0.24	0.03	8.0	0.3	0.04	0.02	0.06	0.06	1.30	2.31	7.8	4.7
Musang (Plinthic Kandiudults)												
A	1.76	0.12	14.7	2.5	0.10	0.10	0.14	0.12	2.99	6.42	7.2	3.9
AB	1.03	0.08	12.9	1.6	0.03	0.03	0.03	0.02	3.42	8.01	1.4	4.1
B21	0.35	0.04	8.8	0.9	0.02	0.03	0.04	0.03	3.31	5.18	2.3	4.5
B22	0.24	0.03	8.0	0.3	0.03	0.02	0.00	0.08	3.46	5.08	2.6	4.5
B23	0.24	0.03	8.0	0.4	0.03	0.04	0.04	0.15	3.70	6.00	4.3	4.7
B24	0.16	0.03	5.3	0.3	0.02	0.04	0.01	0.04	4.38	10.38	1.1	4.7
BC	0.13	0.02	6.5	0.5	0.01	0.01	0.00	0.02	5.02	8.95	0.4	4.8
Awang (Aquic Paleudults)												
A	1.63	0.15	10.9	4.2	0.16	0.18	0.14	0.19	2.33	2.86	23.4	4.0
AB	0.91	0.09	10.1	2.2	0.12	0.13	0.09	0.10	2.15	2.94	15.0	4.5
B21	0.57	0.05	11.4	1.4	0.04	0.06	0.06	0.08	2.18	2.84	8.5	4.3
B22	0.43	0.05	8.6	1.2	0.02	0.02	0.06	0.12	2.84	2.73	8.1	4.5
B23	0.36	0.05	7.2	0.8	0.01	0.02	0.06	0.12	1.96	2.83	7.4	4.6
B24	0.28	0.04	7.0	1.1	0.02	0.03	0.04	0.09	1.86	2.61	6.9	4.5

source of soil nitrogen, phosphorus and sulfur. Organic matter content is evaluated from the total C and N.

Some chemical properties, including C, N and macro-nutrients, of four soil series from natural stands of primary forest, not from either the 50-ha Plot or Plot 1, are shown in Table 4. Unfortunately, no data is available from regenerating forest. Total C and N contents range from 0.13 to 3.37% and 0.02 to 0.26%, respectively. Available P is in the range 0.3 to 4.2 mg kg^{-1}. The sum of exchangeable cations (K + Na + Mg + Ca) varies between 0.11 to 1.54 cmol (+) kg^{-1} and CEC from 2.31 to 11.56 cmol (+) kg^{-1}. Exchangeable Al is 1.30 to 5.02 cmol(+) kg^{-1}. Base saturation results, between 0.4 and 29.8%, decrease abruptly from the upper to deeper horizons. The pH (H_2O), range 3.9 to 4.8, increased with depth. This last observation may reflect the increase in Fe and Al oxide contents and the decrease in organic matter content (Sanchez 1976).

Kauffman et al. (1998) reported a spread of organic C contents from 0.8 to 2.3% from a range of humid tropical forests. Yamashita et al. (1999) reported similar values in the Sakaerat Forest and much higher values in the Kog Ma Forest of Thailand. Some top soils in a dipterocarp forest in East Kalimantan (Ohta & Effendi 1992b) and a hill dipterocarp forest in Peninsular Malaysia (Tange et al. 1998) also have higher C content than those of the Pasoh FR. The trends for N content are similar to those of C.

In the Sakaerat Forest, available P in top soils ranges from 11.9 to 20.8 mg kg^{-1} (Sakurai et al. 1998). A broader variation (about 10 to 40 mg kg^{-1}) has been reported from the top horizons of a hill dipterocarp forest soil (Tange et al. 1998). Ultisols of East Kalimantan have an even wider range of available P, about 4 to 50 mg kg^{-1} (Ohta & Effendi 1992b). In the Pasoh FR, available P in topsoil was at most 4.2 mg kg^{-1}. A soil is usually deemed infertile if the available P falls to 3 mg kg^{-1} or less. By comparison with other tropical rain forest soils in Southeast Asia, the Pasoh FR seems to be deficient in available P.

In Ferralsols or other dominant soils in humid tropics, the ranges of total cations and CEC in topsoil are 1.0 to 2.2 cmol (+) kg^{-1} and 6.6 to 20.0 cmol (+) kg^{-1}, respectively. The exceptions are Luvisols and Cambisols (Kauffman et al. 1998). Values from Pasoh FR are within this range, but located at lower end. The base saturation rate is very low at deeper horizons reflecting low cation content. The soils of East Kalimantan (Ohta et al. 1993), Lambir (Ishizuka et al. 1998), Peninsular Malaysia (Tange et al. 1998), and Sakaerat (Sakurai et al. 1998) Forest show similar values to those presented in Table 4.

The average value of exchangeable Al for a range of tropical forest soils is reported to be less than 2 cmol (+) kg^{-1} (Kauffman et al. 1998). The highest value reported by Sakurai et al. (1998) in a deep horizon soil from the Sakaerat Forest was 2.15 cmol (+) kg^{-1}. Soils of the Pasoh FR contain greater amounts of exchangeable Al (maximum 5.02 cmol (+) kg^{-1} in Musang (MUS) series) than some previously reported values (Kauffman et al. 1998; Sakurai et al. 1998). However, Ohta et al. (1993) found much higher exchangeable Al contents ranging from 2 to 18 cmol (+) kg^{-1} in East Kalimantan. Aluminium or Al in soil fixes phosphorus to eventually form varicite or wavellite, which are unavailable to plants. Furthermore, Al *per se* is harmful to plant life.

4.3 Relationships between physical and chemical properties

Physical and chemical properties affect each other. CEC, available P and exchangeable Al are related to soil particle size (Fig. 6). The changes in CEC are

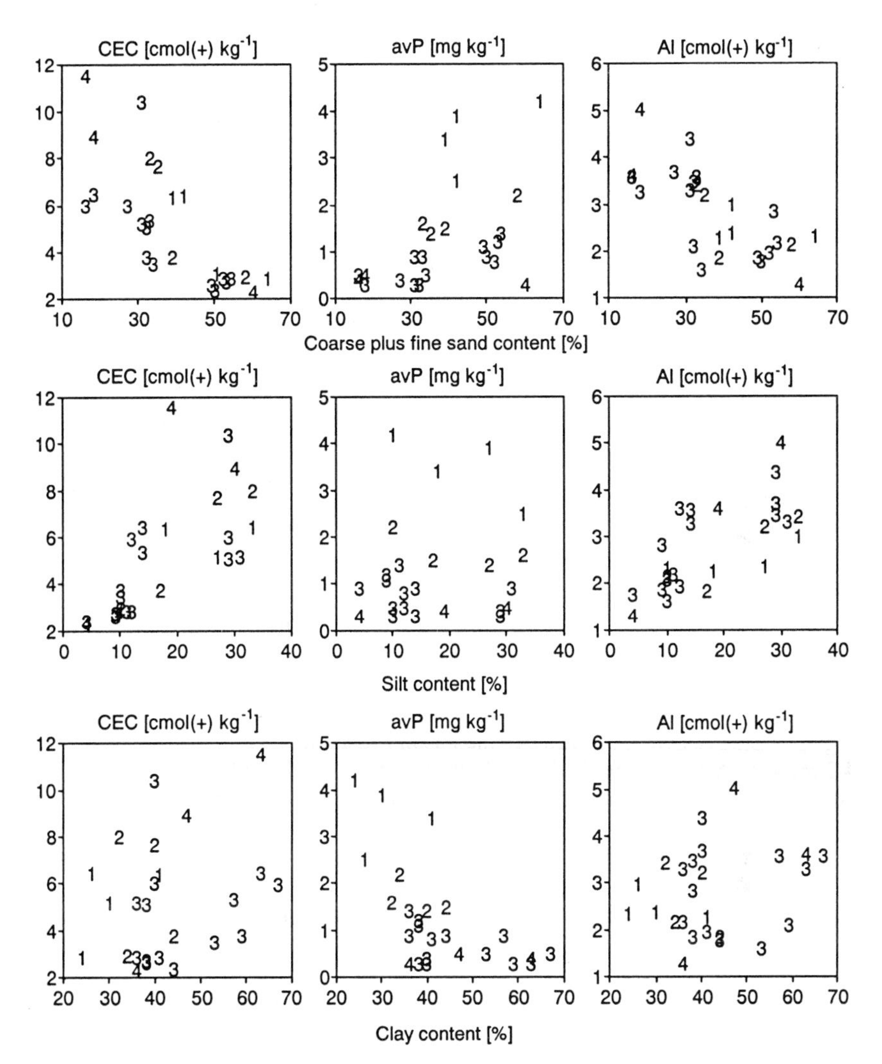

Fig. 6 Relationships between the soil particle sizes and three chemical properties. Upper, middle and bottom indicate the relationships of (coarse plus fine) sand fraction, silt and clay to chemical properties, respectively. The symbols '1', '2', '3' and '4' indicate A horizon, AB horizon, B horizon and BC horizon, respectively.

inversely related to coarse sand content and positively related to silt content. However, the clay fraction, which can function as an effective cation exchange site (Ohta et al. 1993), has no obvious effects on CEC in the Pasoh FR soil. Organic matter has potential for cation retention, but no significant correlations are observed between silt content and total C, nor between CEC and total C. Thus, there is a possibility that the silt fraction, or silt-related properties, might play important roles in cation retention in soils of the Pasoh FR.

Available P shows positive but weak correlation with sand content and negative correlation with clay content. This negative relationship between P availability and clay content is also observed in Ultisols at East Kalimantan (Ohta et al. 1993). A higher degree of P immobilization is observed in clayey Ultisols, than in sandy Ultisols (Woodruff & Kamprath 1965). Available P shows significant correlation with C and N. Sources of available P are weathered parent materials and organic matter. In weathered tropical soils, organic matter may be an important source of available P. Organic matter acts not only as a source of P but also tends to deactivate P-fixing agents (Ohta et al. 1993). Organic matter accumulates on the forest floor at the soil surface and gradually decreases with soil depth. Although the physical properties of soil could affect P availability, the amount of organic matter is the more dominant factor in determining P availability. Since the depth of a soil horizon indirectly determines the amount of organic matter, data points from the same horizon are plotted at approximate levels in a scatter diagram (Fig. 6).

Exchangeable Al indicates a negative relationship with sand content and a positive relationship with silt content. This trend for exchangeable Al is similar to that of CEC. There is a significant correlation between exchangeable Al and CEC. In the Pasoh soils exchangeable Al occupies at least 36% of the available CEC. The Awang series shows greatest Al occupancy at all horizons with more than 70% of CEC taken up by exchangeable Al. Al is widely distributed in the soils, the exchangeable Al content is dependent upon CEC, and Al affects the availability of P. Therefore the CEC and exchangeable Al are both inversely related to available P. The available P content is low in horizons with greater exchangeable Al. Comparatively high exchangeable Al content increases the possibility of P deficiency (Attiwill & Adams 1993).

5. SOIL BIOLOGICAL PROPERTIES

The constituents of soil biota include microbes, soil animals and plants. In this context, the soil N, an important resource for soil animals and plants, cycles between the abiotic and biotic components in soil system and is regulated by the microbial activity. Fine roots are the active interface between plants and the soil system. Both the soil N and fine root biomass are the critical biological factors for sustaining the forest productivity. To date however, they have been little studied in the Pasoh FR.

5.1 Soil N dynamics

Soil N is one of the critical nutrients for plant growth. The N availability limits net primary production in terrestrial ecosystems (Vitousek & Howarth 1991). Traditionally, N availability has been assessed by measuring net N mineralization rates in soils (Binkley & Hart 1989). Ordinarily, net N mineralization rate is defined as the increment in inorganic N during a certain period of incubation, i.e. [net N mineralization rate] = [inorganic N pool after incubation] - [inorganic N pool before incubation]. Reich et al. (1997) showed that the aboveground net primary production increased linearly with annual net N mineralization rate in temperate forest ecosystems.

Recently a new pathway to utilize organic N has been discovered in some ecosystems at higher latitudes where the N availability is low (Chapin et al. 1993; Northup et al. 1995; Read et al. 1989). The increased N content in wet and dry deposition has altered soil processes including soil N dynamics around industrialized areas in temperate zones (Aber et al. 1989; Dise & Wright 1995). The Pasoh FR received 9.6 kg N ha^{-1} yr^{-1} through precipitation in 1973 (estimated by authors using the data presented in Manokaran (1978)). N deposition ranged from 1.7 to 21 kg N ha^{-1} yr^{-1} in the tropics (Vitousek & Sanford 1986). The tropical forest in Monteverde received 7.5 kg N ha^{-1} yr^{-1} from 1991 to 1992 (Clark et al. 1998). Although the N deposition in tropical forest is still low by comparison with that in temperate zones, anthropogenic N deposition could possibly increase in tropical region for decades (Matson et al. 1999). Despite the importance of grasping of soil N dynamics in tropical forest ecosystems, the basic study on net N mineralization rates in soils was insufficient in tropical forests (Vitousek & Sanford 1986). Since then, net N mineralization rate has been measured mainly in the neotropics. Even now, studies on net N mineralization rates are limited for the tropical rain forests of Southeast Asia. We discuss the pool size of inorganic N and net N mineralization rate mainly in neotropics and then compare our results from the Pasoh FR.

5.2 Inorganic N pool

The pool size of inorganic N shows the balance between input to soil system (e.g. mineralization, deposition and fixation) and output (e.g. plant uptake, leaching and immobilization). In boreal coniferous forest, the pool of inorganic N decreases in summer and increases in winter (Bashkin & Kudeyarov 1977). At higher latitudes, the input of inorganic N seems to underperform the N output. Conversely, in a temperate coniferous plantation, the pool size increases in summer and decreases in winter (Yamashita et al. 1992). At this location the input exceeded the output even in summer. In the dry tropical forests of India (Roy & Singh 1995) and the Amazonian forests (Neil et al. 1995), the pool size decreases in the wet (summer) season and increases in the dry (winter) season. Soil systems appear to supply more N in the temperate zone and less in the boreal and tropical region, than plant requirements. Low input of inorganic N in the boreal regions is attributable to low mineralization rates due to low temperature. High output rate due to vigorous growth of plants is one of the causative factors for decrease in inorganic N pool during the summer season in the tropical savanna and forest.

Though the pool size oscillates with ambient environment, temporal variation is often ignored, because it is reported as an average. In the Amazonian forests, the pool of NH_4^+-N at 0 – 5 cm deep soil ranges from 1.83 to 17.59 μg N g^{-1} and that of NO_3^--N from 1.07 to 11.87 μg N g^{-1} (Neil et al. 1997). In the tropical savanna of India, the ranges in 0 – 10 cm deep soil for NH_4^+-N and NO_3^--N are 2.6 to 5.8 μg N g^{-1} and 0.2 to 1.2 μg N g^{-1}, respectively (Singh et al. 1991). The organic layer on the forest floor contains 7.8 to 94.4 μg N g^{-1} of NH_4^+-N and 0 to 2 μg N g^{-1} of NO_3^--N in the neotropics (Vitousek & Matson 1988). On an area basis, the inorganic N pool is <40 kg N ha^{-1} 15 cm^{-1} for NH_4^+-N and <5 kg N ha^{-1} 15 cm^{-1} for NO_3^--N in a secondary forest of Costa Rica (Matson et al. 1987), and 1.6 kg N ha^{-1} 10 cm^{-1} for NH_4^+-N and 6.7 kg N ha^{-1} 10 cm^{-1} for NO_3^--N in an old-growth forest of Costa Rica (Zou et al. 1992).

In the Pasoh FR, we determined the pool sizes of NH_4^+-N and NO_3^--N in 0 – 10 cm deep soil in the Plot 1, Plot 2, logged forest and material from termite mounds (Table 5). Differences in soil types may also affect the pool of inorganic N, but the current condition of the forest canopy could be a more effective determinant. Termite mounds accumulate organic matter and nutrient elements and are thought to be key contributors to nutrient dynamics in the Pasoh FR. Soil cores, 25 cm^2 in cross section and 10 cm long, were collected from the closed canopy site, gap center and gap peripherals of the primary (including Plot 1) and regenerating forests (Plot 2), and from another site in logged forest outside the Pasoh FR. The pool size ranged from 1.4 to 5.5 μg N g^{-1} for NH_4^+-N and from 12.2 to 19.3 μg N g^{-1} for NO_3^--N

Table 5 Pool size of inorganic N in soils of the Pasoh FR. Values are means ± 1SE. The word 'Sum' indicates the sum of NH_4^+-N plus NO_3^--N.

Site	condition	*N*	Concentration per dry weight [μg g^{-1} DW]			Concentration per soil volume [kg ha^{-1}10cm^{-1}]		
			NH_4^+-N	NO_3^--N	Sum	NH_4^+-N	NO_3^--N	Sum
Primary forest	closed forest	10	2.6 ± 0.2	12.2 ± 2.0	14.8 ± 1.9	1.8 ± 0.2	8.3 ± 1.3	10.1 ± 1.2
	gap center	8	1.4 ± 0.2	13.2 ± 1.1	14.6 ± 1.0	1.0 ± 0.2	9.7 ± 1.0	10.7 ± 0.9
	gap peripheral	8	1.8 ± 0.2	15.1 ± 3.7	16.9 ± 3.6	1.3 ± 0.1	11.2 ± 2.8	12.5 ± 2.7
Regenerating forest	closed forest	10	3.3 ± 0.7	16.9 ± 5.0	20.2 ± 5.3	2.5 ± 0.6	13.0 ± 4.2	15.6 ± 4.5
	gap center	8	2.8 ± 0.3	19.4 ± 3.4	22.3 ± 3.6	2.3 ± 0.2	15.9 ± 2.7	18.3 ± 2.9
	gap peripheral	8	6.1 ± 1.5	17.7 ± 3.6	23.9 ± 4.5	4.6 ± 1.1	13.5 ± 2.7	18.1 ± 3.3
Logged forest	not specified	10	4.7 ± 0.8	17.1 ± 1.9	21.8 ± 1.4	4.7 ± 0.9	17.0 ± 2.3	21.6 ± 2.2
Termites mounds	black coloured	1	65.8	67.7	134	–	–	–
	brown coloured	1	377	25.3	402	–	–	–

at the forest sites. Materials from termite mounds showed much higher contents of inorganic N reflecting higher biological activity. On an area basis, NH_4^+-N pool ranged from 1.0 to 4.7 kg N ha^{-1} 10 cm^{-1} and NO_3^--N ranged from 8.3 to 17.0 kg N ha^{-1} 10 cm^{-1}.

The NO_3^--N pool exceeded that of NH_4^+-N at all sites except one of the termite mounds (brown colored). But many of the published reports mentioned earlier, show the pool size of NH_4^+-N exceeding that of NO_3^--N. Negatively charged NO_3^--N tends to leach from the soil profile, due to the lack of anion exchange capacity (AEC) on soil particles under ordinary conditions. Positively charged NH_4^+-N is adsorbed onto soil particles that have substantial CEC. Upward transport or the accumulation of NO_3^--N is supposed to occur when the soil becomes drier (Sanchez 1976; Wild 1972). Where enough H^+ is produced in soil, the possibility of generating AEC dependent on zero point charge is suggested (Robertson 1986). AEC conditions (unlikely to occur in the Pasoh FR soils) or drier soil could lead to the NO_3^--N pool size being greater than that of NH_4^+-N in the Pasoh FR. In addition, Matson et al. (1999) pointed out the importance of electrostatic adsorption in controlling losses of excess NO_3^--N to aquatic ecosystems in the tropics.

Since the inherent properties on N dynamics were not described in each forest type before they had been disturbed, we can only compare the current status of N properties among the current forest types. With these restrictions, the effects of forest types and canopy conditions are tested by two-way ANOVA (Table 6). The NO_3^--N pool size shows no significant differences between forest type and conditions of canopy. The pool size of NH_4^+-N in the primary forest is significantly smaller than in other forest types ($P < 0.01$) and also differs between canopy conditions ($P < 0.05$). Soils in the gap peripherals contain more NH_4^+-N than under other canopy conditions. The sum of NH_4^+-N and NO_3^--N differs between the primary and regenerating forest ($P < 0.05$). As a result, the soils of the primary forests contain lesser amounts of inorganic N than in the disturbed forests. Vitousek & Denslow (1986) reported that the pool size of NO_3^--N in the Costa Rican forest showed no increase at the treefall gaps. Similarly, we found no increase in inorganic N pool size at the gap sites as a whole (gap center plus peripheral sites) by comparison with the closed canopy sites. However, in the regenerating forest, a significant increase was observed in the gap peripherals by comparison with the closed canopy sites ($P < 0.05$). Although Vitousek & Denslow (1986) concluded that an increase, not in nutrient, but in light availability seems to represent a more important shift in resources within the treefall gap, our results are not consistent with theirs at least for this specific forest type.

Table 6 Results of an analysis of variance to test for the effects of forest type (primary vs. regenerating) and canopy conditions (closed canopy vs. gap) on the inorganic N pool size at 0–10 cm depth in the soils of the Pasoh FR.

		NH_4-N		NO_3-N		Sum	
	df	*F*	*P*	*F*	*P*	*F*	*P*
Forest type (F)	1	9.715	0.003	2.857	0.097	4.875	0.032
Canopy condition (C)	1	0.021	0.887	0.462	0.500	0.440	0.510
F × C	1	3.294	0.076	0.002	0.965	0.103	0.750

5.3 Net N mineralization rate

The net N mineralization rate is strongly affected by ambient temperature and water regime. The rates are greatest in tropical forest soils and least in arctic or arid ecosystems (Attiwill & Adams 1993). In tropical rain forest, unlike temperate zones, however, seasonal variability in net N mineralization rate is relatively small (Matson et al. 1987; Vitousek & Denslow 1986). Although it is not satisfactory, one-off N mineralization measurements could be representative of annual patterns in the humid tropics (Vitousek & Matson 1988).

In Costa Rican forest, the net N mineralization rate is around 3 $\mu g\ N\ g^{-1}\ day^{-1}$ and most of the mineralized N was nitrified with no time lag (Robertson 1984). Vitousek & Denslow (1986) reported the net N mineralization rate in intact forest and gaps ranged from 0.9 to 2.1 $\mu g\ N\ g^{-1}\ day^{-1}$ in Costa Rican forest on volcanic soil. Two years later, Vitousek & Matson (1988) summarized the net N mineralization rate in a range of neotropical soils. They cited ranges from 0 (on Andept in Hawai'i) to 6.8 $\mu g\ N\ g^{-1}\ day^{-1}$ (at a cleared site on Andept in Costa Rica) at a depth of 0 – 15 cm and from 0 to 8 $\mu g\ N\ g^{-1}\ day^{-1}$ in organic layers. They also mentioned that these values were greater than in the temperate forest soils, except for forests on white sand or in upper montane areas. In the Amazonian forest, the N mineralization rate is also rapid and ranges from 0.93 to 5.47 $\mu g\ N\ g^{-1}\ day^{-1}$ at 0 – 5 cm soil depth and 0.49 to 2.39 $\mu g\ N\ g^{-1}\ day^{-1}$ at depths of 5 – 10 cm (Neil et al. 1997). Annual estimates of the net N mineralization rate in the Costa Rican forest are 588 to 1140 $kg\ N\ ha^{-1}\ yr^{-1}$ (Matson et al. 1987). Nitrogen content in the aboveground fine litterfall is important for plant N requirements, and in various neotropical forests it has a range from 74 to 224 $kg\ N\ ha^{-1}\ yr^{-1}$ (Vitousek & Sanford 1986). Annual estimates for the net N mineralization rate overwhelmingly exceed annual plant requirements. As the net nitrification rate was high at all the above sites, NO_3^--N is thought to be a major N source for plants in these tropical forests. In contrast, Ultisols in East Kalimantan produce NH_4^+-N but not NO_3^--N and the maximum N mineralization rate is 8.3 μg N $g^{-1}\ day^{-1}$ in the A horizon (Ohta et al. 1992b). Even though the production rate of NH_4^+-N is relatively high, nitrification rates remain low in East Kalimantan.

We measured the net N mineralization rate of a primary forest soil from the Pasoh FR, at 25°C for 30 days using a laboratory incubation method. We collected organic layer, black topsoil and light yellow sub-soil. Results are shown in Table 7. The highest rate of 0.98 $\mu g\ N\ g^{-1}\ day^{-1}$ is observed in the topsoil but this is still at the lower end of previously reported values. Assuming these values are representative of annual patterns, this extrapolates to an annual estimate on an area basis of 100 $kg\ N\ ha^{-1}\ yr^{-1}$ in 0 – 10 cm deep soil. The annual litterfall of this site is 8.6 $Mg\ ha^{-1}\ yr^{-1}$ and the N concentration of leaf litter is around 1.2% dry mass (Yamashita & Takeda 1998), the forest canopy needs about 90 $kg\ N\ ha^{-1}\ yr^{-1}$. The soil at 0 – 10 cm deep can produce inorganic N just equal to plant requirements in the Pasoh FR. Of the upper 10 cm of soil, the topsoil (0 – 2 cm) produces half of the N required by plants. As discussed above, more fertile sites produce more inorganic N than plant requirements. Excess N at such sites is lost to the aquatic system or denitrified to the atmosphere. Since inorganic N production is restricted to the topsoil and the mineralization rate is relatively low, N seems to cycle tightly within the terrestrial system in the Pasoh FR.

5.4 Fine root biomass

Fine roots play an important role not only in absorbing water and nutrients, but also in dry matter production by rapid turnover. Fine root biomass (FRB) measurement is the first step to clarifying the role of fine roots at ecosystem level.

In October 1991, fifteen soil cores (5.6 cm in diameter) were taken to the depth of 20 cm to measure the FRB in a primary forest of Pasoh FR. The soil cores were then cut horizontally at 4 cm intervals. The soil samples were sieved using tap water and the collected root samples separated according to diameter at root (d): $d < 1$ mm, $1 \leqq d < 2$ mm, $2 \leqq d < 3$ mm, $3 \leqq d < 5$ mm and $d \geqq 5$ mm. The results show that the FRB for roots less than 2, 3 and 5 mm in diameter are respectively 624, 751, and 970 g m^{-2}. Vogt et al. (1996) compiled published FRB data for the tropics and showed that the average of 24 tropical broad-leaved evergreen forests was 1,020 (170) g m^{-2} (standard error in parenthesis). The average of six tropical broad-leaved evergreen forests growing on the same Ultisols as in the Pasoh FR was 638

Table 7 Net N mineralization rate at the various soil horizons in the natural forest of Pasoh.

	Concentration per dry weight [μg N g^{-1} DW day^{-1}]			Concentration per soil volume [kg N ha^{-1} cm^{-1} yr^{-1}]		
	NH_4^+-N	NO_3^--N	Sum	NH_4^+-N	NO_3^--N	Sum
Organic layer	0.46	0.00	0.46	0.8	0.0	0.8
Top soil	0.01	0.97	0.98	0.3	25.5	25.8
Sub soil	-0.02	0.25	0.23	-0.5	6.5	6.0

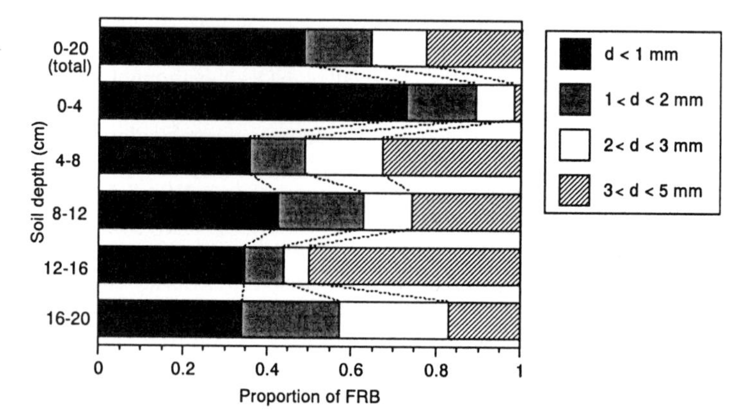

Fig. 7 Proportions of the fine root biomass (FRB) in different diameters at each soil layer. The symbol 'd' shows diameter of fine root.

(306) g m^{-2}. Yoda & Kira (1982) showed that the total FRB for d < 1 cm within 1 m depth was 2,050 g m^{-1} at the Pasoh FR and that the FRB decreased exponentially from surface to deeper horizons. Due to the different limits in the fine root diameters (2–6 mm), the data in Vogt et al. (1996) have been rearranged for comparison. The FRB in d < 2 mm and the FRB in d < 5 mm become 381 (131, N = 7) and 1,342 (263, N = 13) g m^{-2}, respectively. The former value is less than a third lower than the latter. In contrast, the FRB in d < 2 mm occupies 64% of the entire FRB in d < 5 mm in the Pasoh FR (Fig. 7).

Within the top 20 cm, the FRB in d < 1 mm constitutes 73% of total FRB in d < 5 mm at 0 – 4 cm layer and around 40% in all subsequent layers. The proportion of FRB in 3 ≦ d < 5 mm size range fluctuates widely; it is least in the 0 – 4 cm layer and highest at 12 – 16 cm. The vertical distribution differs between the root diameter classes (Fig. 8). The FRB in d < 1 mm peaks between 0 and 4 cm and decreases abruptly with increasing soil depth, by comparison with the gradual decrease in

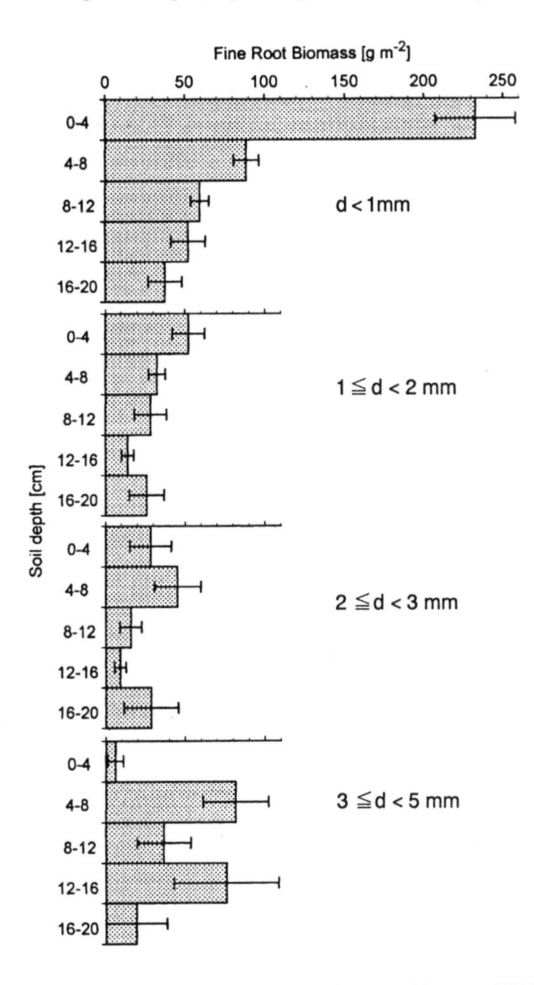

Fig. 8 Vertical distributions of the fine root biomass (FRB) in different diameters. 'd' shows fine diameter of root. Bars indicate one standard error of the mean.

FRB $1 \leqq d < 2$ mm. The size ranges $2 \leqq d < 3$ mm and $3 \leqq d < 5$ mm have peaks below 4 cm in depth. The two smaller classes of FRB are distributed most abundantly in the top layer indicating a lighter texture, while the larger diameter FRB is most abundant somewhere below 4cm depth in the clayey heavy soil in this forest. In the light of these observations, what factors control fine root distribution? From a large-scale perspective Vogt et al. (1996) reported that FRB was positively correlated with annual air temperature, and negatively with the P content of aboveground litterfall. At a more local level, they compared FRB of various diameters with the thickness of the organic layer and A_1, layer at each sampling point, and the organic content in the soil core. This showed that FRB ($d < 1$ mm) correlated with the thickness of A_1 layer, but not with the thickness of the organic layer. This suggests that it is not the accumulation of organic matter, but nutrient release in the A_1 layer that stimulates fine root proliferation.

The variability between the sampling points should be considered to examine the reliability of the data in this study. Bengough et al. (2000) showed that the coefficients of variance of the FRB in core samples from grassland were typically between 30 – 50%. The coefficients of variance in this study are 34, 43, 66 and 90% for $d < 1$ mm, $1 \leqq d < 2$ mm, $2 \leqq d < 3$ mm and $3 \leqq d < 5$ mm fine roots, respectively, indicating relatively sparse distribution of the larger diameter roots. Although the sampling depth (20 cm, in this study) is not enough to collect deep roots, variability of the FRB in $d < 2$ mm is easily detected as it is most abundant in the surface soil layer (just below the organic layer). The core diameter of the soil sampler (5.6 cm) and the number of replications (15) used in this study are within the range commonly adopted procedures (Oliveira et al. 2000). In the future, research on fine root turnover, detailed root distribution, and the interactions of roots of different species is needed, to address the role of roots of various species in the forest ecosystem.

6. CONCLUSION

The Pasoh FR is situated on relatively flat topography. However, there is a wide variation in soil types. Small differences in parent material or drainage cause different types of soils to develop. Although each soil type shows very unique properties, trends in physical properties as a whole can be categorized as low sand content and high silt content. We found that the silt fraction or silt related factors play a more important role than organic matter in cation retention. Critical chemical properties of Pasoh FR soils are extremely high content of Al and low P availability. These appear to act interactively. Where Al is high, P availability decreases, therefore control of Al is required to sustain land productivity in the Pasoh FR.

The inorganic N pool is characterized by a large NO_3^--N pool, which exceeds the NH_4^+-N pool. This might be attributed to the drier conditions or possibly to anion retention. The net N mineralization rate is quite low in comparison with other tropical rain forests. Not only P, but also N, seem to be critical factors for the plant growth.

The soils of the Pasoh FR should be categorized as infertile. The proportion of thinner fine root to total fine root is higher than reported values elsewhere. The proliferation of thinner roots enables the plants to absorb the scarce nutrients in the soil. However, the effective root zone is restricted to the top few cm of soil. N mineralization is also restricted to that zone. As disturbing this zone may cause irreversible damage to the terrestrial ecosystem, the conservation of this zone should be addressed especially in the Pasoh FR.

ACKNOWLEDGEMENT

The authors are grateful to Dr. Amir Husni bin Mohd Shariff for providing basic data on soil survey. They also thank to staff of Soil Chemical Laboratory of FRIM for rendering invaluable laboratory works, to Prof. Hiroshi Takeda for giving us the chance to join the NIES, FRIM and UPM joint research project, and to two anonymous reviewers for critical comments on the manuscript. This paper is partly supported by Japan Environment Agency (Grant No. E-2).

REFERENCES

Aber, J. D., Nadelhoffer, K. J., Streudler, P. & Melillo, J. M. (1989) Nitrogen saturation in northern forest ecosystems. Bioscience 39: 378-386.

Allbrook, R. F. (1972) The soils of Pasoh forest reserve Negeri Sembilan. Malay. For. 36: 22-33.

Attiwill, P. M. & Adams, M. A. (1993) Nutrient cycling in forests. New Phytol. 124: 561-582.

Bashkin, V. N. & Kudeyarov, V. N. (1977) Studying the year round dynamics of mineral nitrogen in a grey forest soil. Sov. Soil Sci. 8: 41-48 (in Russian with English summary).

Bengough, A. G, Castrignano, A., Pages, L. & van Noordwijk, M. (2000) Sampling strategies, scaling, and statistics. In Smit, A. L., Bengough, A. G., Engels, C. & Pellerin, van N. S. (eds). Root methods. Springer-Verlag, Berlin, pp.147-173.

Binkley, D. & Hart, S. C. (1989) The components of nitrogen availability assessments in forest soils. Adv. Soil Sci. 10: 57-112.

Chapin III, F. S., Molianen, L. & Kielland, K. (1993) Preferential use of organic nitrogen for growth by a non-mycorrhizal arctic sedge. Nature 361: 150-153.

Clark, K. L., Nadkarni, N. M., Schaefer, D. & Gholts, H. L. (1998) Atmospheric deposition and net retention of ions by the canopy in a tropical montane forest, Monteverde, Costa Rica. J. Trop. Ecol. 14: 27-45.

Dise, N. B. & Wright, R. F. (1995) Nitrogen leaching from European forests in relation to nitrogen deposition. For. Ecol. Manage. 71: 153-161.

FAO (1995) Forest Resources Assessment 1990, Global synthesis. Forestry Paper No. 124. FAO, Rome.

Ishizuka, S., Tanaka, S., Sakurai, K., Hirai, H., Hirotani, H., Ogino, K., Lee, H. S. & Kendawang, J. J. (1998) Characterization and distribution of soils at Lambir Hills National Park in Sarawak, Malaysia, with special reference to soil hardness and soil texture. Tropics 8: 31-44.

Jabatan Pertanian (1993) Panduan: Mengenali siri-siri tanah utama di Semenanjuang Malaysia. Jabatan Pertanian, Semananjung Malaysia, Kuala Lumpur. 118pp.

Kauffman, S., Sombroek, W. & Mantel, S. (1998) Soils of rainforests: characterization and major constraints of dominant forest soils in the humid tropics. In Schulte, A. & Ruhiyat, D. (eds). Soils of Tropical Forest Ecosystems. Springer, Berlinpp. pp.9-20.

Lal, R. (1987) Tropical Ecology and Physical Edaphology J. Wiley & Sons, New York, 732pp.

Manokaran, N. (1978) Nutrient concentration in precipitation, throughfall and stemflow in a lowland tropical rain forest in Peninsular Malaysia. Malay. Nat. J. 30: 423-432.

Matson, P. A., Vitousek, P. M., Ewel, J. J., Mazzarino, M. J. & Robertson, G. P. (1987) Nitrogen transformations following tropical forest felling and burning on a volcanic soil. Ecology 68: 491-502.

Matson, P. A., McDowell, W. H., Townsend, A. R. & Vitousek, P. M. (1999) The globalization of N deposition: ecosystem consequences in tropical environments. Biogeochemistry 46: 67-83.

Mbagwu, J. S. C., Lal, R. & Scott, T. W. (1983) Physical properties of three soils in southern Nigeria. Soil Sci. 136: 48-55.

Neil, C., Piccolo, M. C., Steudler, P. A., Melillo, J. M., Feigl, B. J. & Cerri, C. C. (1995) Nitrogen dynamics in soils of forests and active pastures in the western Brazilian Amazon basin. Soil Biol. Biochem. 27: 1167-1175.

Neil, C., Piccolo, M. C., Cerri, C. C., Steudler, P. A., Melillo, J. M. & Brito, M. (1997) Net nitrogen mineralization and net nitrification rates in soils following deforestation for

pasture across the southwestern Brazilian Amazon Basin landscape. Oecologia 110: 243-252.
Northup, R. R., Yu, Z. S., Dahlgren, R. A. & Vogt, K. A. (1995) Polyphenol control of nitrogen release from pine litter. Nature 377: 227-229.
Ohta, S. & Effendi, S. (1992a) Ultisols of "lowland dipterocarp forest" in East Kalimantan, Indonesia. 1. Morphology and physical properties. Soil Sci. Plant Nut. 38: 197-206.
Ohta, S. & Effendi, S. (1992b) Ultisols of "lowland dipterocarp forest" in East Kalimantan, Indonesia. 2. Status of carbon, nitrogen, and phosphorus. Soil Sci. Plant Nut. 38: 207-216.
Ohta, S., Effendi, S., Tanaka, N. & Miura, S. (1993) Ultisols of "lowland dipterocarp forest" in East Kalimantan, Indonesia. 3. Clay minerals, free oxides, and exchangeable cations. Soil Sci. Plant Nut. 39: 1-12.
Oliveira, M. R. G., van Noordwijk, M., Gaze, S. R., Brouwer, G., Bona, S., Mosca, G. & Hairiah, K. (2000) Auger sampling, ingrowth cores and pinboard methods. In Smit, A. L., Bengough, A. G., Engels, C. & Pellerin, van N. S. (eds). Root Methods. Springer-Verlag, Berlin, Heidelberg, Germany, pp.175-210.
Paramananthan, S. (1983) Field legend for soil surveys in Malaysia. UPM-PSTM, Kuala Lumpur, 80pp.
Read, D. J., Leake, J. R. & Langdale A. R. (1989) The nitrogen nutrition of mycorrhizal fungi and their host plants. In Boddy, L., Marchant, R. & Read, D. J. (eds). Nitrogen, phosphorus and sulphur utilization by fungi. Cambridge University Press, Cambridge, pp.181-204.
Reich, P. B., Grigal, D. F., Aber, J. D. & Gower, S. T. (1997) Nitrogen mineralization and productivity in 50 hardwood and conifer stands on diverse soils. Ecology 78: 335-347.
Richter, D. D. & Babbar, L. I. (1991) Soil diversity in the tropics. Adv. Ecol. Res. 21: 315-389.
Robertson, G. P. (1984) Nitrification and nitrogen mineralization in a lowland rainforest succession in Costa Rica, Central America. Oecologia 61: 99-104.
Robertson, G. P. (1986) Nitrification and denitrification in humid tropical ecosystems: potential controls on nitrogen retention. In Proctor, J. (ed). Mineral Nutrients in Tropical Forest and Savanna Ecosystems. Blackwell Scientific Publications, Oxford, pp.55-69.
Roy, S. & Singh, J. S. (1995) Seasonal and spatial dynamics of plant-available N and P pools and N-mineralization in relation to fine roots in a dry tropical forest habitat. Soil Biol. Biochem. 27: 33-40.
Sakurai, K., Tanaka, S., Ishizuka, S. & Kanzaki, M. (1998) Differences in soil properties of dry evergreen and dry deciduous forests in the Sakaerat Environmental Research Station. Tropics 8: 61-80.
Sanchez, P. A. (1976) Properties and management of soils in the tropics. J. Wiley & Sons, New York, 618pp.
Singh, R. S., Raghubanshi, A. S. & Singh, J. S. (1991) Nitrogen-mineralization in dry tropical savanna: effects of burning and grazing. Soil Biol. Biochem. 23: 269-273.
Sulaiman, S., Abdul Rahim, N. & LaFrankie, J. V. (1994) Pasoh Climatic Summary (1991-1993). FRIM Research Data No. 3. FRIM, Kepong.
Soil Survey Staff (1998) Keys to Soil Taxonomy. Eighth edition. USDA-NRCS, Washington D.C., 328pp.
Tange, K., Yagi, H., Sasaki, S., Niiyama, K. & Abd. Rahman, K. (1998) Relationship between topography and soil properties in a hill dipterocarp forest dominated by *Shorea curtsii* at Semangkok Forest Reserve, Peninsular Malaysia. J. Trop. For. Sci. 10: 398-409.
Vitousek, P. M. & Denslow, J. S. (1986) Nitrogen and phosphorus availability in treefall gaps of a lowland tropical rainforest. J. Ecol. 74: 1167-1178.
Vitousek, P. M. & Howarth, R. W. (1991) Nitrogen limitation on land and in the sea: How can it occur? Biogeochemistry 13: 87-115.
Vitousek, P. M. & Matson, P. A. (1988) Nitrogen transformations in a range of tropical forest soils. Soil Biol. Biochem. 20: 361-367.
Vitousek, P. M. & Sanford Jr, R. L. (1986) Nutrient cycling in moist tropical forest. Annu. Rev. Ecol. Syst. 17: 137-167.
Vogt, K. A., Vogt, D. J., Palmiotto, P. A., Boon, P., O'hara, J. & Asbjorsen, H. (1996) Reviews

of root dynamics in forest ecosystems grouped by climate, climatic forest types and species. Plant Soil 187: 159-219.

Wild, A. (1972) Nitrate leaching under bare fallow at a site in northern Nigeria. J. Soil Sci. 23: 315-324.

Woodruff, J. R. & Kamprath, E. J. (1965) Phosphorus adsorption maximum as measured by the Langmuir isotherm and its relationship to phosphorus availability. Soil Sci. Soc. Am. Proc. 29: 148-150.

Yamashita, T. & Takeda, H. (1998) Decomposition and nutrient dynamics of leaf litter in litter bags of two mesh sizes set in two dipterocarp forest sites in Peninsular Malaysia. Pedobiologia 42: 11-21.

Yamashita, T., Takeda, H. & Watanabe, H. (1992) Spatio-temporal variations of mineral nitrogen in a *Chamaecyparis obstusa* plantation soil. Bull. Kyoto Univ. For. 64: 51-60 (in Japanese with English summary).

Yamashita, T., Nakanishi, A., Tokuchi, N. & Takeda, H. (1999) Chemical properties and nutrient accumulation in two Siamese forest soils. Appl. For. Sci., Kansai 8: 89-94 (in Japanese with English summary).

Yoda, K. & Kira, T. (1982) Accumulation of organic matter, carbon, nitrogen and other nutrient elements in the soils of a lowland rainforest at Pasoh, Peninsular Malaysia. Jpn. J. Ecol. 32: 275-291.

Zou, X., Valentine, D. W., Sanford Jr., R. L. & Binkley, D. (1992) Resin-core and buried bag estimates of nitrogen transformations in Costa Rican lowland rainforests. Plant Soil 139: 275-283.

Part II

Vegetation Structure, Diversity and Dynamics

8 Leaf Phenology of Trees in the Pasoh Forest Reserve

Noriyuki Osada[1], Hiroshi Takeda[2], Akio Furukawa[3], Toshinori Okuda[4] & Muhamad Awang[5]

Abstract: We studied the phenology of leaf emergence in 94 trees and the leaf dynamics of 17 selected trees in the Pasoh Forest Reserve (Pasoh FR), Malaysia. We tested the following hypotheses: (i) the phenology of leaf emergence differs among trees of different heights because of differences in the relative importance of meteorological factors, and (ii) the timings of leaf emergence and leaf fall are synchronized within crowns. Six-month cycles in the phenology of leaf emergence were detected for the trees of various heights, and phenology was synchronized among different height trees. The phenology of leaf emergence was correlated with long-term mean rainfall seasonality, but not with rainfall occurring in the preceding 1-month period. This suggests that, irrespective of height differences, these trees responded to proximate cues that anticipated the seasonality of water availability. Moreover, leaf emergence occurred synchronously with leaf fall in most trees. This indicates that seasonal change in leaf number is small for most trees, and that standing leaf biomass does not change seasonally in this site. This is contrary to tropical dry forests, where standing leaf biomass is smaller in dry seasons than in rainy seasons.

Key words: El Niño, forest structure, irradiance, leaf emergence, leaf fall, Malaysia, meteorological conditions, rainfall seasonality, tree height.

1. INTRODUCTION

Tropical rain forests develop in regions with high temperatures and abundant year-round rainfall (Whitmore 1998). Western Peninsular Malaysia is typical of such regions; there are constant high temperatures and bimodal rainfall peaks each year, but dry seasons are not severe (Dale 1959). Leaf phenology has been studied intensively in this area (Gong & Ong 1983; Medway 1972; Ng 1981; Ogawa 1978; Putz 1979; Wong 1983; Yamashita et al. 1995). Despite the similar climate, however, some of these studies found different patterns in phenology. For example, Gong & Ong (1983) related leaf fall seasonality to drought stress, but Ogawa (1978) and Yamashita et al. (1995) related it to rainy conditions. Medway (1972) and Ng (1981) related phenology of leaf emergence to rainfall seasonality, but Putz (1979) and Wong (1983) did not find such a relationship.

Most of these investigators examined only tall trees (Medway 1972; Ng 1981; Putz 1979) or shrubs (Wong 1983), or they did not distinguish the canopy from understory trees (Gong & Ong 1983; Ogawa 1978; Yamashita et al. 1995).

---

[1] Graduate School of Agriculture, Kyoto University. Present address: Nikko Botanical Garden, Graduate School of Science, The University of Tokyo, Nikko 321-1435, Japan.
E-mail: ss29326@mail.ecc.u-tokyo.ac.jp
[2] Kyoto University, Japan.
[3] Nara Women's University, Japan.
[4] National Institute for Environmental Studies (NIES), Japan.
[5] Universiti Putra Malaysia (UPM), Malaysia.

Tropical rain forests are characterized by a complex vertical structure (Ashton & Hall 1992; Richards 1996) that creates heterogeneous environments for trees (Aoki et al. 1978; Yoda 1978). Consequently, canopy trees may be less likely to be limited by water and shade stresses than understory trees, because taller trees have deeper root systems (Cavender-Bares & Bazzaz 2000; Donovan & Ehleringer 1991; Wright 1992) and receive more light energy (Parker 1995; Yoda 1978) than smaller ones. If these stresses are the main factors affecting the leaf phenology of trees (van Schaik et al. 1993; Wright 1996), then leaf phenology may differ between canopy and understory species, as well as between saplings and mature trees of a given canopy species (Osada et al. 2002). Thus, clear differences in leaf phenology may be detected between canopy and understory trees within a stand (e.g. Opler et al. 1980; Shukla & Ramakrishnan 1982; Wright 199; Wright & Cornejo 1990).

In addition, because most trees drop some or all of their leaves during dry seasons, the standing leaf mass is smaller during dry seasons than during wet seasons in tropical dry forests (e.g. Borchert 1994; Bullock & Solis-Magallanes 1990; Frankie et al. 1974; Reich & Borchert 1984; Shukla & Ramakrishnan 1982). This seasonal variation standing for leaf mass influences the carbon budget of tropical forests. In tropical rain forests, however, most studies of leaf phenology have focused on either leaf emergence (Medway 1972; Ng 1981; Putz 1979; Wong 1983) or leaf fall (Gong & Ong 1983; Ogawa 1978; Procter et al. 1983), but the relationship between them has not been studied. If standing leaf mass changes seasonally in tropical rain forests, then the static measurement of leaf biomass (e.g. Kira 1978) will not represent a year-round pattern, and we cannot use it to estimate the carbon budget directly. Therefore, it is important to study the relationship between the phenology of leaf emergence and leaf fall in a tropical rain forest. Yamashita et al. (1995) related phenology of leaf emergence and leaf fall using stipules and leaves. This indirect method can be applied to some species that have a well-defined stipule that is dropped when new leaves are produced, but the relationship between leaf emergence and stipule fall is not always a good index because it varies with species. Rather, in order to examine different species, it is necessary to observe leaf emergence and leaf fall directly.

The objective of our study was to test the following hypotheses: (i) the phenology of leaf emergence differs among trees of different heights in a tropical rain forest stand, because of differences in the relative importance of meteorological factors, and (ii) the timings of leaf emergence and leaf fall are synchronized within crowns. For this purpose, we examined the leaf phenology and leaf dynamics of trees of various heights in a small plot in the Pasoh Forest Reserve (Pasoh FR) using a canopy walkway system.

2. MATERIALS & METHODS

2.1 Study site

The study was carried out in the Pasoh FR, Peninsular Malaysia (2°59' N, 102°18' E). The Pasoh FR is a lowland dipterocarp forest that belongs to the Red Meranti-Keruing type, and is dominated by *Shorea* spp. (Red Meranti group) and *Dipterocarpus* spp. (Keruing; Manokaran et al. 1992). The emergent layer averages 46 m and the height of the main canopy is 20–30 m (Manokaran & Swaine 1994).

The reserve is located in the Jelebu district of Negeri Sembilan state, where the annual rainfall is lowest in Peninsular Malaysia (Dale 1959). Annual rainfall at Kuala Pilah (37 km south of the Pasoh FR) averages 1,850.2 mm/year, with two peaks in April–May and November–December (Manokaran & Swaine 1994). The mean monthly temperature ranges from 26.0°C to 27.7°C (Manokaran & Swaine 1994).

A canopy walkway system was built in April 1992. It consisted of three towers (two 32 m and one 52 m tall), which were joined by 20 m walkways 32 m above the ground. Since the system is situated on a hill, the top of the 52 m tower is the highest place in the reserve, except for the eastern boundary.

2.2 Qualitative measurements of the phenology of leaf emergence in the stand

A triangular plot of ≒ 1,087 m^2 area was established around the tower system, with a 10-m buffer zone enclosing the triangle formed by the three towers. Tree heights were measured for the trees ≧ 5 cm DBH (Diameter at Breast Height) inside the plot by climbing the towers and estimating tree heights relative to the heights of the three towers (Osada et al. 2002).

For most of these trees (N = 94), the phenology of leaf emergence was observed monthly from May 1997 to May 1999. Some trees were excluded from this observation because of difficulties in observing the whole crowns or in distinguishing new leaves from old ones. Intensity of leaf emergence was assigned to the following four grades: 0 = no new leaves within the crown; 1 = less than one-fourth of the crown was covered with new leaves; 2 = one-fourth to three-fourths of the crown was covered with new leaves; 3 = more than three-fourths of the crown was covered with new leaves.

2.3 Quantitative measurement of leaf dynamics in selected species

Seventeen trees of 16 shade-tolerant species were selected adjacent to the canopy walkway system for measuring leaf demography (Table 1; Osada et al. 2001). For each tree, two sample 'branch-units' of about 1 m^3 in volume were chosen from the upper and lower parts of the crown. There were no leaves proximal to the sample branch units.

All sample branch units were tagged and sketched in September 1995 to analyze the number and position of leaves. The number and position of missing

Table 1 List of sample trees (Osada et al. 2001). Nomenclature follows Kochummen (1997). Stature class follows Manokaran et al. (1992).

Abbreviation	Sample tree	Family	DBH (cm)	Height (m)	Lowest Leaf Height (m)	Height of Branch Unit (m) Upper	Height of Branch Unit (m) Lower	Study Period (months)	Stature Class
Et	*Elateriospermum tapos* Bl.	Euphorbiaceae	54.1	35.0	18.0	32.5	19.5	44	Canopy
Xs 1	*Xanthophyllum stipitatum* Benn.	Polygalaceae	48.7	34.0	14.5	31.0	16.0	29	Canopy
Ds	*Dipterocarpus sublamellatus* Foxw.	Dipterocarpaceae	37.7	33.0	18.0	32.5	21.5	38	Emergent
Pc	*Ptychopyxis caput-medusae* (Hk.f.) Ridl.	Euphorbiaceae	36.0	33.0	21.0	32.5	23.5	44	Canopy
Gs	*Ganua* sp.1	Sapotaceae	27.2	26.0	17.0	24.0	18.5	44	Canopy
Er	*Eugenia rugosa* (Korth.) Merr.	Myrtaceae	18.9	19.0	11.0	18.5	13.5	44	Canopy
Mf	*Mangifera foetida* Lour.	Anacardiaceae	16.1	20.5	14.5	18.5	16.0	44	Canopy
Cs	*Chionanthus* sp.1	Oleaceae	16.0	22.0	16.0	19.0	17.0	44	Understory
Xs 2	*Xanthophyllum stipitatum* Benn.	Polygalaceae	13.3	18.5	10.0	16.5	13.5	44	Canopy
Dm	*Diplospora malaccensis* Hk.f.	Rubiaceae	11.8	15.0	11.0	14.5	12.5	44	Understory
Ml	*Macaranga lowii* King ex Hk.f.	Euphorbiaceae	9.3	12.0	9.0	12.0	9.0	44	Understory
Mm	*Monocarpia marginalis* (Scheff.) Sinclair	Annonaceae	7.5	10.5	3.5	10.0	5.5	44	Canopy
So	*Santiria oblongifolia* Bl.	Burseraceae	6.4	8.5	5.5	7.5	6.5	44	Canopy
Hd	*Homalium dictyoneurum* (Hance) Warb.	Flacourtiaceae	5.5	9.5	6.0	9.0	6.0	44	Canopy
As	*Actinodaphne sesquipedalis* Hk.f. & Thoms. ex Meisn.	Lauraceae	3.6	5.5	3.5	5.0	3.5	44	Understory
Mg	*Mallotus griffithianus* Hk.f.	Euphorbiaceae	3.0	6.0	2.0	5.0	3.0	39	Treelet
Ae	*Alangium ebenaceum* (Clarke) Harms	Alangiaceae	2.5	5.0	3.5	4.5	3.5	44	Understory

leaves and newly emerged leaves were recorded monthly from October 1995 through May 1999. Leaf demography was investigated in detail and is described in a related study (Osada et al. 2001).

2.4 Meteorological factors

Rainfall, net radiation, air temperature, and relative humidity were monitored at the top of the tallest tower (52 m). Net radiation was measured with a net radiometer (MF-40, Eko Instruments Trading), and air temperature and relative humidity were measured with a temperature and relative humidity meter (HMP 35C, Vaisala). These data were collected every minute, and their 30-minute averages were automatically recorded (CR10, Campbell). Rainfall data were collected every 30 minutes with a rain gauge (B-011-00, Yokogawa Weathac Corporation) (Tani & Abdul Rahim, unpublished data). In addition, monthly rainfall data measured at Kuala Klawang, about 24 km west of the Pasoh FR, were used for the long-term mean trends (Malaysian Meteorological Services 1947–1996).

2.5 Data analysis of the phenology of leaf emergence in the stand

The phenology of leaf emergence was compared among trees in 5 height classes: 5–10 m, 10–15 m, 15–20 m, 20–30 m, and > 30 m. The phenologies of leaf emergence of these height classes were compared using Kendall's coefficients of concordance. The phenological patterns of leaf emergence were analyzed using circular statistics (Zar 1996). The date of phenological events were converted to angles of 0–360°, where 0° corresponds to 1 January. The mean angle (a) and vector length (r) were then calculated:

$$a = \arctan(Y/X), \text{ if } X > 0 \text{ or } a = 180 + \arctan(Y/X), \text{ if } X < 0$$

$$r = \sqrt{X^2 + Y^2}$$

where $X = (\sum f_i \cos a_i)/n$, $Y = (\sum f_i \sin a_i)/n$, $n = \sum f_i$, a_i = the midpoint of the measurement interval and f_i = the frequency of occurrence of the data within that interval. Some previous studies detected 6-month cycles corresponding to rainfall seasonality, but others did not. We therefore used the Rayleigh test to examine whether phenological activity had a 1-year or a 6-month cycle.

Cumulative daily net radiation and rainfall were calculated, and daily maximum and minimum temperatures and relative humidities were averaged over a 1-month period, from the census date each month to the census date the following month. Since we studied leaf phenology at monthly intervals, it is reasonable to explain phenology in relation to meteorological factors occurring in the preceding 1-month period. Therefore, the relationship between monthly patterns of leaf emergence and meteorological factors of the preceding 1-month period were investigated using the Kendall rank correlation.

2.6 Data analysis of the timings of leaf emergence and leaf fall within crowns

The sampled units from the upper and lower crown of the same tree consisted of different numbers of branches, so values could not be compared directly. The numbers of leaves that emerged or fell in each month were therefore standardized by dividing by the number of branches originally present. The timing of leaf emergence was compared with that of leaf fall to examine whether leaf emergence and leaf fall were

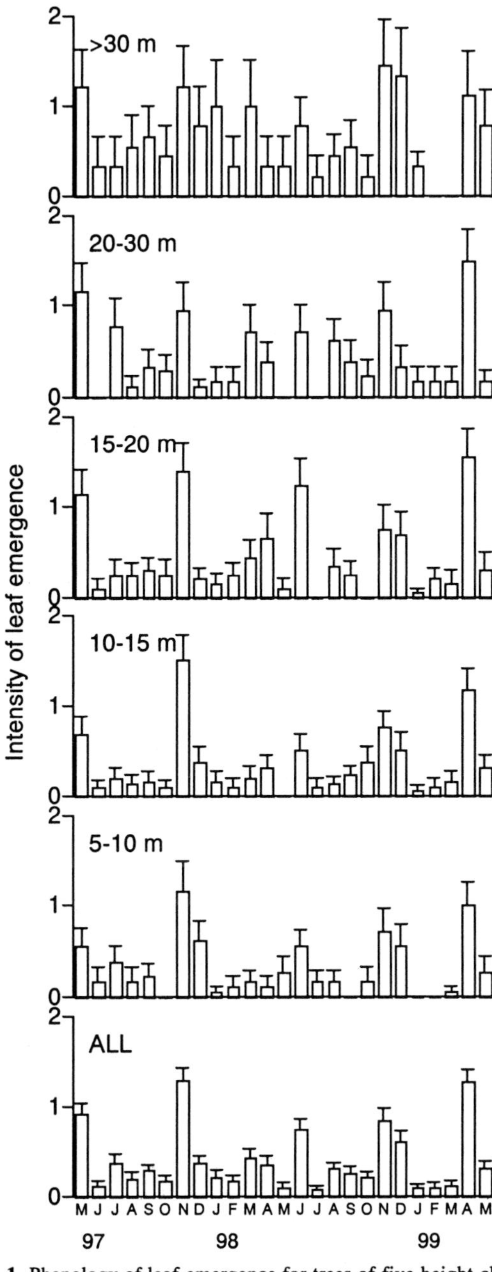

Fig. 1 Phenology of leaf emergence for trees of five height classes, and for all trees inside the plot (Mean ± *SD*). Sample sizes were 18, 29, 20, 18 and 9 for 5–10 m, 10–15 m, 15–20 m, 20–30 m and > 30 m in height, respectively. For each tree, intensity of leaf emergence was assigned into the following four grades: 0 = no new leaves within the crown; 1 = less than one-fourth of crown was covered with new leaves; 2 = one-fourth to three-fourths of the crown was covered with new leaves; 3 = more than three-fourths of the crown was covered with new leaves.

synchronized in each crown position. In trees that produced new leaves after dropping old ones, the monthly observation inevitably resulted in the following two patterns: (i) leaf emergence and leaf fall occurred within the same month in some instances, and (ii) leaf emergence occurred 1 month after leaf fall in other instances. A similar situation occurred in species that produced new leaves before dropping the old ones. To exclude this effect, 3-month moving averages were compared by cross correlation to assess whether leaf emergence occurred before or after leaf fall.

3. RESULTS

3.1 Phenology of leaf emergence within the stand

Fig. 1 shows the phenology of leaf emergence for trees of 5 height classes. For all height classes, there were 5 peaks in leaf emergence during the 2 years: May and November 1997, June and November–December 1998 and April 1999. Using the Rayleigh test, a 1-year cycle was detected only for the 15–20 m height class, but a 6-month cycle was detected for all classes (Table 2). Kendall's coefficients of concordance among the phenologies of the 5 height classes were highly significant ($W = 0.718, P < 0.001$). Moreover, the phenologies of all these height classes were highly correlated with each other (Kendall rank correlation, $\tau = 0.316$–0.669, $P < 0.05$ for all pair-wise combinations; all were significant at $P < 0.05$ after sequential Bonferroni correction). Thus, the phenology of leaf emergence was well synchronized among the height classes. The mean intensities of leaf emergence during the 2 years were $0.631 \pm 0.417, 0.422 \pm 0.398, 0.442 \pm 0.447, 0.343 \pm 0.361$ and 0.307 ± 0.315 ($\pm$ *SD*) for the trees with > 30 m, 20–30 m, 15–20 m, 10–15 m and 5–10 m tall, respectively, and they were greater in taller trees (Kruskal-Wallis test, $\chi^2 = 12.78, P = 0.012$).

There was little rainfall from spring 1997 to summer 1998, when a strong El Niño persisted, and rainfall seasonality during this period was quite different from the long-term mean (Fig. 2). Weak positive correlations were detected between the timing of leaf emergence and net radiation during the preceding 1-month period for trees >15 m tall and for all trees (Table 3). Similarly, weak positive correlations were detected with long-term mean rainfall seasonality for all height classes, but not with rainfall of the preceding 1-month period. There were no correlations between the phenology of leaf emergence and other meteorological factors (maximum and minimum temperature, and maximum and minimum relative humidity) for all height classes.

3.2 The timings of leaf emergence and leaf fall within crowns

The timings of leaf emergence and leaf fall were synchronized in 15 out of 17 trees in one or two crown parts ($P = 0.0012$, Table 4). Clear patterns were observed in two trees (*Et* and *Cs*), which dropped most of their leaves before producing new ones, but this was not detected statistically for *Cs* because of its irregular leaf fall patterns. Leaf fall occurred before the production of new leaves in three trees (*Et*, *Cs* and *As*). In contrast, leaf fall occurred after the production of new leaves in seven trees (*Gs*, *Er*, *Mf*, *Xs*2, *Mm*, *Hd* and *Ae*) and leaf fall occurred simultaneously with emergence in 5 trees (*Xs*1, *Ds*, *Pc*, *Dm* and *So*). Thus, leaf emergence and leaf fall were synchronized in most trees, and trees that produced new leaves before dropping the old ones were more abundant than those that produced new leaves after dropping the old ones.

Table 2 Result of Rayleigh test for 1-year cycle and 6-month cycle in phenology of leaf emergence (+$P < 0.10$, **$P < 0.01$, ***$P < 0.001$, others $P > 0.10$).

Height class (m)	1-year cycle Rayleigh test *r*	*z*		6-month cycle Rayleigh test *r*	*z*		1st peak Mean date	95% confidence interval	2nd peak Mean date	95% confidence interval
>30	0.09	1.15		0.25	8.85	***	14-May	30-Apr – 28-May	13-Nov	30-Oct – 27-Nov
20-30	0.10	1.95		0.20	7.82	***	18-Apr	3-Apr – 3-May	18-Oct	3-Oct – 2-Nov
15-20	0.15	4.68	**	0.32	22.09	***	29-Apr	20-Apr – 7-May	28-Oct	20-Oct – 6-Nov
10-15	0.10	2.73	+	0.40	40.66	***	28-Apr	22-Apr – 4-May	28-Oct	22-Oct – 3-Nov
5-10	0.02	0.06		0.45	27.53	***	9-May	2-May – 17-May	8-Nov	31-Oct – 15-Nov
ALL	0.05	2.05		0.31	93.03	***	1-May	27-Apr – 5-May	31-Oct	27-Oct – 4-Nov

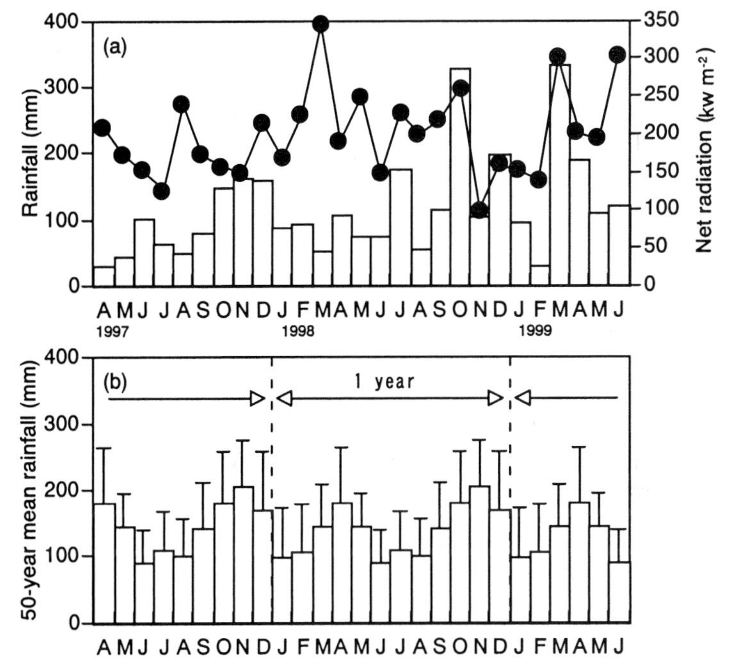

Fig. 2 Seasonality of rainfall (bars) and net radiation (circles with line) in Pasoh FR (a: M. Tani & Abdul Rahim N. unpubl. data) and 50-year mean rainfall seasonality at Kuala Klawang, about 24 km from Pasoh FR (Mean ± *SD*) (b: data source from Malaysian Meteorological Services 1947–1996).

Table 3 Kendall's rank correlations (τ) between phenology of leaf emergence and meteorological factors of a preceding 1-month period (+$P < 0.10$, *$P < 0.05$, **$P < 0.01$, others $P > 0.10$).

Height class	Rainfall	Mean rainfall	Net radiation	Max temp.	Min temp.	Max RH	Min RH
>30	0.21	0.26 +	0.24 +	0.03	0.02	0.20	0.15
20-30	0.11	0.30 *	0.41 **	0.11	0.02	-0.07	-0.03
15-20	0.06	0.30 *	0.32 *	0.23	0.09	0.00	0.00
10-15	0.12	0.47 **	0.23	0.10	0.03	0.00	0.03
5-10	0.23	0.35 *	0.14	0.21	0.14	0.02	-0.02
ALL	0.15	0.36 *	0.31 *	0.15	0.01	0.03	0.03

Table 4 Timings of leaf emergence and leaf fall. Two branch units of about 1 m^3 in volume were selected from the upper and lower crowns and were analyzed separately. Three-month moving averages were compared by cross correlation (Kendall's τ). N indicates the number of months analyzed (+ $P < 0.10$, * $P < 0.05$, ** $P < 0.01$, *** $P < 0.001$, others $P > 0.10$).

Sample tree		Upper crown		Lower crown	
		N	τ	N	τ
Et	F → E	39	0.47 ***	42	0.30 *
	=	40	0.44 ***	43	0.50 ***
	E → F	39	0.05	42	0.43 ***
Xs 1	F → E	27	-0.14	27	0.01
	=	28	-0.05	28	0.40 **
	E → F	27	-0.01	27	0.28 +
Ds	F → E	37	0.13	37	-0.12
	=	38	0.28 *	38	0.07
	E → F	37	0.21 +	37	0.12
Pc	F → E	42	0.42 ***	42	0.19
	=	43	0.55 ***	43	0.22 +
	E → F	42	0.25 *	42	0.15
Gs	F → E	42	0.14	42	0.12
	=	43	0.36 **	43	0.31 *
	E → F	42	0.45 ***	42	0.44 ***
Er	F → E	42	0.15	42	-0.07
	=	43	0.41 ***	43	0.15
	E → F	42	0.70 ***	42	0.39 ***
Mf	F → E	42	0.10	29	0.23
	=	43	0.50 ***	30	0.45 **
	E → F	42	0.32 **	29	0.28 +
Cs	F → E	42	0.11	33	0.07
	=	43	0.04	34	0.41 **
	E → F	42	0.00	33	0.63 ***
Xs 2	F → E	42	0.22 +	42	0.00
	=	43	0.15	43	0.27 *
	E → F	42	0.16	42	0.44 ***
Dm	F → E	42	0.30 **	16	-0.38 +
	=	43	0.33 **	17	-0.40 *
	E → F	42	0.31 **	16	-0.37 +
Ml	F → E	42	0.17	42	0.02
	=	43	0.21 +	43	0.02
	E → F	42	0.13	42	0.05
Mm	F → E	42	0.14	18	0.29
	=	43	0.32 **	19	0.36 +
	E → F	42	0.28 *	18	0.40 *
So	F → E	42	0.62 ***	42	0.66 ***
	=	43	0.41 **	43	0.54 ***
	E → F	42	0.39 **	42	0.40 **
Hd	F → E	42	0.23 +	36	-0.27 *
	=	43	0.42 ***	37	0.09
	E → F	42	0.43 ***	36	0.43 **
As	F → E	42	0.33 *	42	0.26 +
	=	43	0.42 **	43	0.17
	E → F	42	0.24 +	42	-0.03
Mg	F → E	37	-0.08	37	-0.03
	=	38	-0.12	38	-0.02
	E → F	37	-0.11	37	-0.03
Ae	F → E	42	0.02	42	-0.10
	=	43	0.23 +	43	-0.05
	E → F	42	0.34 **	42	-0.04

F → E, Leaf fall occurred before producing new leaves;
= , Leaf fall and leaf emergence occurred at the same time;
E → F, Leaf fall occurred after producing new leaves.

4. DISCUSSION

4.1 Phenology of leaf emergence within the stand

In this study, we found clear synchronous 6-month cycles in the phenology of leaf emergence for trees in all height classes (Table 2). Since the mean intensity of leaf emergence was greater in taller trees, the absolute intensity of leaf emergence may be influenced by height, although seasonality may not be. This result is contrary to our expectations. The phenology of leaf emergence was not correlated to rainfall that occurred during the preceding 1-month period, but it was correlated to long-term mean rainfall seasonality. There was little rainfall from spring 1997 to summer 1998, and rainfall seasonality in this period was quite different from the long-term mean (Fig. 2). In irrigation experiments conducted in a tropical wet forest in Panama, Wright (1991) found that the phenologies of several shrub species were not limited directly by current water deficits, but were controlled by proximate cues that anticipated water deficits. Similarly, the phenology of leaf emergence in our site may be controlled by proximate cues that anticipate the seasonality of water availability, and therefore correlations may not be detected between the phenology of leaf emergence and rainfall seasonality of the preceding 1-month period. This result contrasts that of Osada et al. (2002), who found that leaf production was limited by light resources available to each individual, but that leaf fall was enhanced by drought stress in saplings of the canopy species *Elateriospermum tapos* Bl. (Euphorbiaceae) in the same site. Responses to drought stress may differ across species in a forest stand (e.g. Borchert 1994), and therefore more intensive study is necessary to investigate the intra- and interspecific differences in leaf phenology.

In addition, the phenology of leaf emergence at greater heights was also related to net radiation in the preceding 1-month period (Table 3). Thus, irradiation may enhance leaf production at higher positions in the canopy. However, these are just correlations, and physiological studies of trees are required to determine which factors are more important in stand-level leaf phenology.

4.2 The timings of leaf emergence and leaf fall within crowns

Leaf emergence occurred synchronously with leaf fall in most trees, and trees produced new leaves within 1 month before or after dropping old leaves. Trees that dropped old leaves after producing new leaves were more abundant than those that dropped old leaves before producing new ones. However, two individuals of the same species (*Xs*) had different patterns, and even the two crown positions of the same tree also showed different patterns (Table 4). This is probably because we studied leaf phenology at monthly intervals, and more frequent observations are necessary to determine the relationships between the phenologies of leaf emergence and leaf fall. Nonetheless, since most trees produced and dropped their leaves synchronously, seasonal variation in leaf number is relatively small in most trees, and standing leaf mass will not change seasonally in this site. Therefore, the year-round patterns of leaf biomass are well represented by static measurements of leaf biomass (Kira 1978; Osada et al. 2001).

ACKNOWLEDGEMENTS

We thank Drs. M. Yasuda, N. Osawa, A. Makmom, Mr. J. Shamsuddin and the members of the Laboratory of Forest Ecology, Kyoto University, for their valuable suggestions. Drs. M. Tani and Abdul Rahim Nik allowed us to use the unpublished meteorological data. Dr. M. Yasuda provided the rainfall data of Kuala Klawang. The present study is a part of a Joint Research Project between the NIES, FRIM and UPM (Global Environment Research Program granted by Japan Environment Agency, Grant No. E-1). This study was partly supported by Japan Society for the Promotion of Science Research Fellowships for Young Scientists for N. Osada.

REFERENCES

Aoki, M., Yabuki, K. & Koyama, H. (1978) Micrometeorology of Pasoh forest. Malay. Nat. J. 30: 149-159.

Ashton, P. S. & Hall, P. (1992) Comparisons of structure among mixed dipterocarp forests of northwestern Borneo. J. Ecol. 80: 459-481.

Borchert, R. (1994) Soil and stem water storage determine phenology and distribution of tropical dry forest trees. Ecology 75: 1437-1449.

Bullock, S. H. & Solis-Magallanes, J. A. (1990) Phenology of canopy trees of a tropical deciduous forest in Mexico. Biotropica 22: 22-35.

Cavender-Bares, J. & Bazzaz, F. A. (2000) Changes in drought response strategies with ontogeny in *Quercus rubra*: implications for scaling from seedlings to mature trees. Oecologia 124: 8-18.
Dale, W. L. (1959) The rainfall of Malaya I. J. Trop. Geogr. 13: 23-37.
Donovan, L. A. & Ehleringer, J. R. (1991) Ecophysiological differences among juvenile and reproductive plants of several woody species. Oecologia 86: 594-597.
Frankie, G. W., Baker, H. G. & Opler, P. A. (1974) Comparative phenological studies of trees in tropical wet and dry forests in the lowlands of Costa Rica. J. Ecol. 62: 881-919.
Gong, W. K. & Ong, J. E. (1983) Litter production and decomposition in a coastal hill dipterocarp forest. In Sutton, S. L., Whitmore, T. C. & Chadwick, A. C. (eds). Tropical rain forest: ecology and management, Blackwell, Oxford, pp.275-285.
Kira, T. (1978) Community architecture and organic matter dynamics in tropical lowland rain forests of Southeast Asia with special reference to Pasoh Forest, West Malaysia. In Tomlinson, P. B. & Zimmermann, M. H. (eds). Tropical trees as living systems, Cambridge University Press, Cambridge, pp.561-590.
Kochummen, K. M. (1997) Tree flora of Pasoh forest. Forest Research Institute Malaysia, Kepong, Malaysia.
Malaysian Meteorological Service (1947-1996). Monthly abstract of meteorological observations. Malaysian Meteorological Service, Kuala Lumpur.
Manokaran, N. & Swaine, M. D. (1994) Population dynamics of trees in dipterocarp forests of Peninsular Malaysia. Malayan Forest Records No. 40. Forest Research Institute Malaysia, Kepong, Malaysia, 173pp.
Manokaran, N., LaFrankie, J. V., Kochummen, K. M., Quah, E. S., Klahn, J. E., Ashton, P. S. & Hubbell, S. P. (1992) Stand table and distribution of species in the 50 ha research plot at Pasoh Forest Reserve. Forest Research Institute Malaysia, Kepong, Malaysia.
Medway, L. (1972) Phenology of a tropical rainforest in Malaya. Biol. J. Linn. Soc. 4: 117-146.
Ng, F. S. P. (1981) Vegetative and reproductive phenology of dipterocarps. Malay. For. 44: 197-221.
Ogawa, H. (1978) Litter production and carbon cycling in Pasoh forest. Malay. Nat. J. 30: 367-373.
Opler, P. A., Frankie, G. W. & Baker, H. G. (1980) Comparative phenological studies of treelet and shrub species in tropical wet and dry forests in the lowlands of Costa Rica. J. Ecol. 68: 167-188.
Osada, N., Takeda, H., Furukawa, A. & Muhamad, A. (2001) Leaf dynamics and maintenance of tree crowns in a Malaysian rain forest stand. J. Ecol. 89: 774-782.
Osada, N., Takeda, H., Furukawa, A. & Muhamad, A. (2002) Ontogenetic changes in leaf phenology of a canopy species, *Elateriospermum tapos* (Euphorbiaceae), in a Malaysian rain forest. J. Trop. Ecol. 18: 91-105.
Parker, G. G. (1995) Structure and microclimate of forest canopies. In Lowman, M. D. & Nadkarni, N. M. (eds). Forest Canopies. Academic Press, San Diego, pp.73-106.
Procter, J., Anderson, J. M., Fogden, S. C. L. & Vallack, H. W. (1983) Ecological studies in four contrasting lowland rain forests in Gunung Mulu National Park, Sarawak. J. Ecol. 71: 261-283.
Putz, F. E. (1979) A seasonality in Malaysian tree phenology. Malay. For. 42: 1-24.
Reich, P. B. & Borchert, R. (1984) Water stress and tree phenology in a tropical dry forest in the lowlands of Costa Rica. J. Ecol. 72: 61-74.
Richards, P. W. (1996) Tropical rain forest. Cambridge University Press, Cambridge.
Shukla, R. P. & Ramakrishnan, P. S. (1982) Phenology of trees in a sub-tropical humid forest in northeastern India. Vegetatio 49: 103-109.
Van Schaik, C. P., Terborgh, J. W. & Wright, S. J. (1993) The phenology of tropical forests: adaptive significance and consequences for primary consumers. Annu. Rev. Ecol. Syst. 24: 353-377.
Whitmore, T. C. (1998) An introduction to tropical rain forests. Oxford University Press, Oxford.
Wong, M. (1983) Understory phenology of the virgin and regenerating habitats in Pasoh Forest

Reserve, Negeri Sembilan, west Malaysia. Malay. For. 46: 197-223.
Wright, S. J. (1991) Seasonal drought and the phenology of understory shrubs in a tropical moist forest. Ecology 72: 1643-1657.
Wright, S. J. (1992) Seasonal drought, soil fertility and the species density of tropical forest plant communities. Trends Ecol. Evol. 7: 260-263.
Wright, S. J. (1996) Phenological responses to seasonality in tropical forest plants. In Mulkey, S. S., Chazdon, R. L. & Smith, A. P. (eds). Tropical forest plant ecophysiology. Chapman & Hall, New York, pp.440-460.
Wright, S. J. & Cornejo, F. H. (1990) Seasonal drought and leaf fall in a tropical forest. Ecology 71: 1165-1175.
Yamashita, T., Takeda, H. & Kirton, L. G (1995) Litter production and phenological patterns of *Dipterocarpus baudii* in a plantation forest. Tropics 5: 57-68.
Yoda, K. (1978) Three-dimensional distribution of light intensity in a tropical rain forest of West Malaysia. Malay. Nat. J. 30: 161-177.
Zar, J. H. (1996) Biostatistical analysis. Prentice Hall, New Jersey.

9 Eco-Morphological Grouping of Non-Dipterocarp Tree Species in a Tropical Rain Forest Based on Seed and Fruit Attributes

Mamoru Kanzaki[1], Song Kheong Yap[2], Yuka Okauchi[3], Katsuhiko Kimura[4] & Takuo Yamakura[3]

Abstract: Eight ecological and morphological attributes of 39 non-dipterocarp tree species were examined in the Pasoh Forest Reserve (Pasoh FR), Peninsular Malaysia. The attributes included three related to seed germination (seed longevity in the soil, responsiveness of germination to gaps and germination type), three morphological attributes of seed or fruit (fruit morphology type, seed coat character, and seed weight), and two attributes related to niche preference (successional status and life-form of adult plant). These attributes were treated as categorical data, and the dissimilarity matrix between species was subjected to clustering analysis using the UPGMA algorithm. Three cluster were recognized. The first cluster consists of 17 primary species that have short-lived seeds. These species also have large seeds, no special seed coat, and hypogeal germination that is not responsive to gaps. The second group consists of 12 gap-secondary species; most have extra long-lived seeds with epigeal germination in response to gaps, a capsule and an oily seed coat.The third cluster consists of 10 primary species, but differs from the first group in seed longevity; they have intermediate to long seed longevity (mean life ≧ 3 months and < 10 years). This group is also characterized by epigeal germination and a woody seed coat, which are common in gap-secondary species, and large seeds like those of primary species with short-lived seed. Our research suggests the existence of a third functional group, primary species with long-lived seed, which is distinct from primary and gap-secondary species in the classical meaning. The attributes that characterize this group seem to be adaptations to mammalian seed herbivores or dispersers.

Key words: buried seed, cluster analysis, Detrended Correspondence Analysis (DCA), functional group, germination, seed and fruit attributes, tropical rain forest.

1. INTRODUCTION

Seed and seedling stages are quite important in the life cycle of plants, and seed attributes are thought to largely represent species life-history traits (Ng 1978; Salisbury 1942; Swaine & Whitmore 1988). Among seed attributes, seed size has a significant relationship with various ecological traits of plants, such as shade-tolerance (Foster & Janson 1985; Grubb & Metcalfe 1996; Hewitt 1998; Metcalfe & Grubb 1995; Osunkoya 1996), seedling growth (Seiwa 2000; Seiwa & Kikuzawa 1989), growth form of the adult plant (Mazer 1989; Metcalfe & Grubb 1995) and

[1] Division of Forestry and Biomaterials Science, Graduate School of Agriculture, Kyoto Univer sity, Kyoto 606-8502, Japan. E-mail: mkanzaki@kais.kyoto-u.ac.jp
[2] Forest Reseach Institute Malaysia (FRIM), Malaysia.
[3] Graduate School of Science, Osaka City University, Japan.
[4] Faculty of Education, Fukushima University, Japan.

Table 1 List of 39 species examined and their attributes. There is one replicate sample for *Macaranga gigantea*.

Species	Family	Fruit type	Seed coat	Seed weight	Seed longevity	Responsiveness to gap	Germination type	Life form	Successional status
Antidesma cuspidatum M. A.	Euphorbiaceae	D	W	S	L	N	E	ST	P
Calophyllum macrocarpum Hk. f.	Guttiferae	D	W	L	I	N	H	LT	P
Canarium littorale Bl.	Burseraceae	D	W	M	I	N	E	LT	P
Cassia nodosa Buch.-Ham. ex Roxb.	Leguminosae	F	W	M	L	N	E	MT	P
Commersonia bartramia (L.) Merr.	Sterculiaceae	C	W	S	EL	R	E	ST	S
Cryptocarya griffithiana Wight	Lauraceae	B	N	M	S	N	H	MT	P
Daemonorops geniculata (Griff.) Mart.	Palmae	B	W	M	L	N	E	Rattan	P
Dracontomelon dao (Blanco) Merr. Rolfe	Anacardiaceae	D	W	M	EL	N	E	LT	P
Dysoxylum acutangulum Miq.	Meliaceae	C	N	L	S	N	H	MT	P
Eugenia duthieana King	Myrtaceae	B	N	L	S	N	D	MT	P
Eugenia sp.	Myrtaceae	B	N	L	S	N	SH	MT	P
Ficus glandulifera (Wall. ex Miq.)	Moraceae	Fig	N	S	L	R	E	ST	P
Garcinia parvifolia (Miq.) Miq.	Guttiferae	B	N	M	S	N	H	MT	P
Glochidion sericeum Hk. f.	Euphorbiaceae	C	O	S	EL	R	E	ST	S
Gmelina arborea Roxb.	Verbenaceae	D	W	M	EL	R	E	LT	S
Gmelina elliptica J. E. Smith	Verbenaceae	D	W	M	I	R	SH	Shrub	S
Grewia miqueliana Kurz	Tiliaceae	B	N	M	I	N	E	ST	P
Hibiscus macrophyllus Roxb. ex Hornem.	Malvaceae	C	N	S	I	R	E	MT	S
Knema scortechinii (King) Sinclair	Myrisinaceae	F	N	L	S	N	H	LT	P
Lithocarpus encleisacarpus (Korth.) A. Camus	Fagaceae	N	N	L	I	N	H	LT	P
Litsea castanea Hk. f.	Lauraceae	B	N	M	S	N	H	LT	P
Litsea cordata (Jack) Hk. f.	Lauraceae	B	N	M	S	N	H	LT	S
Macaranga gigantea (Rchb. f. & Zoll.) M. A.	Euphorbiaceae	C	O	M	I	N	E	ST	S
Macaranga gigantea (Rchb. f. & Zoll.) M. A.	Euphorbiaceae	C	O	M	I	R	E	ST	S
Mallotus macrostachyus (Miq.) M. A.	Euphorbiaceae	C	O	M	EL	R	E	ST	S
Mallotus paniculatus (Miq.) M. A.	Euphorbiaceae	C	O	M	EL	R	E	ST	S
Memecylon excelsum Bl.	Melastomataceae	B	N	M	S	N	E	LT	P
Milletia atropurpurea (Wall.) Benth.	Leguminosae	F	N	L	S	N	H	MT	P
Ormosia macrodisca Baker	Leguminosae	F	N	L	S	N	SH	MT	P
Ormosia venosa Baker	Leguminosae	F	N	L	S	N	SH	MT	P
Phyllanthus emblica L.	Euphorbiaceae	D	W	M	EL	R	E	MT	S
Polyalthia rumphii (Bl.) Merr.	Annonaceae	B	N	M	S	N	D	MT	P
Pyrenaria acuminata Planch.	Theaceae	B	W	M	I	N	E	ST	P
Quercus argentata Korth.	Fagaceae	N	N	L	S	N	H	LT	P
Santiria griffithii (Hk. f.) Engl.	Burseraceae	D	N	M	S	N	H	MT	P
Sapium baccatum Roxb.	Euphorbiaceae	C	N	M	L	R	E	MT	S
Styrax benzoin Dryand	Styracaceae	B	W	L	I	N	E	LT	P
Trema orientalis (L.) Bl.	Ulmaceae	D	W	S	EL	R	E	MT	S
Vitex pinnata L.	Verbenaceae	D	W	S	L	R	E	MT	S
Xylopia fusca Maing. ex Hk. f. & Thoms	Annonaceae	F	N	M	S	N	H	MT	P

Fruit type: B, berry; C, capsule; D, drupe; F, follicle; N, nut.
Sead coat: W, woody coat; O, oily coat; N, no special coat.
Seed weight: S, small seed (< 0.01 g); M, medium seed (0.01–1 g); L, large seed (> 1 g).
Seed longevity: S, short-lived (< 3 mo); I, intermediate-lived (3 mo–1 yr); L, long-lived (1–10 yr); E, extra-long-lived (> 10 yr).
Gap responsiveness: R, responsive ; N, nonresponsive.
Germination type: E, epigeal; H, hypogeal; SH, semi-hypogeal; D, durian.
Life-form: LT, large tree (> 30 m in height); MT, medium tree (15 m to 30 m); ST, small tree (< 15 m).
Successional status: P, primary species; S, gap-secondary species.

seed dispersal mode (Leishman & Westoby 1994). The morphological attributes of fruit are also significantly associated with seed dispersal mode (Gautier-Hion et al. 1985; Pijl 1982; Roth 1987).

In the previous study (Kanzaki et al. 1997), we demonstrated variation in the seed dormancy and germination traits of Malaysian tropical rain forest trees, using seed burial and germination experiments. Sixteen percent of primary forest species produce seed that germinates quite slowly in forest soil, with a mean life span exceeding one year. These results suggest the existence of a third ecological group, with seed and seedling traits that differ from those of the pioneer and non-pioneer categories of Swaine & Whitmore (1988). This study tests this prediction using numerical classification methods to compare the eco-morphological attributes of seeds, fruit, seedlings, and adult plants.

2. MATERIALS

Seeds were collected for a burial test to examine seed population dynamics and assess seed longevity in the soil (Kanzaki et al. 1997). All the seed samples were collected between December 1991 and February 1995, either from a primary forest in or from the secondary vegetation next to the primary forest of Pasoh Forest Reserve (Pasoh FR) (Table 1). Of the 45 species examined in Kanzaki et al. (1997), six species were excluded from this study because data on some of the attributes analyzed in the present study were unavailable. The 39 species examined represent 34 genera and 21 families. Since no general flowering occurred during the seed-collection period, all of the species examined are non-dipterocarp species. Except for *Gmelina arborea*, which was introduced from India/Myanmar, all plant species are native to Peninsular Malaysia. One woody liana (rattan), *Daemonorops geniculata*, was included in the study.

One duplicate for *Macaranga gigantea* was included in the data set because two burial experiments conducted for the species produced different seed response to gaps. Therefore, the total number of samples was 40.

3. DATA SET

Eight attributes based on the previous study by Kanzaki et al. (1997) and other references were examined for the 39 species. The attributes were categorized as follows (also see Table 2).

Table 2 Attributes examined. Attributes of 39 species were classified into two to six categories as shown below.

Attribute	No. categorie	Categories
Fruit type	6	berry, drupe, capsule, follicle, nut and fig
Seed coat	3	woody, oily, and no specieal coat
Seed weight	3	small (<0.01 g), medium (>0.01g and <1g), and large seeds (>1 g)
Seed longevity	4	short (< 3 mo), intermediate (≥3 mo & <1 yr), long (≥1 yr & <10 yr), and extra-long (>10 yr)
Responsiveness of germination to gap	2	gap responsive and nonresponsive
Germination type	4	epigeal, hypogean, semi-hypogeal, and durian
Life-form	5	large tree, medium tree, small tree, shrub and rattan
Successional status	2	primary and gap-secondary

3.1 Fruit type

Fruit morphology was classified into six categories: berry, drupe, capsule, follicle, nut and fig fruit, based on the descriptions in Ng (1991, 1992) and Whitmore (1972a, b).

3.2 Seed coat

Seeds with a thick hard coat and seeds surrounded by thick woody endocarp were grouped into the "woody coat" category. Some of the seeds with a hard seed coat are surrounded by oily sarcotesta. These seeds were placed in a distinct group, the "oily coat" category. Fifteen species belonged to the woody coat category and *Macaranga gigantea*, *Glochidion sericeum* and two species of *Mallotus* belonged to the oily coat category. The remaining 20 species were grouped in a "no special coat" category. *Santiria griffithii*, a species with fruit categorized as drupe, was included in the no special coat category because the endocarp was very thin and weak.

3.3 Seed weight

Fresh seed weight was obtained by Kanzaki et al. (1997). Each value is the average weight of at least 20 seeds for each of the species examined. Seed was grouped into three categories by weight: small seeds (< 0.01 g), medium seeds (≧ 0.01 and < 1 g) and large seeds (≧ 1 g).

3.4 Seed longevity

Kanzaki et al. (1997) estimated the mean longevity of buried seed by monitoring a seed population artificially buried in the surface soil layer of the closed forest stand at Pasoh FR for 3 years. In this prior study, the survival of ungerminated seeds was regressed using the Weibull distribution model, and the mean life of the ungerminated seed in the soil was estimated. The estimated mean life was classified into four categories: short-lived (< 3 months), intermediate-lived (3 months to 1 year), long-lived (1 year to 10 years)and extra-long-lived (over 10 years). These values adequately represent differences in the germination speed of the species examined under shaded conditions in the dense forest.

3.5 Responsiveness of germination to gap conditions

Kanzaki et al. (1997) conducted two germination tests. In one, an artificial soil seed bank was buried in forest soil under the closed rain forest canopy and in the other, seed was sown in a germination bed in a glass house to simulate open habitat conditions. Seeds that germinated rapidly under open site conditions but never germinated under the forest soil conditions were defined as "gap responsive", while the seeds that germinated in both of the germination tests were defined as "non-responsive". The species examined included 12 gap responsive and 27 non-responsive species.

3.6 Germination type

Ng (1978b, 1991, 1992) categorized the germination type of Malaysian tree seed into four types: epigeal, hypogeal, semi-hypogeal and durian. In semi-hypogeal germination the cotyledons are exposed, unlike hypogeal germination, but the hypocotyl is not elongated. In durian germination, the cotyledons remain within the seed coat, like hypogeal germination, and the hypocotyl is elongated. The germination type of the examined species was determined by direct observation during our germination test or by following the description in Ng (1991, 1992). The species examined consisted of 21 epigeal, 2 durian, 11 hypogeal and 5 semi-hypogeal species.

3.7 Life-form

The life-form of the adult plant was classified into 5 categories: large tree (> 30 m high), medium tree (15 m to 30 m), small tree (< 15 m), shrub and liana (rattan) based on the floristic description by Ng (1978a, 1989) and Whitmore (1972a,b).

3.8 Successional status

The species were classified in two categories: primary forest species and gap-secondary forest species. The assignment was based on the floristic description by Ng (1978a, 1989) and Whitmore (1972a,b) or our observations. If a species was common in both secondary and primary forest, it was included in the primary category.

4. STATISTICAL ANALYSIS

4.1 Test of association

Correlation between each pair of attributes was examined by the test of association based on G statistics subjected to William's correction (Sokal 1973).

4.2 Cluster analysis

The eco-morphological similarity of species was expressed by the Jaccard Index of Similarity (S):

$$S = a/(a + b + c),$$

where a represents the number of characters common to two species, and b and c represent the number of characters unique to each species. The dissimilarity between species, D, was then defined by

$$D = 1 - S,$$

and the matrix of dissimilarity was subjected to a clustering analysis using the unweighted pair group method with arithmatic mean (UPGMA) algorithm (Sneath & Sokal 1973). In addition to the clustering analysis of the species, the attributes were also subjected to UPGMA clustering. In this case, the similarity between attributes was also defined by the Jaccard index. In the cluster analyses, the order of species or attributes in the distance matrix was randomized ten times and the matrices were re-analyzed to confirm the consistency of the dendrograms produced.

A program package, SPSS ver. 9.0 (SPSS Inc., Chicago, USA), was used for these clustering analyses.

4.3 Ordination of species

To extract the representative axes that reflect most of the eco-morphological variation between species, Detrended Correspondence Analysis (DCA) (Hill & Gauch 1980) was employed. Each species was characterized by 29 dummy variables, which corresponded to 29 categories mentioned above. When a species possessed a character, the corresponding dummy variable was given a value of 1; otherwise, the variable was given a value of 0. The data matrix, consisting of 40 samples 29 dummy variables, was subjected to DCA using the program package PC-ORD ver. 3.0 (MJM Software, Gleneden, USA).

5. RESULTS

5.1 Clustering of species

One of the dendrograms produced by 10 iterations is shown in Fig. 1. The cophenetic correlation (Romesburg 1989) was 0.865, indicating that the obtained dendrogram reflected the structure of the original dissimilarity matrix well. The ten dendrograms obtained by changing the order of species in the distance matrix are not identical, but three clusters were always recognized when the dendrograms were truncated at a dissimilarity level of 0.6 to 0.7. The members of each cluster also never changed between iterations, and Cluster-1, 2 and 3 in Fig. 1 consisted of 17, 10 and 12 species, respectively. Cluster-3 included two *Macaranga gigantea* samples as shown in section 2.

5.2 Association and clustering of attributes

Of 28 combinations of attributes, 17 combinations showed significant associations (Table 3). Only one attribute, the life-form of the adult plant, had no significant

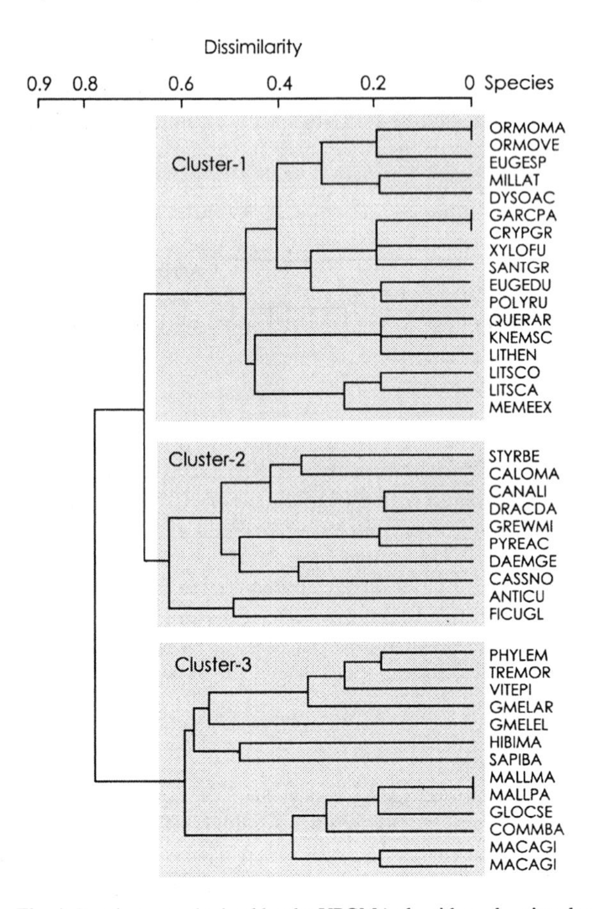

Fig. 1 Dendrogram obtained by the UPGMA algorithm, showing three eco-morphological species groups. The species name code consists of the initial 4 letters of the genus name and the initial 2 letters of the specific name. See Table 1 for the species.

association with the remaining attributes. In contrast, successional status and responsiveness of germination to gap were significantly associated with all the other attributes except life-form. Thus, most of the attributes have strong non-random associations with each other.

A cluster analysis was conducted for the attributes. As mentioned, the life-form of the adult plant had no association with the other attributes. Therefore, the

Table 3 Results of the test of the association among fruit and seed attributes. The data for 39 tropical rain forest species were examined and subjected to the *G*-test. Asterisks indicate significant associations between attributes. *** $P < 0.005$; ** $P < 0.01$; * $P < 0.05$. See Table 2 for details of the attributes.

	Attribute	2	3	4	5	6	7	8
1	Fruit type	***	***	***	ns	ns	ns	ns
2	Seed coat	-	***	**	**	***	ns	ns
3	Successional status	-	-	***	*	***	***	ns
4	Responsiveness of germination	-	-	-	***	***	***	ns
5	Germination type	-	-	-	-	***	**	ns
6	Seed longivity	-	-	-	-	-	**	ns
7	Seed weight	-	-	-	-	-	-	ns
8	Life-form	-	-	-	-	-	-	-

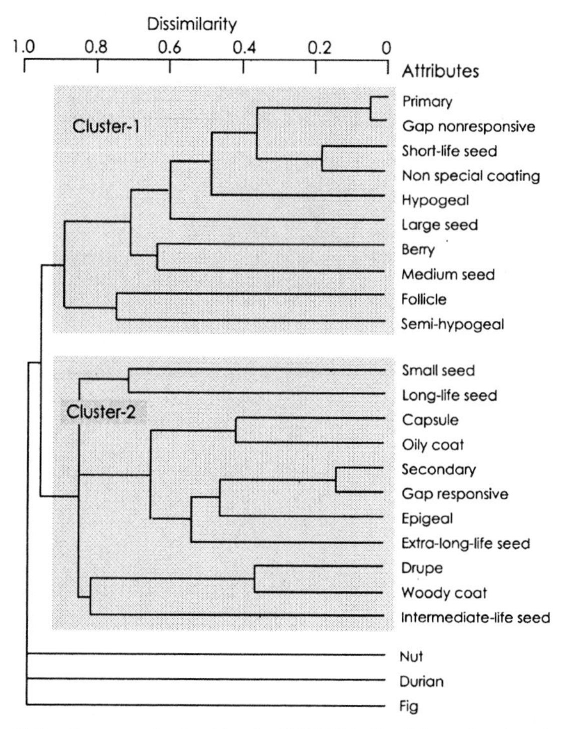

Fig. 2 Dendrogram obtained by the UPGMA algorithm, showing the two character sets observed in the species examined. See Table 2 for an explanation of the characters.

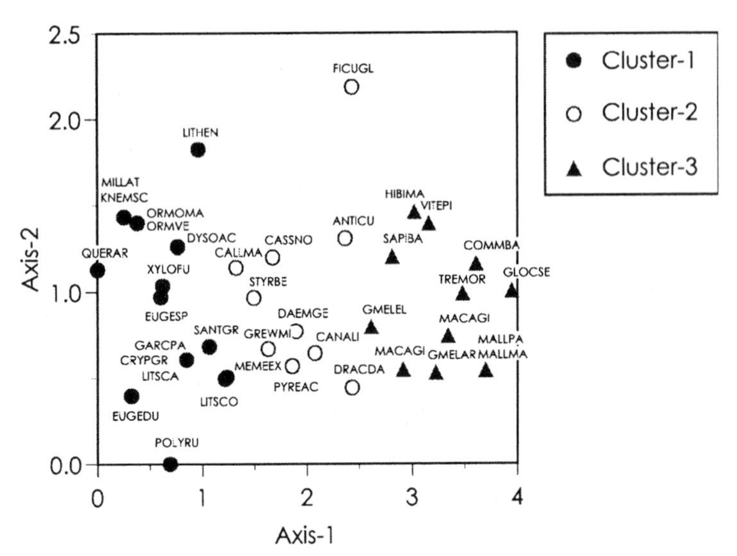

Fig. 3 Ordination of species using eight attributes. Axes 1 and 2 were obtained by the DCA. Clusters 1, 2 and 3 are those obtained in Fig. 1.

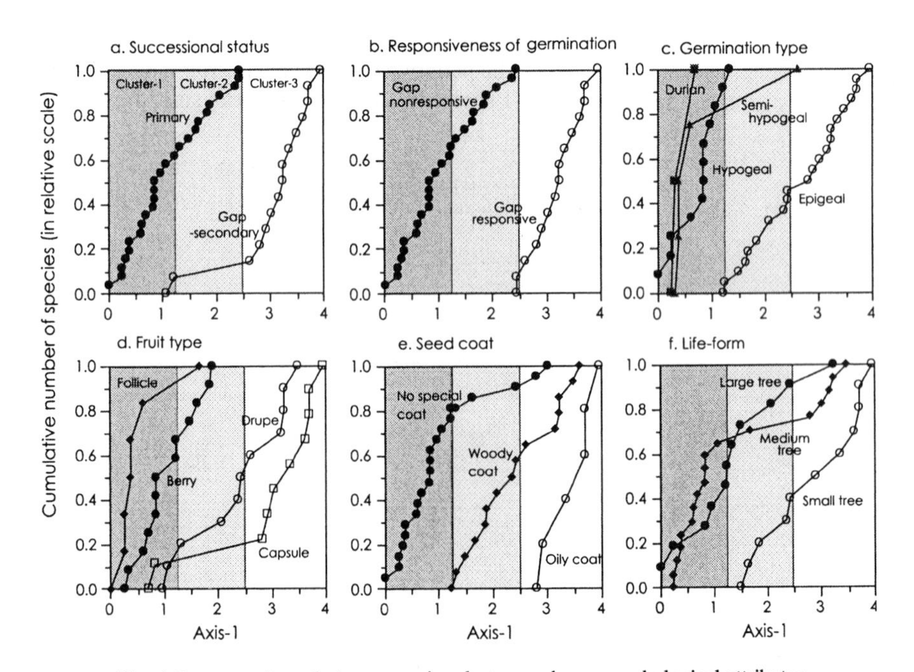

Fig. 4 Correspondence between species clusters and eco-morphological attributes. The cumulative number of species classified into a given category was plotted in relative scale against the species score on Axis obtained by DCA (see Fig. 3). The domains in the diagrams correspond to the three species clusters in the dendrogram shown in Fig. 1.

categories under the life-form were excluded from the data matrix and the dissimilarity matrix for the remaining 22 categories was subjected to UPGMA clustering analysis. Two clusters were recognized, as shown in Fig. 2. The first cluster consisted of 10 characters, including primary species, non-responsive germination, short-lived seed and large seed. The second cluster consisted of 11 characters, including secondary species, extra-long-lived seed, gap responsive germination and small seed.

5.3 Ordination of species

Axis-1 extracted by DCA explained 70% of the original variance of the multivariate distance among species. The second and third axes explained less than 6% of the variance. Therefore, Axis-1 accounts for most of the eco-morphological variation among species. The scores of the species on the first and second axis are plotted in Fig. 3. As expected, the three clusters of species obtained by the cluster analysis were well discriminated by the first axis score. The three groups were differentiated by species scores of 1.25 and 2.5. The scores on Axis-2 did not contribute to discriminating the three clusters.

In Fig. 4, the cumulative number of species classified into a given category was plotted against the Axis-1 score of species to examine the correspondence between the three clusters of species and attributes. In terms of successional status, all the primary species fell into Cluster-1 or 2 and all but one gap-secondary species fell into Cluster-3 (Fig. 4a). A similar good correspondence between the species clusters and their attributes was also observed for responsiveness of germination (Fig. 4b) and germination type (Fig. 4c). All the non-responsive seeds fell into Cluster-1 or 2 and most of the gap-responsive seeds fell into Cluster-3. Similarly, most of the hypogeal seeds fell into Cluster-1, and most of the epigeal seeds fell into Cluster-2 or 3. While the pattern was not as clear, fruit type and seed coat also corresponded to the Axis-1 score. In terms of life-form, small trees only corresponded to Cluster-2 or 3, but large and medium trees were not associated with a particular species cluster.

In Fig. 5, seed weight and estimated seed life span are plotted against the Axis-1 score of the species. Both the original numerical seed weight and the estimated seed longevity had significant correlations with the Axis-1 score (Spearman's rank correlation coefficient $R = -0.82$, $P < 0.01$, for seed weight and $R = 0.91$, $P < 0.01$, for seed longevity).

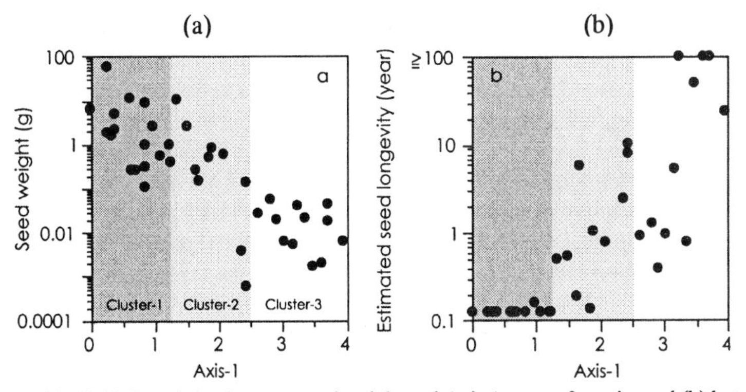

Fig. 5 (a) Correlation between seed weight and Axis 1 score of species and (b) between seed longevity and Axis 1 score. The domains in the diagrams correspond to the three species groups in the dendrogram shown in Fig. 1.

6. DISCUSSION

6.1 Three species groups in a tropical rain forest

Species Clusters-1 and 2 consisted of primary forest species with large seeds that did not germinate in response to gap conditions. On the other hand, species Cluster-3 consisted of gap-secondary species with small seeds surrounded by oily or thick woody organs. These contrasting character sets correspond well to the character sets adopted to discriminate pioneer and non-pioneer species (Foster & Janson 1985; Swaine & Whitmore 1988).

This study, however, distinguished two groups within the non-pioneer species. Species Cluster-1 was characterized by hypogeal germination and short-lived seed, while Cluster-2 was characterized by epigeal germination, longer-lived seed, and woody organs surrounding the seed. The Cluster-1 species had bigger seeds than the Cluster-2 species. These syndromes of eco-morphological attributes found in non-pioneer species suggest the existence of some selective force that separates non-pioneer species. The most outstanding functional difference between the two species groups was germination rate. The species of Cluster-1 germinated within three months after burial, and no seed survived more than three months even if it had not germinated. On the other hand, seeds of Cluster-2 species germinated slowly, and the estimated mean seed life was from 3 months to 10 years. Ng (1978b, 1980) also distinguished two contrasting types of germination, named rapid and delayed germination, in 350 Malaysian tree species. As Ng (1978b) pointed out, the rapid germination species produce carpets of seedlings after their seed fall, so the predator satiation hypothesis (Janzen 1971) could be applied to these species when fruiting is infrequent. The delayed germination species tend to produce a few seedlings at a time over an extended period (Ng 1978b). During their dormant phase, seeds are under predator pressure, but simultaneously have the opportunity to be dispersed by animal dispersers (Hopkins & Graham 1987; Ng 1978b). Most of the seeds of Cluster-2 species are surrounded by thick woody organs, such as the endocarp of *Calophyllum macrocarpum*, *Canarium littorale* and *Dracontomelon dao* and the thick woody seed coat of *Styrax benzoin* and *Pyrenaria acuminata*. The woody organs surrounding the seed seem to function as protection against predators and may be evolutionary adaptations to mammal seed dispersers.

Gautier-Hion et al. (1985) analyzed dispersal syndromes in African tropical rain forest. One of their syndromes is the ruminant-rodent-elephant syndrome, which includes plant species with large, dull-colored drupe type fruit with dry fibrous flesh and well-protected seeds. This syndrome contrasts with the bird-monkey syndrome, which includes plant species characterized by brightly colored berries or drupe-type fruit without a protective seed cover. Their result suggests that the well-protected seeds surrounded by woody endocarp are adaptive to ground inhabiting mammal dispersers. Although our data set dose not cover dispersal mode, the morphological character sets peculicar to each of species Cluster-1 and 2 are concordant with the two character sets, the bird-monkey and the ruminant-rodent-elephant syndromes in Gautier-Hion et al. (1985).

The present study suggests the existence of two functional groups, characterized by different seed dispersal and germination patterns, within the primary rainforest species. Further studies on the seed dispersal and post-dispersal consequences of the seeds are required.

6.2 Associations among attributes

Although most of the attributes examined here had significant associations with each other, these attributes formed two clusters. One of the character sets was common to primary species groups (species Clusters-1 and 2) and the other character set was common to the gap-secondary group (species Cluster-3). Osunkoya (1996) showed that fruit size, seed size, hardness of seed coat, seedling cotyledon function and fruit maturation could predict the light requirement of Australian rain forest trees. Hewitt (1998) showed a significant relationship between seed size and shade-tolerance for American temperate angiosperms. Metcalfe & Grubb (1995) also found the significance relationships between seed mass and shade tolerance by the comparison of confamilial species. Their results are concordant with our results and suggest the importance of seed and seedling attributes in predicting a regeneration niche (Swaine & Whitmore 1988).

In our analysis, the life-form of the adult plant was not significantly associated with other attributes and did not effectively discriminate any species groups. This is reasonable, because pioneer species have life-forms ranging from shrub to large tree (Swaine & Whitmore 1988). However, Leishman & Westoby (1994) and Metcalfe & Grubb (1995) found that plant life-form was significantly associated with seed size; larger life-forms tend to have a larger seed mass. Suzuki & Ashton (1996) and Yamada & Suzuki (1999) also showed that seed weight was significantly associated with wing size (and hence with dispersal mode) and the height of the adult tree for the two families that include many wind dispersed species, the Dipterocarpaceae and Sterculiaceae. However, they found that the smaller tree species have a larger seed mass. This discrepancy is probably caused by differences in the dispersal mechanism of the examined species and implies different evolutionary selection on seed size between wind-dispersed seeds and other seeds, such as animal-dispersed seeds.

6.3 Phylogenetic consideration

As pointed out by Mazer (1989), Metcalfe & Grubb (1995) and Osunkoya (1996), phylogenetic considerations are necessary for comparative ecological study. In our study, the number of species examined was not sufficient to test the effect of phylogeny on the eco-morphological attributes. However, more than half of the members of species Cluster-3 were from the Euphorbiaceae. The eco-morphological attributes peculiar to this group probably reflect characters phylogenetically fixed in the Euphorbiaceae. Furthermore, our study did not include Dipterocarpaceae, the dominant family in the Southeast Asian tropics. Further collection of sample species and comprehensive data analysis including phylogenetic considerations is therefore required to validate the present study.

ACKNOWLEDGEMENTS

We wish to thank the Forest Research Institute Malaysia (FRIM) and the staffs of FRIM for permission to use the Pasoh FR and their various supports including the identification of plants. We also thank Drs. A. Furukawa, Y. Morikawa, N. Kachi, T. Okuda, Y. Maruyama and M. Yasuda for their kind support and suggestions for this study. Thanks are also due to Dr. J. V. LaFrankie for his help with identification of plants. This study was funded by a Global Environmental Research Program grant from the Japan Environment Agency (Grant No. E-2) and partly by a Grant-in-Aid (No. 07304080) from the Ministry of Education, Science and Culture, Japan.

REFERENCES

Foster, S. A. & Janson, C. H. (1985) The relationship between seed size and establishment conditions in tropical woody plants. Ecology 66: 773-780.

Gautier-Hion, A., Duplantier, J. M., Quris, R., Feer, F., Sourd, C., Decoux, J. P., Dubost, G., Emmons, L., Erard, C., Ilecketsweiler, P., Moungazi, A., Roussilhon, C. & Thiollay, J. M. (1985) Fruit characters as a basis of fruit choice and seed dispersal in a tropical forest vertebrate community. Oecologia 65: 324-337.

Grubb, P. J. & Metcalfe, D. J. (1996) Adaptation and inertia in the Australian tropical lowland rain-forest flora: contradictory trends in intergeneric and intrageneric comparisons of seed size in relation to light demand. Funct. Ecol. 10: 512-520.

Hewitt, N. (1998) Seed size and shade-tolerance: a comparative analysis of North American temperate trees. Oecologia 114: 432-440.

Hill, M. O. & Gauch, H. G (1980) Detrended correspondence analyseis: an improved ordination technique. Vegetatio 42: 47-58.

Hopkins, M. S. & Graham, A. W. (1987) The viability of seeds of rainforest species after experimental soil burials under tropical wet lowland forest in north-eastern Australia. Aust. J. Ecol. 12: 97-108.

Janzen, D. H. (1971) Seed predation by animals. Annu. Rev. Ecol. Syst. 2: 465-492.

Kanzaki, M., Yap, S. K., Okauchi, Y., Kimura, K. & Yamakura, T. (1997) Survival and germination of buried seeds of non-dipterocarp species in a tropical rain forest at Pasoh, West Malaysia. Tropics 7: 9-20.

Leishman, M. R. & Westoby, M. (1994) Hypotheses on seed size: tests using the semiarid flora of western New South Wales, Australia. Am. Nat. 143: 890-906.

Mazer, S. J. (1989) Ecological, taxonomic, and life history correlates of seed mass among indiana dune angiosperms. Ecol. Monogr. 59: 153-175.

Metcalfe D. J. & Grubb, P. J. (1995) Seed mass and light requirements for regeneration in Southeast Asian rain forest. Can. J. Bot. 73: 817-826.

Ng, F. S. P. (ed). (1978a) Tree Flora of Malaya vol. 3. Longman Malaysia, Kuala Lumpur.

Ng, F. S. P. (1978b) Strategies of establishment in Malayan forest trees. In Tomlinson, P. B. & Zimmerman, M. H. (eds). Tropical trees as living systems, Cambridge Universty Press, Cambridge, pp.129-162.

Ng, F. S. P. (1980) Germination ecology of Malaysian woody plants. Malay. For. 43: 406-437.

Ng, F. S. P. (ed). (1989) Tree flora of Malaya vol. 4. Longman Malaysia, Kuala Lumpur.

Ng, F. S. P. (1991) Mannual of forest fruits, seeds and seedlings vol. 1. Forest Research Institute Malaysia, Kuala Lumpur.

Ng, F. S. P. (1992) Mannual of forest fruits, seeds and seedlings vol. 2. Forest Research Institute Malaysia, Kuala Lumpur.

Osunkoya, O. O. (1996) Light requirements for regeneration in tropical forest plants: Taxon-level and ecological attribute effects. Aust. J. Ecol. 21: 429-441.

Pijl, L. van der (1982) Principles of dispersal in higher plants (3rd ed). Springer-Verlag, Berlin.

Romesburg, H. C. (1989) Cluster analysis for researchers. Robert E. Krieger Publishing Company, Florida.

Roth, I. (1987) Stratification of a tropical forest as seen in dispersal types. W. Junk, Dordrecht.

Salisbury, E. J. (1942) The Reproductive capacity of plants. George Bell, London.

Seiwa, K. (2000) Effects of seed size and emergence time on tree seedling establishment: importance of developmental constraints. Oecologia 123: 208-215.

Seiwa, K. & Kikuzawa, K. (1989) Seasonal growth patterns of seedling height in relation to seed mass in deciduous broad-leaved tree species. Jpn. J. Ecol. 39: 5-15 (in Japanese with English summary).

Sneath, P. H. A. & Sokal, R. R. (1973) Numerical taxonomy. W. H. Freeman & Company, San Francisco.

Sokal, R. R. (1973) Introcuction to biostatistics. W. H. Freeman & Company, San Francisco.

Suzuki, E. & Ashton, P. S. (1996) Sepal and nut size of fruits of Asian Dipterocarpaceae and its implications for dispersal. J. Trop. Ecol. 12: 853-870.

Swaine, M. D. & Whitmore, T. C. (1988) On the definition of ecological species groups in

tropical rain forest. Vegetatio 75: 81-86.
Whitmore, T. C. (ed) (1972a) Tree flora of Malaya vol. 1. Longman Malaysia, Kuala Lumpur.
Whitmore, T. C. (ed) (1972b) Tree flora of Malaya vol. 2. Longman Malaysia, Kuala Lumpur.
Yamada, T. & Suzuki, E. (1999) Comparative morphology and allometry of winged diaspores among the Asian Sterculiaceae. J. Trop. Ecol. 15: 619-635.

10 Microhabitat Preference of Two Sympatric *Scaphium* Species in a 50 ha Plot in Pasoh

Toshihiro Yamada[1], Toshinori Okuda[2] & N. Manokaran[3]

Abstract: A remarkable feature of tropical woody biodiversity is the presence of a large number of closely related, sympatric species in some genera. Such species are most likely to have relatively similar niches, sharing the same ecological and physiological heritage via their common ancestral lineage. How these sympatric species coexist is a central to understanding of the maintenance of high biodiversity in tropical rain forests. The mode of the coexistence can be divided into equilibrium and non-equilibrium. The equilibrium coexistence poses that each species occupies a different niche, which results from and reduces direct competition. For the equilibrium coexistence of plant species, the diversification of a microhabitat in relation to topography and shading conditions is among the most important. On the other hand, the non-equilibrium postulates that equilibrium forces at work are weak and species coexist by means of a stochastic non-equilibrium factor. To detect how sympatric species coexist in a tropical rain forest, we compared tree forms and local spatial distribution patterns between two sympatric *Scaphium* species (Sterculiaceae) of a tropical canopy tree, *S. macropodum* and *S. linearicarpum* using data obtained from a 50 ha plot in Pasoh Forest Reserve (Pasoh FR). The total numbers of *S. linearicarpum* and *S. macropodum* larger than 1 cm in DBH (Diameter at Breast Height) in the plot were 148 and 726, respectively. Frequency distributions of DBH followed L-shaped distributions in both cases, suggesting that they have near-equilibrium population structures. Both species were distributed unevenly and were aggregated in some clumps in the plot. Furthermore, their distributions were strongly associated with topographic variables. Their habitat preferences in relation to topography were similar each other; they were rare on valleys and flat areas and abundant on slopes, although the range of *S. macropodum*'s distribution is larger than that of *S. linearicarpum.* An index of spatial association shows a positive spatial association between two species. These results suggest that habitat niche segregation in relation to topography does not exist between these species and therefore they must be under a severe direct competition for occupying a restricted area where is suitable for their regeneration. For the equilibrium coexist of these species, differentiation in regeneration niches associated with gap regimes is required. But the shading conditions of saplings and pole size trees do not differ between these species. Thus the regeneration niche differentiation appears unlikely to exist between them. Consequently, we can hypothesize that a stochastic non-equilibrium factor is important for the coexistence of these two *Scaphium* species.

Key words: coexistence, equilibrium, habitat niche, non-equilibrium, tree form.

---

[1] Faculty of Environmental and Symbiotic Sciences, Prefectural University of Kumamoto, Kumamoto 862-8502, Japan. E-mail: tyamada@pu-kumamoto.ac.jp
[2] National Institute for Environmental Studies (NIES), Japan.
[3] Forest Research Institute Malaysia (FRIM), Malaysia.

1. INTRODUCTION

Tropical rain forests are the most biologically diverse terrestrial biome. Many hypotheses how the biodiversity is maintained have been proposed (cf. Chesson & Warner 1981; Connell 1978; Grubb 1977; Janzen 1970; Kohyama 1993; Tilman 1982). These have been developed in two different directions, equilibrium and non-equilibrium hypotheses. The former category postulates an equilibrium community and thus the principle of competitive exclusion. In a community at equilibrium each species occupies a different niche that results from and reduces direct competition (Whittaker 1975). Therefore, they are sometimes called the niche-diversification-hypothesis. On the contrary, the latter posits that equilibrium forces at work are weak. By this argument, species diversity is decreased through random local extinction in a stochastic dynamic system. However, species diversity can be maintained when the rates of local extinction are low and balanced by local immigration and speciation (mass effects). Whether or not tropical rain forests are equilibrium and how strong are equilibrium forces in tropical rain forests are still obscure and open to discussion.

An important niche gradient for tree species is a microhabitat related to topography and/or edaphic conditions. Many scientists reported that the species-specific correlations between tree distribution and soil and/or topography in tropical rain forests (Ashton 1969, 1988; Austin et al. 1972; Lescure & Boulet 1985; Richards 1952; Rogstad 1990; Svenning 1999). These results are the evidence of the existence of an equilibrium force in tropical rain forests. On the contrary, Hubbell & Foster (1986), who studied the distribution of trees in a 50-ha plot in Barro Colorado Island (BCI) showed that most of the common trees in this neotropical rain forest belonged to a large habitat generalist guild. They emphasized that it is impossible to explain species richness by niche differentiation, and treated non-equilibrium forces as the working mechanism of species richness. Furthermore, Duivenvoorden (1995) showed that most of Colombian tropical rain forest trees inhibiting in the well-drained upland habitat would be soil generalists, which limits the importance of habitat specialization for maintaining tree species richness.

Another important niche for tree species is the specialization to regeneration conditions along a gap-understudy gradient; that is regeneration niche (Grubb 1977). Forests are recognized as the mosaic of different phases of the growth cycle from canopy gaps to mature, closed stands (Whitmore 1984). Two distinctive ecological species groups, pioneer and non-pioneer have been defined along this forest growth cycle (Swain & Whitmore 1988). The pioneers entirely depend either their seed germination or seedling establishment on forest gaps. On the other hand, the non-pioneers are those whose seeds can germinate and seedlings can establish and survive under forest shade. Even under the closed canopies, light conditions are highly heterogeneous with respect to the complex, heterogeneous canopy structure (Lieberman et al. 1989). Some scientists revealed the existence of difference in life-historical traits in terms of growth response to shading conditions among the non-pioneers (Clark & Clark 1992; Clark et al. 1993; Itoh 1995; Lieberman et al. 1995).

One of the remarkable features of tropical woody biodiversity is the presence of a series of sympatric congeneric species in some genera (cf. Ashton 1982; Fedorov 1966; Itoh 1995; Kochummen et al. 1990; Rogstad 1990; Whitmore 1984; Yamada et al. 2000a). Such species are most likely to have relatively similar niches, sharing the same ecological and physiological heritage via their common ancestral lineage (Rogstad 1989). The maintenance mechanisms may therefore appear to be most critical for the coexistence of such species, and thus is a central to understanding the maintenance of high biodiversity in tropical rain forests. Moreover, monophyletic groups are more informative for evolutionary processes of tropical trees because they differ at fewer gene loci and therefore in fewer life history characteristics. Several studies on the coexistence mechanism of the sympatric species have so far been made (Itoh 1995; Rogstad 1990; Yamada et al. 2000a). Rogstad (1990) studied the coexistence mechanism of six distinct tree species of the monopheric *Polyalthia hypoleuca* complex in the Far Eastern tropics and showed that they have clear divergence in auto-ecological characteristics that affect their spatial distribution. Itoh (1995) showed that *Dryobalanops lanceolata* Burck and *D. aromatica* Gaertn. f., which are predominant species in Bornean tropical lowland rain forests have antagonistic patterns of microhabitat preference related to edaphic conditions.

Scaphium Schott and Endl. (Sterculiaceae) is a small genus of large briefly deciduous trees comprising six species. Yamada et al. (2000a) showed that all the six species but *S. linearicarpum* (Mast.) Pierre. has divergent habitats at various spatial scales and coexist because they reduce direct competition by habitat niche differentiation. However, they did not refer to *S. linearicarpum* due to the lack of its enough information. Here, we report the patterns of distribution of two sympatric *Scaphium* species, *S. macropodum* (Miq) Beumee ex Heyne and *S. linearicarpum* in relationships to topography and shading conditions using data derived from a 50-ha plot set in the Pasoh Forest Reserve (Pasoh FR), Malaysia. We tested three predictions, which follow from the niche-diversification-hypothesis. The first prediction is that the distribution of the *Scaphium* species will be influenced by microhabitat heterogeneity that is related to topography. If the *Scaphium* species are specialized on particular topographic variables, it will be recognized as potential evidence of existence of the equilibrium force in a community. The second is that the *Scaphium* species should show antagonistic microhabitat relationships. If they coexist by microhabitat specialization related to topography, then antagonistic patterns of microhabitat preferences should be present. The last is that the *Scaphium* species have been diversified the regeneration niches. If they have diversified the regeneration conditions, it can be detected by the analysis of their spatial distributional in relation to local shading conditions. With them we reviewed the importance of equilibrium and non-equilibrium forces on their coexistence.

2. MATERIALS AND METHODS

2.1 Study site and species

The study was performed in an equatorial foothill forest in Pasoh FR, Negeri Sembilan, West Malaysia. The annual rainfalls are ca. 170 cm. A 50-ha plot was set in the forest in Pasoh FR by Forest Research Institute Malaysia (FRIM) and Smithsonian Institute and all individual woody plants larger than 1 cm in DBH within the plot were identified, measured by DBH, tagged, and mapped in 1985 (Kochummen & LaFrankie 1990). They were recensused twice in 1990 and 1995. *S. linearicarpum* and *S. macropodum* are found in the 50-ha plot (Manokaran et al. 1990). *S. macropodum* is the 13th most abundant species in trees larger than 30 cm in DBH in the 50-ha plot (Kochummen et al. 1990). They are the members of trees that make up the main continuous canopy to emergent layers in the forest. Details of the forests and plots such as climates, soils, vegetation and forest physiognomies are described in Chaps.1, 2, 3, 4 and 7.

Scaphium is endemic to the South-East-Asiatic region defined by Good (1974). The distributional records of *S. linearicarpum* are rare in Borneo and common in the Malaya and Sumatra, and *S. macropodum* is the most widespread and common throughout Indochina, Malaya, Borneo and Sumatra (Yamada et al. 2000a). The stature of *S. macropodum* is over 45 m in Borneo (Yamada & Suzuki 1996). The trees flower on bare twigs after leaf-fall (Kostermans 1953) and produce gyration or wind-dispersed fruits with a boat shaped wing derived from a dehiscing follicle. The dispersal distance of the fruit seldom exceeds 50 m from the base of parent trees (Ridley 1930; Yamada & Suzuki 1997). The species' saplings are shade tolerant and are abundant under the closed canopies (Yamada & Suzuki 1997).

2.2 Field methods

This study used data set derived from the 50-ha plot set in the forest in Pasoh FR by FRIM and Smithsonian Institute. All trees identified as *Scaphium* at the first forest inventory were re-identified to correct identification error and measured by DBH in August 1999.

2.3 Data analysis

The spatial distribution patterns of *Scaphium* trees were analyzed by Morishita's I_δ index (Morishita 1959). In this method, I_δ are calculated for various quadrat sizes. For any quadrat size where species are aggregated, random and uniformly distributed, the values of the I_δ index take more than, equal to and less than 1.0, respectively. A statistical difference between the observed distribution and the Poisson model of random distribution was tested by the F-test (Morishita 1959). Furthermore, this method provides mean clump size when species are aggregated. A spatial association between the two *Scaphium* species was computed by Iwao's ω index of the degree of habitat overlapping (Iwao 1977). The ω index take a maximum of 1.0 for complete overlapping, through 0.0 for independent occurrence, to a minimum of -1.0 for completely exclusion. The dispersion patterns and spatial associations were examined over a range of quadrat sizes from 61 m^2 to 250,000 m^2.

For the analysis of microhabitat preference of the *Scaphium* species related to topography, we subdivide the 50-ha plot into 5,000 10 m × 10 m subplots. Then we categorized the subplots into six topographic types; ridge top, higher parts of a slope, middle parts of a slope, lower parts of a slope, flatland and valley using the subplot's degree ratio and index of convexity (Yamakura et al. 1995) calculating from the ground elevation height of the four corners of the subplot. Details of the calculation procedure and definition of the topographic types can be seen in Okuda et al. (in press). The dependency of distribution of the *Scaphium* species on topography was analyzed by chi-square test for observed frequency over the topography types.

For the quantification of local shading conditions of each individual *Scaphium* tree, we used a canopy closure index, G (Lieberman et al. 1989, 1995). The index estimates the shading at a given individual tree based on the density and heights of neighboring trees and uses the following measures: the horizontal distance (d) between the focal tree and each taller neighbor within a given radius, the vertical distance (Δh) between the focal tree and each taller neighbor, and the hypotenuse distance (h) between the focal tree and each taller neighbor, calculated from d and Δh. The ratio $\Delta h/h$ is the sine of the include angle θ. The G is defined as the sum of these ratios for all taller neighbors within the specified radius:

$$G = \Sigma \sin \theta$$

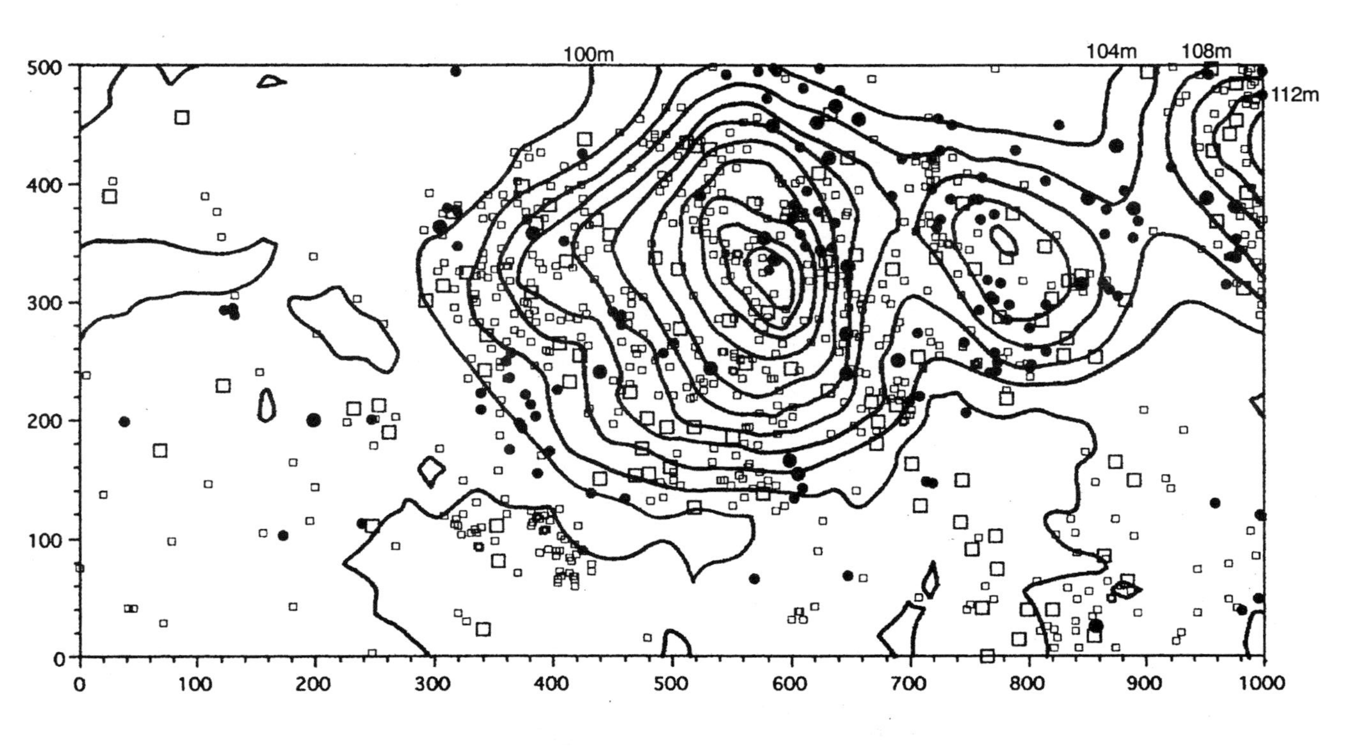

Fig. 1 Spatial distributions of *Scaphium linearicarpum* (●) and *S. macropodum* (□) in a 50 ha plot in Pasoh FR. Larger and smaller symbols represent trees larger and smaller than 20 cm in DBH, respectively. The contour interval is 2 m shown in italic letter.

Tree height (H) was estimated independently from species using the following extended allometry empirically determined in the same forest (Kato et al. 1978).

$$1 / H = 1 / (1.66\, DBH^{1.085}) + 1 / 60$$

We used the data of tree positions and DBH of all trees larger than 1 cm in the plot for the calculation G. The specified radius for the calculation of G cannot be determined a priori. However, we used 10 m as the specified radius because this corresponded to the crown area of emergent trees in a Bornean tropical rain forest (Kohyama et al. 2001). We thus excluded trees located where the distance from the nearest margin was less than the radius, 10 m from the calculation to avoid the margin effect in the fixed plot.

3. RESULTS

3.1 Size structure of the *Scaphium* populations

The DBH frequency distributions were the classical L-shaped for both species (Table 1), suggesting that some big trees occurred with many small trees and they belong to an ecological species group of non-pioneers. The abundance of *S. macropodum* was the higher, with 726 individuals than *S. linearicarpum* which had 148 trees. The maximum size of DBH of *S. macropodum* was 75.3 cm, and that of *S. linearicarpum* was 68.9 cm.

3.2 Spatial distribution of *Scaphium*

The spatial distribution of *Scaphium* trees in the 50-ha plot was shown in Fig. 1.

Table 1 Population structures of two *Scaphium* species in a 50 ha plot in Pasoh FR.

Species	Number of individuals (/50 ha)	Maximum DBH (cm)	Skewness of DBH distribution	Kurtosis of DBH distribution
Scaphium linearicarpum	148	68.9	1.83	2.86
Scaphium macropodum	726	75.3	2.32	5.60

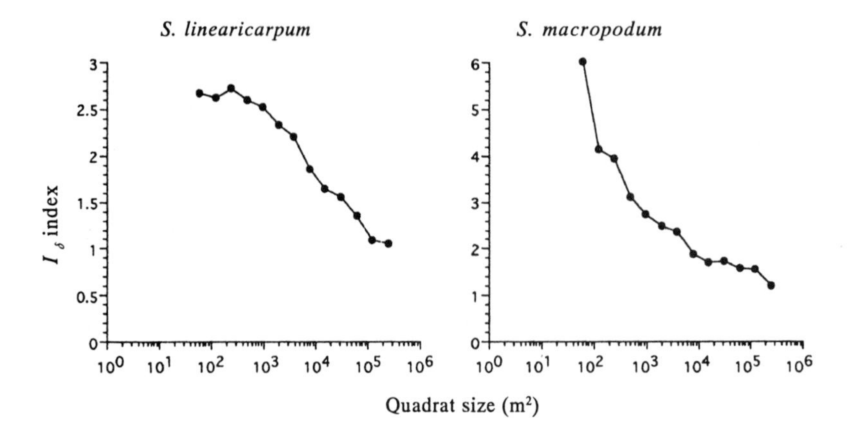

Fig. 2 Changes in I_δ index with quadrat size for *S. linearicarpum* and *S. macropodum*.

Both species were distributed unevenly in the plot. The Morishita's index of dispersion I_δ revealed that both *Scaphium* species had aggregated distributions which were significantly different from Random distribution in any quadrat sizes examined (F-test, $P < 0.01$, Fig. 2). Large clump sizes of 62,500 m^2 for *S. linearicarpum* and of 31,250 m^2 for *S. macropodum* were detected.

3.3 Microhabitat preference in relation to topography

The distributions of the *Scaphium* species were highly affected by topography (Table 2). The chi-square test over topographic types reveals that the observed frequency in topography types of both *Scaphium* species significantly skewed into lower, middle and higher parts of a slops ($P < 0.01$). It follows that both species prefers lower, middle and higher parts of slops for their microhabitats in relation to topography. The difference in observed frequencies over topography types between the *Scaphium* species was tested by chi-square test and was not significant ($P > 0.3$). Therefore, microhabitat preference in relation to topography is similar between the *Scaphium* species.

3.4 Spatial association between two *Scaphium* species

The index of spatial association, ω was always positive throughout any quadrat

Table 2 Number of *Scaphium* trees over six topographic types.

Topographic types	*S. linearicarpum*		*S. macropodum*	
	Observed N	Expected N*	Observed N	Expected N*
Flat alluvial areas	19	56	150	305
Ridgetop	1	1	8	4
Valley and reverie areas	6	10	22	54
Lower parts of a slope	42	23	180	123
Middle parts of a slope	33	20	178	107
Higher parts of a slope	14	6	90	35

*Expected N shows the tree number expected if trees distributed randomly independent from topography.

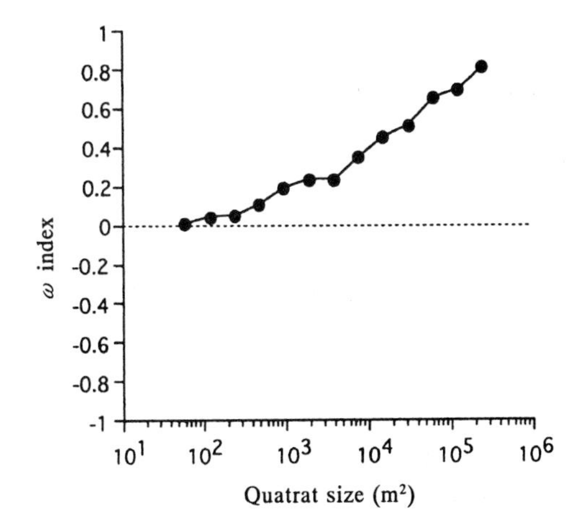

Fig. 3 Changes in the ω index between *S. linearicarpum* and *S. macropodum*.

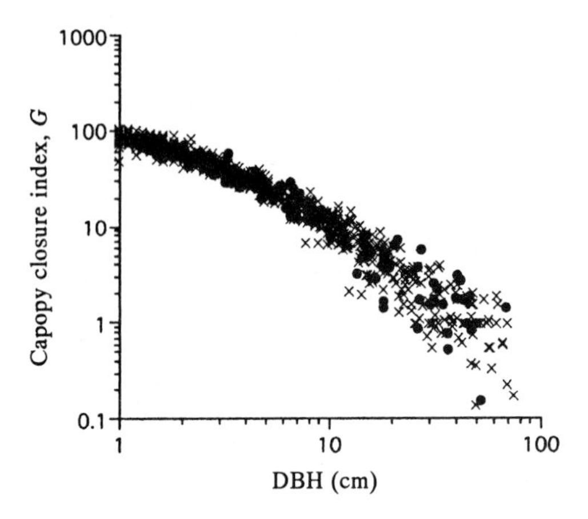

Fig. 4 Relationship between DBH and canopy closure index G for *Scaphium linearicarpum* (×) and *S. macropodum* (●).

Table 3 Shading conditions of saplings and poles of *Scaphium linearicarpum* and *S. macropodum*. Shading conditions are evaluated by a canopy closure index G.

Size class	*S. linearicarpum*				*S. macropodum*			
	Max.	Median	Min.	N	Max.	Median	Min.	N
Saplings (1≦DBH<2 cm)	88.0	68.8	28.5	28	105.5	72.0	41.7	198
Poles (10≦DBH<15 cm)	12.2	8.4	3.3	9	16.7	8.4	2.0	78

sizes examined (Fig. 3), suggesting a strong positive spatial association between the species. The value of ω was larger than 0.2 in quadrats larger than 2,000 m^2. This implies that the habitats of these species overlap well in larger spatial scales than 2,000 m^2.

3.5 Microhabitat preference in relation to shading conditions

The value of G decreased with increasing tree size for both species (Fig. 4). To factor out the bias of tree size on G, we compared the G between the *Scaphium* species in trees of similar sizes. We analyzed the interspecific difference in G in tree sizes from 1 to 2 cm in DBH (saplings) and 10 to 15 cm in DBH (poles) using Mann-Whitney U-test. For both saplings and pole size trees, medians of G were not significantly different between each species (Table 3). This suggests that the shading conditions of saplings and poles of both species are similar each other, and thus microhabitat in relation to shading conditions is similar between the *Scaphium* species.

4. DISCUSSION

The first prediction, the distribution of the *Scaphium* species are strongly influenced by microhabitat heterogeneity, which is related to topography, can be acceptable. Both species were distributed unevenly in the plot (Fig. 1) and aggregated in some clumps (Fig. 2). The distributions of both species were skewed into lower, middle and higher parts of slopes, and rare in valleys (Table 2). This would be a potential evidence of the existence of an equilibrium force in tropical rain forest communities. The degree to which the equilibrium force contribute to determining the structure and species composition of Pasoh forest community is discussed in Okuda et al. (1997).

The second prediction, the sympatric *Scaphium* species show an antagonistic microhabitat relationship cannot be accepted. The microhabitat preferences in relation to topography were similar each other (Table 2) and their distributions were overlaps well in the plot (Fig. 3). These results suggest that the sympatric *Scaphium* species are under a severe competition for occupying a restricted site where is suitable topography for their regeneration. This is in contradiction to an earlier study of *Scaphium* species (Yamada et al. 2000a). Yamada et al. (2000a) compared local and geographical distributions among *Scaphium* species and showed that all but *S. linearicarpum* have divergent habitats at various spatial scales and coexist because they reduce direct competition by habitat niche differentiation. They finally concluded that the equilibrium force in relation to habitats might play the predominant role that permits their coexistence. But the results of the study do not support the idea. We cannot find the habitat segregation between *S. linearicarpum* and *S. macropodum*. Therefore, the results give a limit of the importance of equilibrium force in relation to habitat on the coexistence of *Scaphium* species.

For the equilibrium coexistence of *S. linearicarpum* and *S. macropodum* in the plot, the diversification of regeneration niches is required. Yamada et al. (2000b) showed that three sympatric *Scaphium* species, *S. borneense* (Merr.) Kosterm., *S. longipetiolatum* (Kosterm.) Kosterm. and *S. macropodum* have diversified above ground architectures and allometries which leads to the equilibrium coexistence by promoting regeneration niche differentiation in Lambir Hills National Park, Sarawak. However, we cannot find any evidences of regeneration niche diversification between two *Scaphium* species, which is the third prediction. Frequency distributions of DBH followed the L-shaped distribution for both species (Table 1), suggesting that they all belong to the guild of non-pioneer species (Swaine & Whitmore 1988). Moreover, shading condition evaluated by a canopy closure index, *G* did not differ between the sympatric *Scaphium* species (Table 3).

Our results suggest that there is no microhabitat specialization between two sympatric *Scaphium* species. It is concluded that a stochastic non-equilibrium force may be a predominant factor for the coexistence of the species, although there still exist some possibilities that the density or distance dependent predator of seeds and/or seedlings (Connell 1971; Janzen 1970) and the diversification of reproductive schedules (Chesson & Warner 1981), which lead to the equilibrium coexistence underlie their coexistence.

This study dwells on only one genus among many comprising tropical rain forests. We should analyze the existence of niche divergence in many other genera as well as in many other places. Moreover the distribution of *Scaphium* trees is likely to be influenced not only by congeners but all other trees in the community. The community level analysis of niche divergence would be also needed for the further understanding of the maintenance of high biodiversity in the tropical rain forest.

ACKNOWLEDGEMENTS

This study is a part of a joint research project between the NIES, FRIM, and UPM (Grant No. is E-2 (3) from the Global Environment Research Program, Japanese

Environment Agency). The large-scale forest plot at the Pasoh FR are on going project of the Malaysian Government, and were initiated by FRIM through its former Director-General, Dato' Dr. Salleh Mohd. Mor, and under the leadership of Drs. N. Manokaran, P. S. Ashton and Stephen P. Hubbell. Supplementary funding was provided by the National Science Foundation, USA; the United Nations, though it is Man and the Biosphere (MAB) program; UNESCO-MAB grants; UNESCO-ROSTSEA; and the continuing supports of the Smithsonian Tropical Research Institute (STRI) and National Institute for Environment Studies (NIES).

REFERENCES

Ashton, P. S. (1969) Speciation among tropical forest trees; some deductions in the light of recent evidence. Biol. J. Linn. Soc. 1: 155-196.

Ashton, P. S. (1982) Dipterocarpaceae. Flora Malaysiana. vol. 9 Part 2: 237-552.

Ashton, P. S. (1988) Dipterocarp biology as a window to the understanding of tropical forest structure. Annu. Rev. Ecol. Syst. 19: 347-370.

Austin, M. P., Ashton, P. S. & Greig-Smith, P. (1972) The application of quantitative methods to vegetation survey III. A reexamination of rain forest data from Brunei. J. Ecol. 60: 305-324.

Chesson, P. L. & Warner, R. R. (1981) Environmental variability promotes coexistence in lottery competitive systems. Am. Nat. 113: 923-943.

Clark, D. A. & Clark D. B. (1992) Life history diversity of canopy and emergent trees in a neotropical rain forest. Ecol. Monogr. 62: 315-344.

Clark, D. A., Clark D. B. & Rich, P. M. (1993) Comparative analysis of microhabitat utilization by saplings of nine tree species in neotropical rain forest. Biotropica 25: 397-407.

Connell, J. H. (1971) On the role of natural enemies in preventing competitive exclusion in some marine animals and in rain forest trees. In Boer, P. J. & Gradwell, G. R. (eds). Dynamics of Population. Center for Agricultural Publishing and Documentation, Wageningen. pp.361-381.

Connell, J. H. (1978) Diversity in tropical rainforest and coral reefs. Science 199: 1302-1310.

Duivenvoorden, J. F. (1995) Tree species composition and rain forest-environment relationships in the middle Caqueta area. Colombia, NW Amazonia. Vegetatio 120: 91-113.

Fedorov, A. A. (1966) The structure of tropical rain forest and speciation in the humid tropics. J. Ecol. 54: 1-11.

Good, R. (1974) The geography of flowering plants (4th ed). Longman, Kuala Lumpur.

Grubb, P. J. (1977) The maintenance of species richness in plant communities. The important role of the regeneration niche. Biol. Rev. 52: 107-145.

Hubbell, S. P. & Foster, R. B. (1986) Commonness and rarity in a neotropical forest: implications for tropical tree conservation. In Soule, M. E. (ed). Conservation Biology. The Science Scarcity and Diversity. Sinauer, Sunderland, Massachusetts, pp.205-231.

Itoh, A. (1995) Regeneration processes and coexistence mechanism of two Bornean emergent dipterocarp species. Ph.D. diss., Kyoto Univ.

Iwao, S. (1977) Analysis of spatial association between two species based on the interspecies mean crowding. Res. Popul. Ecol. 8: 243-260.

Janzen, D. H. (1970) Herbivores and the number of tree species in tropical forests. Am. Nat. 104: 501-528.

Kato, R., Tadaki, Y. & Ogawa, H. (1978) Plant biomass and growth increment studies in Pasoh forest. Malay. Nat. J. 30: 211-224.

Kochummen, K. M., LaFrankie, J. V., & Manokaran, N. (1990) Floristic composition of Pasoh Forest Reserve, a lowland rain forest in Peninsular Malaysia. J. Trop. For. Sci. 3: 1-13.

Kohyama, T. (1993) Size structured tree populations in gap-dynamics forest - the forest architecture hypothesis for the coexistence of species. J. Ecol. 81: 131-143.

Kohyama, T. Suzuki, E., Partomihardjo, T. & Yamada, T. (2001) Dynamic steady state of patchmosaic tree size structure of a mixed dipterocarp forest regulated by local crowding.

Ecol. Res. 16: 85-98.
Kostermans, A. J. G. H. (1953) The genera *Scaphium* Schott & Endl. and *Hildegrdia* Schott & Endl. (Sterculiaceae). J. Sci. Res. Indonesia 2: 13-23.
Lescure, J-P. & Boulet, R. (1985) Relationship between soil and vegetation in a tropical rain forest in French Guiana. Biotropica 17: 155-164.
Lieberman, M., Lieberman, M. & Peralta, R. (1989) Forest are not just Swiss cheese: canopy stereogeometry of non-gaps in tropical forests. Ecology 70: 550-552.
Lieberman, M., Lieberman, M., Peralta, R. & Hartshorn, G. S. (1995) Canopy closure and the distribution of tropical forest tree species at La Selva, Costa Rica. J. Trop. Ecol. 11: 161-178.
Manokaran, N. & LaFrankie, J. V. (1990) Stand structure of Pasoh Forest Reserve, a lowland rain forest in Peninsular Malaysia. J. Trop. For. Sci. 3: 14-24.
Morishita, M. (1959) Measuring of the distribution of individuals and analysis of the distribution patterns. Memories of Faculty of Science, Kyushu University, Series of Entomology (Biology) 2: 215-235.
Okuda, T., Kachi, N., Yap, S. K. & Manokaran, N. (1997) Tree distribution pattern and fate of juveniles in a lowland tropical rain forest - implications for regeneration and maintenance of species diversity. Plant Ecol. 131: 155-171.
Okuda, T., Nor Azman, H., Manokaran, N., Saw, L. Q., Amir, H. M. S. & Ashton, P. S. Local variation of canopy structure in relation to soils and topography and the implications for species diversity in a rain forest of Peninsular Malaysia. In Losos, E. C. & Leigh, E. G. J. (eds). Forest diversity and dynamism: Results from the global network of large-scale demographic plots, University Chicago Press, Chicago (in press).
Richards, P. W. (1952) The Tropical Rain Forest. Cambridge University Press, Cambridge.
Ridley, H. N. (1930) Dispersal of Plants Throughout the World. L. Reeve, Kent.
Rogstad, H. S. (1989) The biosystematics and evolution of the *Polyalthia hypoleuca* species complex (Annonaceae) of Malesia. I. Systematic treatment. J. Arnold Arboretum 70: 153-246.
Rogstad, H. S. (1990) The biosystematics and evolution of the *Polyalthia hypoleuca* species complex (Annonaceae) of Malesia. II. Comparative distributional ecology. J. Trop. Ecol. 6: 387-408.
Svenning, J. (1999) Microhabitat speciation in a species-rich palm community in Amazonian Equador. J. Ecol. 87: 55-65.
Swaine, M. D. & Whitmore, T. C. (1988) On the definition of ecological species group in tropical rain forests. Vegetatio 75: 107-117.
Tilman, D. (1982) Resource Competition and Community Structure. Princeton University Press, Princeton. NJ.
Whitmore, T. C. (1984) Tropical Rain Forest of the Far East. Clarendon Press, Oxford.
Whittaker, R. H. (1975) Communities and Ecosystems. MacMillan, New York.
Yamada, T. & Suzuki, E. (1996) Ontogenetic change in leaf shape and crown form of a tropical tree, *Scaphium macropodum* (Sterculiaceae) in Borneo. J. Plant Res. 109: 211-217.
Yamada, T. & Suzuki, E. (1997) Change in spatial distribution during life history of a tropical tree, *Scaphium macropodum* (Sterculiaceae) in Borneo. J. Plant Res. 110: 179-186.
Yamada, T., Itoh, A., Kanzaki, M., Yamakura, T., Suzuki, E. & Ashton, P. S. (2000a) Local and geographical distribution for a tropical tree genus, *Scaphium* (Sterculiaceae) in the Far East. Plant Ecol. 148:23-30
Yamada, T., Yamakura, T. & Lee, H. S. (2000b) Architectural and allometric differences among *Scaphium* species are related to microhabitat preferences. Funct. Ecol. 14:731-737.
Yamakura, T., Kanzaki, M., Itoh, A., Ohkubo, T., Ogino, K., Chai, E. O. K., Lee, H. S. & Ashton, P. S. (1995) Topography of a large-scale research plot established within a tropical rain forest at Lambir, Sarawak. Tropics 5: 41-56.

11 Diversity of Putative Ectomycorrhizal Fungi in Pasoh Forest Reserve

Lee Su See[1], Roy Watling[2] & Evelyn Turnbull[2]

Abstract: The tropical rainforests of Peninsular Malaysia are dominated by trees of the ectomycorrhizal family, the Dipterocarpaceae. Between 1992 and 1997 collections of putative ectomycorrhizal fungi were made in a lowland rainforest at Pasoh Forest Reserve (Pasoh FR). Collections were made during March in each year and additionally in early September 1995 and late August 1996, to coincide with the fungal fruiting seasons which occur at the end of a prolonged dry spell. During each visit of about three days duration, collections were made beside the major trails and in the Arboretum. In 1995 and 1996, collections were extended to the newly established regeneration plots, A–E where the forest had been logged in the 1950's. A total of 296 species distributed in 19 families were recorded with many of the collections being new to science. The most frequently collected fungi were members of the family Russulaceae; a total of 114 species of *Russula* and 35 species of *Lactarius* were provisionally identified. This was followed in order of decreasing frequency by members of the Boletaceae (45 species), Amanitaceae (34 species), Cantharellaceae (13 species), Entolomataceae (13 species including possible saprophytes), Tricholomataceae (10 species), Cortinariaceae (9 species), Sclerodermataceae (8 species), Gautieriaceae (3 species), Hymenogastraceae and Secotiaceae (2 species each), and Chamonixiaceae, Clavulinaceae, Elasmomycetaceae, Gomphaceae, Hydnaceae, Hymenochaetaceae, Pisolithaceae and Thelephoraceae (1 species each). Two hundred and thirteen species, or about 72% of the collections, were single collections. Only 102 species could be placed in previously described taxa; 66% of the taxa found are apparently new to science and could only be assigned to the proximity of a known species consortium. Very few species were collected successively in two or more consecutive years, and only two species, *Tylopilus maculatus* and *Cantharellus ianthinus* were collected every year. Our collections also show that hypogeous fungi are present in the tropical rainforest. Enumeration of just one group, i.e. the putative ectomycorrhizal fungi, and data from other studies on wood decomposer fungi in Pasoh FR demonstrate that fungal biodiversity in a Malaysian lowland rainforest is very high and that more research of a long-term nature is required.

Key words: dipterocarp forest, fungal diversity, tropical macrofungi.

1. INTRODUCTION

The tropical rainforests of Malaysia are dominated by trees of the ectomycorrhizal family, the Dipterocarpaceae (see Lee 1998). Members of this family are some of the most important sources of timber in Malaysia and South-East Asia, making up to about 80% of the timber exported from the region. In 1996 dipterocarps made up 33% of log production and 35% of sawn timber exported from Peninsular Malaysia

[1] Forest Research Institute Malaysia (FRIM), 52109 Kuala Lumpur, Malaysia.
E-mail: leess@frim.gov.my

[2] Royal Botanic Garden, Edinburph EH3 5LR, UK.

(Anonymous 1996). As with other mycorrhizal symbioses, the dipterocarp ectomycorrhizal association has been demonstrated to enhance uptake of phosphorus and improve seedling growth in some dipterocarp species (Lee & Alexander 1994; Yazid et al. 1994). Other tree families in the tropical lowland rainforest which are also known to contain members capable of forming ectomycorrhizas are the Fagaceae, Leguminosae and Myrtaceae. However, little is known of these associations in Malaysia.

The taxonomy of ectomycorrhizal fungi in many parts of the world, particularly in the tropics, is poorly known. It is estimated that more than 70% of the Malaysian fungal flora remain to be described (E. J. H. Corner, pers. comm.) and therefore it is not surprising that until recently little was known of the ectomycorrhizal fungi in lowland rainforests. Ectomycorrhizal fungi are predominantly members of the ascomycetes and the basidiomycetes and in a few cases members of the zygomycetes. A list of all the known ectomycorrhizal fungi is given in Brundrette et al. (1996).

Fungi mainly from the families Amanitaceae, Boletaceae, Russulaceae and Sclerodermataceae have been reported to be associated with dipterocarps in South-East Asia (Becker 1983; Chalermpongse 1987; Hadi & Santoso 1988; Lee 1992; Ogawa 1992; Smits 1994; Watling & Lee 1995; Yasman 1994; Zarate et al. 1994), India (Bakshi 1974) and Sri Lanka (De Alwis & Abeyanake 1980), but the identities of many of these fungi need to be confirmed. In addition, a range of dipterocarp ectomycorrhizal morphotypes has been observed and described (Becker 1983; Berriman 1986; Lee 1988; Lee et al. 1997), indicating the diversity of the associated fungi.

Here we present the results of collections of putative ectomycorrhizal basidiomata made between 1992 and 1997 in Pasoh Forest Reserve (Pasoh FR) to give an insight of the diversity of this group of fungi in a lowland rainforest. The major part of the results reported here was obtained from work carried out under Project 5–The Frequency of Ectomycorrhizal Fungi in Logged and Unlogged Lowland Dipterocarp Forest, as part of a collaborative research programme funded by the bilateral Malaysia-United Kingdom Programme for the Conservation, Management and Development of Forest Resources (Lee et al. 1998).

2. MATERIALS AND METHODS

Collections of basidiomata of putative ectomycorrhizal fungi (see Watling & Lee 1995) were made in Pasoh FR from beside the major trails, the Arboretum and in regenerating forest which had been selectively logged in the 1950's using the Malayan Uniform System (MUS, Wyatt-Smith 1963, see Chap. 2). Collections were made during March each year from 1992 until 1997 and additionally in early September 1995 and late August 1996, to coincide with the fungal fruiting seasons which occur at the end of a prolonged dry spell. Corner (1935) observed that there are two general fruiting seasons in the Malay Peninsula, the first in March and the second about August or September. In 1995 and 1996, collections were extended to the regenerating forest where five 1 ha plots, A–E, had been established. Each collecting period was of about three days duration. Methods of collecting and processing specimens are described in Watling & Lee (1995) and Watling et al. (1998).

The mycelial connections of some basidiomata were traced to roots of host trees to confirm their mycorrhizal status (Watling & Lee 1995). However, this is a very time consuming activity and could not be embarked on as a routine part of this study. There are always difficulties in linking an identified basidioma with the roots of a specific tree. The identity of trees with which basidiomata appeared to be closely associated were noted whenever possible but in some cases this was less obvious than others.

3. RESULTS AND DISCUSSION

Most of the fungi were found along trails and in natural clearings in the primary forest and in disturbed areas in the regenerating forest. Corner (1935) states that agaric fungi tend to fruit more frequently in open places because this habitat is more exposed to alternate drying and wetting.

A total of 296 taxa from 19 families of putative ectomycorrhizal fungi were collected from Pasoh FR between 1992 and 1997 (Table 1). However, of these only 102 could be placed in previously described taxa (Appendix), leaving approximately 66% to be assigned to the proximity of a known species consortium. Many of these taxa are undoubtedly new to science. Confirmation of the identity of many of the fungi to species level, in particular in the Russulaceae, will remain problematical unless monographic treatments are published.

Members of the Russulaceae dominated the collections with 114 and 35 provisionally identified species of *Russula* and *Lactarius*, respectively. However, of these, only about 17% of the *Lactarius* and less than 10% of the *Russula* could be assigned to known species. Although the majority of the species found agree with the familiar species-facies exhibited in temperate taxa, there are some very distinctive differences in morphology and life strategies. Differences between temperate and West African species in the Russulaceae have been noted (Buyck et al. 1994) but these are different from those noted in Malaysia. It is clear that the Malaysian Russulaceae require a large future taxonomic input.

The next most numerous species were members of the Boletaceae followed by the Amanitaceae and Cantharellaceae. In contrast to the Russulaceae, about 62% of the Amanitaceae, 64% of the Boletaceae and 38% of the Cantharellaceae could be identified to known species. The identification of these groups of fungi was greatly aided by the publications of Corner for the boletes (Corner 1970, 1972) and Cantharellaceae (Corner 1966), and Corner & Bas (1962) for the genus *Amanita*.

Members of other families which were collected included 13 species of Entolomataceae including possible saprophytes, 10 species of Tricholomataceae, 9 species of Cortinariaceae (five of which were confirmed as species of *Inocybe*), 8 species in Sclerodermataceae, 3 unidentified species in Gautieriaceae, 2 species each in Hymenogastraceae and Secotiaceae, and one species each in Chamonixiaceae, Clavulinaceae, Elasmomycetaceae, Gomphaceae, Hydnaceae, Hymenochaetaceae, Pisolithaceae and Thelephoraceae. The Entolomataceae have been included here as there are indications that some members of this family can form ectomycorrhizas (Brundrette et al. 1996; Singer 1986).

The results of this study show that at least 10 species of hypogeous fungi are also present (Table 1), not supporting tentative suggestions from classical herbarium studies that such fungi are absent in the tropical rainforest. Corner & Hawker (1953) published the only substantive paper on South-East Asian hypogeous fungi, the results of which have been extended more recently (Watling 2001).

Approximately 76% of the collections which could be identified to species were first described by Corner & Bas (1962) and Corner (1970, 1972) confirming both the magnitude of his contribution to tropical mycology and the absence of subsequent monographic studies.

All the *Amanita* spp. listed here have been previously recorded for

Table 1 Families and genera of putative ectomycorrhizal fungi collected between 1992 and 1997 in Pasoh FR, Peninsular Malaysia (*including possible saprophytes).

Family/ Genus	No. of species	No. of species identified to known taxa
Amanitaceae		
Amanita spp.	34	21
Boletaceae		
Boletus spp.	24	12
Mucilopilus sp.	1	0
Pulveroboletus spp.	3	2
Tylopilus spp.	2	2
Boletellus sp.	1	1
Phylloporus spp.	7	7
Rubinoboletus spp.	3	2
Hiemiella spp.	2	1
Strobilomyces spp.	2	2
Cantharellaceae		
Cantharellus spp.	12	4
Craterellus sp.	1	1
Chamonixiaceae		
Chamonixia sp.	1	1
Clavulinaceae		
Clavulina sp.	1	1
Cortinariaceae		
Cortinarius spp.	2	0
Inocybe spp.	5	5
Phaeocollybia spp.	2	0
Elasmomycetaceae		
Zelleromyces sp.	1	1
* Entolomataceae		
Entoloma spp.	8	4
Leptonia spp.	2	2
Nolanea spp.	3	2
Gautieriaceae		
Gautieria spp.	3	0
Gomphaceae		
Gloeocantharellus sp.	1	1
Hydnaceae		
Hydnum sp.	1	1
Hymenochaetaceae		
Coltricia sp.	1	1
Hymenogastraceae		
Dendrogaster spp.	2	1
Russulaceae		
Lactarius spp.	35	6
Russula spp.	114	10
Pisolithacèae		
Pisolithus sp.	1	1
Sclerodermataceae		
Horakiella sp.	1	0
Scleroderma spp.	7	4
Secotiaceae		
Secotium spp.	2	0
Thelephoraceae		
Sarcodon sp.	1	1
Tricholomataceae		
Laccaria spp.	7	2
Tricholoma spp.	3	3
Total	**296**	**102**

Peninsular Malaysia by either Watling & Lee (1995, 1998) or Turnbull & Watling (1999). Fourteen of the 20 species were only known previously from Singapore as was the single species of *Limacella*. Twelve of the boletes have also been recorded by Watling & Lee (1995, 1998) and of those which were not, one was described from Singapore and one from Borneo (Sabah) (Corner 1972). Of the 7 taxa in *Phylloporus* two were previously known only from Singapore. All 16 Russulaceae identified to species have not previously been recorded from Peninsular Malaysia: unlike *Amanita* (Corner & Bas 1962), the boletes (Corner 1972) or Cantharellaceae (Corner 1966), there have been no monographic treatments for Malesian Russulaceae. Our determinations are tentative until type material has been examined. Three Cantharellaceae, one Australian and two from Sabah and the Pacific Islands, are new to Peninsular Malaysia. *Inocybe* has been dealt with by Turnbull (1995) and of the species recorded, two were previously known only from Japan. In all amongst the collections so far analysed, 33 (30%) are new to Peninsular Malaysia. Prior to the present study and recent publications of Watling et al. (1998), over 90% of the taxa would have been new to Peninsular Malaysia.

In his book, *Boletus* in Malaysia, Corner (1972) stated that the richness of the boleti came from the diversity of the Fagaceae. In a later publication, he was of the opinion that the apparent absence of *Amanita*, *Russula* and *Boletus* from the forests of the Solomon Islands was due to the absence of the Fagaceae (Corner 1993). We would like to suggest that the high diversity of the putative ectomycorrhizal fungi in a lowland rainforest in Peninsular Malaysia could be a reflection of the diversity of the Dipterocarpaceae; there being 168 species of dipterocarps in Peninsular Malaysia (Symington 1974). However, we presently do

Table 2 Number of species of putative ectomycorrhizal fungi collected each year between 1992 and 1997. Figures in parentheses indicate number of additional species collected only from the Regenerating Plots A–E (* including possible saprophytes).

Family	Year							
	1992	1993	1994	1995		1996		1997
				Spring	Autumn	Spring	Autumn	
Amanitaceae	5	7	9	8 (0)	3 (0)	1 (0)	14 (5)	11
Boletaceae	7	18	7	6 (1)	1 (0)	3 (1)	10 (4)	21
Cantharellaceae	4	2	5	1 (0)	3 (1)	1 (1)	1 (1)	6
Chamonixiaceae	0	0	0	0	1 (0)	0	0	0
Clavulinaceae	0	0	0	0	0	0	0	1
Cortinariaceae	4	0	2	0 (2)	4 (0)	0	1 (0)	1
Elasmomycetaceae	0	0	0	0	0	1 (0)	0	0
* Entolomataceae	2	4	1	1 (0)	1 (0)	0	3 (0)	1
Gautieriaceae	1	0	0	0 (1)	0	1 (0)	1 (1)	0
Gomphaceae	0	0	1	0	0	0	0	0
Hydnaceae	0	0	0	0	1 (0)	1 (0)	0	1
Hymenochaetaceae	0	0	0	0	0	1 (0)	0	0
Hymenogastraceae	0	0	0	1 (0)	0 (1)	0	2 (0)	1
Russulaceae	59	12	22	29 (9)	33 (7)	4 (3)	3 (2)	4
Pisolithaceae	1	0	0	1 (0)	0	0	0 (1)	0
Sclerodermataceae	2	0	0	3 (0)	6 (0)	1 (0)	3 (0)	5
Secotiaceae	0	0	0	0 (1)	0	0 (1)	0	0
Thelephoraceae	1	0	0	0	0	0	0	0
Tricholomataceae	1	2	1	2 (2)	2 (0)	0	1 (0)	1
Total	87	45	48	52 (16)	55 (9)	14 (6)	39 (14)	53

not have sufficient data or information about the association of particular ectomycorrhizal fungi with specific dipterocarp species except for *Shorea leprosula* (Lee et al. 1997). Other families of trees which are known to contain ectomycorrhizal members are Celastraceae, Leguminosae, Myristicaceae and Ulmaceae (Smits 1994), and there are indications that trees in other families may also do so (Watling & Lee, unpubl. data).

Since collections were also made in the autumn of 1995 and 1996, in these two years figures from only the spring collections were used when comparing the number of fungus species collected over the five years. The number of species collected each year was variable but the highest and lowest number of species were collected in 1992 and spring 1996, respectively (Table 2). Apart from that, the numbers were rather similar for the other four years. Spring 1996 was probably not a good season for fungi in Pasoh FR because of the low rainfall received in both February and March of that year (Fig. 1). In other years, rainfall was generally low in January and/or February followed by higher rainfall in March. The number of species collected in autumn 1996 was higher than in spring of the same year but still below the number reported for autumn of the preceding year. The lower collection of autumn 1996 was probably due to the low rainfall in the months of July, August and September of that year. Corner (1935) has noted that rain in itself is not stimulation but that dry weather followed by ensuing heavy rain causes the development of fungal basidiomata in the Malay Peninsula.

The data also showed that in years where many collections of members of the Russulaceae were made, there were few collections of species of *Amanita*, for example in 1992, 1994 and 1995 (Table 2). On the other hand when numerous species of *Amanita* were collected, there were few members of the Russulaceae, e.g. autumn 1996 and 1997. This could imply that these groups of fungi have different

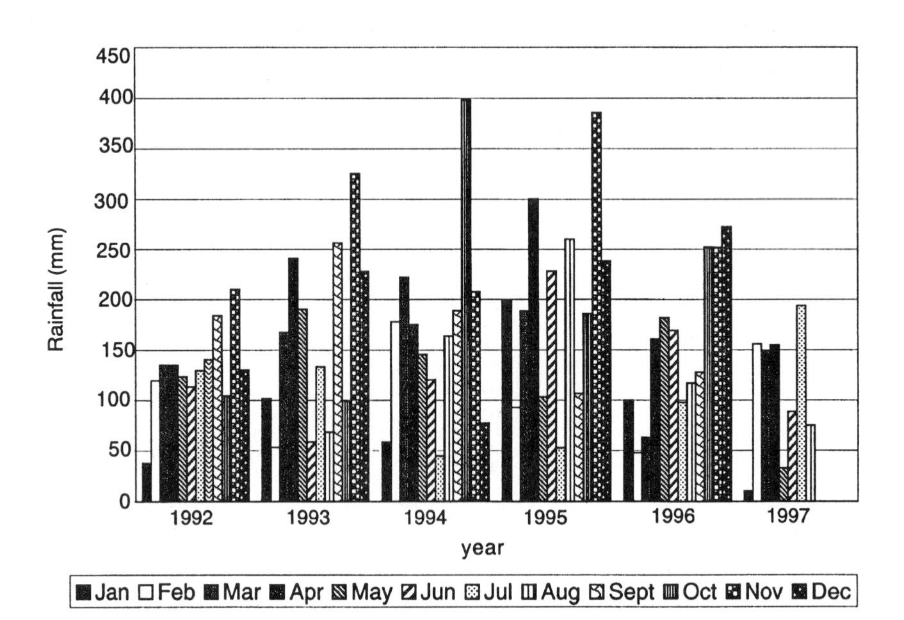

Fig. 1 Monthly rainfall record for FELDA (Federal Land Development Authority) Pasoh Dua Station, the nearest weather station to Pasoh FR, from January 1992 to August 1997.

Table 3 Fungi collected successively in two or more consecutive years between 1992–1997 in Pasoh FR.

Family	Species	Years collected
Amanitaceae	*Amanita* sp. 6 Corner & Bas	1995, 1996
	Amanita alauda	1996, 1997
	Amanita demissa	1996, 1997
	Amanita fritillaria f. *malayensis*	1996, 1997
	Amanita obsita	1995, 1996, 1997
	Amanita privigna	1993, 1994
	Amanita sychnopyramis	1994, 1995, 1996
	Amanita tjibodensis	1996, 1997
	Amanita sp. (Gymnopodae)	1995, 1996
Boletaceae	*Boletus peltatus*	1992, 1993, 1994, 1995
	Boletus peltatus var. *decolorans*	1996, 1997
	Boletus tristiculus	1993, 1994
	Phylloporus bellus var. *cyanescens*	1992, 1993
	Phylloporus rubescens	1995, 1996
	Pulveroboletus frians	1993, 1994, 1996, 1997
	Tylopilus maculatus	1992, 1993, 1994, 1995, 1996, 1997
	Rubinoboletus ballouii	1992, 1993
	Rubinoboletus ballouii var. *fuscatus*	1995, 1996, 1997
	Heimiella retispora	1996, 1997
Cantharellaceae	*Cantharellus ianthinus*	1992, 1993, 1994, 1995, 1996, 1997
	Craterellus cornucopioides var. *mediosporus*	1995, 1996
Cortinariaceae	*Inocybe sphaerospora*	1995, 1996, 1997
Gautieriaceae	*Gautieria* sp. 1	1995, 1996
Hydnaceae	*Hydnum repandum*	1995, 1996, 1997
Hymenogastraceae	*Dendrogaster cambodgensis*	1995, 1996, 1997
Russulaceae	*Lactarius* sp. 1 (Plinthogalli)	1992, 1993
	Lactarius sp. 2 (Plinthogalli)	1995, 1996
	Lactarius sp. 3 (Plinthogalli)	1995, 1996, 1997
	Lactarius sp. (Piperites)	1995, 1996
	Russula alboareolata	1992, 1993, 1994, 1995
	Russula cf. *sororia*	1995, 1996
	Russula subnigricans	1992, 1993, 1994
	Russula sp. (Foetentinae sp. 25)	1994, 1995, 1996
	Russula sp. (Foetentinae sp. 33)	1994, 1995
	Russula sp. (Illota group)	1994, 1995
	Russula sp. (Lepidinae)	1992, 1993
	Russula sp. (Nigricantinae)	1992, 1993
	Russula sp. (red pileus)	1994, 1995
Pisolithaceae	*Pisolithus aurantioscabrosus*	1995, 1996
Sclerodermataceae	*Horakiella* sp.	1995, 1996, 1997
	Scleroderma leptopodium	1995, 1996
Tricholomataceae	*Tricholoma mons-fraseri*	1995, 1996

phenological patterns or that they fruit in response to different triggering mechanisms.

In those years when collections were made both in spring and autumn, only 6% of the species were found in both spring and autumn of 1995 and 4% in both collecting seasons in 1996. Overall about 85% of the fungi were only collected once over the seven-year period, and very few species (about 15%) were collected in two or more consecutive years (Table 3). Corner (1993) has made similar observations in Singapore, the lowland forests of Peninsular Malaysia and Sarawak. Only two species were collected every year, i.e. *Tylopilus maculatus* (Corner) Watling & Lee (Watling & Lee 1999) and *Cantharellus ianthinus* Corner. The evidence is very convincing that the stimulus for fungi to fruit is primarily the dry weather which checks mycelial growth and that the ensuing rain causes the development of basidiomata (Corner 1935). Although this is true when applied to the overall annual fungal fruiting seasons, it would appear that some other mechanism is operating when one considers fruiting of the same species of fungus within a shorter period of several months. Here it is tempting to speculate that the mycelium of the fungus needs to accumulate sufficient reserves after the first fruiting season of the year before it can fruit again during the second fruiting season five to six months later.

4. CONCLUSION

Corner (1993) concluded that the Malesian region (the geographical region stretching from the Malay Peninsula to the Solomon Islands) holds not only the richest angiosperm vegetation but, also, the richest fungus biota. On the other hand, the Malaysia: Country Study on Biological Diversity (Anonymous 1997) states that there are only about 400 species of known fungi in the country. This only reinforces Corner's (1993) point that the Malesian fungus flora is the least known of fungus biotas and the most difficult to elucidate. Our study has shown that by enumerating just one group of fungi, i.e. the putative ectomycorrhizal fungi, the presence of at least over 200 species were found in the lowland rain forest of Pasoh FR alone. The very high fungal biodiversity in Pasoh FR and other similar lowland rain forest areas is confirmed by a separate study on wood inhabiting polypores in Pasoh FR and other lowland rain forests in Peninsular Malaysia where many previously undescribed fungi have been discovered (T. Hattori, pers. comm.). We do not know whether other groups of fungi, e.g. the litter decomposers, also demonstrate such high diversity but this is likely as has been demonstrated by studies in tropical lowland forests in South America (Singer & Araujo 1979). There is thus clearly a need for more long-term research on tropical fungi, and surveys and taxonomic studies are essential for progress in all aspects of fungal biology in the region.

ACKNOWLEDGEMENTS

We would like to thank the Forest Research Institute Malaysia and the Overseas Development Authority, UK for their support. We would also like to thank Mrs. Noraini Sikin Yahya of FRIM for technical assistance.

REFERENCES

Anonymous (1996) Forestry Statistics Peninsular Malaysia 1995. Forestry Department, Peninsular Malaysia, Kuala Lumpur, 132pp.

Anonymous (1997) Malaysia: Country Study on Biological Diversity. Assessment of Biological Diversity in Malaysia. Ministry of Science, Technology and the Environment, Malaysia, Kuala Lumpur, 186pp.

Bakshi, B. K. (1974) Mycorrhiza and its Role in Forestry. Forest Research Institute and Colleges, Dehra Dun, India, 89pp.

Becker, P. (1983) Mycorrhizas of *Shorea* (Dipterocarpaceae) seedlings in a lowland Malaysian rainforest. Malay. For. 46: 146-170.

Berriman, C. P. (1986) Mycorrhizas of *Shorea* (Dipterocarpaceae) in relation to host specificity and soil phosphorus status. B. thesis, Univ. Aberdeen.

Brundrette, M., Bougher, N., Dell, B., Grove, T. & Malajczuk, N. (1996) Working with Mycorrhizas in Forestry and Agriculture. ACIAR Monograph 32. Canberra, Australia, 374pp.

Buyck, B., Thoen, D. & Watling, R. (1994) The Guinea-Congolian domain: Ectomycorrhizal fungi - a case study. In: Lowland rain forest of the Guinea-Congo domain. Proc. R. Soc. Edin. B 104: 313-334.

Chalermpongse, A. (1987) Mycorrhizal survey of dry-deciduous and semi-evergreen dipterocarp forest ecosystems in Thailand. In Kostermans, A. J. C. H. (ed). Proceedings of the Third Round Table Conference on Dipterocarps. 16-20 April 1985, Mulawarman University, Samarinda, Kalimantan. UNESCO, Indonesia, pp.81-103.

Corner, E. J. H. (1935) The seasonal fruiting of agarics in Malaya. Gardens' Bulletin, Straits Settlements 9: 79-88.

Corner, E. J. H. (1966) A Monograph of Cantharelloid Fungi. Oxford University Press, Oxford. 255pp.

Corner, E. J. H. (1970) *Phylloporus* Quél. and *Paxillus* Fr. in Malaya and Borneo. Nova Hedwigia 20: 793-822.

Corner, E. J. H. (1972) *Boletus* in Malaysia. Government Printing Office, Singapore. 263 pp.

Corner, E. J. H. (1993) 'I am a part of all that I have met' (Tennyson's Ullysses). In Isaac, S., Frankland, J., Watling, R. & Whalley, T. (eds). Aspects of Tropical Mycology. Cambridge University Press, pp. 1-13.

Corner, E. J. H. & Bas, C. (1962) The Genus *Amanita* in Singapore and Malaya. Persoonia 2: 241-304.

Corner, E. J. H. & Hawker, L. E. (1953) Hypogeous fungi from Malaya. Trans. Br. Mycol. Soc. 36: 125-137.

De Alwis, D. P. & Abeyanake, K. (1980) A survey of mycorrhizae in some tropical forest trees of Sri Lanka. In Mikola, P. (ed). Tropical Mycorrhiza Research. Clarendon Press, Oxford, pp.146-153.

Hadi, S. & Santoso, E. (1988) Effect of *Russula* spp., *Scleroderma* sp. and *Boletus* sp. on the mycorrhizal development and on the growth of five dipterocarp species. In Mohinder, M. S. (ed). Agricultural and Biological Research Priorities in Asia. International Foundation for Science and Malaysian Scientific Association, pp.183-185.

Lee, S. S. (1988) The ectomycorrhizas of *Shorea leprosula* Miq. (Dipterocarpaceae). In Ng. F. S. P. (ed). Trees and Mycorrhizas. Proceedings of the Asian Seminar. 13-17 April 1987, Forest Research Institute Malaysia, Kepong, pp.189-209.

Lee, S. S. (1992) Some aspects of the biology of mycorrhizas of the Dipterocarpaeae. Ph.D. diss., Univ. Aberdeen.

Lee, S. S. (1998) Root symbiosis and nutrition. In Appanah, S. & Turnbull, J. M. (eds). A Review of Dipterocarps: Taxonomy, ecology and silviculture. CIFOR, Bogor, pp.99-114.

Lee, S. S. & Alexander, I. K. (1994) The response of two dipterocarp species to nutrient additions and ectomycorrhizal infection. Plant Soil 163: 299-306.

Lee, S. S., Alexander, I. J. & Watling, R. (1997) Ectomycorrhizas and putative ectomycorrhizal fungi of *Shorea leprosula* Miq. (Dipterocarpaceae). Mycorrhiza 7: 63-81.

Lee, S. S., Dan, Y. M., Gauld, I. D. & Bishop, J. (eds). (1998) Conservation, Management and Development of Forest Resources. Proceedings of the Malaysia-United Kingdom Programme Workshop. FRIM, Kepong, 21-24th October 1996, 392pp.

Ogawa, M. (1992) Mycorrhiza of dipterocarps. In Proceedings of the BIO-REFOR Tsukuba Workshop, 19-21 May 1992. BIO-REFOR, IUFRO/SPDC, Tsukuba, Japan, pp.55-58.

Singer, R. (1986) The Agaricales in Modern Taxonomy (4th ed.). Koeltz Scientific Books, Koenigstein. 981pp.

Singer, R. & Araujo, I. (1979) Litter decomposition and ectomycorrhiza in Amazonian forests. I. A comparison of litter decomposing and ectomycorrhizal basidiomycetes in latosol-terra-firme rain forest and white sand podzol campinarana. Acta Amazonica 9: 25-41.

Smits, W. T. M. (1994) Dipterocarpaceae: mycorrhizae and regeneration. Thesis, Agricultural University Wageningen. Tropenbos Series 9. The Tropenbos Foundation III, Wageningen, 243pp.

Symington, C. F. (1974) Foresters' Manual of Dipterocarps. Malayan Forest Records No. 16, Forest Research Institute Malaysia, Kuala Lumpur, 244pp.

Turnbull, E. (1995) *Inocybe* in Peninsular Malaysia. Edinburgh Journal of Botany 52: 351-359.

Turnbull, E. & Watling, R. (1999) Taxonomic and floristic notes on Malaysian higher fungi III. Malay. Nat. J. 53: 189-200.

Watling, R. (2001) An investigation of Thai mushrooms. Department of Agriculture Newsletter of Plant Pathology and Microbiology 10: 9-20 (in Thai).

Watling, R. & Lee, S. S. (1995) Ectomycorrhizal fungi associated with members of the Dipterocarpaceae in Peninsular Malaysia-I. J. Trop. For. Sci. 7: 657-669.

Watling, R. & Lee, S. S. (1998) Ectomycorrhizal fungi associated with members of the Dipterocarpaceae in Peninsular Malaysia- II. J. Trop. For. Sci. 10: 421-430.

Watling, R. & Lee, S. S. (1999) Some larger fungi of Semangkok Forest Reserve, Selangor. Malay. Nat. J. 53: 311-318.

Watling, R., Lee, S. S. & Turnbull, E. (1998) Putative ectomycorrhizal fungi of Pasoh Forest Reserve, Negri Sembilan, Malaysia. In Lee, S. S., Dan, Y. M., Gauld, I. D. & Bishop, J. (eds). Conservation, Management and Development of Forest Resources. Proceedings of the Malaysia-United Kingdom Programme Workshop, 21-24 October 1996, Kuala Lumpur. FRIM, Kepong, pp.96-104.

Wyatt-Smith, J. (1963) Manual of Malayan Silviculture for Inland Forests. vol. I. & II. Malayan Forest Records No. 23 (reprinted in 1995 with two additional chapters and an overall index), Forest Research Institute, Malaysia, Kepong.

Yasman, I. (1994) Ectomycorrhizal sporocarp appearance in a dipterocarp forest, East Kalimantan, Indonesia. In Suzuki, K, Sakurai, S. & Ishii, I. (eds). Proceedings of the International Workshop of BIO-REFOR, 20-23 September 1993, Yogyakarta, Indonesia. BIO-REFOR, IUFRO/SPDC, pp.179-181.

Yazid, S. M., Lee, S. S. & Lapeyrie, F. (1994) Growth stimulation of *Hopea* spp. (Dipterocarpaceae) seedlings following inoculation with an exotic strain of *Pisolithus tinctorius*. For. Ecol. Manage. 67: 339-343.

Zarate, J. T., Watling, R., Jeffries, P., Dodd, J. C., Pampolina, N. M., Sims, K., Lorilla, E. B. & de la Cruz, R. (1994) Survey of ectomycorrhizal fungi associated with pines and dipterocarps in the Philippines. In Suzuki, K., Sakurai, S. & Ishii, I. (eds). Proceedings of the International Workshop of BIO-REFOR, 20-23 September 1993, Yogyakarta, Indonesia. BIO-REFOR, IUFRO/SPDC, pp.182-185.

Appendix Putative ectomycorrhizal fungi identified to known taxa (*including possible saprophytes).

Amanitaceae
- *Amanita* sp. 2 Corner & Bas
- *Amanita* sp. 6 Corner & Bas
- *Amanita alauda* Corner & Bas
- *Amanita centunculus* Corner & Bas
- *Amanita cinctipes* Corner & Bas
- *Amanita demissa* Corner & Bas
- *Amanita elata* (Massee) Corner & Bas
- *Amanita hemibapha* ssp. *similis* (Boedijn) Corner & Bas
- *Amanita gymnopus* Corner & Bas
- *Amanita mira* Corner & Bas
- *Amanita modesta* Corner & Bas
- *Amanita obsita* Corner & Bas
- *Amanita pilosella* Corner & Bas
- *Amanita privigna* Corner & Bas
- *Amanita sychnopyramis* Corner & Bas
- *Amanita tjibodensis* Boedijn
- *Amanita tristis* Corner & Bas
- *Amanita vestita* Corner & Bas
- *Amanita virginea* Massee
- *Amanita xanthomargaros* Corner & Bas
- *Limacella singaporeana* Corner

Boletaceae
- *Boletus destitutus* Corner
- *Boletus fumosipes* Peck
- *Boletus nigroviolaceus* Heim
- *Boletus peltatus* Corner & Watling
- *Boletus peltatus* var. *decolorans* nom. prov.
- *Boletus pernanas* Pat. & Baker
- *Boletus phaeocephalus* Pat. & Baker
- *Boletus polychrous* Corner
- *Boletus tristior* Corner
- *Boletus tristis* Pat. & Baker
- *Boletus tristiculus* Massee
- *Boletus valens* Corner
- *Boletellus emodensis* (Berk.) Singer
- *Heimiella retispora* (Pat. & Baker) Boedijn
- *Phylloporus bellus* var. *cyanescens* (Massee) Corner
- *Phylloporus bogoriensis* Hoehn.
- *Phylloporus brunneolus* Corner
- *Phylloporus orientalis* var. *brevisporus* Corner
- *Phylloporus orientalis* var. *orientalis* Corner
- *Phylloporus parvisporus* Corner
- *Phylloporus rufescens* Corner
- *Pulveroboletus frians* (Corner) Horak
- *Pulveroboletus icterinus* (Pat. & Baker) Watling
- *Rubinoboletus ballouii* (Peck) Heinemann & Ram.
- *R. ballouii* var. *fuscatus* (Corner) Hongo
- *Strobilomyces mollis* Corner
- *Strobilomyces velutipes* Cooke & Massee
- *Tylopilus maculatus* (Corner) Watling & Lee
- *Tylopilus spinifer* (Pat. & Baker) Watling & Lee

Cantharellaceae
- *Cantharellus ianthinus* Corner
- *Cantharellus lilacinus* Cleland & Cheel
- *Cantharellus odoratus* (Schw.) Fr.
- *Cantharellus omphalinoides* Corner
- *Craterellus cornucopioides* var. *mediosporus* Corner

Chamonixiaceae
- *Chamonixia mucosa* (Petri) Corner & Hawker

Clavulinaceae
- *Clavulina cartilaginea* (B. & Per.) Corner

Cortinariaceae
- *Inocybe aequalis* (Horak) Turnbull & Watling
- *Inocybe angustifolia* Corner & Horak
- *Inocybe aurantiocystidiata* Turnbull & Watling
- *Inocybe avellanea* Kobayasi
- *Inocybe sphaerospora* Kobayasi

Elasmomycetaceae
- *Zelleromyces malaiensis* (Corner & Hawker) A.H. Smith

*Entolomataceae
- *Entoloma caeruleoviride* Corner & Horak
- *Entoloma corneri* Horak
- *Entoloma flavidum* (Massee) Corner & Horak
- *Entoloma pallido-flavum* (Hennings & Nyman) Horak
- *Leptonia decolorans* var. *decolorans* (Horak) Largent
- *Leptonia mougeotti* (Fr.) P.D. Orton
- *Nolanea cystopus* (Berk.) Pegler
- *Nolanea maderaspatana* Pegler

Gomphaceae
- *Gloeocantharellus okapaensis* Corner

Hydnaceae
- *Hydnum repandum* L. : Fr.

Hymenochaetaceae
- *Coltricia oblectans* (Berk.) Cunn.

Hymenogastraceae
- *Dendrogaster cambodgensis* Pat.

Russulaceae
- *Lactarius gerardii* Peck
- *Lactarius hygrophoroides* Berk. & Curt.
- *Lactarius subplinthogalus* Coker
- *Lactarius subserifluus* Longyear
- *Lactarius sumstinei* Peck
- *Lactarius vellereus* (Fr.) Fr.
- *Russula alboareolata* Hongo
- *Russula crustosa* Peck
- *Russula cyanoxantha* (Schaeff.) Fr.
- *Russula heterophylla* (Fr.) Fr.
- *Russula japonica* Hongo
- *Russula nauseosa* f. *japonica* Hongo
- *Russula pallidospora* (Bl. in Romagn.) Romagn.
- *Russula subnigricans* Hongo
- *Russula violeipes* Quél.
- *Russula virescens* (Schaeff.) Fr.

Pisolithaceae
- *Pisolithus aurantioscabrosus* Watling

Sclerodermataceae
- *Scleroderma columnare* Berk. & Br.
- *Scleroderma echinatum* (Petri) Guzmán
- *Scleroderma leptopodium* Har. & Pat.
- *Scleroderma sinnamariense* Mont.

Thelephoraceae
- *Sarcodon thwaitsei* (B. & Br.) Maas Geest.

Tricholomataceae
- *Laccaria murina* Imai
- *Laccaria vinaceoavellanea* Hongo
- *Tricholoma mons-fraseri* Corner
- *Tricholoma rhizophoreti* Corner
- *Tricholoma termitomycoides* Corner

12 Community Structure of Wood-Decaying Basidiomycetes in Pasoh

Tsutomu Hattori[1] & Lee Su See[2]

Abstract: More than 200 species of polypores, the most important group of wood-decaying basidiomycetes have been found in Pasoh Forest Reserve (Pasoh FR), Malaysia. Among them, about 35 species are pantropical, 30 species are paleotropical and only 7 species are cosmopolitan or widespread in the Northern Hemisphere. Many of the others are probably tropical Asiatic species. We analyzed niche differentiation of wood-decaying basidiomycetes according to the diameter of the substrata. Among the common species in Pasoh FR, *Coriolopsis retropicta*, *Microporus xanthopus* etc., were restricted to small substrata such as fallen branches and twigs while *Erythromyces crocicreas*, *Ganoderma australe* etc., were found mostly on large substrata. *Earliella scabrosa*, *Stereum ostrea* etc., occurred on both large and small substrata. We examined host specificity and preference of wood-decaying basidiomycetes in Pasoh FR. Many species such as *G. australe* and *Nigroporus vinosus* occurred in various tree families, but some species were restricted to or were most frequently found in one or a few tree families. For example, *Erythromyces crocicreas*, *Fomitopsis dochmia* etc. were restricted to dipterocarp trees. We also examined the effects of decomposition stage of substrata on species of wood-decaying basidiomycetes. Some species such as *G. australe*, *M. affinis* etc. were mainly found on newly fallen trees while other species such as *Nigroporus vinosus*, etc., were on well-decomposed trees. Species richness of wood-decaying basidiomycetes was higher in a primary forest plot than in a regenerating forest plot. This suggests that a low frequency of treefall in the regenerating forest reduced the species richness of wood-decaying basidiomycetes.

Key words: biodiversity, biogeography, host specificity, niche, polypores.

1. INTRODUCTION

Polypores and other wood-decaying basidiomycetes are important decomposers of woody debris in forest ecosystems. In tropical rainforests, termites are known to be important decomposers, but the ecological importance of wood-decaying basidiomycetes in the forest should not be overlooked. Fungi play a very important role in recycling the organic matter of the forest.

It should also be noted that some of the wood-decaying basidiomycetes might even be utilized by humans in the future. Some species such as *Ganoderma* spp., *Wolfiporia cocos* (Schw.) Ryv. & Gilb., *Phellinus linteus* (Berk. & Curt.) Teng have histories of medical uses while others such as *Lentinula edodes* (Berk.) Pegler, *Pleurotus ostreatus* (Jacq.:Fr.) Kummer and *Grifola frondosa* (Dicks.:Fr.) S. F. Gray are cultivated as foods. Furthermore, their enzymatic activities focus on the degradation of phenolic compounds that are harmful to humans and other organisms.

---

[1]Forestry and Forest Products Research Institute (FFPRI), Tsukuba 305-8687, Japan.
E-mail: hattori@affrc.go.jp
[2]Forest Research Institute Malaysia (FRIM), Malaysia.

It is difficult to access the biodiversity of macrofungi, because of the pronounced periodicity and yearly fluctuations of fruitbody production and the short life span of most macrofungi. However, polypores are classified as durables or perennials because of the longevity of their sporocarps (Arnolds 1995) and fruit, which is less seasonal than other macrofungi.

Recently, there have been some studies on niche differentiation and biodiversity of polypores, mainly in coniferous forests of northern Europe. Many of wood-decaying basidiomycetes show specificity or preference for substrata having certain characteristics, for example, substratum size. In boreal areas, niches of wood-decaying basidiomycetes were differentiated according to substratum conditions and their effects on the diversity of decay fungi discussed (Bader et al. 1995; Renvall 1995). Lodge (1996) also suggested the preference of several wood-decaying basidiomycetes for logs, branches and twigs in Puerto Rico.

In temperate regions, many wood-decaying basidiomycetes only occur on specific tree genera or families (Ryvarden & Gilbertson 1993, 1994). On the other hand, Lodge (1997) reported that strong host-specificity is rare among wood decomposer agarics though it is common among leaf decomposer fungi. However, there is no similar information on host-specificity among polypores in Tropical Asia.

Here, we show niche differentiation of polypores and other wood-decaying basidiomycetes in Pasoh, and discuss the effects of logging on the diversity of wood-decaying basidiomycetes as it relates to niche differentiation.

2. MATERIALS AND METHODS

2.1 Biogeographical characters

We visited Pasoh Forest Reserve (Pasoh FR) 10 times from 1991 to 1999, and made intensive collections of polypores within the lowland areas, mainly within Plot 1 (IBP plot) and along the trails (see color plate 2 for location). Collected specimens were determined in the laboratory. The total number of species and number of new species collected were recorded after each visit.

2.2 Effect of the substratum size

In Pasoh FR, we examined the effect of substratum size on the occurrence of wood-decaying basidiomycetes, mainly polypores. We selected Plot 1 as the research plot. This 2 ha (200 m × 100 m) plot is situated in a primary forest. We also established 40 belt plots (each 50 m × 10 m) mainly within regenerating forests along the Main Trail and the Nature Trail. We examined woody debris including fallen trees, branches and twigs in the established plots. We recorded the occurrence of wood-decaying basidiomycetes on these substrata and measured the mean diameter of each substratum where the basidiocarps were attached.

2.3 Effect of the substratum species

In order to reveal the effect of the substratum species on the occurrence of wood-decaying basidiomycetes in Pasoh FR, we recorded wood-decaying basidiomycetes within and around Plot 1 found on fallen trees whose scientific names were known. Additionally, wood samples were taken from the substrata on which some wood-decaying basidiomycetes were growing. In the laboratory, the wood samples were sliced with a microtome and examined with a light microscope, then determined by genus or family.

2.4 Effect of the decomposition stage of the substrata

We marked 29 trees that fell in 1992–1995 mainly inside the Plot 1. Wood-decaying basidiomycetes occurring on these trees were recorded once a year for 3–6 years after the trees fell.

2.5 Effect of logging on decay fungal community

Primary Forest Plot (PFP, 100 m × 100 m) was established within Plot 1. Regenerating Forest Plot (RFP, 100 m × 100 m) was established within the area where intensive logging had been conducted in the 1950s, 300 m north of the PFP.

Fallen trees and branches of more than 10 cm diameter and 2 m in length were numbered and mapped within the plots. Diameter, length and occurrence of basidiocarps of polypores were recorded for each fallen tree. Fallen branches and twigs more than 2 cm were also examined for occurrence of basidiocarps of polypores. Each plot was divided into 100 subplots (10 m × 10 m) and the occurrence of each species of polypores was recorded in each subplot.

3. RESULTS AND DISCUSSION

3.1 Biogeographical characters

We have collected more than 200 species of polypores and several other decay fungi in Pasoh FR. Still, quite a number of new species were recorded after each visit (Fig. 1). We suggest that Pasoh FR is relatively rich in polypore species because nearly 300 species of polypores might be expected from this single research site compared to only about 330 species recorded from the whole of Europe where most of the species have already been listed (Ryvarden & Gilbertson 1993, 1994).

Among the species recorded from Pasoh FR, only the following are cosmopolitan or widespread in the Northern Hemisphere: *Ceriporia reticulata*

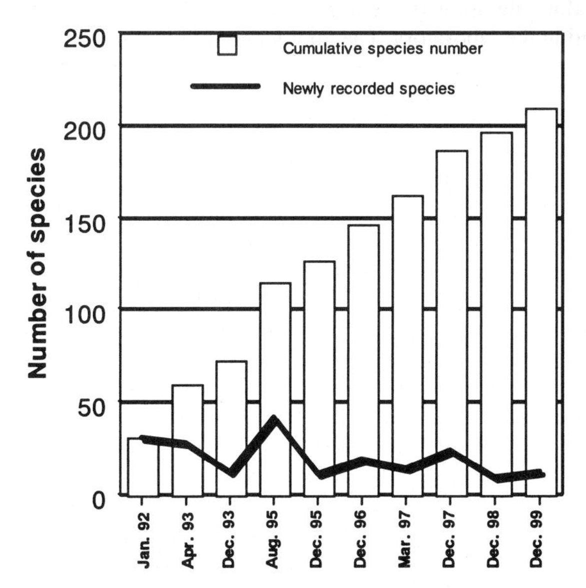

Fig. 1 Newly recorded species at each visit and cumulative species number of polypores in Pasoh FR.

(Pers.:Fr.) Dom., *C. viridans* (Berk. & Br.) Donk, *Coltricia cinnamomea* (Pers.) Murr., *Phellinus gilvus* (Schw.) Pat., *Pycnoporus sanguineus* (L.:Fr.) Murr., *Schizopora paradoxa* (Fr.) Donk and *Tyromyces chioneus* (Fr.) Karst.

Several species are pantropical: *Antrodiella liebmanii* (Fr.) Ryv., *Coriolopsis caperata* (Berk.) Ryv., *Cyclomyces setiporus* (Berk.) Pat., *Earliella scabrosa* (Pers.) Gilb. & Ryv., *Flabellophora obovata* (Jungh.) Corner, *Fomitopsis carnea* (Bl. & Nees:Fr.) Imaz., *F. dochmia* (Berk. & Br.) Ryv., *Ganoderma australe* (Fr.) Pat., *Gloeoporus thelephoroides* (Hook.) Cunn., *Grammothele fuligo* (Berk. & Br.) Ryv., *G. lineata* Berk. & Curt., *Hexagonia tenuis* (Hook.) Fr., *Loweporus tephroporus* (Mont.) Ryv., *Megasporoporia cavernulosa* (Berk.) Ryv., *M. setulosa* (Henn.) Ryv. & Wright, *Nigrofomes melanoporus* (Mont.) Murr., *Nigroporus vinosus* (Berk.) Murr., *Oxyporus molissimus* (Pat.) Reid, *Phellinus noxius* (Corner) Cunn., *P. senex* (Nees & Mont.) Imaz., *P. setulosus* (Lloyd) Imaz., *P. umbrinellus* (Bres.) Ryv., *Phylloporia chrysita* (Berk.) Ryv., *P. spathulata* (Hook.) Ryv., *Polyporus brasiliensis* (Berk.) Corner, *P. dictyopus* Mont., *P. tenuiculus* (Beauv.) Fr., *Rigidoporus lineatus* (Pers.) Ryv., *R. microporus* (Fr.) Overeen, *R. vinctus* (Berk.) Ryv., *Tinctoporellus epimiltinus* (Berk. & Br.) Ryv., *Trametes elegans* (Spreng.:Fr.) Fr., *Trametes modesta* (Kunt.:Fr.) Ryv.

Several other species are paleotropical: *Amauroderma rugosum* (Bl. & Nees:Fr.) Torr., *Aurificaria indica* (Massee) Reid, *Coriolopsis strumosa* (Fr.) Ryv., *Cyclomyces tabacinus* (Mont.) Pat., *Echinochaete russiceps* (Berk. & Br.) Reid, *Flavodon flavus* (Kl.) Ryv., *Fomitopsis feei* (Fr.) Kreisel, *Gloeoporus croceopallens* Bres., *Loweporus fuscopurpureus* (Pers.) Ryv., *L. roseoalbus* (Jungh.) Ryv., *Microporus affinis* (Bl. & Nees:Fr.) Kl., *M.* cf. *vernicipes* (Berk.) Kl., *M. xanthopus* (Beauv.:Fr.) Kunt., *Perenniporia ochroleuca* (Berk.) Ryv., *Phellinus glaucescens* (Petch) Ryv., *P. lamaensis* (Murr.) Heim, *P. pachyphloeus* (Pat.) Pat., *Pyrofomes albomarginatus* (Lév.) Ryv., *P. tricolor* (Murr.) Corner, *Rigidoporus hypobrunneus* (Petch) Corner, *Theleporus calcicolor* (Sacc. & Syd.) Ryv., *Trametes menziezii* (Berk.) Ryv., *T. meyenii* (Kl.) Lloyd, *Trichaptum durum* (Jungh.) Corner.

Many other species are probably endemic to tropical areas of Asia. A number of species described by Corner (1983, 1984, 1987, 1989a, b, 1991, 1992) from tropical Asia were also collected in Pasoh FR, but many of them are still undetermined and several undescribed species will likely be found.

3.2 Effect of the substratum size

Among the frequently occurring species, the following were found mainly on larger substrata of at least 10 cm diameter: *Erythromyces crocicreas* (Berk. & Br.) Hjorts. & Ryv., *Fomitopsis carnea*, *Ganoderma australe*, *Ganoderma* cf. *chalceum* (Cke.) Stey., *Phellinus lamaensis*, *Rigidoporus microporus* etc. (Fig. 2). Species often seen on both large and small substrata, but usually not on small branches of less than 5 cm diameter, are as follows: *Earliella scabrosa*, *Microporus affinis*, *Nigroporus vinosus* and *Stereum ostrea* (Fig. 3). There are several species whose occurrence is restricted to small branches and twigs of usually less than 5 cm diameter. These are *Coriolopsis retropicta*, *Megasporoporia* sp. No. 1 and *Microporus xanthopus* etc. (Fig.4).

Decay fungi occurring on larger trunks were different from those on branches and twigs. This suggests that the abundance of fallen trunks and/or branches will likely affect the species richness and composition of wood-decaying basidiomycetes. A small number of species would be expected in a stand where few fallen trees of large diameter are present.

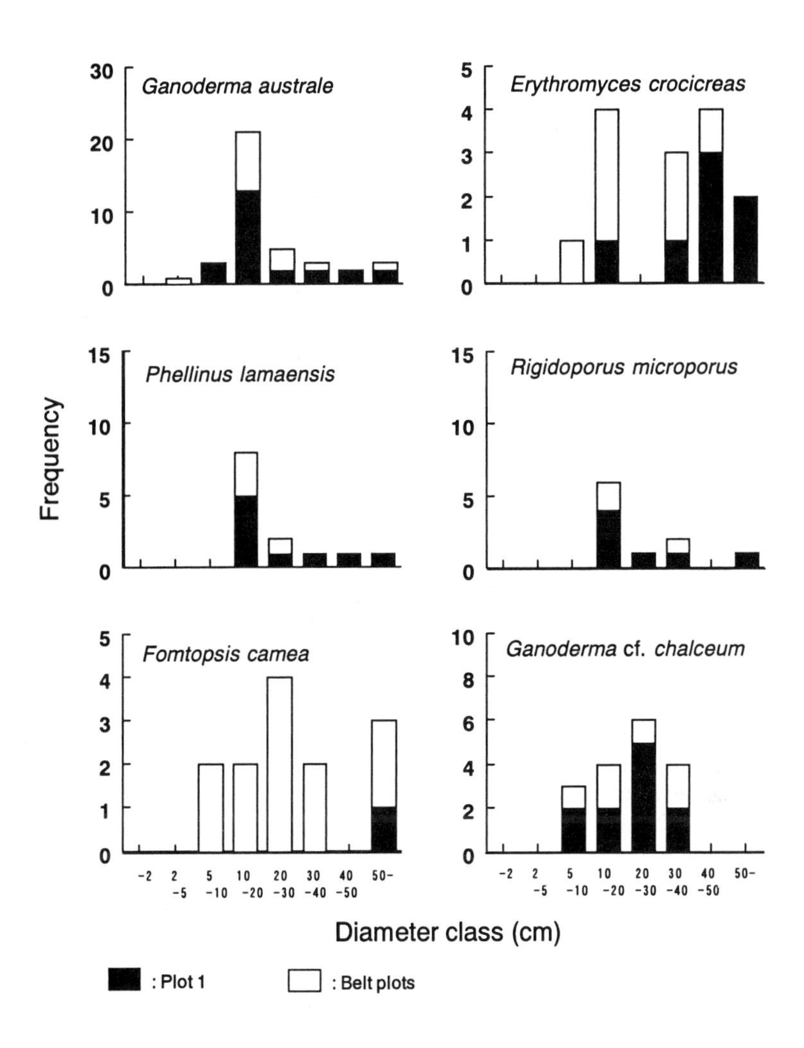

Fig. 2 Frequencies of wood-decaying basidiomycetes on substrata of different diameter classes: species occured frequently on larger substrata.

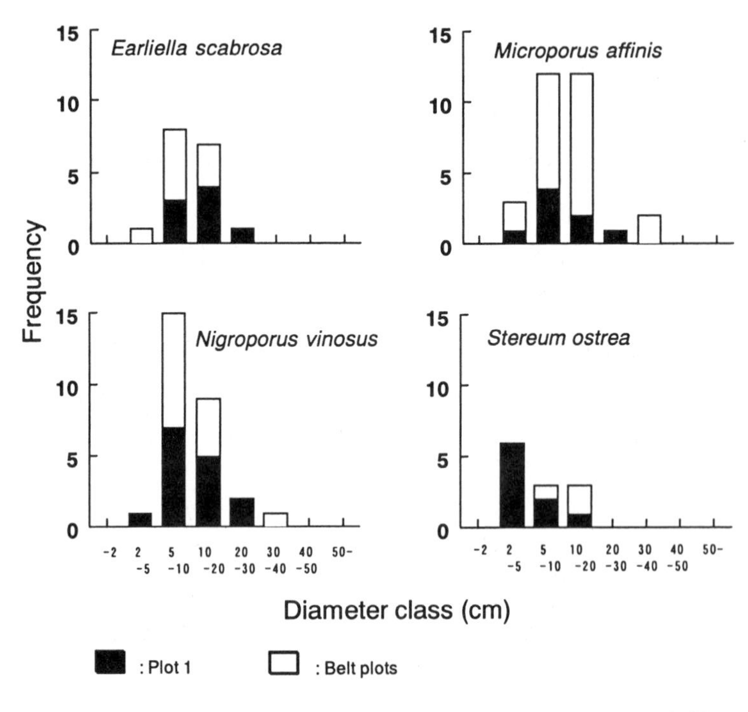

Fig. 3 Frequencies of wood-decaying basidiomycetes on substrata of different diameter classes: species occurred both on larger and smaller substrata.

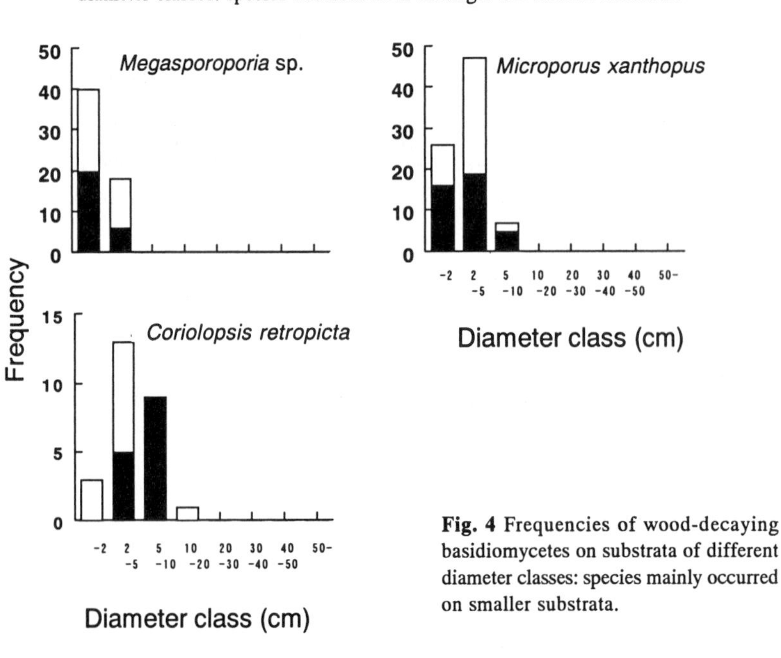

Fig. 4 Frequencies of wood-decaying basidiomycetes on substrata of different diameter classes: species mainly occurred on smaller substrata.

3.3 Effect of the substratum species

Some species were restricted to certain tree families and are thus defined as specialists. Some other species were not restricted to certain tree families, but mainly occurred on them. These fungi are defined as species showing host preference.

Several species were specific to dipterocarp trees. *Perenniporia dipterocarpicola* Hattori & Lee was always recorded on standing or fallen trunks of dipterocarp trees (Hattori & Lee 1999). *Erythromyces crocicreas* is another species specific to fallen trunks of dipterocarp trees, including newly fallen trees. The following species were also found on dipterocarp trees but on old timbers: *Daedalea aurora* (Ces.) Aoshima, *Fomitopsis carnea*, *F. dochmia*, *Phellinus* cf. *fastuosus* (Lév.) Ryv. etc.

Perenniporia hexagonoides Hattori & Lee was hitherto found only on the bases of living *Scaphium macropodum* (Miq.) Beumee ex Heyne (Sterculiaceae) (Hattori & Lee 1999). This species may be specific to the genus *Scaphium* or members of the Sterculiaceae.

Nigrofomes melanoporus and *Pyrofomes albomarginatus* were mostly found on dipterocarp trees, but they were also found on other hosts. *Microporus affinis* was frequent on members of the Dipterocarpaceae and Fagaceae, but was also found on other hosts. *Earliella scabrosa* and *Trametes elegans* showed some preference for Leguminosae and Euphorbiaceae, respectively, but their preference is probably not so strong.

Most of the other frequently occurring species did not show any preference for any particular tree family. For example, *Ganoderma australe*, one of the most frequent species in Pasoh FR, was found on 15 tree families, *Nigroporus vinosus* on 13 families and *Rigidoporus microporus* on 10 families.

In tropical areas, it is believed that wood-decaying basidiomycetes have little specificity or preference because in general density of each tree species, genera and families are low in the stands (Lindblad 2000). However, there are several species that show specificity to or preference for dipterocarp trees in Pasoh FR. This may be attributed to the presence of a large mass of fallen dipterocarp trees in Pasoh FR where the huge emergent trees are dominated by Dipterocarpaceae whose wood is usually not easily decomposed and thus persists for a long time on the forest floor.

Here, host ranges are revealed only for some common wood-decaying basidiomycetes. Among rarer wood-decaying basidiomycetes, there may be other additional specialists such as *Phellinus scorodocarpi* Corner specific to *Scorodocarpus borneensis* (Baill.) Becc. (Olacaceae) described by Corner (1991).

3.4 Effect of the decomposition stage of the substrata

Ganoderma australe was the most frequent species in the early stage (around one to two years after treefall) and was recorded on 14 out of 29 trees examined. Other species frequently seen in the early stage were *Schizopora* cf. *flavipora* (Cke.) Ryv., *Rigidoporus microporus*, *Erythromyces crocicreas* (all on dipterocarp trees), *Microporus affinis*, *Cyclomyces tabacinus* and *Earliella scabrosa*. Basidiocarp developments of these fungi were often widespread on the trees examined suggesting that colonization occurred at an early stage after treefall and the basidiocarps developed quickly.

Basidiocarps of *Loweporus fuscopurpureus*, *Phellinus lamaensis*, *Tinctoporellus epimiltinus* were also seen on undecomposed trees, but usually

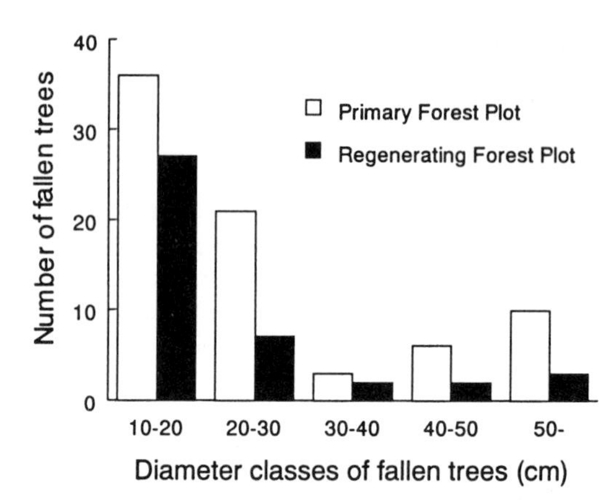

Fig. 5 Number and diameter classes of fallen trees in the Primary Forest Plot and the Regenerating Forest Plot.

around two or more years after treefall. It is unclear as to whether they are later colonizers or early colonizers that produce basidiocarps slowly.

Basidiocarps of *Nigroporus vinosus* were always seen on decayed and softened wood (2 or more years after treefall). Other species also seen during the later stage of decomposition were *Steccherícium seriatum*, *Antrodiella liebmanii* and *Polyporus tenuiculus*. Additionally, *Antrodiella* spp., *Oligoporus* spp., *Tyromyces* spp. and *Trechispora* sp. were also frequently seen in Pasoh FR on other trees in the later stages of decomposition.

These results suggest that the presence of fallen trees under various stages of decomposition increase species richness of wood-decaying basidiomycetes. Therefore, we conclude that forest stands where treefalls occur occasionally are more species rich in wood-decaying basidiomycetes than are those with intensive treefalls in a certain year.

3.5 Effect of logging on decay fungal community

Seventy-six fallen trees and branches were recorded in the Primary Forest Plot (PFP). Among them, 36 (47%) had a diameter of less than 20 cm and 10 (13%) had a diameter larger than 50 cm. Forty-one fallen trees and branches were recorded in the Regenerating Forest Plot (RFP). Twenty-seven (66%) had a diameter less than 20 cm and three (7%) had a diameter larger than 50 cm (Fig. 5). Several newly fallen trees of different diameter classes were observed in the PFP. However, most of the fallen trees of larger diameter seen in the RFP were old dipterocarps, presumably branches cut during logging.

In the RFP, fallen trees of larger diameter were fewer than in the PFP and mostly consisted of old dipterocarp trees. Therefore, the fungal species frequently occurring in this plot were those characteristic of smaller substrata (e.g. *Megasporoporia* sp. No. 1, *Microporus xanthopus* etc.) and those of old dipterocarp trees (e.g. *Fomitopsis carnea* and *F. dochmia*). In the PFP, fungal species of small substrata were also frequently seen but those found on old dipterocarp trees were not recorded. Species characteristic of large substrata, such as *Ganoderma australe* and *Phellinus lamaensis*, were more frequently observed in the PFP than in the RFP.

In comparing these two plots, the total frequency and number of wood-decaying basidiomycete species not specific to either small twigs or old dipterocarp trees were much higher in the PFP. This in fact is the main reason for the reduced diversity of wood-decaying basidiomycetes in the RFP.

In forests where intensive logging had been carried out in the past, few natural treefalls are expected because few large trees are left in the forest (see Chap. 2). In an undisturbed forest, natural treefalls often occur, resulting in the creation of tree fall gaps. Some flushes of wood-decaying basidiomycetes will occur in such areas followed by the occurrence of other fungi in later stages of the succession. There were few obvious tree fall gaps in the RFP but several gaps were observed in the PFP.

This suggests that intensive logging in forests reduces the biodiversity of wood decay fungi although flushes of some fungi may occur temporarily just after logging. In order to conserve the biodiversity of wood decay fungi, less intensive and more sustained logging is needed.

4. CONCLUSION

The diversity of wood-decaying basidiomycetes is obviously very high in Pasoh FR where more than 200 species of polypores have been recorded mainly from a relatively small area of a 2 ha plot and along trails. This is in contrast with Europe where about 330 species of polypores have been recorded by a much larger number of both amateur and professional mycologists over a very much longer time. We would expect more species to be recorded from Pasoh FR if collections were extended over a larger area and over a longer period. If we were to consider the tropical rainforest in the Southeast Asia as a whole, we would then expect the diversity of wood-decaying basidiomycetes to be very high indeed.

Species richness of wood-decaying basidiomycetes appears to be enhanced when fallen trees in different stages of decomposition are present. This suggests that the number of species of wood-decaying basidiomycetes tends to be higher in forest stands where treefalls occur occasionally compared to forests where treefalls are more intensive but temporal. The results also show that wood-decaying basidiomycetes can be used as indicators of the status/condition of a forest as certain fungi are only found in association with trees of a certain diameter and/or species.

Finally, our results suggest that intensive logging in forests reduces the biodiversity of wood-decaying basidiomycetes although flushes of some fungi may occur temporarily just after logging. Thus in order to conserve the biodiversity of wood-decaying basidiomycetes and to ensure efficient and effective nutrient recycling for sustainable forest management, logging must be less intensive and more sustainable.

ACKNOWLEDGEMENT

This study was supported by the Global Environmental Research Fund (Grant No. E-1) by the Japan Environmental Agency.

REFERENCES

Arnolds, E. (1995) Problems in measurements of species diversity of macrofungi. In Allsopp, D., Colwell, R. R. & Hawksworth, D. L. (eds). Microbial Diversity and Ecosystem Function, CAB International, Wallingford, UK, pp.337-353.

Bader, P., Jansson, S. & Jonsson, B. G (1995) Wood-inhabiting fungi and substratum decline in

selectively logged boreal spruce forests. Biol. Conserv. 72: 355-362.
Corner, E. J. H. (1983) Ad Polyporaceas I. *Amauroderma* and *Ganoderma*. Beiheft zur Nova Hedwigia 75: 1-182.
Corner, E. J. H. (1984) Ad Polyporaceas II & III. Beiheft zur Nova Hedwigia 78: 1-222.
Corner, E. J. H. (1987) Ad Polyporaceas IV. The genera *Daedalea*, *Flabellophora*, *Flavodon*, *Gloeophyllum*, *Heteroporus*, *Irpex*, *Lenzites*, *Microporellus*, *Nigrofomes*, *Nigroporus*, *Oxyporus*, *Paratrichaptum*, *Rigidoporus*, *Scenidium*, *Trichaptum*, *Vanderbylia*, and *Steccherinum*. Beiheft zur Nova Hedwigia 86: 1-265.
Corner, E. J. H. (1989a) Ad Polyporaceas V. The genera *Albatrellus*, *Boletopsis*, *Coriolopsis* (dimitic), *Cristelloporia*, *Diacanthodes*, *Elmerina*, *Fomitopsis* (dimitic), *Gloeoporus*, *Grifola*, *Hapalopilus*, *Heterobasidion*, *Hydnopolyporus*, *Ischnoderma*, *Loweporus*, *Parmastomyces*, *Perenniporia*, *Pyrofomes*, *Stecchericium*, *Trechispora*, *Truncospora* and *Tyromyces*. Beiheft zur Nova Hedwigia 96: 1-218.
Corner, E. J. H. (1989b) Ad Polyporaceas VI. The genus *Trametes*. Beiheft zur Nova Hedwigia 97: 1-197.
Corner, E. J. H. (1991) Ad Polyporaceas VII. The xanthochroic polypores. Beiheft zur Nova Hedwigia 101: 1-175.
Corner, E. J. H. (1992) Additional resupinate non-xanthochroic polypores from Brazil and Malesia. Nova Hedwigia 55: 119-152
Hattori, T. & Lee, S. S. (1999) Two new species of *Perenniporia* described from a lowland rainforest of Malaysia. Mycologia 91: 525-531.
Lindblad, I. (2000) Host specificity of some wood-ihnabiting fungi in a tropical forest. Mycologia 92: 399-405.
Lodge, D. J. (1996) Microorganisms. In Reagan, D. P. & Robert, B. W. (eds). The food web of a tropical forest. University of Chicago Press, Chicago, pp.54-108.
Lodge, D. J. (1997) Factors related to diversity of decomposer fungi in tropical forests. Biodiversity Conserv. 6: 681-688.
Renvall, P. (1995) Community structure and dynamics of wood-rotting Basidiomycetes on decomposing conifer trunks in northern Finland. Karstenia 35: 1-51.
Ryvarden, R. & Gilbertson, R. L. (1993) European Polypores, Part 1. Fungiflora, Oslo.
Ryvarden, R. & Gilbertson, R. L. (1994) European Polypores, Part 2. Fungiflora, Oslo.

Part III

Plant Population and Functional Biology

13 Comparative Biology of Tropical Trees: a Perspective from Pasoh

Sean C. Thomas[1]

Abstract: From the late 1960s, tree species at Pasoh Forest Reserve (Pasoh FR) have been the focus of numerous comparative studies. Research to date has examined spatial distributions and edaphic preferences, reproductive phenology, vertical stratification, and interactions with pollinators and seed predators. Recent studies have also begun to address physiological and morphological bases for ecological variation. Pasoh FR is dominated by groups of trees that are phylogenetically closely related. As many as 47 species within a single genus (*Eugenia*) occur in sympatry at the site. Detailed studies of several of these speciose groups indicate that described morphospecies correspond to reproductively-isolated taxa. Closely related species commonly differ in terms of reproductive timing, microhabitat preference, and vertical stratum at maturity. Comparative studies have also emphasized several bases for ecological "niche differentiation" at Pasoh, both within and among phylogenetic groups. Tree species at the site show pronounced differences in terms of vertical stratification, and relative to local patterns of water flow, soil parent material, and light requirements for growth. Recent theoretical studies suggest that niche differentiation may not be necessary to maintain high species richness in tropical forest communities. How, then, should one interpret the large ecological differences found among tree species at Pasoh FR? The existence of large interspecific differences in such characteristics as demographic rate parameters and size at reproduction indicates that one central assumption of "null" models–that species are equivalent–is simply incorrect. Models that fully incorporate interspecific ecological differences as well as dispersal limitation and other central elements of null models of tropical forest communities are needed to further our theoretical understanding of tropical tree diversity, and to better predict tropical forest responses to human and natural disturbance. Comparative studies are essential to parameterize and test such models, and also serve an important role in providing basic "silvics" information essential to forest management and conservation in the region.

Key words: life-history variation, niche differentiation, null models, reproductive isolation, species diversity, vertical stratification.

1. INTRODUCTION

It is fair to say that biologists know only two things about the vast majority of living species of tropical trees: (1) the few details of gross morphology that have served to distinguish that species from others, and (2) taxonomic affiliation. In other words, we know only the name of the species, and the details of morphology that have led it to be given this name. The list of traits for which little or no information exists is, correspondingly, very long indeed. Growth rate, shade tolerance, regeneration pattern, reproductive and vegetative phenology, seed

[1] Faculty of Forestry, University of Toronto, Ontario M5S 3B3, Canada.
E-mail: sc.thomas@utoronto.ca

dispersal distances, dispersal agents, pollinators, germination and seedling establishment requirements, edaphic associations, nutrient requirements, age and size at maturity, reproductive output, drought tolerance, mycorrhizal symbionts, herbivores, seed predators, disease organisms and vectors–these are just a few of the characteristics for which detailed data are generally absent. Such data are of more than "academic interest": they are a requirement for informed forest management, as reflected by the fact that precisely such information forms the core data reported in standard silvics manuals used by temperate-zone foresters (e.g. Burns & Honkala 1990).

The context for studies of ecological differences among rain forest tree species has most often been the maintenance of species diversity. Traditional views, particularly those borrowed from the zoological literature through the 1980s, emphasized "niche" differences as the primary mechanism for the maintenance of species diversity in ecological communities (e.g. Bergh & Braakhekke 1978; Denslow 1980; Dobzhansky 1950; Klopfer & MacArthur 1961; Newman 1982; Richards 1969). The study of niche relationships of trees faces substantial challenges, of both a theoretical and practical nature. First, all plants require essentially the same qualitative kinds of resources (light, water, macro- and micro-nutrients, CO_2, O_2, physical space), a fact which complicates the study of tree "niches" in many ways. A second challenge to studies of niche relationships in tropical trees is their great size and long lifespan. It has often been assumed that tree species differ mainly in terms of characters expressed early in ontogeny: i.e. in terms of their "regeneration niche" sensu Grubb (1977, 1996). However, recent work has documented large interspecific differences in growth and physiological characteristics of adult trees (e.g. Clark & Clark 1992; Thomas 1996a; Thomas & Bazzaz 1999). For example, the fact that tree species which attain greater heights display higher maximal photosynthetic rates and other "sun-plant" characteristics even as saplings in the understory suggests an important role of selection on adult tree characteristics in shaping the overall phenotype of a given species (Reich 2000; Thomas & Bazzaz 1999). Thus, it is important to understand not only resource requirements of saplings, but also how such requirements change throughout tree ontogeny. Finally, the simple fact that tree species occur at such low densities in tropical forests makes all comparative studies of tree ecology in the tropics immensely more challenging than in temperate regions.

In recent years plant "niches" have fallen into disfavor, at least among some schools of ecologists. A variety of mechanisms that may potentially account for tree diversity in tropical rain forests have been proposed that do not require ecological "niche" differences among species. First, the "natural enemies" hypothesis (Becker & Wong 1985; Connell 1971; Janzen 1970; Okuda et al. 1995, 1997; Wills et al. 1997; Wills & Condit 1999) posits that there is specialization of diseases, herbivores, and/or seed predators on a given tree species. While differences between species are necessary for natural enemies to maintain species diversity, such differences do not correspond to "niche differentiation" in the usual sense. More recently, there has been a growing realization of just how slow dynamics of competitive exclusion are likely to be in tropical forest systems characterized by long tree lifespans, local interactions, and limited dispersal (e.g. Brokaw & Busing 2000; Hubbell 1995, 2001; Hubbell & Foster 1986). It seems likely that competitive exclusion among tropical trees occurs only over time scales on the order of thousands to millions of years. Over such a time scale speciation and regional migration must also be considered (Hubbell 2001).

From the onset of the IBP (International Biological Program) era of research in the 1960s and '70s, Pasoh Forest Reserve (Pasoh FR) has been the focus of a large number of studies focusing on the comparative biology of tropical trees. The main purpose of this paper is to review this work, with reference also to studies of comparative tree biology elsewhere in the Sundaland region of SE Asia. I develop three general points: namely, (1) that work at Pasoh FR and elsewhere in SE Asia provides rich documentation of the existence of ecological differences among tropical tree species, even those that are very closely related; (2) that habitat specificity and vertical stratification are especially important as "niche" axes; and (3) that comparative tree biology is not merely of theoretical interest from the perspective of equilibrium theories of species coexistence, but rather is first and foremost essential for sustainable management, biodiversity conservation, and forest restoration.

2. SPECIES RICHNESS AND THE PREVALENCE OF SPECIOSE TAXA IN SE ASIAN FORESTS

The forests of SE Asia are among the most species-rich in the world. There has been a long-standing debate in the literature as to whether the highest local species richness is found in the forests of Amazonia or Sundaland (Ashton 1977; Gentry 1988b; Whitmore 1984). Gentry (1982, 1988a,b) contended that the world's most diverse forests were restricted to upper Amazonia, and indeed, the record 1-ha sample (of trees ≧ 10 cm DBH) has since been recorded from Amazonian Ecuador (Valencia et al. 1994). Comparably high species richness has also been reported recently from forests in central Amazonia on poorer soils with lower rainfall (De Oliveira & Mori 1999). Gentry's own sampling work in SE Asia suggested that Sundaland forests were less diverse than the forests of upper Amazonia (Gentry

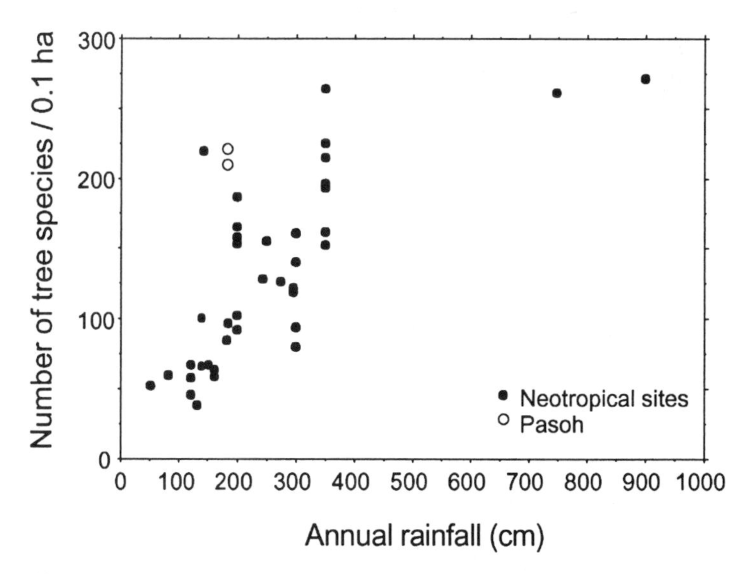

Fig. 1 Tree species richness (trees ≥ 2.5 cm in 0.1 ha) as a function of annual rainfall, with two plots at Pasoh FR compared to 40 neotropical sites. Data are from the archive of A. Gentry (http://www.mobot.org/MOBOT/research/applied_research/gentry.html), with climate information from Gentry, 1988b.

Table 1 Relative proportion of species-rich taxa in the Pasoh FR 50-ha plot compared with similar plots in Central America (Barro Colorado Island, Panama), and Africa (Korup Forest Reserve, Cameroon). Based on data from the first plot census in each case.

	Pasoh	BCI	Korup
Total number of species	820	313	420
Total number of genera	295	187	231
Number of genera with > 5 species	37	5	8
% of tree flora in genera with >5 species	49%	17%	17%
Five most speciose genera (Number of species / genus)	*Eugenia* (47)	*Inga* (14)	*Cola* (23)
	Diospyros (23)	*Psychotria* (13)	*Diospyros* (14)
	Aglaia (22)	*Ficus* (12)	*Trichoscypha*
	Garcinia (18)	*Piper* (8)	*Garcinia* (10)
	Litsea (14)	*Miconia* (7)	*Psychotria* (7)

1988b). However, his conclusion was based on a very limited number of SE Asian samples. Interestingly, Gentry's data for Pasoh FR indicate a remarkably high species richness, given the site's relatively low rainfall (Fig. 1). The 50 ha plot at Pasoh FR contains ~820 species of trees (Kochummen et al. 1991; Manokaran et al. 1990), compared to ~1200 for similar plots at Lambir Forest Reserve in Sarawak, Borneo, and Yasuni in Amazonian Ecuador (Condit et al. 2000; Valencia 1998). Comparative data thus indicates that Pasoh FR, while not approaching the tree species richness found in the richest areas of either the neotropics or Borneo, is still remarkably species-rich, particularly given the low rainfall at the site.

One intriguing characteristic of the tree flora of the Sundaland region of Semenanjung, Sumatra and Borneo, is the prevalence of highly species-rich tree taxa found growing in close sympatry (Ashton 1969; Van Steenis 1969). The large-scale plot data allow for some preliminary quantitative comparisons across major biogeographic regions (Table 1). Within the 50 ha plot sample at Pasoh FR, one finds an astounding 47 species of *Eugenia* (Myrtaceae), 23 species of *Diospyros* (Ebenaceae), and 22 of *Aglaia* (Meliaceae) (Botanical nomenclature follows Kochummen 1997, except as noted). In total, 37 genera have at least 6 species represented in the plot at Pasoh FR. This is markedly higher than the local intra-generic diversity recorded in 50 ha plots at Barro Colorado Island, Panama, and Korup Forest Reserve, Cameroon, where only 5 and 8 tree genera having 5 or more species were represented, respectively. Nearly half of the total tree species count in the 50 ha sample at Pasoh FR is due to genera with > 5 species in the plot, while the comparable figure for BCI and Korup is 17% in each case.

Given this pattern of high species richness in narrow taxonomic groups, comparative studies of sibling species or small monophyletic groups are of particular interest and importance. First, it is necessary to verify that putative species are reproductively isolated, and likewise that morphologically-defined species correspond to single biological species. Second, analyses of such groups provide what has been viewed as a kind of ultimate test of the importance of niche divergence in maintaining species diversity in tropical forests (Ashton 1969, 1977; Rogstad 1989, 1990). It should be emphasized that finding ecological differences between sibling species pairs (or within small monophyletic groups) does not necessarily imply that these differences are sufficient to ensure long-term co-existence. Nevertheless, such studies provide important information regarding the nature of ecological shifts that occur through the process of speciation, and thus a critical link between comparative ecology and evolutionary biology in tropical trees.

3. ARE DESCRIBED TREES AT PASOH FR GOOD BIOLOGICAL SPECIES ?

One reaction to summaries of the astounding diversity of trees in Sundaland and similar areas in the tropics, a reaction not uncommon among temperate forest biologists, is outright skepticism. Are all the described tree taxa really valid biological "species"? As in essentially all areas of the tropics, taxonomists have approached the problem of species delineation by relying almost entirely on morphological characteristics. Problems, both theoretical and practical, with species delineation have certainly not gone unrecognized among tropical plant systematists (e.g. Stevens 1990). However, most named taxa are essentially based on subjective impressions of morphological distinctiveness. How can one be sure that all of the named morphological entities labeled as "species" are in fact reproductively isolated?

There are certainly some "species" present at Pasoh FR for which taxonomic status is debatable. Facultative or obligate apomicts represent an important example. Embryological studies have documented apomixis via adventive embryony in a number of taxa of SE Asian trees (Ha et al. 1988; Kaur et al. 1978, 1986). The best-studied example is the genus *Garcinia*, which includes the apomictic village fruit tree *Garcinia mangostana*. Most *Garcinia* species can produce viable seed via adventive embryony (Richards 1990a,b). Moreover, compared with other dioecious trees at Pasoh FR, which tend to show male-biased sex ratios, *Garcinia* populations show a tendency toward female bias (Thomas 1997). This trend is consistent with facultative apomixis in the genus as a whole. In at least one exceptional case, *Garcinia scortechinii*, no staminate trees have been found in the reserve, though staminate trees do exist elsewhere in Peninsular Malaysia. This remarkable pattern indicates that *G. scortechinii* is in fact a geographic parthenogen, with the asexual form occurring at Pasoh FR (Thomas 1997).

Hybridization, particularly of the variety that so greatly complicates systematy in certain temperate genera such as *Quercus* and *Crataegus*, might also be suspected as a source of apparent species diversity at Pasoh FR. In Peninsular Malaysia, the best documented case of hybridization is between *Shorea leprosula* and *S. curtisii*. Putative hybrids with intermediate morphological characteristics are known from several locations. That such trees are in fact hybrids was confirmed by isozyme analysis and formation of sterile fruits in at least one population (Gan 1976; cited in Ashton 1988). Another instance of a possible sterile hybrid has also recently been reported from Pasoh FR. Trees identified within the 50 ha plot as *Garcinia parvifolia* include a morphologically-distinct subset that continuously produces pistillate flowers, but apparently never form viable seed (Thomas 1997). Anecdotally, I have also found within the 50 ha plot one *Ixora* individual with a phenotype intermediate between *I. pendula* and *I. lobbii*.

Although apomixis and hybridization do occur, most speciose genera at Pasoh FR that have been closely examined have not yielded any evidence for apomictic swarms or hybrids. *Garcinia* and putative hybrids in *Shorea* are thus almost certainly exceptional cases. Speciose genera in which evidence suggests reproductive isolation of member taxa at Pasoh FR include *Polyalthia* (*hypoleuca* complex) (Annonaceae), *Shorea* section Muticae (Diptercarpaceae), *Scaphium* (Sterculiaceae), *Aporosa** (Euphorbiaceae), *Baccaurea* (Euphorbiaceae), *Diospyros* (Ebenaceae), and *Ixora* (Rubiaceae). Studies on these taxa are reviewed below. Three main types of observations have been made that are consistent with reproductive isolation of species in these groups: (1) differences in timing of

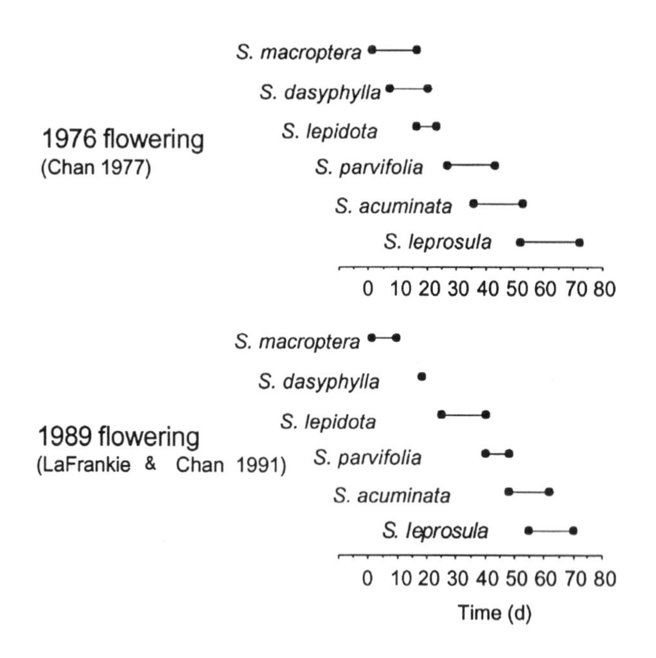

Fig. 2 Confirmation of sequential flowering in *Shorea* section Muticae at Pasoh FR (redrawn from LaFrankie & Chan 1991).

reproduction within genera; (2) differences in spatial location of reproductive display or size at onset of maturity; and (3) differences in spatial distribution and/or edaphic preference.

3.1 Differences in reproductive timing

One well-documented mechanism of reproductive isolation in some trees represented at Pasoh FR is differentiation in flowering phenology. Observations made during the 1976 mast flowering first described a remarkable pattern of sequential flowering in *Shorea* section Muticae (Ashton et al. 1988; Chan & Appanah 1980). The six species in this group were observed to flower sequentially over a 10 week period during a major masting event in 1976, with the peak of flower production and presumed anthesis in each species differing by ~5–10 days (Fig. 2). Differentiation of flowering in this same set of species was again observed during the 1989 masting event (LaFrankie & Chan 1991). Moreover, the same precise sequence of flowering was observed during both events (Fig. 2).

The flowering sequence in *Shorea* section Muticae has been known for more than 20 years. Do closely-related sympatric trees show similar patterns in other taxa? A two and half year study of reproductive phenology of several speciose tree genera at Pasoh FR, revealed several similar patterns of temporal segregation of flowering (Thomas 1993; Fig. 3). In 1990, *Diospyros scortechinii*, *D. apiculata*,

* The genus *Aporosa* has commonly been referred to as *Aporusa* in prior studies. Schot (1995), by reference to the original manuscripts describing the genus, found that the former name is correct under the international rules of botanical nomenclature.

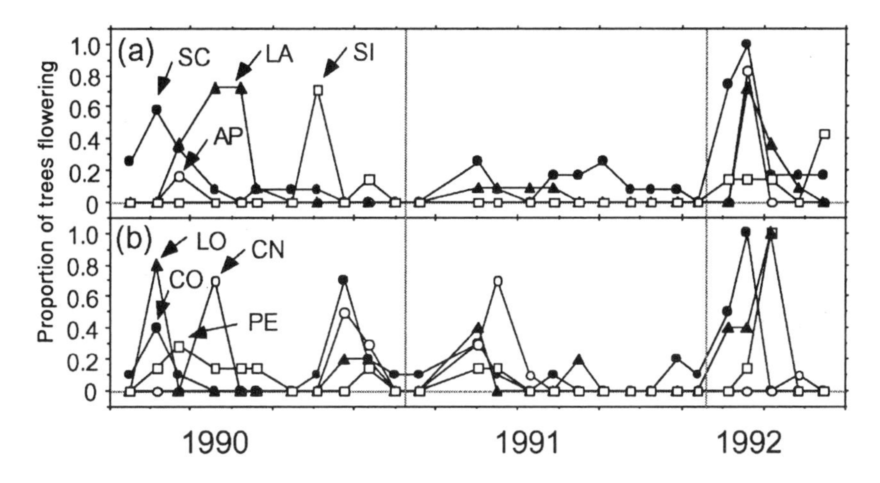

Fig. 3 Flowering phenology in sympatric species of (a) *Diospyros*, and (b) *Ixora* at Pasoh FR, based on monthly surveys. *Diospyros* species include *D. scortechinii* (abbreviated SC), *D. apiculata* (AP), *D. latisepala* (LA), and *D. singaporensis* (SI). *Ixora* species include *I. concinna* (CO), *I. lobbii* (LO), *I. pendula* (PE), and *I. congesta* (CN). In each case, the set of congeneric species show evidence of temporal segregation of flowering, and a similar sequence of flowering among species between flowering events. Data from Thomas (1993).

D. latisepala and *D. singaporensis* flowered sequentially over a 7-month period. In 1991, only a low level of flowering occurred in these taxa, but in 1992 the same sequence of flowering was also observed, though with a greater degree of overlap among the first 3 species in the sequence. *Ixora* species, which flower frequently and conspicuously in the understory at Pasoh FR, also exhibited temporal segregation of flowering. *Ixora congesta* and *I. lobbii* share morphologically similar showy orange inflorescences. During 3 consecutive flowerings in Feb.–May of 1990, 1991 and 1992, the flowering peak in *I. lobbii* preceded that of *I. congesta* by ~1 month (*I. concinna* may precede *I. lobbii* in this sequence, but more frequent observations are needed to confirm such a pattern). Two other white-flowered *Ixora* species, presently classified as varieties of *I. grandifolia* (*I. grandifolia grandifolia* Zoll. & Mor. and *I. grandifolia lancifolia* Corner), showed temporally distinct flowering peaks as well. *I. grandifolia grandifolia* flowered in Dec. 1989 and Jan. 1991, while *I. grandifolia lancifolia* showed flowering peaks in Mar. 1990, Sep. 1990, Mar. 1991 and Aug. 1991.

3.2 Vertical stratification and reproductive display

Another mechanism by which species may maintain reproductive isolation is through reliance on pollinators that are active at different locations in the forest canopy. Vertical stratification of pollinators has been observed in many tropical forests, both among major taxonomic groups (Basset et al. 1992; Kato 1996; Kato et al. 1995) and within taxa, including butterflies (Beccaloni 1997), and ambrosia beetles (Chap. 31), among others. In SE Asia there appears to generally be strong divergence in pollinators between the upper canopy and emergent strata versus sub-canopy and understory trees (Appanah 1990; Chan & Appanah 1980; Momose et al. 1998;

Yap 1976, 1982). There is evidence that the small-bodied insect group thysanaptera (thrips) are important pollinators of some dipterocaps (Chan & Appanah 1980); however, Hemiptera (Appanah 1990) and small beetles (Sakai et al. 1999) are also involved. Large-bodied bees (*Apis dorsalis* and *Xylocopa* spp.) are important pollinators in other tree taxa, such as emergent legumes in the genera *Koompassia* and *Sindora*. A wide variety of insect species have been implicated as pollinators of sub-canopy and understory tree taxa (Kato 1996). Some particularly important mid- and understory pollinators include bees of the genera *Trigona* and *Amegilla*, and also nectarivorous birds such as the abundant purple-naped sunbird (*Hypogramma hypogrammicum*).

Because of this vertical differentiation of pollinators, we should expect species which differ in terms of size at reproductive onset and/or vertical location of reproductive display to be effectively reproductively isolated. There are, in fact, dramatic examples of sibling species that differ dramatically in terms of size and vertical location of the reproductive display. The two species presently classified as varieties of *Ixora grandifolia* differ greatly in size as well as in the timing of flowering and fruiting. *I. grandifolia grandifolia* attains heights of ~24 m and flowers at ~12 cm DBH, while *I. grandifolia lancifolia* is a shrub reaching only ~5 m in height that flowers at <1 cm DBH (Thomas 1996a,b). Another example of interest is closely-related species that differ in terms of reproductive display. Numerous trees in the Sundaland region are cauliflorous, bearing reproductive structures on the stem or base of the tree. The species *Baccaurea parviflora* and *B. racemosa* are very similar in leaf structure and other aspects of morphology, but the flowers and fruits of the smaller *B. parviflora* are displayed near the base of the tree, while reproductive structures of *B. racemosa* are displayed in the crown and upper parts of the stem (Thomas & LaFrankie 1993). Other tree genera at Pasoh FR which include both cauliflorous and non-cauliflorous members include *Diospyros* (Ebenaceae), *Polyalthia* (Annonaceae), *Saraca* (Leguminosae), and *Ficus* (Moraceae). That cauliflory evolved as a means to better attract understory pollinators is an idea dating back to early observations of Wallace (1878).

3.3 Differences in spatial/edaphic distributions

A third important potential mechanism for reproductive isolation is via habitat segregation. The importance of habitat differentiation with respect to edaphic and hydrological characteristics has been emphasized in prior discussions of speciose taxa in SE Asia (Ashton 1969, 1977; Rogstad 1989, 1990; Van Steenis 1969). Members of the monophyletic *Polyalthia hypoleuca* complex show clear differentiation with respect to edaphic variables, though secondarily through differences in stature (Rogstad 1990: Fig. 4). Direct observations on pollinators and experimental cross-pollination studies have also demonstrated reproductive isolation of species pairs within this group (Rogstad 1994).

Detailed studies of closely-related species at Pasoh FR have, in at least a couple instances, detected additional biological species that were incorrectly "lumped" under the same morphospecies. *Aporosa falcifera* as identified during the initial 50 ha plot census is in fact comprised of 2 distinct types: *Aporosa falcifera* sensu lato, and a new taxon (labeled *Aporosa* "dark" sp. nov. in prior publications) not matching prior collections in Peninsular Malaysia. The new form is found essentially only in riparian areas with restricted drainage, and differs dramatically in bark characteristics (*Aporosa falcifera* sensu lato has whitish bark; *Aporosa* "dark" sp. nov. has dark brown, almost black bark on mature individuals).

Collections of the new form also indicate substantial differences in floral morphology (*Aporosa* "dark" sp. nov. possesses highly lobed petals on pistillate flowers, and larger flowers overall). The leaves of the two species are very similar, but leaves on saplings of *Aporosa* "dark" sp. nov. possess extrafloral nectaries that *Aporosa falcifera* sensu lato lacks). Other morphologically-defined taxa that appear to include more than one biological species include members of the genus *Garcinia* (*G. parvifolia* [see above]; *Garcinia* "small" sp. nov., mis-identified as *Garcinia eugeniifolia*, also includes some individuals of *G. opaca*).

To summarize, studies at Pasoh FR strongly suggest that most tree species described on a morphological basis appear to correspond to good biological species. More specifically, closely-related species pairs tend to be differentiated in terms of reproductive timing, vertical stratification and display of floral parts, and/or with respect to soil variables. The detailed investigations of speciose tree genera at Pasoh FR also provide some evidence that there may be a greater number of biological species present than have been detected using morphological characteristics alone. Detailed studies of tree population ecology within narrow taxonomic groups have lead to the discovery of putative "cryptic" species that were not detected on the basis of morphological assessments. It is almost certainly the case that additional studies of other groups would result in discovery of additional cryptic taxa that would otherwise go undetected and undescribed.

4. ECOLOGICAL DIFFERENTIATION AT PASOH FR

A number of authors have argued that most tropical trees are ecological generalists, showing relatively little differences among species (e.g. Hubbell & Foster 1986; Pitman et al. 1999). Such a pattern would be consistent with "null models" of tropical forest dynamics, which assume that all species are ecologically equivalent (Hubbell 1997, 2001; Leigh et al. 1993; Vandermeer 1996). In direct contradiction, many empirical studies have documented differences in habitat associations, life-history characteristics, and other traits of tropical trees. In a general sense, "ecological differences" constitute a potentially infinite set of contrasting

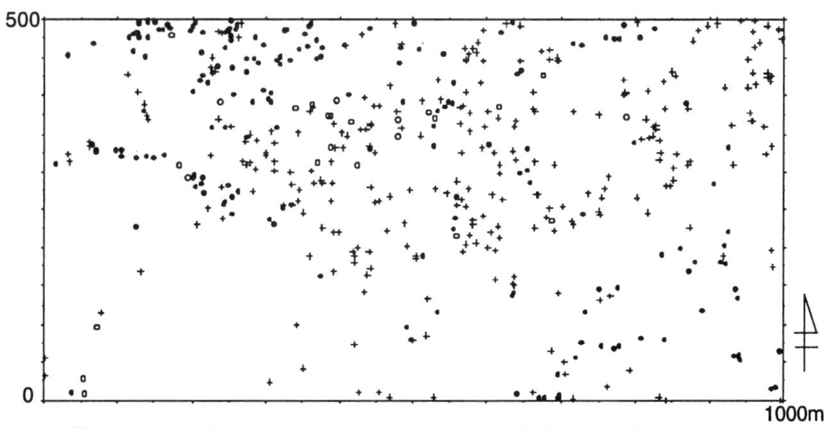

Fig. 4 Spatial distributions of tree species in the *Polyalthia hypoleuca* complex in the Pasoh FR 50 ha plot (500 m × 1000 m), confirming earlier observations of Rogstad (1989, 1990). Filled circles = *P. glauca*; Open circles = *P. hypoleuca*; Crosses = *P. sumatrana*. Data from the first census (1985–1987).

morphological and physiological traits, as well as biotic interactions. To effectively compare species and understand the potential importance of observed differences, we need at the very least a sensible classification scheme for "kinds" of tree niche differences that may be of importance.

In a widely-cited review, Grubb (1977) described four different kinds of niches in plant communities. The "habitat niche" constitutes the range of physical and chemical conditions tolerated by the mature phase of a given species. Familiar examples in tropical forests would thus include differences in distribution related to soil type or elevation. The "life-form niche" (sometimes referred to as the "structural niche") refers to the potential for physical separation of species in space, particularly in the vertical dimension or as a consequence of space or other resources generated by other plants. Examples of structural niche differentiation may thus include differences in vertical stratum occupied by mature trees, or in substrate requirements related to the presence of other plants, as in the case of epiphytes or saprophytes. The "phenological niche" is related to differing temporal patterns of resource use, with spring ephemeral vs. later-emerging understory species in temperate deciduous forests often mentioned as an example. Finally, Grubb's notion of the "regeneration niche" concerns all phenotypic characters related to replacement of adult individuals through the plant's life cycle. Gap partitioning related to differences in species' shade-tolerance is the paradigmatic example of differentiation of the regeneration niche. In the following sections I make use of Grubb's categorization to briefly review prior work at Pasoh FR and elsewhere in the Sundaland region of SE Asia on each of these aspects of ecological variation.

4.1 The habitat niche

The lowland forest area encompassed by Pasoh FR lies on a gently rolling topography, and includes soils derived from both sedimentary rocks (Triassic shales with interbedded sandstone and quartzite), and from fluviatile granitic alluvium in low-lying areas (Allbrook 1973; Ashton 1976; Manokaran & Swaine 1994). The large hill (Bukit Palong) at the east side of the reserve is composed of acidic granite, and possesses a markedly different red ultisolic soil. Areas within the eastern part of the lowland area of the reserve show some influence of the acidic granite parent material, notably in the far northeast corner of the 50 ha plot.

Quantitative analyses of tree spatial distributions in relation to site characteristics include early studies based on a set of 10 small (0.4–2 ha) plots enumerating trees > 30 cm DBH (Ashton 1976; Wong & Whitmore 1970). Several of these plots were later incorporated into the 50 ha CTFS (Center for Tropical Forest Science) forest dynamics plots at Pasoh FR. On the basis of a Principle Components Analysis (PCA) of 10, 20 m × 200 m plots, Wong & Whitmore (1970) concluded there was no evidence for correlations between species distributions and soil types within the reserve, and that local floristic differences were driven mainly by clumping due to limited dispersal. This conclusion was also consistent with similar studies elsewhere in Peninsular Malaysia around this time (Lee 1968; Poore 1968). Ashton (1976) extended a subset of the plots used by Wong & Whitmore, and conducted a set of association analyses using sub-divisions of the plots. In contrast to Wong & Whitmore (1970), these analyses supported the existence of consistent variations in floristic composition with soil type. In particular, a number of common understory trees emerged as the best indicators of soil type in the data. For example, *Aporosa globifera* (mis-identified as *Aporusa pseudoficifolia* in Ashton 1976), was the most consistent indicator of alluvial soils within the data set.

The 50 ha mapped forest plot at Pasoh FR provides a vast amount of data to examine segregation of species distributions relative to edaphic variation (Manokaran et al. 1992b). Quantitative analyses have previously been presented examining spatial aggregation within individual species (Condit et al. 2000; He et al. 1997), and spatial patterning of tree diversity in the plot (He et al. 1994, 1996). Quantitative analyses of habitat correlations of individual species have been conducted in a few cases (I. Debski, pers. comm.), and a more comprehensive analysis using randomization methods is planned. Here, as a preliminary step, I present a categorization of species based on subjective interpretations of the mapped 50 ha distributions, and also drawing on personal knowledge of local distributions outside the plot area (Table 2). Although not a quantitatively rigorous evaluation, the subjective evaluation presented takes advantage extensive field experience in the area.

Among the commoner species, roughly 60% appear to be "habitat generalists", found throughout the plot area, while ~40% are "habitat specialists" occurring predominantly in areas that correspond closely to soil or hydrological characteristics. The two most common kinds of edaphic specialization are related to riparian habitat and soil drainage. About 17% of the commoner tree species are restricted to riparian areas, and a similar proportion of species are restricted to upland areas with little or no individuals found in riparian zones. A much smaller proportion of the tree flora appears to show strong specialization related to soil type. About 4% of species were mainly or entirely restricted to sandy alluvial soils that predominant in the southwestern part of the plot. Only ~1% of species were restricted to the distinctive red granitic ultisolic soils found in the far northeastern corner of the plot. Examples of each of these patterns are shown in Fig. 5.

Several more subtle aspects of relationships between species distributions and edaphic characteristics are worthy of comment. First, the classification presented was based on overall distributions of all trees > 1 cm DBH. However, there is evidence that adult trees commonly show a higher degree of habitat specificity than do saplings (Webb & Peart 2000). Size-class specific analyses would therefore likely indicate a higher proportion of species showing strong edaphic preferences. Second, a wide variety of kinds of distribution are included in the each of the categories. In particular, the "upland" species are a very heterogeneous group. Some "upland" species are absent from riparian areas only (e.g. *Rinorea anguifera* in Fig. 5); however, many others also show some bias with

Group	Count	%	%*
Generalists	354	42	61
Edaphic specialistss			
Upland	98	12	17
Riparian/swamp	97	12	17
Alluvial soils	22	3	4
Granitic ultisols	7	1	1
Rare species	256	31	
Known problem taxa	3	0	

* Excluding rare species and known problem taxa.

Table 2 Categorization of habitat preferences of tree species enumerated within the 50 ha plot at Pasoh FR. Categories are based on a subjective evaluation of the mapped plot data (Manokaran et al. 1992b), aided by personal knowledge of spatial distributions elsewhere in the reserve and in nearby forests. Examples of representative distribution types are given in Fig. 5.

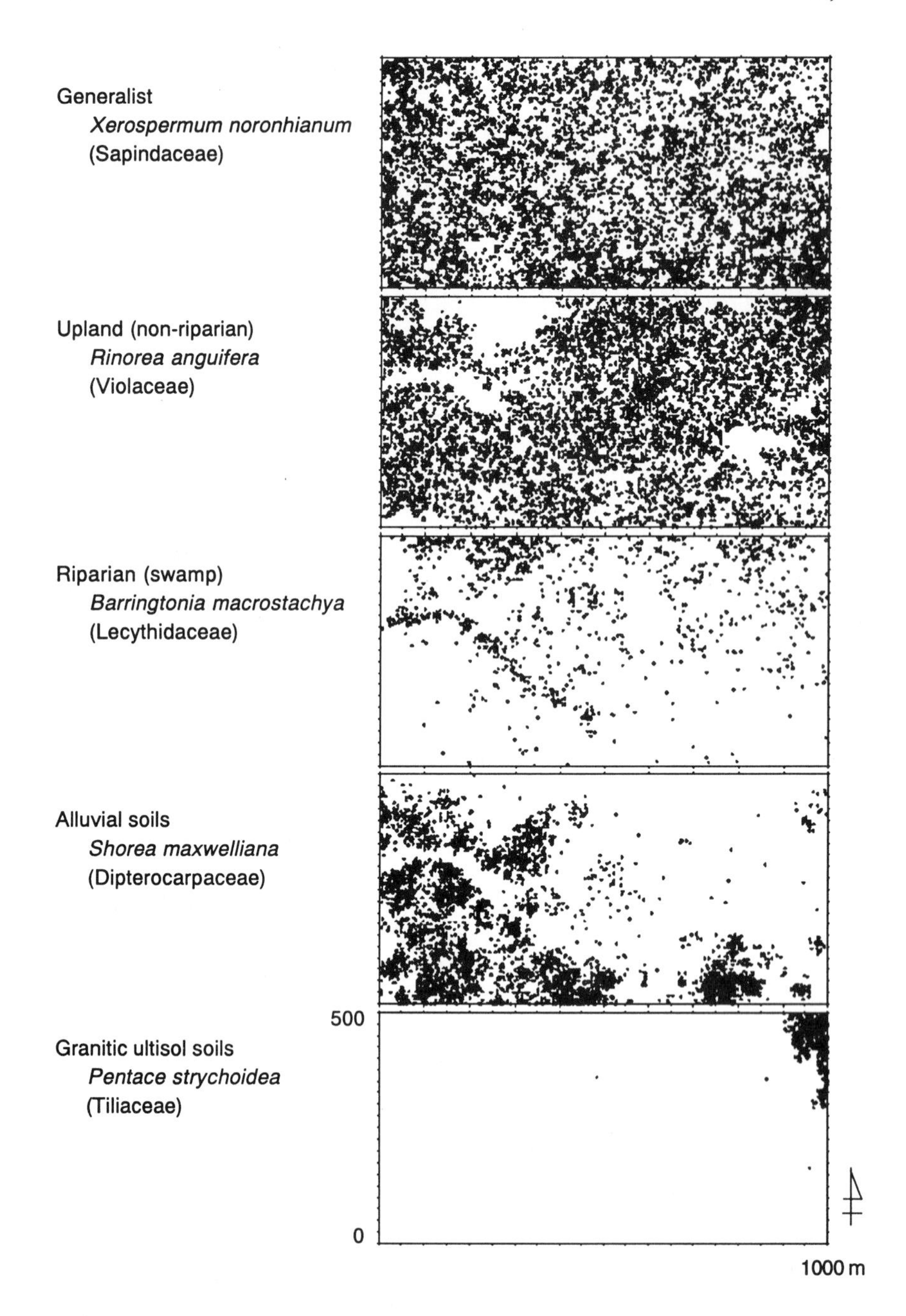

Fig. 5 Examples of the kinds of relationships between edaphic variables and species distributions found within the 50 ha plot at Pasoh FR, based on first (1985–1987) census.

respect to soil type and/or topography, such as on slopes or on hilltops. Likewise, "riparian" species include some that are restricted to larger periodically-inundated floodplains, others that occur only in or near intermittent streams, and some species that occur mainly around the margin of riparian zones (e.g. *Garcinia scortechinii*– see Thomas 1997). Many riparian species include a few outlying individuals that occur mainly on hilltops, possibly indicating the importance of soil oxygen content as a physiological mechanism determining distributions in many riparian species.

The figure of ~40% of the flora showing strong edaphic specialization is remarkably similar to results of quantitative analyses previously reported. Hubbell & Foster (1986) found significant habitat associations in 36% (15 of 41) common tree species at Barro Colorado Island. In another Sundaland forest (Gunung Palung Park, Indonesian Borneo), Webb & Peart (2000) report significant habitat associations in 42% (21 of 49) species examined.

4.2 The life-form (or structural) niche

Ecological differences related to life-form are obviously of critical importance in any analysis of total plant diversity in any forest ecosystem. The non-tree flora of Pasoh FR has received considerable attention, including studies of the lianas (Appanah et al. 1993), herbaceous plants (Wong 1981, 1983), pteridopytes (Sato et al. 2000) and cryptogams (Wolseley et al. 1996).

Tree species size can be quantitatively evaluated as the asymptotic maximum height attained by a given species (Thomas 1996a). In theory, this would be best accomplished by following height growth of a large cohort of trees through time, and estimating the average maximum height attained prior to senescence. In practice, a nearly unbiased estimate of this value may be calculated using static allometric relationships between tree height and diameter, taking advantage of the fact that while height growth virtually ceases late in ontogeny, diameter increment continues to increase (Thomas 1996a). Such relationships were used to estimate the "asymptotic" maximum height for a total of 42 tree species at Pasoh FR (Thomas 1996a; Thomas, unpubl. data). These estimates were in turn compared with species-specific diameter distributions in the 50 ha plot to generate a regression function to

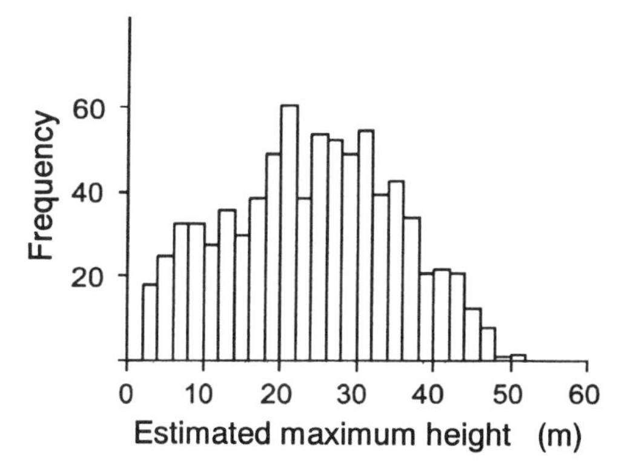

Fig. 6 Distribution of estimated asymptotic height (H_{max}) values for species occurring within the 50 ha plot at Pasoh FR. H_{max} was estimated for each species as a function of the 97th percentile of diameter (using the 50 ha plot data). The relationship used was derived on the basis of allometric estimates of H_{max} for 42 tree species (Thomas 1993, 1996a).

estimate asymptotic height for all of the species represented in the plot (Fig. 6). Asymptotic species heights show wide variation among trees represented in the 50 ha plot at Pasoh FR, with a range of ~3–52 m (note that some individual trees at Pasoh FR attain heights in excess of 60 m). There is no clear peak at what would be considered canopy height (30–40 m); rather, the modal species is an understory tree with a maximum height near 20 m.

There are two major hypotheses as to how differences in tree species size may facilitate species coexistence. Kohyama's "forest architecture hypothesis" posits that small understory trees may coexist with larger species by taking advantage of canopy gaps and completing their life cycles before large species are able to displace them from a given gap (Kohyama 1993). A more traditional view implicit in many early studies of forest stratification is that understory trees are simply able to persist and successfully reproduce under the shade of canopy trees, and are not dependent on canopy gaps. This idea is consistent with physiological studies that have emphasized high efficiency of carbon gain of understory trees under low light conditions (Givnish 1988; Thomas & Bazzaz 1999).

Comparative studies at Pasoh FR have also shown that many aspects of growth, physiology, and life history are correlated with differences in maximal tree height (Table 3). In essentially every case, smaller-statured understory trees show characteristics associated with physiological efficiency under low light conditions, rather than an ability to rapidly grow and reproduce in transient light gaps. In particular, smaller understory trees tend to have lower growth rates, lower mortality rates, lower maximum photosynthetic rates, but a higher efficiency of carbon gain under low light levels (Table 3; Thomas & Bazzaz 1999). That these trends represent "adaptations" that have independently evolved in multiple clases has also been supported by phylogenetically-corrected analyses (Thomas 1996a,b,c; Thomas & Bazzaz 1999).

Table 3 Some characteristics associated with asymptotic height in species of *Aporosa*, *Baccaurea*, *Diospyros*, *Garcinia*, and *Ixora* studied at Pasoh FR (in some cases only a subset of these groups are included). Coefficient of determination (R^2) values are given for linear regressions where species are treated as variates: * $P < 0.05$; ** $P < 0.01$; *** $P < 0.001$. All relationships are positive in slope (increase in character with increase in asymptotic height) with the exception of wood density and initial slope of H-D allometric relationships, which are higher in smaller- than larger-statured trees.

Character	R^2	reference
Growth rate (trees 1-2 cm)	0.202**	1
Growth rate (adult trees)	0.565***	1
Wood density	0.330*	1
Initial slope of H-D allometric relationship	0.454**	1
Size (dbh) at onset of maturity	0.582***	2
Strength of correlation between staminate flower production and size (DBH)	0.452**	3
Degree of ontogenetic change in leaf size	0.336***	4
Mortality rate (trees < 2 cm)	0.696**	5
Photosynthetic capacity (area basis)	0.556***	6

References: 1 (Thomas 1996a), 2 (Thomas 1996b), 3 (Thomas 1996c), 4 (Thomas & Ickes 1995), 5 (Thomas 1993), 6 (Thomas & Bazzaz 1999).

4.3 The phenological niche

The examples most commonly given for differentiation of the phenological niche in plants are from the temperate zone, namely spring ephemeral understory plants or winter vs. summer annuals. In the more nearly aseasonal environment at Pasoh FR, there would appear to be little scope for such extreme forms of phenological niche separation, in which a given species is physiologically active during time periods when competing species are quiescent. However, more subtle differences in ecological timing may be important. One influential hypothesis for the maintenance of species diversity in tropical forests and similar systems has been the "storage effect" (Chesson & Warner 1981; Warner & Chesson 1985). Briefly, this hypothesis postulates that tree species differ in the timing of reproduction, such that there is no temporal correlation in reproductive output among species, and that the first-arriving individual in a given location is able to monopolize that space. In addition, reproductive output is considered an attribute of a species (not of an individual), implying that a single surviving member of a given species can produce a number of propagules equal to the most common species in the system. The lack of correlation in reproduction among species must be driven by some difference in species responses to environmental variation (if species were ecologically identical, they should show a temporal correlation in reproduction that would approach unity). For this reason, the "storage effect" mechanism may be considered a special type of species coexistence resulting from differences in the phenological niche.

The forests of SE Asia, Pasoh FR included, are renowned for the phenomenon of mass flowering and mast fruiting at multi-year intervals (Ashton et al. 1988; Curran et al. 1999; Janzen 1974; Toy 1991; Toy et al. 1991). What is almost certainly the longest-term data on flowering and fruiting patterns has been collected at Pasoh FR and several other forest reserves in the south-central area of the Malay peninsula (Yap & Chan 1990). While these data show a strong interannual masting signal, there are numerous species, even among the dipterocarps, that flower and fruit between masting events. Some even show annual or near-annual fruiting (such as *Dipterocarpus crinitus* and *Neobalanocarpus heimii* - see also Chap. 15). Understory trees, on the other hand, commonly show relatively frequent reproduction that is not synchronized with masting cycles (Fig. 3; Thomas 1993; Wong 1983; Yap 1976). While some understory taxa, such as many members of the Euphorbiceae, show a strong annual cycle (e.g. *Aporosa* and *Baccaurea*: Thomas & LaFrankie 1993), other understory trees and shrubs lack a clear annual pattern (Fig. 3). The most obvious of the frequent-flowering taxa include members of the Rubiaceae, such as *Ixora* spp. and the nearly continuous-flowering *Porterandia anisophylla*. However, there are many other taxa with less showy floral displays that also show this pattern.

In sum, phenological studies at Pasoh FR suggest that temporal segregation of reproduction is common, in spite of the existence of the mast flowering cycle. There is thus some potential importance of a "storage effect" mechanism in the system, though its relative importance is very much unclear.

4.4 The regeneration niche

A relatively small set of common "pioneer" species are found in roadside regenerating forests near Pasoh FR, the most typical of which include *Arthrophyllum diversifolium*, *Fagraea racemosa*, *Ficus fistulosa*, *F. glossularioides*, *Macaranga gigantea*, *M. hypoleuca*, *M. triloba*, *Mallotus paniculatus*, *Melastoma malabathricum*, *Trema tomentosa* and *Vitex pinnata*. All of these species are either absent or rare within the 50 ha plot of primary forest at Pasoh FR, occurring as saplings mainly in large gaps. In total, only ~1.4% of stems and ~3% of species enumerated in the 50 ha plot represent pioneer species (Thomas 2003).

Although the vast majority of species at Pasoh FR fall into the category of "primary forest trees", there are certainly differences in light requirements for growth and other components of "shade tolerance". Timber trees in the region, in particular the dipterocarps, have long been categorized into shade tolerance classes using wood density as a proxy measure (cf. Appanah & Weinland 1993). The most important "heavy hardwood" species at the site, characterized by very high shade tolerance and slow growth, include *Neobalanocarpus heimii*, and *Shorea maxwelliana*. Common species of "medium hardwoods" include *Dipterocarpus* species (such as *D. cornutus*, *D. costulatus* and *D. crinitus*). The silviculturally-important red merantis (including *Shorea leprosula*, *S. lepidota*, *S. parviflora* and *S. acuminata*) are among the most abundant light hardwoods at the site, and are characterized by a high demand for light but very rapid early growth. These pronounced ecological differences among the Dipterocarps are very well known, and form an essential part of the working knowledge of foresters in Peninsular Malaysia (Appanah & Weinland 1993).

Moad (1993), working at Pasoh FR and Sepilok Forest Reserve (Sepilok FR) in Sabah, provides a considerably more precise description of differences in shade tolerance and related characteristics of common Dipterocarps at the two sites. This study documents that the light-demanding species show lower survivorship rates in the understory and a higher minimum light requirement for growth (the latter analyzed on the basis of relationships between seedling stem extension and incident light, evaluated using hemispherical photograph analysis). More light-demanding species also showed a steeper increase in growth rate in response to increasing light levels than did more shade-tolerant species. These differences in growth patterns corresponded closely to differences in photosynthetic light responses, with light-demanding species showing higher maximum photosynthetic rates, but also higher respiration rates and a higher leaf-level light compensation point than did more shade-tolerant species (Fig. 7; Moad 1993).

Although differences in light requirements for growth have been quantified, there has been relatively little work to date on gap-phase regeneration at Pasoh FR. It is clear that physiological differences exist, as do differences in mortality and growth rates (Manokaran et al. 1992a), and that these differences correspond well with the traditional species groupings based on wood density.

5. DISCUSSION

Both historically (e.g. Ashton 1969; Klopfer & MacArthur 1961; Richards 1969) and in the recent literature (e.g. Clark et al. 1998; Davies et al. 1998; Kobe 1999; Svenning 2000; Yamada et al. 2000), many researchers have sought to explain species diversity in tropical forests in terms of "niche" differences in patterns of resource use and habitat partitioning. The studies reviewed here broadly support the idea that tree species at Pasoh FR tend to show pronounced differences in a wide variety of ecological characteristics. In particular, differences in maximum tree size and in spatial distribution relative to soil hydrology appear to be of considerable importance. Current evidence on "regeneration niche" differences is less strong, with, for example, only a few "pioneer" trees present that recruit only in gaps. Also, while differences in light requirements for growth have been demonstrated in some groups, such as the dominant dipterocarps (Moad 1993), the significance of such differences to community structure has received little attention.

Demonstrations of ecological differences among tropical tree species suggest that niche differences may contribute to the maintenance of diversity, but not that they actually do (Tilman & Pacala 1993). There is thus an obvious need for modeling studies that would take into account the kind of ecological differences found at Pasoh FR, and make projections of the long-term consequences of these differences to forest community structure and dynamics. Empirical data from Pasoh FR has been the basis of several important forest simulation models (Appanah et al. 1990; Bossel & Krieger 1991, 1994; Huth & Ditzer 2001; Liu & Ashton 1998, 1999; Oikawa 1985). However, to date such models have lumped species into broad categories. Accordingly, these models have not addressed issues of species coexistence directly. Although a strong modeling effort examining community processes and species diversity is currently lacking, it is still instructive to speculate how ecological variation described here may affect long-term dynamics, specifically in comparison to null models of tropical forest dynamics (Hubbell 2001).

Species-neutral models generally make two main assumptions. First, that community dynamics are a stochastic, zero-sum game; and second, that species are equivalent in terms of the main parameters describing this game–namely, rates of birth, death, and dispersal (Hubbell 2001). It is obvious that the second assumption does not strictly hold at Pasoh FR. Trees at Pasoh FR are not equivalent, but rather diverge markedly in terms of vital rates and dispersal patterns. As

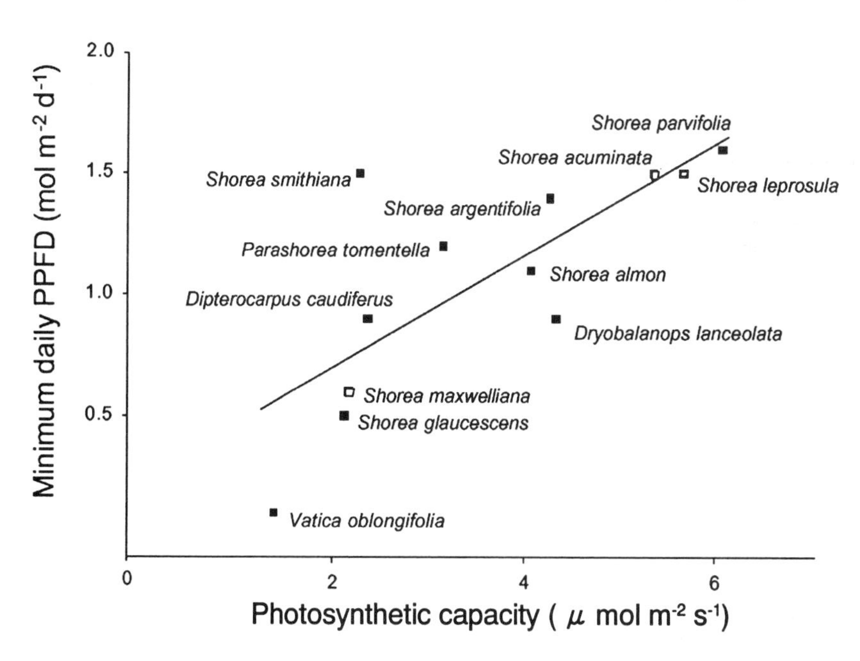

Fig. 7 Relationship between minimum daily light requirement for growth and photosynthetic capacity for dipterocarp species studied at Pasoh FR (open symbols), and Sepilok FR (closed symbols). Minimum daily photosynthetic photon flux density for growth was estimated on the basis of regressions between stem extension growth and light levels estimated using hemispherical photograph analyses. Photosynthetic capacity of shade-acclimated seedlings were measured in situ at the two study sites. Linear regression is significant at $P = 0.004$, $R^2 = 0.584$. Data from Moad (1993).

illustrated above, a high proportion of species at Pasoh FR (> 40%) are largely or entirely restricted to one or more habitat types (Table 3, Fig. 5), and species vary widely in stature (Fig. 6), in timing of reproduction (Fig. 3), and in shade tolerance (Fig. 7). In fact, recent analyses have suggested that the Pasoh FR flora is especially variable in terms of demographic parameters shown by individual species (Condit et al. 1999). It has been conjectured (cf. Hubbell 2001), that some patterns of ecological variation, such as correlated variation along an axis of shade-tolerance to shade-intolerance, result in an effective equivalence of vital rates across species. However, a formal test of this conjecture has not been made. An alternative possibility is that variation in vital rates will essentially add additional stochasticity to the zero-sum random walk shown by a given species. Rather than enhancing species diversity, as per the traditional view, variation among species could thus potentially result in a reduction in species diversity over the long term. The jury would appear to be out on this fundamental point.

A less obvious contradiction between the patterns reviewed here and recently-developed null models concerns species size. The idea of the "zero-sum random walk" presupposes a strong form of local competition in which one and only one tree can occupy a given point in space (Hubbell 2001). However, one important result of comparative studies at Pasoh FR is that tree species show very wide variation in maximal size (Thomas 1996a; Fig. 6), and in size at onset of reproduction (Thomas 1996b). The modal tree species at Pasoh FR is an understory tree that attains a height of ~20 m, far below the crowns of the large emergent species that may reach 50–60 m. Viewed in terms of crown area occupied by an individual, reproductively mature tree, one could pack ~32 individuals of the largest trees occurring at Pasoh FR in 1 ha, several hundred more typical "canopy" trees, or several thousand sub-canopy trees and treelets (Thomas 2003). Of course, most understory taxa are able to complete their life-cycles in the absence of canopy gaps, so more than one tree species can and does effectively occupy the same horizontal space. It is not clear how a "zero-sum random walk" model could be effectively modified to incorporate the important biological phenomenon of understory trees. It may be premature to state that these basic observations effectively falsify Hubbell's model. However, I do conclude that some of the most important kinds of ecological variation observed at Pasoh FR present very serious "problems" for null models of tropical forest dynamics.

Beyond an "academic" interest in understanding how tropical forests work, studies in comparative ecology of tropical trees are essential to inform practical matters of forest management and conservation. This becomes especially clear when one considers specific environmental changes that are now or have already been experienced by the Pasoh FR tree flora. Such changes include impacts of selective logging (Chap. 2), forest edge formation and fragmentation (Liu et al. 1999; Thomas 2003), effects of the regional haze phenomenon, including reductions in light levels (Tang et al. 1996), and increased densities of wild pigs (*Sus scrofa*), with consequent physical disturbance of understory vegetation (Ickes 2001; Ickes et al. 2001; Chap. 35). Other impacts are likely to also include changes in local hydrology and climate related to local and regional land-use patterns, increased deposition of nitrogen and acid rain resulting from regional anthropogenic pollution, and impacts of globally rising levels of carbon dioxide. To understand the impacts of any of these changes, the first thing one would wish to know is how variable species responses to this impact are, and which species are likely to suffer. It is instructive that analyses of both increased pig densities, and of edge and fragmentation effects, suggest that the dominant dipterocarps at Pasoh FR, as a group, exhibit some of the strongest negative impacts (Thomas 2003; Chap. 35).

More than 20 years ago, P. W. Richards emphasized the importance of Pasoh FR as a remnant natural forest area and unique source of scientific information (Richards 1978). While the early work at Pasoh FR during the IBP era emphasized ecosystem processes and productivity, the present review illustrates the fact that comparative studies of tree biology have been a major focus of research since then. It is to be hoped that this long history of work by many scientists will ultimately result in on-the-ground improvements in forest management and conservation. Certainly any attempt to understand and predict the future of Pasoh FR, and similar remnants of lowland forest throughout the tropics, must be based on an understanding of ecological differences among the tree species that comprise the main structural elements of the forest. Even if we were satisfied that a neutral model of equivalent species was sufficient to "explain" tree species diversity in the primeval forest, there is little reason to think that such a null model would usefully predict how the forest is likely to change in future in the face of anthropogenic change. It's also incumbent on tropical forest biologists to ultimately consider how tropical forest areas can be increased in the future. Once, on a long walk through the core area of Pasoh FR, I heard it said that "each tree has its story". The descriptive biology of exploring such "tree stories" may not be fashionable biology; however, I suggest that the body of descriptive silvics data yielded by such studies may ultimately be among the most valuable and practical contributions to emerge from the forest at Pasoh FR.

ACKNOWLEDGEMENTS

The large-scale forest plot at the Pasoh FR are on going project of the Malaysian Government, and were initiated by the Forest Research Institute Malaysia (FRIM) through its former Director-General, Dato' Dr. Salleh Mohd. Mor, and under the leadership of Drs. N. Manokaran, P. S. Ashton and Stephen P. Hubbell. Supplementary funding was provided by the National Science Foundation (USA); the United Nations, though it is Man and the Biosphere (MAB) program; UNESCO-MAB grants; UNESCO-ROSTSEA; and the continuing supports of the Smithsonian Tropical Research Institute, Panama and National Institute for environmental Studies (NIES). D. Kenfack, R. Condit, and A. Moad are thanked for permission to present unpublished data.

REFERENCES

Allbrook, R. F. (1973) The soils of Pasoh Forest Reserve, Negeri Sembilan. Malay. For. 36: 22-33.

Appanah, S. (1990) Plant-pollinator interactions in Malaysian rain forests. In Bawa, K. S. & Hadley, M. (eds). Reproductive Ecology of Tropical Forest Plants. UNESCO, Paris, pp.85-102.

Appanah, S. & Weinland, G. (1993) Planting quality timber trees in Malaysia. Malaysian Forest Record No. 38. Forest Research Institute Malaysia, Kepong, Malaysia.

Appanah, S., Gentry, A. H. & LaFrankie, J. V. (1993) Liana diversity and species richness of Malaysian rain forests. J. Trop. For. Sci. 6: 116-123.

Appanah, S., Weinland, G., Bossel, H. & Krieger, H. (1990) Are tropical rain forests non-renewable? An enquiry through modelling. J. Trop. For. Sci. 2: 331-348.

Ashton, P. S. (1969) Speciation among tropical forest trees: some deductions in the light of recent evidence. Biol. J. Linn. Soc. 1: 155-196.

Ashton, P. S. (1976) Mixed dipterocarp forest and its variation with habitat in the Malayan

lowlands: a re-evaluation at Pasoh. Malay. For. 39: 56-72.
Ashton, P. S. (1977) A contribution of rainforest research to evolutionary theory. Ann. Mo. Bot. Gard. 64: 694-705.
Ashton, P. S. (1988) Dipterocarp biology as a window to the understanding of tropical forest structure. Annu. Rev. Ecol. Syst. 19: 347-370.
Ashton, P. S., Givnish, T. J. & Appanah, S. (1988) Staggered flowering in the Dipterocarpaceae: New insights into floral induction and the evolution of mast fruiting in the aseasonal tropics. Am. Nat. 132: 44-66.
Basset, Y., Aberlenc, H. & Delvare, G. (1992) Abundance and stratification of foliage arthopods in a lowland rain-forest of Cameroon. Ecol. Entomol. 17: 310-318.
Beccaloni, G. (1997) Vertical stratification of ithomiine butterfly (Nymphalidae: Ithomiinae) mimicry complexes: the relationship between adult flight height and larval host-plant height. Biol. J. Linn. Soc. 62: 313-341.
Becker, P. & Wong, M. (1985) Seed dispersal, seed predation, and juvenile mortality of *Aglaia* sp. (Meliaceae) in lowland dipterocarp rainforest. Biotropica 17: 230-237.
Bergh, J. P., van den & Braakhekke, W. G. (1978) Coexistence of plant species by niche differentiation. In Freysen, A. H. J. & Woldendorp, J. W. (eds). Structure and functioning of plant populations. North-Holland, Amsterdam, pp.125-138.
Bossel, H. & Krieger, H. (1991) Simulation model of natural tropical forest dynamics. Ecol. Model. 59: 37-71.
Bossel, H. & Krieger, H. (1994) Simulation of multi-species tropical forest dynamics using a vertically and horizontally structured model. For. Ecol. Manage. 69: 123-144.
Brokaw, N. & Busing, R. T. (2000) Niche versus chance and tree diversity in forest gaps. Trend. Ecol. Evol. 15: 183-188.
Burns, R. M. & Honkala, B. H. (1990) Silvics of North American Trees. vol. 2. Hardwoods. US Department of Agriculture, Washington, DC, USA.
Chan, H. T. & Appanah, S. (1980) Reproductive biology of some Malaysian dipterocarps. I. Flowering biology. Malay. For. 43: 132-143.
Chesson, P. L. & Warner, R. R. (1981) Environmental variability promotes coexistence in lottery competitive systems. Am. Nat. 117: 923-943.
Clark, D. A. & Clark, D. B. (1992) Life history diversity of canopy and emergent trees in a neotropical rain forest. Ecol. Monogr. 62: 315-344.
Clark , D. B., Clark, D. A. & Read, J. M. (1998) Edaphic variation and the mesoscale distribution of tree species in a neotropical rain forest. J. Ecol. 86: 101-112.
Condit R, Ashton, P. S., Manokaran, N., LaFrankie, J. V., Hubbell, S. P. & Foster, R. B. (1999) Dynamics of the forest communities at Pasoh and Barro Colorado: comparing two 50 ha plots. Philos. Trans. R. Soc. London Ser. B 354: 1739-1748.
Condit, R., Ashton, P., Baker, P., Bunyavejchewin, S., Gunatilleke, S., Gunatilleke, N., Hubbell, S., Foster, R., Itoh, A., LaFrankie, J. V., Lee, H., Losos, E., Manokaran, N., Sukumar, R. & Yamakura T. (2000) Spatial patterns in the distribution of tropical tree species. Science 288: 1414-1418.
Connell, J. H. (1971) On the role of natural enemies in preventing competitive exclusion in some marine animals and in rain forest trees. In Den Boer, P. J. & Gradwell, G. (eds). Dynamics of Populations. PUDOC, Wageningen, the Netherlands, pp.298-312
Curran L. M., Caniago, I., Paoli, G. D., Astianti, D., Kusneti, M., Leighton, M., Nirarita, C. E. & Haeruman, H. (1999) Impact of El Niño and logging on canopy tree recruitment in Borneo. Science 286: 2184-2188.
Davies, S. J., Palmiotto, P. A., Ashton, P. S., Lee, H. S. & LaFrankie, J. V. (1999) Comparative ecology of 11 sympatric species of *Macaranga* in Borneo: tree distribution in relation to horizontal and vertical resource heterogeneity. J. Ecol. 86: 662-673.
Denslow, J. S. (1980) Gap partitioning among tropical rainforest trees. Biotropica 12: 47-55.
De Oliveira, A. & Mori, S. (1999) A central Amazonian terra firme forest. I. High tree species richness on poor soils. Biodiversity Conserv. 8: 1219-1244.
Dobzhansky, T. (1950) Evolution in the tropics. Am. Sci. 38: 209-221.
Gentry, A. H. (1982) Neotropical floristic diversity: phytogeogrpahical connections between

Central and South America, Pleistocene climatic fluctuations or an accident of Andean orogeny? Ann. Mo. Bot. Gard. 69: 557-593.
Gentry, A. H. (1988a) Tree species richness of upper Amazonian forests. Proc. Natl. Acad. Sci. USA 85: 159-159.
Gentry, A. H. (1988b) Changes in plant community diversity and floristic composition on environmental and geographical gradients. Ann. Mo. Bot. Gard. 75: 1-34.
Givnish, T. J. (1988) Adaptation to sun and shade: a whole plant perspective. Aust. J. Plant Physiol. 15: 63-92.
Grubb, P. J. (1977) The maintenance of species richness in plant communities: the importance of the regeneration niche. Biol. Rev. 52: 107-145.
Grubb, P. J. (1996) Rainforest dynamics: the need for new paradigms. In Edwards, D. S. (ed). Tropical rainforest research–current issues. Kluwer, Amsterdam, the Netherlands, pp.215-233.
Ha, C. O., Sands, V. E., Soepadmo, E. & Jong, K. (1988) Reproductive patterns of selected understorey trees in the Malaysian rain forest: the apomictic species. Bot. J. Linn. Soc. 97: 317-331.
He, F., Legendre, P., Bellehumeur, C. & LaFrankie, J. V. (1994) Diversity pattern and spatial scale: A study of a tropical rain forest of Malaysia. Environ. Ecol. Stat. 1: 265-286.
He, F., Legendre, P. & LaFrankie, J. V. (1996) Spatial pattern of diversity in a tropical rain forest in Malaysia. J. Biogeogr. 23: 57-74.
He, F., Legendre, P. & LaFrankie, J. V. (1997) Distribution patterns of tree species in a Malaysian tropical rain forest. J. Veg. Sci. 8: 105-114.
Hubbell, S. P. (1995) Towards a theory of biodiversity and biogeography on continuous landscapes. In Carmichael, G. R., Folk, G. E. & Schnoor, J. L. (eds). Preparing for Global Change: A Midwestern Perspective. SPB Academic Publishing, Amsterdam, pp.171-199.
Hubbell, S. P. (1997) A unified theory of biogeography and relative species abundance and its application to tropical rain forests and coral reefs. Coral Reefs 16: S9-S21.
Hubbell, S. P. (2001) The unified neutral theory of biodiversity and biogeography. Princeton University Press, Princeton, NJ, USA, 448pp.
Hubbell, S. P. & Foster, R. B. (1986) Biology, chance, and history and the structure of tropical rainforest tree communities. In Diamond, J. & Case, T. J. (eds). Community Ecology. Harper & Row, New York, pp.314-329.
Huth, A. & Ditzer, T. (2001) Long-term impacts of logging in a tropical rain forest– a simulation study. For. Ecol. Manage. 142: 33-51.
Ickes, K. (2001) Density of native wild pigs (*Sus scrofa*) in a lowland Dipterocarp rain forest of Peninsular Malaysia. Biotropica 33: 682-690.
Ickes, K., De Walt, S. J. & Appanah, S. (2001) Effects of native pigs (*Sus scrofa*) on woody understorey vegetation in a Malaysian lowland rain forest. J. Trop. Ecol. 17: 191-206.
Janzen, D. H. (1970) Herbivores and the number of tree species in tropical forests. Am. Nat. 104: 501-529.
Janzen, D. H. (1974) Tropical black-water rivers, animals, and mast-fruiting in the Dipterocarpaceae. Biotropica 6: 69-103.
Kato, M. (1996) Plant-pollinator interactions in the understory of a lowland mixed dipterocarp forest in Sarawak. Am. J. Bot. 83: 732-743.
Kato, M., Inoue, T., Hamid, A., Nagamitsu, T., Merdek, M., Nona, A., Itino, T., Yamane, S. & Yumoto, T. (1995) Seasonality and vertical structure of light-attracted insect communities in a dipterocarp forest in Sarawak. Res. Popul. Ecol. 37: 59-79.
Kaur, A., Ha, C. O., Hong, K., Sands, V. E., Chan, H. T., Soepadmo, E. & Ashton, P. S. (1978) Apomixis may be widespread among trees of the climax rain forest. Nature 271: 75-88.
Kaur, A., Jong, K., Sands, V. E. & Soepadmo, E. (1986) Cytoembryology of some Malaysian dipterocarps, with some evidence of apomixis. Bot. J. Linn. Soc. 92: 75-88.
Klopfer, P. & MacArthur, R. (1961) On the causes of tropical species diversity: niche overlap. Am. Nat. 95: 223-226.
Kobe, R. K. (1999) Light gradient partitioning among tropical tree species through differential

seedling mortality and growth. Ecology 80: 187-201.
Kochummen, K. M. (1997) Tree Flora of Pasoh Forest. Kuala Lumpur, Malaysia: Forest Research Institute Malaysia, 462pp.
Kochummen, K. M., LaFrankie, J. V. & Manokaran, N. (1991) Floristic composition of Pasoh Forest Reserve, a lowland rain forest in Peninsular Malaysia. J. Trop. For. Sci. 3: 1-13.
Kohyama, T. (1993) Size-structured tree populations in gap-dynamic forest - the forest architecture hypothesis for the stable coexistence of species. J. Ecol. 81: 131-144.
LaFrankie, J. V. & Chan, H. T. (1991) Confirmation of sequential flowering in *Shorea* (Dipterocarpaceae). Biotropica 23: 200-203.
Lee, P. C. (1968) Studies on *Dryobalanops aromatica*. Ph.D. diss., Univ. Malaya.
Leigh, E. G, Wright, S. J., Herre, E. A. & Putz, F. E. (1993) The decline of tree diversity on newly isolated tropical islands - a test of a null hypothesis and some implications. Evol. Ecol. 7: 76-102.
Liu, J. & Ashton, P. S. (1998) FORMOSAIC: an individual-based spatially explicit model for simulating forest dynamics in landscape mosaics. Ecol. Model. 106: 177-200.
Liu, J. & Ashton, P. S. (1999) Simulating effects of landscape context and timber harvesting on tree species richness of a tropical forest. Ecol. Appl. 9: 186-201.
Liu, J., Ickes, K., Ashton, P. S., LaFrankie, J. V. & Manokaran, N. (1999) Spatial and temporal impacts of adjacent areas on the dynamics of species diversity in a primary forest. In Mladenoff, D. J. & Baker, W. L. (eds). Spatial modeling of forest landscape change: approaches and applications. Cambridge University Press, Cambridge, UK, pp.42-69.
Manokaran, N., LaFrankie, J. V., Kochummen, K. M., Quah, E. S., Klahn, J. E., Ashton, P. S. & Hubbell, S. P. (1990) Methodology for the fifty-hectare research plot at Pasoh Forest Reserve. Research Pamphlet No. 104. Forest Research Institute Malaysia. Kepong, Malaysia, 69pp.
Manokaran, N., Kassim, A. R., Hassan, A., Quah, E. S. & Chong, P. F. (1992a) Short-term population dynamics of dipterocarp trees in a lowland rain forest in Peninsular Malaysia. J. Trop. For. Sci. 5: 97-112.
Manokaran, N., LaFrankie, J. V., Kotchummen, K. M., Quah, E. S., Klahn, J. E., Ashton, P. S. & Hubbell, S. P. (1992b) Stand tables and distribution of species in the fifty-hectare research plot at Pasoh Forest Reserve. FRIM Research Data Series 1. Forest Research Institute Malaysia, Kepong, Malaysia, 454pp.
Manokaran, N. & Swaine, M. D. (1994) Population dynamics of trees in dipterocarp forests of Peninsular Malaysia. Malayan Forest Record No. 40, Forest Research Intitute Malaysia, Kepong, Malaysia, 173pp.
Moad, A. S. (1993) Dipterocarp sapling growth and understory light availability in tropical lowland forest, Malaysia. Ph.D. diss., Harvard Univ., 340pp.
Momose, K., Yumoto, T., Nagamitsu, T., Kato, M., Nagamasu, H., Sakai, S., Harrison, R. D., Itioka, T., Hamid, A. A. & Inoue, T. (1998) Pollination biology in a lowland dipterocarp forest in Sarawak, Malaysia. I. Characteristics of the plant-pollinator community in a lowland dipterocarp forest. Am. J. Bot. 85: 1477-1501.
Newman, E. I. (1982) Niche separation and species diversity in terrestrial vegetation. In Newman, E. I. (ed.) The plant community as a working mechanism. Special Publications No. 1 of the British Ecological Society. Blackwell Scientific Publications, Oxford, pp.61-78.
Oikawa, T. (1985) Simulation of forest carbon dynamics based on a dry-matter production model. I. Fundamental model structure of a tropical rainforest ecosystem. Bot. Mag. Tokyo 98: 225-238.
Okuda, T., Kachi, N., Yap, S. K. & Manokaran, N. (1995) Spatial pattern of adult trees and seedling survivorship in *Pentaspadon motleyi* in a lowland rain forest in Peninsular Malaysia. J. Trop. For. Sci. 7: 475-489.
Okuda, T., Kachi, N., Yap, S. K. & Manokaran, N. (1997) Tree distribution pattern and fate of juveniles in a lowland tropical rain forest– implications for regeneration and maintenance of species diversity. Plant Ecol. 131: 155-171.
Pitman, N. C. A., Terborgh, J., Silman, M. R. & Nuez, P. (1999) Tree species distributions in an upper Amazonian forest. Ecology 80: 2651-2661.

Poore, M. E. D. (1968) Studies in Malaysian rain forest. I. The forest on triassic sediments in Jengka Forest Reserve. J. Ecol. 56: 143-196.
Reich, P. B. (2000) Do tall trees scale physiological heights? Trends Ecol. Evol. 15: 41-42.
Richards, A. J. (1990a) Studies in *Garcinia*, dieocious tropical forest trees: agamospermy. Bot. J. Linn. Soc. 103: 233-350.
Richards, A. J. (1990b) Studies in *Garcinia*, dioecious tropical forest trees: the origin of the mangosteen (*G mangostana* L.). Bot. J. Linn. Soc. 103: 301-308.
Richards, P. W. (1969) Speciation in the tropical rain forest and the concept of the niche. Biol. J. Linn. Soc. 1: 149-153.
Richards, P. W. (1978) Pasoh in perspective. Malay. Nat. J. 30: 145-148.
Rogstad, S. H. (1989) The biosystematics and evolution of the *Polyalthia hypoleuca* species complex (Annonaceae) of Malesia. I. Systematics. J. Arnold Arboretum 70: 153-246.
Rogstad, S. H. (1990) The biosystematics and evolution of the *Polyalthia hypoleuca* species complex (Annonaceae) of Malesia. II. Comparative distributional ecology. J. Trop. Ecol. 6: 387-408.
Rogstad, S. H. (1994) Biosystematics and evolution of the *Polyalthia hypoleuca* species complex (Annonaceae) of Malesia. III. Floral ontogeny and breeding systems. Am. J. Bot. 81: 145-154.
Sakai, S., Momose, K., Yumoto, T., Kato, M. & Inoue, T. (1999) Beetle pollination of *Shorea parvifolia* (section Mutica, Dipterocarpaceae) in a general flowering period in Sarawak, Malaysia. Am. J. Bot. 86: 62-69.
Sato, T., Saw, L. G. & Furukawa, A. (2000) A quantitative comparison of pteridophytes diversity in small scales among different climatic regions in eastern Asia. Tropics 9: 83-90.
Schot, A. M. (1995) A synopsis of taxonomic changes in *Aporosa* Blume (Euphorbiaceae). Blumea 40: 449-460.
Stevens, P. F. (1990) Nomenclatural stability, taxonomic instinct, and flora writing– a recipe for disaster? In Baas, P., Kalkman, K. & Geesink, R. (eds). The plant diversity of Malesia. Kluwer Academic Publisher, the Netherlands, pp.387-410.
Svenning, J. C. (2000) Small canopy gaps influence plant distributions in the rain forest understory. Biotropica 32: 252-261.
Tang, Y. H., Kachi, N., Fukuyama, A. & Awang, M. (1996) Light reduction by regional haze and its effect on simulated leaf photosynthesis in a tropical forest of Malaysia. For. Ecol. Manage. 89: 205-211.
Thomas, S. C. (1993) Interspecific allometry in Malaysian rainforest trees. Ph.D. diss., Harvard Univ., 291pp.
Thomas, S. C. (1996a) Asymptotic height as a predictor of growth and allometric characteristics in Malaysian rain forest trees. Am. J. Bot. 83: 556-566.
Thomas, S. C. (1996b) Relative size at reproductive onset in rain forest trees: a comparative analysis of 37 Malaysian species. Oikos 76: 145-154.
Thomas, S. C. (1996c) Reproductive allometry in Malaysian rain forest trees: biomechanics vs. optimal allocation. Evol. Ecol. 10: 517-530.
Thomas, S. C. (1997) Geographic parthenogenesis in a tropical rain forest tree. Am. J. Bot. 84: 1012-1015.
Thomas, S. C. (2003) Ecological correlates of tree species persistence in tropical forest fragments. In Losos, E. C. & Leigh, E. G Jr. (eds). Forest diversity and dynamism: Findings from a network of large-scale tropical forest plots. University of Chicago Press, Chicago (in press).
Thomas, S. C. & LaFrankie, J. V. (1993) Sex, size, and interyear variation in flowering among dioecious trees of the Malayan rain forest. Ecology 74: 1529-1537.
Thomas, S. C. & Bazzaz, F. A. (1999) Asymptotic height as a predictor of photosynthetic characteristics in Malaysian rain forest trees. Ecology 80: 1607-1622.
Thomas, S. C. & Ickes, K. (1995) Ontogenetic changes in leaf size in Malaysian rain forest trees. Biotropica 27: 427-434.
Tilman, D. & Pacala, S. W. (1993) The maintenance of species richness in plant communities. In

Ricklefs, R. E. & Schluter, D. (eds). Species diversity in ecological communities: historical and geographical perspectives. University of Chicago Press, Chicago, USA, pp.13-25.
Toy, R. J. (1991) Interspecific flowering patterns in the Dipterocarpaceae in West Malaysia: implications for predator satiation. J. Trop. Ecol. 7: 49-57.
Toy, R., Marshall, A. & Pong, T. (1991) Fruiting phenology and the survival of insect fruit predators–a case-study from the south-east Asian Dipterocarpaceae. Philos. Trans. R. Soc. London Ser. B 335: 417-423.
Valencia, R. (1998) Preliminary comparisons between Yasuni and BCI plots. Inside CTFS Summer 1998: 3.
Valencia, R., Balslev, H. & Paz y Miño, C. G. (1994) High tree alpha-diversity in Amazonian Ecuador. Biodiversity Conserv. 3: 21-28.
Vandermeer, J. (1996) Disturbance and neutral competition theory in rain forest dynamics. Ecol. Model. 85: 99-111.
Van Steenis, C. G. G. J. (1969) Plant speciation in Malesia, with special reference to the theory of non-adaptive saltatory evolution. Biol. J. Linn. Soc. 1: 97-133.
Wallace, A. R. (1878) Tropical nature and other essays. MacMillan, London, UK.
Warner, R. R. & Chesson, P. L. (1985) Coexistence mediated by recruitment fluctuations: a field guide to the storage effect. Am. Nat. 125: 769-787.
Webb, C. O. & Peart, D. R. (2000) Habitat associations of trees and seedlings in a Bornean rain forest. J. Ecol. 88: 464-478.
Whitmore, T. C. (1984) Tropical rain forests of the far east (2nd ed.). Clarendon Press, Oxford.
Wills, C. & Condit, R. (1999) Similar non-random processes maintain diversity in two tropical rainforests. Proc. R. Soc. London Ser. B 266: 1445-1452.
Wills, C., Condit, R., Foster, R. B. & Hubbell, S. P. (1997) Strong density- and diversity-related effects help to maintain tree species diversity in a neotropical forest. Proc. Natl. Acad. Sci. USA 94: 1252-1257.
Wolseley, P., Ellis, L., Harrington, A. & Moncrieff, C. (1998) Epiphytic cryptogams at Pasoh Forest Reserve, Negri Sembilan, Malaysia–Quantitative and qualitative sampling in logged and unlogged plots. In Lee, S. S., Dan, Y. M., Gauld, I. D. & Bishop, J. (eds). Conservation, management and development of forest resources: proceedings of the Malaysia-United Kingdom Programme Workshop, 21-24 October 1996, Kuala Lumpur, Malaysia. Forest Research Institute Malaysia, Kepong, Malaysia, pp.61-83.
Wong, M. (1981) Impact of dipterocarp seedlings on the vegetative and reproductive characteristics of *Labisia pumila* (Myrsiniaceae) in the understory. Malay. For. 44: 370-376.
Wong, M. (1983) Understory phenology of the virgin and regenerating habitats in Pasoh forest reserve, Negeri Sembilan, West Malaysia. Malay. For. 46: 197-223.
Wong, Y. K. & Whitmore, T. C. (1970) On the influence of soil properties on species distribution in a Malayan lowland dipterocarp forest. Malay. For. 33: 42-54.
Yamada, T., Itoh, A., Kanzaki, M., Yamakura, T., Suzuki, E. & Ashton, P. S. (2000) Local and geographical distributions for a tropical tree genus, *Scaphium* (Sterculiaceae) in the Far East. Plant Ecol. 148: 23-30.
Yap, S. K. (1976) The reproductive biology of some understorey fruit tree species in the lowland dipterocarp forest of West Malaysia. Ph.D. diss., Univ. Malaya, 146pp.
Yap, S. K. (1982) The phenology of some fruit tree species in a lowland dipterocarp forest. Malay. For. 45: 21-35.
Yap, S. K. & Chan, H. T. (1990) Phenological behavior of some *Shorea* species in Peninsular Malaysia. In Bawa, K. S. & Hadley, M. (eds). Reproductive Ecology of Tropical Forest Plants. UNESCO, Paris, pp.21-35.

14 Palms of Pasoh

Nur Supardi Md. Noor[1]

Abstract: The diverse biological factors of palms and their economic importance make palms a particularly interesting group for biodiversity studies. The objective of the present study was to analyze species diversity in relation to soil types within the primary forest and the impact of logging on the palm communities in Pasoh Forest Reserve (Pasoh FR). The study has shown that there were distinct differences in palm species diversity between dry alluvial and wet alluvial soils in the primary forest of Pasoh FR. Both these soil types differ in palm species diversity and density from that on lateritic and shale-derived soils. Both soil types have large numbers of palm individuals. The greatest number of species was on wet alluvial soils. Dry alluvial soils were generally poor in terms of both species and abundance of palms. Higher moisture content is an important component in growth of plants in tropical rain forest. Low water retention of dry alluvial soils that contain a large amount of coarse sand may be the limiting factor for the successful germination, survival and the growth of palm species that avoids this soil. The diversity of palms can be associated with a few factors besides edaphic ones. The impact of *Sus scrofa* (pigs/wild boars) activities is particularly significant at Pasoh FR. The species richness and density, and the recovery of palms at Pasoh FR were affected by tree harvesting operation. The impact of logging on the palm communities at Pasoh FR is evidently clear even 42 years after logging. The effect of light intensity was not clearly determined and requires further investigation.

Key words: impact of logging, palms, rattans, soils, species diversity.

1. INTRODUCTION

Palms are a group of plant species of significant ecological importance. They belong to a clearly defined and easily identified angiosperm family, Arecaceae (syn. Palmae). A characteristic of most forest in Peninsular Malaysia is the abundance of palms in the undergrowth. Climbing palms or rattans are abundant at the forest edges and in canopy gaps. This feature discriminates the forests in South-East Asia from the rain forests of Amazonia, where there are no rattans (Appanah & Salleh 1991).

Palms in Peninsular Malaysia grow from just above sea level to the montane forests (Dransfield 1988; Whitmore 1973). However, most of palm species are found in the lowland dipterocarp forest. Dransfield (1988) estimated about 140 species or 70% of palm species in Peninsular Malaysia are found in the lowland dipterocarp forest, 100 species (c. 50%) in the hill dipterocarp forest, and 20 species (c. 10%) in montane forest, though not all species are confined to the different forest types. This shows the importance of lowland dipterocarp forests as habitats of palms.

As most lowland dipterocarp forests in Peninsular Malaysia have been cleared or exploited, the distribution of palm species in these forests is far from original condition. Thus, it is important to evaluate the original palm species in the remaining forest types, such as the forests of Pasoh Forest Reserve (Pasoh FR). In their natural habitats, palms grow as shrub palms (either short-stemmed or

---

[1]Forest Research Institute Malaysia (FRIM), 52109 Kuala Lumpur, Malaysia.
E-mail: supardi@frim.gov.my

acaulescent), straight-bole tree-like (arborescent or palm tree) species, or climbers (Dransfield 1978). Species in these different habit groups may be affected by logging in different extents. Palms are either single-stemmed (solitary) or loosely or tightly clumped (clustering). In terms of their floristic biology, some are dioecious, some are monoecious, while others are hermaphrodite. These diverse factors make palms a particularly interesting group for biodiversity studies.

Palms are of great importance to the forest dwellers and rural communities residing close to a forest. Most species are put to some use. Some species are used for construction of dwellings, basketry, twine, food, ornament, ritual and medicine. In Peninsular Malaysia, the stems (canes) of some rattan species are collected for a well-developed industry (Abdul Latif & Shukri 1989). Rattan stems, skins or cores are used in the manufacturing of furniture and handicrafts. Pressure from uncontrolled collection of rattan has resulted in the scarcity of some rattan species (Kiew 1989) and depletion of their genetic resources (Nur Supardi 1997).

Efforts to distinguish palm communities with respect to their habitat preference have been conducted mainly in Central and South America (e.g. Kahn & de Granville 1992). There, the floristic composition, diversity, density and structure of forest ecosystems can be characterised by the species of palms present. Such studies have not yet been carried out in Peninsular Malaysia or in much of the Asia-Pacific region. A study at Pasoh FR revealed that the distribution of some groups of tree species on a local scale is determined by response to canopy disturbance, dispersability of fruits and small local differences in relief, drainage and soil (Whitmore 1988) as shown in maps of the 50 ha demographic plot at Pasoh FR (Manokaran et al. 1992).

Studies were carried out in 1996 and 1997 to investigate the impact of logging on the species diversity of palms at Pasoh FR. The forest logged more than 40 years ago was compared with the 50 ha plot which is in the pristine state (primary forest; Manokaran 1998; Chap. 2). It was assumed that before logging the regenerating forest was similar to the surviving primary forest. As there are local differences in relief, drainage and soil, it is therefore crucial to study the variability of palm species diversity within the primary forest. The aims of the study are:

1) To determine the species richness and abundance of palms in the primary forest of Pasoh FR.

2) To examine the impact of logging on species richness and abundance of palms at Pasoh FR.

2. STUDY SITE AND METHODS

2.1 Study Site

The study sites are the primary forest in the 50-ha demographic plot, within Compartments 32 and 33 and the regenerating forest in Compartment 25 that was felled in 1955.

Four plots (1 to 4) of size 0.12 ha (60 m × 20 m) were demarcated each on four major soil types in the primary forest of the 50 ha plot. The soils of the 50 ha plot have been mapped by Adzmi & Suhaimi (unpublished). There are 11 soil series grouped into four broad soil types:

Soil Type I: Soils developed on sub-recent alluvium (riverine alluvium), poorly drained or dry alluvium;

Soil Type II: Soils developed on sub-recent alluvium (riverine alluvium), well-drained or wet alluvium;

Soil type III: Soils developed on sedentary soils (parent material: shale);

Soil Type IV: Soils developed on reworked material (lateritic soils).

Nine 0.12 ha plots (A4, A5, B2, B3, B4, B5, C1, D1, E1) were placed in the regenerating forest which dominated by Soil Type IV.

2.2 Methods

The number smaller sub-plots (quadrats of 5 m × 5 m within the plots) having individual palm species were compared for the regenerating and primary forests. The number of quadrats with individuals of a species was calculated as the proportion of the total number of quadrats and is presented as percentage occupancy of the species.

Censuses were carried out from May 1996 till October 1997. All palm individuals with their bases found within the palm sub-plots or on the northern and eastern boundaries of these sub-plots were included in the census. Identifications were made on newly germinated seeds to mature palms were based on work by Dransfield (1979), Saw (1995), Whitmore (1973) and identification of seedlings and juveniles through thorough observation.

An individual palm may be single-stemmed or clustering. All individuals were tagged with plastic ribbons that were numbered. The abundance of palms or their densities may be expressed as the number of individuals or the number of stems per unit area.

Items that were recorded during the census were species name, individual number, number of stems per individual, sex (if inflorescence or infructescence were present), availability of flowers or fruits, damage caused by animals and occurrence of wind throws.

Each stem was classified into one of four growth stages: seedling, juvenile, intermediate and mature (adapted from Stockdale, 1994). A seedling was defined as having no stem and a leaf with few leaflets (seedling leaf). A juvenile has a short stem or is stemless, has more leaves and a broader base of the plant. It has the leaf morphology of more mature plants. A palm at the intermediate stage of growth has a base that is sometime as broad as mature palm, particularly if the leaf sheath has not been shed. The stem and the leaves are much shorter than those on mature individuals. Climbing palms at the intermediate stage are equipped with climbing organs. A mature acaulescent palm has large base, a developed crown or has produced suckers or on inflorescence. A mature arborescent palm has a tall stem, while a mature climbing palm has a long stem with about one fifth of its length near the base of the plant exposed (leaf sheath removed).

The light ratings give a rough estimate of light reaching the forest floor. Light ratings were recorded for each individual. For mature climbing palms, observations were made at the base. They are a visual estimation of light penetration through the tree canopies. The ratings as adapted from Dawkins (1958) are:

1. No direct light;
2. Some side light through a gap or edge of overhead canopy;
3. Some overhead light from partial opening of tree crown;
4. Full overhead light at noon;
5. Full overhead light for half of the day.

Observations were also made on fruiting/flowering female individuals, effect of wind throws and wild pig activities in the sub-plots. Comparisons between soils types and forest types were using *t*-test. The differences in species number were tested using the analysis of variance (ANOVA).

Table 1 Palm species found in the primary and regenerating forests of Pasoh FR.

Palm Species	Unlogged Forest						Logged Forest	
	Growth Form¶	Soil Type I	Soil Type II	Soil Type III	Soil Type IV	Outside subplots	Sub-plots	Outside subplots
a. Rattan Species								
1 *Calamus castaneus*	Ac	X	X	X	X	X		
2 *Calamus densiflorus*	C	X	X	X	X	X	X	X
3 *Calamus diepenhorstii**	C	X	X	X	X	X	X	X
4 *Calamus insignis* var. *insignis**	C	X	X	X	X	X	X	X
5 *Calamus javensis*	C	X	X	X	X	X	X	X
6 *Calamus laevigatus* var. *laevigatus**	C	X	X	X	X	X	X	X
7 *Calamus ornatus**	C					X		
8 *Calamus oxleyanus*	C					X		
9 *Calamus scipionum*	C					X		
10 *Calamus speciosissimus*	C	X	X	X	X	X	X	X
11 *Calamus tumidus*	C	X	X	X	X	X	X	X
12 *Daemonorops angustifolia*	C					X		
13 *Daemonorops calicarpa*	Ar	X	X	X	X	X	X	X
14 *Daemonorops didymophylla*	C		X	X	X	X		X
15 *Daemonorops geniculata*	C	X	X	X	X	X	X	X
16 *Daemonorops grandis*	C					X		
17 *Daemonorops hystrix* var. *hystrix**	C	X	X	X	X	X	X	X
18 *Daemonorops kunstleri*	C		X	X	X	X		X
19 *Daemonorops leptopus*	C	X	X		X	X		X
20 *Daemonorops macrophylla*	C	X	X		X	X		X
21 *Daemonorops micracantha*	C	X	X	X	X	X	X	X
22 *Daemonorops verticillaris*	C	X	X	X	X	X	X	X
23 *Korthalsia flagellaris*	C					X		
24 *Korthalsia laciniosa*	C		X	X	X	X		
25 *Korthalsia rigida*	C	X	X	X	X	X		X
26 *Korthalsia scortechinii*	C		X	X	X	X		
27 *Myrialepis paradoxa*	C					X		
28 *Plectocomiopsis geminiflora*	C					X		
b. Other palm species:								
29 *Arenga obtusifolia*	Ar						X	X
30 *Caryota mitis*	Ar		X			X		
31 *Eleiodoxa conferta*	Ac					X		
32 *Eugeissona tristis*	Ac							X
33 *Iguanura wallichiana* subsp. *wallichiana**	Ac		X			X		
34 *Licuala ferruginea*	Ac	X	X	X	X	X	X	X
35 *Licuala kunstleri*	Ac	X	X	X	X	X	X	X
36 *Licuala longipes*	Ac	X	X	X	X	X	X	X
37 *Licuala triphylla*	Ac		X	X	X	X	X	X
38 *Nenga pumila* var. *pachystachya**	Ar		X			X		
39 *Oncosperma horridum*	Ar	X	X		X	X		X
40 *Orania sylvicola*	Ar					X		
41 *Pholidocarpus macrocarpus*	Ar					X		
42 *Pinanga disticha*	Ac	X	X	X	X	X		
43 *Pinanga malaiana*	Ar			X		X		
44 *Pinanga simplicifrons*	Ac	X	X	X	X	X	X	X
45 *Salacca affinis*	Ac		X			X		

* subspecies and varities included

X = species present

¶ Ac=acaulescent and/or short-stemmed palm; Ar=arborescent (tall/short) palm tree; C=climbing palm.

3. RESULTS AND DISCUSSION

3.1 Species richness

Pasoh FR: In general, there were 45 species of palms in 17 genera in Pasoh FR (Table 1). The most common are the rattans, *Calamus diepenhorstii, C. insignis, Daemonorops verticillaris* and *D. micracantha* (Nur Supardi et al. 1995). All these species are of commercial value. They are small to medium sized canes that are split and used as cores and skins for the production of high quality rattan furniture and handicrafts. A few other species of commercial importance are *C. tumidus, C. ornatus, C. scipionum, C. insignis, C. laevigatus* var. *laevigatus* and *C. speciosissimus*.

Several other palm species found in Pasoh FR are also of some use. Fruits of *Salacca affinis* and *Eleiodoxa conferta* are edible. The folded leaves of *Licuala longipes* are collected for making a local food, 'ketupat'. The trunks of *Oncosperma horridum* are used to make poles and can be turned into beautiful (but heavy) furniture or handicrafts. Its terminal bud ('umbut") is used in local dishes. Other species, though of less or no commercial value are breeding grounds for insects and birds. Squirrels and birds eat their fruits. Some mammals (particularly monkeys) feed on the apices of palms. A few palm species found in Pasoh FR have the potential to be planted as ornamental plants. Examples are *Licuala* spp. (acaulescent species) and *Pholidocarpus macrocarpus* (acaulescent or palm-tree species).

Primary forest: Thirty-two species of palm belonging to 10 genera were found in the study area (16 plots, 1.92 ha). Twenty of these species in three genera were rattans (Table 1). The remaining (12) species in seven genera were other palms. Among the rattans, there were two acaulescent species (*Calamus castaneus* and *Daemonorops calicarpa*). Eleven of the 18 climbing rattans are of commercial importance. There were four arborescent palms, *Caryota mitis, Nenga pumila* var. *pachystachya, Oncosperma horridum* and *Pinanga malaiana*. The other species were acaulescent or short-stemmed species.

Ten species present in the 50 ha demographic plot but not included in the plots used in this study were *Calamus ornatus, C. oxleyanus, Daemonorops angustifolia, D. grandis, Eleiodoxa conferta, Korthalsia flagellaris, Orania sylvicola, Pholidocarpus macrocarpus, Plectocomiopsis geminiflora* and *Myrialepis paradoxa*. Among the species not recorded in the sub-plots, *D. angustifolia, E. conferta* and *K. flagellaris* were known to grow on wet alluvium close to rivers or fresh water swamps.

Regenerating forest: Within the study area (9 plots, 1.08 ha) on lateritic soils there were 18 palm species in six genera. These include twelve species of rattan and six other palm species. The smaller study area of lateritic soils in the primary forest (4 plots, 0.48 ha) had 27 species. There was at least a 33% decrease in the number of palm species in the regenerating forest as opposed to the primary forest.

Ten species found on lateritic soils in the primary forest were absent from the assessed plots of the regenerating forest. Four of the ten absent species were not observed anywhere in the regenerating forest. These species were *C. castaneus, K. laciniosa, K. scortechinii* and *P. disticha*. *C. castaneus* and *P. disticha* were clustering acaulescent or short-stemmed species that require very little light during their early stages of growth. These two species were not abundant in the primary forest. *C. castaneus* was abundant only on wet alluvium. Only a few individuals of *P. disticha* were found on lateritic and shale-derived soils in the primary forest.

Logging could have removed adults of these species and created unfavorable conditions for the growth of juveniles.

The two *Korthalsia* species are both multi-stemmed, although they differ morphologically and have different climbing habits. *K. laciniosa* is a commercial large-diameter rattan that climbs high in the forest canopy and has aerial branches. It seldom forms large clusters. *K. scortechinii* is a smaller-diameter rattan. In the primary forest of Pasoh FR only a few stems of this species climb above 5 m. Under intense light in canopy gaps, it sometimes forms large thickets. Openings created during logging could have favored the development of this species, especially at the preliminary stage after logging. Its absence from the regenerating forest is interesting. This species was not abundant on any soil in the primary forest. The few individuals likely to be present before logging may not have survived disturbance or the dense canopy cover in the regenerating forest may have suppressed the growth of survival stems.

Six other species were found only outside the plots. These are *K. rigida, D. didymophylla, D. kunstleri, D. leptopus, D. macrophylla* and *Oncosperma horridum. D. macrophylla* was found in a pocket of shale-derived soils. Its presence there could be a matter of chance since it was not found on shale in the primary forest.

All the ten species absent in the regenerating forest were present in the primary forest at very low densities. They made up about 1.4% of the total number of individuals in the primary forest, where they were not well distributed among the assessed plots. The most widespread amongst them, *K. rigida*, was found in only 7.3% of the sub-plots, while the others occur in less than 4% of the sub-plots (see "percentage of occupancy", Table 6). Logging could have removed the few adults of these species and created unfavorable conditions for seedlings establishment.

There was only one species present in the regenerating forest but not found in the primary forest. This species, *Arenga obtusifolia*, is common on hill slopes east of Pasoh FR forest. It occurs in regenerating forest, as well as primary forest, in lowland as well as up to 950 m altitude. It sometimes dominates disturbed forest area (J. Dransfield, pers. comm.). However, it was not found in large numbers in the regenerating forest of Pasoh FR. The individuals found may have originated from seeds dispersed after logging.

Out of the eleven commercial rattan species found in the primary forest, only eight were found within the surveyed area of the regenerating forest. Two other commercial species, *K. rigida* and *D. didymophylla* were outside the plots. The only commercial species absent from the regenerating forest was *K. laciniosa*.

3.2 Species richness on different soil types

Table 2(a) shows that the wet alluvial soil at Pasoh FR was the most species rich. There were 30 palm species on the wet alluvium soil. The dry alluvial soils were the most species poor, with only 20 species. The numbers of species for two other soil types were intermediate. The result from ANOVA, as in Table 2(b), shows that species richness between the soil types differs significantly ($P < 0.01$) for all growth stages or when seedlings were excluded. The results from t-tests made (Table 2(c)), show that difference was significant ($P < 0.01$) between the palm communities on dry alluvium and the other soil types. However, variations of species richness between palm communities on wet alluvium and lateritic soils, and that between lateritic and shale-derived soils were not significantly different ($P < 0.05$).

A high species richness of palms on wet alluvial was due partly to the fact

Table 2 The number of species of rattans and other palms in different plots within soil types of the primary forest of Pasoh FR.

(a)

Soil Types		Plot 1		Plot 2		Plot 3		Plot 4			
		all growth stages	excluding seedling	all growth stages	excluding seedlings	all growth stages	excluding seedlings	all growth stages	excluding seedlings	Mean	CV
I (Dry Alluvium)	Rattan	13	12	9	9	8	7	11	11	10.3	22
	Other palms	4	4	4	3	1	1	4	3	3.3	46
	All palms	17	16	13	12	9	8	15	14	13.5	25
II (Wet Alluvium)	Rattan	15	14	17	16	17	16	19	16	17	9.6
	Other palms	6	6	9	7	8	7	8	6	7.8	16
	All palms	21	20	26	23	25	23	27	22	24.8	11
III (Shale-Derived)	Rattan	13	12	15	15	14	13	16	16	14.5	8.9
	Other palms	6	6	5	5	4	4	5	5	5	16
	All palms	19	18	20	20	18	17	21	21	19.5	6.6
IV (Lateritic)	Rattan	12	11	14	13	16	13	17	15	14.8	15
	Other palms	4	4	7	7	5	4	6	6	5.5	24
	All palms	16	15	21	20	21	17	23	21	20.3	15

(b)

			df	Sum of Square	Mean square	F value	Pr (F)
Results of Analysis of Variance	All growth stages	Soil	3	256.5	85.5	13.267	0.002 **
		Subplot	3	29.5	9.833	1.526	0.273 ns
		Residuals	9	58	6.444		
			df	Sum of Square	Mean square	F value	Pr (F)
	Excluding seedlings	Soil	3	189.19	63.062	11.809	0.002 **
		Subplot	3	25.687	8.562	1.603	0.256 ns
		Residuals	9	48.062	5.34		

(c)

		t	df	P-value
Results of t-tests between Soils (all growth stages)	I vs II	-5.219	6	0.002 **
	I vs III	-3.286	6	0.017 *
	I vs IV	-2.976	6	0.025 *
	II vs III	3.584	6	0.012 *
	II vs IV	2.262	6	0.064 ns
	IV vs III	0.461	6	0.661 ns

that four species (none of them rattans) occurred only on this soil type. These species were *Caryota mitis, Iguanura wallichiana* subsp. *wallichiana, Nenga pumila* var. *pachystachya* and *Salacca affinis.* These four species are known to be associated with the wetter soil.

The relative species poverty of the dry alluvial soils occurs because, aside from the four species associated with the wet soil, another eight species were absent from all plots on the dry alluvial soil. Among these species were two commercial rattans, *D. didymophylla* and *K. laciniosa.* Three other rattans, not of commercial value, were also absent from the dry alluvium. These were *D. kunstleri, D. leptopus* and *K. scortechinii.* Amongst the non-rattans, *Licuala triphylla, P. disticha* and *P. malaiana* were three other palm species not found on dry alluvium. These eight species were not abundant on any soil type at Pasoh FR. Their absence from the sub-plots on dry alluvium could simply be due to sampling error.

Eighteen out of 32 species, including nine of the 11 commercially valuable species of rattan, occurred on all soil types. This suggests that some species are generalists (widely distributed) but their abundance may vary according to soil types.

Table 3 The number of palm species per 0.12 ha sub-plots within the regenerating and primary forests on lateritic soils at Pasoh FR.

	Plots	Number of species at different				Total	
		Seedlings	juveniles	interme-diates	matures	all growth stages	excluding seedlings
Regenerating forest (R)	A4	6	7	4	1	8	8
	A5	5	8	1	2	9	8
	B2	7	11	2	0	11	11
	B3	6	8	5	1	10	9
	B4	3	8	7	2	10	10
	B5	5	11	4	2	13	12
	C1	10	11	3	0	14	13
	D1	7	6	3	3	10	8
	E1	8	9	0	0	11	9
	Mean	**6**	**9**	**3**	**1**	**11**	**10**
	CV %	32	21	65	89	18	19
Primary forest (P)	1	12	15	9	8	16	15
	2	11	16	13	11	21	20
	3	18	17	7	5	21	17
	4	18	20	13	7	23	21
	Mean	**15**	**17**	**11**	**8**	**20**	**18**
	CV %	26	13	29	32	15	15
	t-values	11.64	15.25	11	14.67	14.28	15.48
	(R vs P)	$P<0.001$	$P<0.001$	$P<0.001$	$P<0.001$	$P<0.001$	$P<0.001$

3.3 Species richness according to growth stages

Primary forest: Not all species were represented by all growth stages at Pasoh FR. For example, seedlings were represented in only 29 out of 32 species. These include all rattan species. The three species that lack seedlings were *Nenga pumila* var. *pachystachya, P. malaiana* and *Salacca affinis*. There was only one mature individual (on wet alluvium) of *N. pumila* var. *pachystachya* in the primary forest plots. This species is very rare throughout Pasoh FR. The absence of any seedlings in the sub-plots may be a reflection of the fact that the inflorescence of this monoecious species is strongly protandrous, so that self-pollination and effective fruit-set are extremely unlikely. Infrequent flowering of *P. malaiana* and *S. affinis* may be the reason for the absence of their seedlings. Data on the phenology and seedling survival of these species might unravel on this problem.

Two species represented only by seedlings were *O. horridum,* an arborescent palm normally found on hill slopes and abundant on nearby hills in the north-eastern part of Pasoh FR, and *C. tumidus*, a rattan of superior cane quality. There were only 86 individuals (1.7 ha^{-1}) of *O. horridum* in the 50 ha plot (Manokaran et al. 1992). The data from a rattan inventory made in a 5 ha area located westernmost of the 50 ha plot between 1992 and 1993 (Nur Supardi et al. 1995) revealed that there were two mature individuals of *C. tumidus.* (Data source: FRIM Rattan Silviculture Unit).

Caryota mitis, an arborescent palm confined to the wet alluvium, was represented in the plots only by 65 seedlings and two juvenile plants. No mature individual of this species was observed in the 50 ha plot. Both *C. tumidus* and *Caryota mitis* must have seeds that are widely dispersed, or able to remain for some years in the seed bank, or both.

Mature individuals were presented in 23 out of the 32 species present in the primary forest. Five rattan species have no mature plants in the sub-plots some were observed elsewhere in the 50 ha demographic plot. Among the non-rattans, mature individuals of *L. triphylla, O. horridum* and *P. malaiana* were only found

Table 4 The number of individuals and stems of palms in plots (each 0.12 ha) of regenerating and primary forests on lateritic soils at Pasoh FR.

Forest Types	Subplots	Number of individuals						Number of stems					
		seedlings	juveniles	inter-mediates	matures	all growth stages	excluding seedlings	Seedlings	juveniles	inter-mediates	matures	all growth stages	excluding seedlings
Regenerating forest (P)	A4	56	47	7	4	114	58	61	52	7	4	124	63
	A5	88	86	1	2	177	89	88	91	1	2	182	94
	B2	100	108	8	0	216	116	102	108	8	0	218	116
	B3	22	47	7	1	77	55	27	48	7	1	83	56
	B4	21	45	8	2	76	55	28	48	8	2	86	58
	B5	11	55	12	4	82	71	13	56	12	4	85	72
	C1	65	130	3	0	198	133	70	135	4	0	209	139
	D1	30	28	10	4	72	42	41	47	6	5	99	58
	E1	29	42	0	0	71	42	32	46	0	0	78	46
	Mean	47	65	6	2	120	73	51	70	6	2	129	78
	CV %	68	53	65	93	50	44	59	47	63	97	45	41
Primary forest (P)	1	994	366	36	27	1423	429	1034	394	54	29	1511	477
	2	472	233	27	29	761	289	493	267	36	51	847	354
	3	890	143	22	13	1068	178	914	153	23	14	1104	190
	4	556	284	59	23	922	366	611	304	61	27	1003	392
	Mean	728	257	36	23	1044	316	763	280	44	30	1116	353
	CV %	35	36	46	31	27	34	33	36	40	51	25	34
	t-values	18.2	12.1	11.6	19	21.6	13.9	19	12.8	14.2	12.5	22.8	14.5
	(R vs P)	$P<0.001$	$P<0.001$	$P<0.001$	$P<0.001$	$P<0.001$	$P<0.001$	$P<0.001$	$P<0.001$	$P<0.001$	$P<0.001$	$P<0.001$	$P<0.001$

outside the sub-plots. The results suggest that, as in other plant groups, survival amongst seedlings and juveniles of palms is generally low.

Regenerating forest: If seedlings were excluded, the mean number of species in the regenerating forest was similar to the primary forest (Table 3). The species only represented by a seedling was *L. triphylla*. Most plots have fewer species when seedlings were excluded. Differences in the number of species for all growth stages between the regenerating and primary forests are obvious (t-test, $P < 0.01$, Table 3).

The consequence of logging is the low abundance of palms in younger stages that can survive and grow to maturity. Mature individuals in the regenerating forest were represented by eight species. Except for *C. laevigatus* var. *laevigatus* and *L. longipes,* which are normally solitary, the other species are clustering in habit. This growth strategy could be an advantage to the survival of these species.

3.4 Abundance of palms

Primary forest: When seedlings were included, there were 12,830 palm individuals in 1.92 ha of the primary lowland forest of Pasoh FR (more than 6,600 individuals ha^{-1}). There were 4,219 non-seedling individuals (> 2,100 ha^{-1}) out of which 85% or 3,599 individuals (> 1,750 ha^{-1}) were rattans. Thus, rattans dominate the palm community in the primary forest.

Some individuals were multi-stemmed; thus the number of stems at different growth stages exceeded the number of individuals. The average number of stems per individual on all soil types did not exceed two, as most individuals were single-stemmed, though a high percentage of the species produced sucker shoots.

Regenerating forest: There were 1,008 palm individuals ha^{-1} in the regenerating forest, as opposed to 8,691 individuals ha^{-1} in the primary forest. Only 12% of the palm individuals were found more than 40 years after logging. A similar difference was observed for the number of stems. There were 1,085 and 9,300 stems ha^{-1} in the regenerating and primary forests, respectively. The impact of logging is shown here by the reduction in the density of palms. This is also expressed by the high t-

values as shown in Table 4.

Eight commercial species were represented by 814 individuals ha^{-1} as opposed to 7,355 individuals ha^{-1} in the primary forest. This is a difference of 89%. The proportion of commercial individuals to the density of all palm decreased from 81% in the primary forest to 76% in the regenerating forest.

3.5 Density of palms on different soil types

The number of palm individuals differs significantly ($P<0.01$) between the dry alluvium and the other soil types. The variations between the other soil types were not significant ($P<0.01$; Table 5(c)). The density of palms on different soil types ranges from 961 individuals or 1,143 stems on 0.48 ha of dry alluvium to 4,561 individuals or 4,892 stems on shale (Tabe 5(a)). More than 90 % of the individuals were rattans. Rattans were most abundant on shale-derived soil, where there were 1,316 non-seedling individuals in 0.48 ha (2,742 ha^{-1}). There were more individuals of non-rattans on lateritic soils than other soil types, i.e. 296 individuals (617 ha^{-1}) of which most were non-seedlings. A large number (117) of the 240 non-seedling individuals on the lateritic soils were *P. simplicifrons*. This short-stemmed palm species was found on all soil types.

The dry alluvium had the fewest number of palms, with only 326 non-seedling

Table 5 The variation in density of rattans and other palm species between plots within soil types at Pasoh FR (Note: figures represent numbers of individuals per 0.12 ha sub-plot).

(a)

Soil Types		Sub-plot 1 all growth stages	Sub-plot 1 excluding seedlings	Sub-plot 2 all growth stages	Sub-plot 2 excluding seedlings	Sub-plot 3 all growth stages	Sub-plot 3 excluding seedlings	Sub-plot 4 all growth stages	Sub-plot 4 excluding seedlings	Mean	CV
I (Dry Alluvium)	Rattan	391	103	168	73	189	51	158	62	226.5	49
	Other palms	25	17	10	7	5	4	15	9	13.8	62
	All palms	416	120	178	80	194	55	173	71	240.3	49
II (Wet Alluvium)	Rattan	611	177	869	343	775	272	703	180	739.5	15
	Other palms	26	24	64	40	46	34	40	25	44.0	36
	All palms	637	201	933	383	821	306	743	205	783.5	16
III (Shale-Derived)	Rattan	679	223	1243	419	1156	265	1229	409	1076.8	25
	Other palms	64	59	59	51	62	50	69	60	63.5	6.6
	All palms	743	282	1302	470	1218	315	1298	469	1140.3	24
IV (Lateritic)	Rattan	1377	387	712	247	1038	155	751	233	969.5	32
	Other palms	46	42	49	42	30	23	171	133	74.0	88
	All palms	1423	429	761	289	1068	178	922	366	1043.5	27

(b)

Results of Analysis of Variance			df	Sum of Square	Mean square	F value	Pr (F)
	All growth Stages	Soil	3	3777	654521	10.887	0.002 **
		Sub-plot	3	1954562	1259.1	0.021	0.996 ns
		Residuals	9	538596	59844		
			df	Sum of Square	Mean square	F value	Pr (F)
	Excluding Seedlings	Soil	3	201812.7	67270.9	8.394	0.006 **
		Sub-plot	3	17988.7	5996.2	0.748	0.55 ns
		Residuals	9	72130.1	8014.5		

(c)

Results of *t*-test Between four soil types (all growth stages, all species)		*t*	df	*P*-value
	I vs II	-6.334	6	0.001 **
	I vs IV	-5.253	6	0.002 **
	I vs III	-6.158	6	0.008 **
	II vs IV	-1.684	6	0.143 ns
	II vs III	-2.416	6	0.052 ns
	IV vs III	0.497	6	0.78 ns

individuals in 0.48 ha (Table 5(a)) The wet alluvium and lateritic soils have 1,095 and 1,262 non- seedling individuals in 0.48 ha (2,281 and 2,629 ha^{-1}), respectively. The patterns were similar, though the densities increased greatly, when seedlings were included.

There were more palm individuals and fruiting individuals on shale-derived soils than on any other soil types. These were mainly *D. verticillaris.* I observed that individuals of this species have longer stems on this soil type. They may have established earlier or they could have grown faster on shale-derived soils.

3.6 Differences in density of individual species on different soil types

D. verticillaris had the highest number of individuals at seedling and juvenile growth stages on all soil types. The greatest density of this species was on shale-derived soils, having 6,577 individuals ha^{-1}. Seedlings of *D. verticillaris* made up more than 75 % of the palm seedlings on wet alluvium, lateritic and shale-derived soils. However, on the dry alluvium it contributes slightly less than 50% of the palm seedlings.

There were only 650 seedlings ha^{-1} of *D. verticillaris* on the dry alluvium, but this figure was nevertheless more than the number of seedlings of any other species on any soil type. The number of mature individuals of this species varied between 17 and 44 individuals ha^{-1}, according to soil type. Results from the rattan inventory made in 1992 and 1993 (Nur Supardi et al. 1995) showed that there were only between six to 29 mature individuals ha^{-1} of *D. verticillaris* (Data source: FRIM Rattan Silviculture Unit). The inventory was made mostly on dry alluvium and partly on wet alluvium. Like other species, with the exception of *C. densiflorus* and *C. insignis, D. verticillaris* was less abundant on dry alluvium soils. *C. densiflorus* and *C. insignis* were found in greater abundance on dry alluvium than other soil types. These commercial species made up about 39% of the non-mature individuals on dry alluvium. They occurred at higher density on this soil than on other soil types. Four species were most abundant on shale: *D. micracantha* (a commercial rattan), *D. geniculata* and *D. hystrix* (two other rattans) and *L. kunstleri* (a non-rattan). Two species were most abundant on wet alluvium: *C. javensis* (a climbing rattan) and *C. castaneus* (an acaulescent rattan). One species, *P. simplicifrons* (a short-stemmed non-rattan), was most abundant on lateritic soils.

3.7 Species densities among growth stages

Primary forest: Only a few species had large numbers of mature individuals. Four species had more than 30 mature individuals ha^{-1}. The species were *D. verticillaris* (on wet alluvium and shale-derived soils), *L. ferruginea* (except on dry alluvium), *L. longipes* (lateritic and shale-derived soils) and *P. simplicifrons* (lateritic soils).

Most species had more seedling and juvenile individuals than intermediate and mature individuals. The exceptions were *L. ferruginea, L. longipes, L. kunstleri* and *I. wallichiana* subsp. *wallichiana*. *L. ferruginea* had 50 and 46 intermediate and mature individual ha^{-1} on wet alluvium. During the 1985 tree census of the 50 ha plot, Manokaran et al. (1992) recorded 448 individuals (9 ha^{-1}) of this species with diameter of $\geqq$ 1 cm. Most of the individuals were found on wet alluvium. Pig activities might have reduced the number of younger plants of *Licuala* species in the plots. Most of the young individuals of *Iguanura wallichiana* subsp. *wallichiana* were observed as sucker shoots that resprout on old stems that die off as a result of herbivory. No germinating seeds were observed and none of the mature individuals was producing inflorescences. Shoot damage by insects may be one reason why these mature individuals were not able to flower.

Regenerating forest: There were only 17 mature individuals ha^{-1} in the regenerating forest, none of which were observed to be flowering or fruiting during the census. This may be due to sampling errors. There is a possibility that there were few or inadequate numbers of mature male and female individuals of dioecious or functionally dioecious monoecious species for the production of fruits within the sub-plots or they were too far apart.

The number of mature individuals in the regenerating forest to the primary forest was less by 91%. Fewer mature individuals mean fewer seeds produced. The number of seedlings in the regenerating forest was less by 94% (393 vs. 6067). This difference is also shown by the drop in the ratio of seedlings to mature from 32:1 in the primary forest to 23:1 in the regenerating forest.

The ratio of juveniles and intermediates to mature individuals in the regenerating forest was greater than the primary forest. The ratios for juvenile and intermediate were 32:1 and 3:1, while in the primary forest the ratios were 11:1 and 1.6:1, respectively. There were more juveniles than seedlings in the regenerating forest. They may have originated from seeds or seedlings that survived logging damage or they were a collection of individuals that were dispersed after logging over the years. This suggests that the growth of the palms is very slow.

No pig activity was observed within sub-plots of the regenerating forest. Traces of pig activity were found outside the sub-plots of the regenerating forest but the damage was not as severe as in the primary forest. Two reasons for this were suggested: i) the top soil of lateritic soils which dominates the regenerating forest is shallow; the band of laterite may hinder pigs from churning deep into the soils in search of earthworms; ii) hunting pressure in the regenerating forest.

A majority of the palm species is clustering in habit, but very few form large clumps. The most abundant species in both forest types was *D. verticillaris*. 56 percent of the total number of seedlings and 40% of the juveniles belong to this species. These percentages were, however, less than those in the primary forest (83% and 53%). This is an indication that forest disturbance has reduced the relative abundance of this species amongst the palm community at Pasoh FR.

On the contrary, logging has increased the relative abundance of some other palm species. Another commercial rattan species, *C. diepenhorstii,* makes up 16% of the palm community in the regenerating forest compared to 3.5% in the primary forest. *D. hystrix* (non-commercial rattan species) accounted for almost 7.4% of the total number of palms in the regenerating forest, an increase of 5.5%. In general, there were fewer individuals of most species (except for *C. javensis*) but the percentages for most species in the regenerating forest were slightly higher. This was due to a large drop in the numbers of individuals of *D. verticillaris* (from 72% in the primary forest to 44% in the regenerating forest).

One acaulescent and one dwarf undergrowth species had lower densities. These were *D. calicarpa* and *P. simplicifrons.* The reduction was obvious in *P. simplicifrons,* which has the second largest number of individuals amongst the non-rattan in the primary forest. It had only one juvenile in the regenerating forest. Hence, logging is detrimental to the survival and growth of this short-stemmed species. Another short-stemmed species that was absent in regenerating forest is *P. disticha.*

3.8 Light and the distribution of palm species in regenerating forest

The average estimated light for all palm species in the regenerating forest is presented in Table 6. The figures for *L. triphylla* and *P. simplicifrons* were, however, not representative as there was only one individual of each of the species. Though the average light rating for the two forests is not apparent (mean figures between 2.0 to 2.8 for regenerating forest; 2.5 to 2.9 for the primary forest), I observed that most of the regenerating forest was more shaded than the primary forest. There were fewer gaps between the crowns of smaller, shorter but more numerous trees of dbh above 10 cm in the regenerating forest, average of 708 trees ha^{-1} in the regenerating forest, 530 trees ha^{-1} in the primary forest). This is clearly shown by the averaged figures of the light ratings for individual species found in the sub-plots (Table 6) . The values for most species were higher for the primary forest. All palm individuals were growing under some sidelight or overhead light from the partial opening or gap of a tree crown (light rating between 2 and 3).

4. GENERAL DISCUSSION

The forest of Pasoh FR is uniform in relief but has differing geology and soil types. I therefore investigated the variation between the palm communities on different soil types in the primary forest and examined other factors that influence the diversity

Table 6 The light ratings and percentage of occupancy for palm species in the regenerating and primary forests on lateritic soils at Pasoh FR.

	Regenerating			Primary		
Species	Av. light rating	No. of quadrats	% occupancy	Av.light rating	No. of quadrats	% occupanc
Calamus castaneus		0	0	2.8	7	3.6
C. densiflorus	2.6	63	14.6	2.7	56	29.2
C. diepenhorstii	2.3	77	17.8	2.8	67	34.9
C. insignis var. *insignis*	2.6	42	9.7	2.7	47	24.5
C. javensis	2.5	12	2.8	2.7	9	4.7
C. laevigatus	2.6	30	6.9	2.7	63	32.8
C. speciosissimus	2.6	18	4.2	2.6	34	17.7
C. tumidus	2.8	4	0.9	2.6	13	6.8
Daemonorops calicarpa	2.7	4	0.9	2.7	56	29.2
D. didymophylla		0	0	3	2	1
D. geniculata	2.8	41	9.5	2.8	70	36.5
D. hystrix	2.7	60	13.9	2.5	41	21.4
D. kunstleri		0	0	2.8	5	2.6
D. leptopus		0	0	3	1	0.5
D. macrophylla		0	0	3	3	1.6
D. micracantha	3	3	0.7	2.9	10	5.2
D. verticillaris	2.6	195	45.1	2.6	183	95.3
Korthalsia laciniosa		0	0	3	3	1.6
K. rigida		0	0	2.7	14	7.3
K. scortechinii		0	0	2.9	4	2.1
Arenga obtusifolia	3	3	0.7		0	0
Licuala ferruginea	2.5	25	5.8	2.8	49	25.5
L. kunstleri	2.7	18	4.2	2.7	18	9.4
L. longipes	2.7	9	2.1	2.6	31	16.1
L. triphylla	2	1	0.2	3	1	0.5
Oncosperma horridum		0	0	3	5	2.6
Pinanga disticha		0	0	3	4	2.1
P. simplicifrons	3	1	0.2	2.7	52	27.1

of palms. Forest on wet alluvium was the most species rich (32 species in 10 genera within 0.48 ha). Lateritic and shale-derived soils were intermediate in species richness, but had dense communities of palms (8,691 and 9,514 individuals ha^{-1}). Dry alluvial soils were poor in both species number and density of palms. This soil type has low water retention capacity. It was also subjected to disturbance by wild pigs that reduced the palm population. Wind throws were observed to be initially damaging, but appear to enhance regeneration of palms within the gaps created.

Logging reduced species richness of palms. There were only 18 species (5 genera) within 1.08 ha as opposed to 27 species (6 genera) in 0.48 ha in the primary forest on a similar soil type. The reduction in palm density is even greater (1,008 individuals ha^{-1} in the regenerating plots vs. 8,691 individuals ha^{-1} in the primary plots). Variation within the regenerating forest seemed to be dependent on the degree of damage caused during logging.

Some of the tree species found at Pasoh FR were characteristic of hill forest (Kochummen et al. 1990). Northeast of Pasoh FR is the hill dipterocarp forest which is connected to the mountainous Main Range of Peninsular Malaysia. The hill dipterocarp element is not evident in the palms at Pasoh FR. None of the palm species found at Pasoh FR is species only found in the hill forests. However, many species found in the lowland forest of Pasoh FR are common in hill as well as lowland forests. Species found in lowland dipterocarp forest of Pasoh FR but not in hill dipterocarp forest were *C. tumidus, I. wallichiana* subsp. *wallichiana, L. ferruginea, L. kunstleri, L. longipes, L. triphylla* and *S. affinis.*

The primary lowland forest of Pasoh FR was not particularly rich in palm species. However, it is not poor in term of rattan species (20 in 3 genera). The lowland evergreen forest of Endau-Rompin, about 180 km southeast of Pasoh FR is diverse in its palm flora (Dransfield & Kiew 1987). But within this area, there were 26 rattan species in four genera and 27 rattan species in five genera in two continuous plots of 3.5 and 8ha (Tan et al. 1990). They surveyed rattans along a topographical gradient from 30–450 m a.s.l., hence capturing species found within this altitudinal range. Stockdale (1994) registered 21 rattan species in four genera in Brunei Darussalam in the island of Borneo. Borneo is known to be very rich in palms (Dransfield, pers. comm.). She inventoried a 1.5 ha plot measuring 50 m × 300 m across a topographical gradient from 60 to 210 m a.s.l. In East Kalimatan, a part of Borneo known to be poor in palms, Stockdale (1994) recorded 20 species of rattan in four genera within sub-plots totaling 0.75ha. She surveyed fifteen 50 m × 10 m sub-plots within 200 ha at 100 m altitude. These figures suggested that the remaining lowland dipterocarp forests of Peninsular Malaysia and Borneo house about the same number (at least 20) of rattan species. Thus, differences in species richness of palms in these forests reflect differences in the numbers of other groups of palms except rattan.

The obvious difference between palm communities of the Malesia floristic region and those of the Amazonia is the abundance of rattans. In the terra firme forests (tropical forest on well-drained soils) of Amazonia most (70% or more) palm species were small or medium-sized, under story species (Kahn et al. 1988; Kahn & de Granville 1992).

The diversity of palms in Amazonia has been intensively studied within plots of different sizes. There were nine palm species in six genera in a 0.18 ha plot in the lower Waki River valley (130–150 m altitude) in French Guiana. In Surinam, nine species (7 genera) were found on six 0.25 ha sub-plots (total 0.75 ha) in terra firme forest which include some areas on waterlogged soils. Along the changing topography of the terra firme in the lower Tocantins valley in Tucurui, Brazil, there were 12 species (8 genera) in 3.84 ha. These were parts of Eastern Amazonia. The central part of Amazonia is much richer in palm species. In terra firme forest on the lower Rio Negro valley, near Manaus, Brazil, there were 26 species (9 genera) in 0.72 ha (Kahn & de Granville 1992). Western Amazonia has the richest palm assemblage. In the lower Ucayali River Basin, Peru, there were 29 species (16 genera) and 34 species (21 genera) in 0.71 ha and 0.5 ha (Kahn & Meija 1991). The number of species is comparable to Pasoh FR (32 species, 10 genera), but the number of genera is greater.

The area surveyed at Pasoh FR has 14,107 stems and 12,830 individuals in 1.92 ha. Rattans made up 13,151 stems, while 956 stems belong to other palm species. This is about 7350 palm stems ha^{-1} (6,850 rattans ha^{-1} and 500 other palms ha^{-1}). This is rather lower than the densities between 7,676 and 9,864 stems ha^{-1} reported for palms in terra firme forests of Amazonia (Kahn et al. 1988; Kahn & de Granville 1992). There were more palm stems in Amazonia compared to Pasoh FR because the number of multi-stemmed species is dominated. Some of these species were dominant in the forest under story.

Dry alluvial soil at Pasoh FR was poor in palm species and low in densities of palms (20 species, 961 individuals) compared to other soil types (between 25 and 32 species, 3134 and 4561 individuals). Wild pigs were noticeably more active on the dry alluvial soils than on the other soil types. The impacts of wild pigs on the vegetation on dry alluvium were great and have been observed by several authors (Manokaran & LaFrankie 1990; Manokaran & Swaine 1994; Saw 1995). Pigs uproot palm seedlings as they forage in the leaf litter and surface soil, churning up the soil to a considerable depth (Chap.35). They utilize leaves of palms and other growing plants as bedding in the construction of "nests". A male wild pig was observed to bite and collect most of undergrowth vegetation within an area of more than 25 m^2 at Pasoh FR. This was done just before the female gave birth (Samad Latif, a Pasoh FR staff, pers. comm.). Pigs' 'nests' were mould-like structures that may reach a size of 3 m × 2 m × 0.5 (height) m. Medway (1963) found that out of 174 plants used to construct a nest, there were seven thorny rattan stems between 26 and 120 cm tall.

Pigs may effectively defoliate some species, particularly acaulescent individuals of *L. ferruginea* that have large leaves. Wild pigs were sometimes selective in their diet and habitats (Diong 1973, see Chap. 35), but most of the time they were generalists in terms of food (omnivorous).

Only seven of the 15 rattan species present on dry alluvial soil had reached maturity, although 12 species had seedlings and juveniles. Pigs may be one factor preventing those established individuals from reaching maturity. Their nesting habit which include snapping and uprooting woody understory vegetation reduces the number of juvenile and intermediate palm individuals, hence a lesser number of mature individuals.

Light requirements differ with species, especially rattans, which require different light regimes at different stages of life (Aminuddin 1992; Manokaran 1985). Chazdon (1986) made similar observations for understory palms in Costa Rica.

The field staff at Pasoh FR observed that gaps caused by wind throws have become common since the clearing of surrounding forests for oil-palm plantations. These result in small gaps. There were five light gaps with estimated size between 25 to 75 m^2 caused by wind throws in the palm plots (1.92 ha). This is about 0.025

ha or 1.3% of the total area surveyed (1.92 ha). In two cases, the gaps appear to reduce the number of palm species. This was observed in plot 3 on dry alluvium and plot 1 on wet alluvium, which lacked four species compared to the average of the other sub-plots on the same soil type. Plot 1 on dry alluvium and plots 2 and 4 on lateritic soils were also affected by wind throws.

I observed that the wind throws in these plots were quite recent, possibly occurring about three to six months before enumeration of the sub-plots. A fallen tree or a huge branch is damaging to the undergrowth vegetation. The numbers of palm individuals in the affected plots may have been reduced. There is very little recruitment in these plots. The damage by wind throws in plot 1 on dry alluvium was observed to have occurred more than a year before enumeration of the plot. This indicates that the effect of wind throws is initially damaging to the palm community, but it is anticipated that over time the chances of regeneration will increase. This requires further investigation.

The various soil types differ in composition, acidity and texture (Adzmi & Suhaimi, unpubl.). Dry alluvial soils were rich in carbon, high in exchangeable calcium and potassium and have more coarse sand than other types of soils. But, they were low in fine sand and silt. *C. insignis* and *C. densiflorus* have the highest density on dry alluvium and may prefer the extra amounts of carbon and exchangeable calcium. This is yet to be proven. Apparently, dry alluvium has low water retention due to the larger content of coarse sand. This may not suit most palm species at Pasoh FR.

On the contrary, the "palm species rich" wet alluvial soil contains high percentages of fine sand and silts, hence is able to retain more moisture in the soils. It has the highest level of available phosphorus and exchangeable magnesium, is the least acidic and is low in nitrogen. The figures obtained from the soil analyses of the wet alluvium did not differ so much from dry alluvium as from other soil types. Hence, one reason for the higher number of palms on wet alluvium may be its ability to store water for the vegetation. In the drier months its water table is much higher than that on other soils (Adzmi & Suhaimi, unpubl.). The variations between shale-derived and lateritic soil were apparent, but further studies were needed to determine their relationships to the presence and absence of palm species on these soils.

If a forest is disturbed, the number of species is reduced. The number of species lost depends on the degree of disturbance. The subsequent development of a disturbed forest depends on what is left and what comes in after disturbance. Regeneration takes time, and the time taken for a forest to recover or come to a state of equilibrium, similar to that before it was logged, depends on the above-mentioned factors. Thirty-four years after logging, the forest of Pasoh FR had not yet reached the state it was prior to logging (Manokaran 1998). The palm species diversity continues to be much affected even 42 years after logging. Rare species or uncommon species in the undisturbed forest were those most affected. They either totally disappeared or were reduced in number. The densities of surviving species were very low, even when seedlings were excluded. The number of mature individuals is particularly low, thus limiting regeneration.

Most of the rattan species, including the commercial ones, were found in the regenerating forest. The numbers of rattans for Pasoh FR were however much higher than the number of rattans in a one year-old regenerating forest in Berau, East Kalimantan, as reported by Stockdale (1994). Her study recorded 408 rattan individuals, which is less by 28% to an adjacent primary forest. However, the number of rattan stems (shoots) in the one year logged sub-plots (1,775 stems ha^{-1}) in Berau is more abundant than at Pasoh FR (1,075 ha^{-1} or 1,161 in 1.08 ha).

There were two species that were particularly sensitive to logging and clear felling. The species were *C. castaneus* and *P. simplicifrons*. Other acaulescent (or short-stemmed) species, which were not found in abundance, were likely to have disappeared as a result of disturbance. The level of disturbance is an important factor in the survival of this group of species. Other factors contributing to the species diversity of palms in a disturbed site were past distribution, topography, dispersal agents, fruiting frequency, light and secondary damage caused by animals such as pigs.

Most of the climbing palms found in Pasoh FR were commercially important. They were able to survive and regenerate following logging, but their production is much affected. There is a lack of studies on the growth of these species in the wild. Casual observations showed that growth of these species is moderately slow. It will take a few more decades for the regenerating and the clear-felled forests to produce harvestable rattan stems.

5. CONCLUSION

The present study has shown that there were distinct differences in palm species diversity between dry alluvial and wet alluvial soils in the primary lowland forest of Pasoh FR. Both these soil types differ from lateritic and shale-derived soils in terms of species richness and density. Lateritic and shale-derived soils varied but were not distinctively different. Both had large numbers of palm individuals. The greatest number of species was found on wet alluvial soil. Dry alluvial soil was generally poor in terms of both species and abundance of palms. Moisture was an important component in growth of plants in tropical rain forest. Low water retention of dry alluvial soils that contain a large amount of coarse sand may be the limiting factor for the successful germination, survival and growth of some palm species.

The diversity of palms can be associated with a few factors besides edaphic ones. The impact of pig activities is particularly significant at Pasoh FR. The effect of light was not demonstrated and requires further investigation. The species richness and density of palms at Pasoh FR were much affected by forest disturbance. The degree of their absence depends on the level or size of disturbance. Logging regime practiced at Pasoh FR in 1950s greatly reduced the density of palms.

ACKNOWLEDGEMENT

This work was made possible by the financial support of the Overseas Development Administration of the United Kingdom who financed the study under the Conservation Management and Development of Forest Resources Project (CMDFR), and FRIM and the Government of Malaysia, who pay costs of the fieldwork. The author express his deepest gratitude to Dr. Barbara Pickersgill of Reading University and Dr. John Dransfield of the Royal Botanic Gardens, Kew, for their constant support, advice and encouragement throughout the study as supervisors.

REFERENCES

Abdul Latif, M., & Shukri Mohamad (1989) The rattan industries in Peninsular Malaysia. Part I: Distribution and current status of rattan manufacturers in Peninsular Malaysia. RIC Occasional Paper No.16, 14pp.

Aminuddin, M. (1992) Aspects of the physiology of rattans. In Wan Razali, M., Manokaran,

N. & Dransfield (eds). A Guide To The Planting of Rattan. Malayan Forest Records No. 35. Forest Research Institute Malaysia, Kuala Lumpur, pp.35-38.
Appanah, S. & Salleh, M. N. (1991) Natural regeneration and its implications for forest management in the dipterocarp forest of Peninsular Malaysia. In Gomez-Pompa, Whitmore, T. C. & Hardley, M. Rain Forest Regeneration and Management. Man and the Biosphere Series vol. 6. UNESCO, Paris and The Parthenon Publishing Group. pp.361-369.
Chazdon, R. L. (1986) Physiological and morphological basis of shade tolerance in rain forest understory palms. Principes 30: 92-99.
Dawkins, H. C. (1958) The management of natural tropical high forest with special reference to Uganda. Institute Paper 34, Imperial Forestry Institute, University of Oxford.
Diong, C. H. (1973) Studies of Malayan wild pig in Perak and Johore. Mal. Nat. J. 26: 120-151.
Dransfield, J. (1978) Growth forms of rain forest palms. In Tomlinson, P. B. & Zimmermann, M. H. (eds). *Tropical Trees as Living Systems.* Cambridge University Press, Cambridge. pp.247-268.
Dransfield, J. (1979) A manual of rattans of the Malay Peninsula. Malayan Forest Records No. 26. Forest Research Institute Malaysia, Kuala Lumpur, 270pp.
Dransfield, J. (1988) Forest palms. In Cranbrook, E. (ed). Key Environments Malaysia. International Union for Conservation of Nature and Natural Resources, Pergamon Press, pp.37-48.
Dransfield, J. (1992) The taxonomy of rattans. In Wan Razali, M., Dransfield, J. & Manokaran, N. (eds). Guide to cultivation of rattan. Mal. For. Rec. No 30. Forest Research Institute Malaysia, Kepong, pp.1-10.
Dransfield, J. & Kiew, R. (1987) Annotated checklist of the palms at Ulu Endau, Johore, Malaysia. Malay. Nat. J. 41: 257-265.
Kahn, F. & de Granville, J-J. (1992) Palms in forest ecosystems of Amazonia. Ecological Series No. 95. Springer-Verlag, 226pp.
Kahn, F. & Meija, K. (1991) The palm communities of two 'terra firme' forests in Peruvian Amazonia. Principes 35: 22-26.
Kahn, F., Meija, K. & de Castro, A. (1988) Species richness and density of palms in Terre Firme forests of Amazonia. Biotropica 20: 266-269.
Kiew, R. (1989) Utilization of palms in Peninsular Malaysia. Malay. Nat. J. 43: 26-50.
Kochummen, K. M., LaFrankie, J. & Manokaran, N. (1990) Floristic composition of Pasoh Forest Reserve, a lowland rain forest in Peninsular Malaysia. J. Trop. For. Sci. 3: 1-13.
Manokaran, N. (1985) Biological and ecological considerations pertinent to the silviculture of rattans. In Wong, K. M. & Manokaran, N. (eds). Proceedings of the Rattan Seminar, Kuala Lumpur 2-4 October, 1994. The Rattan Information Centre, Forest Research Institute Malaysia, Kepong, pp.95-105.
Manokaran, N. (1998) Effect, 34 years later, of selective logging in the lowland dipterocarp forest at Pasoh, Peninsular Malaysia, and implications on present day logging in the hill forests. In Lee, S. S., Dan, Y. M., Gauld, I. D. & Bishop, J. (eds). Conservation, management and development of forest resources. Proceedings of the Malaysia-United Kingdom Programme Workshop 21-24th October 1996. Forest Research Institute Malaysia, Kepong. pp.41-60.
Manokaran, N. & LaFrankie, J. (1990) Stand structure of Pasoh Forest Reserve, a lowland rain forest in Peninsular Malaysia. J. Trop. For. Sci. 3: 14-24.
Manokaran, N. & Swaine, M. D. (1994) Population dynamics of trees in dipterocarp forests of Peninsular Malaysia. Malayan Forest Records No. 40. Forest Research Institute Malaysia, Kuala Lumpur, 173pp.
Manokaran, N., LaFrankie, J. V., Kochummen, K. M., Quah, E. S., Klahn, J. E., Ashton, P. S. & Hubbell, S. P. (1992) Stand table and distribution of species in the 50-ha research plot at Pasoh Forest Reserve. FRIM Research Data, No. 1.
Medway, L. (1963). Pigs' nest. Malay. Nat. J. 17: 41-45.
Nur Supardi, M. N. (1997) The activities on conservation and genetic improvement of rattan in Malaysia. Meeting of Experts on the Conservation and Genetic Improvement of Bamboo

and Rattan. IPGRI Asia Pacific Region. Serdang.
Nur Supardi, M. N., Shalihin, S. & Aminuddin, M. (1995) Rattan inventory techniques. In Proceedings of Conference on Forest and Forest Products Research. Kepong. 3rd-4th October 1995. Forest Research Institute Malaysia, Kepong. pp.79-86.
Saw, L. G (1995). Taxonomy and ecology of Licuala in Peninsular Malaysia. Ph.D. diss., Univ. Reading.
Stockdale, M. C., (1994). Inventory methods and ecological studies relevant to the management of wild populations of rattan. D. thesis, Univ. Oxford.
Tan, C. F., Raja Barizan R. S. & Nur Supardi, M. N. (1990). Rattans from Ulu Kinchin, Pahang, Malaysia. The Rompin-Endau Expedition: 1989. Malay. Nat. J. 43: 250-255.
Whitmore, T. C. (1973). Palms of Malaya. Oxford University Press.
Whitmore, T. C. (1988). Forest types and forest zonation. In Cranbrook, E. (ed). Key Environment Malaysia. International Union for Conservation of Nature and Natural Resources. Pergamon Press, pp.20-30.

15 Flowering Predictions Based on 20 Years of Phenological Observations of *Neobalanocarpus heimii* (King) Ashton

Marzalina Mansor[1], Jayanthi Nadarajan[2] & Ang Khoon Cheng[2]

Abstract: This study focuses on the phenological behavior of *Neobalanocarpus heimii*, a tree species with a high reputation as a 'primary hardwood'. Twenty years of monthly phenological observations were made on this species at Forest Research Institute Malaysia (FRIM), Pasoh Forest Reserve (Pasoh FR), Ulu Gombak Forest Reserve (Ulu Gombak FR) and Ampang Forest Reserve (Ampang FR). The data indicates that there is a high probability (> 95%) of an onset of flowering following the end of a dry period. Flowering occurs in most years (either annually or biannually). The event can either peaks during March–May and maybe followed by a second peak during Sept.–Nov. in all four phenological plots. Based on the percentage of each month that flowers, it can be concluded that the flowering patterns can be predicted for at least in three of the forest areas. This indicates that some dipterocarp species do not show 'masting' behavior and such species can be very useful for reforestation programs.

Key words: chengal, dipterocarp, flowering events, flowering occurrence, phenology.

1. INTRODUCTION

Neobalanocarpus heimii (chengal), a member of the Dipterocarpaceae, produces a heavy hardwood timber, which is highly valued for its durability, strength, treatability and working properties. This species has a high reputation as a 'primary hardwood' and is confined to the Peninsular Malaysia and Pattani, Thailand (Ashton 1982; Symington 1943). In Peninsular Malaysia, *N. heimii* occurs in all states except in Perlis, Penang and Malacca. It is often found on undulating, well-drained areas with soils of average fertility (Ashton 1982; Wyatt-Smith 1987) but its abundance decreases with an increase in altitude. Symington (1943) reported that it was also found at low densities (less than 5 trees/ha) and natural forests with many large trees of thies species usually lack of suitable representation of small and intermediate sizes.

The high endemicity, limited distribution, over-exploitation, and decreasing area of preferred habitats with reproductively mature populations justify the need to undertake conservation measures in order to safeguard existing populations. There is a considerable risk that *N. heimii* will continue to be exploited in the near future due to the high demand for this species. Kiew et al. (1985) once listed this species as one of the ten endangered species in Malaysia. With regards to this, the Malaysian government has taken steps in implementing conservation measures in order to safeguard the timber resources in the remaining forests and to practice sustainable harvesting. Most of the plots for this species occurring in the virgin

[1] Forest Research Institute Malaysia (FRIM), 52109 Kuala Lumpur, Malaysia.
E-mail: mzalina@frim.gov.my
[2] FRIM, Malaysia.

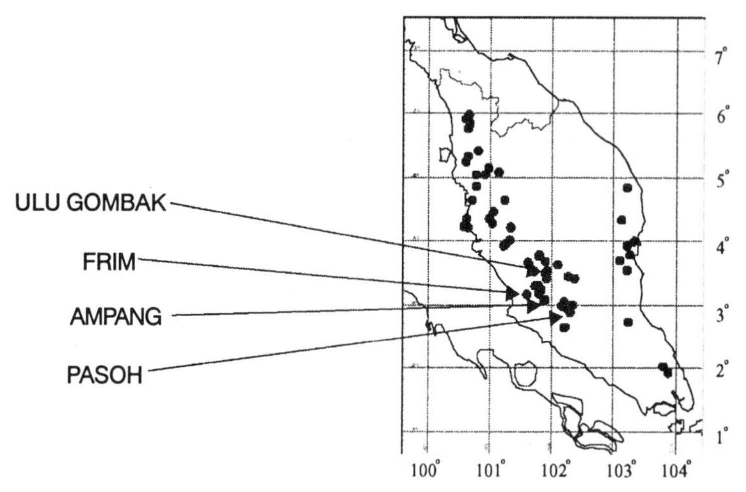

Fig. 1 Natural distribution of *N. heimii* within Peninsular Malaysia (black dots) and the location of study sites (pointed by arrows).

jungle reserve system in Malaysia have been designated as research plots and these are scattered throughout the lowland dipterocarp forests, the largest being in Balok FR, Compartment 8 in Pahang (Saw & Raja Barizan 1991).

Most South-East Asian timber species tend to display erratic flowering-fruiting behavior and this limits the continuous supply of planting materials, especially seeds. In Peninsular Malaysia, major general or mass flowering events have been reported in 1950, 1953, 1957, 1958, 1968, 1976, 1981, 1982 and 1996 with weaker events in 1955, 1960, 1963 and 1983 (Appanah 1993; Numata et al. 1999; Yap & Marzalina 1990). To date, the longest observation period reported was for 12 years (Yap 1985). It has been estimated that in aseasonal zones in South-East Asia, flowering occurs at intervals of 2 to 5 years (Burgess 1972; Krishnapillay & Tompsett 1998) or 3 to 8 years (Cockburn 1975). Study by Ashton et al. (1988) found that a drop in minimum temperature for at least consistently 5–8 days was the cause of flowering events 8 to 9 weeks later. They concluded that rainfall was not directly responsible. But this may not be the case for *N. heimii*. There are also species that flower annually but the occurrence is very low with few mother trees within the population bear it. Meanwhile according to Bawa (1998) most dipterocarps appear to be strongly cross-pollinated and this definitely is attributed to the production of seeds if the number of flowering trees of the same species is low and the occurrence is infrequence. In view of the fact that seeds are the only viable method of propagation and regeneration of *N. heimii*, the remark aspects will significantly affect future timber supply. Therefore, the knowledge on phenological behavior of *N. heimii* is very critical consecutively to obtain the actual timing of flowering and seed collection.

This study focused on the phenological observation of *N. heimii* in order to determine the pattern of flowering and fruiting of this species. The number of trees sampled within the phenological plot was meant to understand the natural flowering-fruiting behavior of change. Therefore the idea was to find out when will the flowering event occur and the possibilities of occurrence for each month of the year. It is hoped that prediction of flowering patterns can assist both better understanding of the natural reproductive system of *N. heimii* and for future production of planting material for applied forestry and conservation purposes.

2. METHOD

The study was carried out in phenological plots located in arboretum and plantation areas of the Forest Research Institute Malaysia (FRIM) Kepong, Pasoh Forest Reserve (Pasoh FR), Ulu Gombak Forest Reserve (Ulu Gombak FR) and Ampang Forest Reserve (Ampang FR)(Fig. 1). Observations were made monthly on 4–10 chengal mother trees that were mainly located near the trail within each phenological plot. Tree selection was based on phenotypic appearance. Trees selected for observation had straight bole form and the majority was above 25 m height and more than 100 cm girth at breast height. The Seed Technology section's phenological team used a pair of binoculars (Leica, lens of 10 mm × 42 mm). When in doubt, samples of branches were brought down for further examination. The flushing of the leaves was not recorded as Burgess (1972) reported that leaf flushes occur in almost equal frequency and are apparently are not affected by weather changes. It can easily be confused that the wingless seed shape were similar to the color and structure of the leaves. Both flower petals and fruit aborted can be found on the ground. Only events of flowering and fruiting were recorded. The data were then used to focus on the flowering patterns as to predict when will the events occur again and in which month it will happen, i.e. termed as the flowering occurrence. This study indicated the onset of the flowering events. It did not reveal the number of trees involved in the pattern since the objective is to find out the possibilities of the event to occur again and whether it related to the rainfall season. It is also important to note that during the 20 years of phenological observations, some of the selected trees were no longer available and the new ones within the phenological plot were selected.

Monthly records of total amount of precipitation (mm) were obtained from the FRIM Nursery section. Precipitation data for Pasoh, Ulu Gombak and Ampang FRs were gathered from the Malaysian Meteorological Station from Pasoh Dua, Ulu Gombak and Ampang Klang Gate respectively, within the time of study. For data analysis, the correlation analysis was applied using Minitab software (Minitab Inc.,USA, website: http://www.minitab.com/contacts). The data mean values of both rainfall and flowering occurrence of 20 years observations for each month were used for data analysis.

3. RESULTS

The flowering and fruiting patterns of *N. heimii* over 20 years of observation were projected in Fig. 2. When no observation was carried out, it was noted as 'n'. The missing data appear only 12% of the total observations made in 240 months. From this figure, it was found that the period of flowering event ranges from only one month to a maximum of six months. This observation confirmed that *N. heimii* tends to flower gregariously almost every year as mentioned by Medway (1972), Appanah (1979) and Yap (1985). Shorter length flowering occurrence may be related to unfavorable environmental conditions. During such periods, it was observed that the flowers failed to set seed, where the majority of fruits were aborted. Meanwhile longer period of six months flowering events were observed to occur on different trees in a synchronized manner. Such a pattern within the same species of dipterocarp has been reported by Chan & Appanah (1980). Appanah (1979) observed that the blooming of individual *N. heimii* trees was short, ranging from 11–20 days.

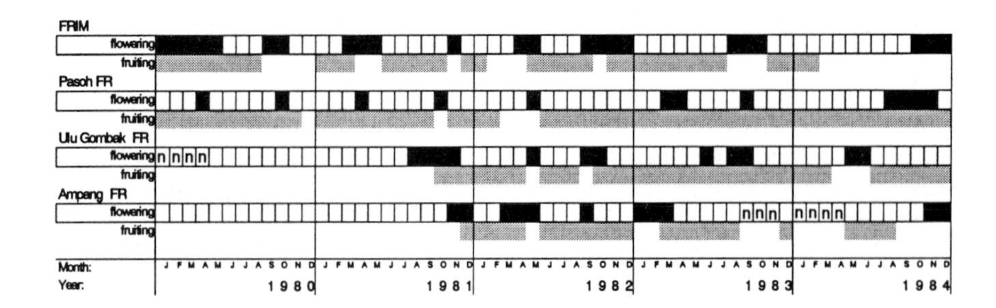

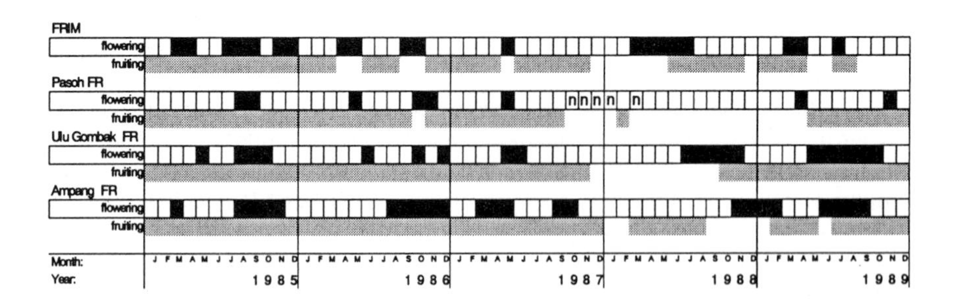

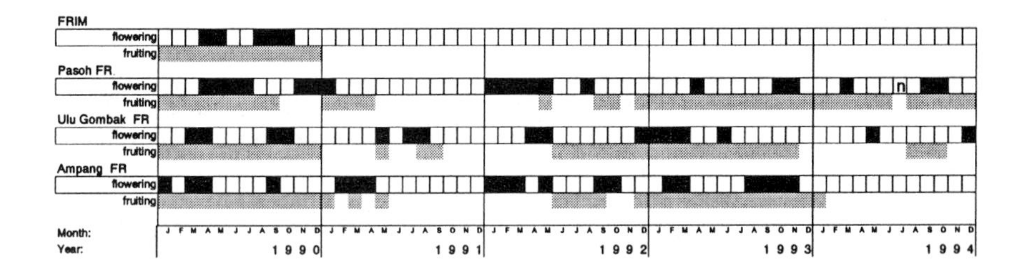

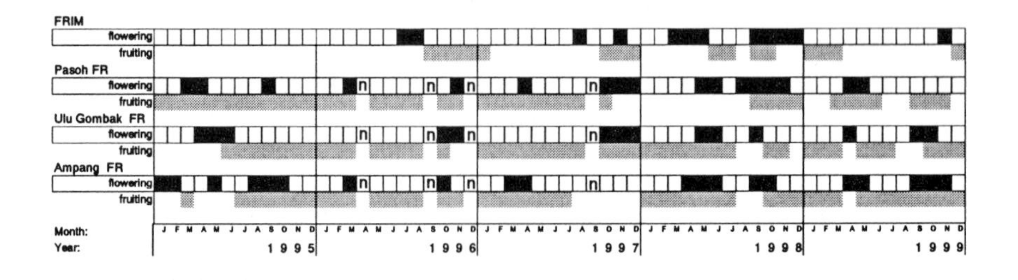

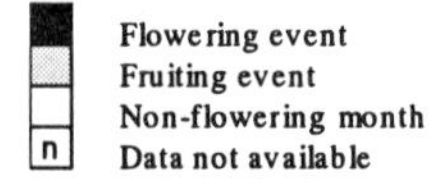

Fig. 2 Flowering and fruiting events within 20 years for *N. heimii* in four FR areas.

Table 1 Annual and biannual occurrence of *N. heimii* flowering seasons from 1980 to 1999.

Study sites	'80		'82		'84		'86		'88		'90		'92		'94		'96		'98	'99
FRIM	2	2	2	1	1	2	2	1	1	2	2	Nil	Nil	Nil	Nil	Nil	1	2	2	1
Pasoh FR	2	2	1	2	1	1	2	1	Nil	2	2	Nil	2	2	2	2	2	2	2	1
U.Gombak FR	Nil	1	2	2	1	2	2	1	1	1	2	2	1	2	1	2	2	1	2	2
Ampang FR	Nil	1	2	1	1	2	1	2	1	2	2	1	2	2	Nil	2	2	1	2	2

3.1 Flowering patterns

Fig. 2 revealed that flowering events of *N. heimii* were either annually or biannually or what is termed as 'supra-annual' by Numata et al. (1999). When tabulated the presence or absence of flowering patterns in Table 1, the occurrence can be deduced to the following prediction:

1) FRIM-the biannual and annual flowering episodes occurred twice in consecutive year from 1981 to 1990. However there were no flowering events detected from 1991 to 1995. We were not sure of the cause of it. In 1997 and 1998, the biannual flowering periods commenced again. From such pattern, one can aspects that the flowering periods of *N. heimii* could take place as twice annually and twice biannually in FRIM.
2) Pasoh FR-flowering events tended to occur very haphazardly with biannual flowering periods present in 65% of the observation periods, and in each year from 1992 to 1998. The intervals between each flowering event varied between 2 months to 12 months. Therefore it is not easy to postulate the flowering event for Pasoh FR.
3) Ulu Gombak FR-There have been flowering events every year since 1981. Medway (1972) reported that this species was found to flower in 1963 and 1968 but none from 1964 to 1967 and 1969 in this same Forest Reserve. But from our data, interestingly enough, the annual flowering period occurred once followed by twice biannual events. Such pattern occurred consistently from the period of 1981–1987, 1989–1992 and 1994–1999. Therefore it can be predicted that the coming year will be an annual flowering event and continued with twice biannual events.
4) Ampang FR-there was no systematic pattern from 1981–1985. Annual flowering followed by twice biannual events occurred from 1988–1993 and from 1995–1999. There was no flowering in 1994. Such a pattern tends to follow the flowering patterns of Ulu Gombak FR.

3.2 Flowering occurrence

When the appearance of flowering events by each month was tabulated, Fig. 3 indicates that the occurrence of flowering in most years (either annually or biannually) can either peak within March–May and or be followed by the second peak occurring within September–November in all four phenological plots. From the summary of analysis (Table 2) attained, all pattern of flowering events were significantly correlated with the rainfall ($P < 0.05$). Hence this shows that the model fits the data well. The critical value for correlation coefficient with sample size equal to 12 is 0.576. When compared both Fig. 3 and Table 2, the following findings can be deducted:

1) Fig. 3(a) FRIM: the percentage of flowering occurrence each month over 20 years indicated that the probability of occurrence were the least in January,

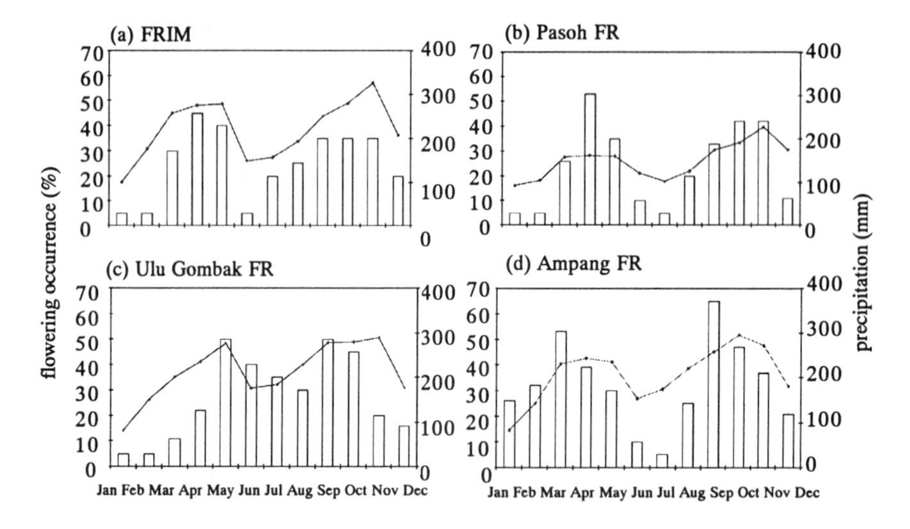

Fig. 3 Monthly rainfall and the percentage of *N. heimii* flowering occurrence in each month for 20 years at respective sites within four FR areas. Bars indicate flowering occurrence. Line indicate precipitation.

Table 2 Summary table of correlation analysis between percentage of flowering occurrence and rainfall.

Location	Correlation coefficient	*P*	*N*	*F*
FRIM	0.878	0.001	12	33.76
Pasoh FR	0.772	0.008	12	14.73
Ulu Gombak FR	0.662	0.019	12	7.8
Ampang FR	0.599	0.04	12	5.59

February and June (only 5%); followed by March, July, August and December (20–30%). 30% of the occurrence can happen from September to November and the highest possibilities were detected in April and May, 45% and 40% respectively. It was found out that the flowering season started at the end of a drying period.

2) Fig. 3(b) Pasoh FR: the probability of flowering to occur for each month were less in January, February, June, July and December (range from 5–10%), followed by August (20%), March (26%), September (33%) and May (35%). The first peak of the flowering period also falls much more frequently in April (53%) and the second peak in October and November (42%).

3) Fig. 3(c) Ulu Gombak FR: the flowering events charted the lowest occurrence in January, February and March (5-11%), followed by December (16%), November (20%), April (22%), August (30%), July (35%), June (40%) and October (45%). The possibilities of occurrence were the highest in May and September (50%).

4) Fig. 3(d) Ampang FR: flowering occurrence was lowest in July and June (5–10%), followed by December, August and January (2–26%); February, April, May and November (30–39%); and October (47%). The occurrence first peaked in March (53%), followed by the second peak in September, which is the highest (65%).

4. DISCUSSION

From the results above, the observation upon 20 years of flowering events offers a prediction that at FRIM, *N. heimii* flowering events can take place twice annually and twice biannually. While in Ulu Gombak and Ampang FRs, such event can be predicted to occur as an annual flowering, continue with twice biannual events. However, it is not easy to postulate the *N. heimii* flowering event in Pasoh FR.

Flowering probabilities for *N. heimii* trees in Ampang FR to flower were higher in January and February as compared to the other three forest areas. While such events in Ulu Gombak occurred more towards the second and third quarter of the year. The majority of flowering events occur first in March at Ampang FR, followed by FRIM and Pasoh FR (both in April) and in May at Ulu Gombak FR. The second flowering event was most likely to take place at Ampang FR and Ulu Gombak FR in September and at Pasoh FR in October and November. At FRIM, the tendency of second flowering event was to occur from September to November. This confirms the prior results reported Yap (1985), which also found two flowering peaks that occurred from March to May and September to November in the same FRs. The results also indicated that the flowering probability increases each month with the increment of the average total precipitation levels and then is reduced as the total levels falls. Dry spells could mean less cloud, greater insulation and higher temperatures with lengthen sunshine hours and increases in photosynthetic active period were known to have a profound effect on the plant physiology. Since there is a highly significant between flowering and rainfall, the increment of precipitation can be the hint for the foresters to start observing for flowering of *N. heimii*.

Our observations indicated that the terminal inflorescences occurred unevenly within the crown. During flowering events, the creamy white petals open fully, exposing the yellow anthers and emitting a sweet fragrance. *N. heimii* strong sweet fragrance and its large size flowers as compared to those of *Shorea*'s, likely indicated an ability to attract more pollinators from a distance. Anthesis is diurnal and according to Appanah (1979), *N. heimii* flowers are mainly pollinated by bees from the genera *Apis* and *Trigona*. His study indicated that the floral production peaks at around fourteenth to fifteenth day and conclude that the peak flowering events for *N. heimii* tend to happen in a single burst similar to *Shorea* species. Such behavior may maximize opportunities for out-crossing and simultaneously increases the size of the pool for potential gene exchange (Chan & Appanah 1980).

From this study we observed that the seeds take approximately 4–5 months to mature from the time of flower initiation. Thus observed pattern of continuous fruiting may be largely due to the effect of staggered flowering prior to fruit development. The amount of seeds produced per tree varies from few to hundreds. Some fallen fruits were found to be attacked by seed beetles. Thus the timing of seed collection is very important, as these seeds readily germinate after seed fall. Since chengal seeds are wingless, majority of the seeds can be found right under the canopy of the mother tree. It was reported that the highest number of seedlings could be found within 5 m of the stem, but here the highest mortality also occurred (Siti Rubiah 1990). This could be due to intense competition and insufficient light for survival and establishment of these species or affects of species-specific pathogens. The report on seed germination and seed storage and seedling growth behavior will be reported in another paper.

5. CONCLUSION

This study revealed some important characteristic of flowering events of *N. heimii*. Based on the 20 years of phenological data, although the results of individual flowering trees were not accounted for, the probability of flowering events and the occurrence in each month of the year were summarized and can be predicted for this species. Flowering events at FRIM was observed to occur as alternate twice annually and twice biannually, while trees of *N. heimii* in Ulu Gombak FR and Ampang FR exhibited an alternate of an annual flowering period occurred once followed by twice biannual patterns. The highest possibilities of the first flowering occurrence in a year was March in Ampang FR, April in both FRIM and Pasoh FR, and May in Ulu Gombak FR. Since the biannual events tend to take place frequently over the past 20 years in all four areas surveyed, the second peak can be expected to occur as follows: September to November in FRIM, September in Ampang FR and Ulu Gombak FR, and October to November in Pasoh FR. Although FRIM, Ampang FR and Ulu Gombak FR areas are closer to on another compared to Pasoh FR, it seems *N. heimii* mother trees in Ampang FR flower first followed by trees in FRIM, Pasoh FR and Ulu Gombak's. This again may be caused by microclimate in the individual areas. Thus, more microclimatic data specifically in the forest areas such as temperature, sunshine hours and humidity should be gathered in order to evaluate the possible factors that triggering flowering events in chengal. The statistical analysis indicates that there is a significant ($P < 0.05$) correlation between flowering and rainfall. At FRIM and Pasoh FR, the onset of flowering events was highly correlated to the onset of the rainy season. The average precipitation immediately before flowering was lowest in the year. Appanah (1993) found that the mass flowering occurred once every 2 to 10 years and had no close correlation with physiography or water availability. Studies have also suggests that gregarious flowering is triggered when the night-time temperatures drop to as low as 19°C for 5–8 days in a row about 8–9 weeks earlier than the actual occurrence of flowering (Appanah 1993; Numata et al. 1999). However, as *N. heimii* tends to flower annually or biannually, we suggest that drought may also trigger flowering occurrence in this species. This indicates that *N. heimii* does not show 'masting' behavior and such species can be very useful for reforestation programs. Therefore foresters will be able to procure the seeds of *N. heimii* by figuring the prediction of flowering events given in this study and thus, they are able to sustain the production of *N. heimii* planting material in the future.

ACKNOWLEDGEMENTS

We wish to thank the Director Generals of FRIM for supporting the continuous 20 years of phenological observation reported in this study. We acknowledge the former study leaders i.e. Drs. F. S. P. Ng , H. T. Chan and S. K. Yap for the initiation and continuance of the phenological studies in FRIM. The budget was maintained by IRPA projects under the Ministry of Science and Technology Malaysia. We are in debt to all the 'orang asli' tree climbers who aided this work especially Juntak So and Ramli Perdus, data entry by Fadzlinah Zolpatah and Hamsinah Hashim, Dr. Toshinori Okuda (NIES) for constant encouragement, Dr. Saw, L. G. for providing Fig. 1 (map of natural distribution of *N. heimii*), Mr. Nor Azman Hussein for helping out with the graphs, and Dr. Chua, L. S. L. for her valuable comments and suggestions. Appreciation also goes to Mr. K. L. Ng from the Meteorological Department, the Nursery and Hydrological section in FRIM for the rainfall data that they provided. We are also grateful to the Forestry Department for providing access to the forest reserve areas within the states involved.

REFERENCES

Appanah, S. (1979) The ecology of insect pollination of some tropical rainforest trees. Ph.D. diss., Univ. Malaya.

Appanah, S. (1993) Mass flowering of dipterocarp forests in the aseasonal tropics. J. Biosci. 18: 457-474.

Ashton, P. S. (1982) Dipterocarpaceae. In Flora Malesiana Serial 1. Spermatophyta 9: 251-552.

Ashton, P. S., Givnish, T. J. & Appanah, S. (1988). Staggered flowering in the dipterocarpaceae: New insights into floral induction and the evolution of mast fruiting in the aseasonal tropics. Am. Nat. 132: 44-66.

Bawa, K. S. (1998) Conservation of genetic resources in Dipterocarpaceae. In Appanah, S. & Turnbull, J. M. (eds). A review of dipterocarps-taxonomy, ecology and silviculture. CIFOR, Indonesia, pp.45-56.

Burgess, P. F. (1972) Studies on the regeneration of the hill forest of the Malay Peninsula- the phenology of dipterocarps. Malay. For. 35: 103-123.

Chan, H. T. & Appanah, S. (1980) Reproductive biology of some Malaysian Dipterocarps. I: Flowering biology. Malay. For. 43: 132-143.

Cockburn, P. F. (1975) Phenology of dipterocarps in Sabah. Malay. For. 38: 160-170.

Kiew, B. H., Kiew, R., Chin, S. C., Davidson, G. & Ng, F. S. P. (1985) Malaysia's 10 most endangered animals, plants and areas. Malay. Nat. J. 38: 2-6.

Krishnapillay, B. & Tompsett, P. (1998) Seed handling. In Appanah, S. & Turnbull, J. M. (eds). A review of dipterocarps-taxonomy, ecology and silviculture. CIFOR, Indonesia, pp.73-88.

Medway, F. L. S. L. (1972) Phenology of a tropical rainforest in Malaya. Biol. J. Linn. Soc. 4: 117-146.

Numata, S., Kachi, N., Okuda, T. & Manokaran, N. (1999) Chemical defences of fruits and mast-fruiting of dipterocarps. J. Trop. Ecol. 15: 695-700.

Saw, L. G. & Raja Barizan, R. S. (1991) Directory of plant genetic resources in Malaysia. Research Pamphlet No. 109, 161pp.

Siti Rubiah, Z. (1990) Studies on germination and seedling growth of *Neobalanocarpus heimii* (King) Ashton. M. thesis, Univ. Pertanian.

Symington, C. F. (1943) Foresters' Manual of Dipterocarps. Malayan Forest Record No.16. Penerbit Universiti Malaya, Kuala Lunpur, 244pp.

Wyatt-Smith, J. (1987) Manual of Malayan Silviculture for inland forest. Part III-Chap. 8: Red Meranti- Keruing Forest. Research Pamphlet No. 101, pp.1-89.

Yap, S. K. (1985) Gregarious flowering of dipterocarps: observation base on fixed tree populations in Selangor and Negri Sembilan, Malay Peninsula. In Kostermans, A. J. G. H. (ed). Proceedings of 3rd Roundtable Conference on Dipterocarps, UNESCO, Jakarta, pp.305-317.

Yap, S. K. & Marzalina, M. (1990) Gregarious flowering- fact or myth. In Proceedings of International Conference on Forest Biology and Conservation in Borneo. Yayasan Sabah, Kota Kinabalu, Sabah, Malaysia, pp.1-4.

16 Leaf Physiological Adjustments to Changing Lights: Partitioning the Heterogeneous Resources across Tree Species

Atsushi Ishida[1], Akira Uemura[1], Naoko Yamashita[2], Michiru Shimizu[3], Takashi Nakano[4] & Ang Lai Hoe[5]

Abstract: Rain forest trees with various light requirements exhibit a high variation in their physiology and morphology as they encounter contrasting light environments, contributing to higher leaf carbon gain in each environment. A desire to understand the species-specific variations of capturing and using of light resources in terms of photosynthetic processes provides the setting for this chapter. Both niche partitioning and chance effects in the distribution of tropical trees are widely recognized. The objective of this chapter is to clarify tree responses at the single leaf and whole plant levels for the spatial and temporal changes of lights, providing an improved background of ideas on rain forest dynamics and coexistence and an improved plan for the sustainable forest management and conservation. Recent evidences that the ecophysiological adjustments of plant to changing lights influence replacement success under the changing environments and tropical forest dynamics, are reviewed briefly.

Key words: canopy gap, chlorophyll fluorescence, leaf gas exchange, leaf initiation rate, ontogenetic change, regeneration niche, successional status.

1. INTRODUCTION

Tropical rain forests are notable for their richness in tree species. Recent topics in the coexistence of species in tropical forests are reviewed (Denslow 1987; Primack 1992). There are two major perspectives regarding species coexistence; equilibrium (niche partitioning) and non-equilibrium (stochastic existence). Some studies in the tropics have raised the possibility that fine-scale differences determine partitioning among tree species with respect to canopy-gap size (Brokaw 1987; Denslow 1980; Popma & Bongers 1988; Schnitzer & Carson 2001), soil texture and water condition (Davies et al. 1998; Niiyama et al. 1999), or maximum tree size and longevity (Swaine & Whitmore 1988). On the other hand, field data over a long term by Hubbell et al. (1999) did not support the hypothesis that partitioning occurs among tropical tree species. Denslow (1980) postulates that a dynamic equilibrium exists in a community between disturbance rates and the rates of competitive displacement among species. Pacala & Rees (1998) suggest that niche partitioning is important if disturbance regimes occur on a small scale, whereas competition

[1] Forestry and Forest Products Research Institute (FFPRI), Tsukuba 305-8687, Japan. E-mail: atto@ffpri.affrc.go.jp
[2] Hokkaido Research Center, FFPRI, Japan.
[3] The University of Tokyo, Japan.
[4] Yamanashi Institute of Environmental Science, Japan.
[5] Forest Research Institute Malaysia (FRIM), Malaysia.

colonization is more important if disturbance regimes occur on a large scale. Nakashizuka (2001) postulates that the major mechanisms of coexistence change throughout the life-development stage, contributing to the coexistence of tree species in a community.

The objectives of the present review represent the species-specific differences of resource use, especially in light in species of lowland rain forest, and provide perspective issues in currently ecophysiological research of rain forest plants. The spatial and temporal changes in lights occurred by disturbance or competitive conditions among plants. Over the past decade, several studies have been shown that the change of resources caused by disturbance and the species-specific physiological response for the changing resources are related to the dynamics of species replacement (Clark & Clark 1992; Loik & Holl 2001) or invasive success in newly canopy-gap sites (Baruch et al. 2000; Durand & Goldstein 2001; Yamashita et al. 2000). The understanding of species-specific differences of resource use provides an improved background of ideas on rain forest dynamics and coexistence, and contributes to the making of an improved plan for the sustainable forest management and rehabilitation of degraded forests in tropical areas.

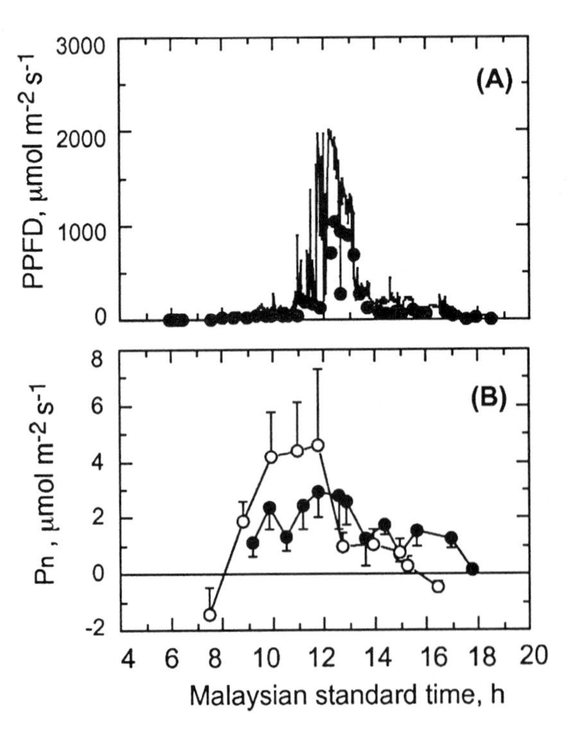

Fig. 1 Diurnal changes of (A) photosynthetic photon flux density (PPFD; solid line) on a horizontal surface and PPFD at the leaf surface of *Neobalanocarpus heimii* (closed circle) and (B) net photosynthetic rate (P_n) in *Shorea leprosula* (open circle) and *Neobalanocarpus heimii* (closed circle) seedlings planted under a canopy gap. Vertical bars are 1 *SD*. Modified from Ishida et al. (1999b).

2. LEAF PHYSIOLOGY OF SAPLINGS GROWING IN SUNLIT AND SHADED SITES

The differences of canopy-gap size can change the daily total amount of quantum and the diurnal time courses of diffuse and direct lights. The amount of light through the canopy layer is an important factor on the amount of leaf carbon gain of understory seedlings. The slight differences in the amount of light in rain forest understory can have a major influence on the growth rates of understory plants (Pearcy 1987) and species richness (Svenning 2000). Some tree species growing under canopy gaps in tropics use mainly diffused light in photosynthesis (Ackerly & Bazzaz 1995b). Ishida et al. (1999b) also found that the seedlings of *Shorea leprosula* Miq. and *Neobalanocarpus heimii* (King) Ashton (Dipterocarpaceae) growing under a canopy gap well used diffused light in the morning for a net photosynthetic rate, rather than direct sunlight at midday or diffused light in the afternoon (Fig. 1). Thus, the effective light for the carbon gain temporally varies in the sapling under the canopy gap.

The dipterocarp species vary in shade tolerance at the seedling stage. Turner (1990) showed that there is the high variation of seedling mortality and height growth among dipterocarp species at a forest understory in Peninsula Malaysia. When comparisons are made across species that included extreme differences in leaf physiology, niche partitioning in light resource is probably found even within climax trees. *N. heimii* and *S. leprosula* are typical canopy trees of Dipterocarpaceae, and these species are commonly found at the Pasoh Forest Reserve (Pasoh FR) from sapling to adult stages (Manokaran et al. 1992). Both species are emergent trees in Malaysian forests, but *N. heimii* is a slow-growing dipterocarp and has a high wood density (980 kg m^{-3}), whereas *S. leprosula* is a fast-growing dipterocarp and has a relatively low wood density (580 kg m^{-3}). Kachi et al. (1995) found that the survival rate was higher in *N. heimii* saplings than in *S. leprosula* saplings in deep shady conditions of understory at the Pasoh FR, and the growth rate was higher in *S. leprosula* saplings than in *N. heimii* saplings at the center of a canopy gap.

Table 1 Leaf physiological and morphological characteristics in sapling leaves of a pioneer tree and two dipterocarps with contrasting growth rate and wood density[a].

	SLA[b] ($m^2 kg^{-1}$)	Fv/Fm at predown	Max. net photosynthetic rate: Area basis ($\mu mol\ m^{-2} s^{-1}$)	Max. net photosynthetic rate: Mass basis ($\mu mol\ kg^{-1} s^{-1}$)	Max. ETR[b] Area basis ($\mu mol\ m^{-2} s^{-1}$)	Max. NPQ[b]
Macaranga gigantea **(a pioneer tree)**						
Sunlit site	19.2 a	0.77 a	8.1 a	155 a	125 a	3.3 a
Small gap site	25.4 b	0.79 a	4.5 b	112 a	85 b	3.5 a
Shorea leprosula **(a fast-growing dipterocarp)**						
Sunlit site	9.4 c	0.78 a	11.0 a	103 a	123 a	2.7 a
Small gap site	11.9 d	0.77 a	5.4 b	64 b	65 b	3.4 a
Neobalanocarpus heimii **(a slow-growing dipterocarp)**						
Sunlit site	12.1 d	0.76 a	1.9 c	23 c	67 b	4.5 b
Small gap site	13.2 e	0.79 a	3.5 d	46 d	71 b	3.4 a

[a] Modified from Ishida et al. (1999b). The same letters (a–d) after the data in each column indicate homogeneous groups according to *t*-test ($P < 0.05$). The wood density is approximately 300 kg m^{-3}, 580 kg m^{-3} and 980 kg m^{-3} in *Macaranga gigantea, Shorea leprosula* and *Neobalanocarpus heimii,* respectively.

[b] SLA is the ratio of leaf area to dry mass; ETR is electron transport rate through PSII; and NPQ is non-photochemical quenching.

In leaf morphology, the difference in SLA (specific leaf area; the ratio of leaf area to leaf dry mass) between the saplings growing at a sunlit site and a small-canopy-gap site (small-gap site) was large in a pioneer *Macaranga gigantea* and a fast-growing *S. leprosula*, whereas the differences was small in a slow-growing *N. heimii* (Table 1). Thus, the plasticity in leaf thickness was lower in *N. heimii*; the term "plasticity" used in this chapter is defined as the differences in leaf characteristics at a constant environment with contrasting light regimes, i.e. the differences between existing sun and shade leaves. Bazzaz & Pickett (1980) hypothesize that sun-adapted plants exhibit greater plasticity than shade-tolerant plants. Thus, the low plasticity in leaf thickness of *N. heimii* saplings would reflect their shade-tolerant character, as a genetic property.

In leaf physiology, though the leaf-area based photosynthetic rate was not necessary higher in a pioneer than in dipterocarps, the leaf-mass based photosynthetic rate was higher in a pioneer (Table 1). The maximum net photosynthetic rate was higher at the sunlit site than at the small-gap site in *M. gigantea* and *S. leprosula*, whereas the relationship was reversed in *N. heimii*. The result that the maximum net photosynthetic rate in shade leaves positioned above those of sun leaves, has been reported in some late-successional, shade-tolerance trees (Bazzaz & Carlson 1982; Kubiske & Pregitzer 1996), but not all shade-tolerant species in tropics (e.g. Yamashita et al. 2000). The low photosynthetic rate at the sunlit site in *N. heimii* saplings will be associated with the low photoinhibition avoidance as a shade-tolerance tree.

Non-photochemical quenching (NPQ) is a measure of the thermal dissipation of excess-absorbed light energy in leaves, as a mechanism of photoinhibition avoidance (Björkman & Demmig-Adams 1994). The value of NPQ at a given photosynthetic photon flux density (PPFD) was higher in *N. heimii* leaves than in *S. leprosula* and *M. gigantea* leaves, especially at the sunlit site (Table 1). This indicates that because *N. heimii* exhibits low light-use properties of leaves (i.e. low photosynthetic capacity), much excess-absorbed light energy needs to dissipate actively to avoid photoinhibition probably associated with a large pool size of xanthophyll pigments of the leaves. In spite of the high NPQ, the maximum value of quantum yield of the photosystem II (PSII) measured in darkness at predawn (Fv/Fm) was lowest in the leaves of *N. heimii* saplings planted at the sunlit site (but this was no significance). The value (0.76) was lower than the values of 0.81–0.83 found in unstressed leaves of higher plants (Björkman & Demmig 1987), indicating that the leaves of *N. heimii* suffered from slight but sustained photoinhibition.

These morphological and physiological traits of canopy leaves indicate that the sapling of *N. heimii* is more shade-tolerant and less favorable in bright environments, in spite of a typical emergent tree. These data at the single leaf level are consistent with other shade-tolerant characteristics, such as demography (high survival rate in the deep shaded site), slow plant growth rate, and heavy woods. Thus, we postulate a hypothesis that leaf physiology is associated with ecological status for light use at a life-development stage. The overall results indicate that the regeneration of *N. heimii* favors sites where a small canopy gap occurs or where a frequent canopy closure occurs until the height of the saplings reaches the forest canopy height, supporting that regeneration niche partitioning exists according to *sensu* Grubb (1977). However, because there is no clear partitioning in topography between *N. heimii* and *S. leprosula* trees in the 50-ha plot in the Pasoh FR (Manokaran et al. 1992), more long-term work is still needed.

3. ONTOGENETIC CHANGE OF LEAF PHYSIOLOGY AS TREE GROWS

Many studies on rain forests have been conducted in the ecophysiological processes on tree seedlings or shrubs that were not tall, because of the difficulty of access to the top canopy of large trees. Most studies on shade tolerance were defined by tropical seedling performance in survival (e.g. Augspurger 1984), in growth (e.g. Kitajima 1994; Popma & Bongers 1988; Sasaki & Mori 1981; Turner 1990) and in leaf physiology (e.g. Ishida et al. 1999b; Liang et al. 2001; Turner & Newton 1990). Although *N. heimii* is a typical emergent tree that is exposed to high light at the adult stage, the saplings exhibit more shade tolerance. Grubb (1996) suggests the possibility that the amount of light needed to maintain growth increases as the tree size increases; therefore there is crossover in the relative ranking of species for shade tolerance between seedlings and adult-tree stages.

To examine the possibility of crossover for shade tolerance, we compared the diurnal leaf gas exchange and the leaf morphology at the top canopy between a sapling and an adult tree of *N. heimii* in Peninsular Malaysia. The results showed no significant differences in the maximum stomatal conductance and net photosynthetic rate and SLA between adult and sapling trees (Fig. 2). On the other hand, we also obtained a contrast result in the top canopy leaves of *Carpinus laxiflora* (Sieb.et Zucc.) Bl., a temperate deciduous tree in Japan; i.e. the stomatal conductance and net photosynthetic rate were significantly higher in an adult tree than in a sapling tree growing in a sunlit site. The value of SLA in the top canopy leaves of *C. laxiflora* was significantly lower (i.e. thicker) in the adult tree than in the sapling tree. Thus, the high net photosynthetic rate in the adult tree of *C. laxiflora* was due to both high stomatal conductance and thick leaf, probably

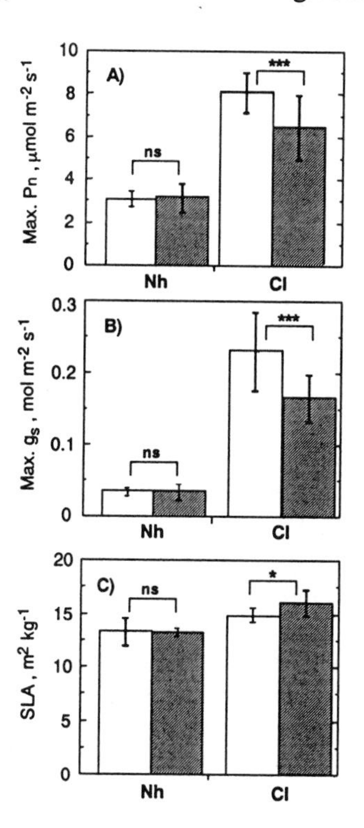

Fig. 2 Comparison of (A) maximum net photosynthetic rate at ambient air (P_n) and (B) maximum water vapor stomatal conductance (g_s) under relatively low leaf-to-air vapor pressure deficit and high light intensity, and (C) specific leaf area (SLA) at the top canopy between an adult tree (open bars) and a sapling (shaded bars) in *Neobalanocarpus heimii* (Nh) and *Carpinus laxiflora* (Cl). Vertical bars are ±1 SD, and ns shows no significant difference ($P > 0.05$; *t*-test); * < 0.05; *** < 0.001. Data in Nh saplings are modified from Ishida et al. (1999b) and the other data are unpublished.

associated with high nitrogen content per unit leaf area.

The ontogenetic changes in leaf physiology and morphology with increasing tree size may generate crossover in relative ranking for shade tolerance among tree species in a community (Grubb 1996). However, the ontogenetic change was still not supported by many published data (Thomas & Bazzaz 1999). Variations in the photosynthetic capacity of individual tropical trees have previously been examined from different viewpoints: 1) direct effect of the amount of light received (Ackerly & Bazzaz 1995a; Pearcy 1987; Zotz & Winter 1993), 2) allocation process of nutrients within leaves (Kitajima et al. 1997a), 3) ontogenetic aging process of leaves (Ishida et al. 1999c), and 4) hydraulic limitation related to tree height (Ryan &Yoder 1997; Yoder et al. 1994). Yoder et al. (1994) compared the diurnal photosynthetic rate and stomatal conductance between a young conifer tree (10 m tall height) and an old tree (32 m tall height) in lodgepole pine (*Pinus contorta* var. *latifolia* Englem.). They showed that the net photosynthetic rate and stomatal conductance in the young tree positioned above those of the old tree, and speculated that the reduced stomatal conductance in the old conifer tree was due to an increased hydraulic resistance resulting from increased path length through trachid caused by the tall height (i.e. hydraulic limitation hypothesis in the maximum tall height of trees). On the other hand, we found the reverse relation in *Carpinus laxiflora* (Fig. 2), probably this was due to the high availability of water and nutrients from soils caused by hydraulic upturn, because of the development of the root system as the tree grows and because of the low hydraulic resistance at a given length in a vessel compared with the trachid of conifer.

Thomas & Bazzaz (1999) found a positive relationship between the leaf-area based maximum photosynthetic capacity of the sapling stage and the maximum height reached by the mature stage within 28 late successional species (excluding *N. heimii*) in the Pasoh FR. Their observation may support a hypothesis that the maximum height of a mature tree limited by the balance of carbon budget between photosynthesis and respiration at the whole plant level rather than by hydraulic limitation, based on a hypothesis that there is no crossover in relative species ranking for shade tolerance throughout the life-development stage. However, the net photosynthetic rate in *N. heimii* was relatively low among dipterocarp trees, in spite of an emergent tree (see Table 1 and Chap. 17). Therefore, more work is needed to evaluate whether there are ontogenetic changes in leaf physiology and morphology as trees grow, and is needed to examine the factor determining the limitation of heights in tropical trees.

High water and nutrient contents in soils and high CO_2 concentrations in the rain forest understory clearly improve the carbon balance at the leaf level (Liang et al. 2001) or the whole plant level (Burslem et al. 1995; Ishida et al. 2000b; Kitajima 1994) in seedlings. Ishida et al. (2000b) showed that the instantaneous growth CO_2 compensation point of *Shorea smithiana* Sym. seedling increased three times (from 100 μ mol mol^{-1} to 300 μ mol mol^{-1}) from the single-leaf level at midday to the whole-plant level for 24h including dark respiration of leaves, stems, and roots, indicating the importance of dry matter allocation between productive (i.e. leaves) and non-productive (i.e. stems and roots) organs within a plant body on the carbon economy at the whole-plant level. Kitajima (1994) showed that high LAR (leaf area ratio; the ratio of total leaf area to total dry biomass) contributed to high whole-plant carbon gain in the seedling shade tolerance of tropical trees. Probably, the change of dry matter allocation as trees grow can influence the carbon economy at the whole-plant level, especially under changing environments. Understanding of the interactive effects of dry matter allocation and resource availability on carbon balance at the whole-plant level is shortage, especially related with the growth of trees.

4. PHYSIOLOGICAL ADJUSTMENTS TO CHANGING LIGHTS

A cycle of suppression (a result of canopy-gap closing) and release (a result of canopy-gap opening) of height growth must be repeated several times until the height of saplings reaches the forest canopy height (Canham 1990). Throughout the process, the canopy leaves experience different light conditions from sun to shade caused by mutual shading or competition and from shade to sun caused by disturbance. The range and speed of the acclimation to changing lights represent an important part in carbon gain, tree growth, reproduction and forest dynamics; the term "acclimation" used in this chapter is defined as the temporary change of leaf characteristics in response to changing lights.

4.1 Changing from sun to shade

When light conditions diminish, an increase in the net photosynthetic rate at low lights is important for maintaining positive carbon gain on the balance of photosynthesis and respiration (Givnish 1988). Because the adult tree of *Dryobalanops aromatica* Gaertn. f. (Dipterocarpaceae) develops a dense canopy (Fig. 3), the apex leaves at the uppermost canopy initiate under sunlit conditions, and these leaves are subject to self-shading as they age (Fig. 4). In contrast, the leaves in the lower canopy are exposed to uniform dim light conditions throughout the leaf life span. The ratio of chlorophyll to nitrogen in the leaf (chl/N ratio) was higher in the lower-canopy leaves than in the uppermost-canopy leaves. The ratio

Fig. 3 The top canopy of 60-year-old *Dryobalanops aromatica* tree at FRIM arboretum and the maximum canopy height is about 35 m.

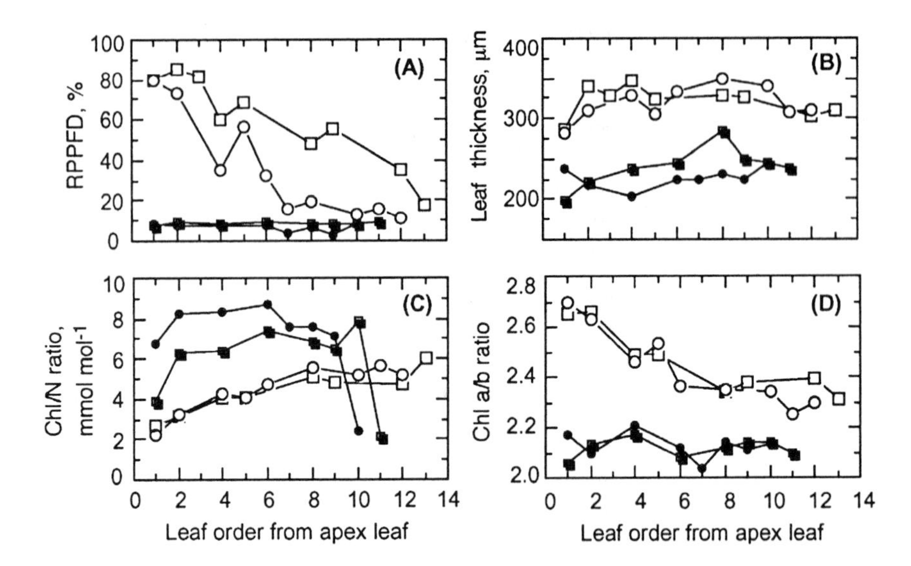

Fig. 4 Changes in (A) photosynthetic photon flux density relative to full sunlight (RPPFD), (B) leaf thickness, (C) chlorophyll/nitrogen ratio and (D) chlorophyll a/b ratio with leaf order from the apex leaf in two uppermost shoots (open symbols) and two lower shoots (closed symbols) of an adult tree of *Dryobalanops aromatica*. Leaf order 1 is the just fully expanded leaves. Modified from Ishida et al. (1999c).

of chlorophyll a to b in the leaf (chl a/b ratio) was lower in the lower-canopy leaves than in the uppermost-canopy leaves. The values of chl/N and chlorophyll a/b ratios remained unchanged in the lower-canopy leaves through their leaf life span. On the other hand, in the uppermost-canopy leaves, the chl/N ratio gradually increased and the chlorophyll a/b ratio gradually decreased as the self-shading progressed, but dry-mass based total nitrogen remained unchanged. These data indicate that a rearrangement of leaf nitrogen between light-harvesting and carbon-fixation components within a leaf occurred in response to self-shading even in mature canopy leaves (Field 1983).

However, no change of leaf thickness was found in the mature leaves at both the uppermost and lower canopies in *D. aromatica* (Fig. 4). Thus, morphological properties to adjust to changing lights were more conservative than the physiological properties. This result will be reasonable for considering the difficulty of rearrangement of cell walls. However, this result is not general in tropical trees, there is a report that the leaf thickness increased even in mature leaves in *Bischofia javanica* Bl. because of the increase in cell length, when transferred from shade to sun (Kamaluddin & Grace 1992).

In spite of the differences in the patterns of the light environments with leaf order between the uppermost-canopy and the lower-canopy leaves of *D. aromatica,* the leaf-mass based maximum photosynthetic rate decreased with increasing leaf age in a similar pattern between both canopy leaves (Fig. 5). Although the nitrogen rearrangement from carbon-fixation components to light-harvesting components with progressing self-shading can partially explain the decrease in the maximum photosynthetic capacity with leaf aging in the uppermost-canopy leaves, the

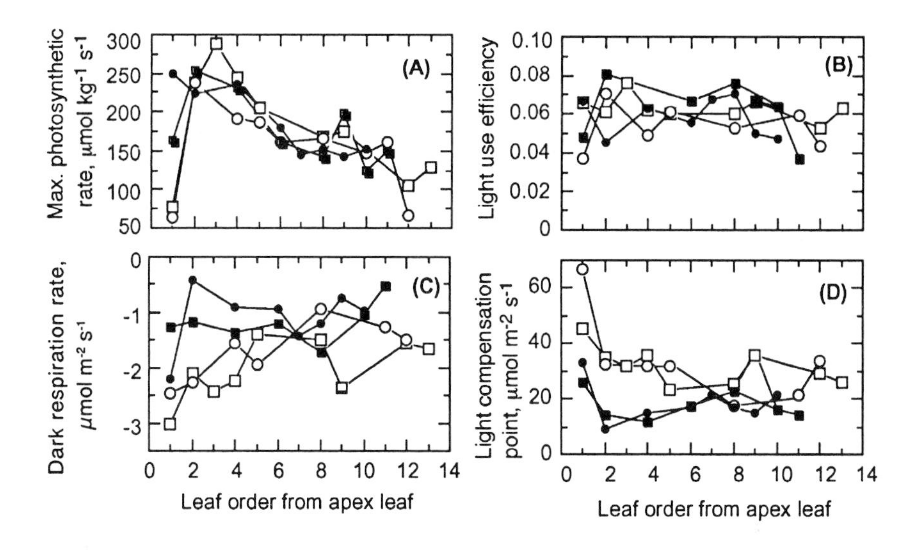

Fig. 5 Changes in (A) mass-based photosynthetic rate under high light and high CO_2 conditions, (B) light use efficiency, (C) dark respiration rate and (D) light compensation points with the leaf order of *Dryobalanops aromatica*. Legend are the same as Fig. 4.

decrease in photosynthetic capacity with leaf aging will be mainly caused by other effects, such as leaf aging rather than self-shading. The changes of photosynthetic capacity with leaf aging were not correlated to the leaf nitrogen content (Ishida et al. 1999c). Although good relationships between leaf nitrogen content and photosynthetic rate have been well demonstrated (e.g. Evans 1989), a decrease in photosynthetic nitrogen-use efficiency with leaf age was also found in other tropical trees (Kitajima et al. 1997b; Sobrado 1994), resulting in no good correlation between leaf nitrogen content and photosynthetic rate. Because interactions of multiple biotic and abiotic factors are relevant to the variations of photosynthetic rate within individual tropical trees, difficulties for interpreting the variations of photosynthetic rate remained (Ishida et al. 1996).

An adjustment in net photosynthetic rate for self-shading was found in the respiration rate, rather than in the maximum photosynthetic rate in *Dryobalanops aromatica* leaves (Fig. 5). Light-use efficiency measured under 0–30 μ mol m^{-2} s^{-1} PPFD exhibited no difference between the uppermost-canopy and the lower-canopy leaves, and no change was noted in leaf order in either canopy leaves (mean ± 1*SD*, 0.076 ± 0.011). The dark respiration rate and the light compensation point remained unchanged in the lower-canopy leaves. In the uppermost-canopy leaves, however, dark respiration rate decreased (toward 0), and the light compensation point decreased (toward 0) as self-shading progressed, though no leaf thickness (Fig.4B) was changed. These data showed that the adjustment for self-shading with respect to carbon balance at the single leaf level was due to a reduction in the amount of dark respiration, instead of adjustments of light-use efficiency or the maximum photosynthetic rate in *D. aromatica*.

4.2 Changing from shade to sun

The availability of light resource changes in space and time by disturbances. A canopy-gap opening improves the light conditions of the understory, but excess light and elevated temperature often bring photoinhibition to the leaves of dipterocarp seedlings (Ishida et al. 2000a; Kitao et al. 2000; Turner & Newton 1990). Even in tropical pioneer trees growing at a sunlit site, photoinhibition was found when the top canopy leaves were horizontally fixed with fine wire and the leaves were exposed to direct sunlight at midday (Ishida et al. 1999a).

Yamashita et al. (2000) indicated that the species replacement success after

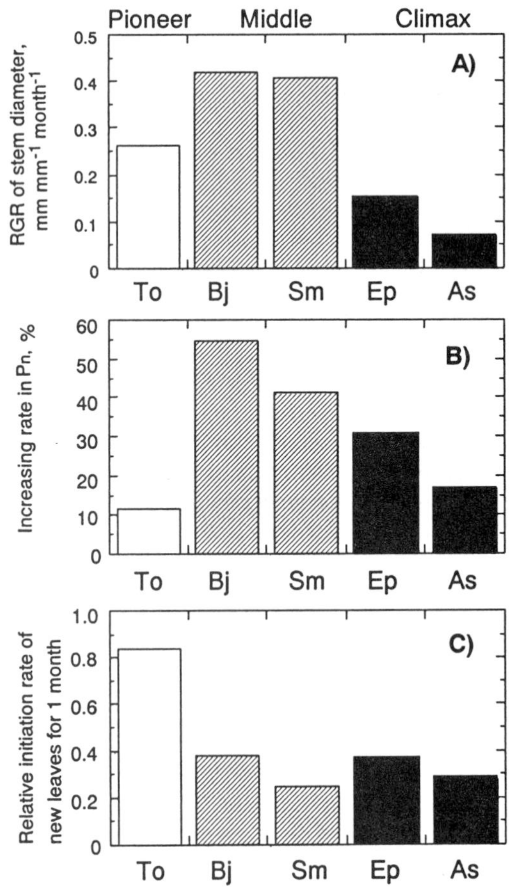

Fig. 6 One month after the saplings were tranfered from shade to sun, (A) relative growth rate (RGR) of stem diameter and (B) increasing rate of net photosynthetic rate (P_n) of the existing shade leaves before transfer. At a shaded condition, (C) relative initiation rate of new leaves (the ratio of the number of new leaves to the number of existing leaves) in the saplings; a pioneer *Trema orientalis* (To), middle successional *Bischofia javanica* and *Schima mertensiana* trees (Bj and Sm, respectively), and late successional *Elaeocarpus photiaefolius* and *Ardisia sieboldii* trees (Ep and As, respectively). Modified from Yamashita et al. (2000).

changing lights from shade to sun was due to the physiological acclimation not only in existing shade leaves by minimizing photoinhibition and enhancing photosynthetic rate but also in the deployment of new sun leaves with high photosynthetic capacity. When the seedlings of four tropical tree species were transferred from shaded to sunlit conditions, the increase in photosynthetic capacity for the mature leaves grown under shaded conditions was higher in middle successional trees (*Bischofia javanica* Blume and *Schima mertensiana* (Sieb. et Zucc.)) than in a pioneer tree (*Trema orientalis* Blume) or a late successional tree (*Elaeocarpus photiaefolius* Hook. Et Arn. and *Ardisia sieboldii* Miquel.) (Fig. 6). The susceptibility of PSII by excess-light energy (i.e. the level of photoinhibition after the sun transfer) was lowest in the mature leaves of the pioneer. The initiation rate of new leaves that adjusted to new conditions was faster in the pioneer tree than in the middle or late successional trees. These data show that the acclimation to increasing light in the pioneer tree was dependent on the initiation of new leaves adjusted to new environments, rather than on the acclimation of old leaves. On the other hand, the acclimation in the late successional tree species was dependent on the acclimation of old leaves that developed under a shaded condition, rather than new-leaf initiation. During the one month after tree saplings were transferred from shade to sun, the relative growth rate of stem diameter as a measure of the growth at whole plant level was higher in the middle successional tree species than in the pioneer or late successional trees. This has important implications for the relative success of middle successional tree species, which have a high acclimation capacity of existing shade leaves and a relatively high initiation rate of new leaves. On the Bonin Islands in the Pacific Ocean, *Bischofia javanica* is a middle successional tree and is one of the most successful invasive species after disturbances by typhoons (Yamashita et al. 2000).

Becker et al. (1999) argued that leaf sizes are generally smaller in heath forests with lower nutrient availability in soils, comparing with lowland dipterocarp forests. However, the individual leaf sizes largely vary among lowland dipterocarp species. When comparing four *Shorea* species coexisting in a lowland rain forest, small individual leaves were associated with high leaf initiation rates (Fig. 7). This

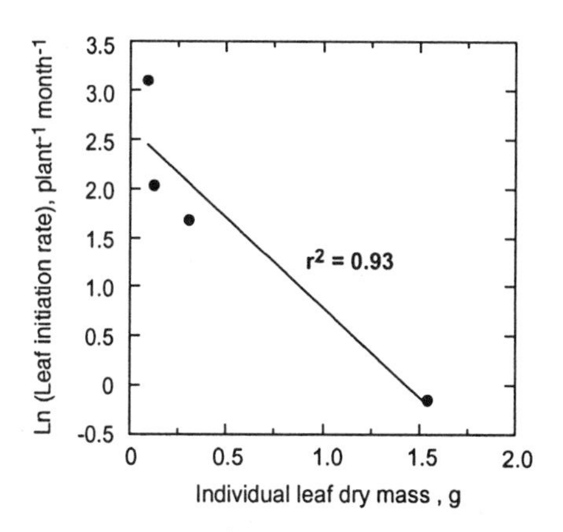

Fig. 7 Relationship between individual leaf dry mass and Ln (leaf initiation rate) of the individual seedlings planted at a sunlit site in *Shorea leprosula, S. pauciflora, S. gratisima* and *S. smithiana* (unpubl. data).

predicts that saplings with small individual leaves can rapidly adjust to the changing light from shade to sun because of the high leaf initiation rate, even if the dipterocarp trees have similar leaf physiological properties under constant light environments. Thus, the variation of leaf size in dipterocarps may be related to regeneration niche in changing lights at lowland dipterocarp forests. Bazzaz (1996) predicts that recent global environmental change increases pioneer plants in various areas. However, we predict that the large-scale disturbances caused by human and fire impacts increase the richness of pioneers or plants with relatively small individual leaves, whereas small-scale disturbances such as a canopy-gap opening caused by a tree fall promotes an increase in middle successional tree species or plants with relatively large individual leaves as invasive or replaceable success species in lowland dipterocarp forests. The interspecific comparison of the use of changing resources in time and space, rather than the use of resources under constant environments, will foster better understanding of forest dynamics, species coexistence, and species replacement.

5. CONCLUDING REMARKS

The species-specific variations of capturing and using changing resources at the single leaf and whole plant levels would influence the whole life history and demography of coexisting tree species in a community, contributing to the understanding of forest dynamics and species replacement. The environment of rain forests without severe climates such as cool or prolonged drought will permit the packing of various plant species that have a specialized suite of physiological and ecological traits at several developmental stages, perhaps contributing to the maintenance of high biodiversity in this area. The increase in limiting factors for carbon gain, growth, and survival will decrease functional and species richness in a community.

Various issues represented in the present chapter still remain in plant ecophysiological research at rain forests. These issues include high biodiversity and the coexistence of species, the height limitation of trees, the relationship between nutrients and photosynthesis in leaf, the interactive effects of dry matter allocation and resource availability on the whole-plant carbon gain, the crossover of relative ranking across species throughout life stages, the resource use under changing and constant environments, and the linkage between leaf properties and regeneration form. The globally environmental change and the recent increase of human influence on tropical forests should promote an increase in invasive-success species, associated with high use capacity of changing resources. The accumulation of ecophysiological data among tropical tree species will contribute to predict the future of forests in the tropics and will make an improved plan for sustainable forest management and the rehabilitation of degraded forests.

ACKNOWLEDGEMENTS

This study was financially supported by the Joint Research Project between the NIES, FRIM and UPM (Global Environmental Research Program, supported by Japan Environmental Agency, Grant No. E-2-3).

REFERENCES

Ackerly, D. D. & Bazzaz, F. A. (1995a) Leaf dynamics, self-shading and carbon gain in seedlings of a tropical pioneer tree. Oecologia 101: 289-298.

Ackerly, D. D. & Bazzaz F. A. (1995b) Seedling crown orientation and interception of diffuse

radiation in tropical forest gaps. Ecology 76: 1134-1146.
Augspurger, C. K. (1984) Light requirements of neotropical tree seedlings: a comparative study of growth and survival. J. Ecol. 72: 777-795.
Baruch, Z., Pattison, R. R. & Goldstein, G. (2000) Responses to light and water availability of four invasive Melastomataceae in the Hawaiian islands. Internatinal J. Plant Sci. 161: 107-118.
Bazzaz, F. A. (1996) Plants in changing environments. Cambridge University Press, Cambridge, UK, 320pp.
Bazzaz, F. A. & Carlson, R.W. (1982) Photosynthetic acclimation to variability in the light environment of early and late successional plants. Oecologia 54: 313-316.
Bazzaz, F. A. & Pickett, S. T. A. (1980) Physiological ecology of tropical succession: a comparative review. Annu. Rev. Ecol. Syst. 11: 287-310.
Becker, P., Davies, S. J., Moksin, M. H. J., Ismail, M. Z. & Simanjuntak, P. M. (1999) Leaf size distributions of understorey plants in mixed dipterocarp and heath forests of Brunei. J. Trop. Ecol. 15: 123-128.
Björkman, O. & Demmig, B. (1987) Photon yield of O_2 evolution and chlorophyll fluorescencecharacteristics at 77K among vascular plants of diverse origins. Planta 170: 489-504.
Björkman, O. & Demmig-Adams, B. (1994) Regulation of photosynthetic light energy capture, conversion, and dissipation in leaves of higher plants. Ecol. Studies 100: 17-47.
Brokaw, N. V. L. (1987) Gap-phase regeneration of three pioneer tree species in a tropical forest. J. Ecol. 75: 9-19.
Burslem, D. F. R. P., Grubb, P. J. & Turner, I. M. (1995) Responses to nutrient addition among shade-tolerant tree seedlings of lowland tropical rain forest in Singapore. J. Ecol. 83: 113-122.
Canham, C. D. (1990) Suppression and release during canopy recruitment in *Fagus grandifolia*. Bulletin of the Torrey Botanical Club 117: 1-7.
Clark, D. A. & Clark, D. B. (1992) Life history diversity of canopy and emergent trees in a neotropical rain forest. Ecol. Monogr. 62: 315-344.
Davies, S. J., Palmiotto, P. A., Ashton, P. S., Lee, H. S. & Lafrankie, J. V. (1998) Comparative ecology of 11 sympatric species of *Macaranga* in Borneo: tree distribution in relation to horizontal and vertical resource heterogeneity. J. Ecol. 86: 662-673.
Denslow, J. S. (1980) Gap partitioning among tropical rainforest trees. Biotropica 12: 47-55.
Denslow, J. S. (1987) Tropical rainforest gaps and tree species diversity. Annu. Rev. Ecol. Syst. 18: 431-451.
Durand, L. Z. & Goldstein, G. (2001) Photosynthesis, photoinhibition, and nitrogen use efficiency in native and invasive tree ferns in Hawaii. Oecologia 126: 345-354.
Evans, J. R. (1989) Photosynthesis and nitrogen relationships in leaves of C_3 plants. Oecologia 78: 9-19.
Field, C. B. (1983) Allocating leaf nitrogen for the maximization of carbon gain: leaf age as a control on the allocation program. Oecologia 56: 341-347.
Givnish, T. J. (1988) Adaptation to sun and shade: a whole- plant perspective. Aust. J. Plant Physiol. 15: 63-92.
Grubb, P. J. (1977) The maintenance of species-richness in plant communities: the importance of the regeneration niche. Biol. Rev. 52: 107-145.
Grubb, P. J. (1996) Rainforest dynamics: the need for new paradigms. In Edwards, D. S., Booth, W. E. & Choy, S. C. (eds). Tropical Rainforest Research–Current Issues. Kluwer Academic publishers, the Netherlands, pp.215-233.
Hubbell, S. P., Foster, R. B., O'Brien, S. T., Harms, K. E., Condit, R., Wechsler, B., Wright, S. J. & Loo de Lao, S. (1999) Light-gap disturbances, recruitment limitation, and tree diversity in a neotropical forest. Science 283: 554-557.
Ishida, A., Toma, T., Matsumoto, Y., Yap, S. K. & Maruyama, Y. (1996) Diurnal changes in leaf gas exchange characteristics in the uppermost canopy of a rain forest tree, *Dryobalanops aromatica* Gaertn.f. Tree Physiol. 16: 779-785.
Ishida, A., Toma, T. & Marjenah (1999a) Leaf gas exchange and chlorophyll fluorescence in

relation to leaf angle, azimuth, and canopy position in the tropical pioneer tree, *Macaranga conifera*. Tree Physiol. 19: 117-124.

Ishida, A., Nakano, T., Matsumoto, Y., Sakoda, M. & Ang, L. H. (1999b) Diurnal changes in leaf gas exchange and chlorophyll fluorescence in tropical tree species with contrasting light requirements. Ecol. Res. 14: 77-88.

Ishida, A., Uemura, A., Koike, N., Matsumoto, Y. & Ang, L. H. (1999c) Interactive effects of leaf age and self-shading on leaf structure, photosynthetic capacity and chlorophyll fluorescence in the rain forest tree, *Dryobalanops aromatica*. Tree Physiol. 19: 741-747.

Ishida, A., Toma, T., Ghozali, D. I. & Marjenah (2000a) In situ study of the effects of elevated temperature on photoinhibition in climax and pioneer species. Ecol. Studies 140: 269-280.

Ishida, A., Toma, T., Mori, S. & Marjenah (2000b) Effects of foliar nitrogen and water deficit on the carbon economy of *Shorea smithiana* Sym. seedlings. Biotropica 32: 351-358.

Kachi, N., Okuda, T., Yap, S. K. & Manokaran, N. (1995) Biodiversity and regeneration of canopy tree species in a tropical rain forest in Southeast Asia. J. Environ. Sci. 9: 17-36.

Kamaluddin, M. & Grace, J. (1992) Photoinhibition and light acclimation in seedlings of *Bischofia javanica*, a tropical forest tree from Asia. Ann. Bot. 69: 47-52.

Kitajima, K. (1994) Relative importance of photosynthetic traits and allocation patterns as correlated of seedlings shade tolerance of 13 tropical trees. Oecologia 98: 419-428.

Kitajima, K., Mulkey, S. S. & Wright, S. J. (1997a) Seasonal leaf phenotypes in the canopy of a tropical dry forest: photosynthetic characteristics and associated traits. Oecologia 109: 490-498.

Kitajima, K., Mulkey, S. S. & Wright, S. J. (1997b) Decline of photosynthetic capacity with leaf age in relation to leaf longevities for five tropical canopy tree species. Am. J. Bot. 84: 702-708.

Kitao, M., Lei, T. T., Koike, T., Tobita, H., Maruyama, Y., Matsumoto, Y. & Ang, L. H. (2000) Temperature response and photoinhibition investigated by chlorophyll fluorescence measurements for four distinct species of dipterocarp trees. Physiol. Plant. 109: 284-290.

Kubiske, M. E. & Pregitzer, K. S. (1996) Effects of elevated CO_2 and light availability on the photosynthetic light response of trees of contrasting shade tolerance. Tree Physiol. 16: 351-358

Liang, N., Tang, Y. & Okuda, T. (2001) Is elevation of carbon dioxide concentration beneficial to seedling photosynthesis in the understory of tropical rain forest? Tree Physiol. 21: 1047-1055.

Loik, M. E. & Holl, K. D. (2001) Photosynthetic responses of tree seedlings in grass and under shrubs in early-successional tropical old fields, Costa Rica. Oecologia 127: 40-50.

Manokaran, N., LaFrankie, J. V., Kochummen, K. M., Quah, E. S., Klahn, J. E., Ashton, P. S. & Hubbell, S. P. (1992) Stand table and distribution of species in the Fifty Hectare Research Plot at Pasoh Forest Reserve. FRIM Research Data, Forest Research Institute Malaysia (FRIM), Kuala Lumpur, Malaysia.

Nakashizuka, T. (2001) Species coexistence in temperate, mixed deciduous forests. Trends Ecol. Evol. 16: 205-210.

Niiyama, K., Rahman, K. Abd, Iida S., Kimura, K., Azizi, R. & Appanah, S. (1999) Spatial patterns of common tree species relating to topography, canopy gaps and understorey vegetation in a hill dipterocarp forest at Semangkok forest reserve, Peninsular Malaysia. J. Trop. For. Sci. 11: 731-745.

Pacala, S.W. & Rees, M. (1998) Models suggesting field experiments to test two hypotheses explaining successional diversity. Am. Nat. 152: 729-737.

Pearcy, R. W. (1987) Photosynthetic gas exchange responses of Australian tropical forest trees in canopy, gap and understory micro-environments. Funct. Ecol. 1: 169-178.

Popma, J. & Bongers, F. (1988) The effect of canopy gaps on growth and morphology of seedlings of rain forest species. Oecologia 75: 625-632.

Primack, R. B. (1992) Tropical community dynamics and conservation biology. Bioscience 42:

818-821.
Ryan, M. G. & Yoder, B. J. (1997) Hydraulic limits to tree height and tree growth. Bioscience 47: 235-242.
Sasaki, S. & Mori, T. (1981) Growth responses of dipterocarp seedlings to light. Malay. For. 44: 319-345.
Schnitzer, S. A. & Carson, W. P. (2001) Treefall gaps and the maintenance of species diversity in a tropical forest. Ecology 82: 913-919.
Sobrado, M. A. (1994) Leaf age effects on photosynthetic rate, transpiration rate and nitrogen content in a tropical dry forest. Physiol. Plant. 90: 210-215.
Svenning, J.-C. (2000) Small canopy gaps influence plant distributions in the rain forest understory. Biotropica 32: 252-261.
Swaine, M. D. & Whitmore, T. C. (1988) On the definition of ecological species groups in tropical rain forests. Vegetatio 75: 81-86.
Thomas, S. C. & Bazzaz, F. A. (1999) Asymptotic height as a predictor of photosynthetic characteristics in Malaysian rain forest trees. Ecology 80: 1607-1622.
Turner, I. M. (1990) Tree seedling growth and survival in a Malaysian rain forest. Biotropica 22: 146-154.
Turner, I. M. & Newton, A. C. (1990) The initial responses of some tropical rain forest tree seedlings to a large gap environment. J. Appl. Ecol. 27: 605-608.
Yamashita, N., Ishida, A., Kushima, H. & Tanaka, N. (2000) Acclimation to sudden increase in light favoring an invasive over native trees in subtropical islands, Japan. Oecologia 125: 412-419.
Yoder, B. J., Ryan, M. G., Waring, R. H., Schoettle, A. W. & Kaufmann, M. R. (1994) Evidence of reduced photosynthetic rates in old trees. For. Sci. 40: 513-527.
Zotz, G. & Winter, K. (1993) Short-term photosynthesis measurements predict leaf carbon balance in tropical rain-forest canopy plants. Planta 191: 409-412.

17 Gas Exchange and Turgor Maintenance of Tropical Tree Species in Pasoh Forest Reserve

Yoosuke Matsumoto[1], Yutaka Maruyama[2], Akira Uemura[1], Hidetoshi Shigenaga[1], Shiro Okuda[3], Hisanori Harayama[1], Satoko Kawarasaki[4], Ang Lai Hoe[5] & Son Kheong Yap[5]

Abstract: Net photosynthetic rate (Pn), stomatal conductance (Gw), water use efficiency (WUE), and osmotic potential (Ψs_0) were studied for 46 tropical tree species, including 24 tree species found in Pasoh Forest Reserve, in order to clarify their ecophysiological traits. The maximum value of Pn (Pn_{max}) varied from 2.5 to 24.2 $\mu molCO_2\ m^{-2}\ s^{-1}$ with an average of 9.1 $\mu molCO_2\ m^{-2}\ s^{-1}$. The maximum value of Gw (Gw_{max}) varied from 30 to 1,300 $mmolH_2O\ m^{-2}\ s^{-1}$ with an average of 340 $mmolH_2O\ m^{-2}\ s^{-1}$. These values were relatively low compared to those of temperate tree species. Intrinsic water-use efficiency (IWUE, Pn_{max}/Gw_{max}) of the tropical species was also relatively low compared to the temperate tree species. Ψs_0 of the tropical species was relatively high (less negative) compared to the temperate tree species, suggesting that the capacity of leaves to maintain positive turgor is relatively low in the tropical species. The lower photosynthetic rate in the tropical species was due not only to lower stomatal conductance, but also to lower photosynthetic efficiency, compared to the temperate species. The rapid growth of tropical tree species may be due to favorable environmental factors such as relatively constant temperature and moisture levels, which permit photosynthesis all the year-round.

Key words: leaf water osmotic potential, maximum net photosynthetic rate, maximum water vapor conductance, temperate tree species, tropical tree species, water-use efficiency.

1. INTRODUCTION

The photosynthetic efficiency, or maximum net photosynthetic rate, of a leaf would not always be correlated with its growth rate of trees. However, photosynthetic efficiency is generally higher in fast-growing pioneer species than in slow-growing climax species among broad-leaved tree species in temperate forests (Reigh et al. 1998). Maruyama et al. (1997) reported that net photosynthetic rates of dipterocarp trees, which are known as relatively slow-growing species, were low compared to those of fast-growing trees such as acacia and teak in the tropics.

Photosynthetic rate is limited by CO_2 diffusion through stomata and decreases with decreasing stomatal conductance. The opening of stomata to uptake CO_2 would release water vapor from stomata at the same time. A higher stomatal conductance would increase assimilation but at the same time would decrease leaf

[1] Forestry and Forest Products Research Institute (FFPRI), Tsukuba 305-8687, Japan. E-mail: ymat@ffpri.affrc.go.jp
[2] Hokkaido Research Center, FFPRI, Japan.
[3] Shikoku Research Center, FFPRI, Japan.
[4] Ibaraki University, Japan
[5] Forest Research Institute Malaysia (FRIM), Malaysia.

water content through transpiration. Transpiration also increases linearly with vapor pressure deficit (VPD). In the humid tropics, ambient air is almost saturated with water vapor early in the morning. As a result of extremely high ambient air temperatures, VPD often reaches above 30 hPa at midday in the tropics, which is comparable to the highest VPD in temperate forests during a hot and dry summer. Thus, water loss via transpiration become large in tropical trees during daytime. Decline in leaf water content would cause a lowering of leaf water potential and turgor, and thus, may affect photosynthesis. For the canopy leaves, water uptake from soil is rather difficult because of large hydraulic resistance and low gravitational potential due to the height. Thus, the maintenance of positive turgor plays an important role in the physiological processes of canopy leaves, especially in tropical rain forests where canopy trees grow extremely high.

In order to evaluate net photosynthetic rate, water use efficiency, and the capacity to maintain positive turgor of tropical tree species in the Pasoh Forest Reserve (Pasoh FR), we summarized information on stomatal gas exchange, specifically the balance between photosynthesis and stomatal conductance, and osmotic potential of leaves, which we have studied in the last decade (Ishida et al. 1996; Maruyama et al. 1997, 1998; Matsumoto et al. 1994, 1996, 2000).

2. MATERIALS AND METHODS

Forty six tropical tree species, including 25 dipterocarp and 21 non-dipterocarp species, were used for the study (Table 1). Among these, 25 tree species have been found in Pasoh FR. Measurements were carried out in the Pasoh FR, the campus of the Forest Research Institute Malaysia (FRIM), the Bidor Research Substation of FRIM and the Cikus Experimental Site of JICA (Japan International Cooperation Agency).

Various sized trees from small potted seedlings to emergent ones were used for the measurement. Well-exposed sun leaves were chosen from these trees and photosynthetic rates were measured with portable photosynthesis and transpiration measurement systems (SPB-H2, SPB-H3, and SPB-H4, ADC (UK)-Shimadzu, Japan). Transpiration rate and stomatal conductance were measured using a steady-state porometer (LI-1600, Li-Cor, USA).

For the measurements of osmotic potential, leaves were collected in the morning when air humidity was high and sealed immediately in a plastic bag with wet filter paper in order to prevent transpirational water loss. Then the leaves were deeply frozen for at least two days and melted under room temperature. Cell sap was collected from these leaves and the osmotic potential of the cell sap was determined using osmometers (Model 5100 and 5100B, Wescor Inc., USA).

3. PHOTOSYNTHESIS AND STOMATAL CONDUCTANCE

We compared the values of net photosynthetic rate (Pn) and stomatal conductance (Gw) of tropical tree species and defined the maximum value of Pn and Gw of each species as Pn_{max} and Gw_{max}, respectively. Figure 1 shows the maximum net photosynthesis rate (Pn_{max}) of tropical tree species. Pn_{max} varied from 2.5 to 24.2 μmol CO_2 m^{-2} s^{-1} with an average value of 9.1 μmol CO_2 m^{-2} s^{-1}. Larcher (1995) summarized photosynthetic rates of tropical trees to be 10–16 and 5–7 μmol CO_2 m^{-2} s^{-1} for sun and shade leaves, respectively. Zotz & Winter (1996) reported photosynthetic rates of 10 to 20 μmol CO_2 m^{-2} s^{-1} for tropical trees. Compared to these values, the variation in Pn_{max} obtained in this section was relatively large. Pn_{max} was highest in *Acacia mangium* and *A. auriculiformis*, which are well known

Table 1 Species used in this study.

Species Name	Distribution in Pasoh FR*	Measurements**		
		Pn	Gw	Ψ_{So}
Dipterocarp species				
Dipterocarpus cornutus	Locally common in non-swampy areas	○	○	○
Dipterocarpus kerrii	-	○	○	○
Dipterocarpus oblongifolius	-	○	○	○
Dipterocarpus sublamellatus	Very rare	○	○	-
Dryobalanops aromatica	-	○	○	○
Hopea nervosa	-	○	○	-
Hopea odorata	-	○	○	○
Hopea pubescens	-	-	-	○
Neobalanocarps heimii	Very common	○	○	○
Parashorea densiflora	Widely distributed but uncommon	-	-	○
Shorea accuminata	Common and widely distributed	○	-	○
Shorea assamica	-	○	○	○
Shorea bracteolata	Common and widely distributed	-	-	○
Shorea curtisii	-	○	○	○
Shorea glauca	-	-	-	○
Shorea guiso	Of patchy distribution mainly in swampy areas	-	-	○
Shorea leprosula	Very common and widely distributed	○	○	○
Shorea macroptera	Common and widely distributed	○	○	○
Shorea multiflora	Very rare, No stem above 10cm in DBH	-	-	○
Shorea ovalis	Common throughout	○	○	○
Shorea ovata	-	-	-	○
Shorea parvifolia	Common and widely distributed	○	○	○
Shorea pauciflora	Common and widely distributed	○	○	-
Shorea platyclados	-	○	○	-
Shorea roxburghii	-	-	-	○
Non dipterocarp species				
Acacia auriculiformis	-	○	○	○
Acacia mangium	-	○	○	○
Alstonia angustiloba	Uncommon, of scattered distribution	○	-	○
Azadirachta excelsa	-	○	○	○
Cinnamomum iners	Uncommon	○	-	-
Durio sp.	-	-	-	○
Endospernum malaccensis	Very rare	○	-	○
Gonystylus affine	-	○	○	○
Heritiera sp.	Common and widely distributed in non-swampy areas	-	-	○
Hevea brasiliensis	-	○	○	-
Intsia palembanica	Common and widely distributed except on higher ground	○	-	-
Khaya ivoreusis	-	○	-	-
Melastoma malabathricum	Very rare	-	-	○
Palaquium sp.	Rare to common	-	-	○
Pentaspadon motleyi	Common	-	-	○
Ptychopyxis caput-medusae	Very rare	○	○	-
Scaphium macropodum	Common in non-swampy areas	○	○	○
Sindora coriacea	Uncommon	-	-	○
Tectona grandis	-	○	○	○
Terminalia catappa	-	○	○	○
Xanthophyllum amoneum	Rare	○	○	-

*According to Kochummen (1997) and Manokaran et al. (1992). **Pn, net photosynthetic rate; Gw, stomatal conductance; Ψs_0, osmotic potential at full turgidity; ○, previously measured data.

as fast growing pioneers. Other fast growing species like *Tectona grandis, Cinnamomum iners,* and *Azadirachta excelsa* also showed a relatively high Pn_{max}. Among dipterocarps, *Hopea odorata* and *Shorea assamica*, which are known as relatively fast-growing dipterocarp species, had a relatively high Pn_{max}. The Pn_{max} of those found in the Pasoh FR varied from 2.5 to 14.4 μmol CO_2 m^{-2} s^{-1}, and was within the same range as the other species except for the fast growing ones (Fig. 2).

Among species found in the Pasoh FR, the distribution of those with a relatively high Pn_{max} such as *Cinnamomum iners, Alstonia angustiloba, Dipterocarpus sublamellatus,* and *Endospernum malaccensis* is uncommon or very rare (Table 1). On the other hand, species commonly found in Pasoh FR such as *Shorea leprosula, S. parvifolia, S. accuminata, S. pauciflora, S. macroptera,*

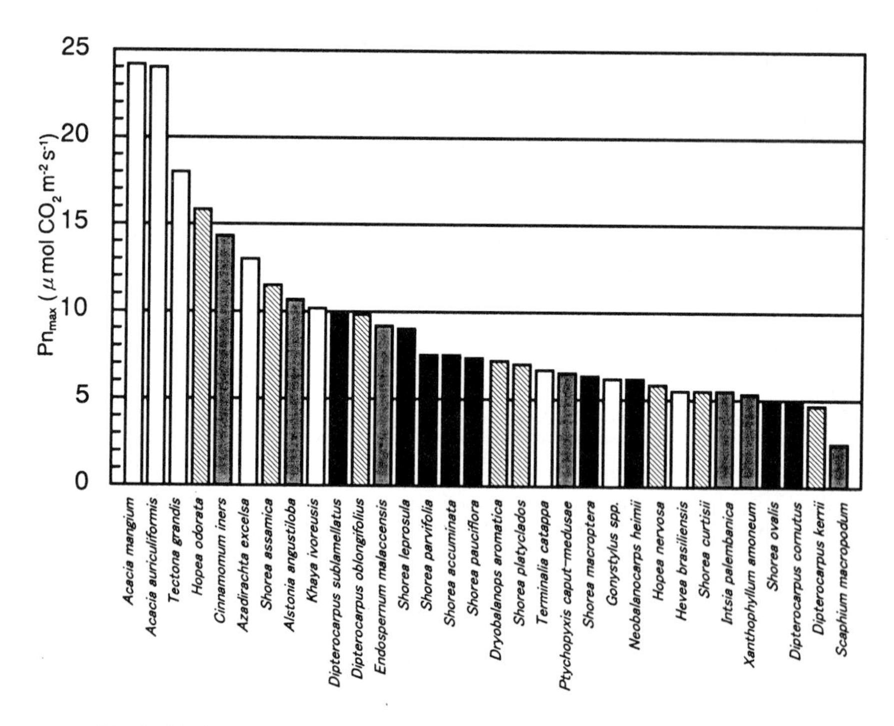

Fig. 1 Maximum net photosynthetic rate (Pn_{max}) of tropical tree species (after Matsumoto et al. 2000). Closed and hatched column, dipterocarp species; open and gray, non-dipterocarp species; closed and gray, species distributed in Pasoh FR.

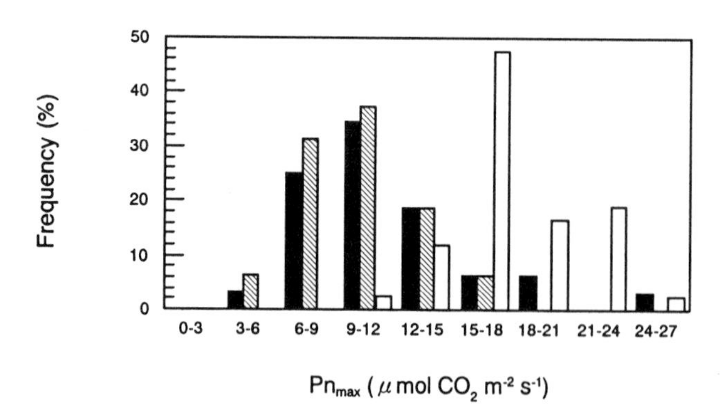

Fig. 2 Relative frequency distribution of maximum net photosynthetic rate (Pn_{max}) of tropical tree species and temperate tree species (after Matsumoto & Maruyama 2002; Matsumoto et al. 1999). Closed, tropical tree species; open, temperate tree species; hatched, species distributed in Pasoh FR.

Neobalanocarpus heimii, *Intsia palembanica*, and *S. ovalis* showed a relatively low Pn_{max}. When compared with temperate tree species (Matsumoto et al. 1999), the average value of Pn_{max} of tropical tree species was significantly low (*t*-test, $P < 0.01$, Fig. 2). Therefore, the high primary productivity and rapid growth generally observed in tropical species would be due not to photosynthetic efficiency but to favorable environmental factors such as relatively constant temperature and moisture levels, which permit photosynthesis year-round.

The maximum value of stomatal conductance (Gw_{max}) varied from 30 to 1,300 mmol H_2O m^{-2} s^{-1} with an average of 340 mmol H_2O m^{-2} s^{-1} (Fig. 3). As photosynthesis is limited by CO_2 diffusion through stomata, species with a high Pn_{max} like *A. mangium*, *A. auriculiformis*, *T. grandis*, *S. assamica* and *H. odorata* showed a relatively high Gw_{max}. Except for these species, the frequency distribution of Gw_{max} was almost the same in those found in Pasoh FR and all the tropical tree species (Figs. 3 and 4). However, like the Pn_{max}, the Gw_{max} of those commonly found in the Pasoh FR was relatively low.

Compared to temperate species, the Gw_{max} of tropical species was relatively low (Fig. 4) but the difference was not significant. The difference in Pn_{max} between tropical species and temperate species was more remarkable than that in Gw_{max} (Figs. 2 and 4), and Pn_{max} for the same Gw_{max} was lower in tropical species, especially those found in the Pasoh FR, than in temperate species (Fig. 5). These results suggest that the lower photosynthetic rate of the species found in the Pasoh FR would be due to a smaller capacity for both CO_2 diffusion and CO_2 fixation compared to temperate species.

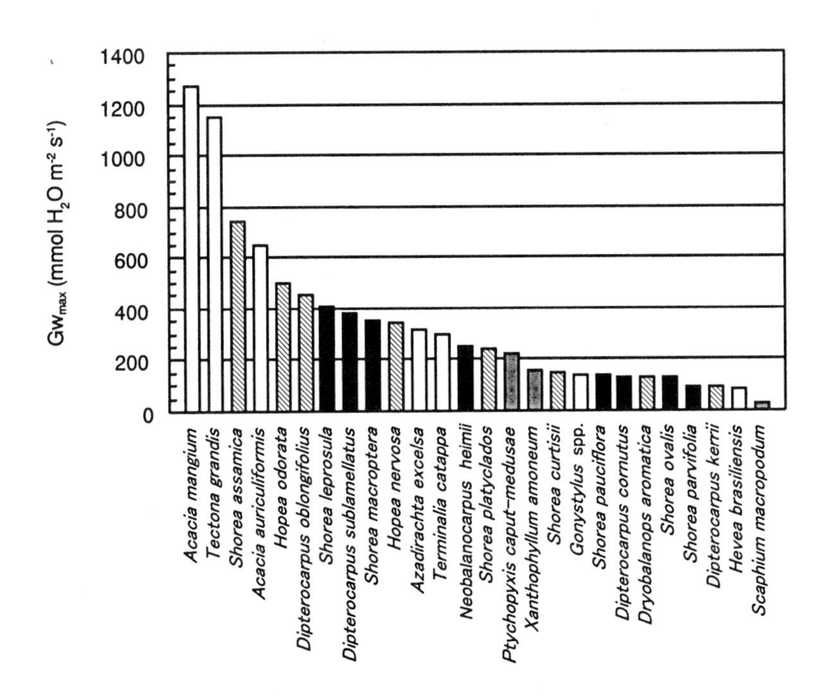

Fig. 3 Maximum water vapor conductance (Gw_{max}) of tropical tree species. Legends are the same as in Fig. 1.

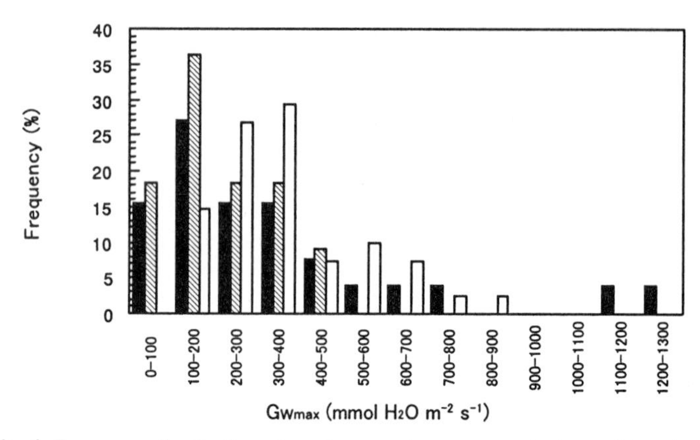

Fig. 4 Frequency distribution of maximum water vapor conductance (Gw_{max}) of tropical tree species and temperate tree species. Legends are the same as in Fig. 2.

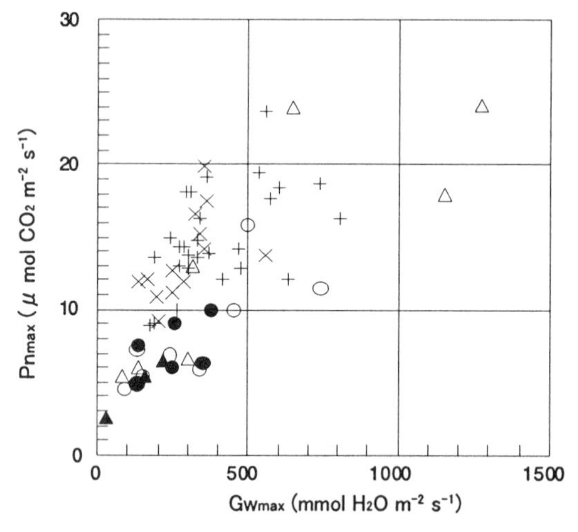

Fig. 5 Maximum water vapor conductance (Gw_{max}) and maximum net photosynthetic rate (Pn_{max}) of tropical and temperate tree species.
Circles, dipterocarp species; triangles, non-dipterocarp species; closed symbols, species distributed in Pasoh FR; inclined crosses, evergreen broadleaved tree species in temperate tree species; crosses, deciduous broadleaved tree species in temperate tree species

4. WATER ECONOMY AND TURGOR MAINTENANCE

In order to evaluate the water economy of stomatal gas exchange and maintenace of turgor, intrinsic water use efficiency (IWUE, IWUE = Pn_{max}/Gw_{max}) and leaf osmotic potential were compared among species. The IWUE of potentials tropical tree species varied from 15 to 79×10^{-6} mol CO_2/mol H_2O (Fig. 6). When stomatal conductance is high, the change in photosynthesis with changing stomatal conductance (dPn/dGw) becomes marginal, as the increase in Ci (intercellular CO_2 partial pressure) with stomatal conductance becomes marginal. Therefore, IWUE generally declines at a higher stomatal conductance, and thus, species with a high Pn_{max} are expected to have a low IWUE. The relatively low IWUE of species with a high Pn_{max} like *A. mangium*, *T. grandis*, and *S. assamica* would be due to their high Gw_{max}. These species may require large amounts of water to maintain their high photosynthetic rate. However, *A. auriculiformis*, *H. odorata*, and *A. excelsa*, which also had a relatively high Pn_{max}, showed a moderate IWUE, suggesting their superior gas exchange characteristics. *A. auriculiformis* is known as an extremely drought-hardy species and has been used for the rehabilitation of deeply degraded land (Ang & Maruyama; 1995, Ang et al. unpublished). *H. odorata* also was reported to have a large capacity to cope with environmental stress and was recommended for planting purposes (Ang et al. 1992). *A. excelsa* occurs naturally in tropical seasonal forests in Thailand. Therefore, these species are expected to have a comparatively high IWUE. The IWUE of the species distributed in the Pasoh FR was almost within the range of that of the other tropical species. As mentioned previously, species commonly found in Pasoh FR showed relatively low Pn_{max} and Gw_{max} values, but the IWUE of those species was within the same range as the other tropical species.

Though the Gw_{max} of tropical trees was relatively low compared to temperate trees (Fig. 4), the large VPD during daytime would cause increasing transpirational water loss. Furthermore, the Pn_{max} for same Gw_{max} was lower in tropical trees than in temperate trees (Fig. 5). These results suggest that tropical trees would require greater amounts of water for the same photosynthetic production than temperate trees. Water uptake from soil is rather difficult for canopy leaves because of the large hydraulic resistance and low gravitational potential due to the height. This

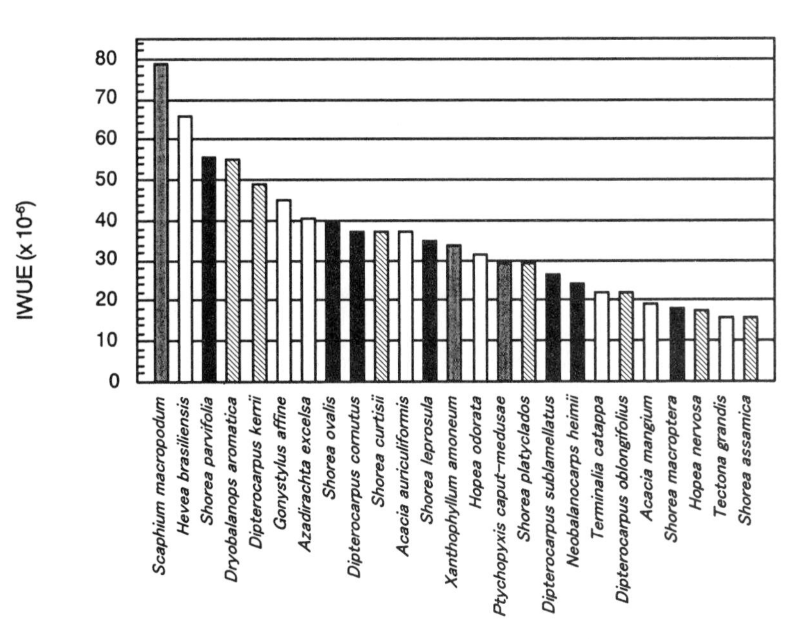

Fig. 6 Intrinsic water use efficiency (IWUE) of tropical tree species. Legends are the same as in Fig. 1.

would cause a decline in leaf water content and, as a result, would cause a lowering leaf water potential and turgor. Therefore, turgor maintenance is of great importance in physiological processes of canopy leaves, especially in tropical rain forests where canopy trees grow extremely high.

Leaf osmotic potential at full turgidity (Ψs_0) varied from –2.1 to –0.8 MPa (Fig. 7). Ψs_0 was lowest (absolute value was largest), and thus, the capacity to maintain positive turgor is largest, in *A. excelsa*. This species also had a relatively high IWUE among fast growing species with a high Pn_{max} (Figs. 1 and 6). The large capacity for turgor maintenance together with superior gas exchange characteristics of this species may permit successful establishment and rapid growth in drier forests. We expected a relatively low Ψs_0 for species commonly found in the Pasoh FR, most of which grow extremely high. However, Ψs_0 of these species was within the same range as the other tropical species. As reported by Maruyama et al. (1998) and Furukawa et al. (2001), canopy leaves may avoid water stress by reducing transpirational water loss through stomatal closure during the daytime, rather than by maintaining positive turgor through osmotic adjustment, i.e. lowering osmotic potential through solute accumulation and/or solute concentration in leaves.

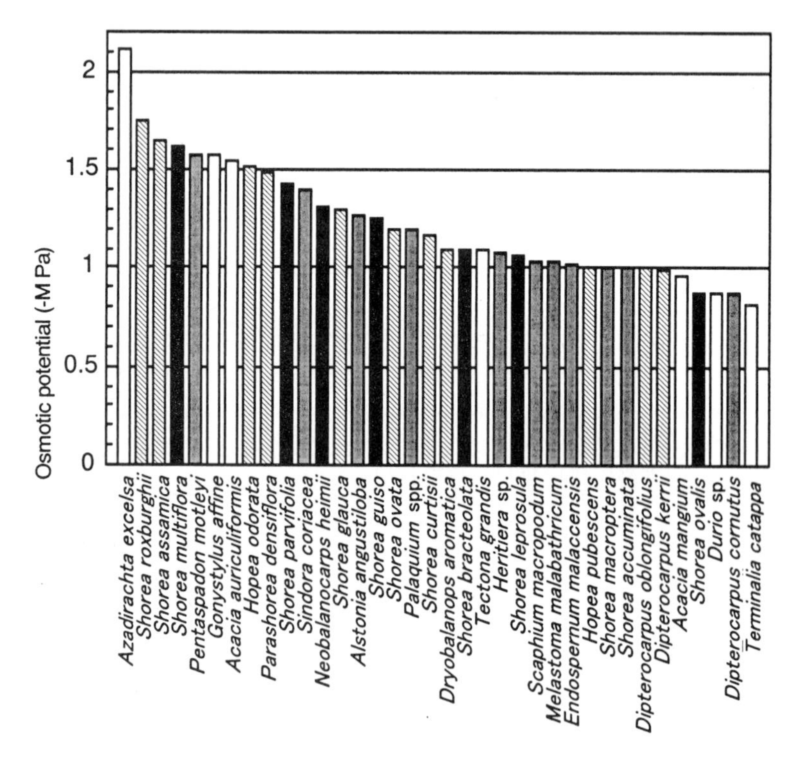

Fig. 7 Osmotic potential at water saturation (Ψs_0) of tropical tree species. Legends are the same as in Fig. 1.

ACKNOWLEDGEMENT

The authors are grateful to Drs. Y. Morikawa and A. Furukawa for providing them the chance to join the NIES/FRIM/UPM joint project. This study was supported by the Global Environment Research Fund of the Japan Environment Agency.

REFERENCES

Ang, L. H. & Maruyama, Y. (1995) Growth and photosynthesis of *Acacia auriculiformis* and *Acacia mangium* seedlings planted on sand tailings. Proceedings of the International Workshop of Bio-Refor, Kangar, Malaysia, pp.67-71.

Ang, L. H., Maruyama, Y., Wan Razali, W. M. & Abd. Rahman, K. (1992) The early growth and survival of three commercial dipterocarps planted on decking sites of a logged-over hill forest. Proceedings of the International Symposium on Rehabilitation of Tropical Rainforest Ecosystems: Research and Developement Priorities, Kuching, Malaysia, pp.2-4.

Ang, L. H., Matsumoto, Y., Maruyama, Y., Uemura, A. & Ang, T. B., (in preparation) Growth and gas exchange characteristics of four timber species grown on sand tailings.

Furukawa, A., Toma, T., Maruyama, Y., Matsumoto, Y., Uemura, A., Ahmad Makmom, A. & Muhamad, A. (2001) Photosynthetic rates of four tree species in the upper canopy of a tropical rain forest in the Pasoh Forest Reserve in Peninsular Malaysia. Tropics 10: 519-527.

Ishida, A., Toma, T., Matsumoto, Y., Yap, S. K. & Maruyama, Y. (1996) Diurnal changes in leaf gas exchange characteristic in the uppermost canopy of a rain forest tree, *Dryobalanops aromatica* Gaertn. f.. Tree Physiol. 16: 779-785.

Kochummen, K. M. (1997) Tree flora of Pasoh forest. Malayan Forest Records No. 44. Forest Research Institute Malaysia, Kepong, 461pp.

Larcher, W. (1995) Physiological plant ecology (3rd ed). Springer, Berlin, 506pp.

Manokaran, N., Lafrankie, J. V., Kochummen, K. M., Quah, E. S., Klahn, J. E., Ashton, P. S. & Hubbell, S. P. (1992) Stand table and distribution of species in the 50-hectare research plot at Pasoh Forest Reserve. Research data No.1, Forest Reserch Institute Malaysia, 454pp.

Maruyama, Y., Toma, T., Ishida, A., Matsumoto, Y., Morikawa, Y., Ang, L. H., Yap, S. K. & Iwasa, M. (1997) Photosynthesis and water use efficiency of 19 tropical tree species. J. Trop. For. Sci. 9: 434-438.

Maruyama, Y., Uemura, A., Ishida, A., Shigenaga, H., Ang, L. H. & Matsumoto, Y. (1998) Photosynthesis, transpiration, stomatal conductance and leaf water potential of several tree species. Proceedings of the Second International Symposium on Asian Tropical Forest Management, Samarinda, Indonesia, pp.263-275.

Matsumoto, Y. & Maruyama, Y. (2002) Gas exchange characteristics of major tree species in Ogawa Forest Reserve. In Nakashizuka, T. & Matsumoto, Y. (eds). Diversity and interaction in a temperate forest community: Ogawa Forest Reserve of Japan. Ecological Studies vol. 158, Springer, Tokyo, pp.201-213.

Matsumoto, Y., Uemura, A., Okuda, S., Morikawa, Y., Maruyama, Y., Yap, S. K. & Ang, L. H. (1994) Ecophysiological research on adaptation of tropical tree species to environmental stress. Research report of FRIM/UPM/NIES joint research project for 1993, pp.72-77.

Matsumoto, Y., Ishida, A., Shigenaga, H., Uemura, A., Toma, T., Maruyama, Y., Osumi, Y., Morikawa, Y., Yap, S. K. & Ang, L. H. (1996) Ecophysiological properties on tropical tree species. Tropical Rain Forest Ecosystem and Biodiversity in Peninsular Malaysia (Research report of the NIES/FRIM/UPM joint research project 1993-1995), National Institute for Environmental Studies and others, pp.43-53.

Matsumoto, Y., Tanaka, H., Kosuge, S., Tanbara, T., Uemura, A., Shigenaga, H., Ishida, A., Okuda, S., Maruyama, Y. & Morikawa, Y. (1999) Maximum gas exchange rates in current sun leaves of deciduous and evergreen broad-leaved 41 tree species in Japan. Jpn. J. For. Environ. 41: 113-121 (in Japanese with English summary).

Matsumoto, Y., Maruyama, Y. & Ang, L. H. (2000) Maximum gas exchange rates and osmotic potential in sun leaves of tropical tree species. Tropics 9: 195-209 (in Japanese with English summary).

Reigh, P. B., Walters, M. B., Tjoelker, M. G., Vanderklein, D. & Buschena, C. (1998) Photosynthesis and respiration rates depend on leaf and root morphology and nitrogen concentration in nine boreal tree species differing in relative growth rate. Funct. Ecol. 12: 395-405.

Zotz, G. & Winter, K. (1996) Diurnal patterns of CO_2 exchange in rainforest canopy plants. In Mulkey, S. S., Chazdon, R. L. & Smith, A. P. (eds). Tropical forest plant ecophysiology, Chapman & Hall, New York, pp.89-113.

18 Sunfleck Contribution to Leaf Carbon Gain in Gap and Understory Tree Seedlings of *Shorea macrophylla*

Yanhong Tang[1], Toshinori Okuda[1], Muhamad Awang[2], Abd Rahim Nik[3]& Makoto Tani[4]

Abstract: Dynamic gas exchange in response to sunflecks has been studied extensively through laboratory experiments and by modeling. However, model estimates of photosynthetic changes in response to sunflecks vary widely depending on the local light environment and on species-specific differences. Moreover, it remains unclear how much sunflecks contribute to leaf carbon gain in natural forests because of a lack of field measurements of dynamic photosynthetic responses. Since sunflecks may increase photosynthetic carbon gain by providing photosynthetic energy and by decreasing limitations on the induction of photosynthesis, we hypothesized that a large proportion of photosynthetic photon flux density (PPFD) received from sunflecks would increase the proportion of carbon gain contributed by sunfleck PPFD. We measured dynamic changes in photosynthesis of *Shorea macrophylla* tree seedlings experimentaly planted under a canopy gap microsite and under the forest canopy in the Pasoh Forest Reserve (Pasoh FR) of Malaysia. The total proportion of daily leaf carbon gain that resulted from sunfleck utilization varied from 26% to 83%, depending on the microsite, during 7 days of measurements. Sunfleck utilization efficiency decreased with increasing threshold values for sunfleck PPFD. The results indicated that total daily carbon gain was closely related to total daily PPFD and total sunfleck PPFD, but was weakly related to diffuse background PPFD.

Key words: dynamic photosynthesis, induction limitation, *Shorea macrophylla*, sunfleck utilization, understory plants.

1. INTRODUCTION

Light is a limiting resource for understory plants in tropical rain forests (Chazdon et al. 1996; Tang et al. 1999). Sunflecks, which consist mainly of direct light that penetrates small gaps in the forest canopy, often contribute a majority of the photosynthetic photon flux density (PPFD) available for photosynthesis by understory plants. Sunflecks can contribute between 20% and 80% of an understory plant's total daily carbon gain (Chazdon 1988; Gross 1982; Pearcy & Calkin 1983; Pearcy 1987; Pearcy 1990) and from 9% to 44% of total annual carbon gain (Pearcy & Pfitsch 1991). The large differences in the contribution of sunflecks to leaf carbon gain are believed to arise from variations in the plant's PPFD environment and from dynamic species-specific photosynthetic responses.

[1] National Institute for Environmental Studies (NIES), Tsukuba 305-8506, Japan.
E-mail: tangyh@nies.go.jp
[2] Universiti Putra Malaysia (UPM), Malaysia
[3] Forest Research Insitite Malaysia (FRIM), Malaysia
[4] Kyoto University, Japan

Sunflecks are usually present for only a small fraction of the day and arise only under clear skies. Moreover, the PPFD in sunflecks is often much higher than the saturated light intensity for photosynthesis in understory plants. These facts suggest that the proportion of leaf carbon gain contributed by sunflecks might be much lower than the proportion of total daily PPFD attributable to sunflecks. On the other hand, sunfleck activity may indirectly increase leaf carbon gain by increasing the induction of photosynthesis or post-illumination CO_2 fixation (Pearcy 1990). It remains unclear how sunfleck utilization contributes quantitatively to total daily carbon gain in understory plants. Understanding the contribution of sunflecks to leaf carbon gain would improve our understanding of the importance of light heterogeneity to understory plants in tropical forests.

So far, few data sets are available for determination of the daily contribution of sunfleck utilization to the leaf's carbon balance during the diurnal course of photosynthesis (Pearcy & Calkin 1983; Pearcy 1987; Pfitsch & Pearcy 1989). In these studies, the contribution of sunflecks to the total daily carbon gain was found to be roughly proportional to the contribution of sunflecks to the total daily PPFD. That is, a rough linear relationship may exist between the proportion of total daily PPFD contributed by sunflecks and the proportion of total daily carbon gain that results from sunfleck PPFD. If this is the case, we would expect that the utilization efficiency of sunfleck PPFD must be close to that of diffuse PPFD, and that the proportion of sunfleck PPFD that exceeds the saturating level of PPFD for the plant would be very small in understory environments. However, little detailed data is available to support or reject these hypotheses. Understanding the contribution of sunfleck PPFD to total daily PPFD would significantly increase our understanding of the mechanisms of photosynthetic sunfleck utilization in natural environments.

In the present study, we measured diurnal variations in the photosynthetic utilization of sunflecks in seedlings of *Shorea macrophylla* transplanted to a microsite under a closed canopy and a second microsite under a canopy gap in a regenerating forest in Malaysia's Pasoh Forest Reserve (Pasoh FR). *S. macrophylla* is not native to the Pasoh forests, but has been widely studied because of its ecological and economic importance among the dipterocarps of the tropical forests of southeast Asia (Appanah & Turnbull 1998). Our major objective was to determine the contribution of sunflecks to total daily carbon gain by the leaves of understory plants by using this species as a model for other dipterocarps growing as understory plants.

2. MATERIALS AND METHODS

2.1 Study site

The Pasoh FR (2° 59' N, 102° 18' E) consists of a core of virgin dipterocarp forest (600 ha) surrounded by a buffer zone that was selectively logged from 1955 to 1959 (Wong 1983; Chap. 2). Meteorological data collected at the Pasoh FR from 1971 to 1974 (Soepadmo 1978) reveal that the mean monthly temperature at ground level is 24℃, with a range from 21 to 26℃; at the top of a 52-m tower in the Pasoh FR, the mean value is 22℃, with a range between 17 and 32℃. Relative humidity measured 2.5 m above the ground is invariably greater than 96% during the wet months (April, May, November and December) and ranges from 61% to 93% during the dry months (February, March, July and August; details are presented in Chaps. 4, 6).

A gap microsite and an closed understory microsite were selected in the regenerating forest (Chap. 2) for the present study. The gap area was 159 m^2 (9.3 m × 17.1 m). Seedlings of *S. macrophylla* were transplanted to both sites in 1994. Diurnal variations in PPFD were measured at 2-second intervals by using a quantum sensor (model LI-190S, Li-Cor Inc., Lincoln, NE, USA) attached to a portable gas-exchange system (model LI-6400, Li-Cor Inc., Lincoln, NE, USA) during photosynthesis measurements. The relative PPFD for the two microsites (expressed as a percentage of the ambient level measured above the canopy) was also measured; it ranged from 5% to 11% for the gap microsite versus 0.8% to 3.3% for the understory microsite during the period when photosynthesis was measured in the present study (from day 114 to day 120 in 1996).

2.2 Gas-exchange measurements

We measured photosynthesis with a portable gas-exchange system. The LI-6400 system has a small leaf chamber closely attached to the sensor head. This design makes it possible to measure dynamic changes in photosynthesis in real time because the return tubing between the leaf chamber and the console that was present in earlier models has been eliminated. Thus, there are almost no time delays to confound correlations between gas-exchange measurements and changes in environmental variables such as light, the CO_2 mole fraction, leaf temperature, etc. However, volume-related time constraints could still exist. The total volume for our leaf chamber and gas analysis cell ranged between 80 and 90 cm^3, depending on the size of the leaves being measured. To examine the effect of this volume on instantaneous photosynthetic responses, we calculated an equilibrium CO_2 concentration based on model (1) in Pearcy et al. (1985). We found that even when the effective chamber volume was five times the actual chamber volume, the corrected assimilation rate differed by less than 1.1% from the measured assimilation rate if the time interval was 2.08 s and the flow rate was 500 cm^3 min^{-1}. On the basis of this estimate, we concluded that the effects of the instrumental time lag were negligible in our data analysis.To examine the sunfleck contribution to leaf carbon gain, we measured photosynthesis under naturally varying light conditions. PPFD was measured in the system's standard leaf chamber near the plane of the leaves by using a miniature GaAsP sensor. The external ambient PPFD was measured with a quantum sensor (model LI-190S) to calibrate the miniature sensor.

To obtain photosynthetic responses that matched those in the natural understory environment, we used ambient CO_2 concentrations and air humidity as far as possible during measurements. A 20-*l* plastic tank and three 1-*l* poly-bottles were used as air buffers to reduce fluctuations in the levels of CO_2 and H_2O in the air entering the chamber. During our measurements, we adjusted humidity levels using the LI-6400 when environmental humidity was higher than 85%. Temperatures in the leaf chamber were only controlled for measurements done under the canopy gap and around noon to avoid heat damage to the leaf from the extremely high temperatures that could have developed within the chamber. During these times, the temperature inside chamber was set to 30℃.

For each measurement, a healthy, fully expanded leaf of *S. macrophylla* was selected. The leaf area used for the photosynthetic measurements was 2 cm × 3 cm. To reduce shading by the chamber walls when the sun was low in the sky, we adjusted the major axis of the chamber so that it ran east-west. Because the seal material on the top side of the chamber is white and the remainder of the wall is a polished silvery color, light could be reflected quite well within the chamber. We therefore made no further adjustments to compensate for the spatial heterogeneity of light within the chamber, though some heterogeneity undoubtedly still existed due to shading of the chamber wall at low solar elevations or due to sunfleck activity (Tang 1997).

An auto-logging program sent data from the LI-6400 system to a computer (HP 200LX, Hewlett-Packard Co. CA, USA) at intervals of 1 to 3 s. The nominal data-logging interval was set to 1 s, but the actual interval differed because of delays imposed by the computer in the instrument. The mean interval was 2.08 s during the measurements, but we used the actual intervals during the data analysis.

In addition to the measurement of the diurnal course of photosynthesis, we also measured photosynthetic response to different levels of PPFD by artificially changing the PPFD. An LED light with balanced red and blue components was used for these measurements.

Since a matching mode was used quite frequently (once every 600 readings) to reduce the magnitude of the error generated by the analyzer, we recalculated the gas-exchange parameters using the equations described by von Caemmerer & Farquhar (1981) instead of directly using the data calculated by the LI-6400. We used a linear model to smooth the differences between CO_2 and H_2O levels in two adjacent matches and compared the means using ANOVA (Abacus Concepts, NC, USA, 1992).

3. RESULTS

3.1 Variation in the physical environments of the seedlings

To examine sunfleck utilization by tree seedlings in tropical rain forests, we measured dynamic CO_2 uptake in *S. macrophylla* leaves for 7 days, from day 114 to day 120

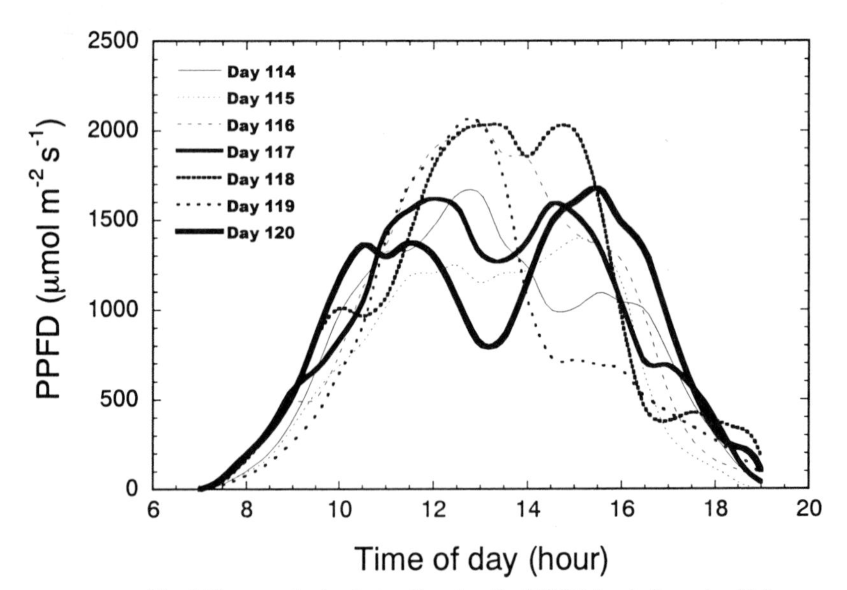

Fig. 1 Photosynthetic photon flux density (PPFD) levels from day 114 to 120 in 1996 while the photosynthetic measurements were made. Original solar radiation was measured 52.6 m above the ground. Data represent the averages for 30 min of measurement.

in 1996. The skies were either very clear or relatively clear during the measurement period (Fig. 1), and this provided potentially high sunfleck activity each day. The physical environment for the *S. macrophylla* leaves enclosed in the measurement chamber exhibited marked diurnal variation (Fig. 2). Environmental CO_2 concentrations were higher at the closed understory microsite, reaching 520 μmol mol^{-1} in the early morning, compared to a maximum of about 480 μmol mol^{-1} at the gap microsite. CO_2 concentrations decreased rapidly thereafter and reached their lowest level around noon. In the afternoon, environmental CO_2 levels increased again, probably as a result of low photosynthetic rates and high respiration rates in the forest (Fig. 2). Sunflecks were much more frequent at the gap microsite than in the closed understory. In the gap, diffuse PPFD varied from 20 to 50 μmol m^{-2} s^{-1}, and most sunflecks occurred between 11:00 and 13:00 local time. In contrast, most diffuse background PPFD levels were lower than 10 μmol m^{-2} s^{-1} in the closed understory. Sunfleck activity varied greatly in intensity and occurrence between these two microsites. Water concentrations in the air were higher at the gap microsite than in the understory.

Sunfleck activity differed markedly between the gap and understory microsites. As shown for two representative days, leaves at the closed understory microsite received PPFD lower than 10 μmol m^{-2} s^{-1} during most of the day (>86%, Figs. 2a, 4a), whereas sunflecks with PPFD greater than 10 μmol m^{-2} s^{-1} occurred almost continuously from 9:30 to 15:00 at the gap microsite (Figs. 2c, 4c). On average, PPFD was greater than 10 μmol m^{-2} s^{-1} for about 18% of the day at the closed understory microsite, versus about 80% of the day at the gap microsite (Fig. 3).

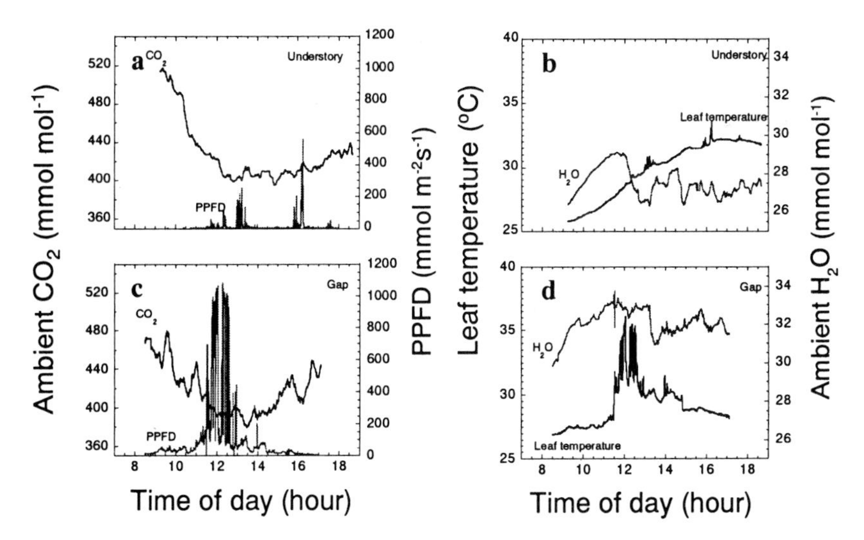

Fig. 2 Characteristics of the physical environment of the *S. macrophylla* leaves enclosed in the chamber used for photosynthesis measurements. Graphs a and b represent values for the understory microsite (day 114, 1996) and graphs c an d represent values for the gap microsite (day 116, 1996). Photosynthetic photon flux density (PPFD) and leaf temperature were measured by sensors inside the leaf-measurement chamber. Ambient CO_2 and H_2O concentrations were measured in the inlet gas in the reference cell of the in LI-6400 photosynthesis system. All data were recorded at a mean interval of 2.03 s.

Sunflecks with PPFD greater than 10 mol m^{-2} s^{-1} contributed the majority of total daily PPFD at both the closed understory and the gap microsites (Fig. 4 a, c). Under clear skies, the frequency distribution for PPFD tended to exhibit two peaks on almost all measurement days during this experiment (Figs. 4a, 4c and data not shown). The lower peak was between 1 and 8 μmol m^{-2} s^{-1} in the closed understory, versus values between 16 and 64 μmol m^{-2} s^{-1} in the gap. The higher peak, however, varied more strongly between the two microsites (data not shown). In the day shown in Figs. 4a and 4c, the peak values ranged between 128 and 256 μmol m^{-2} s^{-1} in the understory and between 512 and 1,024 μmol m^{-2} s^{-1} in the gap.

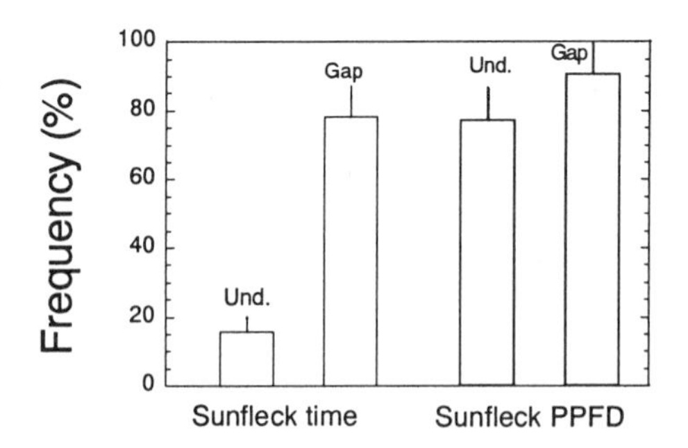

Fig. 3 Percentage of sunflecks with PPFD values greater than 10 μmol m^{-2} s^{-1} as a function of total daily sunfleck time (leftmost two bars) and as a function of total daily PPFD (rightmost two bars) in the understory and gap microsites.

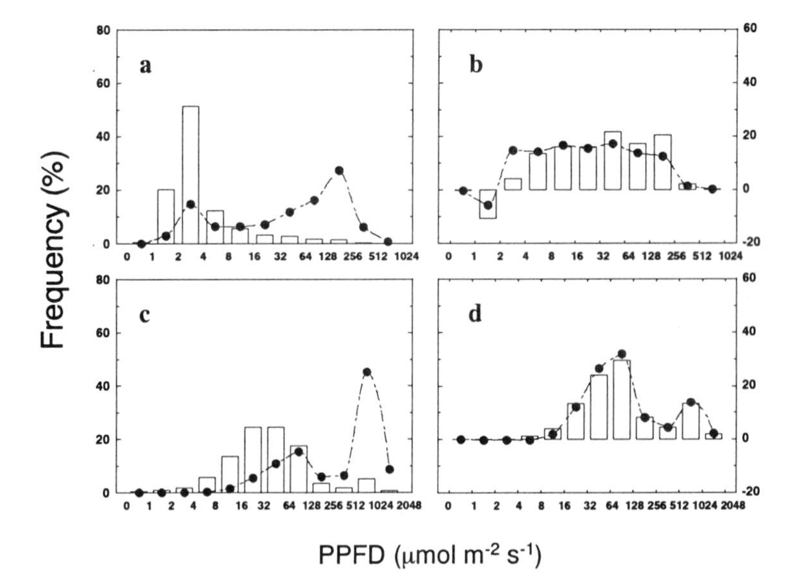

Fig. 4 Percentage of sunfleck time (bars in graphs a and c), sunfleck PPFD (closed circles in graphs a and c), measured carbon gain (bars in graphs b and d), and estimated carbon gain (closed circles in graphs b and d) per each daily total for *S. macrophylla* seedlings in the closed understory (a, b) and gap (c, d) microsites.

In the understory, the frequency distribution of carbon gain contributed by different sunflecks differed greatly from the frequency distribution of PPFD itself (Fig. 4b). At PPFD levels below 2 μmol m^{-2} s^{-1}, net CO_2 uptake was negative, and the majority of the total daily carbon gain occurred at PPFD levels from 4 to 256 μmol m^{-2} s^{-1}. In the gap, more than half of the daily carbon gain resulted from sunflecks with PPFD levels from 16 to 256 μmol m^{-2} s^{-1}. At both microsites, sunflecks with PPFD levels above 10 μmol m^{-2} s^{-1} contributed more than 80% of total daily leaf carbon gain (Fig. 4).

3.2 Photosynthetic utilization of sunflecks

Diurnal variation in the dynamic CO_2 uptake rate in *S. macrophylla* generally followed diurnal changes in PPFD (Fig. 5). In the closed understory, the net CO_2 uptake rate averaged 0 μmol m^{-2} s^{-1} for most of the day, with the highest values (6 μmol m^{-2} s^{-1}) under sunflecks (Fig. 5a, c). Frequent heavy shading resulted in negative CO_2 uptake rates. In the gap, CO_2 uptake rates varied markedly in response to fluctuations in PPFD (Fig 5c). The maximum net CO_2 uptake rate reached 11.2 μmol m^{-2} s^{-1} at noon. Background levels of stomatal conductance (g_s) were similar for leaves in the gap and the understory sites, though the maximum g_s values in the gap (Fig. 5d) increased by nearly 800% under sunfleck conditions in the morning compared with background levels. In contrast, the magnitude of the variation in g_s was relatively weak in the understory (Fig. 5b, c).

The total daily carbon gain that resulted from sunflecks depended almost linearly on total daily PPFD (Fig. 6). The proportion of carbon gain due to sunflecks was closely correlated with the proportion of PPFD contributed by sunflecks. The correlation coefficient was lower at higher sunfleck threshold values. However, total daily PPFD showed a very low correlation with diffuse background PPFD (data not shown).

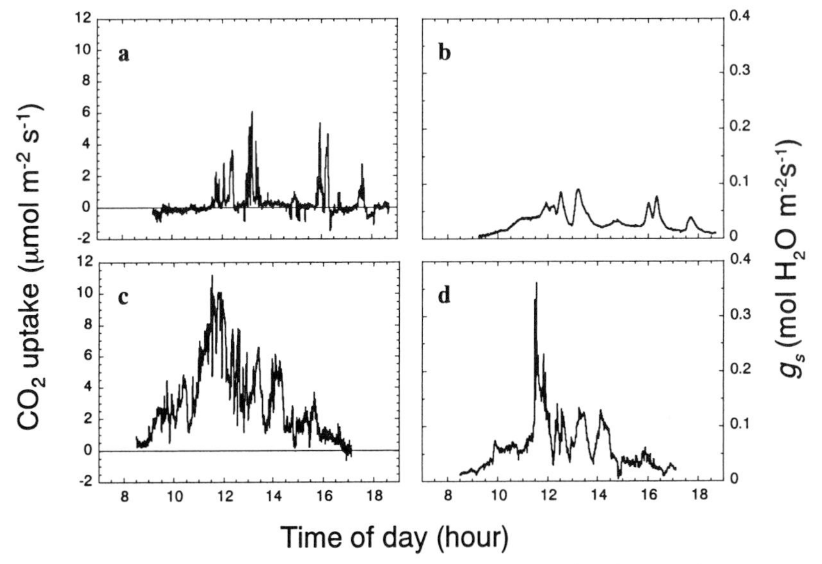

Fig. 5 Diurnal course of CO_2 uptake (graphs a and c) and stomatal conductance (g_s, graphs b and d) for *S. macrophylla* seedlings in the closed understory (a, b) and gap (c, d) microsites. All data were recorded at a mean interval of 2 seconds.

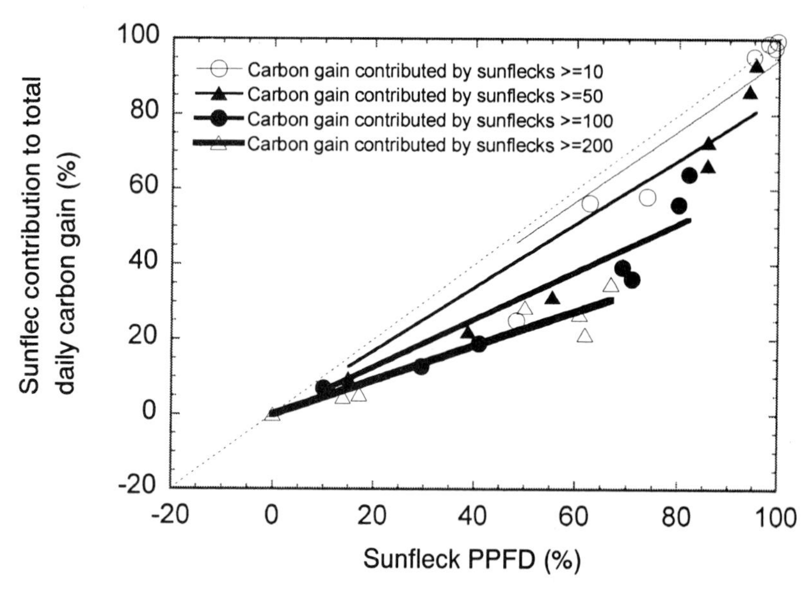

Fig. 6 Correlation between sunfleck contribution to total daily carbon gain and the proportion of total daily PPFD received in the form of sunflecks. Sunflecks were defined by threshold values from 10 to 200 μmol m^{-2} s^{-1} for *S. macrophylla* seedlings.

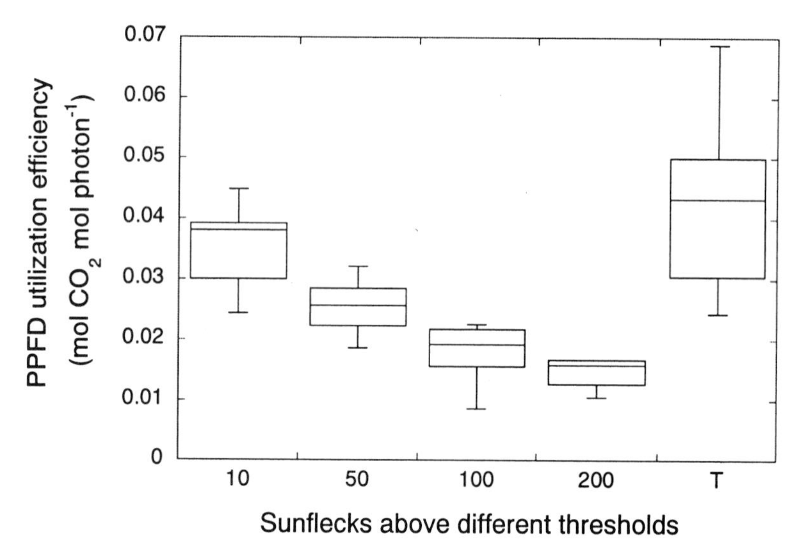

Fig. 7 Photosynthetic utilization efficiency for sunflecks with PPFD values greater than thresholds of 10, 50, 100, and 200 μmol m^{-2} s^{-1}. T: photosynthetic utilization efficiency based on total daily PPFD. Each box encloses 50% of the measurements over the 7-day period; the middle line within the box represents the median value. The top and bottom of the box mark the upper quartile (75%) and the lower quartile (25%) of each data set, respectively. The lines extending from the top and bottom of each box represent standard error bars for the data.

Sunfleck utilization efficiency (moles CO_2 produced per mole photons) decreased with an increasing proportion of sunflecks above a specified threshold value (defined in terms of PPFD, Fig. 7). The daily quantum efficiency was higher than sunflecks at any level. In the understory, the daily quantum efficiency was 0.07 mol CO_2 per mol photons.

4. DISCUSSION

4.1 Sunfleck activity

Few measurements of understory light regimes have been reported for regenerating tropical forests. Ellsworth & Reich (1996) reported that average PPFD decreased from 815 μmol m^{-2} s^{-1} in 1-year-old second-growth forest to less than 400 μmol m^{-2} s^{-1} in 4-year-old second growth forest after abandonment in sites near San Carlos de Rio Negro, Venezuela. Mean PPFD decreased further, to 57 μmol m^{-2} s^{-1}, in a 9-year-old tropical secondary forest, and these values were similar to readings in a 200-m^2 gap in nearby mature forest (Chazdon et al. 1996). Mean PPFD (about 20 μmol m^{-2} s^{-1}) was much lower in the regenerating forest at Pasoh FR than in any of the above-mentioned forests. Since selective logging occurred nearly 40 years before our study, the average PPFD was close to that in the primary forest (Tang et al. 1999).

Even fewer detailed studies have been carried out on sunfleck activity in regenerating lowland tropical forests. Chazdon (1988) has compiled values from the literature which show that in tropical evergreen forests, the mean contribution of sunfleck PPFD to total daily PPFD ranges from 50% to 70%. The contribution of sunflecks to total PPFD in the secondary forest was very high in our study (Fig. 3) and was always above 60% during the seven days measurement period.

4.2 Total contribution of sunfleck utilization to daily carbon assimilation

Only a few data sets are available that report the total contribution of sunfleck utilization under natural light regimes (Björkman et al. 1972; Pearcy & Calkin 1983; Pfitsch & Pearcy 1989). In these studies, 30% to 60% of leaf carbon gain was contributed by sunflecks with PPFD levels greater than 50 μmol m^{-2} s^{-1}. The total contribution of sunfleck utilization to daily leaf carbon gain in *S. macrophylla* varied from 26% to 83% in the present study, depending on the microsite, during seven days of measurements. With a threshold value of 10 μmol m^{-2} s^{-1} for sunflecks, a value close to the light compensation point for most tree seedlings, the contribution of sunflecks to total daily carbon gain was always above 72%. Since diffuse background PPFD in the closed understory of tropical rain forests has frequently been found to be around 10 μmol m^{-2} s^{-1}, the results confirmed that sunfleck PPFD is an important light resource for understory plants (Chazdon et al. 1996). In a deciduous temperate forest, sunfleck contribution to total daily leaf carbon gain was only 10% to 20% (Schulze 1972; Weber et al. 1985).

Pfitsch & Pearcy (1989) found that total daily carbon gain was more closely related to sunfleck PPFD than to diffuse PPFD at different microsites. In this study, we found that total daily carbon gain was closely related to total daily PPFD and total sunfleck PPFD, but was weakly related to diffuse background PPFD during 7 days of measurements (Fig. 6). Daily photosynthetic utilization efficiency was very high, with a maximum total close to 0.07 (Fig. 7). Björkman et al. (1972) found that on an overcast day, the daily efficiency was 0.07 mol CO_2 per mol of incident photons, which is close to the maximum efficiency obtained in our short-term photosynthetic measurements.

ACKNOWLEDGEMENTS

This study was supported by a joint research project between the NIES, FRIM and UPM under global Environment Research Programme granted by Ministry of the Environment, Japan. The authors also thank Ms. Makiko Hirota, Dr. Satoko Kawarasaki and Ms. Soo Woon Kuen for their patience in processing the manuscript.

REFERENCES

Appanah, S. & Turnbull, J. M. (1998) A review of dipterocarps taxonomy, ecology and silviculture. Center for International Forestry Research. Bogor, Indonesia.

Björkman, O., Ludlow, M. M. & Morrow, P. A. (1972) Photosynthetic performance of two rainforests species in their native habitat and analysis of their gap exchange. Carnegie Inst. Washington Yearb. 71: 94-102.

Chazdon, R. L. (1988) Sunflecks and their importance to forest understory plants. Adv. Ecol. Res. 18: 1-63.

Chazdon, R., Pearcy, R., Lee, D. & Fetcher, N. (1996) Photosynthetic responses of tropical forest plants to contrasting light environments. In Mulkey, S. S., Chazdon, R., Smith, A. P., Chapman & Hall (eds). Tropical forest plant ecophysiology, New York, pp.5-55.

Ellsworth, D. S. & Reich, P. B. (1996) Photosynthesis and leaf nitrogen in five Amazonian tree species during early secondary succession. Ecology 77: 581-594.

Gross, L. J. (1982) Photosynthetic dynamics in varying light environments: a model and its application to whole leaf carbon gain. Ecology 63: 84-93.

Pearcy, R. W. (1987) Photosynthetic gas exchange responses of Australian tropical forest trees in canopy, gap and understorey micro-environments. Funct. Ecol. 1: 169-178.

Pearcy, R. W. (1990) Sunflecks and photosynthesis in plant canopies. Annu. Rev. Plant Physiol. Plant Mol. Biol. 41: 421-453.

Pearcy, R. W. & Calkin, H. (1983) Carbon dioxide exchange of C_3 and C_4 tree species in the understory of a Hawaiian forest. Oecologia 58: 26-32.

Pearcy, R. W. & Pfitsch, W. A. (1995) The consequences of sunflecks for photosynthesis and growth of forest understory plants. In Schulze, E. D. & Caldwell, M. M. (eds). Ecophysiology of Photosynthesis. New York, Springer-Verlag. pp.343-359.

Pearcy, R. W., Katherine, O. & Calkin, H. W. (1985) Photosynthetic responses to dynamic light environments by Hawaiian trees. Plant Physiol. 79: 896-902.

Piftsch, W. A. & Pearcy, R. W. (1989) Daily carbon gain by *Adenocaulon bicolor*, a red wood forest understory herb, in relation to its light environment. Oecologia 80: 465-470.

Schulze, E. D. (1972) Die Wirkung von Licht und Temperatur auf den CO_2-Gaswechel verschiedender Lebensformen aus der Krautschicht eines montanen Buchenwaldes. Oecologia 9: 235-258.

Soepadmo, E. (1978) Introduction to the Malaysian IBP Synthesis Meeting. Malay. Nat. 30: 119-124.

Tang, Y. (1997) Light. In Prasad, M. N. V. (ed). Plant Ecophysiology. John Wiley & Sons, Inc., New York, pp.1-40.

Tang, Y. H., Kachi, N., Furukawa, A. & Awang, M. (1999) Heterogeneity of light availability and its effects on simulated carbon gain in tree leaves in a small gap and the understory in a tropical rain forest. Biotropica 31: 268-278.

von Caemmerer, S. & Farquhar, G. D. (1981) Some relationships between the biochemistry of photosynthesis and the gas exchange of leaves. Planta 153: 376-387.

Weber, J. A., Jurik, T.W., Tenhunen, J. D. & Gates, D. M. (1985) Analysis of gas exchange in seedlings of *Acer saccharum*: integration of field and laboratory studies. Oecologia 65: 338-347.

Wong, M. (1983) Understorey phenology of the virgin and regenerating habitats in Pasoh Forest Reserve, Negeri Sembilan, West Malaysia. Malay. For. 46: 197-223.

19 Molecular Phylogeny of Dipterocarpaceae

Tadashi Kajita[1] & Yoshihiko Tsumura[2]

Abstract: The Dipterocarpaceae is distributed mainly in Southeast Asia and its species represent a major canopy component in tropical forests. In Southeast Asia's tropical rain forests in paticular, the Dipterocarpaceae is one of the region's most ecologically and economically significant tree families. The family displays extreme species richness in many tropical rain forests, including Pasoh Forest Reserve (Pasoh FR) – one of key tropical field ecological centres in the world. Dipterocarp species are abundant in the reserve and consequently represent major target species for the studies in this forest. In this paper, recent progress in molecular phylogenetics of Dipterocarpaceae is reviewed, with reference to representatives of all the genera found in the Pasoh FR. A familial relationship of Dipterocarpaceae, Sarcolaenaceae and Cistaceae was re-analyzed using maximum parsimony, neighbor joining, and maximum likelihood methods. A hypothetical phylogenetic tree of genera of Dipterocarpaceae combining all the published phylogenetic relationships is proposed.

Key words: chloroplast DNA, Cistaceae, Sarcolaenaceae.

1. INTRODUCTION

The tree family Dipterocarpaceae is distributed throughout the tropical regions. While species in Africa and South America are relatively few (most of these being shrubs), Asian species are remarkably diverse, with more than 400 species (Ashton 1982). Most of them are emergent trees that dominate Asian lowland tropical rain forest such as Pasoh Forest Reserve (Pasoh FR). Such a remarkable success of the family in Asian tropical forest has been discussed with the acquisition of symbiosis with ectotrophic mycorrhiza (Ashton 1988). As dipterocarp trees are of great value in the timber market, diverse species have been studied taxonomically, primarly using morphological, biochemical and ecological data (Ashton 1982; Symington 1943). However, phylogenetic relationships among taxa of Dipterocarpaceae have not been well characterized. In ecological research, the phylogenetic relationship between species and genera is fundamental to the understanding of the other relationships and interactions between organisms dependent on the dipterocarps as primary producers. In Pasoh FR, many scientists have conducted extensive ecological studies on the dipterocarps and associated organisms. The addition of molecular phylogenetic data to this ecological data will significantly enhance our understanding and knowledge of tropical ecology.

The progress of molecular systematics in the last two decades has tremendously increased our knowledge of the phylogenetic relationships of land plants (Chase et al. 1993; Soltis et al. 1997). Because of its relative simplicity and quantity, molecular data is generally considered more reliable than morphological data for constructing phylogenetic trees. As with many other families of land plants, phylogenetic relationships of the Dipterocarpaceae have recently come to be studied

[1] Botanical Gardens, Graduate School of Science, University of Tokyo, Tokyo 112-0001, Japan. E-mail: tkaji@bg.s.u-tokyo.ac.jp
[2] Forestry and Forest Product Research Institute (FFPRI), Japan.

at various taxonomic levels. In this review, we make a brief survey of the progress of molecular phylogenetic studies of Dipterocarpaceae and present new phylogenetic trees of familial and intrafamilial relationships of the group, based on existing sequence data.

2. MATERIALS AND METHODS

In order to conduct a number of phylogenetic and molecular genetical analyses of Dipterocarpaceae and related families, published sequence data of chloroplast *rbc*L and *atp*B genes of the following species were used. Genbank accession numbers are in parenthesis (*rbc*L and *atp*B, respectively): Dipterocarpaceae: *Anisoptera marginata* Korth. (Y15144, AF035918), *Pseudomonotes tropenbosii* Londoño (*rbc*L: AF030238); Sarcolaenaceae: *Sarcolaena* sp. (Y15147, AJ233070); Cistaceae: *Cistus revolii* Coste & Soulie (Y15140, AF035902), *Helianthemum grandiflorum* DC. (Y15141, AF035907); Neuradaceae: *Neurada procumbens* Linn. (U06814, AJ233069); Sphaerosepalaceae: *Rhopalocarpus* sp. (Y15148, AJ233071); Thymelaeaceae: *Dais cotinifolia* Linn. (AJ233144, AJ233094), *Gonystylus macrophyllus* (Miq.) Airy Shaw (Y15150, AJ233095). The sequences were aligned visually. Phylogenetic analyses were done using PAUP* 4.0 beta (Swofford 1999). Maximum parsimony (MP), neighbor-joining (NJ), and Maximum Likelihood (ML) methods were used. A Branch-and-bound search was done for the MP method. Genetic distances were calculated using Jukes-Cantor method (Jukes & Cantor 1969) for the NJ tree. In both the MP and NJ methods, bootstrap values were calculated from 1,000 replications. "QPS values" (roughly equivalent to the bootstrap values) were calculated with 1,000 "puzzling steps" (Strimmer & Von Haeseler 1996).

Nucleotide substitutions at synonymous sites are computed using the method of Nei and Gojobori (1986). Estimates of the average numbers of substitutions between groups and their variances were calculated using the method of Nei and Jin (1989). For these analyses, the ODEN (Ina 1994) and SEnj (kindly provided by Dr. Y. Ina, Biomolecular Engineering Research Institute, Suita, 565, Japan; yina@beri.co.jp) were used. Results are presented in Fig. 1 and Table 1 and discussed in 3., below.

3. FAMILIAL RELATIONSHIPS

There have been several different opinions regarding the systematic position of Dipterocarpaceae in Angiosperm families. Dipterocarpaceae have generally been placed in the order Theales (Bessey 1915; Cronquist 1981; Hutchinson 1926; Kostermans 1992; Wettstein 1935) or Malvales (Ashton 1982; Dahlgren 1983; Takhtajan 1997; Thorne 1992). According to the global *rbc*L analysis of Angiosperm phylogeny (Chase et al. 1993), Dipterocarpaceae was included in a clade allied to Malvalean genera. Recently, several molecular works on phylogenetic relationships of the Malvales, including Dipterocarpaceae, have been published. Alverson et al. (1998) used the *rbc*L gene for an extensive sample set of Malvales and its allies, and suggest well supported sister relationships of Dipterocarpaceae and Sarcolaenaceae (bootstrap value = 80). The clade consisting of the two families was sister to the Cistaceae (bootstrap value = 100). Nandi et al. (1998) also suggested the close relationships between Dipterocarpaceae and Cistaceae, from a cladistic analysis of Angiosperms using *rbc*L and non-molecular data. In this study, Sarcolaenaceae is placed near Dipterocarpaceae and Cistaceae, however no molecular data for Sarcolaenaceae was included in their analyses. The close relationships of Dipterocarpaceae and Sarcolaenaceae are also suggested by

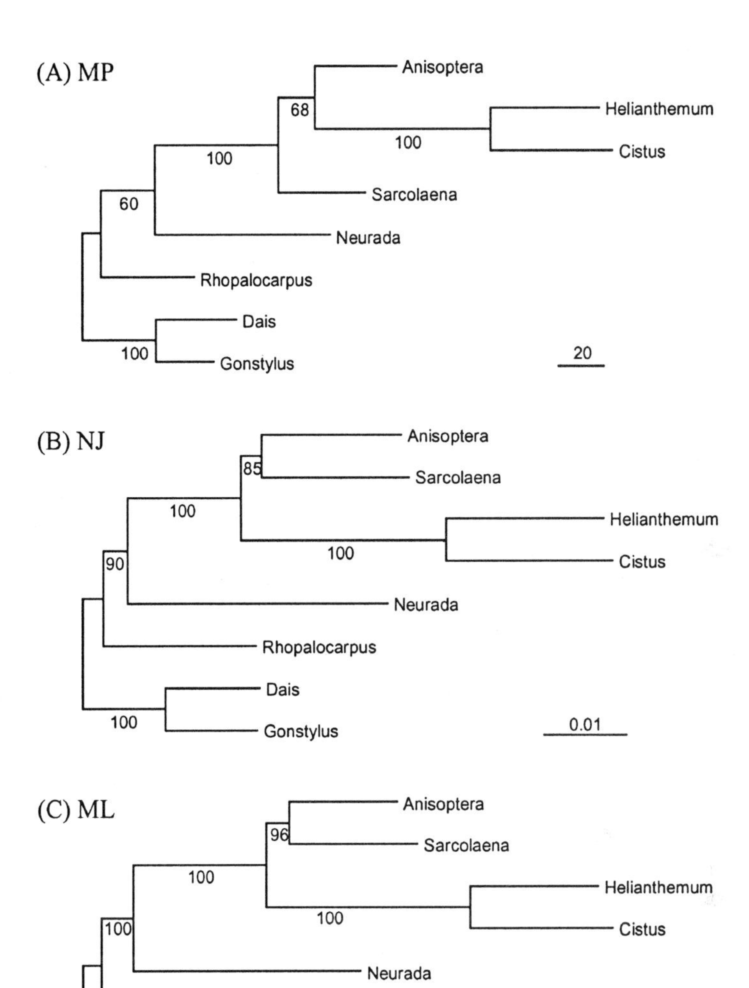

Fig. 1 Familial relationships of Dipterocarpaceae. (A) Maximum parsimonious (MP) tree constructed by branch-and-bound search. Lengths of branches are proportional to the bar of 20 changes indicated in the figure. Bootstrap value of 1,000 replicates is below branches. Tree length = 583, Consistency index (CI) = 0.8113, CI excluding uninformative characters = 0.6986, Retention index (RI) = 0.6919, Re-scaled consistency index (RC) = 0.5613. (B) Neighbor-joining (NJ) tree. Lengths of branches are proportional to the genetic distance of 0.01 indicated in the figure. Genetic distance is calculated by Jukes - Cantor methods. Bootstrap values of 1,000 replicates are below branches. (C) Maximum Likelihood (ML) tree constructed using Quartet puzzling analysis. Number of quartets to examine = 70. Quartets evaluated using approximate likelihood calculations with 1,000 puzzling step. QPS values are under branches. Assumed nucleotide frequencies are, A = 0.283, C = 0.198, G = 0.239, T = 0.281. All sites are assumed to evolve at same rate. HKY model has been applied.

Table 1 Synonymous and nonsynonymous nucleotide substitution between Neuradaceae and others. Nucleotide substitutions of synonymous (Ks) and nonsynonymous (Ka) sites are calculated by the method of Nei & Gojobori (1986).

	rbc L	1371bp		*atp* B	1338bp	
	Ks	Ka	Ks/Ka	Ks	Ka	Ks/Ka
Anisoptera - *Neurada*	0.208	0.018	11.9	0.245	0.021	11.8
	(±0.028)	(±0.004)		(±0.031)	(±0.005)	
Sarcolaena - *Neurada*	0.228	0.016	14.6	0.228	0.026	8.9
	(±0.030)	(±0.004)		(±0.030)	(±0.005)	
Cistus - *Neurada*	0.273	0.021	13.3	0.37	0.029	12.7
	(±0.033)	(±0.004)		(±0.041)	(±0.005)	
Helianthemum - *Neurada*	0.273	0.019	14.7	0.382	0.038	10.2
	(±0.033)	(±0.004)		(±0.042)	(±0.006)	

Morton et al. (1999) and Dayanandan et al. (1999), although their sample set from Malvales were rather restricted and Cistaceae was not included. Bayer et al. (1999) used the chloroplast *atp*B gene in addition to *rbc*L, and analyzed phylogenetic relationships of Malvales. In their phylogenetic tree, which included weighted characters, sister relationships between Dipterocarpaceae (*Anisoptera*) and Cistaceae (*Helianthemum*, *Tuberaria* and *Cistus*) were very well supported (bootstrap value = 99), and Sarcolaenaceae (*Sarcolaena*) was the sister to the clade (bootstrap value = 99). The incongruence between the two *rbc*L phylogenetic trees (Alverson et al. 1998 ; Bayer et al. 1999) was not discussed in the latter paper.

To compare the differences in the sample set, we performed phylogenetic analysis using sequences taken from the sample set of Bayer et al. (1999). We performed three different kinds of analysis. The MP analysis produced the same topology as Bayer et al. (1999), but the NJ and ML analyses suggested sister relationships between Dipterocarpaceae and Sarcolaenaceae (Fig. 1). The branch length of the representatives of Cistaceae is almost twice of the Dipterocarpaceae and Sarcolaenaceae, which may have caused the difficulty in constructing phylogenetic trees in the MP analysis. We also compared the number of nucleotide substitutions at synonymous and nonsynonymous site, between the Neuradaceae and each of Dipterocarpaceae, Sarcolaenaceae and Cistaceae (Table 1). The number of nucleotide substitutions in *rbc*L between *Neurada* and *Cistus* is a little higher than those between *Neurada* and *Anisoptera*, and *Neurada* and *Sarcolaena*. However, the values for *atp*B are considerably higher between *Neurada* and *Cistaceae* genera, which suggests an inconsist molecular clock (i.e. variable rate) in *atp*B gene, which may also explain the different topology given by the MP method.

4. GENERIC RELATIONSHIPS OF DIPTEROCARPACEAE

The family Dipterocarpaceae consists of three subfamilies, namely, Monotoideae (three genera: *Monotes*, *Marquesia* and *Pseudomonotes*), Pakaraimoideae (one genus: *Pakaraimaea*), and Dipterocarpoideae (thirteen genera). The Dipterocarpoideae consists of two tribes; Dipterocarpeae (eight genera: *Dipterocarpus*, *Anisoptera*, *Upuna*, *Cotylelobium*, *Vatica*, *Vateria*, *Vateriopsis* and *Stemonoporus*) and Shoreae (five genera: *Dryobalanops*, *Parashorea*, *Neobalanocarpus*, *Hopea* and *Shorea*) (Londoño et al. 1995). Many authors have been interested in the generic relationships.

Tsumura et al. (1996) studied phylogenetic relationships of Southeast Asian genera of Dipterocarpaceae using PCR-RFLP of chloroplast DNA. Using eleven different genes, they detected 141 site changes among ten genera and 30 species. In their phylogenetic tree (using *Upuna* as an outgroup), monophyly of tribe Shoreae, paraphyly of *Shorea*, monophyly of *Vatica* and *Cotylelobium* and so on were shown. They also compared the sequence divergences of the eleven studied cpDNA genes for Dipterocarpaceae, and showed that *rpo*C, *rbc*L, *pet*B and *trn*K were suitable for molecular phylogenetic analysis.

Kajita et al. (1998) used all the genera but only a portion of the species used in Tsumura et al. (1996), and constructed a phylogenetic tree using nucleotide sequences of *mat*K, the intron of *trn*L and the intergenic spacer (IGS) region between *trn*L and *trn*F. They used *Tilia* (Tiliaceae) as an outgroup according to the phylogenetic relationships proposed by the global *rbc*L analysis of Chase et al. (1993). In the resultant phylogenetic tree, ten genera were divided into two well-supported clades (bootstrap values > 95). One consisted of *Anisoptera*, *Upuna*, *Cotylelobium* and *Vatica*; the other *Dipterocarpus*, *Dryobalanops*, *Shorea*, *Parashorea*, *Neobalanocarpus* and *Hopea*. Genera included in tribe Shoreae comprised a clade, but tribe Dipterocarpeae was paraphyletic because of sister relationships of *Dipterocarpus* to the Shoreae clade. Most bootstrap supports for branches in their phylogenetic tree were more than 95% except for branches within the Dipterocarpeae clade (excluding *Dipterocarpus*), and at the base of the Shoreae clade.

Kamiya et al. (1998) used nucleotide sequences of the intron of *trn*L and IGS between *trn*L and *trn*F to infer phylogeny of 53 dipterocarp species belonging to ten genera. Topology of the generic relationships in the resultant trees is almost identical with those of Kajita et al. (1998). Morton et al. (1999) used nucleotide sequences of *rbc*L to infer phylogenetic relationships of *Pseudomonotes* with other Dipterocarpaceae genera. They surveyed phylogenetic relationships of seven genera of 11 species, which include all the three subfamilies. The relationships of two Monotoideae genera, *Monotes* and *Pseudomonotes*, which are distributed in Africa and South America respectively, are also compared. They obtained a single most parsimonious tree and the sister relationships between Dipterocarpoideae and Pakaraimoideae, although bootstrap value of these clade are not high (51 and 60 respectively). Two genera of Monotoideae formed a highly supported clade (bootstrap value = 98) that is sister to the remaining Dipterocarpaceae clade with bootstrap value less than 50.

Dayanandan et al. (1999) also used *rbc*L gene sequences from 35 species of 13 genera. Nine sequences were the same as those used by Morton et al. (1999), and the topology of the three subfamilies are the same, although those clade are not well supported (Dipterocarpoideae - Pakaraimoideae: bootstrap value < 50; Monotoideae and other Dipterocarpaceae: bootstrap value < 50). Lower resolution of relationships is also seen in most branches, but the monophyly of *Vateria* and *Stemonoporus*, that have *Upuna* as sister, was suggested. Paraphyly of *Shorea* is also suggested.

We already have, as discussed above, molecular phylogenetic trees of Dipterocarpaceae, but molecular markers and taxa surveyed are not completely the same in all trees, and the resolution of some relationships are not highly enough for truly robust conclusion to be drawn. There are still some problems as follows:

1. Monophyly of Dipterocarpaceae: The monophyly of Sarcolaenaceae and Dipterocarpaceae are well supported (see Fig. 1 and section 3 in this paper), but

the monophyly of Dipterocarpaceae is not quite so well supported (bootstrap value < 50 in Dayanandan et al. 1999).

2. Relationships of three subfamilies: The phylogenetic relationships of Monotoideae, Pakaraimoideae and Dipterocarpaceae are not very well supported (bootstrap values < 50 in Dayanandan et al. 1999).
3. The sister relationships of *Dryobalanops* to the other members of the Shoreae clade is not very well supported (bootstrap value = 63 in Kajita et al. 1998).
4. Relationships among *Anisoptera*, *Cotylelobium*, *Vatica*, *Upuna* and *Stemonoporus* are not quite clear (bootstrap value < 80 in Kajita et al. 1998).
5. We still have no information on *Marquesia* and *Vateriopsis*.

To elucidate the whole phylogenetic relationships of Dipterocarpaceae genera, further study is necessary, including *Marquesia* and *Vateriopsis*. It is also desirable to integrate DNA sequence data that have already been published. As the combined analysis of *mat*K, *trn*L intron and the *trn*L-*trn*F IGS regions have the highest bootstrap support among present studies, to combine these data with *rbc*L or other sequence data, such as *pet*B and *rpo*C (the potential of which as a phylogenetic marker was suggested in Tsumura et al. (1996)), will provide reliable phylogenetic relationships.

Although resolution, which is mostly represented by bootstrap values, varies greatly, research has obtained most of the generic relationships by molecular data, except for *Marquesia* and *Vateria*. For this review, we made a hypothetical phylogenetic tree of Dipterocarpaceae (Fig. 2), combining all the published phylogenetical relationships of the family. In this tree, the phylogenetic relationships within Dipterocarpoideae are obtained from Kajita et al. (1998) except for the position of *Stemonoporus* and *Vateria* which is proposed by Dayanandan et al. (1999). The position of *Pakaraimaea*, *Pseudomonotes* and *Monotes* are as suggested by Morton et al. (1999) and Dayanandan et al. (1999). To see the distribution of morphological characters over genera of Dipterocarpaceae, we combined the data matrix of Dayanandan et al. (1999) with others (Goldblatt & Dorr 1986; Nandi 1998; Nandi et al. 1998; Randrianasolo & Miller 1999) (Fig. 2). The direction of some character evolution was not clearly shown in Dayanandan et al. (1999) because of the polytomy in their phylogenetic tree. In our hypothetical tree, all of the branches are dichotomous, and very consistent with the distribution of morphological characters. For example, fruits sepals appear to have evolved from non-thick to thick after *Dryobalanops* branched.

5. PHYTOGEOGRAPHY AND NUCLEAR SUBSTITUTION RATE OF DIPTEROCARPACEAE

The disjunctive distribution of genera of Dipterocarpaceae, phytogeographical perspective of the family is one of the topics that can be addressed using molecular phylogeny. In the *rbc*L tree of Morton et al. (1999) and Dayanandan et al. (1999), Monotoideae which includes both African and South American genera, arose first within Dipterocarpaceae, following the diversification of South American subfamily Pakaraimoideae. The fact that both of these subfamilies diversified basally to the family is consistent with the hypothesis of a Gondwana origin for the family. As there was no discussion of molecular substitution rates for the *rbc*L gene in their analysis, we compared nuclear substitution rate using the available sequence data. Procedures are the same as in section 2.

We compared sequence data of *rbc*L (1311bp) and *atp*B (1338bp) of *Anisoptera*, *Pseudomonotes* and *Sarcolaena*, and calculated nucleotide

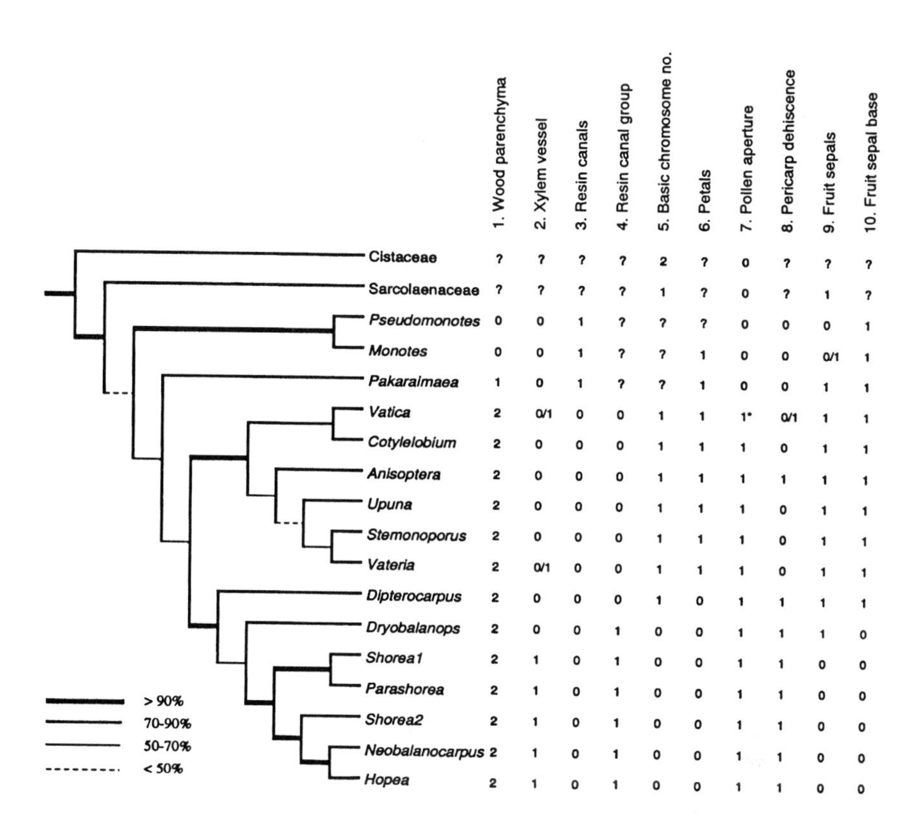

Fig. 2 Hypothetical phylogenetic relationships in Dipterocarpaceae and distribution of morphological characters. The tree topology is a summary of published molecular phylogenies of Dipterocarpaceae (see detail in text). *Shorea* 1 represents most species of *Shorea*, and *Shorea* 2 several species of *Shorea* which is monophyletic to *Hopea*. Bootstrap values in original trees are indicated by thickness of lines of internode as shown in the figure. Distribution of morphological character is according to previous studies (see detail in text). Characters and character states are as follows: 1. Wood parenchyma (0 = uniseriate, 1 = biseriate, 2 = multiseriate); 2. Xylem vessels (0 = solitary, 1 = grouped); 3. Resin canals (0 = present, 1 = absent); 4. Resin canal groups (0 = solitary, 1 = series); 5. Basic chromosome number (0 = seven, 1 = eleven, 2 = others); 6. Petals (0 = connate, 1 = free); 7. Pollen aperture (0 = tricolporate, 1 = tricolpate); 8. Pericarp dehiscence (0 = loculicidal, 1 = non-loculicidal); 9. Fruit sepal aestivation (0 = imbricate; 1 = valvate); 10. Fruit sepal base (0 = thick, 1 = non-thick). Character states which can not be determined or have no information are shown by "?".

Table 2 Average number of substitutions and variances in *rbc*L and *atp*B. Values are calculated in the same way explained in the legend of Table 1.

	rbc L 1311bp		*atp* B 1338bp	
	Ks	Ka	Ks	Ka
Anisoptera - Sarcolaena	0.079 (± 0.016)	0.014 (± 0.004)	0.096 (± 0.018)	0.02 (± 0.004)
Pseudomonotes - Sarcolaena	0.094 (± 0.018)	0.015 (± 0.004)	-	-
Anisoptera - Pseudomonotes	0.086 (± 0.016)	0.015 (± 0.004)	-	-

substitution rate at synonymous and nonsynonymous sites (Table 2). Unfortunately, as the *rbc*L sequences used in Chase et al. (1993), Alverson et al. (1998), Morton et al. (1999) and Dayanandan et al. (1999) have not been made available in DNA databases with the exception of *Pseudomonotes* (Morton et al. 1999), we were unable to use other Dipterocarpaceae *rbc*L data. The Sarcolaenaceae is thought to have originated from the Theales during the Cretaceous period (from 141 to 65 MYA), a hypothesis supported by finds of Xyloolaena (Sarcolaenaceae) type pollen from South Africa (Randrianasolo& Miller 1999). As the number of nucleotide substitution between *Anisoptera* and *Pseudomonotes* is almost as same as that of between *Sarcolaena* and the Dipterocarpaceae, the divergence of these two genera may also have happened during th Cretaceous. As the average nucleotide substitution between Sarcolaenaceae and Dipterocarpaceae is 0.086 (± 0.016), the rate of nucleotide substitution can be calculated as $r = Ks/2T = 3.05x10^{-10} - 6.62x10^{-10}$ (substitutions per site per year). These values are slower than other herbaceous plants, but faster than values proposed in the Betulaceae (Bousquet et al. 1992). According to the *rbc*L-based phylogenetic tree produced by Morton et al. (1999), numbers of site changes from the base of Dipterocarpaceae clade to *Pakaraimaea*, *Monotes* and *Pseudomonotes* are 31, 22 and 30, respectively. Terminal branch lengths of *Monotes* and *Pseudomonotes* are five and thirteen, respectively, and the length of the shared internode is twelve. If the molecular clock is consistent in the Dipterocarpaceae clade, Monotes and Pseudomonotes might branch in the middle of the age of Dipterocarpaceae. The large differences in terminal branch lengths of *Monotes* and *Pseudomonotes* should be examined in detail.

6. SPECIES PHYLOGENY

Many scientists have studied the species relationships of *Shorea* and *Hopea*, which include important dipterocarp timber trees. Ashton et al. (1984) used enzyme electrophoresis to discuss the relationships of ten species of *Shorea*. Tsumura et al. (1996) used eleven species of *Shorea* and seven species of *Hopea*. In their phylogenetic tree, *Shorea singkawang* Buck and *Shorea bracteolata* Dyer made a clade which is sister to the *Hopea* and *Neobalanocarpus* clade. The former is placed in *Shorea* sect. Anthoshorea and the latter in sect. Mutica in the present classification. Other species of sect. Mutica were included in the *Shorea - Parashorea* clade.

In Kamiya et al. (1998) 31 species of *Shorea* (which represented eight sections), four species each from *Hopea* and *Dipterocarpus*, and three species each from *Dryobalanops*, *Vatica* and *Anisoptera*, were included. The relationships of the *Hopea* clade and the clade which includes *Shorea bracteolata* are not clear because of a paucity of information. In the clade, four species of *Shorea* sect.

Anthoshorea (white meranti group) and one species of sect. Pentacne were included. In addition, *S. roxburghii* G. Don and *S. siamensis* Miq., both of which are distributed in monsoon Asia formed a clear clade, but *S. obtusa* Wall. which has same habit as *S. siamensis* was included in the Selangan Batu group. In the other major *Shorea* clade, four clades are recognized. The groups used in forestry – the Selangan Batu, Yellow Meranti, and White Meranti were monophyletic, while the Red Meranti was divided into three clades. This group showed the most nucleotide diversity in the genus. Dayanandan et al. (1999) used 10 species of *Shorea* representing five sections one of which, Doona, was not included in Kamiya et al. (1998). Their *rbc*L tree showed that sect. Doona is a monophyletic group which may form a clade with *Hopea*. All the species appeared in phylogenetic trees of the previous works are shown in Appendix. All the species appeared in phylogenetic trees of the previous works are shown in Appendix.

7. CONCLUSION

Molecular data of Dipterocarpaceae have been accumulated and phylogenetic relationships come to be known in various analyses. However, those data obtained to date have not been integrated, and several taxa have not been included because of the difficulty involved in obtaining materials. It is important to combine all the data and materials to produce an integrated phylogenetic tree. An integrated phylognetic tree will be extremely valuable, not only for understanding the evolutionary relationships but also for comparative studies of ecology and physiology.

Large numbers of scientists are studying Asian tropical forest, in which dipterocarp trees play a central role in the ecosystem. One of the most interesting topics is the co-evolution between *Shorea* species and other organisms such as insects and fungi. Reliable phylogenetic relationships of *Shorea* species are required. As the non-coding region of chloroplast DNA do not give sufficiently high resolution of *Shorea* phylogeny, efforts to survey other genes not only from chloroplast but also from nuclear genome will be necessary.

More than two hundred species of *Shorea* exist in Asian dipterocarp forests. The process of speciation of these taxa has not yet been studied in depth. Further research on genetic variations within and among populations of those species would provide key data to solve these problems.

ACKNOWLEDGEMENTS

We would like to show our gratefulness to Dr. Tsuneyuki Yamazaki of Kyushu University, who is the leading researcher of Dipterocarp phylogeny in Japan. Great thanks are also to Mr. Gregory Kenicer (Botanical Gardens, University of Tokyo) for reading through the manuscript of this paper. The authors thank Norwati Muhammad of the Forest Research Institute of Malaysia (FRIM) for providing materials from the FRIM dipterocarp arboretum.

REFERENCES

Alverson, W. S., Karol, K. G., Baum, D. A., Chase, M. W., Swensen, S. M., McCourt, R. & Sytsma, K. J. (1998) Circumscription of the Malvales and relationships to other Rosidae Evidence from rbcL sequence data. Am. J. Bot. 85: 876-887.

Ashton, P. S. (1982) Dipterocarpaceae. In Van Steenis, C. G. G. J. (ed). Flora Malesiana, Series 1, Spermatophyta, vol. 9, Martinus Nijhoff Pulishers, The Hague, pp.237-552.

Ashton, P. S. (1988) Dipterocarp biology as a window to the understanding of tropical forest structure. Annu. Rev. Ecol. Syst. 19: 347-70.

Ashton, P. S., Gan, Y. Y. & Robertson, F. M. (1984) Electrophoretic and morphological comoparisons in ten rain forest species of Shorea (Dipterocarpaceae). Bot. J. Linn. Soc. 89: 293-304.

Bayer, C., Fay, M. F., De Bruijn, P. Y., Savolainen, V., Morton, C. M., Kubitzki, K., Alverson, W. S. & Chase, M. W. (1999) Support for an expanded family concept of Malvaceae within a recircumscribed order Malvales a combined analysis of plastid atpB and rbcL DNA sequences. Bot. J. Linn. Soc. 129: 267-303.

Bessey, C. (1915) The phylogenetic taxonomy of flowering plants. Ann. Mo. Bot. Gard. 2: 109-164.

Bousquet, J., Strauss, S. H. & Li, P. (1992) Complete congruence between morphological and rbcL-based molecular phylogenies in birches and related species (Betulaceae). Mol. Biol. Evol. 9: 1076-1088.

Chase, M. W., Soltis, D. E., Olmstead, R. G, Morgan, D., Les, D. H., Mishler, B. D., Duvall, M. R., Price, R. A., Hills, H. G, Qiu, Y. L., Kron, K. A., Rettig, J. H., Conti, E., Palmer, J. D., Manhart, J. R., Sytsma, K. J., Michaels, H. J., Kress, W. J., Karol, K. G, Clark, W. D., Hedrén, M., Gaut, B. S., Jansen, R. K., Kim, K. J., Wimpee, C. F., Smith, J. F., Furnier, G. R., Strauss, S. H., Xiang, Q. Y., Plunkett, G. M., Soltis, P. S., Swensen, S. M., Williams, S. E., Gadek, P. A., Quinn, C. J., Eguiarte, L. E., Golenberg, E., Learn, J., G. H., Graham, S. W., Barrett, S. C. H., Dayanandan, S. & Albert, V. A. (1993) Phylogenetics of seed plants: An analysis of nucleotide sequences from the plastid gene rbcL. Ann. Mo. Bot. Gard. 80: 528-580.

Cronquist, A. (1981) An Integrated System of Classification of Flowering Plants. Columbia University Press, New York.

Dahlgren, R. (1983) General aspects of Angiosperm evolution and macrosystematics. Nord. J. Bot. 3: 119-149.

Dayanandan, S., Ashton, P. S., Williams, S. M. & Primack, R. B. (1999) Phylogeny of the tropical tree family Dipterocarpaceae based on nucleotide sequences of the chloroplast rbcL gene. Am. J. Bot. 86: 1182-1190.

Goldblatt, P. & Dorr, L. J. (1986) Chromosome number in Sarcolaenaceae. Ann. Mo. Bot. Gard. 73: 828-829.

Hutchinson, J. (1926) The Families of Flowering Plants. Macmillan, London.

Ina, Y. (1994) ODEN: a program package for molecular evolutionary analysis and database search of DNA and amino acid sequences. CABIOS 10: 11-12.

Jukes, T. H. & Cantor, C. R. (1969) Evolution of protein molecules. In Munro, H. N. (ed.) Mammalian Protein Metabolism, Academic Press, New York, pp.21-132.

Kajita, T., Kamiya, K., Nakamura, K., Tachida, H., Wickneswari, R., Tsumura, Y., Yoshimaru, H. & Yamazaki, T. (1998) Molecular phylogeny of Diptrocarpaceae in Southeast Asia based on nucleotide sequences of matK, trnL intron, and trnL-trnF intergenic spacer region in chloroplast DNA. Mol. Phylogene. Evol. 10 : 202-209.

Kamiya, K., Harada, K., Ogino, K., Kajita, K., Yamazaki, T., Lee, H. S. & Ashton, P. S. (1998) Molecular phylogeny of dipterocarp species using nucleotide sequences of two non-coding regions in chloroplast DNA. Tropics 7: 197-207.

Kostermans, A. J. G. H. (1992) A handbook of the Dipterocarpaceae of Sri Lanka. Wildlife Heritage Trust of Sri Lanka, Colombo.

Londoño, A. C., Alvarez, E., Forero, E. & Morton, C. M. (1995) A new genus and species of Dipterocarpaceae from the Neotropics. I. Introduction, Taxonomy, ecology, and distribution. Brittonia 47: 225-236.

Morton, C. M., Dayanandan, S. & Dissanayake, D. (1999) Phylogeny and biosystematics of Pseudomonotes (Dipterocarpaceae) based on molecular and morphological data. Plant Syst. Evol. 216 : 197-205.

Nandi, O. I. (1998) Ovule and seed anatomy of Cistaceae and related Malvanae. Plant Syst. Evol. 209 : 239-264.

Nandi, O. I., Chase, M. W. & Endress, P. K. (1998) A combined cladistic analysis of Angiosperms using rbcL and non-molecular data sets. Ann. Mo. Bot. Gard. 85: 137-212.

Nei, M. & Gojobori, T. (1986) Simple methods for estimating the numbers of synonymous and nonsynonymous nucleotide substitutions. Mol. Biol. Evol. 3: 418-426.

Nei, M. & Jin, L. (1989) Variances of the average number of nucleotide substitutions within and between populations. Mol. Biol. Evol. 6: 290-300.

Randrianasolo, A. & Miller, J. S. (1999) Taxonomic revision of the genus Sarcolaena (Sarcolaenaceae). Ann. Mo. Bot. Gard. 86: 702-722.

Soltis, D. E., Soltis, P. S., Nickrent, D. L., Johnson, L. A., Hahn, W. J., Hoot, S. B., Sweere, J. A., Kuzoff, R. K., Kron, K. A., Chase, M. W., Swensen, S. M., Zimmer, E. A., Chaw, S. M., Gillespie, L. J., Kress, W. J. & Sytsma, K. J. (1997) Angiosperm phylogeny inferred from 18S ribosomal DNA sequences. Ann. Mo. Bot. Gard. 84: 1-49.

Strimmer, K. & Von Haeseler, A. (1996) Quartet puzzling: a quartet maximum-likelihood method for reconstructing tree topologies. Mol. Biol. Evol. 13: 964-969.

Swofford, D. L. (1999) PAUP*. Phylogenetic Analysis Using Parsimony (*and Other Methods). Version 4. Sinauer Associates, Sunderland, Massachusetts.

Symington, C. F. (1943) Foresters'Manual of Dipterocarps. Malayan Forest Records No. 16. University of Malaya Press, Kuala Lumpur (Reprinted with plates and historical introduction).

Takhtajan, A. (1997) Diversity and Classification of Flowering Plants. Columbia University Press, New York.

Thorne, R. F. (1992) An updated phylogenetic classification of the flowering plants. Aliso 13: 365-389.

Tsumura, Y., Kawahara, T., Wickneswari, R. & Yoshimura, K. (1996) Molecular phylogeny of Dipterocarpaceae in Southeast Asia using RFLP of PCR-amplified chloroplast genes. Theor. Appl. Genet. 93 : 22-29.

Wettstein, R. (1935) Handbuch der Systematischen Botanik. Leipzig.

Appendix: Species appeared in phylogenetic trees of previous works. Species with * are found in the Pasoh FR 50 ha Research Plot.

Tsumura et al. (1996)	Kajita et al. (1998)	Kamiya et al. (1998)	Morton et al. (1999)	Dayanandan et al. (1999)
Anisoptera laevis *	*Anisoptera laevis* *	*Anisoptera costata* *	*Dipterocarpus insignis*	*Anisoptera sp.*
Cotylelobium lanceolatum	*Anisoptera oblonga*	*Anisoptera laevis* *	*Dipterocarpus zeylanicus*	*Cotylolobium scabriusculum*
Dipterocarpus baudii	*Cotylelobium lanceolatum*	*Anisoptera thurifera*	*Monotes sp.*	*Dipterocarpus insignis*
Dipterocarpus kerrii	*Dipterocarpus baudii*	*Cotylelobium lanceolatum*	*Pseudomonotes tropenbosii*	*Dipterocarpus zeylanicus*
Dipterocarpus oblongifolius	*Dipterocarpus kerrii*	*Dipterocarpus alatus*	*Shorea affinis*	*Dryobalanops aromatica*
Dryobalanops aromatica	*Dryobalanops aromatica*	*Dipterocarpus baudii*	*Shorea curtisii*	*Hopea brevipetiolaris*
Hopea apiculata	*Dryobalanops oblongifolia*	*Dipterocarpus kerrii*	*Shorea dyeri*	*Hopea dryobalanoides* *
Hopea dyeri	*Hopea nervosa*	*Dipterocarpus palmebanicus*	*Shorea zeylanica*	*Hopea jucunda*
Hopea helferi	*Hopea odorata*	*ssp. Borneensis*	*Vateria copallifera*	*Monotes sp.*
Hopea latifolia	*Neobalanocarpus heimii*	*Dryobalanops aromatica*	*Vatica affinis*	*Neobalanocarpus heimii*
Hopea nervosa	*Parashorea lucida*	*Dryobalanops lanceolata*		*Pakaraimaea*
Hopea odorata	*Shorea bracteolata*	*Dryobalanops oblongifolia*		*Shorea affinis*
Hopea sangal *	*Shorea macroptera* *	*Hopea dryobalanoides* *		*Shorea assamica*
Hopea subalata	*Shorea ovalis*	*Hopea grifithii*		*Shorea curtisii*
Hopea wightiana	*Upuna borneensis*	*Hopea nervosa*		*Shorea dyeri*
Neobalanocarpus heimii *	*Vatica odorata*	*Hopea odorata*		*Shorea lissophylla*
Parashorea lucida		*Neobalanocarpus heimii* *		*Shorea ovalifolia*
Shorea atrinervosa		*Parashorea lucida*		*Shorea ovalis* *
Shorea bracteolata *		*Shorea acuta*		*Shorea robusta*
Shorea kunstleri		*Shorea argentifolia*		*Shorea stipularis*
Shorea lepidota *		*Shorea beccariana*		*Shorea zeylanica*
Shorea macrophylla		*Shorea biawak*		*Stemonoporus canaliculatus*
Shorea macroptera *		*Shorea bracteolata* *		*Stemonoporus gilimalensis*
Shorea multiflora *		*Shorea bullata*		*Stemonoporus sp.*
Shorea ovalis *		*Shorea curtisii*		*Upuna borneensis*
Shorea parvifolia *		*Shorea exelliptica*		*Vateria copalifera*
Shorea scaberrima		*Shorea faguetiana*		*Vatica affinis*
Shorea singkawang		*Shorea falciferoides*		*Vatica cinerea*
Upuna borneensis		*Shorea fallax*		
Vatica odorata		*Shorea ferruginea*		
		Shorea geniculata		
		Shorea havilandii		
		Shorea laxa		
		Shorea macroptera *		
		Shorea obtusa		
		Shorea ochracea		
		Shorea ovalis *		
		Shorea ovalis *		
		Shorea ovata		
		Shorea parvifolia *		
		Shorea patoiensis		
		Shorea pauciflora *		
		Shorea pilosa		
		Shorea quadriervis		
		Shorea roxburghii		
		Shorea rubra		
		Shorea siamensis		
		Shorea superba		
		Shorea xanthophylla		
		Upuna borneensis		
		Vatica micrantha		
		Vatica oblongifolia		
		Vatica odorata		
		Vatica sarawakensis		

20 Regeneration Strategy and Spatial Distribution Pattern of *Neobalanocarpus heimii* in the Lowland Dipterocarp Forest of Pasoh, Peninsular Malaysia

Claire Elouard[1] **& Lilian Blanc**[2]

Abstract: *Neobalanocarpus heimii* (King) Ashton (chengal) is one of the most dominant species in the Pasoh Forest Reserve (Pasoh FR). This tree presents an uncommon phenological behaviour from the other Dipterocarpaceae species of this region. In general, this species fruits every year with population synchrony unlike other species having mast fruiting behaviour. Fruit production and seedling dynamics were monitored during 34 weeks in plots established under four mother trees. Fruit production was continued throughout the survey, and fruiting patterns varied among the studied mother trees. Fruit dispersal did not extend >16 m from the mother tree trunk; most of the fruits, heavy and wingless, fall close to the trunk. Fruit predation was high and mainly related to insect infestation, causing losses of 23 to 37% of the crop. Pre-dispersal mortality by infestation before fruit fall was considerable through fruit dispersal. Seedling mortality, identified as post-dispersal mortality, ranged from 31 to 52%. Cause of mortality was due to abiotic factors, mainly compaction of the soil (probably related to water stress). Additional losses of seedlings came from fungal infection (20–28%) and insect predation (21–32%). Seedling and sapling dynamics were monitored using another approach. Spatial distributions of seedlings and saplings in four height classes, were compared around ten mother-trees. Survival was quantified in relation to the distance to the mother tree. A spacing-distance mechanism was identified, with a higher number of seedlings and saplings at the edge of the mother tree canopy, confirming the Janzen-Connell theory. Tree spatial distribution was analyzed at a larger scale. Immature trees were aggregated whereas mothers trees exhibited a random distribution. The changes in the spatial structure appears consistent with the sapling spacing-distance mechanism.

Key words: Dipterocarpaceae, dispersal, fruit, regeneration, sapling, seedling, spatial distribution.

1. INTRODUCTION

Long-term management of dipterocarp forests has been on-going for many years, particularly in Malaysia (Appanah & Weinland 1993a). Several selective logging systems (e.g. Selective Management System) introduced for the management of hill forests, are based on advanced regeneration in forms of saplings and of small trees. However, these model have faced several failures and the question of sustainability of this forest management system is still open to question (Appanah et al. 1990).

---

[1] French Institute of Pondicherry, Pondicherry 605 001, India.Present address: Laboratoire de Biologie Evolutive et Biometrie, University C. Bernard Lyon I, Villeurbanne, France.
E-mail: clairel@satyam.net.in
[2] University Claude Bernard, France.

The stock of regeneration for desirable species is supposed to grow rapidly when the canopy is opened. However, species-specific differences in regeneration patterns lead to confuse trials of silvicultural treatments. Therefore, there is an urgent need to identify the reasons of this failure to improve silvicultural techniques that rely on natural regeneration processes.

Lowland dipterocarp forests of the West Malesian region of South-East Asia showed an irregular fruiting cycle (Appanah 1985). General flowering by Dipterocarpaceae and some other botanical families occurs every 2 to 10 years. During mast fruiting years, huge quantities of fruits are dispersed and consequently lead to abundance of seedlings on the ground. This high variability leads to a need to improve our knowledge based on the natural regeneration stages, particularly concerning seedling and sapling dynamics. Studies on the establishment requirements of seedlings and saplings and growth to reproductive stages need to be conducted on species of silvicultural importance.

This article reviews several studies undertaken on aspects of the natural regeneration of *Neobalanocarpus heimii*. We focused on fruit production and dispersal, seedling and sapling recruitment, and survival, with special emphasis on the causes of mortality before and after dispersal by small-scale analyses. The spatial structure of trees is also examined in relation to the processes and factors acting on natural regeneration by large-scale analyses.

2. MATERIALS AND METHODS

2.1 Species characteristics

Studies were conducted in the Pasoh Forest Reserve (Pasoh FR). The studied species is *Neobalanocarpus heimii* (King) Ashton, belonging to the Dipterocarpaceae family. This species has very important ecological role and high economic value. *N. heimii* is found only in Peninsular Malaysia and in South Thailand in lowland as well as in hill forests up to 1000 meters (Symington 1943). The species has suffered severe logging pressure in the past and is now restricted to undisturbed protected or inaccessible forests. In the 50 ha permanent plot of Pasoh FR, this species is the second most abundant in terms of the number of stems (diameter > 10 cm) and the third most important in terms of basal area among the dipterocarp species (Appanah & Weinland 1993b). Its heavy hardwood provides the best-known timber in Peninsular Malaysia. The species also produces a very high quality resin ("damar" in the local name).

N. heimii exhibits reproductive characteristics different from most other dipterocarp species. It flowers almost annually (see Chap. 15) and the fruits, heavy and without wings, fall straight under the mother trees, limiting the dispersal area. This regular production and limited spatial distribution of fruits on the ground makes it easier to study seedling and sapling dynamics and their interactions with the mother tree.

2.2 Methods

Studies on fruits, seedlings and saplings were conducted in a logged-over area, at the external boundary of the 50 ha plot (Chaps. 1, 2; see Color plates). The monitoring on the fruits and seedlings lasted from October 1993 to June 1994, from the beginning of the fruit production up to 34 weeks (244 days). The study on saplings was conducted in May 1994. Two rainfall peaks were observed during the survey, in November 1993 and in February 1994, and a dry period was noticed in January

1994. Tree spatial patterns were analysed from data collected at the 50 ha plot (see Chap. 1).

2.3 Small-scale analyses

Three life-history stages, i.e. fruits, seedlings and saplings, were investigated for four large mother trees spared in the selective logging area. Seedlings were defined as small plants still dependent on the cotyledon reserve, not exceeding 15–20 cm. Once autotrophic, the plants were considered saplings.

The causes of mortality were categorized as insect predation, fungal infection and abiotic factors (drought, inability to establish in the compact soil, physical damages). Two types of mortality were observed: pre-dispersal and post-dispersal. Pre-dispersal mortality was determined as mature non-germinated fruits on the ground and the fruits destroyed by insect predation. Post-dispersal mortality was identified as germinants and seedlings killed by insect predation, fungal infection and abiotic factors.

2.3.1 Fruit production and seedling mortality assessment

Plots were established under four mother trees, producing fruits in abundance and free of conspecific trees in a 70–80 m radius to ensure that the recorded fruits, seedlings and saplings were produced by the putative mother trees. Segments of 30° were surveyed, following the crown shape up to 14 m from the mother tree trunk (Fig. 1a). The segments were divided into sub-segments of 2 meters length each, to study fruit dispersal and seedling survival. Sloping areas (> 20°) that affected seed dispersal were avoided. From October 1993 to June 1994, fruits on the ground were labeled according to the sub-segments to which they belonged, and germination and seedling development were surveyed. After the 34 weeks of our survey, seedlings rarely had more than five leaves.

Mortality rates were calculated as the ratio of fallen dead fruits to total dispersed fruits (pre-dispersal mortality) and the ratio between dead seedlings to the total dispersed fruits (post-dispersal mortality).

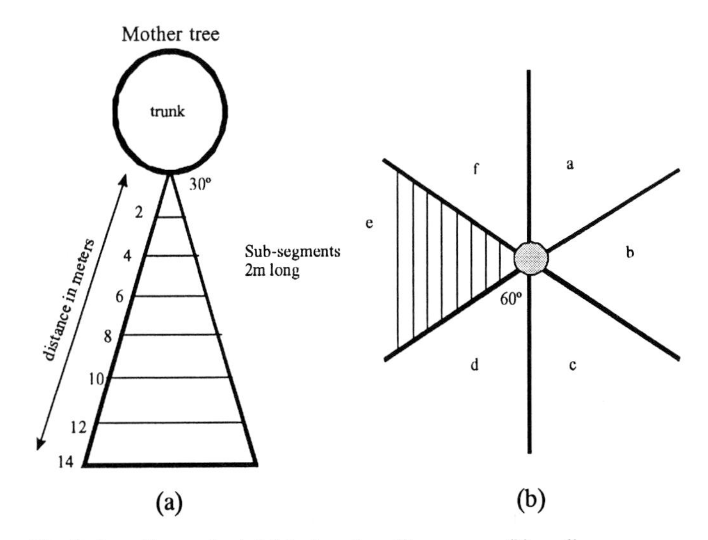

Fig. 1 Sampling method. (a) fruit and seedling survey, (b) sapling survey.

2.3.2 Sapling dynamics

Sapling dynamics were monitored under 10 additional trees. As for the fruit and seedling survey, measurements were made on isolated trees, lacking any conspecific trees in the neighbourhood of 70–80 m radius. The sampling area was partitioned into six segments of 60° each, up to 16 m from the mother tree trunk and partitioned into sub-segments of 2 m length each (Fig. 1b). Saplings in each sub-segment were grouped into 4 height classes: < 50 cm, 50 to 100 cm, 100 to 150 cm and >150 cm. For some trees, class 1(< 50 cm) comprises germinants, seedlings and young saplings. Two trees fruited in 1994 and still had fruits and germinants at the time of the census. All the other trees fruited before 1994 and only two trees had germinants at this time.

2.4 Large-scale analyses and statistical analyses

To study tree spatial pattern in the 50 ha plot, only trees with a diameter over 5 cm were considered. Three classes were defined: mother trees, considered in primary forest as trees with a diameter over 60 cm DBH (Diameter at Breast Height); young trees defined arbitrarily with a diameter between 10 and 60 cm; and juveniles, with a diameter between 5 and 10 cm.

To study tree spatial pattern, we performed a second-order neighbourhood analysis based on a distance-based method (Ripley 1976, 1977; Sterner et al. 1986; Pélissier 1998). From the inter-tree distances, this analysis involves counting all pairs of neighbours in a circle which are below or equal to distance r apart and this number is compared to the one obtained from random distribution of the points. The function $L(r)$ proposed by Besag (1977) enables an interpretation of the spatial pattern: $L(r) = 0$ for a completely random process, $L(r) > 0$ for an aggregated pattern and $L(r) < 0$ for a regular pattern. In an aggregated pattern, the number of observed trees in the neighbourhood of any arbitrary tree is greater (lower) than expected for a completely random process. Ripley analyses were performed with the ADE software package (http://pbil.univ-lyonl.fr/ADE-4/).

Other statistical analyses (ANOVA, c^2) were performed with the software packages Statistica (v. 5.5, Ed. 99, StatSoft) and XLSTAT (v.4.2).

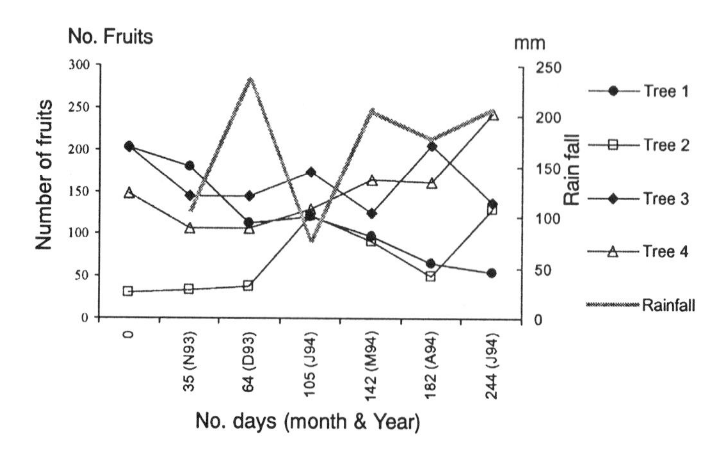

Fig. 2 Fruit production of chengal and rainfall during the 244 days survey.

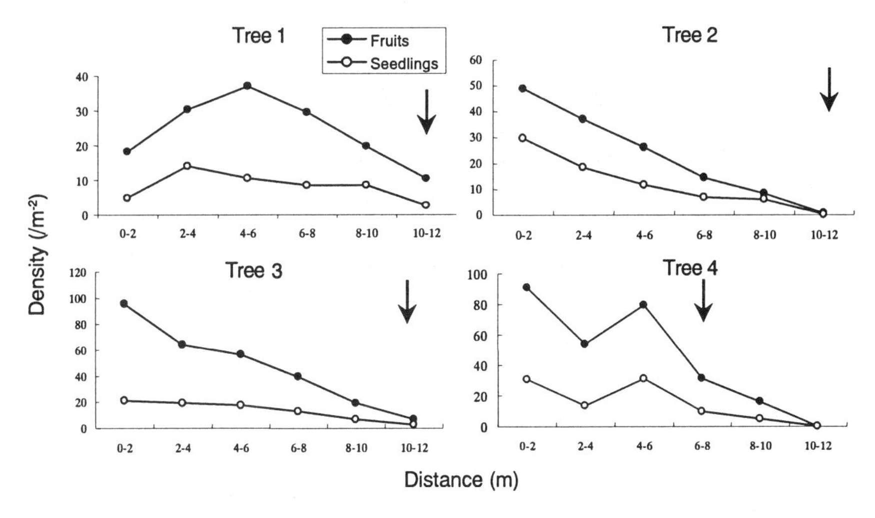

Fig. 3 Distribution of the density (/m^{-2}) of fruits and alive seedlings after 244 days with distance (m) from the trunk of the mother tree. Arrow: crown edge.

3. RESULTS

3.1 Fruit production

3.1.1 Temporal changes in fruit production

Fruit production was continuous throughout the survey, which lasted 244 days (from October 1993 to June 1994), for the 4 trees sampled (Fig. 2). Significant differences of fruit distribution during the survey were noted between the four trees (c^2 =363; df = 18; $P < 0.001$) particularly at the beginning of the survey and at the end (Fig. 2). Tree No. 2 had a very low and irregular production, which started later than for the three other trees. Tree No. 1 had a decreasing production, whereas tree No. 4 showed increasing fruit production. Tree No. 3 had quite constant production throughout the survey. Total production of seeds varied among the 4 trees, and ranged from 497 (Tree No. 2), 828 (Tree No. 1), 1058 (Tree No. 4) to 1132 (Tree No. 3) fruits per 30° segment.

3.1.2 Spatial distribution of the fruit production

Due to the heaviness of the fruit and the absence of wings, primary dispersal was by gravity. The fruits fell straight from the crown to the ground and moved downslope only in relatively steep terrain. Animal dispersal is rarely expected, due to the fact that the fruits contain resin with alkaloids (Appanah & Weinland 1993a). No fruit removal by animals was observed during the survey as can be observed for other barochorous species dispersed by rodents (Forget 1992a,b). Fruit density decreased with increasing distance from the trunk of the mother tree (Fig. 3). However, tree No. 1 showed a different pattern: the 4–6 m class distance from the trunk presented higher fruit distribution than the other classes.

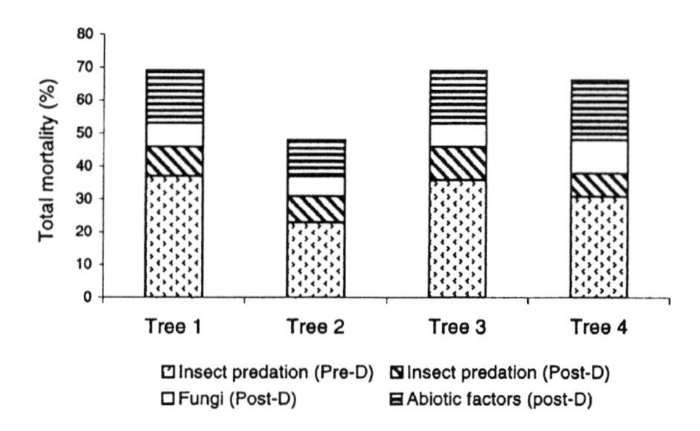

Fig. 4 Pre- and post-dispersal mortality rates for the four trees.

3.2 Fruit and seedling mortality

3.2.1 Pre-dispersal mortality

Fruit mortality ranged from 23 to 37% among sampled trees (Fig. 4), leaving two-third of the fruit production viable. Tree No. 2 had a far lower pre-dispersal mortality rate (23%) than the other trees, which were close to 35%. A high correlation ($R^2 = 0.85$, $P < 0.001$) existed between fruit density and pre-dispersal mortality (Blanc & Elouard unpubl.). Pre-dispersal mortality was mainly due to insect predation. Fruits of *N. heimii* were known to be predated by *Andriplecta* cf *pulverula* (Tortricidae), *Caterema albicostalis* (Pyralidae) and the beetles *Coccotrypes graniceps* and *Poecilips gedeanus* (Scolytidae) (Browne 1961; Daljeet-Singh 1974; Toy 1988). During this survey, *P. gedeanus* was observed destroying fruits: the adults lay their eggs in the fruits on the tree, and the larvae develop inside the fruits, feeding on the cotyledons. However, the main fruit predator was a moth (cf. *Assara*, Pyralidae, Phycitinae; J. Intachat pers. comm.): the adults lay their eggs on the young fruits on the tree. The larvae hatch and burrow into the fruit, develop within and remain as pupae, and emerge as adults after the fruit, completely destroyed, falls (Daljeet-Singh 1974).

3.2.2 Post-dispersal mortality

Mortality of fruits on the ground is insignificant. Few predation (less than 1%) by vertebrates were observed during the survey and in most cases, the fruits taken by rodents were left after a few bites. A small number (less than 0.5%) of fruits died from abiotic factors.

Germination occurred within few days after dispersal. Seedlings suffered three main causes of mortality: insect attack, fungal infection and death due to abiotic factors (Fig. 4).

Leaf herbivory by insects was observed, starting soon after the production of the seedling's first leaves: Microlepidoptera lay their eggs on the young leaves, and the larvae feed on the leaves until pupation. Large leaf areas are ravaged, often the whole foliar system for the young stages, eventually leading to seedling's death. Fungal attacks observed on the seedlings consisted of root-rot and wilt caused by three fungi: *Cylindrocladium camelliae*, *Fusarium* sp. and *Lasiodiplodia theobromae*. However, *C. camelliae* was the most commonly fungus isolated from the diseased seedlings. The main abiotic factor identified as cause of

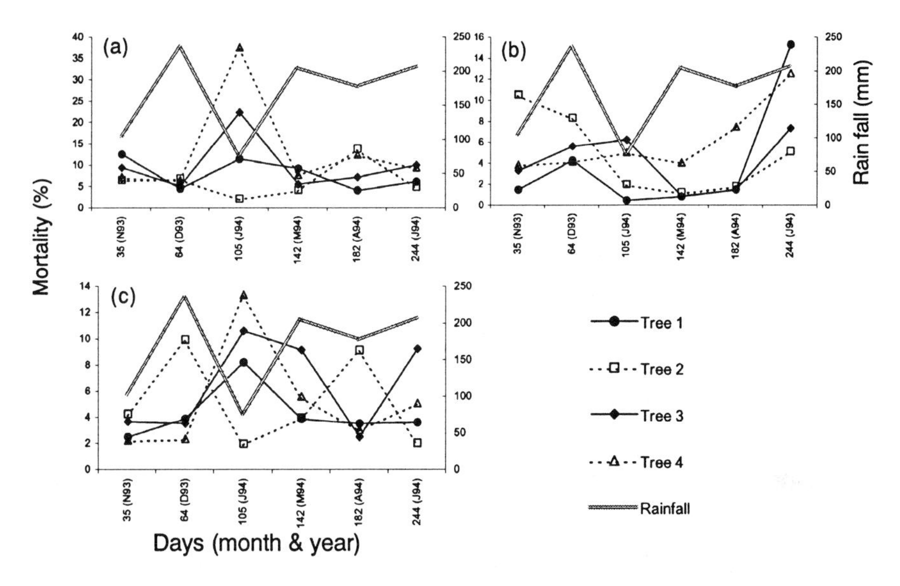

Fig. 5 Post-mortality rates due to abiaotic factores (a), fungal infection (b) and insect attack (c).

death of seedlings was the inability of the seedling hypocotyles to penetrate the soil and develop a viable root system. The soil was at some periods dry and compact, due to low rainfall.

Mortality ranged from 11 to 18% for abiotic factors, 7 to 10% for insect predation and 6 to 11% for fungal infection (Fig. 4). Thus, while insect predation and fungal infection presented similar rates, abiotic factors were responsible for half of the death of the seedlings. The number of living seedlings at the end of the survey was remarkably stable for the trees Nos. 1, 3 and 4, respectively 31%, 32% and 34% of dispersed fruits, and higher, 52%, for the tree No. 2.

Mortality rates due to abiotic factors were quite stable during the survey, usually between 0.02 and 0.15 (Fig. 5a). However higher values were recorded for the third census (105 days) for trees Nos. 3 and 4. It was due to low precipitation compacting the soil. It did not affect the tree No. 2, which had just started fruit dispersal at that period (Fig. 2). Due to this pattern, tree No. 2 avoided the dry period, which can largely explain the low mortality rate recorded for this tree (Fig. 4).

Higher rates of fungal infection were recorded at the end of the survey for three of the trees (Fig. 5b). No clear pattern appeared for insect predation (Fig. 5c), compared to other mortality causes, though the higher rate is recorded at day 105 for three of the trees.

3.2.3 Spatial distribution patterns

There is no difference between the spatial distribution of fruits and living seedlings after 244 days (Fig. 4). The mean distance (from the trunk of the mother tree) of dispersed fruits and living seedlings after 244 days was not significantly different.

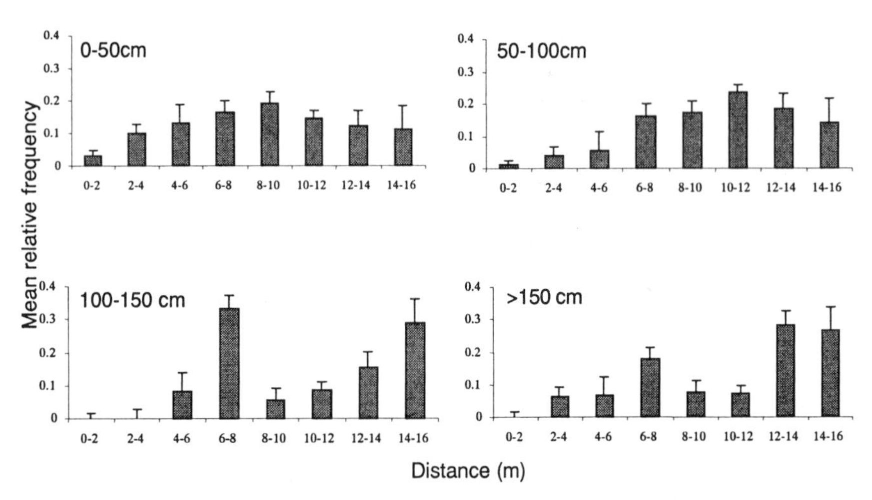

Fig. 6 Mean relative frequency of saplings with distance from ten mother trees. Vertival bars indicate 1*SD*.

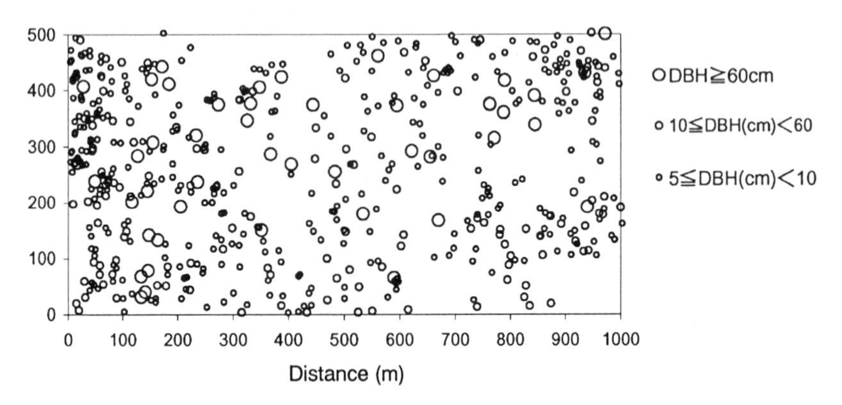

Fig. 7 Tree spatial distribution of *Neobalanocarpus heimii* in the 50 ha plot of Pasoh FR.

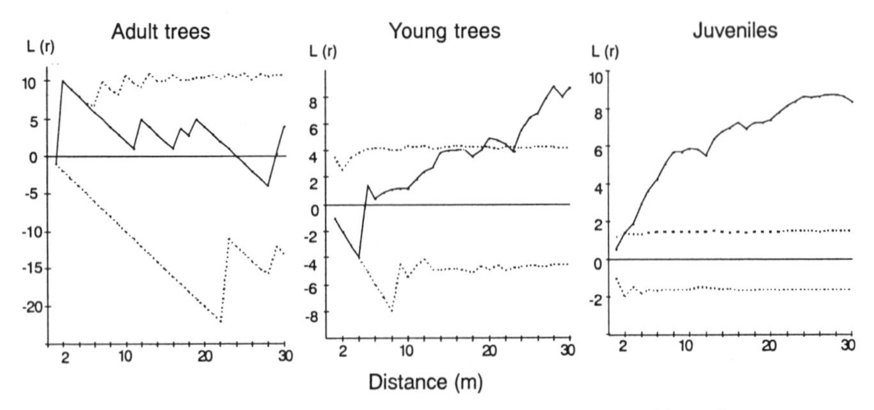

Fig. 8 Ripley' k analyses for adult trees, young trees and juveniles (Confidence envelops for the statistics are generated using Monte Carlo methods).

3.2.4 Saplings

A comparison between fruit production and seedling survival, and sapling survival cannot be done in the present study because the surveys were not conducted on the same trees and were not synchronous in time.

The total number of saplings observed for the 10 trees sampled ranged from 92 to 5,872. The higher number of saplings was found in trees that fruited in 1993 and 1994. More saplings were observed in the lower height class (0–50 cm height) than in the three other height classes sampled. The distribution pattern of the saplings showed that there was an increase of the frequency of saplings at greater distance with increasing height, indicating a spacing mechanism extending from the trunk to the crown's edge of the mother tree (Fig. 6). This suggested that the survival of the saplings is higher beyond the crown's edge of the mother tree than below the canopy.

3.2.5 Tree spatial patterns (large-scale analyses)

On the permanent 50 ha plot, 3,185 individuals (over 1 cm DBH) were recorded in 1990 (Manokaran et al. 1990). The number of trees was 47, 124 and 368 for the mother trees, young trees and juvenile trees, respectively. The density of mother trees was then close to 1 tree/ha. The spatial distribution of trees belonging to the three classes is shown in Fig. 7.

Ripley analyses show that the mother trees are significantly randomly distributed at scales up to 5 meters (Fig. 8). This pattern is different from the ones observed for young and juvenile trees. Their spatial structure were aggregated over scales of 20 and 2 meters, respectively. This analysis shows that tree mortality modifies strongly the spatial distribution during the life cycle. From a clumped distribution at a small scale for juveniles, spatial structure remains aggregated but at higher distance for young trees and then evolves toward a random distribution for mother trees.

4. DISCUSSION

N.heimii is substantially different in regeneration patterns from most other Malaysian dipterocarp species, in its fruiting production behaviour as well as in its dispersal behaviour. The species shows an annual fruiting habit (Appanah & Weinland 1993a; Chap. 15), with small fruit production compared to mast fruit production for other dipterocarp species (Appanah 1979, 1985, 1987, 1998), with fruiting lasting for a longer period. Fruiting was quite constant during the 34 weeks of the survey (Elouard et al. 1996), whereas even during the mast fruiting years, most other dipterocarp species produce fruit over a two weeks to one month period (Krishnapillai & Tompsett 1998). The monitored trees showed individualistic behaviour within a larger fruiting pattern for the species. The fruit dispersal also proved different from most dipterocarp species, which can have substantial wind dispersal of their winged fruits, as in the case of *Shorea leprosula*. The distance reached by *N. heimii* fruits hardly exceeds 16 m from the mother tree trunk. The heavy fruits fall by gravity from the canopy to the ground and there is no additional dispersal mechanism by animals (Blanc & Elouard unpubl.). The spatial distribution of the fruits, presenting a decrease of fruit density with increasing distance from the mother tree trunk, proved conform to the Janzen model (Janzen 1970) of fruit dispersal (Blanc & Elouard unpubl.).

Pre-dispersal mortality, mainly due to insect infestation in immature fruits, accounted for one-third of the fruit production. Other causes of fruit mortality were inability of the fruits to germinate (either from intrinsic factors or from abiotic factors such as drought).

Post-dispersal mortality ranged from 25 to 35% of fruit production and 32 to 51% of viable fruits. This mortality is however more related to abiotic factors than to the two other biotic causes identified during the study. Rainfall during the study period appeared to have been less than during other years (Manokaran & Swaine 1994). Low rainfall was observed at 105 days (June 1994) along with an increase of mortality by abiotic factors, which can then be related to water stress for the germinants and seedlings. Dry compact soil appeared to prevent the penetration of hypocotyles in the soil and the development of the root system. However, tree No. 2 appeared to avoid this dry period with a late fruit dispersal.

Three fungi infecting the seedlings have also been identified as causing seedling mortality on various dipterocarp species in Malaysia, including *Dryobalanops aromatica*, *Hopea odorata*, *H. helferi*, *Shorea acuminata*, *S. glauca*, *S. leprosula*, *S. macroptera*, *S. siamensis* and *Vatica cinerea* (Elouard, unpubl. data). Tree No. 2 presented a low fungal infection pattern, correlated to its low fruit production. The higher rates of fungal infection recorded at the end of the survey for trees Nos. 1, 3 and 4 compared to tree No. 2 may be related to a higher humidity favouring the development of fungi, or correspond to the time needed for seedlings to die from fungal infection.

This study showed no significant changes in the spatial distribution between dispersed fruits and seedlings after 244 days. However the mean distance of living seedlings is always higher than the mean distance of dispersed fruits for each tree (Blanc & Elouard, unpubl.) suggesting a distance-spacing mechanism. This mechanism is not changing the spatial distribution of the seedlings but has a significant effect on the sapling distribution.

Tree spatial distribution appeared to be consistent with conclusions obtained from small-scale analyses on seedlings and saplings. The overall spatial distribution likely results from two processes: an aggregation process at distances that increase from juveniles to young trees and an inhibition process at short distances. For *Neobalanocarpus heimii*, trees appeared close to the mother tree due to short dispersal distances. At the same time, juveniles and young trees appeared at the edge of the crown of the mother tree, as deducted from the sapling study. The poor dispersal and the escape from the crown edge of the mother tree result in clusters at a distance of 20 m for young trees. This is in agreement with the study conducted by Okuda et al. (1997) in Pasoh FR, mentioning that saplings are clumped (aggregated) and adults trees are randomly or regularly distributed. The inhibition process recorded for young trees for small distances (< 4 m) may be also explained by the light availability as conspecific trees can prevent the development of neighbours. *N. heimii* being an emergent species, it grows first in height to reach the canopy strata (Oldeman 1974) then expands in girth. Young trees, before reaching the canopy strata, may have to wait for an opportunity of opening such as a gap due to branch-fall, tree death or tree-fall. The opportunity for such an aperture is higher at the edge of the mother tree, unless the latter dies. This may correspond to the mean distance between regenerating trees around a mother tree and be related to the distance-spacing mechanism and can be related to the circle structure proposed by Pascal (1995).

Forest management needs to rely on knowledge of the phenological behaviour and regeneration strategies of crop tree species. *N. heimii*, as demonstrated before, presents a reproductive behaviour uncommon from other dipterocarp species of the region. Unlike other species that have mast fruiting years producing a large amount of fruits, *N. heimii* may have chosen another strategy to ensure its regeneration. Although only about one third of the fruit production survives as seedlings, the annual fruiting guarantees regular regeneration. However, management of this species needs to take into account that the young stages have a high sensitivity to drought. To ensure the regeneration of this species, it is therefore necessary to consider the importance of the sensitivity of the young stages to changes of the micro-bioclimatic conditions induced by disturbance of the canopy cover.

ACKNOWLEDGEMENTS

This work was undertaken within a European Union Project (CE-STD2) "Dipterocarp domestication in Malaysia: Factors controlling the establishment of valuable trees and reconstitution of dipterocarp forest", co-ordinated by Dr. G. Maury-Lechon, whom the authors are thankful to, for having allowed this work to happen and supporting it. The authors are grateful to the Forest Research Institute Malaysia (FRIM) which hosted them during this study. Gratitude is expressed to Drs. S. Appanah and M. Manokaran who kindly advised the authors during the studies. Thanks are due to Drs. H. T. Chan, Lee Su See, B. Krishnapillay, E. Philip and Mr. Chua who rendered their help during the study, and to Dr. J. Intachat who helped for the identification of insects predating the fruits and seedlings. We thank Dr. M. Manokaran and FRIM for allowing the laboratory of Biometry of the University of Lyon I access to the Pasoh database. Many thanks to Drs. M.-A. Moravie, D. Lee and G. Vasanthy for helpful comments on an earlier version of the manuscript.

REFERENCES

Appanah, S. (1979) The ecology of insect pollination of some tropical rain forest trees. Ph.D. diss, University of Malaya, 213pp.

Appanah, S. (1985) General flowering in the climax rain forests of south-east Asia. J. Trop. Ecol. 1: 225-240.

Appanah, S. (1987) Insect pollinators and the diversity of dipterocarps. In Kostermans, A. J. G. H. (ed). Proceedings of the Third Round Table Conference on Dipterocarps. 16-20 April 1987, Indonesia, UNESCO, pp.277-291.

Appanah, S. (1998) Management of natural forests. In Appanah, S. & Turnbull, J. M. (eds). A review of dipterocarps. Taxonomy, ecology and silviculture. CIFOR, Bogor, Indonesia, pp.133-149.

Appanah, S. & Weinland, G. (1993a) Planting quality timber trees in peninsular Malaysia. Malayan Forest Record No. 38, 221pp.

Appanah, S. & Weinland, G. (1993b) A preliminary study of the 50-hectare Pasoh demography plot: I. Dipterocarpaceae. Forest Research Institute of Malaysia, Research Pamphlet, No. 112.

Appanah, S., Weinland, G., Bossel, H. & Krieger, H. (1990) Are tropical rain forests non-renewable? An enquiry through modelling. J. Trop. For. Sci. 2: 331-348.

Besag, J. (1977) Contribution to the discussion of Dr Ripley's paper. J. R. Stat. Soci. B 39: 193-195.

Browne, F. G. (1961) The Biology of Malayan Scolytidae and Platypodidae. Malayan Forest Records No. 22. Longmans Malysia Sdn. Bhd., pp.1-255.

Daljeet-Singh, K. (1974) The seed pests of some dipterocarps. Malay. For. 37: 24-36.

Elouard, C., Blanc, L. & Appanah, S. (1996) Fruiting and seedling survival of *Neobalanocarpus heimii* in Peninsular Malaysia. In Appanah, S. & Khoo, K. C. (eds). Proceedings of the Fourth Round Table Conference on Dipterocarps, 7-10 Nov. 1994, Chiang Mai, Thailand, pp.520-527.

Forget, P.-M. (1992a) Regeneration ecology of *Eperua grandiflora* (Caesalpiniaceae), a large-seeded tree in French Guiana. Biotropica 24: 146-156.

Forget, P.-M. (1992b) Seed removal and seed fate in *Gustavia superba* (Lecythidaceae). Biotropica 24: 408-414.

Janzen, D. H. (1970). Herbivores and the number of tree species in tropical forests. Am. Nat. 104: 501-528.

Krishnapillay, B. & Tompsett, P. B. (1998) Seed handling. In Appanah, S. & Turnbull, J. M. (eds.) A review of dipterocarps. Taxonomy, ecology and silviculture, CIFOR, Bogor, Indonesia, pp.73-88.

Manokaran, N. & Swaine, M. D. (1994) Population dynamics of trees in dipterocarp forests of Peninsular Malaysia. Malayan Forest Records No. 40. Forest Research Institute Malaysia, Kuala Lumpur, 173pp.

Manokaran, N., LaFrankie, J. V., Kochumen, K. M., Quah, E. S., Klahn, J. E., Ashton, P. S. & Hubbell, S. P. (1990) Methodology for the fifty hectare research plot at Pasoh Forest Reserve. Research Pamphlet, No. 104. Forest Research Institute Malaysia, Kepong.

Okuda, T., Kachi, N., Yap, S. K. & Manokaran, N. (1997). Tree distribution pattern and fate of juveniles in a lowland tropical rain forest–implications for regeneration and maintenance of species diversity. Plant Ecol. 131: 155-171.

Oldeman, R. A. A. (1974) L'architecture de la forêt guyanaise. Mémoires ORSTOM, Paris, 204pp.

Pascal, J.-P. (1995) Quelques exemples de problèmes posés à l'analyste et au modélisateur par la complexité de la forêt tropicale humide. Revue d'Ecologie (Terre et Vie) 50: 237-249.

Pélissier, R. (1998) Tree spatial patterns in three contrasting plots of a southern Indian tropical moist evergreen forest. J. Trop. Ecol. 14: 1-16.

Ripley, B. D. (1976) The second-order analysis of stationary point process. J. Appl. Probability 13: 55-266.

Ripley, B. D. (1977) Modelling spatial patterns (with discussion). J. R. Stat. Soc. Ser. B 39: 172-212.

Sterner, R. W., Ribic, C. A. & Schatz, G. E. (1986) Testing for life historical changes in spatial patterns of four tropical tree species. J. Ecol. 74: 621-633.

Symington (1943) Foresters' manual of dipterocarps. Malayan Forest Reccprds No. 16, Forest Research Institute Malaysia, Kuala Lumpur, 244pp.

Toy, R. J. (1988) The pre-dispersal insect fruit-predators of Dipterocarpaceae in Malaysian rain forest. Ph.D. diss., Univ. Aberdeen, 248pp.

Wyatt-Smith, J. (1966) Ecological Studies on Malayan Forests. I. Composition of and Dynamic Studies in Lowland Evergreen Rain-Forest in two 5-acre Plots in Bukit Lagong and Sungei Menyala Forest Reserves and in two Half-acre Plots in Sungei Menyala Forest Reserve, 1947-59. Research Pamphlet, No. 52. Forestry Department, Malaysia.

21 Mating System and Gene Flow of Dipterocarps Revealed by Genetic Markers

Yoshihiko Tsumura[1], Tokuko Ujino-Ihara[1], Kyoko Obayashi[2], Akihiro Konuma[3] & Teruyoshi Nagamitsu[4]

Abstract: Tropical rain forests are becoming fragmented into small patches and simplified in structure and composition due to their exploitation for timber. Consequently, genetic variation may be decreased and the genetic differentiation between the patches may rise. Developing sustainable forestry regimes that retain genetic diversity is one of the major goals in tropical forestry today. Tropical tree species have predominantly allogamous mating systems, mediated by many kinds of pollinators, so the gene flow through pollen and seeds is one of the most important factors influencing genetic diversity in tropical forests. Recently, molecular markers have been used to determine the genetic diversity and differentiation of forest tree populations. In particular, microsatellite DNA markers, which are highly polymorphic, have been used to evaluate the gene flow through pollen and seeds within forests. We have also developed microsatellite DNA markers in dipterocarp species, and used these markers to analyze mating systems, genetic structures and gene flow in several dipterocarp species at ecological parmanent plots such as those in Pasoh Forest Reserve (Pasoh FR). We discuss, here, characteristics of the pollination and breeding systems in dipterocarps that have been revealed by microsatellite analysis, and the factors that influence them. Finally, the potential value and limitations of using molecular approaches to develop genetic criteria for gene conservation in sustainable forestry are discussed.

Key words: Dipterocarpaceae, inbreeding depression, logging effect, microsatellite, paternity analysis, pollinator, SSR.

1. INTRODUCTION

Knowledge of gene flow and mating systems is essential for understanding the reproductive processes of outcrossing plants, as well as for developing methods to conserve plant populations. Theoretical studies suggest that restricted gene flow probably reduces effective population size, and causes inbreeding depression (Slatkin 1985). In general, restricted gene flow, due for instance to selfing or mating with relatives, causes serious inbreeding depression in outcrossing species (Wang et al. 1999). Thus, restricted gene flow presents a threat to the viability of populations of outcrossing plants.

The Asian tropical rain forests contain a great variation of dipterocarp species, which are predominantly allagomous species pollinated by insects (Bawa 1998). These commercially important timber species are some of the prime targets of selective logging in Southeast Asia. The selective logging has resulted in intense

[1] Forestry and Forest Products Research Institute (FFPRI), Tsukuba 305-8687, Japan.
E-mail: ytsumu@ffpri.affrc.go.jp
[2] University of Tsukuba, Japan.
[3] Niigata University, Japan.
[4] Hokkaido Reserch Center, FFPRI, Japan.

fragmentation of the forests, and low-density populations in the remaining forests. The restricted gene flow caused by selective logging, may also reduce the population viability of tropical trees. Outcrossing rates in some tropical tree species have been shown to be positively correlated to flowering tree density in self-compatible species (Murawski & Hamrick 1991). Thus, to conserve viable populations in tropical tree species, it is important to obtain good estimates of outcrossing rates and pollen flow. To maintain the sustainability of selectively logged forests, we have to know how much genetic variation is retained in the forest after logging.

Several types of DNA marker have been developed in recent years, including microsatellite DNA markers, which have been frequently used to study genetic variation and natural mating systems of plants and animals. This is because they are very sensitive to genetic polymorphism, and can be used to determine parental individuals with a high probability of correct assignation (Chase et al. 1996; Dow & Ashley 1996). We report here on recent progress in dipterocarp study using molecular markers such as microsatellite DNA markers and discuss the kinds of information needed to promote sustainable forestry in Southeast Asia.

2. MICROSATELLITE MARKER

For development of microsatellite marker, enrichment methods have been frequently used presently (Fischer & Backmann 1998; Lench et al. 1996). The developing is not labor and time-consuming than before, and thus, the informations of microsatellite marker have been accumulating in many plant species. Recently, for wild plant species, microsatellites have been used for studies on gene flow, genetic structure and mating systems because of highly polymorphisms (Chase et al. 1996; Dow & Ashley 1996).

Microsatellite markers are especially useful for determining the pollen donors of plants and thus analyzing the effective pollen flow and seed dispersal within populations. Microsatellite DNA markers of tropical species have been reported in several species, including *Pithecellobium elegans* (Chase et al. 1996), *Carapa guianensis* (Dayanandan et al. 1999), *Shorea curtisii* (Ujino et al. 1998), and *Neobalanocarpus heimii* (Iwata et al. 2000). Markers developed in one species can also generally be applied to other closely related species (Dayanandan et al. 1997; Ujino et al. 1998). Ujino et al. (1998) have applied the microsatellite markers derived from *S. curtisii* to the other dipterocarps species. According to the results, the tendency is not completely followed to the molecular phylogeny tree (Kajita et al. 1998; Tsumura et al. 1996), but more than a half of them can be used to the other dipterocarps species.

3. MATING SYSTEM AND GENETIC VARIATION

Using isozyme data, Murawski & Hamrick (1994) demonstrated that the outcrossing rate of *Stemonoporus oblongifolious*, another species of the Dipterocarpaceae, was 84%, while Kitamura et al. (1994) estimated the outcrossing rate of the dipterocarp *Dryobalanops aromatica* to range from 79.4 to 85.6% in primary and regenerating forests, which were not significantly different. The average outcrossing rate of *Shorea curtisii* in primary forest was found to be 96.3% using microsatellite markers (Obayashi et al. 2002), which is within the range of estimates for tropical tree species reviewed by Loveless (1992) and those reported by other authors (Table 1). However, in a selectively logged plot, the average outcrossing rate was found to be only 52.2% (Obayashi et al. 2002): about half the average value in the undisturbed plot and the values reported in previous studies. Reductions of

Table 1 Outcrossing rates in dipterocarps.

Species	Forest type	Outcrossing rate	Reference
Stemonoporus oblongifolious	primary	89.8%	Murawski & Bawa 1993
Dryobalanops aromatica	primary	85.6%	Kitamura et al. 1994
	logged	79.4%	Kitamura et al. 1994
Dryobalanops aromatica	primary	92.3%	Lee 2000
	logged	76.6%	Lee 2000
Shorea congestiflora	primary	87.4%	Murawski et al. 1994
Shorea megistophylla	Primary	86.6%	Murawski et al. 1994
	logged	71.3%	
Shorea trapezifolia	primary	61.7%	Murawski et al. 1994
Shorea leprosula	primary	84.0%	Nagamitsu et al. 2001
Shorea leprosula	primary	83.7%	Lee et al. 2000
Shorea curtisii	primary	96.3%	Obayashi et al. 2002
	logged	52.2%	Obayashi et al. 2002

Table 2 Outcrossing rate and the number of alleles received from paternal trees by outcrossing in each mother tree of *S. curtisii* in undisturbed forest and a selectively logged plot (Obayashi et al. 2002).

Forest type	Investigated No. of mother tree	Average no. of seeds analyzed per mother tree	No. of detected alleles *Shc09*	*Shc07*	*Shc11*	Mean	Outcrossing rate (%)
Undisturbed plot	10	35. 1	4.5	5.9	2.5	4.3	96.3
Logged plot	5	37. 0	3.8	4.6	1.8	3.4	52.2

outcrossing rate in logged forests than in primary forests have been reported for several tropical tree species; *Shorea megistophylla* (Murawski et al. 1994), *Pithecellobium elegans* (Hall et al. 1996), *Carapa procera* (Doligez & Joly 1997) and *Dryobalanops aromatica* (Lee 2000). In *Dryobalanops aromatica*, Lee (2000) documented that outcrossing rate was significantly greater in primary forest than in regenerating forest in Peninsular Malaysia but Kitamura et al. (1994) found no significantly difference between two types of forests in Brunei. Lee (2000) discussed the reason for difference that local microclimate and microenvironment for pollinators might be different from the region where Kitamura et al. (1994) studied. The high inbreeding rate at the selectively logged plot suggested that this species might not have a self-incompatibility mechanism, although some dipterocarp species do have such systems (Soepadmo 1989).

Outcrossing rates may be strongly influenced by flowering-tree density (Murawski & Hamrick 1992) and the types and behavior of pollinators governing the pollen movement (Ghazoul et al. 1998). Both of these factors may be partly responsible for the low outcrossing rate of *S. curtisii* observed at the selectively logged site, since tree density at the site was much lower than at the undisturbed site, and the principal pollinators for the genus *Shorea* are thought to be thrips and/or small beetles with low mobility (Appanah & Chan 1981; Ashton et al. 1988; Momose et al. 1998). The outcrossing rate may not be influenced by the tree density if the species has a wide-ranging pollinator such as *Apis* and/or *Trigona* (Kitamura et al. 1994; Konuma et al. 2000). However, as the outcrossing rate of *Dryobalanops aromatica* with high mobility pollinators such as *Apis* reduced by selectively logging, the microclimate and microenvironment may also influence the pollinator behavior (Lee 2000).

The outcrossing rate varied greatly among individual trees within both plots, especially in the selectively logged plot of *S. curtisii* (Obayashi et al. 2002).

The outcrossing rate of individuals should depend on the neighborhood density of conspecific flowering trees and be lower in isolated individuals than in clusters of trees (Murawski et al. 1990). Nagamitsu et al. (2001) also suggested that *Shorea leprosula*, a tree species with a low outcrossing rate and a similar pollinator, might be isolated from reproductive neighbors in Pasoh Forest Reserve (Pasoh FR). The degree of flowering synchrony with surrounding conspecific trees could also cause variations in outcrossing rate among these trees, since individual *Shorea* trees have relatively short flowering periods of 16 to 26 days (Ashton et al. 1988).

The number of alleles per locus (N_a) and heterozygosity (H_e) of the adult *S. curtisii* population in the selectively logged plot, in which 30% of the wood volume was removed and the density of mature trees was decreased by approximately three quarters about ten years ago, showed similar values to those from the undisturbed plot (Obayashi et al. 2002). This suggests that the disturbance caused by such selective logging events exerts little influence on the population genetic parameters N_a and H_e, as evaluated by highly variable DNA markers like microsatellites. Expected heterozygosity, H_e, is less sensitive to population bottlenecks than the proportion of polymorphic loci or the number of alleles per locus (Barrett & Kohn 1991). In contrast, the mean number of alleles per locus received from a pollen donor in selectively logged plot was significantly smaller than that in the undisturbed plot of *S. curtisii* (Table 2; Obayashi et al. 2002). Disturbances such as selective logging are thought to cause a strong bottleneck, and thus reduce the population size in the following generation. This suggests that some degree of genetic variation might be lost in the next generation of selectively logged forest trees that germinate from newly produced seeds, because the total number of alleles per unit area of the future generation will be decreased if new alleles are not supplied through pollen flow and/or seed dispersal from outside of the plot. Consistent with this hypothesis, Dayanandan et al. (1999) reported that the allelic richness was lower in the sapling cohort of an isolated fragment of *Carapa guianensis* in La Selva forest in Costa Rica than in a more extensive stand. In allogamous species such as dipterocarps, the low outcrossing rate may influence the genetic variation of the future forest. Thus, knowledge of pollinator behavior and flowering tree density for each species must be obtained. A more suitable parameter than heterozygosity to evaluate differences in the genetic diversity between forests would be also needed.

4. GENE FLOW

Konuma et al. (2000) observed long distance pollen flow among flowering *Neobalanocarpus heimii* at a 36 ha study plot in Pasoh FR using microsatellite DNA markers (Fig. 1). Pollen from outside the study plot sired 35% of the seedling and sapling in their study, and the estimated average mating distance was 524 m. *N. heimii* is mainly pollinated by stingless bees or honey bees (Appanah 1985) and the honey bees are known to be long distance pollinators (Dayanandan et al. 1990). Thus, this long distance pollen flow most probably depends on the performance of the pollinator. Pollinator behavior is probably affected by inter-flowering-tree distance (Bawa 1998), thus, flowering tree density can also influence the mating distance of tropical trees. For example, Ghazoul et al. (1998) reported that a pollinator (bee, *Trigona fimbriata*) of *Shorea siamensis* flew much further at a site where there were long distances between flowering host trees than it did at sites where the distances between flowering host trees were shorter. However, there must always be limits on the inter-flowering-tree movements of pollinators.

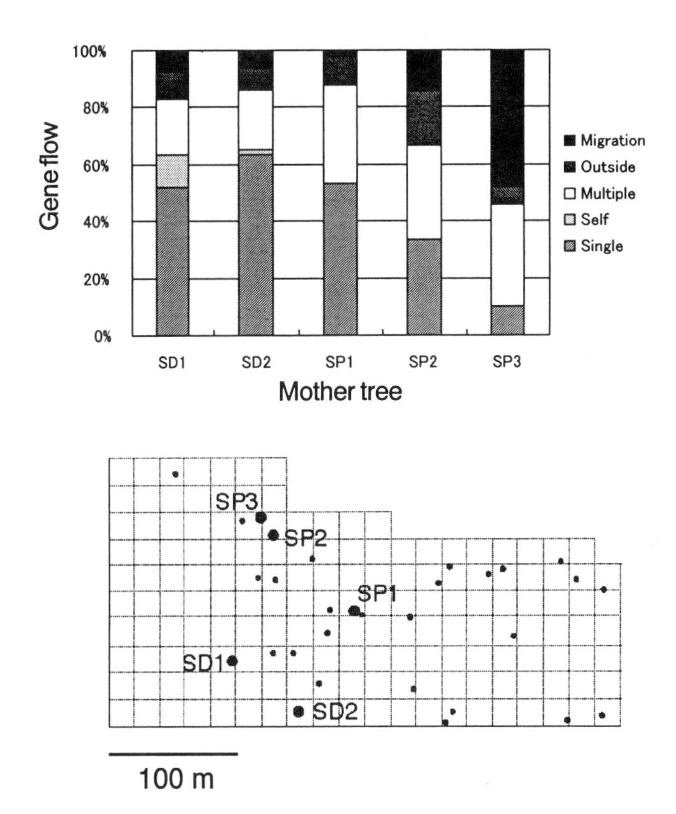

Fig. 1 Profiles of the genotyped offspring. The underlined numbers are numbers of offspring analyzed. Each number around the pie graph indicates the percentge of the following: "Single", which indicates offspring that were assigned to a single pollen parent; "Self", which indicates offspring judged to selfed; "Multiple", indicating offspring that showed more than one pollen parent; "Outside", indicating offspring that did not have a compatible pollen parent within the study site; and "Migration", which indicates seedlings or saplings whose haplotypes were incompatible with that of the nearest reproductive tree (Konuma et al. 2000).

Seed migrations of *N. heimii* have also been inferred, from the observation of offspring with haplotypes that were incompatible with those of neighboring reproductive trees (Konuma et al. 2000). Clearly, fruits or seeds of *N. heimii* had been carried by some dispersal vector such as small mammals. In tropical forests, squirrels disperse fruits of some tree species (Becker et al. 1985). Nagamitsu et al. (2001) reported similar phenomena in *S. leprosula* using microsatellite analysis. Thus, microsatellite analysis is a very sensitive method for observing gene flow within forests, and it can be used to determine parents of seeds, seedlings and saplings at the study site. Such analyses will give us understanding of the range and scope of gene flow, including pollen flow and seed dispersal parameters, for important forest tree species, and thus provide essential information to help design sustainable forestry regimes.

5. INBREEDING DEPRESSION

If seeds are produced by consanguineous matings in allogamous plants, inbreeding depression may be manifested at many subsequent developmental stages, such as fruit maturation (Nagamitsu et al. 2001), germination (Konuma et al. unpubl. data), seedling growth (Konuma et al. 2000) or sapling development. Dipterocarps are also predominantly outcrossing species, so inbreeding depression may be expected in selfed seeds. In *S. leprosula*, the inbreeding rate in one mother tree singnificantly decreased from 0.63 in the immature fruit stage to 0.23 in the mature fruit stage, which suggested that inbreeding depression and/or self-incompatibility might be occurred during fruit maturation (Nagamitsu et al. 2001). Konuma et al. (2000) found selfed offspring among seedlings but not saplings of *N. heimii*, and suggested that inbreeding depression may occur in surviving seedlings. Self-incompatibility is a system that effectively avoids inbreeding, and some dipterocarpaceous species are known to have this system (Soepadmo 1989). However, in low-density populations caused by selective logging, the probability of mating via foreign pollen might decline if the main pollinator does not fly far. Lee (2000) suggested that selective logging activity might reduce the seeds produced through consanguineous mating in the study of *D. aromatica*. He compared the values of (multilocus outcrossing rate (t_m) − single locus outcrossing rate (t_s)) between primary forest in Bukit Sai and logged forest in Lesong and found (t_m - t_s) value of primary forest was higher than of logged forest, which suggest biparental inbreeding. Thus, we have to maintain a suitable flowering tree density and environment for pollinators for genetically sustainable forestry, according to the ecological characters of the trees concerned.

6. GENETIC CRITERIA FOR SUSTAINABLE FORESTRY

To conserve genetic diversity, both within and between population diversity should be considered. For between population diversity, the forest fragmentation caused by timber exploitation presents critical challenges for genetic conservation (Hall et al. 1996; Nason & Hamlick 1996). Microsatellite markers can be used to test the effects of fragmentation on genetic diversity, but organellar DNA markers such as chloroplast DNA might be much more suitable for this purpose (Hamilton 1999). To gauge within-population diversity, microsatellite analysis can be used to obtain detailed information on mating system, genetic structure and, especially, gene flows through pollen and seed. Using the gene flow data in some forest, we can estimate breeding unit parameter that was defined by Nason et al. (1998). This parameter may useful to understand not only a range of gene flow but also *ex situ* conservation. To understand fully the genetic features affecting the within population diversity of species, their ecological characteristics must be thoroughly known, including characters such as flowering phenology and synchrony, pollinator behavior, self-incompatibility systems and apomixis (Kaur et al. 1978). Finally, we have to combine these data in appropriate models to construct valid genetic criteria if we are to maintain genetic diversity for conservation and sustainable forestry.

ACKNOWLEDGEMENT

This study was supported by the Global Environment Research Program by Japan Environment Agency (Grant No. E-1).

REFERENCES

Appanah, S. (1985) General flowering on the climax rain forests of South-east Asia. J. Trop. Ecol. 1: 225-240.

Appanah, S. & Chan, H. T. (1981) Thrips: the pollinators of some dipterocarps. Malay. For. 44: 234-252.

Ashton, P. S., Givnish, T. J. & Appanah, S. (1988) Staggered flowering in the Dipterocarpaceae: new insights into floral induction and the evolution of mast fruiting in the aseasonal tropics. Am. Nat. 132: 44-66.

Barrett, S.H & Kohn, J.R. (1991) Genetic and evolutionary consequences of small population size in plants: implications for conservation. In Falk, D. A. & Holsinger, K. E. (eds). Genetics and Conservation of Rare Plants. Oxford University Press, New York, pp.3-30.

Bawa, K. S. (1992) Mating system, genetic differentiation and speciation in tropical rain forest plants. Biotropica 24: 250-255.

Bawa, K. S. (1998) Conservation of genetic resources in the Dipterocarpaceae. In Appanah, A. & Turnbull, J. M. (eds). A Review of Dipterocarps: Taxonomy, Ecology and Silviculture. Center for International Forestry Research, Bogor, Indonesia, pp.45-55.

Becker, P., Leighton, M., Payne, J. B. (1995) Why tropical squirrels carry seeds out of source crowns. J. Trop. Ecol. 1: 183-186.

Chase, M. R., Moller, C., Kesseli, R. & Bawa, K. S. (1996) Distant gene flow in tropical trees. Nature 383: 398-399.

Dayanandan, S., Attygalla, D. N. C., Abeygunaskera, A. W. W. L., Gunatilleke, I. A. U. N. & Gunatilleke, C. V. S. (1990) Phenology and floral morphology in relation to pollination of some Sri Lankan dipterocarps. In Bawa, K. S. & Hadley, M. (eds). Reproductive ecology of tropical forest plants, Man and the biosphere series, vol. 7. UNESCO, Paris and Parthenon Publishing Group, Carnforth, pp.103-133.

Dayanandan, S., Bawa, K. S. & Kesseli, R. (1997) Conservation of microsatellite among tropical trees (Leguminosae). Am. J. Bot. 84: 1658-1663.

Dayanandan, S., Dole, J., Bawa, K. S. & Kesseli, R. (1999) Population structure delineated with microsatellite markers in fragmented populations of a tropical tree, *Carapa guianensis* (Meliaceae). Mol. Ecol. 8: 1585-1592.

Doligez, A. & Joly, H. I. (1997) Mating system of *Carapa procera* (Meliaceae) in the French Guiana tropical forest. Am. J. Bot. 84: 461-470.

Dow, B.D. & Ashley, M. V. (1996) Microsatellite analysis of seed dispersal and parentable of sapling in bur oak, *Quercus macroptera*. Mol. Ecol. 5: 615-627.

Fisher, D. & Bachmann, K. (1998) Microsatellite enrichment in organisms with large genomes (*Allium cepa* L.). Biotropica 24: 796-804.

Ghazoul, J., Liston, K. A. & Boyle, T. J. B. (1998) Disturbance-induced density-dependent seed set in *Shorea siamensis* (Dipterocarpaceae), a tropical forest tree. J. Ecol. 86: 462-473.

Hall, P., Walker, S. & Bawa, K. S. (1996) Effect of forest fragmentation on genetic diversity and mating system in tropical tree, *Pithecellobium elegans*. Conserv. Biol. 10: 757-768.

Hamilton, M. B. (1999) Tropical tree gene flow and seed dispersal. Nature 401: 129-130.

Iwata, H., Konuma, A. & Tsumura, Y. (2000) Development of microsatellite markers in the tropical tree *Neobalanocarpus heimii* (Dipterocarpaceae). Mol. Ecol. 9: 1684-1685

Kajita, T., Kamiya, K., Nakamura, K., Tachida, H., Wickneswari, R., Tsumura, Y., Yoshimaru, H. & Yamazaki, T. (1998) Molecular phylogeny of Dipetrocarpaceae in Southeast Asia based on nucleotide sequences of matK, trnL intron, and trnL-trnF intergenic spacer region in chloroplast DNA. Mol. Phylogenet. Evol. 10: 202-209.

Kaur, A., Ha, C. O., Jong, K., Sands, V. E., Chan, H. T., Soepadmo, E. & Ashton, P. S. (1978) Apomixis may be widespread among trees of the climax rain forest. Nature 271: 440-442.

Kitamura, K., Rahman, M. Y. B. A., Ochiai, Y. & Yoshimaru, H. (1994) Estimation of the outcrossing rate on *Dryobalanops aromatica* Gaertn. f. in primary and secondary forests in Brunei, Borneo, Southeast Asia. Plant Species Biol. 9: 37-41.

Konuma, A., Tsumura, Y., Lee, C. T., Lee, S. L. & Okuda, T. (2000) Estimation of gene flow in the tropical-rainforest tree *Neobalanocarpus heimii* (Dipterocarpaceae), inferred from paternity analysis. Mol. Ecol. 9: 1843-1852.

Lee, S. L. (2000) Mating system parameters of *Dryobalanops aromatica* Gaertn. F. (Dipterocarpaceae) in three different forest types and a seed orchard. Heredity 85: 338-345.

Lee, S. L., Wickneswar, R., Mahani, M. C. & Zakari, A. H. (2000) Mating system paramerters in a tropical tree species, *Shorea leprosula* Miq. (Dipterocarpaceae) from Malaysian lowland dipterocarp forest. Biotropica 32: 693-702.

Lench, N. J., Norris, A., Bailey, A., Booth, A. & Markham, A.F. (1996) Vectorette PCR isolation of microsatellite repeat sequences using anchored dinucleotide repeat primers. Nucl. Acid Res. 24: 2190-2191.

Loveless, M. D. (1992) Isozyme variation in tropical trees: patterns of genetic organization. In Adams, W. T., Strauss, S. H., Copes, D. L. & Griffin, A. R. (eds). Population genetics of forest trees, Kluwer Academic Publishers, Netherlands, pp.67-94.

Momose, K., Yumoto, T., Nagamitsu, T., Kato, M., Nagamasu, H., Sakai, S., Harrison, R. D., Itioka, T., Hamid, A. A. & Inoue, T. (1998) Pollination biology in a lowland dipterocarp forest in Sarawak, Malaysia. I. Characteristics of the plant-pollinator community in a lowland dipterocarp forest. Am. J. Bot. 85: 1477-1501.

Murawski, D. A. & Bawa, K. S. (1994) Genetic structure and mating system of *Stemonoporus oblongifolius* (Dipterocarpaceae) in Sri Lanka. Am. J. Bot. 81: 155-160.

Murawski, D. A. & Hamrick, J. L. (1991) The effect of the density of flowering individuals on the mating systems of nine tropical tree species. Heredity 67: 167-174.

Murawski, D. A. & Hamrick, J. L. (1992) The mating system of *Cavanillesia platanifolia* under extremes of flowering-tree density: a test of predictions. Biotropica 24: 99-101.

Murawski, D. A. & Hamrick, J. L. (1994) Genetic structure and mating system of *Stemonoporus oblongifolius* (Dipterocarpaceae) in Sri Lanka. Am. J. Bot. 81: 155-160.

Murawski, D. A., Hamrick, J. L., Hubell, S. P. & Foster, R. B. (1990) Mating systems of two Bombacaceous trees of a neotropical moist forest. Oecologia 82: 501-506.

Murawski, D. A., Gunatilleke, I. A. U. N. & Bawa, K. S. (1994) The effect of selective logging on inbreeding in *Shorea megistophylla* (Dipterocarpaceae) from Sri Lanka. Conserv. Biol. 8: 997-1002.

Nagamitsu, T., Ichikawa, S., Ozawa, M., Shimamura, R., Kachi, N., Tsumura, Y. & Muhammad, N. (2001) Microsatellite analysis of the breeding system and seed dispersal in *Shorea leprosula* (Dipterocarpaceae) Int. J. Plant Sci. 162: 155-159.

Nason, J. D. & Hamrick, J. L. (1996) Reproductive and genetic consequences of forest fragmentation: two case studies of Neotropical canopy trees. J. Hered. 8: 264-274.

Nason J. D., Herre, E. A. & Hamrick, J. L. (1998) The breeding structure of a tropical keystone plant resource. Nature 391: 687-689.

Obayashi, K., Tsumura, Y., Ihara-Ujino, T., Niiyama, K., Tanoushi, H., Suyama, Y., Washitani, I., Lee, C.-T., Lee, S. N. & Nuhammad, N. (2002) Genetic diversity and outcrossing rate between undisturbed and selectively logged forests of *Shorea curtisii* (Dipterocaropaceae) using microsatellite DNA analylsis. Int. J. Plant Sci. 163: 151-158.

Slatkin, M. (1985) Gene flow in natural populations. Annu. Rev. Ecol. Syst. 16: 393-430.

Soepadmo, E. (1989) Contribution of reproductive biological studies towards the conservation and development of Malaysian plant genetic resources. In Zakri, A. H. (ed). Genetic Resources of Under-utilised Plants in Malaysia. Malaysian National Committee on Plant Genetic Resources. Kuala Lumpur, pp.1-41.

Tsumura, Y., Kawahara, T., Wiekneswari, R. & Yoshimura, H. (1996) Molecular phylogeny of dipterocarpaceae in Southeast Asia using RFLP of PCR-amplified chloroplast genes. Theor. Appl. Genet. 93: 22-29.

Ujino, T., Kawahara, T., Tsumura, Y., Nagamitsu, T., Wickneswari, R. & Yoshimaru, H. (1998) Development and polymorphism of simple sequence repeat DNA markers for *Shorea curtisii* and other Dipterocarpaceae species. Heredity 81: 422-428.

Wang, T., Hagqvist, R., Tigerstedt, P. M. A. (1999) Inbreeding depression in three generations of selfed families of silver birch (*Betula pendula*). Can. J. For. Res. 29: 662-668.

Part IV

Animal Ecology and Biodiversity

22 Aspects of the Diversity of Geometridae (Lepidoptera) in Pasoh Forest Reserve

Jeremy D. Holloway[1] & Jurie Intachat[2]

Abstract: The moth family Geometridae is introduced, and its value as a bioindicator group is reviewed. The recorded geometrid moth fauna at Pasoh Forest Reserve (Pasoh FR) now stands at 413 species. Application of standard richness estimator statistics suggests the true total may be closer to 500. This total is contrasted with data from the much more extensively recorded Bornean fauna, and more subjectively with other forest sites in Peninsular Malaysia. The distribution of this diversity between sites at Pasoh FR and vertically within the forest is discussed. Details of the faunistic composition and biologies of key components are provided, and all species are listed as an Appendix in a checklist that also includes quantitative data for each species across the samples.

Key words: alpha diversity, beta diversity, indicators, lowland rainforest, moths, plantations, richness estimation, selective logging.

1. INTRODUCTION

This chapter reviews work done in Pasoh Forest Reserve (Pasoh FR), Negeri Sembilan, Peninsular Malaysia, and elsewhere in Malaysia to assess herbivorous insect biodiversity in tropical lowland forest, and the effects of various types of human activity on it. The moth family Geometridae is used as an indicator taxon, as it scores well on a variety of criteria applied to determine indicator quality and identify priority groups.

The moths are sampled using light-traps of various types, a method that, because it involves a response to a 'bait' or attractant, can involve bias. However, this criticism could be levelled at all mass sampling methods in some way or another, and the method has been shown to be the single most useful technique for inventorying species in forest habitats (Hammond 1990, 1994).

Human impact in tropical forest systems varies in intensity from selective logging through conversion to simpler arboreal systems (plantation, agroforestry) to clearance for field crops or housing. Sampling at Pasoh FR has only covered the first of these: no samples have been made from the surrounding oil palm plantations. Therefore notes on the impact of other types of management or cultivation refer to examples from elsewhere.

Sampling at Pasoh FR has only been done within the vicinity of the Field Station, or in primary forest nearby in Compartment 22 (Fig. 1), so the diversity measured may only partly be representative of that of the Reserve as a whole, perhaps only that of the particular type of lowland dipterocarp forest (Red Meranti-Keruing type) found in that area. Further species might be expected to occur in the hill dipterocarp forest towards the north-eastern boundary.

[1] Department of Entomology, the Natural History Museum, London SW7 5BD, UK.
E-mail: jdh@nhm.ac.uk
[2] Owley Farm, South Brent, Devon TQ10 9HN, UK.

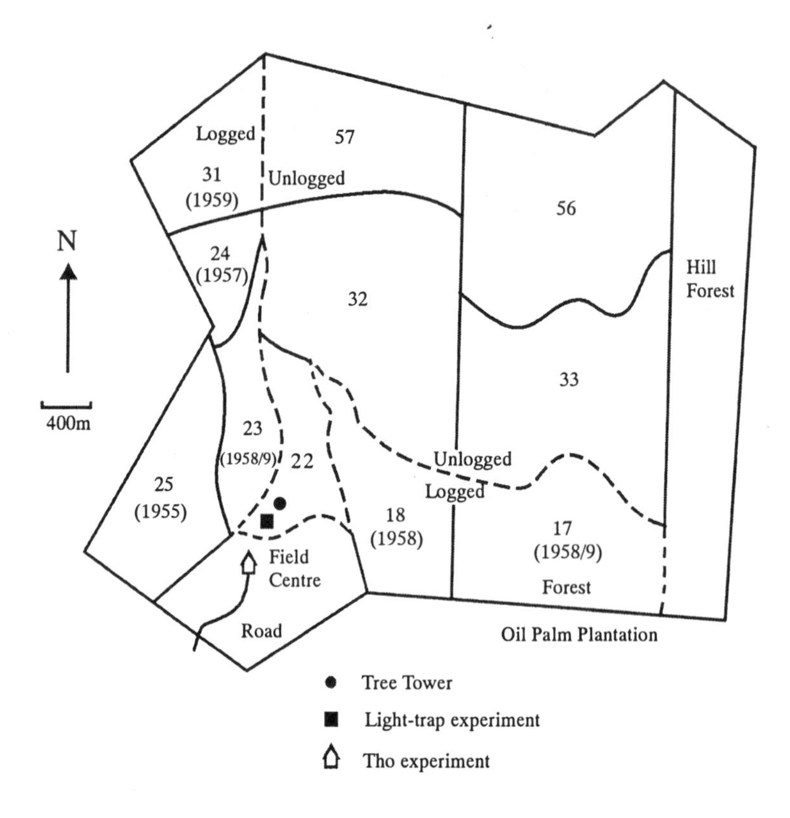

Fig. 1 Trapping sites at the Pasoh FR with the compartment numbers and the year they have been logged in brackets. The Tho experiment site is discussed in section 4 and by Intachat et al. (2001).

Our records from this area of Pasoh FR can be set within the context of a much wider set of surveys conducted in Malaysian Borneo, with some further data from other parts of Peninsular Malaysia. The Bornean fauna is very similar to that of the Peninsula. Indeed, most of the species recorded at Pasoh FR are also known from Borneo, and our records include several that extend the known occurrence of species from Borneo to the Peninsula that were previously thought to be endemic to the former, as indicated on the checklist.

2. THE GEOMETRIDAE

The moth family Geometridae is one of the three largest in the Lepidoptera, the others being the Noctuidae and Pyralidae. The moths are mostly of delicate build but with a large size range from just under 10 mm span to well over 70 mm. They vary in wing pattern from dull, cryptic and bark-like to bright and aposematic such as in the day-flying conifer-feeding genus *Milionia* Walker and the Rhizophoraceae-feeding genus *Dysphania* Hübner. A high proportion of the subfamily Geometrinae is bright green or emerald green. They mostly rest with the wings held flat against the substrate. The known world fauna of about 21,000 species is catalogued in Scoble (1999). This total and that indicated for Borneo (see below) are very similar to current estimates for the butterflies (R. I. Vane-Wright, pers. comm.).

The family is defined on possession of a particular type of hearing organ ventrally at the base of the abdomen, a pair of capsules with a rod-like structure called an ansa that is only found in geometrids. In addition, the larvae of all but a handful of genera have the prolegs (false legs) reduced to a single pair on the sixth abdominal segment in addition to the anal pair. Lepidoptera typically have four pairs, one on each of the third to sixth abdominal segments. This reduction of the prolegs leads to their characteristic looping method of locomotion where they appear to be measuring the surface they walk on, rather like leeches. This is the basis of their various scientific and vernacular names, such as geometers (measurers), inchworms or loopers.

The larvae are external feeders on living plants, mostly defoliators, but with several major groups feeding on the reproductive parts of the plants, particularly flowers, such as many Geometrini and Eupitheciini. The Sterrhini are unusual in feeding on dead foliage or hard fruits. A number feed specifically on particular plant genera or families, but for many species the biology is unknown, so it is difficult to gauge the extent of such specificity accurately. It is reviewed for most Malaysian genera by Holloway (1993 [1994], 1996, 1997).

Further details on the taxonomy and biology of the family may be found in Common (1990), Holloway et al. (2001) and Scoble (1992), as well as in the works by Holloway just cited.

3. INDICATOR SPECIES AND THE GEOMETRIDAE

Various criteria have been proposed for identification of good indicator species. These criteria may be given different priority, depending on whether a given survey is intended to inventory biodiversity or monitor change in it due to various external factors such as climate change or human activity (e.g. Brown 1991; Holloway 1983; Holloway & Stork 1991; New 1998; Pearson 1994). The criteria include: occurrence in moderate diversity over a broad geographical range, including altitude; ease of sampling, enabling accumulation of samples that can provide statistically significant results; taxonomic tractability, with an extensive taxonomic knowledge base; habitat fidelity and specificity; well known biology; patterns shown in common with those of other major taxa; response to change in environmental variables.

Amongst invertebrates, the geometrids score highly for most of these criteria, certainly amongst other Lepidoptera as indicated by Holloway (1984, 1985). Since then, the taxonomic foundation for the group in SE Asia has undergone considerable revision and expansion (Holloway 1993 [1994], 1996, 1997). There are still numerous lacunae in our knowledge of geometrid biology, but this is the case with most invertebrates, even the butterflies. This is more than compensated for by the much higher diversity and finer grained patterns of endemism shown by most invertebrate groups when compared with vertebrates.

Sampling methodology for invertebrates may also be more objective though is still in need of considerable calibration and assessment of bias (New 1998). Light-trapping is prone to such bias, but continued calibration of the method is tending to increase, rather than decrease, confidence in its robustness (e.g. Holloway et al. 2001; Intachat & Woiwod 1999, references cited in both).

4. SAMPLING SITES AND LIGHT-TRAP TYPES

Figure 1 indicates the three sites where geometrids have been sampled intensively (two further sites were sampled over a short period in an experiment to assess the impact of logging, as discussed later and by Intachat et al. (1999a)). Two are in undisturbed primary forest (Compartment 22). The third is within the Field Centre clearing which is sited in forest that was selectively logged between 1955 and 1959, and has since been allowed to regenerating forest (see Chap. 2). There is also a compartment of unlogged forest in the vicinity and an arboretum of 2.2 hectares that contains 267 tree species.

The two sites in undisturbed forest were sampled between June–July 1993 (Intachat & Woiwod 1999) and from July 1994 to July 1995 (Intachat & Holloway 2000). The regenerating forest was sampled over a period of almost three years from November 1982 to October 1985 by the late Dr. Tho Yow Pong and FRIM staff (Intachat 1995; Intachat et al. 2001).

The light-trapping regime used at each site varied. At the first primary forest site, two Rothamsted-type traps (Williams 1948) illuminated by 250 watt clouded lamps were alternated around the corners of a 50m square with two Robinson-type traps (Robinson & Robinson 1950) illuminated by a 125 watt mercury vapour bulb. The latter produces much higher levels of the ultraviolet light to which night-flying moths appear to be particularly attracted. This experiment was conducted every night from dusk to midnight over the 32 days, the traps being rotated round the corners of the square to give a valid comparison of the two models as was done earlier by Taylor & French (1974).

The sampling at the second primary forest site was conducted on a tree tower for five nights each month at new moon, again from dusk to midnight. Three Rothamsted-type traps were run simultaneously on the canopy tower at 1 m, 15 m and 30 m from the ground (Color plate 3). The tree-tower survey was initiated to study vertical distribution of moths and possible stratification within the forest structure.

Dr. Tho used two 100-watt tungsten bulbs hung in front of a white sheet, with all moths on the sheet being collected for four hours after dusk for four or five nights each month around the period of the new moon. The Tho survey was conducted in parallel with monthly climatic records and phenological observations of flushing, flowering and fruiting of a sample of forest trees.

All geometrid moths were collected, pinned, mostly spread, and identified to species as far as possible (almost all). Numbers of individuals of each species in each sample were counted, on an hourly basis for Dr. Tho's survey, and on a nightly basis for the other projects. The data are summarised on a site basis in the checklist.

5. ANALYTICAL METHODS

Comparison of the two trap types (Intachat & Woiwod 1999) was effected by analysis of variance, and diversity was measured by the log-series a measure (Fisher et al. 1943), though tests of the fit of sample data to the log-series α model and the Poisson log-normal model using a Maximum Likelihood Program gave variable results for the two trap types, tending to favour the Poisson log-normal. Nevertheless, the α measure is robust, and the fit is adequate enough for most light trap data to be used generally with confidence (Barlow & Woiwod 1989; Intachat & Woiwod 1999; Taylor et al. 1976).

In the tree tower experiment (Intachat & Holloway 2000), the extent to which the vertical distribution of any species departed from evenness over the three levels was assessed by analysis of variance, the monthly samples having temporal independence though not spatial independence. The extent to which any such height preferences were stratified (coincident) was assessed by cluster analysis, with the more abundant species being compared pairwise according to their proportional representation in the traps at the three levels.

The phenological study of the data collected by the Tho survey has used a generalised linear modelling approach (GLIM) to assess the extent to which moth sample abundance, species richness and diversity can be predicted from the climatic variables and plant phenology. The methods are described in more detail in Intachat et al. (2001).

A measure such as α gives one estimate of diversity that has species richness as a major component. Other estimates of species richness that give an actual predicted species number are frequently applied to this type of sample data. Those most frequently favoured are the Chao estimators (Colwell & Coddington 1994) that are derived simply from the species total and the number of singletons and doubletons in a single sample (Chao 1), or, with multiple samples, from the species total and the numbers of species occurring in only one and only two samples (Chao 2). These will be applied to our data here, though Süssenbach & Fiedler (1999) suggested that a Michaelis-Menten procedure gave a more robust estimate.

6. RESULTS AND DISCUSSION

Major and detailed analyses of all three studies have been or will be published elsewhere as cited below, so the results will be summarised briefly before focusing on general geometrid diversity at Pasoh FR within a south-east Asian context.

The two trap types showed no significant differences in overall diversity measurements, though the samples made by the Rothamsted-type trap were more uniform and consistent. It would therefore appear that the Rothamsted trap performs better in tropical conditions. However, the experiment was performed at ground level in rather dense, closed forest, with each trap invisible to observers at the other three corners of the square. It would be beneficial to repeat the experiment in more open tropical situations before a final judgement on trap merits is passed. Intachat et al. (1999a) used Robinson traps in assessment of logging impact in a variety of forest types, as the portability of these traps can be a logistical advantage in the field (Intachat & Woiwod 1999).

The analysis of the tree tower samples (Intachat & Holloway 2000) showed that a number of species had significantly restricted flight height preferences, and that this effect was stronger in some taxonomic groups, e.g. Geometrinae, than in others, e.g. Ennominae, or was strong in particular genera such as *Zamarada* Moore. However, there was no evidence that these flight height preferences coincided in distinct strata, but species varied in a continuum from broad to narrow flight height preferences over the whole vertical column, with some species turnover from the bottom to the top. There was some indication that abundance and diversity of geometrids in the highest (canopy level) trap was depressed during early months of the survey when low level smoke haze from forest fires was prevalent (Anonymous 1995).

The phenological data indicated that geometrid abundance in samples was positively correlated with high rainfall three months previously and with (possibly consequent on the rainfall) flushing and flowering of trees in the previous month.

High rainfall and humidity during the previous two months had an adverse effect, possibly through encouraging the spread and activity of pathogens (Intachat et al. 2001).

413 geometrid species were recorded in total, and all except five were identified at least to genus (Appendix). About 40 were new records for Peninsular Malaysia, mostly of species previously known only from Borneo. Twelve were species not known from Borneo, though this total is likely to increase when further taxonomic work has been done on the unknowns and the 50 identified to genus only. Nevertheless, the geometrid faunas of lowland forests in Peninsular Malaysia, Borneo and Sumatra are extremely similar, probably a result of their being united as a single land mass, Sundaland, during falls of sea level at the time of the Pleistocene glaciations (Holloway & Jardine 1968).

The total known Bornean fauna of 1079 (Table 1) is more than twice that recorded at Pasoh FR, but of course includes major components of diversity from montane zones and from lowland forest types such as heath forest, mangrove and forest on limestone that are not represented at Pasoh FR. Comparison with specific lowland forest types in Borneo would be fairer, as is discussed below.

Table 1 also breaks down these totals into subfamilies. Most again for Pasoh FR have only half the Bornean total (checklists in Holloway 1993 [1994], 1996, 1997) except for the Larentiinae where the Pasoh FR total is only one tenth of that of Borneo. A high proportion of this subfamily is found in montane zones (and also at high latitudes), and hence its low relative diversity at Pasoh FR is not surprising (Holloway 1986, 1997). The central column of Table 1 gives an estimate of species in Borneo that are known to occur in lowland forest (excluding all strictly montane ones as listed in the references cited), and percentages are more similar to those recorded at Pasoh FR; the Pasoh FR total is between half and two-thirds of this Bornean lowland forest total.

Our Pasoh FR samples also lack the subfamily Oenochrominae. This is restricted in the Oriental tropics to the genus *Sarcinodes* Guenée (e.g. Holloway 1996) that is an outlier from a major Australasian lineage that is a larval feeding specialist on Proteaceae. The small number of host-plant records for *Sarcinodes* are from *Helicia*, one of a few Oriental genera in this mainly southern hemisphere plant family. Most Bornean *Sarcinodes* species also occur in Peninsular Malaysia, so the absence from Pasoh FR may also reflect low levels or absence of Proteaceae. Single species of both *Helicia* and *Heliciopsis* have been recorded in the 50 ha census plot (511 individuals in all) at Pasoh FR (Manokaran et al. 1992). We do not,

Table 1 Numbers of species in different geometrid subfamilies recorded in Borneo and at Pasoh. Percentage are given in brackets.

Subfamily	Borneo				Pasoh	
	(all forest types)		(lowland forest)		(lowland forest)	
Oenochrominae	6	(0.6)	3	(0.4)	0	0
Desmobathrinae	45	(4.2)	35	(4.9)	17	(4.1)
Geometrinae	218	(20.2)	168	(23.4)	106	(25.8)
Ennominae	433	(40.1)	288	(40.2)	191	(46.3)
Orthostixinae	2	(0.2)	2	(0.3)	0	0
Sterrhinae	176	(16.3)	144	(20.1)	74	(17.8)
Larentiinae	199	(18.4)	77	(10.7)	25	(6.0)
TOTAL	1079		717		413	

however, have comparative data from forests where the moth genus occurs.

Table 2 presents data on diversity values and richness estimates for the individual Pasoh FR samples for the Geometridae as a whole, and also for the tree tower samples pooled and the total sample. The raw data for individual species are presented in the checklist (Appendix). Managed forests in Peninsular Malaysia are represented by data from Serendah Forest Reserve, a regenerating lowland dipterocarp forest that was logged in the 1970s, and from a mixed indigenous (mostly dipterocarps) plantation at Kepong. The table also includes data from the Brumas area in the lowlands of Sabah, Borneo, where a similar survey of Lepidoptera in exotic softwood plantations and logged (1984) forest was undertaken over a

Table 2 Numbers of individuals and species of Geometridae taken in the five samples at Pasoh FR and in similar samples from secondary and plantation forests in Peninsular Malaysia and near Brumas in the lowlands of Sabah. Samples from primary and logged-over forest in the Danum Valley, Sabah, and four major lowland forest samples from the G. Mulu National Park, Sarawak, are also included: the latter were accumulated over a much shorter time with a Robinson pattern trap. Figures calculated for diversity (α) and species richness (Chao 1; not Danum, Kepong or Serendah samples) are also presented; those of the former for some Bornean data are rounded to the nearest integer as in the source interactive. Standard errors for α are all within ±4.2.

Site	Individuals	Species	α	Chao 1
PENINSULAR MALAYSIA				
Pasoh total	19077	413	74.8	465
Two-trap	3263	205	67.7	236
Tower: low	1251	172	63.8	247
Tower: middle	1220	173	62.1	276
Tower: high	1157	143	51.3	193
Tower: total	3624	311	81.5	330
Tho phenology	12190	315	64.4	340
Serendah	613	154	66.1	-
Kepong	930	156	53.6	-
BORNEO				
Brumas total	8245	277	55.3	324
Logged-over	1686	192	55.8	261
Eucalyptus	962	165	57.3	216
Paraserianthes	1833	147	37.6	220
Pinus	787	137	47.9	183
Acacia	2326	126	28.6	176
Gmelina	651	118	42.1	166
Danum total	2575	275	83	-
Primary canopy	383	100	44	-
Primary understorey	824	182	72	-
Logged over	332	99	48	-
G. Mulu National Park				
Alluvial forest	616	141	58	176
Hill dipterocarp	1035	172	60	260
Kerangas (heath forest)	400	132	69	253
Forest on limestone	643	196	95	355

year in 1991 (Chey 1994; Chey et al. 1997; Intachat et al. 1999b). The data for Brumas Geometridae only are presented, with values for α and for the Chao 1 richness estimator calculated. Unfortunately it was not possible to obtain comparative samples from undisturbed forest in the Brumas area as it has been mostly cleared (Chey 1994), so control data from the Danum Valley Field Centre area just to the north are presented in the text (Willott 1999, pers. comm.), together with data accumulated over a much shorter time span (four nights rather than several nights each month over a year; Holloway 1984) in four major lowland forest types in the G. Mulu National Park, Sarawak.

Comparative data for Pasoh FR and other lowland dipterocarp forests within Peninsular Malaysia were presented by Intachat et al. (1999a). These sample data were obtained over a much shorter period (three consecutive nights) than most of those in Table 2, and covered paired sites of both logged and undisturbed forest. However, as one pair of samples was from Pasoh FR, they do enable Pasoh FR to be placed in context with the other forest types, though diversity values (for all Geometroidea) are low relative to those for longer runs of data (e.g. α is around 50 for Pasoh FR, rather than 60 or above for most long term samples). Pasoh FR is placed in a second tier of diversity with dipterocarp forests at Tasik Temenggor (another Red Meranti forest) and Endau Rompin (mixed lowland dipterocarp), higher than at Wang Kelian (seasonal White Meranti-Gerutu) and Tanjong Tuan (coastal lowland dipterocarp forest), but lower than at Panti (lowland dipterocarp). Data for logged-over forest and indigenous plantation in the Peninsula are presented by Intachat et al. (1997, 1999b). The biological basis for differences between the faunas of these lowland forest types is as yet unclear, but analyses in progress (Intachat et al. in press) are beginning to enable indicator species for some of them to be identified.

The logged site of the sample pair at Pasoh FR had been logged in the early 1950s, and the site at Panti had also been newly logged, but the logged sites at the other localities had been regenerating for about 20 years (Intachat et al. 1999a). Nevertheless, all showed a similar pattern of no significant loss of diversity as measured by a but a significant drop in overall moth abundance. There were also subtle faunistic changes, including decline or absence of possible dipterocarp indicators (e.g. in the genera *Ornithospila* Warren and *Ectropidia* Warren) from the logged sites.

A quick survey of host-plant records for the 43 species with 100 or more individuals in the Pasoh FR samples includes 11 where there is no information whatever on the species or close relations. Of the others, 12 are probably polyphagous, three may feed on dead foliage, and five may be associated with Dipterocarpaceae and four with Leguminosae. Others involve Fagaceae, Lauraceae, Loranthaceae, Rhamnaceae and Sterculiaceae.

It may be possible to relate geometrid moth diversity to vegetation and floristics with greater precision in the future, when much more information is to hand on the biologies of the moth species. This remains a major challenge for entomologists working in places like Pasoh FR at the start of this new millennium.

Table 2 indicates that there is no significant difference (95% confidence limits) in diversity between four of the Pasoh FR samples, but that from the high trap on the tree tower is lower. However, when the samples for the three tower levels are combined, the diversity value increases significantly, indicating there is species turnover from the ground to the canopy. The Chao 1 values also indicate this, the canopy value being much less than that at middle and low levels, but with pooling of all three leading to a predicted richness very much higher than at any individual level (the Chao 2 value is a comparable prediction of 342 species for the three samples pooled). The two-trap Chao 1 value is close to that for the low trap at the tower site, as might be expected, but the Chao 1 value for the Tho phenology pooled sample is closer to that for the pooled tower samples, despite a diversity value as low as for the two-trap sample and low and middle tower samples. Possibly, as the samples were made in a clearing, a small representation of canopy-flying species was drawn into it from time to time.

In addition to vertical species turnover, there appears to be some element of beta-diversity between sites (see also Robinson & Tuck 1993). When all Pasoh FR samples are pooled, both the α value and Chao 1 estimate rise above values for any individual sample, though the former is still lower than the value for the pooled tower sample, and the latter attains a level over 120 species greater than that for the phenology or pooled tower samples. The Chao 2 value is even greater at 539 species, suggesting that, if Bornean and Peninsular Malaysian geometrid faunas are similar, about half the Peninsular species could occur in the Pasoh FR.

Comparative Bornean data have been accumulated from primary and logged forest at Danum Valley, Sabah, from a wider range of primary forest types in the G. Mulu National Park, Sarawak (now a World Heritage Site), and from an area of softwood plantation and secondary forest at Brumas, Sabah.

Samples from the Danum forest (Willott 1999, pers. comm.; Table 2) were made in the canopy and understorey of primary forest and in forest that was logged over in 1988 (random but frequent sampling at 125 watt mercury vapour lamps over a period from 1993–1995). The four understorey samples gave values of α between 45 and 53, and one of 72 when pooled. The two canopy samples gave values of 32 and 43 that pooled to give a value of 44. Pooling of all these primary forest samples gave a value of 80. The two logged forest samples yielded separate values of 36 and 50, and a value of 48 when pooled. Addition of these logged samples to the primary forest pool increased the overall value of α slightly to 83 (247 species and 1,539 individuals). A few subsidiary primary forest samples were made: these reduced α to 78 but increased the species total to 275 with a total sample of 2,575 moths. Thus, though the individual primary forest sample values are lower than at Pasoh FR, the pooled values are very similar, including the indication of vertical turnover between understorey and canopy. The pooled logged forest value, unlike those for the paired Peninsular Malaysian sites, is somewhat lower than that for the equivalent understorey primary pooled sample, perhaps an effect of the recency of the logging (see also Holloway et al. 1992). The overall number of species recorded is much smaller than at Pasoh FR (275 versus 413), but this probably just reflects the much smaller overall sample (2,575 versus 19,077).

The four Mulu forest types show a wide range of α, with that for alluvial and dipterocarp forest not significantly higher than for the Brumas logged-over forest and, with the kerangas (heath forest) sample, within the range of the individual Pasoh FR samples. The sample from forest on limestone has an α value higher than any other in Table 2, and a Chao 1 value that exceeds all except the totally pooled Pasoh FR sample. All these values are probably not strictly comparable with the others, being obtained over a much shorter timespan. α values at a site tend to increase through time as light-trap sample size accumulates (Barlow & Woiwod 1989; Chey et al. 1997; Robinson & Tuck 1993, periods from 10 nights to 13 months).

At Brumas (samples made at a sheet with a 250 watt mercury-lithium bulb two nights per month through 1991) the higher values of α for logged-over forest and a *Eucalyptus deglupta* plantation approach those for individual Pasoh FR samples, as do those of Chao 1 for the logged-over forest, with *Eucalyptus* and *Paraserianthes falcataria* values a little lower. α values decline from the *Pinus caribaea* plantation through those for *Gmelina arborea* and *Paraserianthes* (values exchanged in error in Intachat et al. (1999b)) to that for *Acacia mangium*. The Chao 1 values are ordered differently: *Paraserianthes* - *Eucalyptus* - *Pinus* - *Acacia* - *Gmelina*. Chey (1994) and Chey et al. (1997) suggested that diversity in these plantations depended mostly on the extent to which components of indigenous, low diversity secondary forest (belukar) were able to develop amongst the mainly exotic crop trees, rather than on the type of crop there. The Chao values for all the samples pooled were 324 for Chao 1 and 342 for Chao 2, equivalent only to the highest single site values at Pasoh FR (pooled tree tower and Tho phenology). Those for plantations alone were lower at around 310. Perhaps this total represents a general degraded secondary forest component that shows much less beta-diversity than does primary forest. Nevertheless, secondary and 'weed'-rich plantation forest would appear to be able to support about one third of Bornean geometrid diversity.

The paired-site results from Peninsular Malaysia mentioned earlier (Intachat et al. 1999a) indicate that logged-over forests can also support high diversity, and that some beta-diversity is also preserved. The intensity of logging may also be important. Strict comparison of floristics between the Brumas logged-over site and those surveyed in Peninsular Malaysia has not been undertaken, but the logging date of the former is more recent than in most of the latter, and the prevalence of lianes, shrubs and *Macaranga* trees was indicative of considerable disturbance (Chey 1994).

Four sites in Peninsular Malaysia were surveyed with Rothamsted-type traps over a similar time-span to that of the Brumas samples (Intachat et al. 1997). Two were in logged-over forest in Serendah Forest Reserve, one of which was clear-felled during the sampling, a third was in an area of poor secondary forest in the same region that had developed in an area that had been clear-felled, with a tin mine, abandoned in the 1970s, and one was in an indigenous plantation, mostly of dipterocarps, established in 1927 at Kepong in an area previously farmed. Plant diversity was twice as high in the logged-over forest as in the plantation, and five times as high as in the secondary forest area. Values of α for geometroids (drepanids and uraniids as well as geometrids) were around 70 for the logged-over and plantation sites, just under 60 for the site logged during the sampling, and below 40 for the secondary forest site. These values are consistent with those obtained elsewhere, allowing a slight reduction for restriction to Geometridae only (values for geometrids in the Serendah site that was not clear-felled and the Kepong plantation site are included in Table 2).

7. CONCLUSIONS

Pasoh FR supports a geometrid moth fauna that appears to be of average diversity amongst Malaysian lowland forests but nevertheless may support over half the geometrid species occurring in the Peninsula and a higher proportion of those restricted to lowland habitats. This diversity has a vertical dimension (canopy vs. understorey) at any individual site, as well as some species turnover between forest sites (beta-diversity).

Comparisons between Pasoh FR and the other forest types within Peninsular

Malaysia (e.g. Intachat et al., in press) indicate that further species turnover (faunistic differences) occurs between one forest type and another, and that this is at least partially retained, albeit with modification, after such forests have been selectively logged over.

Conversion to plantation, or intensive logging, exemplified by sampling in Sabah, appears to reduce diversity to about two thirds of its natural level, with some indication of greater uniformity, less beta-diversity. This retention of diversity may be, for plantations of exotics, more dependent on the development of associated growth of secondary forest species than on the crop species selected. Plantations of indigenous species, however, can over time come to support geometrid diversity approaching that of natural forests.

These data indicate that conservation of insect herbivore diversity in Malaysia requires the retention of significant areas of all indigenous forest types, sensitive methods of timber extraction, and, when plantations are planned, that development of a diverse indigenous understorey is permitted within a landscape mosaic that includes some primary or logged-over forest refuges. The phenological results indicate that rainfall is perhaps the most important factor in the maintenance of current geometrid diversity at Pasoh FR. Natural climate changes or those induced by man such as through increased CO_2 levels or, more locally, through modification of forest cover, will also therefore be likely to influence that diversity.

ACKNOWLEDGEMENTS

Much of the sampling in Peninsular Malaysia was undertaken by Jurie Intachat as part of a doctoral research programme at Oxford University, supervised by Jeremy D. Holloway and Martin Speight, supported by ODA/FRIM Sub-programme 1, Project 5 (The Impact of Forest Development on Faunal Diversity) and IRPA's project RA 103-01-001-DO2 (Biodiversity and Forest Conservation).

We thank Dr. John Willott for providing us with geometrid data from his Danum project in advance of its publication, and the National Institute for Environmental Studies (NIES), Tsukuba, Japan, for permission to use the tree-tower facility at Pasoh FR. We are grateful to Drs. Dick Vane-Wright, Toshinori Okuda and two anonymous referees for their comments on an earlier draft of the text. We thank Ms. Maia Vaswani for keyboarding and text layout, and are also grateful to Mr. Saimas Ariffin and Pasoh Field Station staffs for all their assistance.

REFERENCES

Anonymous (1995) Report on air quality in Malaysia as monitored by the Malaysia Meteorological Service, 1994. Malaysian Meteorological Service, Technical Note 55.

Barlow, H. S. & Woiwod, I. P. (1989) Moth diversity of tropical forest in Peninsular Malaysia. J. Trop. Ecol. 5: 37-50.

Brown, K. S. Jr. (1991) Conservation of Neotropical environments: insects as indicators. In Collins, N. M. & Thomas, J. A. (eds). The Conservation of Insects and their Habitats, London: Academic Press, pp.349-404.

Chey, V. K. (1994) Comparison of biodiversity between rain forest and plantations in Sabah, using moths as indicators. Ph.D diss., Oxford University (unpubl.).

Chey, V. K., Holloway, J. D. & Speight, M. R. (1997) Diversity of moths in forest plantation and natural forests in Sabah. Bull. Entomol. Res. 87: 371-385.

Colwell, R. K. & Coddington, J. A. (1994) Estimating terrestrial biodiversity by extrapolation. Philos. Trans. R. Soc. London Ser. B 345: 101-118.

Common, I. F. B. (1990) Moths of Australia. Melbourne University Press.

Fisher, R. A., Corbet, A. S. & Williams, C. B. (1943) The relation between the number of species and the number of individuals in a random sample of an animal population. J. Anim. Ecol. 12: 42-58.

Hammond, P. M. (1990) Insect abundance and diversity in the Dumoga-Bone National Park, N. Sulawesi, with special reference to the beetle fauna of lowland rainforest in the Toraut region. In Knight, W. J. & Holloway, J. D. (eds). Insects and the Rain Forests of South East Asia (Wallacea), London, Royal Entomological Society, pp.197-254.

Hammond, P. M. (1994) Practical approaches to the estimation of the extent of biodiversity in speciose groups. Philos. Trans. R. Soc. London Ser. B 345: 119-136.

Holloway, J. D. (1983) Insect surveys–an approach to environmental monitoring. Atti XII Congresso Nazionale Italiano di Entomologia, Roma, 1980: 239-261.

Holloway, J. D. (1984) The larger moths of Gunung Mulu National Park; a preliminary assessment of their distribution, ecology and potential as environmental indicators. In Jermy, A. C. & Kavanagh, K. P. (eds). Gunung Mulu National Park, Sarawak, Part II, Sarawak Museum Journal 30, Special Issue 2, pp.149-190.

Holloway, J. D. (1985) Moths as indicator organisms for categorising rain forest and monitoring changes and regenerating processes. In Chadwick, A. C. & Sutton, S. L. (eds). Tropical Rain-Forest, The Leeds Symposium, Special Publication, Leeds Philosophical and Literary Society, pp.235-242.

Holloway, J. D. (1986) Lepidoptera faunas of high mountains in the Indo-Australian tropics. In Vuilleumier, F. & Monasterio, M. (eds). High altitude Tropical Biogeography, Oxford University Press, New York, pp.533-556.

Holloway, J. D. (1993 [1994]) The Moths of Borneo: family Geometridae, subfamily Ennominae. Malay. Nat. J. 47: 1-309.

Holloway, J. D. (1996) The Moths of Borneo: Family Geometridae, subfamilies Oenochrominae, Desmobathrinae and Geometrinae. Malay. Nat. J. 49: 147-326.

Holloway, J. D. (1997) The Moths of Borneo: family Geometridae, subfamilies Sterrhinae and Larentiinae. Malay. Nat. J. 51: 1-242.

Holloway, J. D. & Jardine, N. (1968) Two approaches to zoogeography: a study based on the distributions of butterflies, birds and bats in the Indo-Australian area. Proc. Linn. Soc. London 179: 153-188.

Holloway, J. D. & Stork, N. E. (1991) The dimensions of biodiversity: the use of invertebrates as indicators of man's impact. In Hawksworth, D. L. (eds). The Biodiversity of Microorganisms and Invertebrates: Its Role in Sustainable Agriculture, CAB International, Wallingford, pp.37-62.

Holloway, J. D., Kibby, G & Peggie, D. (2001) The Families of Malesian Moths and Butterflies. Fauna Malesiana Handbook 3. Leiden: E. J. Brill.

Holloway, J. D., Kirk-Spriggs, A. H. & Chey, V. K. (1992) The response of some rain forest insect groups to logging and conversion to plantation. Philos. Trans. R. Soc. London Ser. B 335: 425-436.

Intachat, J. (1995) Assessment of moth diversity in natural and managed forests in Peninsular Malaysia. Ph.D diss., Oxford University (unpubl.).

Intachat, J. & Holloway, J. D. (2000) Is there stratification in diversity or preferred flight height of geometroid moths in Malaysian tropical lowland forest? Biodiversity Conserv. 9: 1417-1439.

Intachat, J. & Woiwod, I. P. (1999) Trap design for monitoring moth biodiversity in tropical rainforests. Bull. Entomol. Res. 89: 153-163.

Intachat, J., Holloway, J. D. & Speight, M. R. (1997) The effects of forest management practices on geometroid moth populations and their diversity in Peninsular Malaysia. J. Trop. For. Sci. 9: 411-430.

Intachat, J., Holloway, J. D. & Speight, M. R. (1999a) The impact of logging on geometroid moth populations and their diversity in lowland forests of Peninsular Malay. J. Trop. For. Sci. 11: 61-78.

Intachat, J., Chey, V. K., Holloway, J. D. & Speight, M. R. (1999b) The impact of forest plantation development on the population and diversity of geometrid moths (Lepidoptera: Geometridae) in Malaysia. J. Trop. For. Sci. 11: 329-336.

Intachat, J., Holloway, J. D. & Staines, H. (2001) Effects of climate and phenology on the abundance and diversity of geometroid moths within a Malaysian natural tropical rainforest environment. J. Trop. Ecol. 17: 411-429.
Intachat, J., Holloway, J. D. & Speight, M. R. (in press) The diversity of geometroid moths with in different types of rain forest in Peninsular Malaysia. Malay. Nat. J.
Manokaran, N., LaFrankie, J. V., Kochummen, K. M., Quah, E. S., Klahn, J. E., Ashton, P. S. & Hubbell, S. P. (1992) Stand table and distribution of species in the 50 ha research plot at Pasoh Forest Reserve. FRIM Research Data No. 1, Forest Research Institute Malaysia, Kepong, pp.210-211.
New, T. R. (1998) Invertebrate Surveys for Conservation. Oxford University Press.
Pearson, D. L. (1994) Selecting indicator taxa for the qualitative assessment of biodiversity. Philos. Trans. R. Soc. London Ser. B 345: 75-79.
Robinson, G. S. & Tuck, K. R. (1993) Diversity and faunistics of small moths (Microlepidoptera) in Bornean rainforest. Ecol. Entomol. 18: 385-393.
Robinson, H. S. & Robinson, P. J. M. (1950) Some notes on the observed behaviour of Lepidoptera in flight in the vicinity of light sources together with description of a light-trap designed to take entomological samples. Entomologists' Gazette 1: 3-15.
Scoble, M. J. (1992) The Lepidoptera, Form, Function and Diversity. Oxford University Press, Oxford.
Scoble, M. J. (ed). (1999) Geometrid Moths of the World, a Catalogue. Melbourne: CSIRO Press.
Süssenbach, D. & Fiedler, K. (1999) Noctuid moths attracted to fruit baits: testing models and methods of estimating species diversity. Nota lepidopterologica 22: 115-154.
Taylor, L. R. & French, R. A. (1974) Effects of light-trap design and illumination on samples of moths in an English woodland. Bull. Entomol. Res. 63: 583-594.
Taylor, L. R., Kempton, R. A. & Woiwod, I. P. (1976) Diversity statistics and the log-series model. J. Anim. Ecol. 45: 255-271.
Williams, C. B. (1948) The Rothamsted light-trap. Proc. R. Entomol. Soc. London Ser. A 23: 80-85.
Willott, S. J. (1999) The effects of selective logging on the distribution of moths in a Bornean rainforest. Philos. Trans. R. Soc. London Ser. B 354: 1783-1790.

Appendix: A checklist of the Geometridae recorded from Pasoh FR. The order of tribes and species follows that in Holloway (1993 [1994], 1996, 1997), but the order of subfamilies is modified to follow the phylogeny of Holloway (1997: Fig. 2). An asterisk (*) indicates a species not shared with Borneo. A dagger (†) indicates a species recorded at Pasoh FR that had previously only been known from Borneo, or Borneo and Sumatra. The first three columns are from the tree tower experiment, the fourth from the phenology experiment and the last from the trap comparison experiment. A question mark (?) indicates that an identification is tentative; voucher material is deposited in the FRIM collection.

	Low	Middle	High	Tho	Two-traps
ENNOMINAE					
Hypochrosini					
Hypochrosis binexata Walker	8	5	6	74	2
Hypochrosis pyrrhophaeata Walker	14	5	2	10	3
Hypochrosis sternaria Guenée	5	4	1	48	11
Celenna callopistes Prout	-	-	-	29	-
Celenna festivaria Fabricius	-	1	-	2	9
Omiza lycoraria Guenée	2	-	-	6	14
Genusa dohertyi Holloway†	1	-	-	-	-
Genusa simplex Warren	3	-	-	-	2
Achrosis alienata Walker	3	-	-	11	1
Achrosis calcicola Holloway†	1	1	-	2	-
Achrosis classeyi Holloway†	1	-	-	1	-
Achrosis fulvifusa Warren	2	3	-	21	3
Achrosis lilacina Warren†	-	-	-	8	1
Achrosis longifurca Holloway†	-	1	-	3	-
Achrosis multidentata Warren	3	-	-	2	4
Fascellina sp. nr. *castanea* Moore	-	-	-	-	11
Fascellina aurifera Warren	-	-	-	2	-
Fascellina castanea Moore	9	-	-	-	26
Fascellina clausaria Walker	7	7	2	49	8
Fascellina inconspicua Warren	-	-	-	-	2
Fascellina quadrata Holloway	-	-	-	-	1
Fascellina viridicosta Holloway	-	-	-	-	1
Corymica arnearia Walker	-	-	-	3	-
Corymica deducta Walker	-	-	-	1	-
Corymica latimarginata Swinhoe	-	1	-	14	-
Scardamiini					
Scardamia iographa Prout	-	-	-	3	1
Aplochlora vivilaca Walker	-	1	-	1	8
Ourapterygini					
Ourapteryx fulvinervis Warren*	-	1	4	6	-
Ourapteryx picticaudata Walker	-	-	-	2	-
Ourapteryx podaliriata Guenée	1	2	3	26	4
Baptini					
Lomographa luciferata Walker	-	2	1	2	1
Tasta micaceata Walker	-	-	-	1	20
Tasta sp. nr. *reflexoides* Holloway	-	15	12	-	-
Hypulia eleuthera Holloway	-	1	-	11	-
Synegia botydaria Guenée	6	-	-	23	3
Synegia camptogrammaria Guenée	-	-	-	1	-
Synegia eumeleata Walker	-	1	-	-	-
Synegia imitaria Walker	-	2	1	3	-
Platycerota vitticostoides Holloway†	-	1	-	-	1
Bulonga schistacearia Walker	-	-	-	6	-
Curbia martiata Guenée	3	18	9	99	77
Borbacha altipardaria Holloway	-	-	-	-	1
Plutodini					
Plutodes sp. (NHM Slide: 13824)†	1	-	-	-	-
Plutodes cyclaria Guenée	6	14	2	210	18
Plutodes evaginata Holloway†	3	9	9	-	-
Plutodes malaysiana Holloway	-	-	-	247	32

Appendix (continued 1)

	Low	Middle	High	Tho	Two-traps
Lithinini					
Nadagara intractata Walker	-	-	-	-	1
Nadagara juventinaria Guenée	1	-	-	-	-
Nadagara scitilineata Walker	-	-	1	-	-
Caberini					
Hyperythra lutea Stoll	-	-	-	5	-
Petelia delostigma Prout	-	1	-	-	-
Petelia medardaria Herrich-Schäffer	1	6	1	60	-
Petelia paroobathra Prout	1	-	-	102	17
Astygisa circularia Swinhoe	2	3	-	37	5
Astygisa metaspila Walker (1200)	-	-	-	16	1
Astygisa stueningi Holloway†	3	2	3	17	2
Astygisa vexillaria Guenée	2	-	-	38	5
Astygisa sp.†	-	1	-	-	-
Thinopterygini					
Pareumelea eugeniata Guenée	1	-	-	-	1
Cassymini					
Cassyma erythrodon Sommerer & Stüning†	-	-	-	24	-
Cassyma chrotadelpha Sommerer & Stü	-	6	6	31	4
Cassyma electrodes Sommerer & Stüning†	-	-	3	-	-
Cassyma quadrinata Guenée	-	-	2	-	-
Cassyma sciticincta Walker	1	9	8	-	4
Cassyma undifasciata Butler†	-	-	4	-	-
Danala laxtaria Walker	-	-	1	1	-
Danala sp. (1460)	-	-	-	9	-
Auzeodes chalybeata Walker	-	1	4	2	1
Peratostega coctata Warren	-	-	2	6	-
Heterostegane contessellata Prout	-	-	-	4	-
Heterostegane subtessellata Walker	-	-	-	2	1
Heterostegane urbica guichardi Holloway	-	-	-	5	-
Heterostegane warreni Prout	5	13	48	31	7
Peratophyga flavomaculata Swinhoe	4	2	1	78	28
Peratophyga spilodesma Prout†	-	-	-	2	-
Peratophyga ?sobrina Prout (1362)	-	-	-	58	-
Peratophyga trigonata Walker	-	1	-	3	1
Peratophyga venetia Swinhoe	1	-	-	7	-
Peratophyga xanthyala Hampson	-	3	6	34	-
Zamarada baliata Felder & Rogenhofer	-	1	28	-	-
Zamarada denticulata Fletcher	1	-	-	490	45
Zamarada eogenaria Snellen	-	-	-	179	73
Zamarada scriptifasciata Walker†	-	16	5	-	-
Zamarada sp. 1(1406)	6	12	7	-	-
Zamarada ucatoides Holloway†	18	27	31	217	45
Sundagrapha lepidata Prout†	-	-	1	-	-
Sundagrapha tenebrosa Swinhoe	6	12	12	99	38
Orthocabera ocernaria Swinhoe	1	-	-	3	-
Eutoeini					
Calletaera foveata Holloway	1	-	-	-	-
Calletaera jotaria Felder & Rogenhofer	-	-	-	20	-
Calletaera sp. (1467)	-	-	-	21	-
Calletaera subexpressa Walker	2	3	-	4	10
Probithia exclusa Walker	1	-	2	-	2
Probithia imprimata Walker	-	1	-	14	1
Eutoea heteroneurata Guenée	-	-	-	15	-
Luxiaria acutaria Snellen	-	-	1	-	-
Luxiaria emphatica Prout	-	4	-	20	-
Luxiaria fictaria Prout	-	1	-	18	2
Luxiaria hyalodela Prout	1	-	-	-	-
Luxiaria phyllosaria Walker	-	-	-	5	2
Luxiaria subrasata Walker	-	-	-	24	8

Appendix (continued 2)

	Low	Middle	High	Tho	Two-traps
Macariini					
Hypephyra subangulata Warren	-	-	-	3	-
Macaria abydata Guenée	-	-	1	-	-
Chiasmia albipuncta Warren	1	-	-	-	-
Chiasmia avitusaria Walker	36	17	10	131	39
Chiasmia bornusaria Holloway†	2	5	2	78	-
Chiasmia hygies Prout†	8	-	-	3	3
Chiasmia mutabilis Warren	1	1	1	5	-
Chiasmia nora Walker	3	6	5	70	4
Chiasmia ozararia Walker	2	1	-	-	-
Chiasmia translineata Walker	5	8	6	57	12
Boarmiini					
Bracca maculosa Walker	2	-	-	1	-
Hyposidra incomptaria Walker	4	-	-	16	4
Hyposidra infixaria Walker	10	17	10	94	25
Hyposidra picaria Walker	1	-	-	-	8
Hyposidra aquilaria Walker	-	-	-	-	3
Hyposidra talaca Walker	3	7	4	36	-
Hyposidra violescens Hampson	-	-	-	3	1
Exeliopsis discipuncta Holloway†	-	-	1	-	-
Exeliopsis macrouncus Holloway	1	-	-	23	4
Krananda semihyalina Moore	-	-	-	6	2
Zanclomenophra subusta Warren	-	-	-	1	-
Racotis boarmiaria Guenée	-	1	-	-	4
Racotis discistigmaria Hampson *	-	-	-	11	1
Acrodontis insularis Holloway	-	-	-	39	-
Chorodna complicataria Walker	1	1	-	16	2
Coremecis maculata Warren	2	-	1	2	40
Coremecis incursaria Walker	-	-	-	16	-
Coremecis sp. *incursaria* Walker	-	-	-	-	7
Amblychia angeronaria Guenée	1	-	-	5	-
Amblychia hymenaria Guenée	-	-	-	110	1
Thoyowpongia nigrodiscus Holloway	1	-	-	-	1
Biston insularis Warren	-	-	-	21	-
Biston pustalata Warren	-	1	-	8	-
Amraica solivagaria Walker	-	3	-	10	-
Iulotrichia decursaria Walker	-	-	-	4	1
Cusiala acutijuxta Holloway	-	-	-	340	-
Cleora contiguata Moore	-	-	-	14	4
Cleora cucullata Butler	-	2	-	-	-
Cleora determinata Walker	-	5	14	499	12
Cleora inoffensa Swinhoe	3	1	1	48	36
Cleora onycha Fletcher	-	2	-	3	-
Cleora propulsaria Walker	-	3	1	27	13
Cleora repetita Butler	-	-	-	1	-
Cleora tenebrata Fletcher	-	-	1	6	8
Ectropis bhurmitra Walker	1	1	-	37	34
Ruttellerona pseudocessaria Holloway	-	1	-	2	-
Ophthalmitis basiscripta Holloway	-	-	-	5	22
Ophthalmitis exemptaria Walker	-	-	-	2	4
Ophthalmitis clararia Walker	4	-	-	-	-
Ophthalmitis viridior Holloway	4	-	1	-	1
Catoria sublavaria Guenée	-	-	-	-	2
Psilalcis bisinuata Hampson	-	1	-	-	-
Psilalcis subfasciata Warren	-	1	-	-	-
Hypomecis costaria Guenée	231	21	-	892	238
Hypomecis dentigerata Warren	7	6	16	9	16
Hypomecis lioptilaria Swinhoe	1	-	-	3	-
Hypomecis separata Walker	13	11	6	130	79
Hypomecis sommereri Sato†	-	-	1	10	5
Hypomecis subdetractaria Prout	-	-	1	36	4
Hypomecis tetragonata Walker	15	19	3	169	61
Hypomecis transcissa Walker	-	-	-	18	8
Abaciscus intractabilis Walker	3	-	-	-	2

Appendix (continued 3)

	Low	Middle	High	Tho	Two-traps
Abaciscus paucisignata Warren	3	-	-	34	64
Abaciscus shaneae Holloway	1	1	-	-	1
Abaciscus sp. (1455)	2	-	2	3	24
Microcalicha punctimarginaria Leech	5	-	-	1	3
Calichodes subrugata Walker	7	2	2	24	67
Myrioblephara pallibasis Holloway	-	-	-	1	-
Ectropidia exprimata Walker	-	1	2	4	4
Ectropidia fimbripedata Warren	2	5	13	61	4
Ectropidia illepidaria Walker	-	-	7	12	2
Ectropidia quasilepidaria Holloway†	1	6	-	2	-
Diplurodes decursaria Walker	14	6	6	15	110
Diplurodes indentata Warren†	1	2	3	-	3
Diplurodes inundata Prout	1	13	12	14	26
Diplurodes kerangatis Holloway	5	4	1	8	5
Diplurodes semicircularis Holloway	5	7	6	1	49
Diplurodes sp. 1 (1266)	-	7	9	51	-
Diplurodes submontana Holloway†	-	7	2	20	11
Diplurodes sugillata Prout†	-	-	1	-	-
Diplurodes triangulata Holloway	9	12	11	126	19
Nigriblephara cheyi Holloway†	-	2	-	-	-
Nigriblephara radula Holloway	2	4	8	11	2
Monocerotesa hypomesta Prout*	-	-	-	3	-
Boarmacaria tenuilinea Warren	-	-	-	-	7
Ennominae unplaced to tribe					
Heteralex rectilineata Guenée	-	-	-	6	3
Pseudocassyma retaka Holloway†	1	-	-	-	-
Pseudocassyma sp. 1 (1228)	1	-	-	-	-
Pseudocassyma sundagraphoides	-	-	-	-	7
DESMOBATHRINAE					
Desmobathrini					
Ozola basisparsata Walker†	-	-	-	4	-
Ozola edui Sommerer	2	-	-	-	-
Ozola pannosa Holloway†	-	-	-	2	-
Ozola sp. (880)	-	-	-	9	5
Noreia achloraria Warren	-	1	-	1	6
Noreia ajaia Walker	1	1	-	2	4
Celerena signata Warren	-	2	-	63	-
Derambila lumenaria Geyer	-	-	1	-	-
Derambila zincaria Guenée	-	1	-	21	-
Derambila sp. (841)	-	-	-	4	4
Derambila ?zancloptera Walker	-	1	-	11	-
Derambila ?nr. *zincaria* Guenée	-	1	-	1	-
Eumeleini					
Eumelea biflavata Warren	1	-	-	1	-
Eumelea rosalia Stoll	-	-	-	5	-
Eumelea ?parda Sommerer	-	-	-	-	3
Eumelea sp.1(852)	-	-	-	1	-
Eumelea sp.2 (854)	-	-	-	2	-
GEOMETRINAE					
Dysphaniini					
Dysphania ?sagana Druce	-	-	-	15	-
Dysphania ?militaris Linnaeus	-	-	-	-	1
Geometrini					
Herochroma flavibasalis Warren	2	1	7	7	-
Herochroma baba Swinhoe*	-	-	-	-	2
Herochroma subtepens Walker	-	-	-	1	1
Epipristis nelearia Guenée	6	15	-	39	28
Epipristis storthophora Prout*	10	-	5	159	17

Appendix (continued 4)

	Low	Middle	High	Tho	Two-traps
Epipristis truncataria Walker	12	6	1	158	40
Pingasa chlora Stoll	-	-	-	2	-
Pingasa rubicunda Warren	1	2	-	96	1
Pingasa ruginaria Guenée	2	-	2	680	12
Pingasa subviridis Warren	-	1	-	1	-
Pingasa tapungkanana Strand	1	-	-	17	2
Pingasa venusta Warren	-	1	-	4	1
Lophophelma rubroviridata Warren	-	-	-	1	-
Lophophelma erionoma Swinhoe	-	-	-	-	1
Lophophelma loncheres Prout	-	-	-	28	2
Lophophelma luteipes Felder &	-	-	-	2	-
Lophophelma funebrosa Warren	-	3	2	-	2
Dindica olivacea Inoue	-	-	-	4	1
Tanaorhinus rafflesii Moore	3	10	5	151	7
Tanaorhinus viridiluleata Walker	2	2	-	-	5
Mixochlora argentifusa Walker	1	3	2	23	2
Mixochlora vittata Moore	1	-	1	19	2
Chloroglyphica xeromeris Prout	-	3	1	5	-
Paramaxates macrocerata Yazaki	1	1	-	24	2
Paramaxates polygrapharia Walker	-	-	-	3	-
Paramaxates yazakii Holloway	1	-	1	-	-
Dooabia puncticostata Prout	-	-	-	1	-
Agathia codina Swinhoe	-	-	-	1	-
Agathia eromenoides Holloway	3	-	-	55	15
Agathia hilarata Guenée*	1	-	-	18	5
Agathia laetata Fabricius	-	-	-	12	3
Agathia laqueifera Prout	1	-	-	-	-
Agathia quinaria Moore	-	3	7	-	-
Ornithospila avicularia Guenée	43	36	6	147	205
Ornithospila bipunctata Prout	2	1	1	25	-
Ornithospila esmeralda Hampson	-	1	1	5	-
Ornithospila lineata Moore	-	-	-	18	-
Ornithospila submonstrans Walker	40	6	2	387	165
Ornithospila succincta Prout	6	11	10	39	7
Ornithospila sundaensis Holloway	3	4	6	19	60
Eucyclodes albisparsa Walker	9	10	8	4	-
Eucyclodes rufimargo Warren	-	-	3	33	4
Eucyclodes semialba Walker	-	16	3	23	-
Rhombocentra semipurpurea Warren	9	1	1	59	-
Spaniocentra lobata Holloway†	19	-	-	-	-
Spaniocentra megaspilaria Guenée†	-	2	4	17	-
Spaniocentra sp. (775)	-	1	-	30	18
Spaniocentra spicata Holloway*	-	17	4	160	54
Comibaena attenuata Warren	-	2	3	86	13
Comibaena biplaga Walker	-	-	-	14	-
Comibaena cassidara Guenée	-	2	-	25	-
Comibaena delicatior Warren*	-	-	1	-	-
Comibaena fuscidorsata Prout	-	-	-	5	-
Comibaena sp. (659)	-	1	-	-	-
Protuliocnemis biplagata Moore	-	1	1	24	10
Protuliocnemis helpsi Holloway	1	-	-	-	-
Protuliocnemis partita Walker	-	-	1	-	2
Comostolodes albicatena Warren	1	-	1	4	-
Comostolodes dialitha West	4	7	2	21	42
Argyrocosma inductaria Guenée	2	2	5	56	-
Aporandria specularia Guenée	-	-	-	19	3
Oenospila flavifusata Walker	2	5	3	21	-
Oenospila sp. nr. *altistrix* Holloway	-	-	-	2	-
Episothalma robustaria Guenée	-	-	-	32	-
Thalassodes dissitoides Holloway	1	3	-	228	5
Thalassodes sp.2 (Geom 10687)	-	-	-	54	-
Orothalassodes hypocrites Prout	1	-	-	5	-
Pelagodes clarifimbria Prout	-	-	-	1	-
Pelagodes semihyalina Walker	2	5	22	44	6
Pelagodes semengok Holloway	-	-	-	17	-

Appendix (continued 5)

	Low	Middle	High	Tho	Two-traps
Hemithea sp. (733)	-	-	-	1	5
Hemithea melalopha Prout	-	11	-	2	-
Pamphlebia rubrolimbraria Guenée	-	-	-	4	-
Idiochlora celataria Walker†	3	3	-	22	7
Idiochlora olivata Warren	2	2	-	14	-
Idiochlora sp. (700/701)	-	-	-	10	-
Albinospila floresaria Walker	-	-	-	12	-
Maxates marculenta Prout†	-	1	-	-	1
Maxates melancholica Prout (708)	-	-	-	28	-
Maxates sp. nr. *acutissima* Walker	-	2	-	-	-
Maxates sp. (789)	-	-	2	-	-
Maxates sp. 1 (750)	-	-	-	49	4
Maxates sp. 2 (758)	-	-	-	4	-
Jodis spumifera Warren	-	-	1	-	-
Jodis subtractata Walker	1	-	-	17	-
Jodis nanda Walker	-	-	-	2	2
Jodis sp. (763)	-	-	-	1	-
Berta chrysolineata Walker	-	-	-	4	-
Berta sp. nr. *digitijuxta* Holloway*	-	-	-	1	5
Berta zygophyxia Prout	-	-	-	2	-
Berta albiplaga Warren	3	-	-	-	-
Berta annulifera Warren	11	5	13	24	68
Comostola cedilla Prout	2	1	6	18	1
Comostola chlorargyra Walker	-	1	-	5	-
Comostola laesaria Walker	2	8	13	24	4
Comostola meritaria Walker	-	-	-	25	15
Comostola orestias Prout	-	-	2	18	1
Comostola pyrrhogona Walker	-	-	2	-	-
Comostola nympha Butler*	-	-	-	11	1
Pseudocomostola cosmetocraspeda Prout	-	3	-	5	2
STERRHINAE					
Cosymbiini					
Chrysocraspeda plumbeofusa Swinhoe	-	-	-	3	2
Chrysocraspeda croceomarginata Warren	-	-	-	2	-
Chrysocraspeda comptaria Swinhoe	-	-	1	30	1
Chrysocraspeda conversata Walker	-	-	-	1	-
Chrysocraspeda remutans Prout†	1	-	1	960	5
Chrysocraspeda juriae Holloway	1	-	-	-	-
Chrysocraspeda argentimacula Holloway	-	-	1	-	-
Chrysocraspeda dracontias Meyrick	1	-	-	3	-
Chrysocraspeda ozophanes Prout	-	-	-	9	2
Chrysocraspeda porphyrochlamys Prout	-	-	-	5	-
Chrysocraspeda sanguinipuncta Swinhoe†	-	-	-	8	1
Chrysocraspeda dramaturgis Holloway†	-	-	-	2	2
Chrysocraspeda ?flavipuncta Warren (956)	-	-	-	4	-
Chrysocraspeda sp. 1 (972)	-	1	1	3	1
Chrysocraspeda sp. 6 (1066)	-	-	-	-	1
Chrysocraspeda sp. 10 (981)	1	-	-	-	-
Cyclophora frenaria Guenée	1	-	-	76	1
Cyclophora indecisa Warren	-	1	-	1	-
Cyclophora rotundata Warren	-	-	-	4	1
"*Cyclophora*" *subdolaria* Swinhoe	2	1	-	38	2
"*Cyclophora*" sp. 1 (927)	3	1	2	51	10
"*Cyclophora*" sp. 3 (911)	2	-	-	41	12
"*Cyclophora*" sp. 4 (917)	4	-	1	1	6
"*Cyclophora*" sp. 6 (914)	-	-	-	5	-
"*Cyclophora*" sp. 7 (915)	-	-	-	6	-
"*Cyclophora*" sp. 9 (1089)	1	-	-	-	-
"*Cyclophora*" sp. 12 (930)	1	1	-	1	-
"*Cyclophora*" sp. 15 (931)	6	1	-	-	1
"*Cyclophora*" sp. 21 (939)	2	-	-	-	-
"*Cyclophora*" sp. J (1088)	-	-	-	3	-
Mesotrophe intortaria Guenée	-	-	-	8	1

Appendix (continued 6)

	Low	Middle	High	Tho	Two-traps
Mesotrophe maximaria Guenée †	1	-	-	2	2
Mesotrophe sp. nr. *nephelospila* Meyrick	-	-	-	1	4
Perixera argyromma Warren	-	1	-	8	1
Perixera clandestina Prout	-	-	-	11	-
Perixera contrariata Walker	2	-	-	9	1
Perixera denticulata Hampson	-	-	-	9	-
Perixera dotilla Swinhoe	-	-	-	11	-
Perixera illepidaria Guenée	-	2	1	1	-
Perixera sarawackaria Guenée	1	-	-	2	1
Perixera thermosaria Walker	-	-	-	3	-
Rhodostrophiini					
Apostegania rectilineata Swinhoe	-	1	-	48	1
Symmacra solidaria Guenée	1	1	-	117	3
Scopulini					
Ignobilia urnaria Guenée	4	1	1	-	2
Zythos obliterata Warren	-	-	1	7	-
Zythos strigata Warren	4	9	2	10	2
Zythos turbata Walker	14	6	4	84	10
Antitrygodes divisaria Walker	6	4	4	6	4
Somatina plynusaria Walker*	2	-	-	-	-
Problepsis delphiaria Guenée	-	1	-	1	-
Problepsis apollinaria Guenée	-	-	-	1	-
Scopula planidisca Bastelberger	7	6	3	-	-
Scopula usticinctaria Walker	-	-	-	1	-
Scopula vacuata Guenée †	8	9	1	31	-
Scopula sp. nr. *albiflava* Warren (1079)	-	-	-	3	-
Scopula sp. 1 (909)	8	1	12	17	3
Scopula sp. 2 (1081)	50	28	19	35	13
Scopula sp. 3 (1082)	-	-	-	35	13
Scopula sp. 4 (1086)	-	-	-	1	-
Scopula sp. 5 (1083)	-	-	-	1	-
Sterrhini					
Idaea mundaria Walker	10	13	10	58	-
Idaea carnearia Warren	1	2	1	2	-
Idaea craspedota Prout	25	84	144	622	82
Idaea phaeocrossa Prout	16	7	10	9	84
Idaea assyria Holloway	1	1	3	7	-
Idaea squamipunctata Warren	114	202	213	266	99
Idaea egenaria Walker	-	-	1	24	-
Idaea violacea Hampson	1	-	1	3	-
Idaea sp. 9 (1001)	-	-	1	-	-
Lophophleps purpurea Hampson	-	1	-	238	-
Lophophleps triangularis Hampson	16	25	40	157	146
Timandrini					
Traminda aventiaria Guenée	1	1	-	-	-
LARENTIINAE					
Trichopterygini					
Sauris interruptata Moore	-	-	-	-	1
Eupitheciini					
Carbia calescens Walker†	-	-	-	3	-
Pomasia vernacularia Guenée complex	18	8	8	272	6
Ziridava kanshireiensis Prout	1	-	-	-	-
Chloroclystis semiscripta Warren	-	-	-	1	-
Chloroclystis sp. (1148)	-	-	-	1	-
Gymnoscelis polyodonta Swinhoe	-	-	-	1	-
Calluga sp. 3 (1151)	1	3	-	-	-
Micrulia catocalaria Moore	1	1	-	6	-
Onagrodes sp. (1153)	-	-	10	-	-

Appendix (continued 7)

	Low	Middle	High	Tho	Two-traps
Eois lunulosa Moore (1138)*	-	-	-	4	-
Eois memorata Walker	8	3	2	66	12
Eois phaneroscia Prout	-	-	-	10	-
Eois sp. nr. *mixosemia* Prout (1121)	-	-	-	46	-
Eois sp. 2 (1136)	4	3	-	2	5
Eois sp. 4 (1139)	-	-	-	10	-
Eois sp. 5 (1140)	-	-	-	16	-
Eois sp. 6 (1142)	-	-	-	7	-
Eois sp. 7 (1134)	1	-	-	-	-
Acolutha pictaria Moore	-	1	1	176	4
Pseudopolynesia amplificata Walker	-	3	-	-	22
UNIDENTIFIED					
sp. 1 (920) (Larentiinae)	2	1	-	-	-
sp. A2 (1324) (Ennominae)	-	1	-	-	-
sp. B1 (1498) (Ennominae)	1	6	-	-	-
sp. level 3 (967) (Sterrhinae)	-	-	2	-	-
sp. 6 (921) (Sterrhinae)	-	-	-	-	2

23 Seasonal Variation and Community Structure of Tropical Bees in a Lowland Tropical Forest of Peninsular Malaysia : the Impact of General Flowering

Naoya Osawa[1] & Yoshitaka Tsubaki[2]

Abstract: All of our experiments were conducted on a canopy tower in Peninsular Malaysia, using an artificial diet. The purpose of our experiment was to clarify the seasonal variation at Pasoh Forest Reserve (Pasoh FR) in and spatial community structure of tropical bees, and to show the impact of general flowering on the community structure of bees. We collected bees over two years at artificial feeders placed at different heights on a canopy tower. Our results showed that there was considerable seasonal variation in the bee community structure: dominant species changed seasonally. From an analysis of the spatial foraging patterns of the bees, we determined that each species might have its own specific searching height. When general flowering occurred, the community structure of bee pollinators changed drastically. Morishita's β was used as a diversity index for an analysis of data from samples collected before, during, and after general flowering. Before general flowering, the value of β was rather high. However, during general flowering, when more than 95% of the bees were the Asian honeybee *Apis cerana*, β decreased. Thereafter, β increased slightly, although 91.6% of individuals were *Trigona peninsularis*. The foraging strategies of tropical bees are discussed in relation to the vertical structure of a tropical rainforest and the impact of general flowering on the community structure of the bees.

Key words: bee, community, foraging strategy, niche differentiation.

1. INTRODUCTION

Lowland forests in the Malaysian Peninsula are characterized by mixed dipterocarp forest (Whitmore 1984). General flowering of most tree species in the Malaysian canopy, especially those in the family Dipterocarpaceae, occurs at intervals of 2 to 10 years (e.g. Ashton et al. 1988; Sakai et al. 1999b; Wood 1956; Yasuda et al. 1999). After general flowering, trees produce large quantities of fruit. One possible explanation for the adaptive significance of general flowering is that the mast fruit produced after flowering escape seed predation by generalists (e.g. Janzen 1970), although some observations contradict the Janzen ideal. There are two peaks in general flowering per year at both Pasoh (Yap & Chan 1990) and Rambir (Sakai et al. 1999b).

Malaysian tropical forests have a high diversity of tree species, implying that few individuals of the same species exist in the same local area. In fact, many dipterocarp species have out-breeding systems that involve animals (e.g. Chan

[1] Japan International Cooperation Agency (JICA).
Present address: Kyoto University, Kyoto 606-8502, Japan.
E-mail: osawa@kais.kyoto-u.ac.jp
[2] National Institute for Environmental Studies (NIES), Japan.

1981; Sakai et al. 1999a). Therefore, pollinators of tropical trees play an important role in the regeneration of the forest. Appanah & Chan (1981) proposed that among dipterocarp species, the flower morphology and flowering behavior of *Shorea* species appear to be highly adapted for pollination by thrips. However, Sakai et al. (1999a) showed that beetles (Chrysomelidae and Curculionidae, Coleoptera) contribute substantially to the pollination of *S. parvifolia* in Sarawak: the beetles accounted for 74% of the flower visitors collected by net-sweeping, and 30% of the beetles carried pollen, whereas thrips accounted for only 16% of the visitors, and 12% of the thrips carried pollen. These results imply that there is considerable geographical variation in pollinators among dipterocarp species, perhaps because dipterocarp flower morphology is less specialized to particular pollinators.

Vertical structures in plants are an important component of tropical rainforests (e.g. Ashton & Hall 1992; Richards 1996). The vertical stratification of a lowland tropical forest imposes a vertical gradient on the light available, which differentially attracts insects and other animals (Young 1982).

During general flowering, a large number of species blooms within a short period of time. This can result in a shortage of pollinators, unless the pollinators can respond quickly (Ashton et al. 1988). General flowering in a dipterocarp tropical rainforest attracts a multitude of pollinators, and results in mast fruiting. In a lowland dipterocarp forest in Sarawak, Momose et al. (1998) recognized 12 categories of pollination systems in 270 plant species; social bees pollinated the greatest number of species (32%), followed by beetles (20%). Thus, social bees (*Trigona, Apis* and *Braunsapis*) played an important role as pollinators. Furthermore, the researchers showed that flowers pollinated by social bees and other diverse insect species have no morphological mechanisms excluding specific flower visitors. Therefore, investigation of vertical and spatial variation in bee communities is required to refine our understanding of pollination systems in tropical rainforests.

Roubik (1992) suggested that many kinds of trees in Southeast Asia are pollinated by a small number of honeybees. Roubik et al. (1995) showed that travel between trees tended to be downward for *Apis dorsata* and upward for *A. koschevnikovi*, while no vertical directional preference was detected within trees.

The present study used an artificial diet to examine the seasonal variation of tropical bees in time and space in a lowland dipterocarp forest and to investigate the impact of general flowering on the community structure of pollinators.

2. MATERIALS AND METHODS

2.1 Experimental methods

All the research was conducted at the Pasoh Forest Reserve (Pasoh FR) in Negeri Sembiran State, Peninsular Malaysia (2°59' N and 102°18' E). The Pasoh FR is lowland dipterocarp forest, typical of the Malaysian Peninsula (Fig. 1). A canopy tower system on the study site was composed of three towers: two were 32 m tall and one was 52 m tall (Color plate 3). The three towers were connected by 20-m long walkways at a height of 30 m, forming a triangle shape. Two petri dishes were placed at 10-m intervals from 0 to 40 m on the 52-m tower; and a sponge saturated with 50 cc of honey (50%) was placed in each dish (Fig. 2). After setting the dishes, 50% honey was also sprayed near the petri dishes for 3 seconds to attract general pollinators. This artificial diet was chosen for all the experiments in order to investigate the vertical foraging patterns of pollinators, since the strong interaction between flowers and pollinators can conceal foraging patterns. One petri dish was

Fig. 1 General flowering at the Pasoh FR. Whitish parts on tree indicate flowers.

Fig. 2 *Apis cerana* at a honey trap.

capped 3 hours after placement and the other was capped 6 hours after placement. All the trapped bees were collected twice a month. The 3- and 6-hour sampling intervals were chosen so as to encompass the early succession of pollinators. Data collected from June 1995 to July 1996 were used for an analysis of the seasonality and foraging area of the bees, and data collected from December 1995 to November 1996 were used for an analysis of the biodiversity of bee pollinators.

2.2 Analysis of biodiversity of bee pollinators

Morishita's β (Morishita 1967) was used to analyze the biodiversity of bee pollinators. When we assign n_i as the total number of individuals in species i, and N as the total number of pollinators, β can be calculated from the following equation:

$$\beta = \frac{N(N-1)}{\sum_i n_i(n_i-1)}$$

In the forest reserve, general flowering occurred from April to July 1996. All the data from December 1995 to November 1996 (12 months) were divided into three four-month categories: before, during, and after general flowering. The combined data for each category were used for the analysis of the impact of general flowering on the community of pollinators.

3. RESULTS

3.1 Seasonal variation in tropical bees

Figure 3 shows the total number of bees collected by the honey traps. In total, ten species of bee (547 individuals), belonging to 2 genera, were trapped: 9 stingless bees in the genus *Trigona*, namely, *T. peninsularis, T. pagdeniformis, T. laeviceps, T. minangkabau, T. canifrons, T. fimbriata, T. geissleri, T. atripes* and *T. nitidiventris*, and one honeybee in the genus *Apis*, namely, *A. cerana*. Among the bees, *T. peninsularis*, *T. pagdeniformis and A. cerana* were the predominant species (Fig. 3).

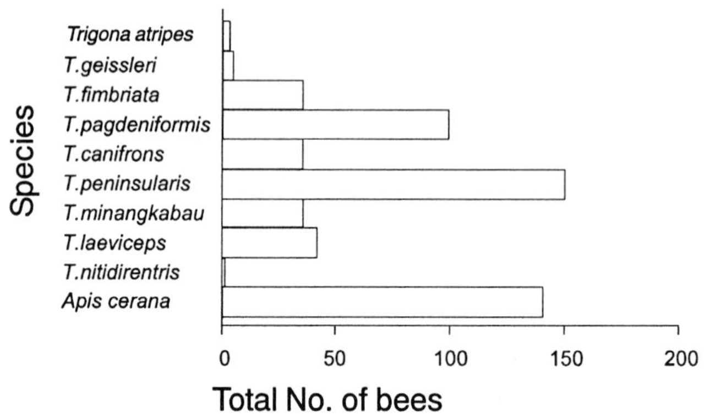

Fig. 3 Total number of bees collected by a honey trap from June 1995 to July 1996 in the Pasoh FR.

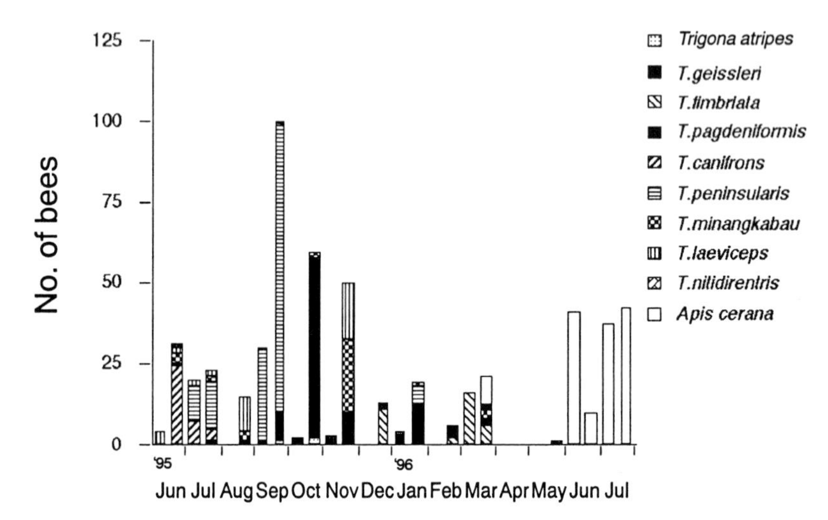

Fig. 4 Seasonal variations of bees from June 1995 to July 1996 in the Pasoh FR.

Figure 4 shows the seasonal variation in bees captured from June 1995 to July 1996. All of the bees collected before March 1996 belonged to the genus *Trigona*, with numbers peaking in September 1995 and decreasing through March 1996. No *Apis* bees were collected before April 1996. General flowering occurred from April to July 1996. Thereafter, the number of bees collected increased through July 1996, but all of the bees caught during this period were *A. cerana*.

The dominant species changed drastically with time during the course of sampling; *T. canifrons* was dominant from June to July 1995, *T. peninsularis* from July to September 1995 *T. pagdeniformis* from October to November 1995, *T. fimbriata* from February to March 1996, and *A. cerana* from June to July 1996 (Fig. 4).

Figure 5 shows the community structure of bees in relation to height. *A. cerana* and *T. laeviceps* were sampled at all heights, while *T. canifrons* was collected only above 30 m. *T. peninsularis* and *T. pagdeniformis* showed considerable variation in collection height, though the former peaked at 30 m and the latter peaked at 10 m. On the other hand, *T. minangkabau* was collected mainly between 0 to 20 m and peaked at 0 m. Thus, there was considerable interspecific variation in the bees' preferred height in early succession.

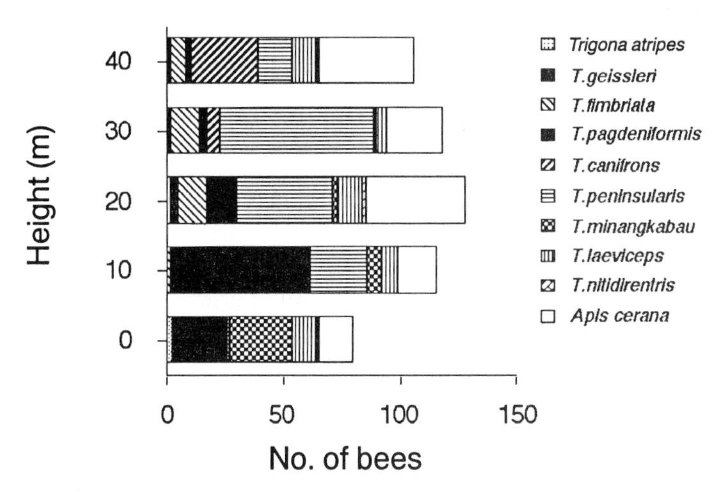

Fig. 5 The community structure of bees by height in the Pasoh FR.

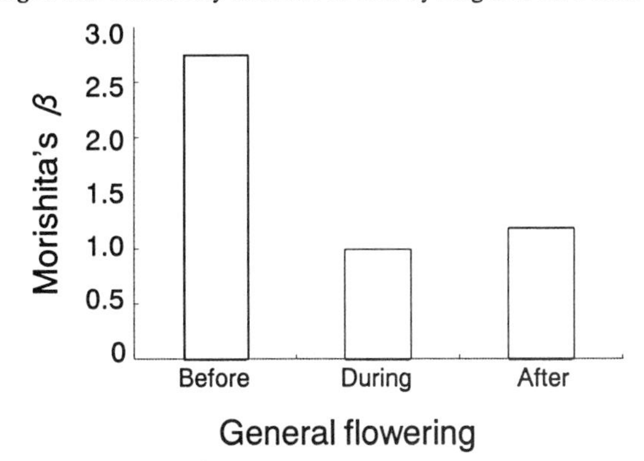

Fig. 6 Morishita's β of bee pollinators at the Pasoh FR before, during and after general flowering.

3.2 Impact of general flowering on the pollinator community

Figure 6 shows the seasonal change of Morishita's β at the Pasoh FR. The value of β before the general flowering was 2.75, but it decreased by 63.6% to 1.0 during general flowering. When the general flowering ended, Morishita's β recovered to 1.19. Before the general flowering, 10 species of bee were collected. However, during general flowering, only 2 species were trapped (96.9% *A. cerana*), and 4 months after general flowering, 6 species were trapped (91.6% *T. peninsularis*).

4. DISCUSSION

In an early study in Peninsular Malaysia, Khoo (1992) sampled bees for three continuous days, and found that the dominant bee species changed drastically with time. Similar results were obtained in this study. Dipterocarp trees started to flower during April 1996. Before general flowering, few species flowered at the same time, implying that there may have been little nectar available. This may have caused intense competition among bees and largely explain the observed seasonal variation in dominant bee species.

It was shown in Sarawak, Malaysia, that *Apis dorsata* and *A. koschevnikovi* have no vertical preference in searching area within trees (Roubik et al. 1995). The same result was observed in *A. cerana* in this study. Furthermore, this study revealed that during general flowering, 96.9% of the individuals trapped were *A. cerana*, which implies that *A. cerana* targets large resources (e.g. general flowering) when foraging. Interestingly, this honeybee was seldom collected until June 1996, and this pattern may also reflect a well-developed searching ability, targeting large resources, in this species. The giant honeybee (*A. dorsata*) may appear slightly before general flowering, since few were observed before February 1996. Detailed observation of *A. dorsata* is still required.

Stingless bees tend to be generalized in their flower-visiting habitats and overlap a great deal in plant species visited (e.g. Hubbell & Johnson 1977). Furthermore, interspecifically aggressive species are large and commonly forage in groups and that they tend to visit and monopolize clumped resources (Johnson & Hubbell 1974, 1975). These results imply that stingless bees are usually resource-limited (e.g. Hubbell & Johnson 1977). In fact, Nagamitsu & Inoue (1997) found in their study in Sarawak that interspecific partitoning of the feeders in stingless bees occurred in foraging time and height but not quality of the food, which may be one of the explanatory mechanisms in many species in stingless bees. The present study also showed the partitioning of the feeder in stingless bees in time and space; stingless bees tended to have their own foraging area in height depend on the species and dominant species in stingless bees drastically changed. We also observed that bees were aggressive toward the other species and they exclude the different species, after stingless bees of the same kind occupied the nectar. A series of these aggressive foraging behaviors in a stingless bee were often observed among large *Trigonas*, especially between *T. canifrons* and *T. peninsularis*. This study showed that a small *Trigona* species, *T. minangkabau* was mainly collected between 0 to 20 m and peaked at 0 m, suggesting that partitioning of the feeder in space is important in determining the foraging area in height in *T. minangkabau.* However, another small *Trigona*, *T. laeviceps* was collected mainly within three hours in setting a trap, but it was collected regardless of height. Therefore, an interaction in partitioning between time and space may be involving in determining the foraging behavior of stingless bees. On the other hand, Khoo (1992) suggested that the quality of the food resource has an influence on the foraging behavior in less aggressive species of stingless bees. Such less aggressive species forage on the resources with low concentration of sugar, which are not favored by larger stingless bees. Therefore, less aggressive species of stingless bees may utilize different food resource from aggressive stingless bees, which might be another mechanism in species coexistence in stingless bees.

This study showed that stingless bees have specific searching heights. Generally, the height of a mature tree is species-specific. The searching image in different stingless bee species may be influenced by interaction between the mature size of each tropical tree and an effective utilization of resources that minimizes competition. Nieh & Roubik (1998) showed in a prior study in Panama that recruitment signals of a stingless bee, *Melipona panamaica*, and can evidently communicate the three-dimensional location of a good food resource in terms of height, distance, and direction. This result implies that foraging behaviors of stingless bees may be involving in the three-dimensional structure of a forest, which is one of the most important characteristics in a lowland tropical rain forest (e.g. Richards 1996). Pollination generalization is predicted when flower rewards are similar across plant species, travel is costly, constraints of behavior and morphology are minor, and/or pollinator lifespan is long relative to the flowering of individual plant species (Waser et al. 1996). However, because specialization of flower morphology for specific pollinators is not common in dipterocarp trees (e.g. Appanah & Chan 1981), the coevolution of pollinators and dipterocarp trees in the tropics may have been different from that in temperate regions. This implies that the time and height differentiation among tropical bees when foraging may have been influenced by the coevolution of pollinators and plants in the Asian tropics, and may result in a few bee species pollinating a large number of plants.

The results of this study indicate that general flowering lowered the biodiversity of bee pollinators, mainly due to the large increase in the numbers of *A. cerana*. However, this does not necessarily imply that general flowering decreases the biodiversity of bee pollinators; rather, it suggests that *A. cerana* has comparatively greater searching ability, and is better able to track the extensive resources generated by general flowering, which may have resulted in the dominance of this species during general flowering. It also suggests that *A. cerana* plays an important role in pollination during general flowering in this lowland dipterocarp forest.

This study also revealed that *T. peninsularis* increased after general flowering. Maeto et al. (2001) showed that true hornets were trapped only in the several months during or just following the mass falling of fruit, and it has been suggested that mast fruiting temporarily increases the population of *Vespa* hornets, which depend on ripe fruits as carbohydrate resources (Maeto et al. 2001). It is known that *Trigona* often harvest sap from diverse sources (Roubik 1989), suggesting that the large amount of carbohydrates generated by mast fruiting may also have favored the increase in the numbers of large *Trigonas*, especially *T. peninsularis*.

ACKNOWLEDGEMENTS

We are much indebted Dr. S. G. Khoo for the identification of stingless bees and his kindly advice for this research. Thanks are also due to Drs. Laurence G. Kirton, Julie Intachat and N. Manokaran, Forest Research Institute Malaysia (FRIM), to conduct this study. We are also thanking Mr. Sharihzal bin Samudin for helping us to conduct the field researches.

REFERENCES

Appanah, S. & Chan, H. T. (1981) Thrips: the pollinators of some dipterocarps. Malay. For. 44: 234-252.

Ashton, P. S., Givnish, T. J. & Appanah, S. (1988) Staggered flowering in the Dipterocarpaceae: new insights into floral induction and evolution of mast fruiting in the aseasonal tropics. Am. Nat. 132: 44-66.

Ashton, P. S. & Hall, P. (1992) Comparisons of structure among mixed dipterocarp forests of north-western Borneo. J. Ecol. 80: 459-481.

Chan, H. T. (1981) Reproductive biology of some Malaysia dipterocarps III breeding systems. Malay. For. 44: 28-36.

Hubbell, S. P. & Johnson, L. K. (1977) Competition and nest spacing in a tropical stingless bee community. Ecology 58: 949-963.

Janzen, H. D. (1970) Herbivores and the number of tree tree species in tropical forests. Am. Nat. 104: 501-529.

Johnson, L. K. & Hubbell, S. P. (1974) Aggressive and competition among stingless bees: field studies. Ecology 55: 120-127.

Johnson, L. K. & Hubbell, S. P. (1975) Contrasting foraging strategies and coexistence of two bee species on a single resource. Ecology 56: 1398-1406.

Khoo, S. G. (1992) Foraging behaviour of stingless bees and honeybees. In Ho, Y. W., Vidyadaran, M. K., Addulla, N., Jainudeen, M. R. & Bahanam, A. R. (eds). Proceedings of the National Intensification of Research in Priority Areas Seminar (Agriculture Section), vol. 2, pp.461-462.

Maeto, K., Osawa, N. & Fukuyama, K. (2001) Impact of mast fruiting on abundance of vespid wasps (Hymenoptera) in a lowland tropical forest of Peninsular Malaysia. Entomol. Sci. 4: 247-250.

Momose, K., Yumoto, T., Nagamitsu, T., Kato, M., Nagamasu, H., Sakai, S., Harrison, R. D., Itioka, T., Hamid, A. A. & Inoue, T. (1998) Pollination biology in a lowland dipterocarp forest in Sarawak, Malaysia. I. Characteristics of the plant-pollinator community in a lowland dipterocarp forest. Am. J. Bot. 85: 1477-1501.

Morishita, M. (1967) The seasonal variation of butterflies in Kyoto. In Morishita, M. & Kira, T. (eds), Nature-Ecological Study Cyuokoronnsha, Tokyo, pp.95-132 (in Japanese).

Nagamitsu, T. & Inoue. T. (1997) Aggressive foraging of social bees as a mechanism of floral resource partitioning in an Asian tropical forest. Oecologia 110: 432-439.

Nieh, J. C. & Roubik, D. W. (1998) Potential mechanisms for the communication of height and distance by a stingless bee, *Melipona panamica*. Behav. Ecol. Sociobiol. 43: 387-399.

Richards, P. W. (1996) The tropical rain forest (2nd ed.). Cambridge University Press, Cambridge.

Roubik, D. W. (1989) Ecology and Natural History of Tropical Bees. Cambridge University Press, Cambridge.

Roubik, D. W. (1992) Loose niches in tropical communities: why are there so few bees and so many trees. In Hunter, M. D., Ohgushi, T. & Price, P. W. (eds). Effects of Resource Distribution on Animal-Plant interaction Academic Press, London, pp.327-354.

Roubik, D. W., Inoue, T. & Hamid, A. A. (1995) Canopy foraging by two tropical honeybees: bee height fidelity and tree genetic neighborhoods. Tropics 5: 81-93.

Sakai, S., Momose, K., Yumoto, T., Kato, M. & Inoue, T. (1999a) Beetle pollination of *Shorea parvifolia* (section *Mutica*, Dipterocarpaceae) in a general flowering period in Sarawak, Malaysia. Am. J. Bot. 86: 62-69.

Sakai, S., Momose, K., Yumoto, T., Nagamitsu, T., Nagamasu, H., Hamid, A., Nakashizuka, T. & Inoue, T. (1999b) Plant reproductive phenology over four years including an episode of general flowering in a lowland dipterocarp forest, Sarawak, Malaysia. Am. J. Bot. 86: 1414-1436.

Waser, N. M., Chittka, L., Price, M. V., Williams, N. M. & Ollerton, J. (1996) Generalization in pollination systems, and why it matters. Ecology 77: 1043-1060.

Whitmore, T. C. (1984) Tropical Rain Forests of the Far East. Clarendon Press, Oxford.

Wood, G. H. S. (1956) Dipterocarp flowering season in Borneo. Malay. For. 19: 193-201.

Yap, S. K. & Chan, H. T. (1990) Phenological behavior of some *Shorea* species in Peninsular

Malaysia. In Bawa, K. S. & Hadley, M. (eds). Reproductive Ecology of Tropical Forest Plants, UNESCO, Paris and The Parthenon Publishing Group, Lancs, pp.21-35.

Yasuda, M., Matsumoto, J., Osada, N., Ichikawa, S., Kachi, N., Tani, M., Okuda, T., Furukawa, A., Rahim Nik, A. & Manokaran, N. (1999) The mechanism of general flowering in Dipterocarpaceae in the Malay Peninsula. J. Trop. Ecol. 15: 437-449.

Young, A. M. (1982) Population biology of tropical insects. Plenum Press, New York and London.

24 Vertical Stratification of Ambrosia Beetle Assemblage in a Lowland Rain Forest at Pasoh, Peninsular Malaysia

Kaoru Maeto[1] & Kenji Fukuyama[2]

Abstract: We report the vertical profile of the species assemblage of ambrosia beetles in a primary forest stand at Pasoh Forest Reserve (Pasoh FR). Ambrosia beetles (xylomycetophagous species of the families Scolytidae and Platypodidae) live in wood, feeding on symbiotic fungi growing on the walls of their galleries. In order to clarify the vertical structure of the assemblage in a tropical rain forest, we used collision traps with ethanol as an attractant set on canopy towers in the forest. In total, 42 species of ambrosia beetles were captured in the ethanol traps placed serially from 1 m to 32 m above the ground. They showed a bi-modal pattern of species distribution in the median height of collection, suggesting that the species are largely divided into canopy species and understorey species. It seems likely that about three quarters of the species are canopy species, probably flying in the main canopy and somewhat shaded subcanopy layers, and most others are understorey species, foraging just above the forest floor. Host taxon selection was not differentiated between the canopy species and the understorey species. However, the canopy species were rather specific to branches of a certain size for each species while the understorey species were mostly generalists in host size selection. Ambrosia beetles are expected to promote early decomposition of wood materials through boring galleries into the wood and infecting them with wood-decaying fungi. Our results suggest that they are rich in species in the forest canopy, where they play an important role in early decomposition and rapid breakdown of standing branches, while some others are confined to the understorey or forest floor, attacking the fallen branches and bigger limbs of trees as well as seedlings.

Key words: alien insects, attractant trap, canopy insects, Coleoptera, decomposer, Platypodidae, Scolytidae, tropical rain forest.

1. INTRODUCTION

A tropical rain forest is a functional ecosystem with a large and multi-layered structure (Richards 1952; Whitmore 1990). Richards (1983) suggested that there are two principal layers of the forest, i.e. the euphotic zone in which the crowns are more or less fully exposed to sunlight, and the shaded oligophotic zone. It may be expected that herbivorous animals are abundant and diverse in the former "productive" zone, while decomposers of decaying plant materials are rich in the latter. However, the actual situation is more complicated.

Two major insect taxa of herbivores in tropical rain forests, Hemiptera and Lepidoptera, show greater abundance of individuals and greater diversity of taxa

[1] Shikoku Research Center, Forestry and Forest Products Research Institute (FFPRI).
Present address: Faculty of Agriculture, Kobe University, Kobe 657-8501, Japan.
E-mail: maeto@kobe-u.ac.jp

[2] Forestry and Forest Products Research Institute (FFPRI), Japan.

in the upper canopy (Rees 1983; Sutton 1983, 1989). Stratification is also distinct in the assemblage of ants, which utilize many types of food resources, in tropical rain forests (Brühl et al. 1998; Itino & Yamane 1995). Soil and litter ants are mostly scavengers and predators, but extrafloral nectar and arthropod exudates may be the primary resources of arboreal ants (e.g. Chap. 32). Maeto et al. (2001) also indicated that vertical distribution is species specific in social wasps, which depend on ripe fruits as well as arthropods as food resources.

According to Hammond et al. (1997), high proportions of fungivore and xylophagous beetles are tree-crown specialists while most herbivores are stratum generalists in a tropical forest in Sulawesi. This may suggest that the euphotic

Fig. 1 A canopy tower in Pasoh FR. Three towers were connected at a height of 30 m by walkways forming a triangle.

zone or upper oligophotic zone is the place where wood materials are decomposed by the beetles, but little information is available on the vertical distribution of wood-inhabiting beetles and how their assemblages are stratified in the forests.

The aim of this paper is to clarify the vertical structure of the ambrosia beetle assemblage in a lowland tropical rain forest in Peninsular Malaysia. Ambrosia beetles, xylomycetophagous species of Scolytidae and Platypodidae, live in wood, feeding on symbiotic fungi growing on the walls of their galleries (Barbosa & Wagner 1989; Beaver 1979; Browne 1961). It is well known that the beetles are predominant early-decomposers of recently dead trees, branches and seedlings. They are very abundant in tropical rain forests, and more than two hundred species have been recorded from Peninsular Malaysia (Beaver & Browne 1978; Browne 1961).

Cachan (1964) described the vertical profile of nine species of ambrosia beetles by collecting them with bait logs in an African tropical forest, but there were too few species to give an overall pattern of the vertical distribution. Since

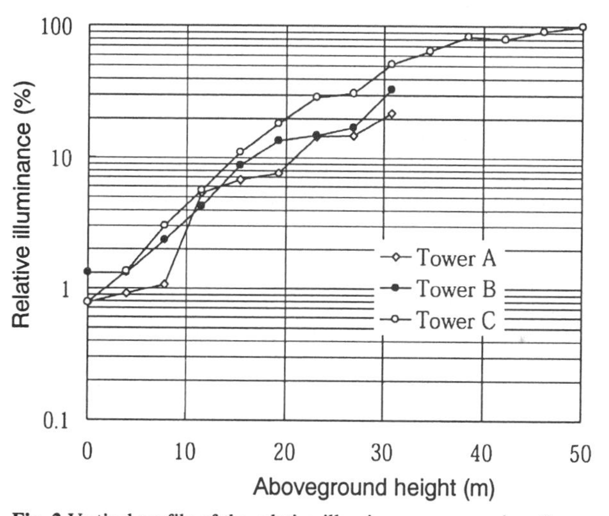

Fig. 2 Vertical profile of the relative illuminance measured on the canopy towers in the study site at mid-day on 2 August 1995.

Fig. 3 A collision trap set on a canopy tower, carrying an ethanol dispenser.

ambrosia beetles are often attracted to ethanol (Barbosa & Wagner 1989; Kinuura & Makihara 1992), we employed the collision trap with ethanol (Maeto et al. 1999) to investigate the vertical profile of ambrosia beetles in flying or foraging activity.

2. MATERIALS AND METHODS

Sampling of ambrosia beetles was conducted on three canopy towers (A and B 32 m tall and C 52 m tall) 20 m apart from each other in a primary stand of Pasoh Forest Reserve (Pasoh FR) in Peninsular Malaysia (Fig. 1) (Lee 1995; Color plate 3). Emergent trees around the tower site were up to about 45 m in height. The light profile measured on the canopy towers at mid-day on 2 August 1995 is shown in Fig. 2. Two digital illuminomaters (Minolta T-1) were used to get relative illuminance. We measured illuminance along the towers simultaneously with the measurement at the top of the tower C (about 50 m above the ground). The relative illuminance tended to decrease exponentially with decreasing height from about 40 m to 0 m above the ground. No distinct layers of the canopy, based on the vertical distribution of light intensity (Yoda 1974), were recognized.

The beetles were sampled with black collision traps (Fig. 3) described in detail by Maeto et al. (1999), with ethanol as an attractant. On each canopy tower, four traps were placed at a height of 1.0 m, 13.0 m, 25.0 m and 32.0 m above the ground in March–April 1993, and nine traps were placed at 0.8 m, 4.6 m, 8.4 m, 12.2 m, 16.0 m, 19.8 m, 23.6 m, 27.4 m and 31.4 m above the ground in November–December 1993 and in July–August 1995. In each case, trapping was carried out for three consecutive weeks. Trapped insects were collected every week, preserved in 70% ethanol and later mounted. Ambrosia beetles of the Scolytidae (Xyleborini and Xyloterini) and the Platypodidae were identified and counted.

Median height of collection for each species was calculated for every trapping period and for every canopy tower, and also by pooling all data together. To confirm the consistency of the median height of collection for species, the rank correlation (Kendall's τ) of the values of co-occurring species was tested between trapping periods and between towers.

Based on biological data given by Browne (1961), the number of host tree families recorded in Malaya, the proportion of the Dipterocarpaceae in the number of host species, and the approximate diameter of host poles or branches were examined in relation to the median height of collection for available species.

3. RESULTS

Forty-two species of ambrosia beetles, i.e. 37 species of the Xyleborini and Xyloterini of the Scolytidae, and five species of the Platypodidae, were collected (Table 1). Vertical profiles of abundance are shown for several dominant species in Fig. 4. *Xylosandrus crassiusculus* (Fig. 5) and *Xyleborus affinis* were collected just above the ground or in the understorey. *Xyleborinus exiguus* and *Indocryphalus* sp. were captured mostly in the middle layer, while *Xyleborus tunggali* and *Euplatypus parallelus* were abundant over 20 m high in the canopy.

The rank correlation of the median height of collection for co-occurred species was always significant both between the trapping periods and between the towers (Tables 2, 3), indicating that the collection height of each species was virtually consistent among the trapping periods and among the towers.

The vertical profile of the median heights of collection after pooling all data for each species shows a bi-modal pattern of species distribution in frequency (Fig. 6). The numbers of species with a median height of collection in the understorey

Table 1 Aboveground height of collection for the species of ambrosia beetles captured in tower traps with ethanol in Pasoh FR, Malaysia (three collections are pooled together).

Species	Aboveground height of collection (m)		
	Median	Range	*N*
Scolytidae - Xyleborini			
Arixyleborus granifer (Eichhoff)	0.8	0.8 — 1.0	3
Arixyleborus granulifer (Eggers)	0.8	0.8 — 1.0	5
Xyleborus affinis Eichhoff	0.8	0.8 — 4.6	16
Arixyleborus medius (Eggers)	1.0	0.8 — 13.0	13
Coptodryas perparvus (Sampson)	1.0	1.0 — 1.0	4
Leptoxyleborus concisus (Blandford)	1.0		1
Xylosandrus crassiusculus (Motschulsky)	1.0	0.8 — 19.8	45
Xyleborus persimilis Eggers	2.7	0.8 — 4.6	2
Cyclorhipidion pruinosum (Blandford)	4.6		1
Cyclorhipidion triangi (Schedl)	4.6		1
Cyclorhipidion subagnatum Wood	8.4		1
Xyleborinus artestriatus (Eichhoff)	8.4		1
Xyleborus adusticollis (Motschulsky)	8.4	0.8 — 31.2	7
Cyclorhipidion agnatum (Eggers)	14.1	8.4 — 19.8	2
Ambrosiodmus aff. *pseudocolossus* (Schedl)	14.5	13.0 — 16.0	2
Xyleborinus exiguus (Walker)	16.0	0.8 — 31.2	53
Xyleborus sp.(M44)	16.0		1
Xyleborinus andrewesi (Blandford)	19.8		1
Eccoptopterus gracilipes (Eichhoff)	23.6		1
Xyleborus sp.(M45)	23.6		1
Coptoborus bicuspis (Browne)	25.0	13.0 — 32.0	9
Coptoborus fragilis (Browne)	25.0	19.8 — 27.4	3
Xyleborus approximatus Schedl	25.0	25.0 — 32.0	3
Xyleborus tunggali Schedl	25.0	1.0 — 32.0	31
Arixyleborus mediosectus (Eggers)	26.2	13.0 — 32.0	10
Eccoptopterus limbus Sampson	27.4	23.6 — 31.2	2
Xyleborus cavulus Browne	27.4		1
Xyleborus sp.(M46)	27.4		1
Ambrosiodmus asperatus (Blandford)	31.2		1
Ambrosiodmus latisulcatus (Eggers)	31.2	25.0 — 31.2	4
Cyclorhipidion sisyrnophorum (Hagendom)	31.2		1
Webbia pabo Sampson	31.2	31.2 — 31.2	2
Xyleborus ciliatoformis Schedl	31.2	25.0 — 31.2	3
Xyleborus haberkorni Eggers	31.2		1
Xylosandrus ater (Eggers)	31.2		1
Xylosandrus mancus (Blandford)	31.2		1
Scolytidae - Xyloterini			
Indocryphalus sp.	25.0	13.0 — 32.0	62
Platypodidae			
Diapus quinquespinatus Chapuis	19.8		1
Euplatypus parallelus (Fabricius)	27.4	8.4 — 31.2	33
Genyocerus diaphanus (Schedl)	27.4	12.2 — 31.2	3
Crossotarsus squamulatus Chapuis	29.3	27.4 — 31.2	2
Platypus quercicola Schedl	31.2	31.2 — 31.2	2

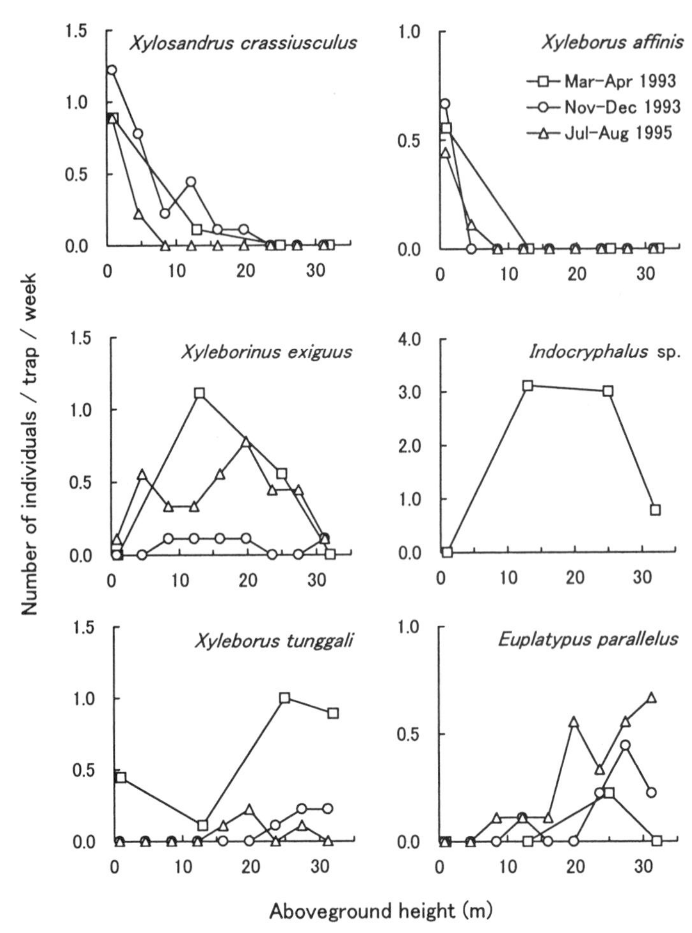

Fig. 4 Vertical abundance of some dominant species of the ambrosia beetles captured in tower traps with ethanol in Pasoh FR, Malaysia.

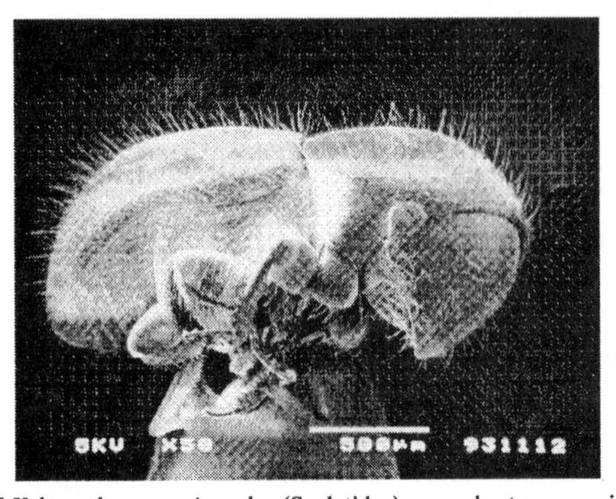

Fig. 5 *Xylosandrus crassiusculus* (Scolytidae), an understorey species of ambrosia beetle.

Table 2 Kendall's coefficient of rank correlation (τ) between towers in the median height of collection for each species (upper triangle) and the number of species treated (lower triangle).

	A	B	C
Tower A	-	0.707**	0.507*
Tower B	15	-	0.566**
Tower C	13	15	-

*P <0.05, **P <0.01

Table 3 Kendall's coefficient of rank correlation (τ) between sampling periods in the median height of collection for each species (upper triangle) and the number of species treated (lower triangle).

	(1)	(2)	(3)
Mar-Apr 1993 (1)	-	0.835**	0.658**
Nov-Dec 1993 (2)	12	-	0.693**
Jul-Aug 1995 (3)	13	12	-

**P <0.01

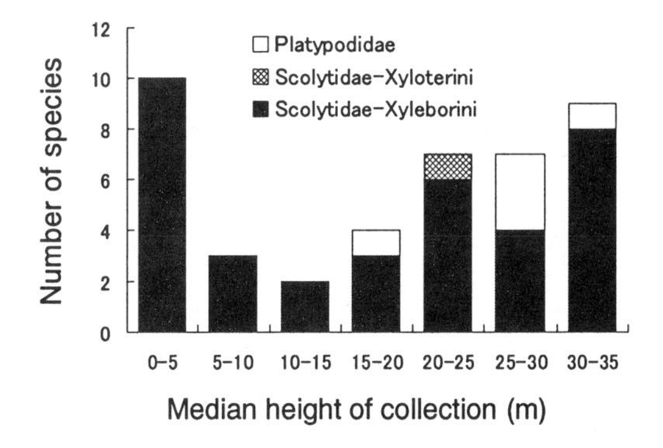

Fig. 6 Frequency distribution of the median height of collection for 42 species of the ambrosia beetles captured in tower traps with ethanol in Pasoh FR.

(0–5 m, 10 spp.), in the middle layer (5–20 m, 9 spp.) and in the canopy (20–35 m, 23 spp.) were significantly different from the expected numbers (6, 18 and 18 spp., respectively) (Chi-square test, $X^2 = 8.556$, $df = 2$, $P < 0.05$). About three quarters of the species were mainly collected over 10 m high above the ground, and most others were just below 5 m high, though a few species, e.g. *X. exiguus* and *Xyleborus adusticollis*, were captured throughout the canopy and the understorey (Table 1).

There was no correlation between the median height of collection and the number of host tree families known to be attacked ($\tau = 0.03$, $N = 28$, $P = 0.85$; Fig. 7), nor between the median height of collection and the dominance of the Dipterocarpaceae among the known hosts ($\tau = -0.12$, $N = 28$, $P = 0.37$; Fig. 8). Meanwhile, 5 out of 6 species with the median height of collection under 5 m high showed a wide range of host size distribution from less than 5 cm to over 20 cm in wood diameter, whereas ony 2 of 11 canopy species over 10 m high showed such a wide host range, 4 species attacked small wood less than 10 cm in diameter and 4 species attacked large wood around 10 cm or more in diameter (Fig. 9).

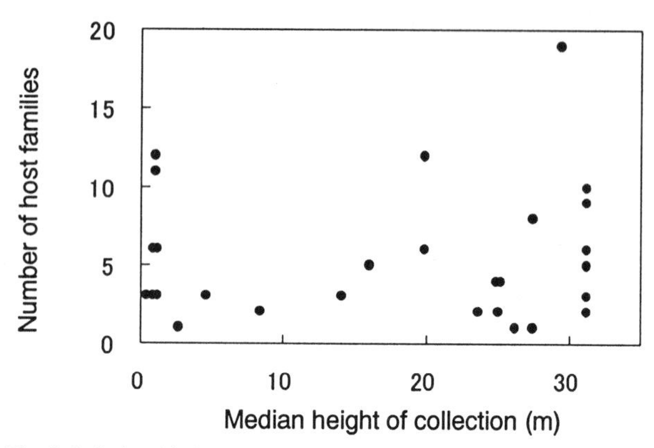

Fig. 7 Relationship between the number of host tree families known to be attacked in Malaya (Browne 1961) and the median height of collection for 26 species of the ambrosia beetles.

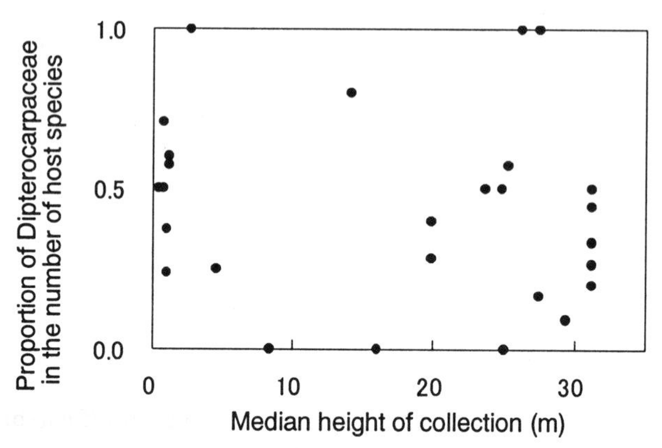

Fig. 8 Relationship between the proportion of the Dipterocarpaceae in the number of host species (Browne 1961) and the median height of collection for 26 species of the ambrosia beetles.

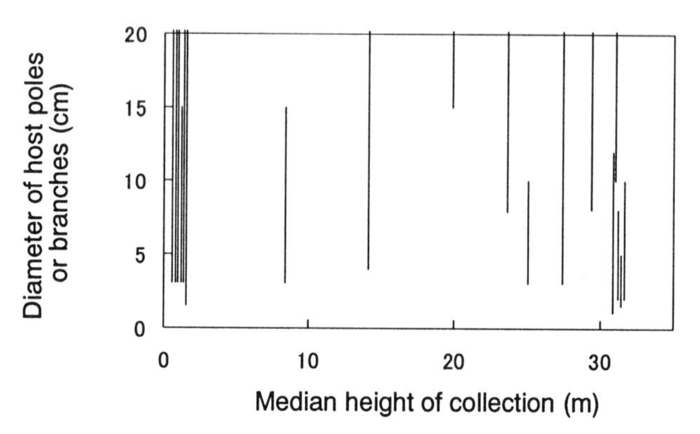

Fig. 9 Relationship between the host size range (Browne 1961) and the median height of collection for 18 species of the ambrosia beetles. Upper ranges are omitted over 20 cm in diameter.

4. DISCUSSION

4.1 Vertical stratification of ambrosia beetle assemblage

Every dominant species of ambrosia beetle collected in Pasoh FR has a specific profile of vertical distribution, as shown by Cachan (1964) for species collected in a humid tropical forest of Côte d'Ivoire, West Africa, and by Weber & McPherson (1983) for *Xylosandrus germanus* in North Carolina. Two of nine African species of the ambrosia beetles reported by Cachan (1964), *Xylosandrus crassiusculus* (as *Xyleborus semigranosus*) (Fig. 5) and *Xyleborus affinis* (as *Xyleborus mascarensis*), were collected in Pasoh FR. They were most abundant at the ground level both in Côte d'Ivoire and Pasoh FR, suggesting that the vertical profile is a consistent feature of the species.

The assemblage of ambrosia beetles showed a bi-modal distribution of species in the median height of collection, suggesting that the species are largely divided into canopy species and understorey species (Fig. 6). It seems likely that about three quarters of the species are canopy species, probably flying in the main canopy and somewhat shaded subcanopy layers, and most others are understorey species, foraging just above the forest floor. In contrast, no such vertical differentiation was shown in the assemblage of ambrosia beetles in temperate forests of northern Japan (Maeto et al. unpubl. data). Vertical stratification of the ambrosia beetles should be developed in tropical rain forests, which are thick in vertical structure and humid through the year. Our study supports the observation by Hammond et al. (1997) that xylophagous and fungivorous beetles are predominantly "canopy specialists" in tropical rain forests.

It might be expected that the canopy species are rather specific to certain tree families such as the Dipterocarpaceae, which are dominant in the upper canopy. However, host taxon selection was not differentiated between canopy species and understorey species (Figs. 7, 8), and this is not surprising since almost all ambrosia beetles are known to be host generalists (Beaver 1979; Browne 1961). At the same time, it seems likely that many canopy species are rather specific to branches of a certain size for each species while most understorey species are generalists in host size selection (Fig. 9). The assemblage of ambrosia beetles is possibly composed of a large group of host-size specialists in the canopy and a small group of host-size generalists in the understorey, though more evidence is necessary to confirm this inference.

As discussed below in detail, the understorey species should depend on fallen trees and branches as well as weakened saplings, and many canopy species probably use standing branches in the canopy. Scarcity of such resources in the middle layer might be a cause of the bi-modal vertical pattern of species richness. Various microclimatological and biological conditions affecting the establishment and growth of symbiotic fungi in wood must be also essential for the explanation of species diversity in ambrosia beetles, but current knowledge is too little to discuss them.

4.2 Ecological role and diversification of ambrosia beetles

Ambrosia beetles are very abundant in humid forests, attacking dying or recently dead trees, branches and saplings (Browne 1961). They are expected to promote early decomposition of wood materials through boring galleries into the wood and infecting them with wood-decaying fungi.

Many fallen trees, branches and bigger limbs of trees are present on the forest floor. The understorey species of ambrosia beetles probably depend on them, but they can use only those dying or recently dead, and such available resources are only a small fraction of the total dead wood on the forest floor. Thus the understorey species would tend to be generalists in host size selection.

Some understorey species also attack healthy or somewhat weakened seedlings and saplings. For example, *Xylosandrus crassiusculus* (as *Xyleborus semiopacus*) and *Xyleborus affinis* have been reported to attack young transplants of *Khaya* (Meliaceae) and *Aucoumea* (Annonaceae) in West Africa (Browne 1963). In temperate forests, an understorey species, *Xylosandrus germanus*, is known to attack apparently healthy plants as well as those dying or recently dead (Weber & McPherson 1983). Such species may affect survival of seedlings and saplings in tropical forests.

There is plenty of branch biomass in the canopy (Kato et al. 1978). Dying or recently dead branches retained in tree crowns are probably suitable resources for the canopy species of ambrosia beetles. For instance, *Ambrosiodmus asperatus* (as *Xyleborus asperatus*), which was collected at 31.2 m high, has been known to have a habit of entering small branches at the tops of the tallest trees (Speyer 1923). Many canopy species, such as *Xyleborus approximatus*, *X. haberkorni* and *Xylosandrus ater*, have been reported to enter twigs or small branches of trees (Browne 1961). Also, we collected *Ambrosiodmus* aff. *pseudocolossus* from standing branches of the Burseraceae in the canopy (Fukuyama et al. unpubl. data). The branches entered by the ambrosia beetles should more easily break off in the wind and fall to the ground (Speyer 1923). It is most likely that the ambrosia beetles play an important role in early decomposition and rapid breakdown of standing branches in the canopy.

However, it should be also noted that some canopy species, such as *Euplatypus parallelus* and *Crossotarsus squamulatus*, often attack newly cut logs in logging sites (e.g. Browne 1961; Ho 1993). The "canopy species" probably include those just flying along the surface of forest canopy to reach large gaps with fallen logs.

4.3 Invasion of exotic species into the primary forest

Although the trap site was located within primary forest of Pasoh FR, some collected species were not indigenous to the forest. One canopy species *Euplatypus parallelus* is of American origin, and seems to have been introduced to the Oriental region recently (Beaver 1999; Browne 1980). It is strongly polyphagous, and is a highly successful species particularly in disturbed forests (Beaver 1999; Ho 1993, as *Platypus parallelus*; Murphy & Meepol 1990, as *P. linearis*).

According to Maeto et al. (1999), one of the understorey species, *X. crassiusculus*, disperses in numbers into the primary forest from neighboring oil-palm plantations. This species has a very wide distribution extending from tropical Africa through the Oriental region to Japan, and to the Pacific islands, but it is scarcely found in deep rain forests (Beaver 1976). It prefers rather disturbed forests or cultivated areas. In Pasoh FR, the population in the primary stand may be maintained or is at least strongly increased by the invasion of individuals reproducing in neighboring oil-palm plantations or young secondary forests (Maeto et al. 1999).

Ambrosia beetles carry various fungi and bacteria, including symbiotic "ambrosia" fungi and also those probably pathogenic to plants (e.g. Beaver 1989).

It is anticipated that some exotic microorganisms will be introduced into the primary forest by the beetles, and then this may affect the original forest ecosystem. Lowland forest reserves are highly exposed to man-modified environments in tropics. Surroundings of the reserves should be properly managed to mitigate the invasion of such alien insects and their associated organisms.

ACKNOWLEDGEMENTS

We are most grateful to Dr. Laurence G. Kirton, Forest Research Institute Malaysia, Kuala Lumpur, for all his assistance during this study. Special thanks are due to Dr. R. A. Beaver, Chiangmai, for identifying the ambrosia beetles and for valuable suggestions on an earlier draft of this manuscript. We also thank two anonymous reviewers for valuable comments on the manuscript. This research was part of a NIES, FRIM and UPM Joint Research Project under the Global Environment Research Programme funded by the Japan Environment Agency (Grant No. E-1).

REFERENCES

Barbosa, P. & Wagner, M. R. (1989) Introduction to forest and shade tree insects. Academic Press, San Diego, 639pp.

Beaver, R. A. (1976) The biology of Samoan bark and ambrosia beetles (Coleoptera, Scolytidae and Platypodidae). Bull. Entomol. Res. 65: 531-548.

Beaver, R. A. (1979) Host specificity of temperate and tropical animals. Nature 281: 139-141.

Beaver, R. A. (1989) Insect-fungus relationships in the bark and ambrosia beetles. In Wilding, N., Collins, N. M., Hammond, P. M. & Webber, J. F. (eds). Insect-fungus interactions, Academic Press, London, pp.121-143.

Beaver, R. A. (1999) New records of ambrosia beetles from Thailand (Coleoptera: Platypodidae). Serangga 4: 29-34.

Beaver, R. A. & Browne, F. G (1978) The Scolytidae and Platypodidae (Coleoptera) of Penang, Malaysia. Oriental Insects 12: 575-624.

Browne, F. G (1961) The biology of Malayan Scolytidae and Platypodidae. Malayan Forest Records 22: 1-255.

Browne, F. G (1963) Notes on the habits and distribution of some Ghanaian bark beetles and ambrosia beetles (Coleoptera: Scolytidae and Platypodidae). Bull. Entomol. Res. 54: 229-266.

Browne, F. G (1980) Bark beetles and ambrosia beetles (Coleoptera, Scolytidae and Platypodidae) intercepted at Japanese ports, with descriptions of new species, IV. Kontyû 48: 490-500.

Brühl, C. A., Gunsalam, G & Linsenmair, K. E. (1998) Stratification of ants (Hymenoptera, Formicidae) in a primary rain forest in Sabah, Borneo. J. Trop. Ecol. 14: 285-297.

Cachan, P. (1964) Analyse statistique des pullulations de Scolytoidea mycétophages en forêt Sempervirente de Côte d'Ivoire. Macroclimat, microclimat, écologie et éthologie. Annales de la Faculté des Sciences, Université de Dakar 14: 5-65.

Hammond, P. M., Stork, N. E. & Brendell, M. J. D. (1997) Tree-crown beetles in context: a comparison of canopy and other ecotone assemblages in a lowland tropical forest in Sulawesi. In Stork, N. E., Adis, J. & Didham, R. K. (eds). Canopy arthropods, Chapman & Hall, London, pp.184-236.

Ho, Y. F. (1993) *Platypus parallelus*, a common ambrosia beetle of timbers. FRIM Technical Information 43: 1-4.

Itino, T. & Yamane, S. (1995) The vertical distribution of ants on canopy trees in a Bornean lowland rain forest. Tropics 4: 277-281.

Kato, R., Tadaki, Y. & Ogawa, H. (1978) Plant biomass and growth increment studies in Pasoh Forest. Malay. Nat. J. 30: 211-224.

Kinuura, H. & Makihara, H. (1992) Scolytid beetles (Coleoptera) captured by hanging traps with attracting chemicals (I) Comparison of traps with different attracting chemicals.

Transactions of Meeting of Tohoku Branch of the Japanese Forestry Society 44: 179-180 (in Japanese).
Lee, S. S. (1995) A guidebook to Pasoh, FRIM Technical Information Handbook No. 3. Forest Research Institute Malaysia, Kuala Lumpur, 73pp.
Maeto, K., Fukuyama, K. & Kirton, L. G. (1999) Edge effects on ambrosia beetle assemblages in a lowland rain forest, bordering oil palm plantations, in Peninsular Malaysia. J. Trop. For. Sci. 11: 537-547.
Maeto, K., Osawa, N.& Fukuyama, K. (2001) Impact of mast fruiting on abundance of vespid wasps (Hymenoptera) in a lowland tropical forest of Peninsular Malaysia. Entomol. Sci. 4: 247-250.
Murphy, D. H. & Meepol, W. (1990) Timber beetles of the Ranong mangrove forests. Mangrove Ecosystem Occasional Papers 7: 5-8.
Rees, C. J. C. (1983) Microclimate and the flying Hemiptera fauna of a primary lowland rain forest in Sulawesi. In Sutton, S. L., Whitmore, T. C. & Chadwick, A. C. (eds). Tropical rain forest: Ecology and management, Blackwell Scientific Publications, Oxford, pp.121-136.
Richards, P. W. (1952) The tropical rain forest. Cambridge University Press, Cambridge, 450pp.
Richards, P. W. (1983) The three-dimensional structure of tropical rain forest. In Sutton, S. L., Whitmore, T. C. & Chadwick, A. C. (eds). Tropical rain forest: Ecology and management, Blackwell Scientific Publications, Oxford, pp.3-10.
Speyer, E. R. (1923) Notes upon the habits of Ceylonese ambrosia-beetles. Bull. Entomol. Res. 14: 11-23.
Sutton, S. L. (1983) The spatial distribution of flying insects in tropical rain forests. In Sutton, S. L., Whitmore, T. C. & Chadwick, A. C. (eds). Tropical rain forest: Ecology and management, Blackwell Scientific Publications, Oxford, pp.77-91.
Sutton, S. L. (1989) The spatial distribution of flying insects. In Leith, H. & Werger, M. J. A. (eds). Tropical rain forest ecosystems, Elsevier Science Publishers, Amsterdam, pp.427-436.
Weber, B. C. & McPherson, J. E. (1983) Life history of the ambrosia beetle *Xylosandrus germanus* (Coleoptera: Scolytidae). Ann. Entomol. Soc. Am. 76: 455-462.
Whitmore, T. C. (1990) An introduction to tropical rain forests. Clarendon Press, Oxford, 226pp.
Yoda, K. (1974) Tree-dimensional distribution of light intensity in a tropical rain forest of West Malaysia. Jpn. J. Ecol. 24: 247-254.

25 The Communal Lifestyle of Web-Building Spiders in Tropical Forests

Toshiya Masumoto[1]

Abstract: Most web-building spiders build their webs solitary, and they usually prevent other individuals from invading their webs. However, some web-builders make their webs communally, and these are found mostly in tropical rain forests. The advantage of communal web building is thought to be improved prey capture. The communal capture of large insect prey has been observed in a number of social web-building spiders and communal spiders have been shown to capture larger prey than do solitary spiders. *Philoponella raffrayi* is a communal web-building uloborid spider that is distributed throughout southeastern Asia. This study describes the colony composition, and prey capture and handling behavior of the uloborid spider *P. raffrayi*. It determines whether the efficiency with which this species captures large insects is higher when spiders hunt cooperatively than when they hunt alone. Although most prey was captured by individual spiders, I occasionally observed two spiders cooperatively wrapping and capturing prey. Cooperation increased prey-capture efficiency when the prey was larger than half of a spider's body length. Furthermore, interspecific associations of *P. raffrayi* in the Pasoh Forest Reserve (Pasoh FR) is described.

Key words: communal web, cooperative prey capture, foraging behavior, *Philoponella raffrayi*.

1. INTRODUCTION

Web-building spiders are generally known to be solitary predators, but some spiders, called social or communal spiders, build their webs communally. To date, 30–40 species of such spiders have been found in the world (Buskirk 1981), mostly in tropical regions of Africa and South or Central America. In South East Asia, *Philoponella raffrayi* (Uloboridae) is reported to be communal.

Members of the genus *Philoponella* (Uloboridae) are found in South and Central America and in tropical Asia and the western Pacific (Lubin 1986; Opell 1979). Although they are known to construct communal webs, but employ only non-cooperative prey capture (see Burgess 1978; Buskirk 1981; Lubin 1986; Smith 1982). However, cooperative prey capture has been reported in a few species of *Philoponella* (Binford & Rypstra 1992; Breitwisch 1989). To understand differences in prey capture and social behavior within the genus *Philoponella*, these behavior of many species need to be investigated and described.

The advantages of communal prey capture have probably played an important role in the evolution of communal or social spiders (Buskirk 1981; Rypstra 1985, 1986; Shear 1970; Uetz 1986, 1989). Communal web building may improve prey capture in two ways: 1) by improving the ability of webs to intercept prey (the "ricochet effect") (Uetz 1989), and 2) by allowing for cooperative immobilization of prey, such that spiders can capture larger and presumably more profitable prey

[1] Graduate School of Agriculture and Life Sciences, University of Tokyo.
Present asddress: D-6, 6, Otani-cho, Otsu 520-0062, Japan. E-mail: tmasumo@pop21.odn.ne.jp

items. The communal capture of large insect prey has been observed in a number of social web-building spiders, such as *Agelena consociata* (Krafft 1969), *Anelosimus eximius* (Christenson 1984; Pasque & Krafft 1992; Vollrath & Rohde-Arndt 1983) and *Mallos gregalis* (Jackson 1979). It has been demonstrated that communal spiders capture larger prey than their solitary relatives (Buskirk 1981; Nentwig 1985; Uetz 1986).

I found colonies of *P. raffrayi* in Pasoh Forest Reserve (Pasoh FR) during 1993–94. This study describes the colony composition and prey capture and handling behavior of the uloborid spider *Philoponella raffrayi*. I focused on answering the question of whether spiders are more efficient in capturing large insects when hunting cooperatively than when hunting alone. I also examined the effect of colony size on prey capture. In addition, in 1991, I observed a *P. raffrayi* communal web located in the FRIM forest in Kepong (Masumoto 1992).

2. MATERIALS AND METHODS

2.1 The spider

Philoponella raffrayi is a communal web-building spider distributed in the undergrowth of tropical rain forests in Peninsular Malaysia (Masumoto 1992; Simon 1891). A *P. raffrayi* colony is composed of individual orb-webs connected to one another by non-adhesive silk. All uloborids lack poison glands and must rely on wrapping to subdue prey (Lubin 1986; Murphy & Murphy 2000).

The average body length of captured individuals was 6.21 mm for females and 3.15 mm for males (Fig. 1). A colony web that was 0.6 m × 0.6 m × 0.7 m when observed on March 9, 1991, grew to 2 m × 2 m × 3.5 m in size by March 23, 1991. This communal web consisted of numerous orb-webs (each about 10 cm in diameter. Fig. 2) surrounding the females of the colony, strong sustainable silks that formed the irregular framework of the colony, and a non-sticky irregular web that formed the center of the colony, where male congregated. The central irregular web was not an orb-web; rather, it was constructed from silk in the form of a radius or frame. Colony size varied according to the number of individuals in the colony. The age of

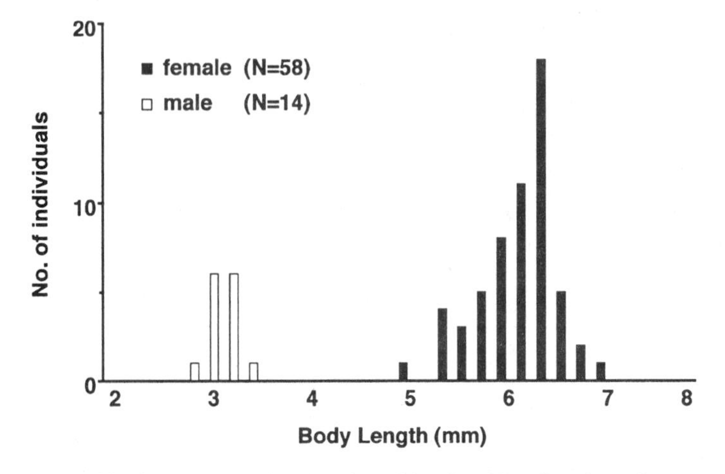

Fig. 1 Body length of adult males and females of *P. raffrayi* in a colony.

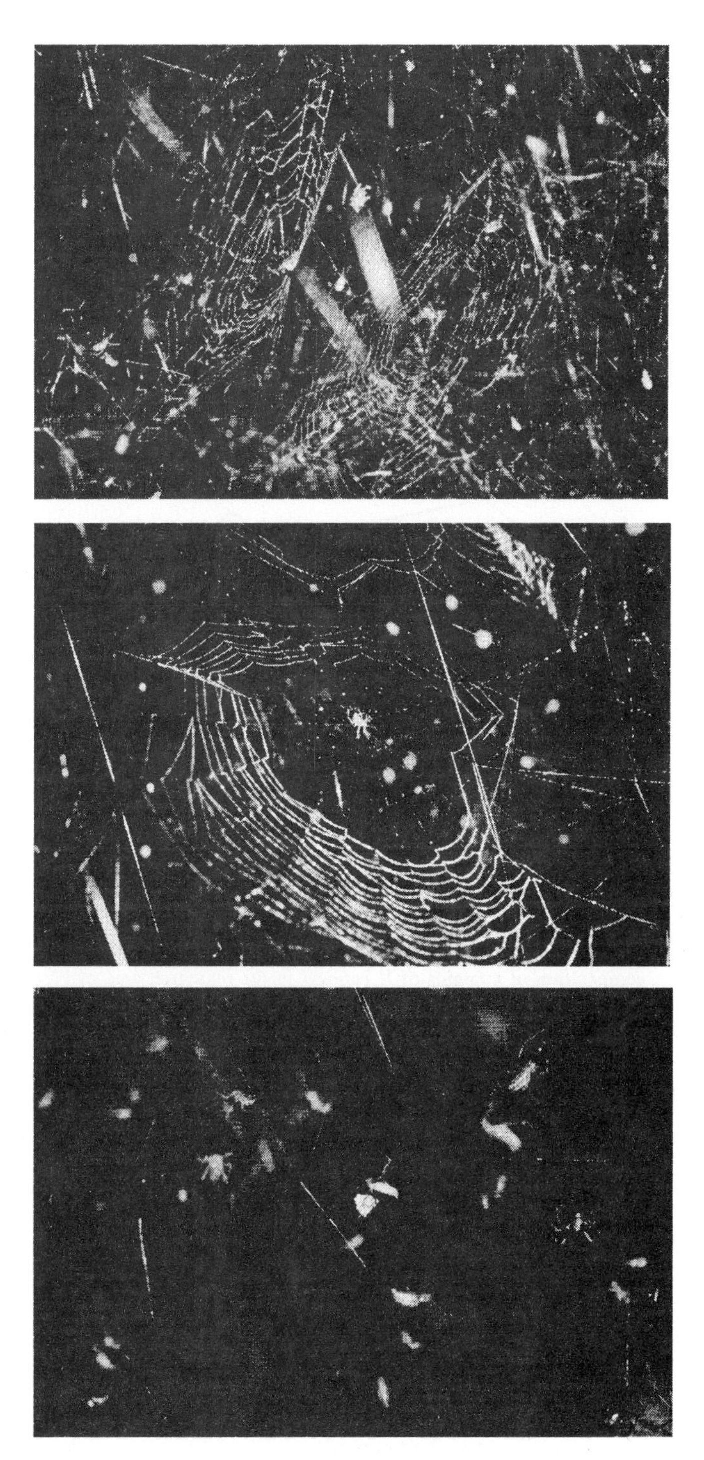

Fig. 2 Pictures of *P. raffrayi*. A part of communal web (upper), an individual orb-web (middle), and prey capture behavior by wrapping (bottom).

adult females was easily identified by their body color: they were orange for at least a week after their final molt, turning black a few weeks later (Masumoto 1992).

2.2 Field observations

I conducted field observations in a two-ha research area (Fig. 3) in the Pasoh FR from February to April 1992, and in March 1993. Twelve colonies found within this area were included in the study. To locate these colonies, I searched for colonies within the study area before observations.

All observations were made during daylight hours, 08:00–18:00. I recorded the number of spiders in the colonies, their stage of maturation, and their behavior, on February 25 and March 17, 1992. Furthermore, I re-visited the same area in April, and checked whether these colonies were present or absent.

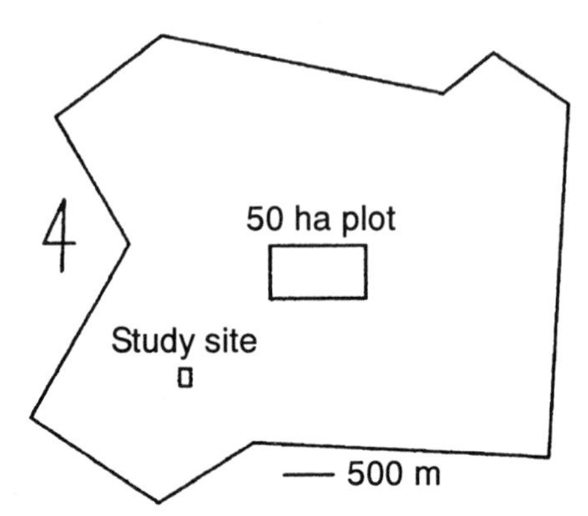

Fig. 3 A map of study site in the Pasoh FR.

Table 1 The composition of colonies of *P. raffrayi* in the reasearch area of Pasoh FR on 25 February and 17 March, 1992. No. 7 and 8 colonies disappeared and No. 9, 10, 11 and 12 colonies appeared on the study area between the dates. The number with * or with** indicates the number of *Argyrodes* spp. or *Portia* spp. respectively. #: of 24 females, 17 females had produced egg sacs on the date (Masumoto 1998).

	25 February, 1992				17 March, 1992			
Colony No.	No. of adult	No. of adult males	No. of juveniles	other species	No. of adult	No. of adult males	No. of juveniles	other species
1	0	0	28	0	20	1	0	0
2	24	1	0	1*	24#	1	0	4*
3	25	7	0	0	23	1	0	1*, 1**
4	0	0	35	0	0	0	25	0
5	44	2	0	0	44	0	0	0
6	6	0	0	0	4	0	0	0
7	15	0	0	0	---	---	---	---
8	15	5	0	0	---	---	---	---
9	---	---	---	---	28	2	0	1**
10	---	---	---	---	4	0	0	0
11	---	---	---	---	3	0	0	2*
12	---	---	---	---	14	1	0	0

On March 16, 1992, I recorded the dimensions of each web, the number of spiders per colony, and the behavior of females, per hour per colony, 9:00 to 17:00. I also examined temporal effects on prey capture.

Since the colony size of *P. raffrayii* was variable, as described below, I also examined the effect of the number of females per colony on the number of females that ate prey.

I collected females and their egg sacs from colony No. 2 (Table 1). After preserving them in 70% alcohol, I counted the number of eggs per sac, and measured female cephalothorax width under a binocular microscope to the nearest 0.1 mm.

In March 1993, I also conducted a total of 17 hours of field observations on the only colony (of 33 females and four males) still present in the study area. For this colony, I recorded stage of maturation per member, and the relative difference in body length between the spider and any insect prey that entered the web. Individuals were not marked, and relative body length between spider and prey was estimated by eye.

3. RESULTS

3.1 Colony composition

In the Pasoh FR, I observed eight colonies in February and ten colonies in March 1992 (Table 1). Between February 25 and March 17 (three weeks), six of the eight colonies remained at the same site, two disappeared from the study area and four new colonies appeared.

In March 1992, females of colony No. 2 (Table 2) produced twig-like egg sacs, hung them from the hub of the web, and began guarding the egg sacs. The mean number of eggs per sac was 118 ± 9.96 *SD* ($N = 15$). The number of eggs per sac was correlated with the cephalothorax width of the respective mother ($R = 0.523, N = 15, P = 0.046$; Spearman's rank correlation). However, eight of 13 females measured had a cephalothorax width of 1.5 mm but produced 101–132 eggs (Masumoto 1998). Oviposition occurred between March and April 1992, the juveniles remained in the same colony in which they had hatched. Each colony consisted of members of a similar developmental stage, exhibiting within-instar size variation. Female developmental stages were synchronous within colony, but non-synchronous among different colonies. Furthermore, the number of spiders per colony never increased, and no fusion of colonies was observed.

Table 2 Composition and relative body size (prey/female spider) of prey insects entering colony of *P. raffrayii*.

Relative size	less than 0.1		0.1 - 0.5		0.5 - 1.0		more than 1.0		total	
	entering web	captured	entering web	captured	entering web	captured	entering web	captured	entering web	captured
Hymenoptera	0	0	3	2	9	0	3	0	15	2
Diptera	32	31	33	30	7	5	3	0	75	66
Coleoptera	0	0	1	1	0	0	0	0	1	1
Lepidoptera	0	0	1	1	0	0	0	0	1	1

In April 1992, no colonies remained at their original sites, and three colonies, each consisting of more than 100 juveniles, appeared at different sites. All adult females had disappeared from these colonies, and I could detect no parent-offspring interaction except for egg sac guarding.

Moreover, *Portia*, a genus of jumping spiders (Salticidae), was found in the study area. *Portia* is known to be a spider killer (Jackson & Wilcox 1990) that enters webs and captures web owners, and I found *Portia* spp. individuals entering a colony of *Philoponella raffrayi* and eating females. The genus *Argyrodes* (Theridiidae) is known to be a kleptoparasitic: it does not weave its own webs, but steals prey from host webs, and some *Argyrodes* spiders also kill web owner spiders. *Argyrodes* species were also found in a *P. raffyrai* colony (Table 1).

3.2 Capture efficiency of communal foraging

During observational periods, I observed 92 insects of four orders entering webs (Table 2). Wrapping was conducted predominantly by individual females. However, when prey was trapped in the periphery of an individual orb web, two cooperating females wrapped seven out of 70 prey items. In such cases, they subdued the prey by throwing silk on it from a distance, and wrapped the prey more tightly as they cooperatively rotated it.

Capture efficiency is defined as the ratio of the number of insects captured and the number of insects entering the webs. The prey capture efficiency of single females was 89–97% when prey length was less than half of the spider body length, but efficiency decreased to mere 8% when the relative prey size was between 0.5 and 1.0 body length; and no insects longer than a spider were captured (Fig. 4). However, the prey capture efficiency of cooperative prey capture by two females was when the relative prey size was between 0.5 and 1.0 spider body length, higher than that achieved by a single female ($P = 0.0027$; Fisher's exact probability).

In all cases in which two females caught prey cooperatively, only one female ate the prey. In six cases, the larger female ate the prey, in one case, the smaller ate prey ($\chi^2 = 3.71$, $0.05 < P < 0.1$). Although there was a tendency for larger females to feed alone on capture prey, it was not clearly significant.

3.3 Effect of colony size on prey eating

The proportion of females eating prey in a colony gradually increased from morning to afternoon (Fig. 5), with 40–50% of females feeding on prey at 17:00 ($N = 81$, $Rs = 0.546$, $P < 0.001$, Spearman's rank correlation). However, the number of females in a colony did not affect the proportion of females eating prey ($N = 81$, $Rs = 0.14$, $P > 0.2$ Spearman's rank correlation; Fig. 6).

4. DISCUSSION

4.1 Advantages of communal prey capture

Communal uloborid spiders, such as *P. oweni*, have long been thought to lack any cooperative prey capture behavior (Buskirk 1981). However, cooperative prey capture has since been reported for a species of *Philoponella* in Cameroon (Breitwisch 1989), and for *P. republicana* in Peru (Binford & Rypstra 1992). Prey capture by *P. raffrayi* is similar to that by *P. republicana*, except that no more than two individuals at a time were observed cooperating. These results indicate that there may be several types of cooperative prey capture in the genus *Philoponella*; some species catch prey alone, some in pairs, and some in group of two or more

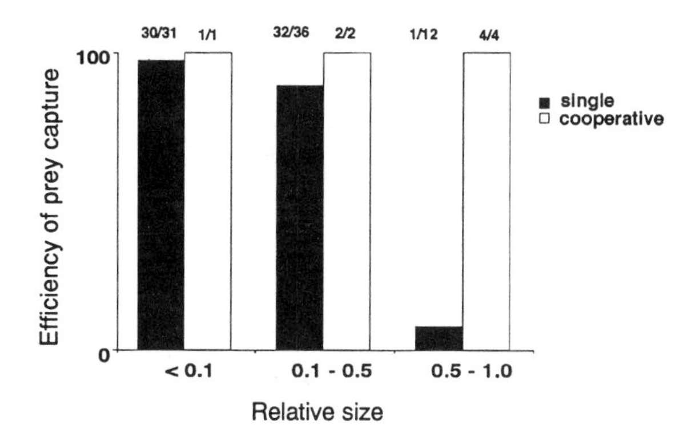

Fig. 4 Prey capture efficiency with the relative prey size.

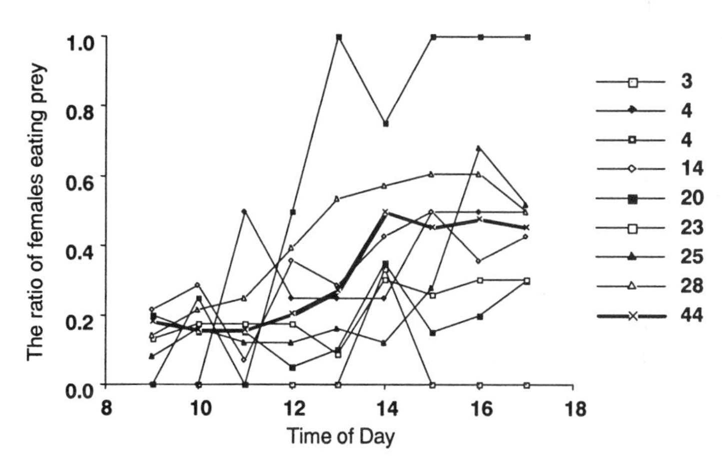

Fig. 5 The proportion of females eating prey in a colony with the time of day. The numbers besides bars indicate the number of females in a colony.

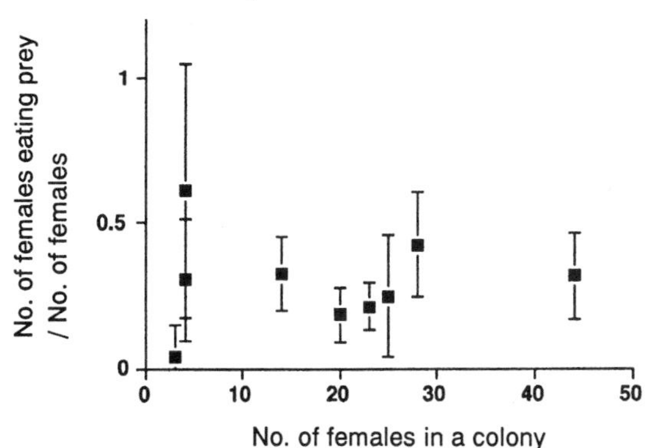

Fig. 6 The proportion of females eating prey with the number of females in a colony. The vertical bars represent $\pm SD$.

individuals.

In this study, cooperative prey capture by *P. raffrayi* is shown to be beneficial for capturing larger prey item. However, individual spiders primarily captured prey, and communal feeding was never observed, though cooperative prey capture sometimes occurred in *P. raffrayi*. These observations indicated that predation related interaction among colony members may be competitive, rather than cooperative.

4.2 Composition of a colony

In social spiders, such as *Agelena consociata* (Roeloffs & Riechert 1988) and *Achaearanea wau* (Lubin & Crozier 1985), genetic variability among same-colony members was reportedly low, and they were highly related. In this study, there are no data on the genetic variability in *P. raffrayi* spiders. Analysis of within-colony genetic variability is needed to describe the kin relationships within colonies in *Philoponella*.

Furthermore, the sex ratio of *P. raffrayi* was biased towards females, similar to many other social spiders (Aviles 1986; Frank 1987).This female biased-sex ratio is indicative of inbreeding, because females in highly inbred systems produce only enough males to mate with existing females (Hamilton 1967). However, if males develop faster and emigrate soon after maturation, the sex ratio would also be biased towards females. Further researches on the life history of *P. raffrayi* are required.

There remain several important unanswered questions in regard to the communal lifestyle of *P. raffrayi*.

1)How are colonies founded?

To resolve relationships among colony members, the process through which a colony is founded must be clarified. The foundation process could affect the genetic relatedness of individuals or female sex bias in a colony, and thus could be important to understand the evolution of communal life living of *P. raffrayi*.

2)What are the costs and benefits of communal living?

In some communal spiders, colony size (the number of spiders in a colony) is positively associated with capture efficiency. However, I could not detect such an association in *P. raffrayi*. Another possible advantage to communal behavior is explained by the "risk-sensitive foraging theory" (Caraco 1981). It predicts that energy gain in tropical rain forest is greater than the energy necessary for individual spiders, variation of prey capture success should be higher in smaller colonies than in larger ones. Future studies could use this theory to test whether communal foraging is advantageous for *P. raffrayi*. On the other hand, Cangialosi (1990) reported that kleptoparasite load was serious in some communal web spiders. However, kleptoparasitism likely had little effects on *P. raffrayi* here, because the number of *Argyrodes* spiders in the study area was small.

Much more research is needed to explain the yet unknown aspects of communal living in the spider, *P. raffrayi*.

ACKNOWLEDGEMENT

I am indebted to Dr. J. Intachat for generous permission to use the laboratory in Forest Research Institute Malaysia (FRIM). I thank Drs. Y. Ono, Y. Tsubaki and A. Furukawa for their encouragement. This study was partly supported by a Grant from Global Environmental Research Program, Environmental Agency, Government of Japan.

REFERENCES

Aviles, L. (1986) Sex-ratio bias and possible group selection in the social spider *Anelosimus eximius*. Am. Nat. 128: 1-12.

Binford, G J. & Rypstra, A. L. (1992) Foraging behavior of the communal spider, *Philoponella republicana* (Araneae: Uloboridae). J. Insect Behav. 5: 321-335.

Breitwisch, R. (1989) Prey capture by a West African social spider (Uloboridae: *Philoponella* sp.). Biotropica 21: 359-363.

Burgess, J. W. (1978) Social behavior in group-living spider species. Symp. Zool. Soc. Lond. 42: 69-78.

Buskirk, R. E. (1981) Sociality in the Arachnida, In Hermann, H. R. (eds). Social Insects. vol. II Academic Press, London New York, pp.281-367.

Cangialosi, K. R. (1990). Social spider defense against kleptoparasitism. Behav. Ecol. Sociobiol. 27: 49-54.

Caraco, T. (1981) Risk-sensitivity and foraging groups. Ecology 62: 527-531.

Christenson, T. E. (1984) Behavior of colonial and solitary spiders of the theridiid species *Anelosimus eximius*. Anim. Behav. 32: 725-734.

Frank, S. A. (1987) Demography and sex ratio in social spiders. Evolution 41: 1267-1281.

Hamilton, W. D. (1967) Extraordinary sex ratios. Science 156: 477-488.

Jackson, R. R. (1979) Predatory behavior of the social spider *Mallos gregalis*: Is it cooperative? Insects Sociaux 26: 300-312.

Jackson, R. R. & Wilcox, R. S. (1990) Aggressive mimicry, prey-specific predatory behavior and predatory behavior and predator-recognition in the predator-prey interactions of *Portia fimbriata* and *Euryattus sp*., jumping spiders from Queensland. Behav. Ecol. Sociobiol. 26: 111-119.

Krafft, B. (1969) Various aspects of the biology of *Agelena consociata* Denis when bred in the laboratory. Am. Zool. 9: 201-210.

Lubin, Y. D. (1986) Web building and prey capture in Uloboridae. In Shear, W. A. (ed). Spiders: webs, behavior, and evolution. Stanford Univ. Press, California, pp.132-170.

Lubin, Y. D. & Crozier, R. H. (1985) Electorophoretic evidence for population differentiation in a social spider *Achaearanea wau* (Theridiidae). Insects Sociaux 32: 297-304.

Masumoto, T. (1992) The composition of a colony of *Philoponella raffrayi* (Uloboridae) in Peninsular Malaysia. Acta Arachnologia 41: 1-4.

Masumoto, T. (1998) Cooperative prey capture in the communal web spider, *Philoponella raffrayi* (Araneae: Uloboridae). J. Arachnologica 26: 392-396.

Murphy, F. & Murphy, J. (2000) An Introduction to the Spiders of South East Asia, With Notes on All the Genera. Malaysian Nature Society, Kuala Lumpur, Malaysia, 625pp.

Nentwig, W. (1985) Social spiders catch larger prey: a study of *Anelosimus eximius* (Araneae: Theridiidae). Behav. Ecol. Sociobiol. 17: 79-85.

Opell, B. D. (1979) Revision of the Genera and Tropical American Species of the Spider Family Uloboridae. Bulletin of Museum Comparative Zoology Harvard University 148: 445-549.

Pasquet, A. & Krafft, B. (1992) Cooperation and prey capture efficiency in a social spider *Anelosimus eximius* (Araneae, Theridiidae). Ethology 90: 121-133

Roeloffs, R. & Riechert, S. E. (1988) Dispersal and population-genetic structure of the cooperative spider, *Agelena consociata*, in west african rainforest. Evolution 42: 173-183.

Rypstra, A. L. (1985) Aggregation of *Nephila clavipes* (L.) (Araneae: Araneidae) in relation to prey availability. J. Arachnol. 13: 71-78.

Rypstra, A. L. (1986) High prey abundance and a reduction in cannibalism: the first step to sociality in spiders (Arachnida). J. Arachnol. 14: 193-200.

Shear, W. A. (1970) The evolution of social phenomena in spiders. Bull. Br. Arachnol. Soc. 1: 65-76

Simon, E. (1891) Observations bilogiques sur les Arachnides. I. Araignees sociables, In Voyage de M.E. Simon au Venezuela (decembre 1881–avril 1888). 11e Memoire Annual Society et France 60: 5-14.

Smith, D. R. (1982) Reproductive success of solitary and communal *Philoponella oweni* (Araneae:

Uloboridae). Behav. Ecol. Sociobiol. 11: 149-154.
Uetz, G. W. (1986) Web-building and prey capture in communal orb weavers, In Shear, W. A. (ed). Spiders: webs, behavior, and evolution. Stanford University Press, California, pp.207-231.
Uetz, G. W. (1989) The "ricochet effect" and prey capture in colonial spiders. Oecologia 81: 154-159.
Vollrath, F. & Rohde-Arndt, D. (1983) Prey capture and feeding in the social spider *Anelosimus eximius*. Z. für Tierpsychol. 61: 334-340

26 The Ant Species Richness and Diversity of a Primary Lowland Rain Forest, the Pasoh Forest Reserve, West-Malaysia

Annette K. F. Malsch[1], Krzysztof Rosciszewski[2] & Ulrich Maschwitz[2]

Abstract: We present results of an inventory of the ant fauna from all strata at Pasoh Forest Reserve (Pasoh FR), Malaysia, with special emphasis on the diversity of leaf-litter inhabiting ants. A variety of sampling methods were used including the "Winkler-Moczarski eclector", the "Berlese-Tullgren method", "pitfall traps" and "hand collection". We collected a total of 489 ant species belonging to 76 genera and 9 subfamilies. The Myrmicinae was the most species rich subfamily with 40.7% of the species collected, followed by Formicinae with 28.2%, Ponerinae 18.0%, Dolichoderinae 5.9%, Cerapachyinae 2.9%; Aenictinae 2.5%, Pseudomyrmecinae 1.4%, Dorylinae 0.2% and Leptanillinae 0.2%. The most species-rich genera were *Camponotus* with 50 species, *Polyrhachis* with 45 species and *Pheidole* with 38 species. Of 475 ant species for which capture sites were recorded approximately 71% of the species seemed to be restricted to one stratum. Nesting sites were recorded for 199 species: 49.7% nested exclusively on the ground, 47.2% in vegetation and only 3% could be found in both. The species accumulation curves showed no tendency toward saturation (e.g. for Berlese and pitfall collections). In regard to all methods, 36% of the species were single captures and 54% single or double captures. In a detailed study on the diversity of ground leaf-litter inhabiting ants we extracted 120 species belonging to 49 genera and 5 subfamilies by using the Winkler-Moczarski eclector. The resulting species accumulation curves of three sample area sizes show no tendency toward saturation, but give a different prognosis of species richness expected at a given sampling site. The ß-diversity values calculated by the Sørensen quotient were 52.7 ± 7.68% comparing nine 25 m^2 plots compared to nine 9 m^2 plots 43.0 ± 8.18%. Our study gave evidence that the sample area size may greatly influence the results on the degree of diversity. Nevertheless, the ground leaf-litter ß-diversity in Pasoh FR is high which is supported by the collection of many rare species and only a small number of species with high abundance. In conclusion the unsaturated species accumulation curves, the large number of rare species and the distinct vertical stratification suggest that there are many more species to be found at Pasoh FR. There are many microhabitats we did not reach. Species foraging only by night are underestimated and the same is probably true for arboreal species. In conclusion our data revealed that the Pasoh FR has to be regarded as one of the most species rich places known in the world, as far as ants are concerned.

Key words: ants, diversity, species richness, vertical stratification, Winkler method.

[1] Johann Wolfgang Goethe-University.
Present address: Faculty of Public Health, University of Bielefeld, 33501 Bielefeld, Germany.
Email: annette.malsch@uni-bielefeld.de
[2] Johann Wolfgang Goethe-University, Germany

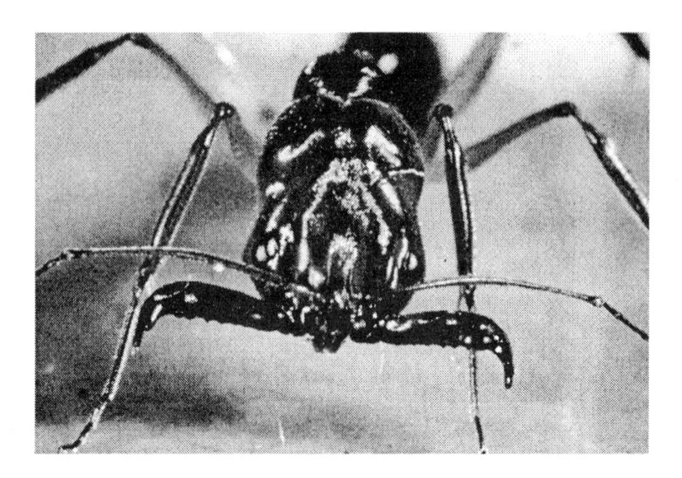

Fig. 1 Frontal view of the head of *Odontomachus* sp. (photograph courtesy of A. Weissflog). The mandibles are linked in 180° position.

1. INTRODUCTION

Tropical habitats are well known for their high species richness and extraordinary diversity. Therefore the evaluation of species richness in tropical regions is of main interest for the assessment of global biodiversity. The estimation of total species number of insects varies from 10 million (Gaston 1991) to 30 million species (Erwin 1982). Although Coleoptera and Lepidoptera are supposed to be the most species-rich insect orders, Hymenoptera is nearly as species-rich, and contains one of the most ecologically important insect groups in tropical rainforests: ants (Formicidae).

Ants are predominant throughout all strata in regard to their biomass, their abundance and their ecological impact on tropical rainforests (Fittkau & Klinge 1973; Stork 1988). In an extensive inventory of arboreal arthropod communities in Sabah, Malaysia, ants contributed more than 50% of all individuals (Floren et al. 1998). Chiba (1978) investigated the soil macrofauna of Pasoh Forest Reserve (Pasoh FR) and found that 45–91% of total individuals were ants.

This omnipresence of ants is one of the first things one notices while walking through a tropical lowland forest. One species is likely to immediately noticed, *Camponotus gigas* (subfamily Formicinae) the "giant forest ant". The workers can reach a lenght of 25–30 mm and forage solitarily on the forest floor by day. They are colored black except their abdomen (gaster) which is reddish brown. Another obvious ant species on the forest floor is the "long-jaw ant" *Odontomachus* spp. (subfamily Ponerinae, Fig. 1). It is brownish colored and its body size is between 8–12 mm long. The most striking, elongate "trap-jaw mandibles" are pulled back by very strong muscles and linked in a 180° position. From this position the mandibles snap shut convulsively and impaling their prey or enemy. This snap is such powerful that one can hear a "clicking" sound.

Also impressive to observe are the epigaeic members of the army ant genus *Aenictus* spp. (subfamily Aenictinae). The brownish-black species can easily be recognized because they are raiding and emigrating in very long trails on the surface of the ground or lower vegetation. Army ants do not build nests like most

Fig. 2 Worker of *Dorylus (Dichthadia) laevigatus* menacing with its sharp mandibles (photograph courtesy of A. Weissflog).

other ants. They dwell in so called "bivouacs", temporary camps in partly sheltered places. When the workers gather to establish the bivouac, they link their legs and bodies together with hooked claws at the tips of their feet. In contrast the south east Asian army ant *Dorylus laevigatus* (Dorylinae) nests and forages completely hypogaeic. Only in very rare incidents when fouraging near to the surface, they can be observed on the ground (Fig. 2).

However, only a few published studies have conducted an inventory of tropical ants across all strata, in order to evaluate total species richness (Brühl et al. 1998; Verhaag 1991). Such extensive surveys of one taxon also allow one to compare taxonomic patterns among regions. For comparability, standardization of methods is needed (Agosti et al. 2000). With this study we surveyed the ant fauna of Pasoh FR throughout all strata by using both standardized and non-standardized methods with an emphasis on the diversity of leaf-litter inhabiting ants.

2. MATERIALS AND METHODS

The main study area was in the centre of the Pasoh FR (Fig. 3). A total of 9.5 months of field work was carried out between 1990 and 1992.

Strata characterization (Fig. 4):

Stratum A: Ground stratum: it comprised the topsoil to 15 cm depth and the substrates lying on the topsoil including the humus layer, leaf-litter and dead wood of all sizes (trunks, branches, twigs, bark, roots and fruits).

Stratum B: Lower vegetation: it comprised all vegetation from the forest floor up to approximately 3 m height including herbs, shrubs, ferns, palms, gingers and small trees.

Stratum C: Arboreal stratum: it comprised all vegetation above 3 m including trees, lianas and epiphytes.

For transition areas in between two strata, the ants were assigned after observation and subjective evaluation. In cases where the species moved freely between two or three strata, it was assigned to each of them.

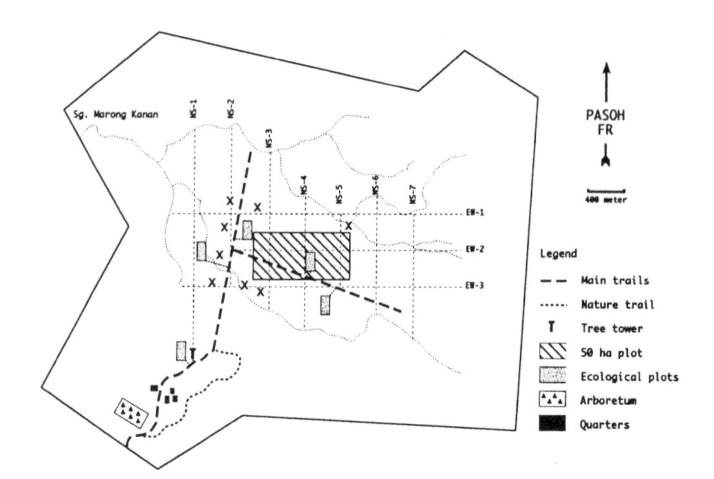

Fig. 3 Main study area in the centre of the Pasoh FR. Inset: × indicates study sites. Map of Pasoh FR modified from Lee (1995).

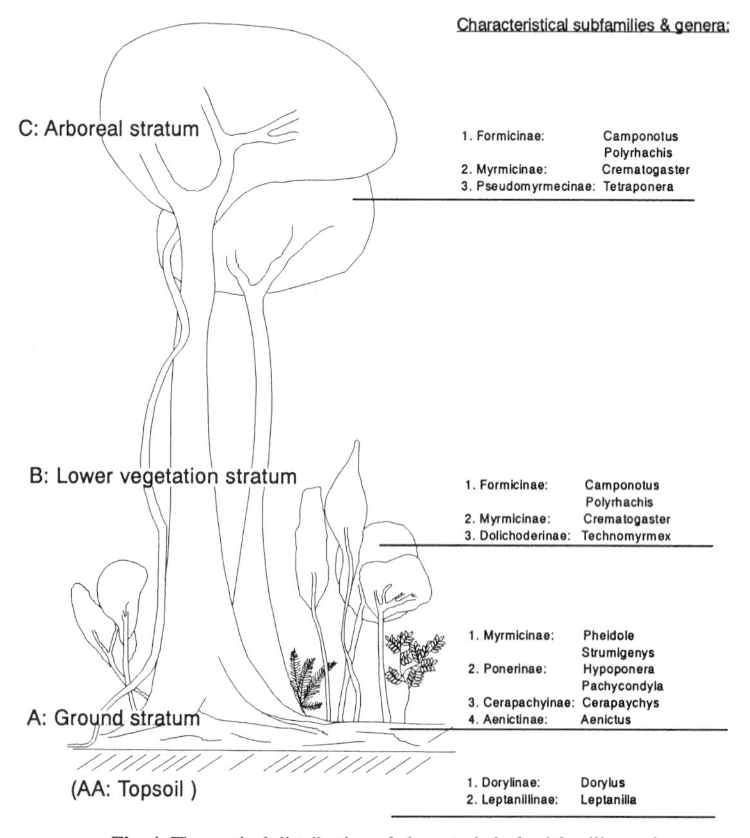

Fig. 4 The vertical distribution of characteristical subfamilies and genera of each stratum (A, B and C) respectively. For better overview the hypogeic stratum (AA) is separated from the ground layer (A).

2.1 Inventory

StratumA:

(1) Pitfall traps: The pitfall traps were filled with 4% formalin and covered with leaves.

a) "100-pitfall": 100 pitfall traps (diam. 1.4 cm, depth 14 cm) were placed on a 10 m × 10 m plot ("plot 2") in the centre of each m² (for 4 weeks)

b) "50-pitfall": 50 pitfall traps (diam. 1.4 cm, depth 14 cm) were placed every meter along a 50 m transect (for 4 weeks)

c) "15-pitfall": 5 pitfall traps (diam. 6.6 cm, depth 10 cm) were placed along three widely separated 5 m transects in equal distances (for 2 weeks).

(2) Berlese-Tullgren method:

a) "40-Berlese": 40 samples of 1/16 m² were collected from plot 2 randomly

b) "10-Berlese": 10 × 1 m² of leaf-litter were taken (see down "collection by hand", stratum A)

(3) Winkler-Moczarski eclector: Five samples of several m² of leaf-litter were taken, each sufficient to fill one standardized Winkler-Moczarski eclector. The places were widely separated and subjectively chosen assuming a high species richness.

The eclector consisted of a white cotton cloth, expanded over two metal frames (33 cm × 27 cm) which were 42 cm apart from each other (Fig. 5). At the top it could be closed over strings for suspension. At the bottom the cotton was forming a tube, in which a plastic beaker, containing a wet leather ribbon, was placed to provide hiding places. Inside the eclector four flat rectangular (22 cm × 39 cm) bags, made from robust nylon mesh of 2 mm width, was hung. These bags were filled with an sifted extract of leaf-litter, humus and peaces of dead wood. The eclectors were installed in shade for two days. For accelerating the desiccation process the mesh bags were emptied and the content mixed and refilled once a day. The cups were checked after 24 hours and then again after additional 24 hours. The live ants could be easily sorted out with forceps (for a more detailed description see Besuchet et al. 1987; Olson 1991).

(4) Collection by hand: The ants were collected with forceps and aspirator. Nest site, colony size and behaviour were recorded (see Appendix).

a) Collection of foraging ants in leaf-litter, in or under dead wood, and on soil surface by day and by night.

b) Sampling of twig nests (diam. ≧ 0.5 cm) in the whole research area.

c) On a 10m x 10 m plot ("plot 1") twigs (diam. ≧ 0.5 cm), leaf-litter and top soil was examined for nest sites.

d) Collecting ant specimens and ant nests on 10 single 1 m² plots. The leaf-litter was extracted with the Berlese method. The soil was dug to a depth of c. 15 cm.

e) Baiting on two widely separated transects with tuna meat, honey and honey

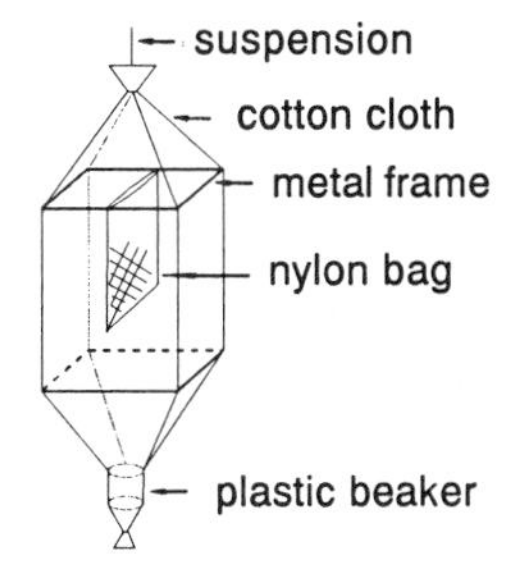

Fig. 5 Winkler-Moszarski-Eclector for the extraction of ants from leaf-litter samples.

water. On each transect 10 bait points were set 5 m apart from each other.

Stratum B:

Collection by hand:

a) Collecting foraging ants in the lower vegetation and at tree stems by day and by night.
b) Sampling under the bark of trees, of silk- and carton nests as well as ants in gangways and under leaves.
c) Examining dead wood of the plants themselves or of fallen dead wood.

Stratum C:

Collection by hand:

a) Investigating fallen parts of the higher arboreal vegetation.
b) Using the climbing rope technique for investigating six trees up to canopy level. A total of 11 arboreal samplings was conducted and a mean height of 25 m could be reached.
c) Sampling of ant specimens from an aluminum framework with a max. height of 40 m by day and by night.

2.2 Diversity study

For the investigation of the diversity of ants in the leaf-litter layer (stratum A) we also used the Winkler-Moczarski eclector. The same procedure as for the inventory was used (see above). Plots selected for sampling met the following criteria: a) more than 80% of ground covered with leaf-litter; b) no permanent wetness of the soil; c) more than 60% overshadowing on the soil surface; d) only moderate growth of shrubs and small trees; e) absence of large trees and logs or similar objects.

A total of 9 plots were investigated. Each plot comprised a 5 m × 5 m area with an additional 3 m × 3 m area nested in the middle of the plot. Each of the two partial areas (9 m^2 and 16 m^2) was examined separately. Therefore three area sizes per plot resulted: 9 m^2, 16 m^2 and 25 m^2. In addition two decomposing logs and their surroundings were sampled. Because of the small sample number the results were only used for completing the faunal survey.

For calculation ß-diversity, the ϕ index ($Cs = 2 \times c / (a + b)$) and the Jaccard index ($C_J = 2 \times c/(a + b - c)$) were used (Magurran 1988): a = species richness of sample A, b = species richness of sample B and c = number of species found in both samples.

2.3 Taxonomic identification:

The genera were identified with the key of Hölldobler & Wilson (1990) and Bolton (1995). Out of 489 morphospecies a total of 152 species (31%) could be determined to species level. The identification was based on workers only. The voucher specimens were deposited in the reference collection of the "Naturhistorisches Museum Karlsruhe", Germany.

3. RESULTS

3.1 Species composition:

We collected a total of 489 ant species belonging to 76 genera and 9 subfamilies (Table 1). In the course of the inventory 468 species representing 75 genera of 9 subfamilies were found. The diversity investigation resulted in 120 species belonging to 49 genera and 5 subfamilies.

The most species-rich genera were *Camponotus* and *Polyrhachis* both belonging to the subfamily Formicinae. Nevertheless, the subfamily Myrmicinae was the most species and genera rich subfamily in this study representing 40% of all species and 40% of all genera. Other important subfamilies included the Formicinae, they were second species-rich (29%), and the Ponerinae which contained the second highest genera number (28%). In addition, one new myrmicine genus represented by one species was discovered. It has been described as *Rostromyrmex pasohensis* by Rosciszewski (1994; Fig. 6).

Ant species composition of stratum A and strata B + C samples differed markedly (Fig. 7). Seven subfamilies were found in the two habitats but preferentially in one (G-test: $G = 121.857, P < 0.001$). Only a few species of the strongly terricolous subfamilies Aenictinae, Cerapachyinae and Ponerinae were found foraging or nesting in stratum B or C. Of the predominantly arboreal Pseudomyrmecinae only one species was also found in the lower vegetation and incidentally on the ground

Table 1 Taxonomic structure of the ant fauna at the Pasoh FR.

Subfamilies	Number of species	% of total species	Number of genera	% of total genera
Myrmicinae	199	40.7	30	39.5
Formicinae	138	28.2	14	18.4
Ponerinae	88	18	22	28.9
Dolichoderinae	29	5.9	5	6.6
Cerapachyinae	14	2.9	1	1.3
Aenictinae	12	2.5	1	1.3
Pseudomyrmecinae	7	1.4	1	1.3
Dorylinae	1	0.2	1	1.3
Leptanillinae	1	0.2	1	1.3
Σ	489	100	77	100

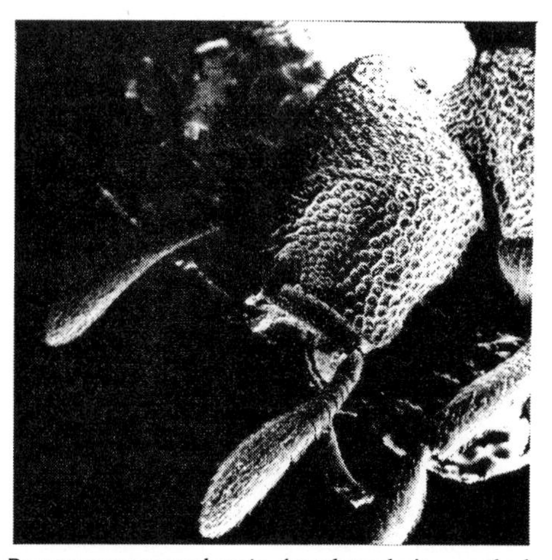

Fig. 6 *Rostromyrmex pasohensis*, dorsolateral view on the head of a worker. One main character is the forward projecting striking prominence ("rostrum") formed by the median portion of clypeus.

(see Appendix). Two subfamilies were restricted to the ground layer: Dorylinae and Leptanillinae. Formicinae and Myrmicinae are well represented across all strata.

Of 102 species found in the high arboreal layer 24 spp. belonged to *Camponotus*, 22 spp. to *Polyrhachis* and 14 spp. to *Crematogaster*. Of 176 species found in stratum B 33 spp. belonged to *Polyrhachis*, 32 spp. to *Camponotus*, 18 spp. to *Crematogaster* and 12 spp. to *Technomyrmex*. Of 300 species found in stratum A 34 spp. belonged to *Pheidole*, 18 spp. to *Strumigenys* and 11 to 14 spp. refer to each of *Aenictus, Camponotus, Cerapachys, Crematogaster, Hypoponera, Leptogenys, Oligomyrmex, Pachycondyla, Paratrechina* and *Tetramorium.*

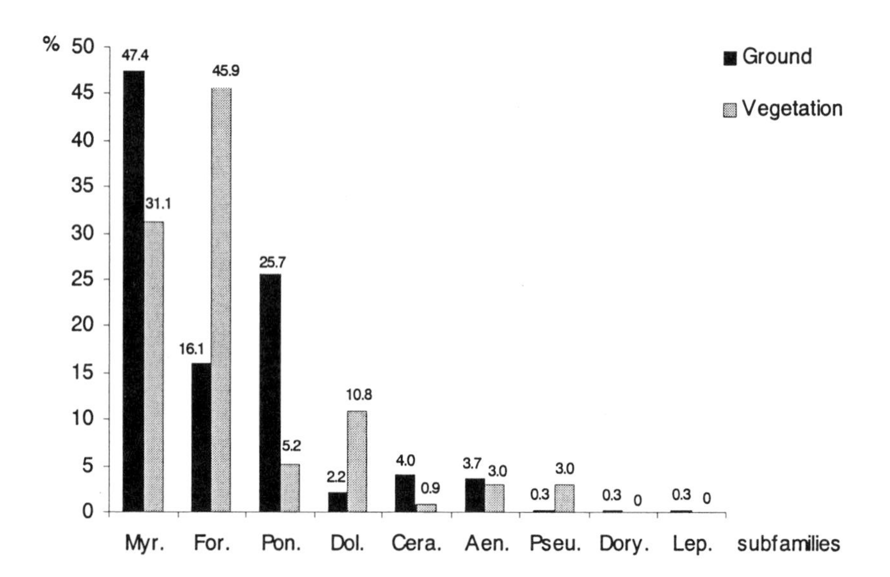

Fig. 7 Taxonomic structure of the subfamily distribution of the vegetation and ground stratum respectively. Myr.= Myrmicinae, For. = Formicinoce, Pon. = Ponentrae, Dol. = Doliehoderinae, Cera. = Cerapachyinae, Aen. = Aenictinae, Pseu. = Pseudomyrmecinae, Dory. = Dorylinae, Lept. = Leptanillinae.

Table 2 List of methods used and achieved species number.

Method	Sample size	Species number
100-pitfall	100 pitfall traps (∅ 1.4 cm) on plot 2 (100 m²)	39
50-pitfall	50 pitfall traps (∅ 1.4 cm) along a 50 m transect	39
15-pitfall	15 pitfall traps (∅ 6.6 cm) 5 traps at 3 sites	34
Inventory-Winkler	5 samples of several m²	103
40-Berlese	40 samples of 1/16 m² on plot 2 (100 m²)	28
10-Berlese	10 x 1 m²	52
Hand collection: Stratum A	10 x 1 m² (incl. Berlese)	75
Hand collection: Stratum A	plot 1 (100 m²)	64
Hand collection: Stratum B	random collection of specimens and nests	175
Hand collection: Stratum C	random collection of specimens and nests	102

3.2 Species richness

For the inventory species were sampled by using various methods. Most of the hand collections were done by subjective search, with no particular sampling structure. Table 2 gives examples of the conducted methods and the achieved species number.

Beside qualitative hand collection the most effective method was the "Inventory -Winkler" method. With 5 fillings of a common Winkler-Moczarski eclector 103 species were extracted. The mean similarity value for the 5 samples calculated by Sørensen was 37.35 ± 5.21%.

The species accumulation curves of the standardized collection methods

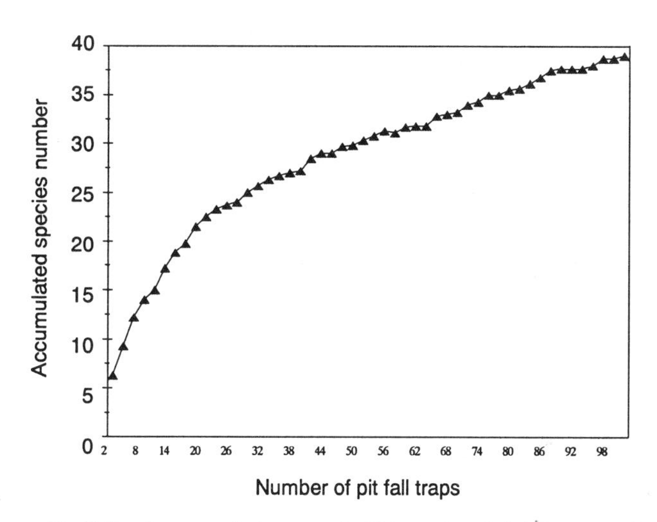

Fig. 8 Species accumulation curve for 100 samples of pit fall traps on plot 2.

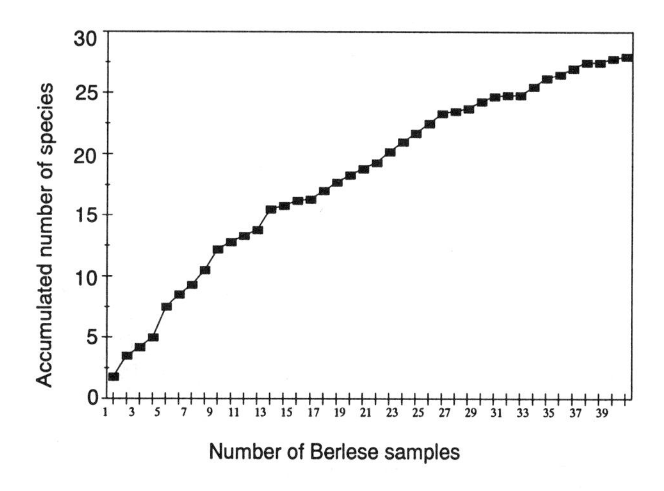

Fig. 9 Species accumulation curve for 40 Berlese samples on plot 2.

showed no tendency toward saturation. The curves of the "100-pitfall" (Fig. 8) and the "40-Berlese" samples are shown as an example (Fig. 9). Both collections were sampled in one area (plot 2) and a total of 55 species were gained. The similarity quotient (ϕ) for these two samples was lower (35.8%) than for collections at two separated sites collected with one method: the "100-pitfall" and the "50-pitfall" samples shared 56% of their species and the "40-Berlese" and the "10-Berlese" shared 45% of the species.

The majority of species were found only rarely: 36% were collected only once, 54% only once or twice. The diversity study similary resulted c. 60% single captures (see below, Fig. 10).

3.3 Vertical stratification

Three hundred species were found in stratum A, 176 species in stratum B, 102 species in stratum C and 230 species in the vegetation in total (accumulated species number of strata B + C).

Of 475 ant species for which capture sites could be safely recorded (Table I, Appendix), 51.4% were found only in stratum A, 32.2% only in the vegetation (strata B + C), and 16.4% were collected in both (Table I, Appendix). Altogether c. 71% of the species were restricted to one stratum. Nesting sites were also recorded for 199 species: 49.7% nested exclusively in the ground stratum, 47.2% in the vegetation and only 3% could be found in both.

47.5 % of the species in the canopy were only found in this layer and 45.8% of the hypogeic species were found nowhere else.

Calculating the similarity of the ant communities of the vertical layers by Sørensen for strata B and C 35.3% (Jaccard Index (C_J): 42.8%), for strata A and B 25.2% (C_J: 28.6%), for strata A and C 9.0% (C_J: 9.4%) and for strata A and (B + C) 29.4% (C_J: 34.5%) results.

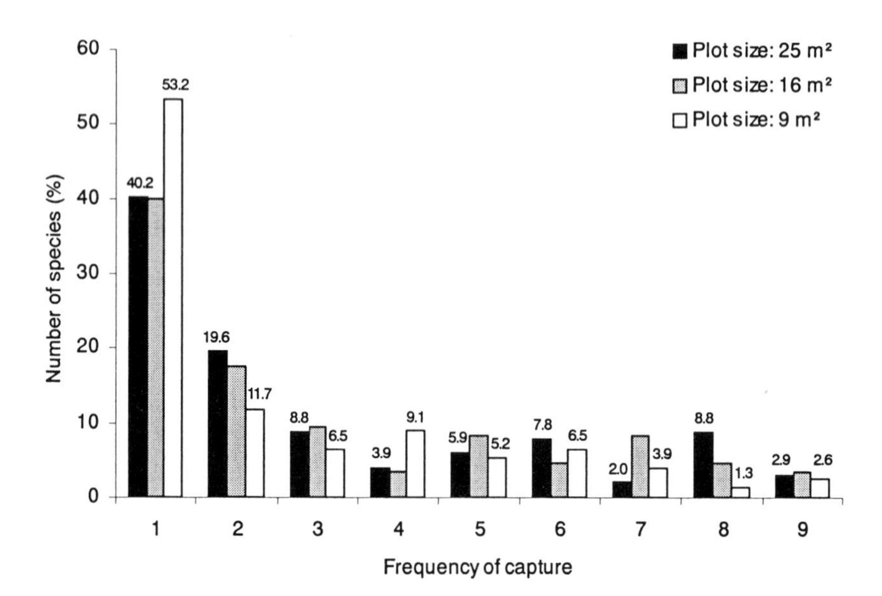

Fig. 10 Frequency of capture for species on nine 9 m², 16 m² and 25 m² respectively

3.4 Nest sites

On the ground a total of 314 twigs and branches (diam. $\geqq$ 0.5 cm) were collected, measured and opened. A total of 58 nests were detected belonging to 28 species. Ant nests could be found in 17.3% of the 255 twigs of plot 1 and in 23.7% of the 59 twigs collected in the investigation area at haphazard. In both collections in medium sized dead wood (2 cm–10 cm in diameter) ant nests were found significantly more often than in small sized wood (< 2 cm in diameter) (*G*-test, plot 1: $G = 20.906$, $P < 0.001$; random collection: $G = 7.603$, $P \leqq 0.01$).

57% of all nest sites of stratum A were in dead wood, while only 36% of the species of stratum B had this nesting preference. Dead wood was not as common in stratum B as in stratum A. Therefore it was impossible to collect twigs with same intensity in both strata. Nevertheless, 40 nests belonging to 36 species were found in dead twigs in stratum B. In addition, 11 species nested under the bark of trees, six of them exclusively: *Cataulacus horridus*, *Crematogaster* sp.5, *Rhopalomastix* sp.1, *Technomyrmex* sp.12, *Paratrechina* sp.6 and *Polyrhachis* cf. *cryptoceroides*.

3.5 The diversity of the ground layer

A total of 120 species were found, belonging to 49 genera. The most species-rich genera were *Pheidole*, *Strumigenys*, *Oligomyrmex*, *Tetramorium* and *Pachycondyla*. Correspondingly the Myrmicinae (58.3%) were the most species rich subfamily followed by the Ponerinae (26.7%), the Formicinae (10.8%), Cerapachyinae (2.5%) and the Dolichoderinae (1.7%).

The distribution of the "species frequency of capture" (Fig. 10) shows that c. 60% of the species were only found once or twice and the columns of the smallest plot size (9 m²) shows a distinct shift in direction to the single captures. This is confirmed by the species accumulation curves of each sample size, none of which shows a tendency toward saturation (Fig. 11).

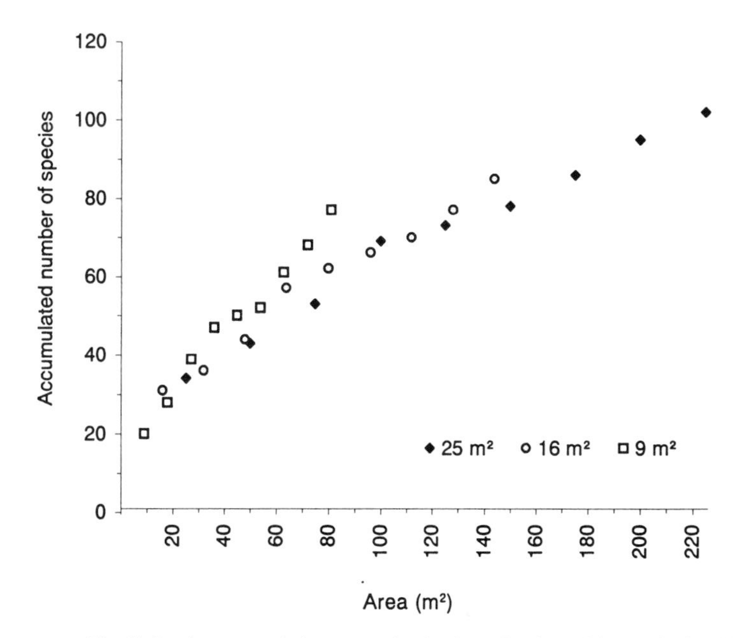

Fig. 11 Species accumulation curves for the three plot sizes of 9 m², 16 m² and 25 m².

The values of CS (mean ± *SD*) are 52.7 ± 7.68% for the 25 m^2 plots and 43.0 ± 8.18% for 9 m^2 plots. The ß-diversity of the nine 9 m^2 plots was significantly higher (Welch-Test (*t*-test): $\alpha/2 = 0.0005, t' = -5.187 < -t = -3.460$) than the ß-diversity of the 25 m^2 plots because more single captures were made (Fig. 10).

On average one more species was found on the 9 m^2 plots per m^2 (2.5 species) compared to the 25 m^2 plots (1.4 species).

4. DISCUSSION

4.1 Species richness

The results presented in this study indicate an extraordinary high ant species richness and states the Pasoh FR as one of the most species rich places known in the world, as far as ants are concerned.

The high percentages of single or double captures of species lead to the characteristic distribution for tropical regions with many rare species and a small number of species with high abundance. Consequently the species accumulation curves (Figs. 8, 9, 11) show no tendency toward saturation. Also a distinct vertical stratification of ant species was found, but it has to be evaluated carefully (see down). These separated ant communities consequently results in a high species turnover between the vertical layers (strata A, B and C) and thereby also contributes to a high species richness.

Most collections were conducted by day, so nocturnal species are doubtless underestimated. The same will be true for species living in hidden microhabitats or strata which were difficult to access, but it has to be assumed that a large portion of the terricolous, epigeic and abundant species were recorded. In addition efficient

Table 3 Comparison of three extensive inventories of ant species richness across all strata.

Study site	Duration	Methods	Total species	Total genera	Total sub-families	Study
East Malaysia: Primary forest 550 m - 800 m a.s.l.	c. 30 months	Winkler method Pitfall traps Fogging method Single-rope technique Tree trunk eclectors Hand collection	524	73	8	Brühl et al. (1998)
Peru: Primary forest Secondary forest Mixed plantation Pastures 220 m a.s.l.	c. 26 months	Berlese method Pitfall traps Baiting Light traps Tree eclectors Single-rope technique Hand collection	520	78	6	Verhaag (1991)
West Malaysia: Primary forest 75 m a.s.l.	c. 10 months	Winkler method Berlese method Pitfall traps Baiting Single-rope technique Hand collection	489	76	9	Malsch et al. (this study)

arboreal sampling methods like Malaise traps or the now well established fogging method, which are reported to give efficient yield of ant species for the canopy layer (Brühl et al. 1998; Longino & Colwell 1997), were not used. Nevertheless, the recorded species number is very high, especially in comparison to other species rich tropical habitats.

A comparison with two inventories yielding a similar high species number by recording ant species across all strata supports the indication of an extraordinary high species richness at Pasoh FR: the sampling effort of the given study was distinctly less than Brühl et al. (1998) and Verhaag (1991; Table 3). Additionally the sampling area comprised no wide altitudinal range of which a high species turnover is known (Brühl et al. 1999; Fisher 1996; Olson 1994) and also no different habitats (e.g. secondary forest, plantation) were included in this study.

However, this analysis lacks the possibility to compare the used methods directly and in detail with each other, because of different handling and samples sizes. Additionally the term "hand collection" very often comprises a variety of methods of which not all are standardized. Especially qualitative hand-collection is very efficient in increasing the species output, but this "experimental design" compromises comparability to other studies and between study sites. Therefore the first standard protocol for monitoring the diversity of ground living ants was developed (Agosti et al. 2000). Concerning the assessment of ant diversity across all strata a standard protocol for sampling ants in the vegetation is still needed. For this investigations comparing and analyzing methods sampling ants in the vegetation for their effectiveness are very useful. Longino & Colwell (1997) conducted a strict quantitative inventory across all strata with Malaise traps, fogging method and Berlese. They suggested a sampling with complementary methods in respect to spatial, temporal and habitat stratification.

Reflecting the given arguments, additional sampling would be very promising to find many more ant species at Pasoh FR.

4.2 Taxonomic Structure

In Pasoh FR we found all nine subfamilies occuring in the Indo Australian region and 76 of 126 genera (Bolton 1995). The most abundant genera found are as well the most abundant and most species rich genera world wide: *Camponotus, Polyrhachis, Pheidole,* and *Crematogaster* (Table II Appendix; Wilson 1976), although *Polyrhachis* occurs only in the old world tropics (Bolton 1973; Maschwitz & Dorow 1993; Wilson 1976). Brühl et al. (1998) obtained similar results from Poring

Table 4 Taxonomic structures of the ant fauna in Pasoh FR, West-Malaysia (this study) and Poring Hot Springs, East-Malaysia (after Brühl et al. 1998).

Subfamilies	% of species at Pasoh FR	% of species at Poring	% of genera at Pasoh FR	% of genera at Poring
Myrmicinae	40.7	39.9	39.5	43.3
Formicinae	28.2	31.5	18.4	20.3
Ponerinae	18.0	11.5	28.9	23.0
Dolichoderinae	5.9	10.2	6.6	6.8
Cerapachyinae	2.9	2.3	1.3	1.4
Aenictinae	2.5	1.5	1.3	1.4
Pseudomyrmecinae	1.4	2.9	1.3	1.4
Dorylinae	0.2	0.2	1.3	1.4
Leptanillinae	0.2	-	1.3	-

Hot Springs, Sabah, East-Malaysia. In contrast, preliminary results recorded in Panguana, Peru (Verhaag 1991) reveals a different ranking of the most species-rich genera. The most species rich genus was *Pheidole* (Myrmicinae) followed by *Camponotus* (Formicinae), *Pseudomyrmex* (Pseudomyrmecinae) and *Gnamptogenys* (Ponerinae). The extensive collections in all strata in Peru were conducted in four different habitats along an disturbance gradient from primary lowland forest to ordinary grass pasture. Therefore comparability is restricted.

The taxonomic structure in Pasoh and Poring Hot Springs is largely similar (Table 4). The differences are probably mainly caused by the ground biased sampling of this study in contrast to the higher sampling effort (fogging method, Floren & Linsenmair 1997) in the arboreal layer in Sabah. This study revealed 22 genera of the mainly terricolous Ponerinae, and the study in Poring only 17 genera. The hypogeic subfamily Leptanillinae (*Leptanilla* sp. 1) was found only at Pasoh. Of the mainly arboreal Dolichoderinae and Formicinae 54 species and 165 species were recorded in Poring while only 29 species of Dolichoderinae and 139 species of Formicinae were collected at Pasoh. A comparison of the taxonomic structures of Pasoh FR and Poring reflects these differences on species and genera level (Table 4).

However, this is the first time that two extensive inventories of ants across all strata of tropical primary lowland forests of one geographical region are available. Even if method-based differences occur, both inventories resulted in a highly concurrent composition of subfamilies and species-rich genera. Therefore the taxonomic structure obtained from these two studies may represent a general pattern of South-East Asian primary lowland forests.

4.3 Vertical stratification

Depending on stratum and the similarity quotient used (CS or C_J) the species turnover ranged between c. 57%–90%. The highest species turnover (90%) occurred between strata A and C (for Pasoh FR as well a similar value (c. 91%) could be extracted from data provided by Bolton 1996). The similarity of the two vegetation strata (B + C) is higher than their respective similarity to the ground stratum (stratum A). They showed a similarity of c. 43% using the Jaccard index. These is a degree of similarity, which had also been found in stratum A between the 9 m² samples of the diversity study. However, different sample sizes are neglected by similarity quotients and thus the results obtained may be biased. A further important aspect is to take the high number of single and double captures of this study into account, which might also be caused by undersampling effects. This contributes to a high species turnover between the strata and results in a strict vertical stratification. These problems must be considered when evaluating the distinct vertical stratification found in this study: 71% of the species seemed to be restricted to one stratum. Following the given arguments, it is likely to assume a less distinct degree of stratification (< 71%). On the other hand Brühl et al. (1998) recorded a similar high vertical stratification with c. 75% of the species restricted to one stratum in Poring Hot Springs (Sabah, East-Malaysia), although both studies used methods with different sampling designs. This may indicate that a high species turnover between the vertical layers could be a common phenomenon in tropical lowland rainforests. In our study the taxonomic composition of the vegetation, strata B and C, is the same on the subfamily level but differs on genera and species level. For example two dolichoderine genera seem to show intergeneric exclusion: only 3 spp. out of 15 *Technomyrmex* species were found in stratum C, but 12 spp. (64.3%) were found in stratum B. Out of 9 *Tapinoma* species 7 spp. (77.8%) occured in the stratum C but only 2 spp. in the stratum B. Moreover six species of the single pseudomyrmecine genus *Tetraponera* were restricted to stratum C and only one species could be found in stratum B and C (Table I, Appendix). For the vegetation layers a stratification is suggested by Brühl et al. (1998) due to plant size, age and species as well as to changing abiotic conditions like temperature, humidity and wind velocity.

The abiotic and biotic differences between vegetation and ground layer are obvious and correspond to strikingly different feeding and nesting habits. The majority of ground-living ants are predators (Hölldobler & Wilson 1990) and nest mainly in dead wood, in leaf-litter and topsoil. Ants in the vegetation feed mainly on food bodies and nectaries offered by plants or on honey dew offered by trophobiotic hemipterans (Blüthgen et al. 2000; Tobin 1991). They have more diverse nesting habits, including carton nests, silk nests, nests in living plants (cavities, epiphytes, myrmecophytes in general), as well as nests in arboreal leaf-litter and in dead wood. Genera which have developed the ability to construct silk or carton nests are the most species-rich in the vegetation (*Polyrhachis, Camponotus* and *Crematogaster*). This is supposed to be one of the main reasons for their successful radiation (Dorow et al. 1990; Maschwitz & Dorow 1993).

Wilson (1959) in New Guinea observed a predominance of arboricolous species foraging down to the ground relative to terricolous species foraging up into the vegetation. We found only 33 species nesting exclusively in one of the two habitats which also foraged in the other. Twenty species had their nest sites in the vegetation and foraged also on the ground and 13 species had their nest sites on the ground and also foraged in the vegetation (strata B + C). Although the data base is small, our observations also indicate a predominance of arboricolous species over terricolous species and their ability to extend their foraging area into a stratum different from their nest stratum.

4.4 Diversity of the ground layer

The CS values (mean ± SD) for leaf-litter ant communities revealed 52.7 ± 7.68% for 25 m² plots and 43.0 ± 8.18% for 9 m² plots. A similar range value (43.6-53.7%) was found in a tropical lowland forest in Costa Rica by Olson (1991) and by Malsch (in prep.) in Sabah, East-Malaysia (51.6 ± 2.7%). These values corresponds to a high number of rare species (Fig. 10) and indicate a high ß-diversity. Additionally, similar range values of the ground ant communities in different tropical lowland forests also argues for equal mechanisms responsible for the concurrent ß-diversity.

The different plot sizes used influenced the measurement of diversity and species richness. The similarity values for the 25 m² and 9 m² plots differed significantly, though the 9 m² plots were subplots of the 25 m². The 9 m² areas revealed strikingly more single captures and consequently in average one more species per 1 m² was found. The species accumulation curve increased more steeply and the diversity values were significantly higher. The results of the 9 m² plots indicating an apparent higher species richness and a more diverse ant community than the 16 m² and the 25 m² plots. This is an important aspect for comparing results of investigations in which different plot sizes were used.

Habitat criteria are also very important for the assessment of diversity. The similarity values for the "Inventory-Winkler" (37.4 ± 5.2%) at subjectively chosen ant species-rich locations are much lower than for the standardized sampling. Moreover the "Inventory-Winkler" resulted (with 5 samplings of only several m²) in as many species (103 spp.) as the "Diversity-Winkler" (102 spp.; 225 m^2 of sifted leaf-litter). This demonstrates the importance and effect of standardized habitat criteria on the resulting data and the high effectivness of qualitative sampling.

5. CONCLUSIONS

Ants are one of the most numerous invertebrate groups in the majority of terrestrial ecosystems, both in terms of number of individuals and species. In addition, ants interact at many trophic levels within an ecosystem. Due to this ecological importance, ants have the potential to deliver reliable biodiversity data.

This inventory using a variety of quantitave and qualitative sampling methods, yielded a high number of ant species (489) and presented the Pasoh FR as one of the most species rich places in SE Asia. This holds true even when the richness of ant species in the Pasoh FR is compared with species rich tropical areas from other continents. Although sampling efforts in Pasoh FR, in terms of time and methods used, were comparatively low, the total number of ant species are similar to those obtained in the Neotropics (e.g. Peru).

At the Pasoh FR, the taxonomic structure of ants seems to represent a general pattern in SE Asian lowland forests. However, it is emphazised that further extensive studies are needed for assessing patterns of biodiversity between different habitat types and biogeographic regions. The most important aspects of these efforts are repeatability and low collection bias. Hence standardized methods are especially needed for the vegetation layer, as qualitative sampling usually compromises the comparability to other studies. Furthermore, it should be of equal interest to work out optimal sample sizes required for an effective and representative sampling of the ant fauna. As this study has shown sample size may greatly influence the species output, thus resulting in a different estimation of species diversity and species richness. More studies with a methodical emphasis are needed to achieve a solid base of comparable applications, and to place rainforest ant faunas in a global context.

ACKNOWLEDGEMENTS

First we wish to thank the Forest Research Institute Malaysia (FRIM), especially Dr. Chang Hung, for permitting and supporting our study at the Pasoh FR. We also like to thank the "Staatliches Museum für Naturkunde, Karlsruhe" and Manfred Verhaag for manifold support. This study was financially supported by the "Deutsche Forschungsgemeinschaft (DFG)" and the "Graduiertenförderung der Universität Frankfurt".

REFERENCES

Agosti, D., Majer, J., Alonso, L. & Schultz, T. (eds.) (2000) Ants: Standard Methods for Measuring and Monitoring Biodiversity. Smithonian Institution Press, 280pp.

Besuchet, C., Burckhardt, D. H. & Löbl, I. (1987) The Winkler-Moczarski eclector as an efficient extractor for fungus and litter coleoptera. Coleopt. Bull. 41: 392-394.

Blüthgen, N., Verhaagh, M., Goitio, W., Jaffé, K., Morawetz, W. & Barthlott, W. (2000) How plants shape the ant community in the Amazonian rainforest canopy: the key role of extrafloral nectarines and homopterans honeydew. Oecologia 125: 229-240.

Bolton, B. (1973). The ant genus *Polyrhachis* F. Smith in the Ethiopian Region (Hymenoptera: Formicidae) Bull. Nat. Hist. Museum London (Ent.) 28: 285-369.

Bolton, B. (1995) Identification guide to the ant genera of the world. Harvard University Press, Cambridge, 222pp.

Bolton, B. (1998) A preliminary analysis of the ants (Formicidae) of Pasoh Forest Reserve. In

Lee, S. S., May, D. Y., Gauld, I. D., & Bishop, J. (eds). Conservation, Management and Development of Forest Resources: Proceedings of the Malaysia United Kingdom Programme Workshop. Forest Research Institute Malaysia, Kuala Lumpur, Malaysia, pp.84-95.

Brühl, C. A., Gunsalam, G. & Linsenmair, K. E. (1998) Stratification of ants (Hymenoptera, Formicidae) in a primary rain forest in Sabah, Borneo. J. Trop. Ecol. 14: 285-297.

Brühl, C. A., Mohamed, M. & Linsenmair, K. E. (1999) Altitudinal distribution of leaf litter ants along a transect in Primary forests on Mount Kinabalu, Sabah, Malaysia. Trop. Ecol. 15: 265-277.

Chiba, S. (1978) Numbers, biomass and metabolism of soil animals in Pasoh Forest Reserve. Malay. Nat. J. 30: 313-324.

Dorow, W. H. O., Maschwitz, U. & Rapp, S. (1990) The natural history of *Polyrhachis (Myrmhopla) muelleri* Forel, 1893 (Formicidae Formicinae), a weaver ant with mimetic larvae and an unusual nesting behaviour. Trop. Zool. 3: 181-190.

Erwin, T. L. (1982) Tropical Forests: their richness in Coleoptera and other arthropod species. Coleopt. Bull. 36: 74-75.

Fisher, B. L. (1996) Ant diversity patterns along an elevational gradient in the Réserve Naturelle Intégrale d´Andringitra, Madagaskar. Fieldiana Zool. 85: 93-108.

Fittkau, E. J. & Klinge, H. (1973) On biomass and trophic structure of the central amazonian rain forest ecosystem. Biotropica 5: 2-14.

Floren, A. & Linsenmair, K. E. (1997) Diversity and recolonization dynamics of selected arthropod groups on different tree species in a lowland rainforest in Sabah, Malaysia with special refernce to Formicidae. In Stork, N. E., Adis, J. & Didham, R. K. (eds). Canopy arthropods. Chapman and Hall, London. pp.344-381.

Floren, A., Linsenmair, K. E. & Biun, A. (1998) Strucure and dynamics of arboreal arthropod communities. Sabah Parks Nat. J. 1: 69-82.

Gaston, K. J. (1991) The magnitude of global species richness. Conservation Biol. 5: 283-293.

Hölldobler, B. & Wilson, E. O. (1990) The Ants. Belknap Press, Cambridge, 732pp.

Lee, S. S. (1995) A guidebook to Pasoh, FRIM Technical Information Handbook No. 3. Forest Research Institute Malaysia, Kuala Lumpur, 73pp.

Longino, J. T. & Colwell, R. K. (1997) Biodiversity assessment using structured inventory: capturing the ant fauna of a tropical rainforest. Ecol. Appl. 7: 1263-1277.

Magurran, A. E. (1988) Ecology diversity and its measurement. University Press, Cambridge, 179pp.

Maschwitz, U. & Dorow, W. H. O. (1993) Nesttarnung bei tropischen Ameisen. Naturwissenschaftliche Rundschau 46: 237-239.

Olson, D. M. (1991) A comparison of the efficiancy of litter sifting and pitfall traps for sampling leaf litter ants (Hymenoptera, Formicidae) in a tropical wet forest, Costa Rica. Biotropica 23: 166-172.

Olson, D. M. (1994) The distribution of leaf litter invertebrates along a neotropical altitudinal gradient. J. Trop. Ecol. 10: 129-150.

Rosciszewski, K. (1994) *Rostromyrmex*, a new genus of myrmicine ants from Peninsular Malaysia (Hymenoptera: Formicinae). Entomol. Scandinavica 25: 159-168.

Stork, N. E. (1988) Insect diversity: facts, fiction and speculation. Biol. J. Linn. Soc. 35: 321-337.

Tobin, J. E. (1991) A neotropical rainforest canopy, ant community: some ecological considersations. In Huxley, C. R. & Cutler, D. F. (eds). Ant plant interactions, Oxford University Press, Oxford, pp.536-538.

Verhaag, M. (1991) Clearing a tropical rainforest–effects on the ant fauna. In Erdelen, W., Ishwaran, N. & Müller, P. (eds). Proceedings of the International and Interdisciplinary Symposium on Tropical Ecosystems. Margraf Scientific Books, Weikersheim, pp.59-68.

Wilson, E. O. (1959) Some ecological characteristics of ants in New Guinea rain forest. Ecology 40: 437-447.

Wilson, E. O. (1976) Which are the most prevalent ant genera? Studia Ent. 19: 187-200.

Appendix: Table I Ant species list of Pasoh FR including: Capture method: Hc = Hand Collection, W = Winkler Method, P = Pitfall Trap, B = Berlese Method. Activity: D = day, N = night, D/N = day and night. Strata (see also methods): A = "ground stratum", with special regard to foraging or nesting sites in top soil ≡ AA; B = lower vegetation (≦ 3 m); C = arboreal stratum (> 3 m). Stratum given in brackets indicates species found predominantly in other strata and found only rarely here. Nests / colony: "+" = hole nest / colony was sampled, "-" = not the hole nest / colony could be sampled, "?" = not sure if the hole nest / colony was collected. Number of workers, queens and alates for nests / colony are given. Alates are signed with * = for female and ** = for male individuals. $_a$ = Sampled with "25-Winkler"; $_b$ = Unaccessible nest sites (e.g. inside trees, deep in the soil) which therefore could not be examined; $_c$ = The given stratum is for temporary nests or assumed temporary nests while nest moving; $_d$ = Biting at the legs of *Aenictus cornutus*; $_e$ = Prey of *Aenictus denatus*; $_f$ = Prey of *Aenictus laeviceps*; $_g$ = in a fallen dead twig from the arboreal stratum; $_h$ = Prey of *Aenictus cornutus*; $_i$ = Prey of *Aenictus* aff. *laeviceps*.

Subfamily *Genus / Species*	Capture method	Activity	Stratum of activity	Nesting stratum	Nest / Colony	Worker	Queen	Alate
Aenictinae								
Aenictus aratus Forel	Hc	D/N	A	A[1]				
Aenictus camposi Wheeler & Chapman	Hc	D/N	A(B)					
Aenictus cornutus Forel	Hc	D	AB	A[1]				
Aenictus dentatus Forel	Hc	D/N	A(B)	A[1]				
Aenictus gracilis Emery	Hc	D/N	AB	A[1]				
Aenictus aff. gracilis Emery	Hc	D	ABC	C[1]				
Aenictus hottai Terayama & Yamane	Hc	D	A(B)	A[1]				
Aenictus laeviceps (F. Smith)	Hc	D/N	AB	A(B)[1]				
Aenictus aff. laeviceps	Hc	D	AB	A[1]				
Aenictus sp.2	Hc,P	D/N	A	A[1]				
Aenictus sp.3	Hc	D	A					
Aenictus sp.5	Hc	D/N	A	A[1]				
Dorylinae								
Dorylus laevigatus (F. Smith)	Hc	N	AA					
Leptanillinae								
Leptanilla sp.1	Hc		AA	AA[17]	- / -	145		
Pseudomyrmecinae								
Tetraponera allaborans (Walker) (s.l.)	Hc		C	C				
Tetraponera attenuata F. Smith	Hc	D	BC(A)	BC	+ / -	34		1**
					- / -	99		7*,9**
					- / -	155		4**
Tetraponera sp.3	Hc	D	C					
Tetraponera sp. PSW-71	Hc	D	C	C				
Tetraponera cf. *modesta* (F. Smith)	Hc		C	C	+ / ?	20	1	
Tetraponera nitida (F. Smith)	Hc	D	C					
Tetraponera sp. PSW-74	Hc		C					
Cerapachyinae								
Cerapachys cf. *antennatus* F. Smith	Hc	D	A					
Cerapachys kodecorum Brown	Hc		A	A				
Cerapachys cf. *suscitatus* (Viehmeyer)	Hc	D	AB					
Cerapachys sp. near *dohertyi* Emery	Hc,B	D/N	A	A	+ / ?	81		
Cerapachys sp. near *sauteri* Forel	Hc	D	A					
Cerapachys sp.1	Hc	D	A					
Cerapachys sp.2	Hc	D	A					
Cerapachys sp.3$_a$	W$_a$		A					
Cerapachys sp.4	Hc		AA					
Cerapachys sp.5	W		A					
Cerapachys sp.6	W		A					
Cerapachys sp.7	Hc		A					
Cerapachys sp.8	Hc	D	A					
Cerapachys sp.10	Hc	D	C					

Appendix: Table I (continued 1)

Subfamily *Genus / Species*	Capture method	Activity	Stratum of activity	Nesting stratum	Nest / Colony	Worker	Queen	Alate
Ponerinae								
Amblyopone sp.1	Hc	D	A	A	- / -	46		3**
Amblyopone sp.2	W		A					
Amblyopone sp.3	Hc		AA	AA	- / -	56	5	
Anochetus graeffei Mayr	Hc,W	D	AA, A	A	? / ?	36	1	
Anochetus rugosus (F. Smith)	Hc	D	A					
Anochetus sp. near *tua* Brown	Hc,P		AA, A	AA				
Anochetus sp.1	Hc							
Anochetus sp.2	Hc,B		AA,A					
Anochetus sp.3	Hc,W		A					
Belonopelta sp.1$_a$	W$_a$		A					
Belonopelta sp.2$_a$	W$_a$		A					
Centromyrmex feae (Emery)	Hc		AA,A					
Cryptopone sp.1	Hc,W		A	A	? / -	26	1	5**
Cryptopone sp.2$_a$	W$_a$							
Diacamma cf. *rugosum* (Le Guillou)	Hc	D	AB	B	+ / -	35		
					+ / -	46		2**
Discothyrea sp.1	Hc,B,W		AA,A	AA				
Discothyrea sp.2	B,W		A					
Gnamptogenys costata (Emery)	Hc	D	AB					
Gnamptogenys menadensis (Mayr)	Hc,P,B	D/N	AB	A	+ / ?	33	1	
					- / -	42		
Gnamptogenys aff. *binghami* (Forel)	Hc,P,B,W	D	AA,A	AA				
Gnamptogenys sp.1	Hc	D	A					
Gnamptogenys sp.2	Hc	N	A					
Gnamptogenys sp.2$_a$	W$_a$		A					
Gnamptogenys sp.3	B		A					
Harpegnathos venator (F. Smith)	Hc		A					
Hypoponera sp.1	Hc		AA					
Hypoponera sp.2	Hc,W		AA,A					
Hypoponera sp.3	Hc,B		AA,A					
Hypoponera sp.4	Hc		AA,A					
Hypoponera sp.5	Hc,B,W		AA,A					
Hypoponera sp.6	Hc,W		A	A				
Hypoponera sp.7	Hc		A	A				
Hypoponera sp.8	Hc		A					
Hypoponera sp.9	Hc,B,W		AA,A					
Hypoponera sp.10	W		A					
Hypoponera sp.11	W		A					
Hypoponera sp.12	W		A					
Hypoponera sp.13	W		A					
Hypoponera sp.14	W		A					
Leptogenys cf. *borneensis* Wheeler	Hc	D/N	A	A$_c$				
Leptogenys crassicornis Emery	Hc	D	A					
Leptogenys diminuta (F. Smith)	Hc,P	D	A	A$_c$				
Leptogenys processionalis (Jerdon)	Hc,P	D/N	A	A$_c$				
Leptogenys mutabilis (F. Smith)	Hc,P	D/N	A	A$_c$				
Leptogenys myops (Emery)	Hc,P	D/N	AA,A					
Leptogenys peuqueti (André)	Hc		AA,A	A$_c$				
Leptogenys sp.5 aff. *assamensis* (Forel)	Hc	D	A	A$_c$				
Leptogenys sp.6 aff. *kraepelini* (Forel)	Hc	D	A					
Leptogenys sp.12 aff. *sagaris* Wilson	P,W		A					
Leptogenys sp.C	Hc	D	A	A$_c$				
Leptogenys sp.E	Hc	D	A					
Myopias sp.1	Hc		A					
Myopone castanea (F. Smith)	Hc		A	A				
Mystrium camillae Emery	Hc		A					
Odontomachus rixosus F. Smith	Hc,P,B,W	D/N	AA,A	AA,A	? / -	15		
					? / -	12		
					? / -	77		
Odontomachus simillimus F. Smith	Hc	N	AA,A	AA				
Odontoponera transversa (F. Smith)	Hc,P,B,W	D/N	AA,A	AA$_b$				
Pachycondyla amblyops (Emery)	Hc,B	D	AA,A					
Pachycondyla darwini (Forel)	Hc		A					
Pachycondyla havilandi Forel	Hc,P	D	A					

Appendix: Table I (continued 2)

Subfamily *Genus / Species*	Capture method	Activity	Stratum of activity	Nesting stratum	Nest / Colony	Worker	Queen	Alate
Pachycondyla insularis (Emery)	Hc,P,B,W		A					
Pachycondyla leeuwenhoeki (Forel)	Hc,B,W	D	AA,A	AA,A	+ / -	61		
Pachycondyla rubra (F. Smith)	Hc,P,W		A	AA				
Pachycondyla sharpi (Forel)	Hc,B		A	AA,A				
Pachycondyla tridentata F. Smith	Hc,P,W	N	A	A	? / ?	73	1	1**
Pachycondyla sp.4	W_a	D	A	A	- / -	421	4	
Pachycondyla sp.4$_a$	Hc,P,B,W		A					
Pachycondyla sp.6	Hc	D	A					
Pachycondyla sp.9	Hc		A					
Pachycondyla sp.11	Hc		A					
Platythyrea cf. *nicobarensis* Forel	Hc		AB					
Platythyrea cf. *quadridenta* Donisthorpe	Hc		A	A				
Platythyrea parallela (F. Smith)	Hc,P	D/N	AB					
Platythyrea tricuspidata Emery	Hc	D	AB					
Platythyrea sp.1	Hc	D	B					
Platythyrea sp.2	Hc	D	B					
Platythyrea sp.3	Hc	D	AB					
Platythyrea sp.5	Hc	D	C					
Platythyrea sp.6	Hc	D	A					
Platythyrea sp.7	Hc,B,W	D	C					
Ponera sp.1		D	AA,A	AA,A	? / ?	7	1	
	Hc,W				+ / ?	9		
Ponera sp.2	Hc,W		A					
Prionopelta sp.1	Hc,B,W		A					
Probolomyrmex aff. *watanabei* Tanaka	W_a		A					
Probolomyrmex sp.1$_a$	Hc,W							
Proceratium sp.1	W		A					
Proceratium sp.2	W		A					
Proceratium sp.3			A					
Myrmicinae								
Acanthomyrmex ferox Emery	Hc	D	AB					
Acanthomyrmex aff. *careoscrobis* Moffett	Hc,B		A	A				
Cardiocondyla nuda (Mayr)	P		A					
Cardiocondyla aff. *wroughtoni* (Forel)	Hc,P		AB	B				
Cardiocondyla sp.1	Hc,B	D	A	A				
Cardiocondyla sp.4	Hc	D	A					
Carebara sp.1	Hc		AB					
Cataulaus horridus F. Smith	Hc	D	AB	B				
Cataulacus hispidulus F. Smith	Hc	D	ABC	BC	+ / +	73	1	
					+ / -	11		1**
Cataulacus latissimus Emery	Hc	D	BC					
Cataulacus praetextus F. Smith	Hc	D	C	C				
Cataulacus aff. *taprobane* F. Smith	Hc	D	C					
Crematogaster difformis F. Smith	Hc	D	ABC	BC				
Crematogaster inflata F. Smith	Hc		B					
Crematogaster longipilosa Forel	Hc	N	A					
Crematogaster sp.1	Hc	D	B					
Crematogaster sp.2	Hc,P	D	AB					
Crematogaster sp.4	Hc	D/N	B					
Crematogaster sp.5	Hc	D/N	B	B				
Crematogaster sp.6	Hc		C					
Crematogaster sp.7	Hc	D/N	AB	B				
Crematogaster sp.8	Hc		BC					
Crematogaster sp.9	Hc,P		ABC	AB				
Crematogaster sp.10	Hc	D	ABC					
Crematogaster sp.12	Hc		B	B	+ / +	576	1	
Crematogaster sp.13	Hc	D	BC	B				
Crematogaster sp.14	Hc	D/N	BC	C				
Crematogaster sp.15	Hc	D/N	BC	B_b				
Crematogaster sp.17	Hc		BC	BC	+ / ?	278	7	1*
					+ / ?	43	3	
Crematogaster sp.18	Hc,B,W	N	A	A	- / -	473		
Crematogaster sp.19	Hc,B,W	D	A					

Appendix: Table I (continued 3)

Subfamily *Genus / Species*	Capture method	Activity	Stratum of activity	Nesting stratum	Nest / Colony	Worker	Queen	Alate
Crematogaster sp.20	Hc		AA	AA				
Crematogaster sp.21	Hc	D	AB					
Crematogaster sp.22	Hc		B	B				
Crematogaster sp.23	Hc	D	C	C	+ / -	2954	1	
Crematogaster sp.24	Hc	D	C					
Crematogaster sp.25	Hc	D	C	C				
Crematogaster sp.26	Hc	D	C					
Crematogaster sp.27	Hc		C	C	+ / -	15		
Crematogaster sp.28	Hc	D	C					
Crematogaster sp.29	P		A					
Crematogaster sp.30	Hc		B	B				
Dilobocondyla borneensis Wheeler	Hc	D	BC					
Dilobocondyla sp.2	Hc	D	C					
Eurhopalothrix heliscata Wilson & Brown	P		A					
Lophomyrmex bedoti Emery	Hc,P,B,W	D/N	AA,A	AA,A	+ / -	28		
					+ / -	129		
					+ / -	144		
					? / -	99		
					+ / -	58		
Lordomyrma sp.1	Hc	D	A					
Mayriella sp.1	W		A					
Meranoplus belli Forel	Hc	D	BC					
Meranoplus bicolor (Guérin-Méneville)	Hc	D	AA	AA				
Meranoplus mucronatus F. Smith	Hc,P	D	AB	A_b				
Meranoplus sp.4	Hc	D	A	A	+ / ?	46	10	4••
					+ / ?	48	1	1••
Monomorium pharaonis (Linné)	Hc,B	D	ABC					
Monomorium aff. *talpa* Emery	W		A					
Monomorium sp.1	Hc	D	AB	B_b				
Monomorium sp.2	Hc	D	BC	BC	+ / ?	75	1	
					? / ?	944	1	
					+ / +	5495	1	
					+ / +	5186	1	1••
					+ / +	4500	1	
					+ / +	7080	1	
					+ / ?	1312		
Monomorium sp.3	Hc	D	A	A	+ / +	146	1	
					+ / +	78	1	
					+ / ?	208	1	
					+ / ?	338		
Monomorium sp.4	Hc,P,B,W	D	AB					
Monomorium sp.4_a	W_a		A					
Monomorium sp.5	Hc,P,B,W		A					
Monomorium sp.7	Hc,P	D	ABC					
Monomorium sp.9	Hc	D	C		+ / ?	97	1	
Monomorium sp.10	Hc,B	D	AB					
Myrmecina sp.1	Hc,P,B,W	D	A	A	? / -	52		1••
Myrmecina sp.2	Hc	D	AA,A	AA	+ / ?	19	2	4•,7••
Myrmecina sp.3	B,W		A	A				
Myrmecina sp.4	Hc,W		A	A				
Myrmecina sp.5	Hc		A					
Myrmecina sp.6	Hc		AA	AA				
Myrmecina sp.7	W		A					
Myrmicaria cf. *arachnoides* (F. Smith)	Hc	N	B	B	+ / ?	73		6•,7••
Myrmicaria cf. *birmana* Forel	Hc	D	B	B	+ / +	2454	12	34••
Oligomyrmex sp.1	Hc,P,B,W	D	A	A	+ / +	91	1	
					+ / ?	18	1	
Oligomyrmex sp.2	Hc		AA	AA	- / -	135		10•,1
Oligomyrmex sp.3	P		A					
Oligomyrmex sp.3_a	W_a		A					
Oligomyrmex sp.4	W		A					
Oligomyrmex sp.5	W		A					
Oligomyrmex sp.5_a	W_a		A					
Oligomyrmex sp.6	Hc	D	B					
Oligomyrmex sp.6_a	W_a		A					

Appendix: Table I (continued 4)

Subfamily *Genus / Species*	Capture method	Activity	Stratum of activity	Nesting stratum	Nest / Colony	Worker	Queen	Alate
Oligomyrmex sp.7	W		A					
Oligomyrmex sp.8	B		A					
Oligomyrmex sp.9	W		A					
Oligomyrmex sp.10	Hc,B		AA,A	AA				
Oligomyrmex sp.11	Hc		d					
Paratopula ankistra Bolton	W		A					
Pheidole aristotelis Forel	Hc,W	D	A	A	+ / ?	47	1	
Pheidole quadricuspis Emery	Hc	D	A					
Pheidole rabo Forel	Hc,P,B,W	D	AA;A(B)	AA	? / ?	44	1	
					? / ?	66	1	
Pheidole (Ischnomyrmex) longipes (F. Smith)	Hc,P,W	D/N	A	A				
Pheidole (Ischnomyrmex) comata F. Smith	Hc,P,B		AA,A	AA				
Pheidole (Ischnomyrmex) sp.1$_a$	W$_a$		A					
Pheidole (Ischnomyrmex) sp.17	Hc		e					
Pheidole sp.3	Hc	D/N	AB	A	- / -	132	1	
Pheidole sp.4	Hc,P,B	D/N	AB					
Pheidole sp.5	Hc,P,W	D	AA,A	AA,A		68	1	
Pheidole sp.6	Hc,P,B		AA,A	AA				
Pheidole sp.7	Hc,P	D/N	AA,A	AA,A	+ / ?	26	1	
					+ / +	89	1	
					+ / ?	23	1	
Pheidole sp.8	Hc,P,B,W	D/N	AA,A(B)	A	? / -	88		
					+ / -	39		
					? / –	164		
					+ / -	189		
					+ / ?	105		
Pheidole sp.9	Hc	D	AA,A					
Pheidole sp.9$_a$	W$_a$		A					
Pheidole sp.10	Hc	D	A					
Pheidole sp.11	Hc		AA,A	AA,A				
Pheidole sp.12	Hc		A	A				
Pheidole sp.13	Hc,P	D	A	A	+ / ?	55		
Pheidole sp.15	Hc,B,W	D	A	A	? / -	375		
					? / -	77		
					+ / ?	104		
					+ / -	73		
					+ / -	46		
					? / ?	23		
					? / -	60		
					+ / -	88		
Pheidole sp.18	Hc		f					
Pheidole sp.19	Hc		f+h					
Pheidole sp.21	W		A					
Pheidole sp.22	Hc		A					
Pheidole sp.23	P		A					
Pheidole sp.24	Hc		A	A				
Pheidole sp.25	P		A					
Pheidole sp.26	P,B		A					
Pheidole sp.27	Hc		AA	AA				
Pheidole sp.28	W		A					
Pheidole sp.29	W		A					
Pheidole sp.30	P		A					
Pheidole sp.31	Hc	D	C					
Pheidole sp.32	Hc	D	C					
Pheidole sp.33	Hc		AA					
Pheidole sp.34	P		A					
Pheidole sp.35	P		A					
Pheidole sp.36	Hc		AA					
Pheidole sp.37	W		A					
Pheidologeton affinis (Jerdon)	Hc,P,B,W	D/N	AA,A					
Pheidologeton silenus (F. Smith)	Hc,P,W	D/N	A(B)					
Pheidologeton sp.3	Hc	D	A					
Pristomyrmex trachylissa (F. Smith)	Hc	D	AB	B$_b$				

Appendix: Table I (continued 5)

Subfamily *Genus / Species*	Capture method	Activity	Stratum of activity	Nesting stratum	Nest / Colony	Worker	Queen	Alate
Pristomyrmex aff. *brevispinosus* Emery	Hc	D	A					
Pristomyrmex sp.1_a	W_a		A					
Pristomyrmex sp.3	Hc		A					
Pristomyrmex sp.4	Hc,B,W		AA,A					
Proatta butteli Forel	Hc,W	D	AA,A,B	AA				
Pyramica (Dysedrognathus) extemenus	W_a		A					
Pyramica (Smithistruma) sp.1	W		A					
Pyramica (Smithistruma) sp.2	P,W		A					
Pyramica (Smithistruma) sp.3	P		A					
Pyramica (Smithistruma) sp.4	B,W		A					
Recurvidris browni Bolton	Hc	D	A				1	
Recurvidris kemneri (Wheeler & Wheeler)	Hc	D	A					
Rhopalomastix sp.1	Hc		B	B	- / -	35	1	4••
Rhoptromyrmex wroughtoni (Forel)	Hc,W		AB				1	
Rostromyrmex pasohensis Rosciszewski	Hc		A	A	+ / ?	5		
					? / ?	8		1••
Solenopsis sp.1	Hc,P,B,W		AA,A	A	+ / +	68	8	
Solenopsis sp.1_a	W_a		A				7	
Solenopsis sp.2	Hc		AA		- / -	821		
Solenopsis sp.2_a	W_a		A					
Solenopsis sp.3	P		A					
Strumigenys doriae Emery	W		A					
Strumigenys juliae Forel	Hc,B,W		AA,A	A	? / -	14		3•
Strumigenys koningsbergeri Forel	Hc,W		A					
Strumigenys signeae Forel	B,W		A					
Strumigenys sublaminata Brown	W		A					
Strumigenys uichancoi Brown	Hc,P,B,W		A	A				
Strumigenys sp.1	Hc,W		A	A				
Strumigenys sp.2	Hc		AA					
Strumigenys sp.3	P,B,W		A					
Strumigenys sp.7	B		A	A				
Strumigenys sp.8	Hc		A	A				
Strumigenys sp.10	W		A					
Strumigenys sp.10_a	W_a		A					
Strumigenys sp.11	W		A					
Strumigenys sp.12	W		A					
Strumigenys sp.14	W		A					
Strumigenys sp.14_a	W_a		A					
Strumigenys sp.16	W		A					
Tetheamyrma subspongia Bolton	W		A					
Tetramorium adelphon Bolton	Hc	D	AB	A	+ / -	23		
Tetramorium aptum Bolton	Hc,B		AA,A					
Tetramorium cf. *curtulum* Emery	Hc,P,B,W	D/N	AA,A	AA	- / -	52		7•
					- / -	56		4•
Tetramorium eleates Forel	Hc		B					
Tetramorium indicum Forel	Hc		B	B	+ / -	62		
Tetramorium insolens (F. Smith)	Hc,B		A	A				
Tetramorium kheppera Bolton	W		A				1	
Tetramorium laparum Bolton	Hc		C	C			1	
Tetramorium meshena Bolton	Hc,W	D/N	AA,A,B	AA,A	? / ?	56	1	2•,9
					? / ?	75		
					+ / ?	16		
Tetramorium obtusidens Viehmeyer	Hc		C?$_g$	C?$_g$				
Tetramorium pacificum Mayr	Hc	D	B	A		81		
Tetramorium seneb Bolton	Hc,B,W		A					
Tetramorium smithi Mayr	Hc	D	AA,A	AA				
Tetramorium sp.1	Hc	D	A,B					
Tetramorium sp.2	Hc	N	A					
Tetramorium sp.9	Hc,P,B		AA,A					
Tetramorium sp.12	W		A	A				
Tetramorium sp.16	W		A					
Tetramorium sp.17	Hc,P,B,W		AA,A	A				
Vollenhovia fridae Forel	Hc,P	D	AA,AB	A				
Vollenhovia rufiventris Forel	Hc	D	B,C?$_g$	C?$_g$				
Vollenhovia sp.2	Hc	N	B				1	

Appendix: Table I (continued 6)

Subfamily *Genus / Species*	Capture method	Activity	Stratum of activity	Nesting stratum	Nest / Colony	Worker	Queen	Alate
Vollenhovia sp.3	Hc	D	B				1	
Vollenhovia sp.4	Hc	D	AB	B	? / ?	12		
Vollenhovia sp.5	Hc,B,W	D	A	A	+ / +	25		
Vollenhovia sp.7	Hc,W	D	A					
Vollenhovia sp.8	Hc		A	A				
Vollenhovia sp.9	B		A					
Vollenhovia sp.10	Hc	D	A					
Vollenhovia sp.11	Hc	D	BC					
Dolichoderinae								
Dolichoderus cuspidatus (F. Smith)	Hc	D	ABC	C_c				
Dolichoderus beccarii Emery	Hc	D	B					
Dolichoderus thoracicus (F. Smith)	Hc,P	D/N	ABC	BC	+ / -	1206	11	
Iridomyrmex sp.1	Hc	D	AA,A	AA_b				
Philidris sp.1	Hc,P	D/N	ABC	ABC	+ / -	85		
Tapinoma sp.1	Hc,B	D	A	AA_b				
Tapinoma sp.2	Hc	D	B	B	- / -	2591	1	
Tapinoma sp.3	Hc	D	BC					
Tapinoma sp.4	Hc	D	C					
Tapinoma sp.5	Hc		C					
Tapinoma sp.6	Hc		C	C	+ / ?	133	1	2••
Tapinoma sp.7	Hc	D	C					
Tapinoma sp.8	Hc	D	C					
Tapinoma sp.9	Hc	D	C					
Technomyrmex strenuus Mayr	Hc		C	C	+ / -	186		5•
					+ / -	345		26•,2••
					+ / -	1780	1	13•
Technomyrmex sp.1	Hc,P,B,W	D/N	A	A	? / -	19		1••
Technomyrmex sp.2	Hc		B	B				
Technomyrmex sp.3	Hc	D/N	AB					
Technomyrmex sp.4	Hc	N	B	B	+ / -	366		152•,7••
Technomyrmex sp.5	Hc	D	BC	B				
Technomyrmex sp.6	Hc		B					
Technomyrmex sp.7	Hc	D	B	B	+ / +	3502	30	1••
Technomyrmex sp.8	Hc		h					
Technomyrmex sp.9	Hc		B					
Technomyrmex sp.11	Hc	D	BC	BC	+ / -	802	1	41•,26••
Technomyrmex sp.12	Hc		B	B				
Technomyrmex sp.13	Hc	D/N	B	B				
Technomyrmex sp.14	Hc	D	B					
Technomyrmex sp.15	Hc		B					
Formicinae								
Acropyga acutiventris Roger	Hc,W		A	A	+ / +	229	1	
Acropyga sp.1$_a$	W$_a$		A					
Acropyga sp.2	Hc		A	A				
Acropyga sp.3	Hc		AA	AA				
Acropyga sp.4	Hc		A	A				
Acropyga sp.5	Hc		AA	A				
Anoplolepis gracilipes (F. Smith)	Hc	D/N	A	AA				
Camponotus (Karavaievia) *asli* Dumpert	Hc	N	BC	BC				
Camponotus (Karavaievia) *melanus* Dumpert	Hc		C	C				
Camponotus (Karavaievia) *striatipes* Dumpert	Hc	N	C	C	+ / -	12		3••
Camponotus badius (F. Smith)					+ / -	11		3••
Camponotus camelinus (F. Smith)					+ / -	14		12••
Camponotus festinus (F. Smith)	Hc	D	BC	B_b				
Camponotus gigas (Latreille)	Hc	D	AA,ABC	AA_b				
Camponotus leonardi Emery	Hc	D/N	ABC					
Camponotus moeschi Forel	Hc,P	D/N	ABC	A_bB_b				
Camponotus trajanus Forel	Hc	D	(A)BC	C_b				
Camponotus sp.2	Hc	N	BC	B				
Camponotus sp.3	Hc	D	C					
Camponotus sp.4	Hc		C	C				

Appendix: Table I (continued 7)

Subfamily *Genus / Species*	Capture method	Activity	Stratum of activity	Nesting stratum	Nest / Colony	Worker	Queen	Alate
Camponotus sp.5	Hc		B					
Camponotus sp.6	Hc	N	AB	A				
	Hc	N	B					
	Hc	N	BC	B	+ / -	90		
					+ / -	69		
Camponotus sp.7					+ / -	37		
Camponotus sp.8					+ / -	39		
Camponotus sp.9	Hc	N	B					
Camponotus sp.10	Hc	D/N	AB					
Camponotus sp.11	Hc,P	D/N	AB	A				
Camponotus sp.13	Hc	D	B					
Camponotus sp.14	Hc	D	B					
Camponotus sp.15	Hc	N	B					
Camponotus sp.16	Hc	N	ABC	C				
Camponotus sp.17	Hc,W		AB	AB				
Camponotus sp.20	Hc	D/N	B	B_b				
Camponotus sp.21	P		A					
Camponotus sp.22	Hc	N	B					
Camponotus sp.23	Hc	D	B					
Camponotus sp.25	Hc	D	B					
Camponotus sp.26	Hc	D	B					
Camponotus sp.27	Hc		f					
Camponotus sp.29	Hc	D	B	B	+ / ?	66	1	8•,14••
Camponotus sp.30	Hc	N	BC					
Camponotus sp.32	Hc		i					
Camponotus sp.33	Hc		C	C	+ / -	20		1•
Camponotus sp.34	Hc	N	B					
Camponotus sp.35	Hc		B	B				
Camponotus sp.36	Hc	D	C					
Camponotus sp.37	Hc		C	C				
Camponotus sp.38	Hc	D	C					
Camponotus sp.39	Hc	D	C					
Camponotus sp.41	Hc		B	B				
Camponotus sp.42	Hc		C					
Camponotus sp.43	Hc		A	A	- / -	2207		460••
Camponotus sp.45	Hc	D	C					
Camponotus sp.46	Hc	D	C					
Camponotus sp.47	Hc	D	C					
Camponotus sp.48	Hc		A	A				
Camponotus sp.49	Hc		B	B				
Cladomyrma maschwitzi Agosti	Hc	D	C	C	+ / -	377		
	Hc		B	B				
Echinopla melanarctos F. Smith	Hc		B	B	+ / +	3341	1	1••
Echinopla striata F. Smith					- / -	3150		60•,70
Echinopla tritschleri Forel	Hc	D	AB	A	+ / ?	59		10••
Echinopla sp.1	Hc	D	(A)BC	B	+ / -	25		3••
Echinopla sp.5	Hc		B	B	+ / -	20		
Echinopla sp.6	Hc	D	B					
Euprenolepis procera Emery	Hc		B					
Euprenolepis sp.2	Hc	D	B					
Gesomyrmex sp.1	Hc, P	D/N	AB	A	? / -	331		
Myrmoteras estrudae Agosti	Hc		AA	AA				
Oecophylla smaragdina (Fabricius)	Hc	D	C	C	+ / -	38		
Paratrechina sp.1	B,W		A					
	Hc,P	D	AB	B	+ / -	1518		
	Hc,P,B,W	D/N	AA,AB	AB	? / ?	54	1	
					+ / ?	118	1	
					+ / -	22		
					+ / -	93		
					? / -	103		
					+ / -	14		
Paratrechina sp.2					+ / -	34		1•,5••
Paratrechina sp.3					+ / -	41		2•
Paratrechina sp.4	Hc,P,B,W	D/N	AA,A					
Paratrechina sp.5	Hc,B		AB	B	+ / +	110	1	2•,1••

Appendix: Table I (continued 8)

Subfamily *Genus / Species*	Capture method	Activity	Stratum of activity	Nesting stratum	Nest / Colony	Worker	Queen	Alate
Paratrechina sp.6	Hc,P	D	AA,A					
Paratrechina sp.7	Hc	D	A	A				
Paratrechina sp.8	Hc	D	AB	B				
Paratrechina sp.9	Hc,P,B,W	D	AB	A				
Paratrechina sp.10	Hc	D	AB	B				
Paratrechina sp.11	Hc	D	A					
Plagiolepis cf. *bicolor* Forel	Hc,P	D	AB					
Plagiolepis sp.2	W		A					
Polyrhachis (*Campomyrma*) sp.39	Hc	D	B					
Polyrhachis (*Cyrtomyrma*) *rastellata* (Latreille)	Hc	D	B					
Polyrhachis (*Hemioptica*) *boltoni* Dorow &	Hc	D	C	C	+ / -	104		
Polyrhachis (*Myrma*) *carbonaria* F. Smith	Hc	D	B	B	+ / -	38		
Polyrhachis (*Myrma*) *illaudata* Walker					+ / ?	33	1	
Polyrhachis (*Myrma*) *striata* Mayr	Hc	D	BC	B				
Polyrhachis (*Myrma*) *sumatrensis* F. Smith								
Polyrhachis (*Myrmatopa*) *lilianae* Forel	Hc	D	AB	AB	+ / -	10		3••
Polyrhachis (*Myrmatopa*) *schang* Forel	Hc	D/N	BC	B				
Polyrhachis (*Myrmatopa*) sp.24	Hc	D	ABC	A				
Polyrhachis (*Myrmatopa*) sp.38	Hc	N	A	A	? / -	130		
Polyrhachis (*Myrmatopa*) *abdominalis* F. Smith					+ / -	91		
Polyrhachis (*Myrmatopa*) *armata* (Le Guillon)	Hc		AB	B	+ / ?	119	1	
Polyrhachis (*Myrmatopa*) *bicolor* F.Smith	Hc	D/N	BC	A	+ / -	23		23•,7••
	Hc		C					
Polyrhachis (*Myrmatopa*) *calypso* Forel	Hc		B					
Polyrhachis (*Myrmatopa*) cf. *cryptoceroides*	Hc	D	BC	C	+ / -	18		
Polyrhachis (*Myrmatopa*) *flavoflagellata*								
Polyrhachis (*Myrmatopa*) *furcata* F.Smith	Hc	D	ABC	B				
Polyrhachis (*Myrmatopa*) *gestroi* Emery	Hc	D	BC	BC	+ / -	49		5•,8••
Polyrhachis (*Myrmatopa*) *hector* F. Smith					+ / -	16		2•,14••
Polyrhachis (*Myrmatopa*) *hippomanes* F. Smith	Hc	D	BC	B	+ / ?	7	1	
Polyrhachis (*Myrmatopa*) cf. *jerdoni* Forel	Hc	D	C	C				
Polyrhachis (*Myrmatopa*) *muelleri* Forel	Hc	D	B					
	Hc	D	B	B	+ / +	2901	1	104••
Polyrhachis (*Myrmatopa*) cf. *mutata* F. Smith					+ / +	934	1	
Polyrhachis (*Myrmatopa*) *ochraceae* Karawajew	Hc	D	B					
Polyrhachis (*Myrmatopa*) *pressa* Mayr	Hc	D	BC	BC	- / -	35		1••
Polyrhachis (*Myrmatopa*) *rufipes* F. Smith	Hc	D	B	B	+ / +	55	1	1•,10••
Polyrhachis (*Myrmatopa*) sp.13					+ / -	68		2••
Polyrhachis (*Myrmatopa*) sp.21	Hc	D	B	B	+ / -	42		
Polyrhachis (*Myrmatopa*) sp.23					+ / -	20		
Polyrhachis (*Myrmatopa*) sp.27	Hc		BC	BC	+ / ?	15		10•,1••
Polyrhachis (*Myrmatopa*) sp.29					+ / -	10		1••
Polyrhachis (*Myrmatopa*) sp.31					+ / -	16		7••
	Hc		B	B				
	Hc		C					
Polyrhachis (*Myrmatopa*) sp.37								
Polyrhachis (*Myrmatopa*) sp.41	Hc	D	C					
Polyrhachis (*Myrmatopa*) sp.42	Hc	D	AB	B	? / -	161		4•,19••
Polyrhachis (*Myrmatopa*) sp.44	Hc	D	BC					
Polyrhachis (*Myrmothrinax*) *frauenfeldi* Mayr	P		A					
Polyrhachis (*Myrmothrinax*) cf. *thrinax* Roger	Hc	D	C					
Polyrhachis (*Myrmothrinax*) sp.40	Hc							
Polyrhachis (*Myrmothrinax*) sp.43	Hc							
Polyrhachis (*Polyrhachis*) bellicosa F. Smith sp.	Hc	D	B	B	+ / -	40		2••
Polyrhachis (*Polyrhachis*) *ypsilon* Emery					+ / -	28		12••
Polyrhachis (*Polyrhachis*) *bihamata* (Drury)					+ / -	32		1•,14••
Polyrhachis (*Polyrhachis*) sp. near *bihamata*					+ / ?	67	1	1•,1••
Prenolepis naoroji Forel	Hc	D	B	B				
Prenolepis sp.2	Hc	D	BC					
Pseudolasius sp.1	Hc	D	C	C				
Pseudolasius sp.1	Hc	D	B					

Appendix: Table I (continued 9)

Subfamily *Genus / Species*	Capture method	Activity	Stratum of activity	Nesting stratum	Nest / Colony	Worker	Queen	Alate
	Hc	N	B	B	+ / ?	64		26•
	Hc	D	B					
	Hc	N	BC	C				
Pseudolasius sp.2	Hc		C					
Pseudolasius sp.3	Hc	D	AB					
Pseudolasius sp.4								
Pseudolasius sp.5	Hc		AB					
Pseudolasius sp.6	Hc	D	BC	A				
Pseudolasius sp.7								
Pseudolasius sp.8	Hc	D	B					
	Hc	N	AB					
	Hc	D	B					
	W_s		A					
	Hc,P,B,W		AA,A	AA,A	+ / -	45		93••
					+ / ?	170		2••
					+ / ?	371	2	223••
					? / -	36	1	
					+ / -	17		24••
					- / -	19	2	10•,1••
	Hc		AA	AA				
	Hc,W		A	A				
	Hc		A					
	Hc		A					
	Hc		A					
	Hc		A					
	Hc		A					

Appendix: Table II Ranking of ant genera by species number / percentage of total species number. [(A) Aenictinae, (C) Cerapachyinae, (Dol) Dolichoderinae, (Dor) Dorylinae, (F) Formicinae, (L) Leptanillinae, (M) Myrmicinae, (Po) Ponerinae, (Ps) Pseudomyrmecinae.]

genus	Species number	%	genus	Species number	%	genus	Species number	%
Camponotus (F)	50	10.2	Pristomyrmex (M)	5	1.0	Diacamma (Po)	1	0.2
Polyrhachis (F)	45	9.2	Solenopsis (M)	5	1.0	Dorylus (Dor)	1	0.2
Pheidole (M)	39	8.0	Cardiocondyla (M)	4	0.8	Dysedrognathus	1	0.2
Crematogaster (M)	30	6.1	Meranoplus (M)	4	0.8	Eurhopalothrix (M)	1	0.2
Tetramorium (M)	19	3.9	Smithistruma (M)	4	0.8	Gesomyrmex (F)	1	0.2
Strumigenys (M)	18	3.7	Amblyopone (Po)	3	0.6	Harpegnathos (Po)	1	0.2
Technomyrmex	15	3.1	Dolichoderus (Dol)	3	0.6	Iridomyrmex (Dol)	1	0.2
Cerapachys (C)	14	2.9	Pheidologeton (M)	3	0.6	Leptanilla (L)	1	0.2
Hypoponera (Po)	14	2.9	Proceratium (Po)	3	0.6	Lophomyrmex (M)	1	0.2
Oligomyrmex (M)	14	2.9	Acanthomyrmex (M)	2	0.4	Lordomyrma (M)	1	0.2
Pachycondyla (Po)	13	2.7	Cryptopone (Po)	2	0.4	Mayriella (M)	1	0.2
Leptogenys (Po)	12	2.5	Dilobocondyla (M)	2	0.4	Myopias (Po)	1	0.2
Aenictus (A)	12	2.5	Discothyrea (Po)	2	0.4	Myopone (Po)	1	0.2
Monomorium (M)	11	2.3	Emeryopone	2	0.4	Myrmoteras (F)	1	0.2
Paratrechina (F)	11	2.3	Euprenolepis (F)	2	0.4	Mystrium (Po)	1	0.2
Vollenhovia (M)	11	2.3	Myrmicaria (M)	2	0.4	Odontoponera (Po)	1	0.2
Platythyrea (Po)	10	2.0	Odontomachus (Po)	2	0.4	Oecophylla (F)	1	0.2
Tapinoma (Dol)	9	1.8	Plagiolepis (F)	2	0.4	Paratopula (M)	1	0.2
Pseudolasius (F)	9	1.8	Ponera (Po)	2	0.4	Philidris (Dol)	1	0.2
Myrmecina (M)	7	1.4	Prenolepis (F)	2	0.4	Prionopelta (Po)	1	0.2
Gnamptogenys (Po)	7	1.4	Probolomyrmex (Po)	2	0.4	Proatta (M)	1	0.2
Tetraponera (Ps)	7	1.4	Recurvidris (M)	2	0.4	Rhopalomastix (M)	1	0.2
Acropyga (F)	6	1.2	Anoplolepis (F)	1	0.2	Rhoptromyrmex (M)	1	0.2
Anochetus (Po)	6	1.2	Carebara (M)	1	0.2	Rostromyrmex (M)	1	0.2
Echinopla (F)	6	1.2	Centromyrmex (Po)	1	0.2	Tetheamyrma (M)	1	0.2
Cataulacus (M)	5	1.0	Cladomyrma (F)	1	0.2			

27 The Bird Community at Pasoh: Composition and Population Dynamics

Charles M. Francis[1] **& David R. Wells**[2]

Abstract: Pasoh Forest Reserve (Pasoh FR) has been the site of several bird population studies, starting in the late 1960s and continuing through the present. So far, at least 220 species of birds have been reported using the primary forest in Pasoh FR, with an additional 13 species in the regenerating forest and edge habitats. Of these, about 184 species are residents that are likely to breed within the primary forest. Estimated population densities for several common babblers were 11–70 individuals per 100 ha suggesting total populations of 200–1,500 individuals within Pasoh FR. However, most species were much less common, many with only a few pairs per 100 ha, or in some cases only a few pairs in Pasoh FR. Such populations are assumed to be too small to be viable in the long term without immigration from outside Pasoh FR. Many species were long-lived, with representatives of 14 species having been captured at least 10 years after initial ringing. The longevity record so far, is held by a Rufous-collared Kingfisher (*Actenoides concretus*) that was ringed as an adult in 1972 and recaptured 21 years later. Apparent survival rates for 14 resident species, after correcting for lower returns of newly ringed birds, averaged 77% per year. As such, populations could persist for many years even if breeding were impaired. Because Pasoh FR is now surrounded on three sides by oil palm plantations, and is too small to support viable populations of many species in isolation, the future of the bird community may be at risk. Further research, to estimate population densities of all species and possible movements to and from the adjoining hill forest, would be valuable to determine which species are most at risk, and to try to identify mitigating measures. Such research would also be of considerable value for understanding the relationships between extinction risk, population movements, population dynamics and rarity.

Key words: avian diversity, density, longevity, population viability, survival rates.

1. INTRODUCTION

Although the overall distribution of rain forest birds in Malaysia is relatively well known (Medway & Wells 1976; Sheldon et al. 2001; Smythies 2000; Wells 1999), details of the community structure, population densities and population dynamics have been studied at relatively few sites. Fogden (1972, 1976) carried out an intensive study between 1964 and 1966 of the population dynamics, seasonality, and densities of birds on 20 ha of Semengo Forest Reserve, 20 km south of Kuching in Sarawak. McClure (1969) estimated densities of birds observed from a canopy platform in lower hill forest near Ulu Gombak in Selangor, Peninsular Malaysia. Medway & Wells (1971) described the bird community composition and estimated densities for selected species at Kuala Lompat, Krau Game Reserve, Pahang, based on a series of short-term surveys. Several studies have compared relative abundance of

[1] Bird Studies Canada, Port Rowan, Ontario, Canada.
Present address: National Wildlife Research Centre, Canadian Wildlife Service, Ottawa, Ontario K1A OH3, Canada. E-mail: Charles.Francis@ec.gc.ca
[2] Serendip, Old Farm, Illington, Thetford, Norfolk IP24 1RP, UK.

birds, as indexed by observation or capture rates, between primary and selectively logged forests in both Peninsular Malaysia and Sabah (Johns 1986, 1996; Lambert 1992; Zakaria & Francis 1999, 2001).

One of the best studied bird communities in Malaysia is at Pasoh Forest Reserve (Pasoh FR). Pasoh FR consists of approximately 600 ha of primary lowland dipterocarp forest surrounded by a buffer area of regenerating forest that was selectively logged in the 1950s and bordered by hill forest to the east (Fig. 1; Soepadmo 1978; Wong 1985; Chap. 2). Some portions in the northern part of Pasoh FR were logged again in the 1980s. A series of ornithological studies has been carried out at Pasoh FR starting in the late 1960s, and continuing intermittently to the present. Collectively, these studies have provided information on bird community composition, differences between regenerating and primary forest, changes in the community over time, population densities, longevity and survival rates of many species.

In this chapter, we review the history of ornithological studies at Pasoh FR and summarize what is known about the bird community composition in the primary forest, based on data from a variety of sources. We then describe some of the key findings from our own long-term mist-netting studies, from 1971 to 1996, on

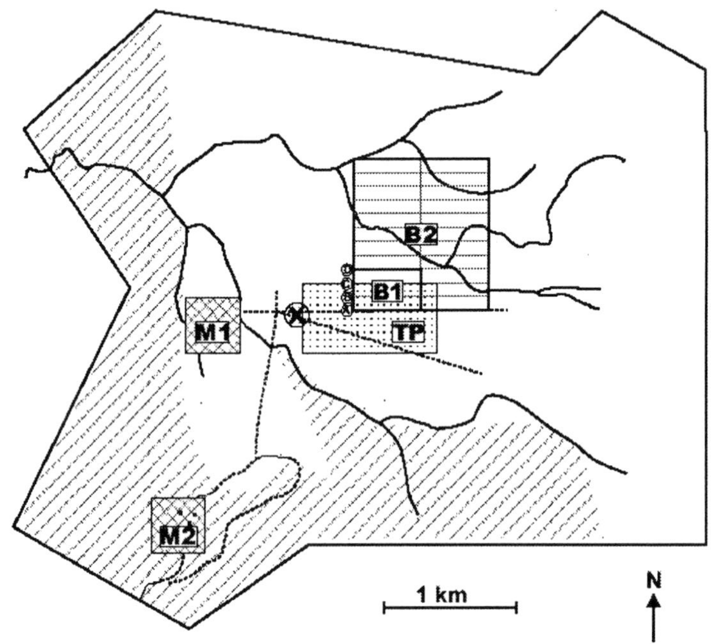

Fig. 1 Map of Pasoh FR, showing study plots referred to in the text, including the 15 ha (**B1**) and 110 ha (**B2**) bird plots established by D. R. Wells and C. M. Francis, the plots in primary (**M1**) and regenerating (**M2**) forest established by M. Wong, the 50 ha tree plot (**TP**), and the former 35m tree tower (**X**). Trails within the 15 ha and 110 ha plots are shown by narrow lines. The 4 E-W trails in the 15 ha plot are labelled A to D from south to north. The main access trails are represented by dashed lines, streams by solid lines, and selectively logged forest by diagonal hatching (boundaries are approximate). The reserve is surrounded by oil palm on all sides except the east where it adjoins a long narrow tract of hill forest (see chap. 2 for details).

population densities, longevities and survival rates of birds in this forest—data important for assessing the viability of bird populations in Pasoh FR. Finally, we close with speculation on what the future may hold for this community, particularly given the relatively small size of Pasoh FR and its relative isolation from other areas of natural forest, and give suggestions for future research on this bird community.

2. ORNITHOLOGICAL STUDIES AT PASOH FR

The first detailed studies of birds at Pasoh FR were by D. R. Wells, who commenced standardized mist-netting and ringing in 1968 in selectively logged forest near the edge of Pasoh FR. That site was subsequently cleared and converted to oil palm plantations. In 1971, he established a new study plot encompassing 15 ha within the primary forest core of Pasoh FR (Fig. 1), in association with the International Biological Programme (Soepadmo 1978). Intensive netting was carried out in that plot at roughly 1–2 month intervals through 1974. In addition, all species of birds heard or seen within the forest during netting sessions or other field work were recorded. Extensive observations were made from a 35 m high wooden tree tower that was erected in 1971 around a large tree about 50 m to the west of the current location of the 50 ha plot (Fig. 1). This tree tower became unsafe for use in the late 1980s when the supporting tree died, and it has since fallen.

Data from that study were used to estimate populations of several understorey species (Wells 1978), incorporated into various compilations of data on Malaysian birds (e.g. Medway & Wells 1976; Wells 1999), and used to understand the dynamics and seasonal activities of certain overwintering northern migrants (Wells 1969, 1990). Data from Pasoh FR were also used as a basis for comparing bird community composition in Malaysia with that in lowland forests of Gabon, Africa (Wells 1988b).

Netting on the original 15 ha plot resumed in 1979 by C. M. Francis and D. R. Wells, and was continued at irregular intervals through 1996, with an expansion of the study area in 1993 to encompass 110 ha. Data from the larger plot were used to estimate revised population densities for selected species of babblers (Zakaria & Francis 1999).

In a separate study, M. Wong established two 16 ha study plots, one in primary forest about 1 km west of Wells' 15 ha plot, and the other in regenerating forest that had been selectively logged in the 1950s (Fig. 1). This study involved intensive netting and observations on both plots from 1978 through 1980 to evaluate community composition and phenology in each area (Wong 1985, 1986). The results indicated that even 25 years after logging, community structure, as measured by relative abundance of species, was not the same as that in the primary forest.

Since 1991, under a NIES-FRIM-UPM project (Chap. 1), H. Nagata at the National Institute for Environmental Studies (NIES), Japan, initiated a study to compare bird communities in the primary forest and edge habitats, using mist-netting and artificial nest experiments. Preliminary results of that research are described in Nagata et al. (1998).

In addition to these long-term projects, a variety of shorter-term studies on birds have been carried out at Pasoh FR. Karr (1980) used results from short term mist-netting to compare bird community composition in Malaysian forests with that elsewhere in the tropics, providing a base for the much more detailed evaluation of trophic structure by Wong (1986). Cooper & Francis (1998) used artificial nests to show that predation rates were significantly higher in the regenerating forests near the edge, than in the primary forest. Styring & Ickes (2001) looked at niche differentiation in woodpecker communities. Many birders visiting the site have contributed observations to enhance understanding of the community composition and populations at this site, and the site continues to be a regular destination for birding tours.

3. BIRD COMMUNITY COMPOSITION

Over the course of these various research projects, at least 220 species have been reported using the primary forest at Pasoh FR (see Appendix). These include 195 species resident in the plot at Pasoh FR, 7 of which usually use edge and scrub habitats but have also been recorded within the primary forest. Many of the records for these "edge" species may have been dispersing individuals that did not remain within the primary forest—for example, the only record of Pied Fantail was a juvenile bird. However, in some areas, even within the primary forest, tree falls have created large gaps, in some cases fairly extensive, which may be suitable habitat for some of these "edge" species.

The residents also include a few species of aerial insectivores that forage over the forest, feeding on insects that originated in the forest, but that are unlikely to breed within Pasoh FR due to a lack of suitable nesting habitat. For example, swiftlets (see Appendix for scientific names) require caves or buildings for nesting, while House Swifts usually nest on cliffs or the sides of buildings. Swifts may regularly forage over an area of 20–30 km or more away from their nesting sites on a daily basis, so it is quite possible that some of these birds were foraging over Pasoh FR while breeding. Waugh & Hails (1983) found that many aerial insectivores selectively feed over forested habitats, presumably because these are a better source of food than open areas.

The remaining species are migratory, and do not breed around Pasoh FR. Most of these are north temperate-tropical migrants that occur in Pasoh FR during the northern winter, although some may use Pasoh FR as a stopover site on their way to wintering areas elsewhere in Malaysia or further south. One species, the Blue-throated Bee-eater, breeds elsewhere in Peninsular Malaysia, and disperses during the non-breeding season. Some of the pigeons, such as Emerald Doves, are also known to move considerable distances within Peninsular Malaysia, though they may breed in Pasoh FR. Of the migrants, 18 occur regularly within the primary forest, while 7 mainly use edge or other habitats and only occasionally use primary forest.

The existing species list is likely to include nearly all of the primary forest species resident within Pasoh FR. Only a few resident species were added in the last 10 years of the study (compare Appendix with Wells 1988a). Apart from the non-breeding aerial insectivores, which were excluded by Wells (1988a), most of these were species that occur mainly in disturbed forest and are unlikely to breed in the primary forest. The exceptions were Chestnut-naped Forktail and Crested Fireback, both of which are usually found in association with streams and appear to be scarce in Pasoh FR, and Spectacled Spiderhunter, a canopy species which may have been overlooked in the past. About 10–15 additional species resident in lowland forest might be expected to occur in Pasoh FR based on range (Medway & Wells 1976), but many of these are usually associated with habitat features such as large rivers that do not occur at Pasoh FR. Several are large conspicuous species that would almost certainly have been detected if they were resident in Pasoh FR. Others, if they do occur, are most likely very rare or localized, and unlikely to have self-sustaining populations.

A number of additional species that live in the oil palm plantations and other disturbed habitats near Pasoh FR regularly make use of the edges of the regenerating forest and the clearings around the FRIM house at Pasoh FR (see footnote to Appendix). Several of these may be expected to enter the primary forest at least occasionally, in the same fashion as some of the other edge species that we have recorded.

The list of migratory species is less likely to be complete. Any forest-dwelling species that migrates to Peninsular Malaysia could potentially stop at Pasoh FR, at least occasionally as a vagrant. However, it seems likely that most of the species that over-winter at Pasoh FR annually, have been recorded.

Few data are available on bird communities from elsewhere in Peninsular Malaysia for comparison with the species list at Pasoh FR. Several additional species have been recorded from primary lowland forests around Kuala Lompat in the Krau Game Reserve or around Taman Negara, both in Pahang about 150–200 km north of Pasoh FR (pers. obs.), although complete species lists have not been prepared for those sites. Some of those additional species are associated with the permanent rivers that occur in both Pahang sites but not Pasoh FR. In addition, the much larger size of those reserves may be sufficient to support populations of some of the larger nomadic hornbills and pigeons that are now rare or absent at Pasoh FR. Nevertheless, the species list, at least for primary forest residents, appears to be only slightly lower than what might have been expected to have occurred in most areas of lowland forests of south-central Peninsular Malaysia, prior to the clearing or logging of much of that forest.

4. DEMOGRAPHIC STUDIES

4.1 Field methods

The most extensive data on demographic methods come from our own mist-netting studies carried out between 1972 and 1996. As many of these results have not been previously published, we first describe in some detail our field methods.

The main study plot, encompassing 15 ha within the primary forest, was about 2 km from the entrance of the forest (Fig. 1). This plot, established in 1971, consisted of four parallel east-west trails, 500 m long and 100 m apart, connected at the ends by north-south trails 300 m long. The lanes were labelled A through D from south to north. Approximately every 50 m along these trails, two alternate mist net lanes were cut at roughly right angles to the trails, for a total of 52 pairs of mist-net lanes. In 1979, when netting resumed, many of these lanes had become overgrown or blocked by tree falls, and only one out of each pair was re-opened. In 1985, a 50 ha plot was established to study tree dynamics (Chap. 1), which overlapped two-thirds of the bird-netting plot. Within this plot, if a net lane became blocked (e.g. due to a tree fall), the net lane was moved to the trail or another area with no obstacles, to minimize damage to any marked trees or saplings that were part of the tree dynamics study.

In 1994, the study area was expanded northwards and eastwards to encompass an area of 110 ha. This expanded area consisted of twelve parallel east-west trails 1 km long, and 100 m apart, with north-south trails at the eastern and western edges and up the centre (Fig. 1). The original 15 ha plot formed the southwestern corner of this larger study area. Some northern portions of this expanded study area were within forest that had been illegally logged in the 1980s, though the boundaries of the logging have not been mapped. Prior to the netting session in 1995, a severe storm felled many trees in the northeastern corner and we were unable to reopen parts of two of the trails, so the effective netting area was reduced by about 5–10 ha.

From 1971–1974, netting took place at 1–2 month intervals (Table 1). On most occasions, after the first few sessions, 52 nets were set along one out of each pair of net lanes. Nets were switched to the other net lane of each pair in alternate netting sessions, to reduce the degree to which birds became accustomed to the net locations. Most nets were 12 m long by 2–2.5 m high with 36 mm mesh and were set continuously (day and night), for 3–5 days. From 1979 to 1993, netting took place

Table 1 Mist-netting sessions at Pasoh FR in the 15 hectare bird study plot (Fig. 1). Most sessions covered the whole plot, except for a few when only the southern 2 (AB) or 3 (ABC) lines were netted. In 1994–1996, all four original lines were covered as well as additional lanes throughout the 110 ha plot. From 25–52 nets were set each day. Ranges of dates include days spent moving nets to new locations if only part of a grid was netted – each net lane was set for 3–5 days per session. Nets were set 24 hours per day (except during heavy rain) until 1987, after which they were usually opened at dawn (07:30–08:00) and closed in the late afternoon (18:00–19:30).

Year	Session	Dates	Number of New Birds	Number of Returns	Notes
1971	1	Jul 14 - Jul 17	90	-	AB only
1972	2	Feb 13 - Feb 17	141	39	AB only
	3	Aug 03 - Aug 09	147	99	ABC only
	4	Sep 12 - Sep 15	103	96	ABC only
	5	Sep 26 - Sep 29	63	83	
	6	Oct 24 - Oct 28	97	106	
	7	Nov 28 - Dec 02	69	115	
1973	8	Jan 15 - Jan 19	75	102	
	9	Feb 15 - Feb 19	79	114	
	10	Mar 19 - Mar 23	92	87	
	11	Apr 21 - Apr 25	94	109	
	12	Jun 04 - Jun 08	83	85	
	13	Jul 11 - Jul 15	77	100	
	14	Aug 14 - Aug 18	78	99	
	15	Sep 18 - Sep 22	59	68	
	16	Oct 16 - Oct 20	53	70	
	17	Nov 20 - Nov 24	48	67	
	18	Dec 18 - Dec 22	43	106	
1974	19	Jan 15 - Jan 19	45	140	
	20	Feb 14 - Feb 18	38	106	
	21	Mar 15 - Mar 19	39	85	
	22	Apr 18 - Apr 21	52	88	
1979	23	Aug 15 - Aug 22	160	23	
1981	24	Aug 29 - Sep 03	126	44	AB only
1982	25	Feb 25 - Mar 01	111	99	ABC only
1986	26	Aug 15 - Aug 19	78	17	AB only
1987	27	Feb 27 - Mar 09	172	56	
	28	Aug 08 - Aug 16	132	99	
1991	29	Feb 04 - Feb 10	108	25	
	30	Aug 23 - Aug 27	69	37	
1993	31	Jun 22 - Jun 24	33	22	AB only
1994	32	Jun 11 - Jul 19	666	81	110 ha
1995	33	Jul 15 - Aug 08	475	239	110 ha
1996	34	Jun 06 - Jun 28	401	310	110 ha

opportunistically, depending upon the availability of personnel (Table 1). On most occasions, nets were set on two trails at a time for 3–4 days (e.g. trails A and B in Fig. 1), then moved to the other two trails (trails C and D in Fig. 1). During a few sessions, only some of the trails were netted (Table 1). Starting in 1991, nets were no longer run 24 hours per day, but instead were opened shortly after dawn and closed about 1 hour before dusk. From 1994–1996, netting took place throughout the 110 ha plot. Nets were set at approximately 50 m intervals along two 1-km trails at a time, and moved to new trails every 3–4 days. Outside of the 15 ha plot, nets were usually set along the trails, to avoid having to cut additional net lanes. Nets were checked at least every two hours throughout the day, with a final check just after dusk (if the nets were being run overnight). Birds were brought to a central location for processing, ringed with a uniquely numbered ring, measured and weighed and then released. The net location of each capture was recorded. Birds captured after dusk were held overnight in a cloth bag and released at dawn. Starting in 1991, many of the birds also received a unique (within species) combination of colour rings, to enable identification of individuals in the field through observations.

4.2 Population density

The only species for which any quantitative information has been obtained on population sizes, are those understorey species that can be regularly captured in mist nets. Over the course of the 25 year study, a total of 2,883 different individuals of 104 species were netted in the 15 ha bird study plot. A further 1,313 individuals were captured from 1994–1996 in the expanded 110 ha plot, outside of the original 15 ha plot, including 16 more species. In addition, there have been 3,016 returns of birds captured in sessions subsequent to those when they were originally ringed (excluding multiple captures within the same netting session). Obviously, total numbers captured do not provide information on population size, owing to turnover in the bird populations over time (mortality, emigration and recruitment) as well as the fact that species vary in their susceptibility to mist nets. However, netting data can be used in various ways to estimate population sizes.

Wells (1978) determined the number of distinct individuals of about 50 species of birds captured within the 15 ha plot between February and September 1973, based on the assumption that there was relatively little population turnover during this period, which is the main breeding season for many species. However, this approach does not consider variation among species in capture probabilities (which can be quite substantial—see section 4.5). Also, with such a small plot, uncertainty in the extent to which territories of birds extend outside of the plot can have a substantial impact on density estimates.

Zakaria & Francis (1999) used open population capture-recapture models (Pollock et al. 1990) to estimate populations of selected babbler species within the 110 ha plot using data from the three capture sessions in 1994–1996. This approach corrects for variation in capture probabilities among species, but requires the assumption that all individuals of a species have equal capture probabilities (Pollock et al. 1990). This may be violated if some birds learn to avoid nets, but because our netting sessions were a year apart during this period (Table 1), this should be a minimal problem.

Converting these numbers to densities still requires assumptions about the area being sampled. If we assume that the territories of these birds were entirely within the plot, population estimates can be converted to densities per 100 ha by dividing by 1.1 (Table 2). If the effective sampling area included an additional area of 100 m around the outside of the plot (i.e. for a total of 156 ha), then estimates are about 30% lower (Table 2). Insufficient data are available on territory sizes to determine the most appropriate value, but 100 m may be reasonable for many species. In addition, the sample probably includes a number of individuals that were not resident within the study areas (see discussion of transients, under survival analyses, below). As such, even the lower values may over-estimate true densities.

Despite these limitations, these data can be used to obtain some preliminary estimates of population sizes within Pasoh FR. Based on the lower estimates, population densities for these 11 babblers ranged from 11–75 individuals per 100 ha with a mean of 29. If we assume uniform densities throughout Pasoh FR, including the selectively logged forests, this suggests a total population of 200–1,500 individuals per species within 2,000 ha. Detailed information on demographic parameters, such as mean values and annual variation in recruitment and survival rates, is required for a detailed population viability analysis, to assess whether these populations are large enough for long-term persistence. In the absence of such information, some conservation biologists have suggested a rule of thumb that 500–1,000 individuals of a species are required to ensure a high probability of long-term persistence of a population (Primack 1993). If these guidelines apply to tropical forest birds, and if the habitat remains intact over time, then only about half of these babbler species have population sizes large enough for a high probability of long-term persistence if Pasoh FR becomes isolated from other populations.

Furthermore, most species of birds in Pasoh FR are much less common than these babblers, which are among the most frequently captured, and hence most abundant of the understorey species within Pasoh FR (Table 3; Wells 1978). Many species occur at much lower densities with only one or a few pairs per 100 ha. Some of the largest species, such as raptors or hornbills, may only have a few family

Table 2 Estimated densities for selected, relatively common species of babblers at Pasoh FR from 1994–1996, based on population estimates from open population capture-recapture models within the 110 ha plot (Zakaria & Francis 1999). First column assumes that all birds remain within the 110 ha plot, while the second assumes their territories include a 100 m buffer around the plot (increasing effective plot area to 156 ha). The latter is probably more realistic, but may still over-estimate densities if some individuals were not resident within the plot, or had territories that extended > 100 m outside of the plot. See Appendix for scientific names.

Species	Estimated density (birds/100 ha)	
	Assumingn plot 110 ha	Assumingn plot 156 ha
Black-capped Babbler	33	23
Black-throated Babbler	29	20
Chestnut-rumped Babbler	41	29
Chestnut-winged Babbler	107	75
Ferruginous Babbler	19	13
Fluffy-backed Tit-Babbler	61	43
Horsfield's Babbler	21	15
Moustached Babbler	51	36
Rufous-crowned Babbler	52	37
Scaly-crowned Babbler	25	18
Short-tailed Babbler	15	11

Table 3 Numbers of recaptures and longevity records for birds ringed within the 15 ha plot at Pasoh FR from 1971–1991. Recaptures are included up until 1996, including some birds that were recaptured outside of the original 15 ha plot. "Max" represents the maximum interval (in years) between original ringing and oldest recapture for that species. See Appendix for scientific names.

Species	Number ringed	Number recaptured	Years between ringing and recapture										
			1	2	3	4	5	6	7	8	9	10	Max
Emerald Dove	10	-	-	-	-	-	-	-	-	-	-	-	-
Gould's Frogmouth	11	2	1	1	-	-	-	-	-	-	-	-	2.6
Cinnamon-rumped Trogon	19	9	1	-	-	-	1	-	-	-	2	-	9.6
Scarlet-rumped Trogon	11	2	-	-	1	-	-	-	-	-	-	-	3.5
Oriental Dwarf Kingfisher	68	15	2	1	-	-	-	-	-	-	-	-	2.1
Rufous-collared Kingfisher	18	9	3	2	-	-	-	-	-	-	-	1	20.8
Rufous Piculet	23	8	2	-	-	1	1	-	-	-	-	-	5.7
Banded Yellownape	9	4	-	-	1	-	2	-	-	-	-	-	5.9
Buff-necked Woodpecker	44	11	4	-	1	-	-	-	-	-	-	-	3.4
Maroon Woodpecker	14	8	2	-	-	-	-	-	-	-	-	-	1.5
Green Broadbill	13	3	1	-	-	-	1	-	-	-	-	-	5
Garnet Pitta	13	3	-	-	-	-	-	-	-	-	-	-	0.9
Hooded Pitta	11	1	-	-	-	-	-	-	-	-	-	-	0.1
Banded Pitta	24	-	-	-	-	-	-	-	-	-	-	-	-
Cream-vented Bulbul	18	5	-	-	-	-	-	-	1	-	-	-	7.8
Grey-cheeked Bulbul	39	21	5	2	-	-	1	1	-	-	-	1	14.5
Yellow-bellied Bulbul	97	50	12	9	-	-	4	1	2	-	1	1	11.5
Hairy-backed Bulbul	79	26	3	2	2	1	2	-	1	-	1	-	9
Greater Racket-tailed Drongo	15	3	1	-	-	-	-	-	-	-	-	-	1.5
Black-capped Babbler	79	28	6	1	-	1	1	-	1	-	1	1	13
Short-tailed Babbler	172	58	4	5	1	-	2	1	3	-	-	-	7.9
White-chested Babbler	11	1	-	-	-	-	-	-	-	-	-	-	0.3
Ferruginous Babbler	77	32	7	2	2	-	2	1	1	-	2	-	9.8
Moustached Babbler	146	61	15	7	3	2	2	1	3	3	1	-	9.3
Sooty-capped Babbler	13	1	-	1	-	-	-	-	-	-	-	-	2.8
Scaly-crowned Babbler	140	69	7	14	1	4	1	-	1	2	-	2	12.5
Rufous-crowned Babbler	51	16	1	2	-	1	-	2	-	-	-	-	6.6
Grey-breasted Babbler	49	19	3	1	-	-	-	-	1	-	-	1	13.6
Striped Wren-Babbler	83	23	4	3	-	1	-	-	-	-	-	-	4
Large Wren-Babbler	29	14	2	-	-	-	1	1	-	1	-	-	8.1
Grey-headed Babbler	38	17	1	-	-	-	1	-	-	-	1	-	9.3
Chestnut-rumped Babbler	113	50	4	3	5	3	-	1	-	1	-	1	11.5
White-necked Babbler	34	14	1	-	-	-	1	-	-	-	-	-	5.6
Black-throated Babbler	81	32	2	2	-	2	2	2	1	-	1	-	9.3
Chestnut-winged Babbler	103	32	5	5	-	-	1	-	-	-	-	-	6
Fluffy-backed Tit-Babbler	57	17	2	-	-	1	2	-	-	-	-	1	13.6
Brown Fulvetta	10	-	-	-	-	-	-	-	-	-	-	-	-
Siberian Blue Robin	112	33	6	-	1	-	-	-	-	-	-	-	3.9
White-rumped Shama	103	27	3	-	1	-	-	-	2	-	-	-	8
Rufous-tailed Shama	44	17	1	2	-	1	1	2	1	-	-	-	7.8
White-crowned Forktail	10	1	-	-	-	-	-	-	-	-	-	-	0.4
Brown-chested Jungle Flycatcher	25	3	2	-	-	-	-	-	-	-	-	-	1.3
Grey-chested Jungle Flycatcher	54	16	1	3	-	-	-	-	-	1	-	1	14.3
Rufous-chested Flycatcher	18	1	-	-	-	-	-	-	-	-	-	-	0.2
Spotted Fantail	15	1	-	-	-	-	-	-	-	-	-	-	0.1
Black-naped Monarch	19	1	-	-	-	-	-	1	-	-	-	-	7
Maroon-breasted Philentoma	7	1	-	-	-	-	-	-	-	-	-	1	13.3
Rufous-winged Philentoma	90	34	4	3	-	-	1	-	1	-	-	3	14.9
Asian Paradise-Flycatcher	65	14	5	2	-	-	-	-	1	-	-	-	7
Purple-naped Sunbird	54	22	3	1	2	-	1	-	-	-	-	-	6
Little Spiderhunter	138	30	6	1	1	-	2	-	2	1	-	-	8.9
Grey-breasted Spiderhunter	11	5	2	1	-	-	-	-	-	-	-	-	2.1
Yellow-breasted Flowerpecker	14	6	2	-	-	-	-	-	-	-	-	-	1

groups in the whole reserve. These species are even less like to persist within Pasoh FR, without movement of individuals from other areas, because small populations are prone to extinction due to stochastic events.

4.3 Changes in the bird community over time

With 25 years of data, the mist-netting data provide some information on changes in the bird community over time. Because the 15 ha plot, as noted above, was not large enough to estimate total population size in all time periods, we instead used contingency analysis to evaluate changes in the relative abundance of species captured within the 15 ha plot over three time periods: 1971–1974, 1979–1987, and 1991–1996. For the 54 species with at least 10 captures, there were highly significant changes in relative abundance ($\chi^2 = 572$, $df = 106$, $P < 0.0001$). Among the more substantial changes, Yellow-bellied Bulbuls decreased from 7.9% to 3.5% of the catch, while Hairy-backed Bulbuls increased from 1.1% to 6.8% and White-rumped Shamas increased from 3.2% to 7.7%. Several babbler species decreased substantially, including Grey-headed (2.7% to 0%) and Scaly-crowned (6.7% to 3.3%), while others increased including Chestnut-rumped (4.4% to 8.8%) and Sooty-capped Babblers (0.1% to 3%), and the Fluffy-backed Tit-Babbler (1.6% to 4.3%). Some caution is required in interpreting these results because they are not independent—an increase in one species leads to a decrease in the relative abundance of all other species. Nevertheless, because no one species comprised >8% of this sample, the differences mentioned here cannot be due entirely to changes in other species.

Unfortunately, these changes are difficult to interpret, because no data are available on natural fluctuations in populations within contiguous forests in Malaysia. Even in the absence of human interference and isolation effects, populations are likely to increase or decrease due to stochastic events, diseases, etc. (Diamond 1975). Furthermore, the netting did not take place in the same months every year and there were changes in the netting protocols, with nets being opened at dawn and closed before dusk in the recent period. Also, those species with the smallest populations, which may be particularly at risk of disappearing from the area due to random chance events, were too rarely caught in mist nets for analysis.

4.4 Longevity

As one of the longest running capture-recapture studies in the Asian tropics, research at Pasoh has provided some of the best available data on longevity for Malaysian birds. Of the 2,748 individuals ringed in the 15 ha study plot in 1991 or earlier, 892 were recaptured at least once in a subsequent netting session. Of these, 433 were recaptured one or more years later, including 120 at least 4 years after ringing, and 14 (of 11 different species) at least 10 years after ringing (Table 3). These recaptures included a few birds recaptured in Marina Wong's primary forest plot (see Fig. 1) as well as several in the 110 ha plot beyond the boundaries of the original 15 ha plot. The record was for a Rufous-collared Kingfisher that was first captured in September 1972 (as an adult), recaptured on two more sessions in the next 4 months, then not seen again until June 1993.

Despite the fact that the record longevity was held by a species for which only 18 individuals were ringed up until 1991, much of the observed variation among species in longevity records was related simply to sample size (correlation of maximum longevity with log of the number of birds ringed for all species: $r = 0.72$, $N = 102$, $P < 0.0001$; only for species with ≥ 10 ringed, $r = 0.47$, $N = 46$, $P = 0.001$).

As such, these longevity records provide only limited information on annual survival rates, the factor most relevant to population dynamics.

4.5 Average survival rates

Estimating annual adult survival rates requires capture-recapture models. We used all available data from both the 15 ha and 110 ha plots, from 1971–1996, to fit such models for all 14 resident species with >100 individuals ringed up until 1995. Survival rates were estimated using the computer program MARK (White & Burnham 1999), with a model that assumed apparent survival and capture probabilities did not vary among sessions, but that apparent survival rates may have differed between the first 12 months after ringing and subsequent years. This provision was made because newly ringed birds may have included transients as well as young birds that were more likely to disperse or die than territorial adults. In contrast, after the first year, most birds in the sample were likely to be resident adults.

Estimates of apparent survival rates (Table 4) were relatively low for all species during the first year after ringing (mean 41%), but much higher in subsequent years (mean 77%). Apparent survival rates, as estimated from capture-recapture models, tend to underestimate true survival rates, because emigration is confounded with mortality. The low apparent survival rates for newly ringed birds do not necessarily mean that these birds had higher mortality. Instead, they were probably due mainly to the presence of "transients" as well as young birds in the initial sample (Johnston et al. 1997). "Transients" refer to individuals that are much less likely to be recaptured within the study area than territorial resident birds. These could be dispersing individuals that were passing through the study area, such as pre-breeders in search of a territory, or they could be birds with established territories elsewhere that had wandered into the plot. Young birds could potentially have been excluded from the analysis if they had all been identified. For some species, the youngest individuals were distinguished by plumage, eye or gape colour, but our knowledge of these criteria improved over the course of the study, and is still incomplete for most species. Furthermore, many species, especially babblers, are believed to remain with their parents in family groups for several months or longer before dispersing to establish their own territories. These older pre-breeders are likely to be indistinguishable from adults, but are presumably just as likely to disperse within their first year as younger birds. As such, survival rates during the first year after ringing represent a mixture of territorial adults, transients that leave the study area, and young birds with higher mortality and dispersal than adults. Estimates of these combined rates provide limited biological information.

However, birds still present in the study area after their first year are likely to be mainly territorial adults. Assuming dispersal of these birds is low, at least for most species, the estimated survival rates after the first year are probably close to the true survival rates of breeding adults.

Assuming constant survival rates over time, and no senescence, the expected mean life span is given by $-1/\ln(S)$ where S represents survival rate. Thus, a survival rate of 77% corresponds to an expected mean lifespan of 3.8 years after a bird becomes an adult. After 10 years, 7% of individuals would still be expected in the population.

Karr et al. (1990) suggested that survival rates of tropical birds were not much different from those of temperate species, and found an average apparent survival rate for birds in a Panama community of only 56% across a range of species. However, they did not use models that differentiated transients or young

Table 4. Apparent annual survival rates (mean ± *SE* with 95% confidence interval) and capture probabilities (mean ± *SE*) for 14 species of resident understorey bird species at Pasoh (*N* = number of distinct individuals ringed). Note that true survival rates are likely higher, because emigration from the study plot is confounded with mortality, especially for the first year after ringing. See Appendix for scientific names.

	Apparent Survival Rates (%)		Capture	
Species	First Year	After First Year	Probability	*N*
Yellow-bellied Bulbul	49 ± 5.9 (37-60)	78 ± 3.6 (70-84)	43 ± 2.7	126
Hairy-backed Bulbul	98 ± 5.9 (17-99)	72 ± 6.2 (59-83)	13 ± 2.1	158
Black-capped Babbler	46 ± 8.5 (30-63)	80 ± 4.5 (69-87)	24 ± 3.1	109
Short-tailed Babbler	20 ± 4.3 (13-30)	77 ± 4.3 (67-84)	29 ± 3.0	188
Ferruginous Babbler	39 ± 6.6 (27-53)	80 ± 4.2 (70-87)	39 ± 3.8	100
Moustached Babbler	49 ± 6.1 (37-60)	77 ± 3.3 (70-83)	30 ± 2.5	201
Scaly-crowned Babbler	39 ± 5.5 (29-51)	79 ± 3.3 (72-85)	32 ± 2.4	162
Chestnut-rumped Babbler	48 ± 6.4 (36-61)	74 ± 4.3 (65-82)	45 ± 3.3	140
Black-throated Babbler	37 ± 7.6 (23-52)	83 ± 4.2 (73-90)	34 ± 4.2	101
Chestnut-winged Babbler	37 ± 6.3 (26-50)	56 ± 8.7 (39-72)	35 ± 3.7	186
Fluffy-backed Tit-Babbler	30 ± 7.8 (17-47)	78 ± 6.4 (63-88)	34 ± 7.0	123
White-rumped Shama	32 ± 5.4 (22-43)	73 ± 6.1 (60-83)	31 ± 4.3	223
Rufous-winged Philentoma	21 ± 5.3 (12-33)	86 ± 3.6 (77-91)	23 ± 2.7	105
Little Spiderhunter	23 ± 5.8 (14-36)	80 ± 4.4 (70-88)	19 ± 2.8	162
Mean	41 ± 1.4	77 ± 0.5	31 ± 0.6	

birds from territorial adults. Several subsequent studies on neotropical birds (Faaborg & Arendt 1995; Francis et al. 1999; Johnston et al. 1997) did use models that distinguish transients, and found significantly higher average survival rates of 66–68%. Nevertheless, those are still substantially lower than the observed value of 77% at Pasoh FR. Fogden (1972) estimated the average survival rate for a range of understorey species in Sarawak, Malaysia at 85–90%, even higher that our estimates. His study was based on a single year, which may or may not have been typical, and he used *ad hoc* statistical methods that may not be strictly comparable to the capture-recapture approach. Thus, it would be premature to conclude that survival rates differ between Sarawak and Pasoh FR. Nevertheless, in combination, these two data sets suggest that survival rates for understorey birds in the Asian tropics are substantially higher than those for understorey birds in areas that have been studied in Central or South America. Whether this is due to lower densities of predators on adults (e.g. bird-eating hawks or falcons), or to other factors, is unknown.

5. FUTURE OF THE BIRD COMMUNITY AT PASOH

Shortly before the start of bird research at Pasoh FR, this forest reserve was contiguous with extensive lowland forests in all directions, although some areas had been selectively logged in the 1950s (Chap. 2). By the early 1970s, Pasoh FR became surrounded on 3 sides with oil palm plantations (Chap. 2), which are hostile habitats for most forest birds. Although still connected to the east with a more extensive tract of hill forest, the avifauna of that tract has not been well studied, and it is, itself, isolated from the main range of Malaysia. The core area of primary lowland forest comprises only 600 hectares, and although the regenerating forest around the core still supports many of the primary forest species, some are absent or occur only at low densities (Wong 1985).

Our data provide only limited information on population changes since isolation. Nevertheless, on theoretical grounds, we can make some predictions about the future of the bird community at Pasoh FR. Isolated forest patches in Singapore (Corlett & Turner 1997) and around Bogor, Java (Diamond et al. 1987) have lost most of their original species. Those that remain are mainly species that also can live in disturbed areas. Although Pasoh FR is many times larger than even the largest forest fragments in Singapore, our population estimates suggest Pasoh FR is probably not large enough to support viable populations of many species in isolation. This prediction finds support in data from offshore islands in Malaysia, where immigration is restricted by water. Even the largest islands, with forested areas of 40–230 km^2 (20–100 times larger than Pasoh FR), support only 13–80 forest species, many fewer than are found at Pasoh (Wells 1976). Although offshore islands differ in many other ways from the mainland, such as reduced vegetation complexity, they do give some indication of what could happen, in the long term, to forest patches that are completely isolated. Research on Barro Colorado Island, which is comparable in size to Pasoh FR, and was formed with the flooding of the Panama Canal in Central America, has shown that over 65 species have disappeared from the island, even though it is separated by only a short stretch of water from the mainland (Robinson 1999).

In addition, there has been some concern about possible changes in the habitat within Pasoh FR. Over the past decade, there have been extensive tree falls in some areas, including parts of our study plots. Such events also occur within contiguous primary forest, and are not necessarily a result of human activities or isolation of Pasoh FR. Nevertheless, they may exacerbate the effects of other disturbances, such as logging, and have enhanced conditions for gap and edge species in some parts of Pasoh FR at the expense of species dependent on the mature forest. Within our 15 ha study plot, this is reflected in the increase in Sooty-capped Babblers and Fluffy-backed Tit-Babblers, both of which thrive in disturbed habitats. Provided that Pasoh FR is protected from further illegal logging, the forest may recover from these disturbances. Nevertheless, given the small size of Pasoh FR, and the proximity of most areas to the edge, we can predict a sustained increase in the relative abundance of species tolerant of edges and disturbance. Also, some of the larger bird species, such as hornbills and pheasants, are subject to additional pressure from illegal hunting, owing to easy access to Pasoh FR from surrounding plantations.

6. FUTURE RESEARCH NEEDS

From a conservation perspective, there is still much that we do not know about the bird community at Pasoh FR. A high priority is to obtain reliable estimates of densities for all of the species in the Pasoh FR community, to estimate remaining population sizes within Pasoh FR, as a basis for measuring future change. This can most effectively be done using a spot-mapping technique (e.g. Robinson et al. 2000) to determine the number of territories of each species. This requires considerable expertise (field workers must be able to identify all species by song) and must be carried out in a plot of at least 100 ha to obtain sufficient sample sizes for reliable estimates. A minimum of 2–3 months of intensive field work would be required to complete such a survey. Ideally, this would be done for more than one plot, to measure variation within Pasoh FR. Supplemental mist-netting and colour-ringing, so that individual birds could be followed, combined with radio-tracking for selected species, would be valuable to validate the spot-mapping approach, to determine how individual birds make use of the forest, and to obtain supplemental information on social structure for each species.

Information on birds in the adjoining hill forest to the east of Pasoh FR would also be useful to determine to what extent this can sustain the community within Pasoh FR. This hill forest is, itself, a relatively narrow tract that does not connect with the main range in Malaysia—details of the extent and condition of that forest are not currently available. Bird communities within that forest may also benefit from interchange of individuals with Pasoh FR. Understanding these dynamics would be valuable for planning of reserves and corridors elsewhere in Malaysia.

It would also be valuable to resume the netting to obtain better information on survival rates and other aspects of population dynamics at Pasoh FR. Further netting within the 110 ha plot, if carried out soon, would be expected to catch many individuals from the earlier studies. If continued for at least 2 sessions, this would greatly improve the precision of survival estimates, and provide data on many additional species. Data on recruitment rates are also necessary to understand population dynamics. This could involve finding and monitoring nests, observing marked pairs to determine the proportion that successfully fledge young, or simply estimating age ratios from mist net captures. Further studies using artificial nests would be valuable to learn more about variation in nest predation rates across Pasoh FR.

Finally, not just from the perspective of conservation at Pasoh FR, but throughout Malaysia, further research would be valuable on the microhabitat requirements of each species. Although the general habitat requirements are known for most species (e.g. which ones are wetland dependent, which live in the understorey and which live in the canopy), few quantitative data are available on microhabitat use. Quantitative studies on the ecology of individual species and their habitat use, such as that by McGowan (1994) on Peacock Pheasants, could help to determine why some species are more sensitive to logging than others, and possibly identify mitigating factors, such as silvicultural treatments, that may improve the quality of logged forest habitats for birds (Zakaria & Francis 2001). Such studies could be particularly effective if combined with existing research studies on the botanical structure at Pasoh FR, such as the 50 ha tree plot, which provides unmatched information on the habitat available within a fixed area.

Because of its long ornithological history, Pasoh FR has the potential to become a natural laboratory for the study of Malaysian bird communities, in the same way as Barro Colorado Island has become a model site for measuring isolation effects on bird communities in Panama (Robinson 1999). In this way, it can make a valuable contribution to the long-term persistence of the rich and diverse avifauna of Malaysia.

ACKNOWLEDGEMENTS

We would like to thank the Economic Planning Unit of the Prime Minister's Department, the Forest Research Institute Malaysia, and the National Parks and Wildlife Department for permission to carry out our bird studies in Peninsular Malaysia. We are grateful to the dozens of people, including many volunteers, who have assisted with the field research over the 26 years of this study, including preparing trails, mist-netting or contributing observations. Financial support was provided by the Zoology Department, Universiti Malaya, and the Wildlife Conservation Society. Drs. Alison Styring, Dennis Yong, Toshinori Okuda and Sean Thomas provided helpful comments on the manuscript.

REFERENCES

Cooper, D. S. & Francis, C. M. (1998) Nest predation in a Malaysian lowland rain forest. Biol. Conserv. 85: 199-202.

Corlett, R. T. & Turner, I. M. (1997) Long-term survival in tropical forest remnants in Singapore and Hong Kong. In Laurence, W. F. & Bierregaard Jr., R. O. (eds). Tropical Forest Remnants. University of Chicago Press, Chicago, USA, pp.333-345.

Diamond, J. M. (1975) Assembly of species communities. In Cody, M. L. & Diamond, J. M. (eds). Ecology and evolution of communities. Harvard University Press, Cambridge, Massachusetts, USA, pp.342-444.

Diamond, J. M., Bishop, K. D. & van Balen, S. (1987) Bird survival in an isolated Javan woodland: island or mirror? Conserv. Biol. 1: 132-142.

Faaborg, J. & Arendt, W. (1995) Survival rates of Puerto Rican birds: are islands really that different? Auk 112: 503-507.

Fogden, M. P. L. (1972) The seasonality and population dynamics of equatorial birds in Sarawak. Ibis 114: 307-343.

Fogden, M. P. L. (1976) A census of a bird community in tropical rain forest in Sarawak. Sarawak Museum J. 24: 251-267.

Francis, C. M., Terborgh, J. S. & Fitzpatrick, J. W. (1999) Survival rates of understorey forest birds in Peru. In Adams, N. J. & Slotow, R. H. (eds). Proceedings of the 22nd International Ornithological Congress, Durban. Bird Life South Africa, Johannesburg, pp.326-335.

Inskipp, T., Lindsey, N. & Duckworth, W. (1996) An annotated checklist of the birds of the Oriental Region. Oriental Bird Club, Sandy, UK.

Johns, A. D. (1986) Effects of selective logging on the ecological organisation of a peninsular Malaysian rain forest avifauna. Forktail 1: 65-79.

Johns, A. G (1996) Bird population persistence in Sabahan logging concessions. Biol. Conserv. 75: 3-10.

Johnston, J. P., Peach, W. J., Gregory, R. D. & White, S. A. (1997) Survival rates of tropical and temperate passerines: a Trinidadian perspective. Am. Nat. 150: 771-789.

Karr, J. R. (1980) Geographical variation in the avifaunas of tropical forest undergrowth. Auk 97: 283-298.

Karr, J. R., Nichols, J. D., Klimkiewicz, M. K. & Brawn, J. D. (1990) Survival rates of birds of tropical and temperate forests: will the dogma survive? Am. Nat. 136: 277-291.

Lambert, F. R. (1992) The consequences of selective logging for Bornean lowland forest birds. Philos. Trans. R. Soc. London Ser. B 335: 443-457.

McClure, H. E. (1969) An estimation of a bird population in the primary forest of Selangor, Malaysia. Malay. Nat. J. 22: 179-183.

McGowan, P. J. K. (1994) Display dispersion and micro-habitat use by the Malaysian peacock-pheasant *Polyplectron malacense* in Peninsular Malaysia. J. Trop. Ecol. 10: 229-244.

Medway, L. & Wells, D. R. (1971) Diversity and density of birds and mammals at Kuala Lompat, Pahang. Malay. Nat. J. 24: 238-247.

Medway, L. & Wells, D. R. (1976) The Birds of the Malay Peninsula. vol. V. Conclusion, and survey of every species. H. F. & G. Witherby Ltd. and Penerbit Universiti Malaya. 448pp.

Nagata, H., Zabaid, A. M. A. & Azarae, H. I. (1998) Edge effects on the nest predation and avian community in Pasoh Nature Reserve. Research report of the NIES/FRIM/UPM joint research project, 1998. National Institute for Environmental Studies, Tsukuba, Japan, 222pp.

Pollock, K. H., Nichols, J. D., Brownie, C. & Hines, J. E. (1990) Statistical inference for capture-recapture experiments. Wildl. Monogr. 107: 1-97.

Primack, R. B. (1993) Essentials of Conservation Biology. Sinauer Associates, Sunderland, Massachusetts, USA.

Robinson, W. D. (1999) Long-term changes in the avifauna of Barro Colorado Island, Panama, a tropical forest isolate. Conserv. Biol. 13: 85-97.

Robinson, W. D., Brawn, J. D. & Robinson, S. K. (2000) Forest bird community structure in central Panama: influence of spatial scale and biogeography. Ecol. Monogr. 70: 209-235.

Sheldon, F. H., Moyle, R. G. & Kennard, J. (2001) Ornithology of Sabah: History, Gazetteer, Annotated Checklist, and Bibliography. Ornithological Monogr. 52.

Smythies, B. E. (2000) The Birds of Borneo (4th ed., revised by Davison, G. W. H.). Natural History Publications, Kota Kinabalu, Sabah.

Soepadmo, E. (1978) Introduction to the Malaysian IBP synthesis meetings. Malay. Nat. J. 30: 119-124.

Styring, A. R. & Ickes, K. (2001) Woodpecker abundance in a logged (40 years ago) vs. unlogged lowland dipterocarp forest in Peninsular Malaysia. J. Trop. Ecol. 17: 261-268.

Waugh, D. R. & Hails, C. J. (1983) Foraging ecology of a tropical aerial feeding bird guild. Ibis 125: 200-217.

Wells, D. R. (1969) A preliminary survey of migration, bodyweight and moult in Siberian Blue Robins, *Erithacus cyane*, wintering in Malaysia. Malay. For. 32: 441-443.

Wells, D. R. (1976) Resident Birds. In Medway, L. & Wells, D. R. (eds). The Birds of the Malay Peninsula. Vol. V. Conclusion, and survey of every species. H. F. & G. Witherby Ltd. and Penerbit Universiti Malaya, pp.1-34.

Wells, D. R. (1978) Numbers and biomass of insectivorous birds in the understorey of rain forest at Pasoh Forest. Malay. Nat. J. 30: 353-362.

Wells, D. R. (1988a) The bird fauna of Pasoh Forest Reserve, Negri Sembilan. Enggang 1: 7-17.

Wells, D. R. (1988b) Comparative evolution of Afro-Asian equatorial-forest avifaunas. In Ouellet, H. (ed). Acta XIX Congressus Internationalis Ornithologicus. University of Ottawa Press, Ottawa, pp.2799-2809.

Wells, D. R. (1990) Migratory birds and tropical forest in the Sunda region. In Keast, A. (ed). Biogeography and ecology of forest bird communities. SPB Academic Publishing, The Hague, pp.357-369.

Wells, D. R. (1999) The Birds of the Thai-Malay Peninsula vol. 1. Non-passerines. Academic Press, London, 648 pp.

White, G. C. & Burnham, K. P. (1999) Program MARK: survival estimation from populations of marked animals. Bird Study 46: S120-S139.

Wong, M. (1985) Understorey birds as indicators of regeneration in a patch of selectively logged West Malaysian rainforest. In Diamond, A. W. & Lovejoy, T. E. (eds). Conservation of tropical forest birds. International Council for Bird Preservation Technical Publication No. 4. Cambridge, UK, pp.249-263.

Wong, M. (1986) Trophic organization of understory birds in a Malaysian dipterocarp forest. Auk 103: 100-116.

Zakaria, M. & Francis, C. M. (1999) Estimating densities of Malaysian forest birds. In Adams, N. J. & Slotow, R. H. (eds). Proceedings of the 22nd International Ornithological Congress, Durban. BirdLife South Africa, Johannesburg, pp.2554-2568.

Zakaria, M. & Francis, C. M. (2001) The effects of logging on birds in tropical forests of Indo-Australia. In Fimbel, R. A., Grajal, A. & Robinson, J. R. (eds). The Cutting Edge: Conserving Wildlife in Logged Tropical Forests, Columbia University Press, New York, pp.193-212.

Appendix: Bird species recorded using the primary forest in Pasoh FR between 1972 and 1996. English and scientific names of species and their taxonomic arrangement and order follow Inskipp et al. (1996) with the exception of a few changes suggested by Wells (1999). Aerial insectivores that forage on insects over the canopy are included, even if they are unlikely to breed within the reserve, but other species only noted flying over the reserve, which were unlikely to land within, or make use of the reserve, are excluded. The following status codes are used: R = resident year-round, M = migrant, E = mainly found in edge habitats, R* = resident species that forages over the reserve but is unlikely to breed within Pasoh because of a lack of suitable nesting habitat.

Species	Status
Phasianidae	
Long-billed Partridge, *Rhizothera longirostris*	R
Black Partridge, *Melanoperdix nigra*	R
Roulroul, *Rollulus rouloul*	R
Red Junglefowl, *Gallus gallus*	RE
Crestless Fireback, *Lophura erythrophthalma*	R
Crested Fireback, *Lophura ignita*	R
Malayan Peacock-Pheasant, *Polyplectron*	R
Great Argus, *Argusianus argus*	R
Ardeidae	
Von Schrenck's Bittern, *Ixobrychus eurhythmus*	M
Accipitridae	
Black Baza, *Aviceda leuphotes*	M
Crested Honey-Buzzard, *Pernis ptilorhynchus*	M
Bat Hawk, *Machaerhampus alcinus*	R
Crested Serpent-Eagle, *Spilornis cheela*	R
Crested Goshawk, *Accipiter trivirgatus*	R
Japanese Sparrowhawk, *Accipiter gularis*	ME
Rufous-bellied Eagle *Hieraaetus kienerii*	R
Changeable Hawk-Eagle, *Spizaetus cirrhatus*	R
Wallace's Hawk-Eagle, *Spizaetus nanus*	R
Falconidae	
Black-thighed Falconet, *Microhierax fringillarius*	R
Columbidae	
Mountain Imperial Pigeon, *Ducula badia*	R
Emerald Dove, *Chalcophaps indica*	R
Little Green Pigeon, *Treron olax*	R
Thick-billed Green Pigeon, *Treron curvirostra*	R
Large Green Pigeon, *Treron capellei*	R
Jambu Fruit Dove, *Ptilinopus jambu*	R
Psittacidae	
Blue-rumped Parrot, *Psittinus cyanurus*	R
Blue-crowned Hanging Parrot, *Loriculus galgulus*	R
Long-tailed Parakeet, *Psittacula longicauda*	R
Cuculidae	
Moustached Hawk-Cuckoo, *Cuculus vagans*	R
Hodgson's Hawk-Cuckoo, *Cuculus fugax*	R
Indian Cuckoo, *Cuculus micropterus*	R
Banded Bay Cuckoo, *Cacomantis sonneratii*	R
Rusty-breasted Cuckoo, *Cacomantis sepulchralis*	R
Drongo Cuckoo, *Surniculus lugubris*	R
Violet Cuckoo, *Chrysococcyx xanthorhynchus*	R
Black-bellied Malkoha, *Phaenicophaeus diardi*	R
Chestnut-bellied Malkoha, *Phaenicophaeus sumatranus*	R
Raffles' Malkoha, *Phaenicophaeus chlorophaeus*	R
Red-billed Malkoha, *Phaenicophaeus javanicus*	R
Chestnut-breasted Malkoha, *Phaenicophaeus curvirostris*	R
Centropodidae	
Short-toed Coucal, *Centropus rectunguis*	R
Strigidae	
White-fronted Scops Owl, *Otus sagittatus*	R
Reddish Scops Owl, *Otus rufescens*	R
Collared Scops Owl, *Otus bakkamoena*	R
Barred Eagle Owl, *Bubo sumatranus*	R
Buffy Fish Owl, *Ketupa ketupu*	R
Brown Wood Owl, *Strix leptogrammica*	R
Brown Boobook, *Ninox scutulata*	R
Tytonidae	
Bay Owl, *Phodilus badius*	R
Podargidae	
Large Frogmouth, *Batrachostomus auritus*	R
Gould's Frogmouth, *Batrachostomus stellatus*	R
Javan Frogmouth, *Batrachostomus javensis*	R
Caprimulgidae	
Malaysian Eared Nightjar, *Eurostopodus temminckii*	R
Apodidae	
Glossy Swiftlet, *Collocalia esculenta*	R
Swiftlets, *Aerodramus* spp.	R
Silver-rumped Spinetail, *Rhaphidura leucopygialis*	R
Brown-backed Needletail, *Hirundapus giganteus*	R
Pacific Swift, *Apus pacificus*	M
Hemiprocnidae	
Grey-rumped Treeswift, *Hemiprocne longipennis*	R
Whiskered Treeswift, *Hemiprocne comata*	R
Trogonidae	
Red-naped Trogon, *Harpactes kasumba*	R
Diard's Trogon, *Harpactes diardii*	R
Cinnamon-rumped Trogon, *Harpactes orrhophaeus*	R
Scarlet-rumped Trogon, *Harpactes duvaucelii*	R
Bucerotidae	
Black Hornbill, *Anthracoceros malayanus*	R
Rhinoceros Hornbill, *Buceros rhinoceros*	R
Helmeted Hornbill, *Buceros vigil*	R
Bushy-crested Hornbill, *Anorrhinus galeritus*	R
White-crested Hornbill, *Berenicornis comatus*	R
Wreathed Hornbill, *Aceros undulatus*	R
Halcyonidae	
Banded Kingfisher, *Lacedo pulchella*	R
Ruddy Kingfisher, *Halcyon coromanda*	ME
White-throated Kingfisher, *Halcyon smyrnensis*	RE
Rufous-collared Kingfisher, *Actenoides concretus*	R
Alcedinidae	
Oriental Dwarf Kingfisher, *Ceyx erithacus*	RM
Blue-eared Kingfisher, *Alcedo meninting*	R
Blue-banded Kingfisher, *Alcedo euryzona*	R

Appendix (continued 1)

Taxon	Status
Meropidae	
Red-bearded Bee-eater, *Nyctyornis amictus*	R
Blue-throated Bee-eater, *Merops viridis*	M
Coraciidae	
Dollarbird, *Eurystomus orientalis*	R
Indacatoridae	
Malaysian Honeyguide, *Indicator archipelagicus*	R
Picidae	
Rufous Piculet, *Sasia abnormis*	R
Grey-capped Woodpecker, *Picoides canicapillus*	R
Rufous Woodpecker, *Celeus brachyurus*	R
White-bellied Woodpecker, *Dryocopus javensis*	R
Banded Yellownape, *Picus miniaceus*	R
Crimson-winged Yellownape, *Picus puniceus*	R
Chequer-throated Yellownape, *Picus mentalis*	R
Olive-backed Woodpecker, *Dinopium rafflesii*	R
Maroon Woodpecker, *Blythipicus rubiginosus*	R
Orange-backed Woodpecker, *Reinwardtipicus validus*	R
Buff-rumped Woodpecker, *Meiglyptes tristis*	R
Buff-necked Woodpecker, *Meiglyptes tukki*	R
Grey-and-Buff Woodpecker, *Hemicircus concretus*	R
Great Slaty Woodpecker, *Mulleripicus pulverulentus*	R
Capitonidae	
Gold-whiskered Barbet, *Megalaima chrysopogon*	R
Red-crowned Barbet, *Megalaima rafflesii*	R
Red-throated Barbet, *Megalaima mystacophanos*	R
Yellow-crowned Barbet, *Megalaima henricii*	R
Blue-eared Barbet, *Melgalaima australis*	R
Brown Barbet, *Calorhamphus fuliginosus*	R
Eurylaimidae	
Dusky Broadbill, *Corydon sumatranus*	R
Banded Broadbill, *Eurylaimus javanicus*	R
Black-and-Yellow Broadbill, *Eurylaimus ochromalus*	R
Green Broadbill, *Calyptomena viridis*	R
Pittidae	
Banded Pitta, *Pitta guajana*	R
Garnet Pitta, *Pitta granatina*	R
Hooded Pitta, *Pitta sordida*	M
Blue-winged Pitta, *Pitta moluccensis*	ME
Pardalotidae	
Golden-bellied Gerygone (Flyeater), *Gerygone sulphurea*	R
Irenidae	
Asian Fairy Bluebird, *Irena puella*	R
Chloropseidae	
Greater Green Leafbird, *Chloropsis sonnerati*	R
Lesser Green Leafbird, *Chloropsis cyanopogon*	R
Blue-winged Leafbird, *Chloropsis cochinchinensis*	R
Laniidae	
Tiger Shrike, *Lanius tigrinus*	ME
Corvidae, Cinclosomatinae	
Malaysian Rail-Babbler, *Eupetes macrocerus*	R
Corvidae, Corvinae, Corvini	
Crested Jay, *Platylophus galericulatus*	R
Black Magpie, *Playsmurus leucopterus*	R
Slender-billed Crow, *Corvus enca*	R
Corvidae, Corvinae, Oriolini	
Dark-throated Oriole, *Oriolus xanthonotus*	R
Corvidae, Corvinae, Campephagini	
Bar-bellied Cuckoo-shrike, *Coracina striata*	R
Lesser Cuckoo-shrike, *Coracina fimbriata*	R
Ashy Minivet, *Pericrocotus divaricatus*	M
Fiery Minivet, *Pericrocotus igneus*	R
Scarlet Minivet, *Pericrocotus flammeus*	R
Bar-winged Flycatcher-shrike, *Hemipus picatus*	R
Black-winged Flycatcher-shrike, *Hemipus hirundinaceus*	R
Corvidae, Dicrurinae, Dicrurini	
Crow-billed Drongo, *Dicrurus annectans*	M
Bronzed Drongo, *Dicrurus aeneus*	R
Greater Racket-tailed Drongo, *Dicrurus paradiseus*	R
Corvidae, Dicrurinae, Rhipidurini	
Spotted Fantail, *Rhipidura perlata*	R
Pied Fantail, *Rhipidura javanica*	RE
Corvidae, Dicrurinae, Monarchini	
Black-naped Monarch, *Hypothymis azurea*	R
Japanese Paradise Flycatcher, *Terpsiphone atrocaudata*	M
Asian Paradise Flycatcher, *Terpsiphone paradisi*	R
Corvidae, Malaconotinae	
Maroon-breasted Philentoma, *Philentoma velatum*	R
Rufous-winged Philentoma, *Philentoma pyrhopterum*	R
Large Wood-shrike, *Tephrodornis gularis*	R
Corvidae,Aegithininae	
Green Iora, *Aegithina viridissima*	R
Great Iora, *Aegithina lafresnayei*	RE
Turdidae	
Eye-browed Thrush, *Turdus obscurus*	ME
Muscicapidae, Saxicolini	
Siberian Blue Robin, *Erithacus cyane*	M
White-rumped Shama, *Copsychus malabaricus*	R
Orange-tailed Shama, *Copsychus pyrropygus*	R
Magpie-Robin, *Copsychus saularis*	RE
Chestnut-naped Forktail, *Enicurus ruficapillus*	R
White-crowned Forktail, *Enicurus leschenaulti*	R
Muscicapidae, Muscicapini	
Rufous-chested Flycatcher, *Ficedula dumetoria*	R
Yellow-rumped Flycatcher, *Ficedula zanthopygia*	M
Narcissus Flycatcher, *Ficedula narcissina*	M
Pale Blue Flycatcher, *Cyornis unicolor*	R
Blue-throated Flycatcher, *Cyornis rubeculoides*	M
Brown-chested Jungle Flycatcher, *Rhinomyias brunneata*	M
Grey-chested JungleFlycatcher, *Rhinomyias*	R
Verditer Flycatcher, *Eumyias thalassina*	R
Ferruginous Flycatcher, *Muscicapa ferruginea*	ME
Asian Brown Flycatcher, *Muscicapa dauurica*	M
Grey-headed Canary Flycatcher, *Culicicapa*	R
Sturnidae	
Hill Myna, *Gracula religiosa*	R
Sittidae	
Velvet-fronted Nuthatch, *Sitta frontalis*	R
Paridae	
Sultan Tit, *Melanochlora sultanea*	R

Appendix (continued 2)

Taxon	Status
Hirundinidae	
Barn Swallow, *Hirundo rustica*	M
Pycnonotidae	
Black-and-White Bulbul, *Pycnonotus melanoleucos*	R
Black-headed Bulbul, *Pycnonotus atriceps*	R
Grey-bellied Bulbul, *Pycnonotus cyaniventris*	R
Puff-backed Bulbul, *Pycnonotus eutilotus*	R
Cream-vented Bulbul, *Pycnonotus simplex*	R
Red-eyed Bulbul, *Pycnonotus brunneus*	R
Spectacled Bulbul, *Pycnonotus erythrophthalmos*	R
Finsch's Bulbul, *Criniger finschii*	R
Grey-cheeked Bulbul, *Criniger bres*	R
Yellow-bellied Bulbul, *Criniger phaeocephalus*	R
Hairy-backed Bulbul, *Hypsipetes criniger*	R
Buff-vented Bulbul, *Hypsipetes charlottae*	R
Streaked Bulbul, *Hypsipetes malaccensis*	R
Sylviidae, Acrocephalinae	
Arctic Warbler, *Phylloscopus borealis*	M
Eastern Crowned Warbler, *Phylloscopus coronatus*	M
Pallas's Grasshopper-Warbler, *Locustella certhiola*	ME
Dark-necked Tailorbird, *Orthotomus atrogularis*	R
Rufous-tailed Tailorbird, *Orthotomus sericeus*	RE
Sylviidae, Timaliini	
White-chested Babbler, *Trichastoma rostratum*	R
Ferruginous Babbler, *Trichastoma bicolor*	R
Horsfield's Babbler, *Malacocincla sepiarium*	R
Short-tailed Babbler, *Malacocincla malaccense*	R
Black-capped Babbler, *Pellorneum capistratum*	R
Moustached Babbler, *Malacopteron magnirostre*	R
Sooty-capped Babbler, *Malacopteron affine*	R
Scaly-crowned Babbler, *Malacopteron cinereum*	R
Rufous-crowned Babbler, *Malacopteron magnum*	R
Grey-breasted Babbler, *Malacopteron albogulare*	R
Chestnut-backed Scimitar-Babbler, *Pomatorhinus montanus*	R
Striped Wren-Babbler, *Kenopia striata*	R
Large Wren-Babbler, *Napothera macrodactyla*	R
Rufous-fronted Babbler, *Stachyris rufifrons*	R
Grey-headed Babbler, *Stachyris poliocephala*	R
White-necked Babbler, *Stachyris leucotis*	R
Black-throated Babbler, *Stachyris nigricollis*	R
Chestnut-rumped Babbler, *Stachyris maculata*	R
Chestnut-winged Babbler, *Stachyris erythroptera*	R
Striped Tit-Babbler, *Macronous gularis*	R
Fluffy-backed Tit-Babbler, *Macronous ptilosus*	R
Brown Fulvetta, *Alcippe brunneicauda*	R
White-bellied Yuhina, *Yuhina xantholeuca*	R
Nectariniidae, Nectariniinae, Dicaeini	
Scarlet-breasted Flowerpecker, *Prionochilus*	R
Yellow-breasted Flowerpecker, *Prionochilus*	R
Crimson-breasted Flowerpecker, *Prionochilus*	R
Plain Flowerpecker, *Dicaeum concolor*	R
Nectariniidae, Nectariniinae, Nectariniini	
Plain Sunbird, *Anthreptes simplex*	R
Red-throated Sunbird, *Anthreptes rhodolaema*	R
Ruby-cheeked Sunbird, *Anthreptes singalensis*	R
Purple-naped Sunbird, *Hypogramma*	R
Purple-throated Sunbird, *Nectarinia sperata*	R
Crimson Sunbird, *Aethopyga siparaja*	R
Little Spiderhunter, *Arachnothera longirostris*	R
Thick-billed Spiderhunter, *Arachnothera*	R
Long-billed Spiderhunter, *Arachnothera robusta*	R
Yellow-eared Spiderhunter, *Arachnothera*	R
Spectacled Spiderhunter, *Arachnothera flavigaster*	R
Grey-breasted Spiderhunter, *Arachnothera affinis*	R
Passeridae, Estrildinae	
White-bellied Munia, *Lonchura leucogastra*	RE

Notes:

1. *Aerodramus* spp. potentially includes Black-nest Swiftlet, *A. maximus*, Edible-nest Swiftlet, *A. fuciphagus* and the migratory Himalayan Swiftlet *A. brevirostris*, but these cannot be reliably distinguished in flight.
2. Additional species reported from within the boundaries of Pasoh FR, in regenerating forest, in clearings and along roads near the edge, or flying over the reserve: Spotted Dove, *Streptopelia chinensis*; Large-tailed Nightjar, *Caprimulgus macrourus*; Blue-tailed Bee-eater, *Merops superciliosus*; Common Flameback, *Dinopium javanense*; Black and Red Broadbill, *Cymbirhynchus macrorhynchus*; Pacific Swallow, *Hirundo tahitica*; Common Iora, *Aegithina tiphia*; Large-billed Crow, *Corvus macrorhynchus*; Asian Glossy Starling, *Aplonis panayensis*; Purple-backed Starling, *Sturnus sturninus*; Common Myna, *Acridotheres tristis*; Orange-bellied Flowerpecker, *Dicaeum trigonostigma*; Yellow-vented Flowerpecker, *Dicaeum chrysorrheum*.

28 Herpetofauna Diversity Survey in Pasoh Forest Reserve, Negeri Sembilan, Peninsular Malaysia

Lim Boo Liat[1] & Norsham Yaakob[2]

Abstract: Herpetofauna surveys were carried out in the primary and regenerating forests at Pasoh Forest Reserve (Pasoh FR) during short periods for each of 1968, 1974, 1987 and 1991. The historical data revealed a total of 75 herpetofauna comprising 26 species of amphibians, 24 species of tortoises, turtles and lizards and 25 species of snakes. Among the 75 species, 62 species were found in primary forest while 54 species from regenerating forest. There was a slight shift in the distribution and composition pattern of this taxon over time between the two forest types. Potential indicator species to environmental changes was established.

Key words: herpetofauna diversity, primary forest, regenerating forest, terrestrial amphibians and reptiles.

1. INTRODUCTION

Among the vertebrate fauna of the Pasoh Forest Reserve (Pasoh FR), the mammals have been fairly well studied by Francis (1989, 1990), Kemper (1988) and Kemper & Bell (1985). Most recent by Yasuda (1998) who worked on the community ecology of terrestrial small mammals in the study area. The avifauna has been very well studied too and over 200 species of birds have been documented in the reserve by Wells (1978) and Wong (1985, 1986). The herpetofauna have been reported by Kiew (1978, 1984) and Kiew et al. (1996) based on occasional and accidental findings. In addition to the herpetofauna, fish is another animal group in the vertebrate taxa that is understudied in the reserve.

Herpetofauna (frogs, lizards, snakes and turtles), as a group receive less attention than mammals and birds in faunal studies. This is most probably due to the nocturnal and cryptic behaviour of most of its members. Amphibians are semi-aquatic and associated with riparian habitat and damp wet places which most people find unpleasant. Land turtles and tortoises, which are less commonly seen than any other fauna groups as they are cryptic in habit, are more difficult to collect. Snakes and lizards on the other hand are encountered more often. Generally, snakes are greatly feared because most people treat them all as poisonous, when in fact only 16 of 142 species in Peninsular Malaysia are poisonous (Tweedie 1983). The lizards, which are generally small, conceal themselves in thick vegetation or forest litter when alarmed.

The survey of herpetofauna in this chapter is based on collection conducted during the small mammal studies carried out by the first and second authors. The collection made was not based on systematic sampling. The checklist provided from this survey however can serve as a baseline for future systematic sampling of this vertebrate fauna in the study area.

---

[1] Department of Wildlife and National Parks (DWNP), 50664 Kuala Lumpur, Malaysia. E-mail : limbooliat@yahoo.co.uk

[2] Forest Research Institute Malaysia (FRIM), Malaysia.

2. STUDY AREA

The description of the study area is well described in Chaps. 1 and 2 in this book.

3. MATERIALS AND METHOD

The survey of the herpetofauna was carried out in the primary forest (PF) and regenerating forest (RF). The survey was done in 1968, 1974, 1987 and 1991 in conjunction with the small mammal studies by the first author (Chap. 29). The survey was conducted five days in each of the PF and RF in each year. This collection was made during the day while checking the small mammal traps. Night sampling with aid of headlamps at the stream and water logged areas in each of the PF and RF was also conducted to survey for frogs. And any nocturnal herpetofauna encountered were also collected. In addition, Orang Asli helpers were also employed to look for these reptilian species in the forest. Time spent for searching herpetofauna was not standardized in each of the RF and PF.

All individuals of herpetofauna collected were given a number and measured.

Table 1 Amphibians recorded from the Pasoh FR, Negeri Sembilan, Peninsular Malaysia.

Amphibian Taxa	Primary Forest (FR)				Regenerating Forest (FR)				Total	Total Species	
	1968	1974	1989	1991	1968	1974	1989	1991	No.	PF	RF
Pelobatidae											
1. *Leptobrachium hendriksoni*	0	1	0	0	1	0	1	1	4	1	1
2. *L.nigrops*	0	0	1	1	0	0	0	1	3	1	1
3. *Megophrys nasuta*	1	0	1	0	0	0	0	0	2	1	0
Bufonidae											
4. *Bufo asper*	1	0	0	1	1	0	1	1	5	1	1
5. *B. melanostictus*	0	0	0	0	1	1	1	1	4	0	1
6. *B. parvus*	1	1	0	1	1	1	1	1	7	1	1
7. *Leptophryne borbonica*	1	1	1	1	0	0	0	1	5	1	1
Ranidae											
8. *Oeidozyga laevis*	1	1	1	1	0	0	0	0	4	1	0
9. *Rana chalconota*	0	0	0	1	1	1	1	1	5	1	1
10. *R. erythraea*	0	0	0	0	1	1	1	2	5	0	1
11. *R. glandulosa*	1	1	1	1	0	0	0	1	5	1	1
12. *R. laticeps*	1	0	0	0	0	1	0	0	2	1	1
13. *R. hosei*	1	1	1	1	0	0	0	1	5	1	1
14. *R. limnocharis*	0	0	0	0	1	1	0	0	2	0	1
15. *R. luctosa*	0	0	0	0	1	1	1	2	5	0	1
16. *R. blythi*	1	1	1	0	1	0	2	3	9	1	1
17. *R. nicobariensis*	1	1	1	0	0	1	0	0	4	1	1
18. *R. paramacrodon*	1	2	0	1	0	1	0	0	5	1	1
Rhacophoridae											
19. *Rhacophorus appendiculatus*	1	0	0	0	0	0	0	0	1	1	0
20. *R. macrotis*	1	0	0	1	1	0	1	1	5	1	1
21. *R. nigropalmatus*	1	0	0	1	0	0	0	0	2	1	0
22. *R. prominanus*	1	1	0	1	0	0	1	0	4	1	1
23. *Polypedates leucomystax*	0	0	0	0	0	1	0	2	3	0	1
24. *P. colletti*	1	0	1	2	0	0	0	0	4	1	0
Microhylidae											
25. *Microhyla berdmorei*	0	0	0	0	1	0	0	1	2	0	1
26. *M. heymonsi*	0	0	0	0	0	0	0	1	1	0	1
No. of individuals	16	11	9	14	11	10	11	21	103	50	53
No. of species	16	10	9	13	11	10	10	16	26	19	21

Most of the specimens collected were deposited either to the wet museum of the Institute of Medical Research (IMR), Kuala Lumpur or University of Malaya, Kuala Lumpur. The identification of these animals was based on publications by various authoritative authors (Berry 1975; Boulenger 1912; Lim 1991; Smith 1931; Tweedie 1983). P. Y. Berry reconfirmed most of the frog species at the time she was with the University of Malaya, Kuala Lumpur.

4. RESULTS AND DISCUSSION

A total of 231 specimens comprised 75 species of herpetofauna were collected in the PF and RF of Pasoh FR. These consisted of 26 species of amphibians (frogs and toads) and 49 species of reptiles (turtles, lizards, snakes). Of the 75 herpetofauna, 132 specimens were collected from PF while 106 specimens found from RF (Tables 1–3).

4.1 Amphibians

A total of 103 specimens of frogs and toads were collected. This comprised 26 species belonging to nine genera of five families (Table 1). Twenty species each were collected from PF and RF. Kiew (1978) identified 20 species of frogs and toads from the PF while one species was opportunistically collected by Lambert, M. R. K. as reported by Kiew et al. (1996). Both these latter collections have added 10 species, which were not encountered in this study. The 10 species were a ranid (*Rana signata*), two rhacophorids species (*Nyctixalus pictus*, *Rhacophorus bimaculatus*) and seven microhylids (*Chaperina fusca*, *Microhyla butleri*, *M. inornata*, *M. supercillaris*, *Kalophrynus pleurostigma*, *K. palmatissimus*, *Kaloula pulchra*). With these additions, the amphibian fauna in the reserve is 36 species authentically recorded to date.

Of the 26 species recorded, seven species (Table 1, Nos. 5, 10, 14, 15, 23, 25 and 26) collected in the regenerating forests are indicator species of disturbed areas (Berry 1975; Kiew et al. 1996). Their absence in the primary forest indicates that the habitat is still in a fairly pristine condition. There are five species (Table 1, Nos. 3, 8, 19, 21 and 24) collected only in the PF and not in the RF. This does not necessarily mean those species were confined to pristine habitats as they have been recorded in disturbed areas too (pers. obs.).

4.2 Reptiles

Tortoises and turtles

Seven specimens of this taxon were examined comprising four species from three families. All four (Table 2, Nos. 1–4) were forest-associated species. Among them, the Malayan flat-shelled turtle (*Notochelys platynota*) is a true primary forest animal (Smith 1931). The spiny turtle (*Heosemys spinosa*) and the Brown tortoise (*Manouria emys*) are fairly common in primary and secondary forests. The Malayan soft-shelled turtle (*Trionyx subplana*), a strictly aquatic species that used to be fairly common in forest streams has now become rare in the past 15 years due to its high demand as food (Pan 1990).

Lizards, monitor lizards, skinks and geckos

Eighty-four individuals comprising 20 species of 11 genera from four families were collected and identified. Of these, nine species were agamid lizards, four varanids, two skinks and five geckos (Table 2, Nos. 5–24).

Agamidae (Tree and Flying lizards)

Of the nine agamid, four were tree lizards (Table 2, Nos. 5–8). The most common were the Green crested lizard (*Bronchocela cristatella*) and Earless lizard (*Aphaniotus fuscus*), which is also associated with disturbed environments. The Horned tree lizard (*Acanthosaura armata*) and the Anglehead lizard (*Gonocephalus borneensis*) were collected from the primary forest.

The remaining five agamid lizards were flying lizards with one species (*Draco quinquefasciatus*) being restricted to primary forest. The other four species (Table 2, Nos. 9–11 and 13) were commonly sighted in the regenerating forest. The most common are *Draco volans* and *D. melanopogon*, both are also associated with more open habitats, such as oil palm estates and around the field station. In this survey, the Spotted-winged flying lizard (*D. melanopogon*) was sighted only in the regenerating forest.

Varanidae (Monitor lizard)

Twelve individuals of four species from a single family were collected (Table 2,

Table 2 Turtles, lizards, monitor lizards and skinks collected from Pasoh FR, Negeri Sembilan, Peninsular Malaysia.

Reptiles Taxa	Primary Forest (FR)				Regenerating Forest (FR)				Total	Total Species	
	1968	1974	1989	1991	1968	1974	1989	1991	No.	PF	RF
TURTLES											
Emydidae											
1. *Heosemys spinosa* – Spiny turtle	1	0	1	1	0	0	1	0	4	1	1
2. *Notochelys platynota* – Malayan flat-shelled turtle	1	1	0	0	0	0	0	0	2	1	0
Testudinidae											
3. *Manouria emys* – Brown tortoise	2	1	0	1	1	1	0	0	6	1	1
Trionycidae											
4. *Trionyx subplana* – Malayan soft-shelled turtle	0	0	0	0	0	0	0	2	2	0	1
LIZARDS, MONITOR LIZARDS & SKINKS											
Agamidae											
5. *Acanthosaura armata* – Horned tree lizard	1	0	1	2	0	0	0	0	4	1	0
6. *Aphaniotis fuscus* –Earless lizard	2	1	1	2	1	1	1	1	10	1	1
7. *Broncocela cristatellus* – Green crested lizard	1	1	1	1	1	1	1	1	8	1	1
8. *Gonocephalus bellii* –Anglehead lizard	1	0	1	1	0	0	0	0	3	1	0
9. *Draco sumatranus* – Common flying lizard	1	2	1	1	1	1	4	6	17	1	1
10. *D. melanopogon* – Spotted-winged flying lizard	0	0	0	0	1	1	0	0	2	0	1
11. *D. fimbriatus* – Flying lizard	0	0	1	0	0	0	1	1	3	1	1
12. *D. quiequefasciatus* – Multicolored wing flying lizard	1	1	1	1	0	0	0	0	4	1	0
13. *D. maximus* – Large flying lizard	1	2	0	0	0	0	1	1	5	1	1
Varanidae											
14. *Varanus salvator* – Water monitor lizard	0	0	0	0	1	1	2	2	6	0	1
15. *V. nebulosus* – Clouded monitor lizard	0	0	0	0	1	0	0	1	2	0	1
16. *V. rudicollis* – Harlequin monitor lizard	0	0	1	2	0	0	0	0	3	1	0
17. *V. dumerilli* - Dumeril's monitor lizard	1	0	0	0	0	0	0	0	1	1	0
Scincidae											
18. *Mabuya multifasciata* – Common sun lizard	0	0	0	0	1	0	2	1	4	0	1
19. *M. longicaudata* – Sun lizard	0	0	1	0	1	0	0	0	2	1	1
Gekkonidae											
20. *Aeluroscalabotes felinus* – Cat-eyed gecko	1	0	0	0	0	0	0	0	1	1	0
21. *Cyrtodactylus pulchellus* – Banded slender-toed gecko	1	0	0	0	0	0	0	0	1	1	0
22. *C. consobrinus* – Peter's slender-toed gecko	1	0	0	1	0	0	0	0	2	1	0
23. *Gecko smithii* – Giant forest gecko	0	0	2	0	0	0	0	0	2	1	0
24. *Ptychozoon kuhlii* – Kuhl's gliding gecko	0	1	1	1	0	0	0	0	3	1	0
No. of individuals	16	10	13	14	9	6	13	16	97	53	44
No. of species	14	8	12	11	9	6	8	9	24	19	13

Nos. 14–17) which represents the entire varanid fauna in Peninsular Malaysia (Harrison & Lim 1957). Two of the species (*Varanus rudicollis*, *V. dumerilli*) were restricted to primary forest habitat and the other two (*V. salvator*, *V. nebulosus*) are human commensal. Of these two, the common were the Water monitor (*V. salvator*) while Dumeril's monitor (*V. dumerilli*) is less common.

Scincidae (Sun lizard)

Six individuals of two species from a single family were examined (Table 2, No. 18–19). Both these lizards (*Mabuya multifasciata*, *M. longicaudata*) were non-habitat specific. Both were human commensal species and the commonest is *M. multifasciata.*

Gekkonidae (Tree and Flying Geckos)

Five species from four genera belonging to a single family were sighted (Table 2, Nos. 20–24). All the species were found only in the primary forests. All five species of geckos are not easily detected because of their ability to change body colour according to light intensity and background colouration causing them to visually merge with the tree trunks upon which they rest.

Table 3 Snakes species collected in Pasoh FR, Negeri Sembilan, Peninsular Malaysia.

Snakes Taxa	Primary Forest (FR)				Regenerating Forest (FR)				Total	Total Species	
	1968	1974	1989	1991	1968	1974	1989	1991	No.	PF	RF
Xenopeltidae											
1. *Xenopeltis unicolor* – Sunbeam snake	0	0	1	0	0	0	0	0	1	1	0
Boidae											
2. *Python reticulatus* – Reticulated python	0	1	0	0	0	0	0	0	1	1	0
Colubridae											
3. *Aplopeltura boa* – Blunt-headed snake	1	0	0	0	0	0	1	0	2	1	1
4. *Gonyosoma oxycephalum* – Red-tailed ratsnake	1	1	0	0	1	0	0	0	3	1	1
5. *Oligodon purpurascens* – Brown kukri snake	1	0	0	0	0	1	0	0	2	1	1
6. *Calamaria lumbricoidea* – Variable reed snake	0	1	0	1	1	0	0	0	3	1	1
7. *C. lovii* – Gimlett's reed snake	1	0	0	0	1	0	0	0	2	1	1
8. *Psuedorhabdion longiceps* – Dwarf reed snake	0	0	0	1	0	1	0	0	2	1	1
9. *Dendrelaphis pictus* – Painted bronze snake	0	1	0	0	0	0	0	0	1	1	0
10. *D. formosus* – Elegant bronze snake	0	0	1	0	0	1	0	0	2	1	1
11. *Chrysopelea paradisi* – Paradise tree snake	1	0	0	0	0	0	0	1	2	1	1
12. *Sibynophis melanocephalus* - Malayan Many-toothed snake	0	0	0	1	0	0	0	1	2	1	1
13. *Boiga dendrophila* – Yellow-ringed cat snake	1	0	0	0	0	1	0	0	2	1	1
14. *B. nigriceps* –Dark-headed cat snake	1	0	0	0	0	0	1	0	2	1	1
15. *B. cynodon* – Dog-toothed cat snake	1	0	0	0	0	0	0	1	2	1	1
16. *Psammodynastes pictus* – Painted mock viper	0	0	1	0	0	0	1	0	2	1	1
17. *Ahaetulla prasina* – Oriental whip snake	0	0	0	1	0	0	1	0	2	1	1
18. *Macropisthodon rhodomelas* – Blue-necked keelback	0	1	0	0	0	0	0	0	1	1	0
19. *M. flaviceps* – Orange-necked keelback	0	0	1	0	0	0	0	1	2	1	1
20. *Enhydris bocourti* – Bocourt's water snake	0	0	0	0	0	0	0	1	1	0	1
Elapidae											
21. *Naja sumatrana* – Common cobra	0	0	0	0	0	0	0	1	1	0	1
22. *Maticora bivirgata* – Blue Malaysian coral snake	0	0	0	1	0	0	0	0	1	1	0
23. *Ophiophagus hannah* – King cobra	0	0	1	0	0	0	0	0	1	1	0
Crotalidae											
24. *Tropidolaemus wagleri* – Wagler's pit viper	1	0	0	1	0	0	0	0	2	1	0
25. *Trimeresurus hageni* –Hagen's pit viper	1	1	1	0	0	0	0	0	3	1	0
No. of individuals	10	6	6	6	3	4	4	6	45	28	17
No. of species	10	6	6	6	3	4	4	6	25	23	17

4.3 Serpents

A total of 44 individuals comprising of 25 species of 19 genera from five families were recorded in the study area. Of the 25 species, 21 species were collected from primary forest while 17 were from the regenerating forest (Table 3).

Xenopeltidae

This family is represented by a single species, the Sunbeam snake (*Xenolpetis unicolor*) and was collected from the RF. This species is non-habitat specific, and has been collected from plantation, lowland to submontane forests (Lim 1955; Tweedie 1983).

Boidae

A single Reticulated python (*Python reticulatus*) was collected in the RF. Scientists working in the area also occasionally saw this species in the primary forest. As the Sunbeam snake, the python is also a non-habitat specific species, ranging from disturbed areas to forests at all elevations. It is particularly common in plantation where prey species (rodents) are more available and common (Lim 1974).

Colubridae

Of the 18 species examined, 9 species are tree snakes (Table 3, Nos. 3-11), 5 species are ground-dwelling forms (Table 3, Nos. 12-16), three species were burrowing snakes (Table 3, Nos. 17-19) and one species freshwater snake (Table 3, No. 20). All 18 species were common snakes and can be found in lowland forests throughout Peninsular Malaysia (Tweedie 1983). Three of these species, the Paradise tree snake (*Chrysopelea paradisi*), the Green whip snake (*Ahaetulla prasina*) and the Elegant bronze back (*Dendrelaphis formosus*) are more commensal in their distribution.

Elapidae

Three species of poisonous snake in this family were identified. One of each of the common cobra (*Naja sumatrana*) and Blue Malaysian coral snake (*Maticora bivirgata*) were collected. The King cobra (*Ophiophagus hannah*) was sighted. Both the King and Common cobras are non-habitat specific in distribution. Their distribution ranged from forest, plantations and occasionally near human habitation. The Blue Malaysian coral snake is strictly a forest species, and the snake is more confined to primary forest, although it is occasionally found in good standing regenerating forests as well (Table 3).

Crotalidae

Five individuals of two species of poisonous pit vipers in this family were examined. Both these vipers, the Wagler's pit viper (*Tropidolaemus wagleri*) and Hagens's pit viper (*Trimeresurus hageni*) were collected from the PF (Table 3). Although both these species have not been encountered in the RF during the surveys, they have been found in secondary forest in other parts of Peninsular Malaysia (Lim et al. 1995).

5. CONCLUSION

It must be noted that the herpetofauna compiled in Tables 1 to 3 was from opportunistic and incidental collections. The 75 species identified at the Pasoh FR constitutes 23.8% of the 315 species recorded from Peninsular Malaysia (Cranbrook 1988). The results showed that the number of species of herpetofauna recorded over the different years between the PF and RF is not similar. It is difficult however to determine the factors that contribute to this observation as the method of collection was not standardised. The different of number of species collected may be reflected from the different workers, the weather of the time of collection and the duration of night sampling in each survey period.

In addition to the different in the number of species collected over time, another observation is on the different in species composition. This survey may indicate that some herpetofauna species are restricted to pristine or disturbed areas. Anecdotal and personal experiences are supporting this observation. For example, the Four-lined tree frog (*Polypedates leucomystax*) is common in disturbed environment. In a study area of a primary hill dipterocarp forest in Kedah, West Malaysia, this tree frog managed to colonise the edge of the 13-km access road (Norsham 2000). They thrived in the water puddles on the road and on *Macaranga* saplings along the road. However, none were sighted out of the road boundary where the forest is still intact. Thus, future studies must determine factors that regulating the distributions and resilience of certain frogs to changes in their environment. This information will be useful in developing conservation measures for management purposes.

The herpetofauna, especially the amphibian taxa show very significant potential as a group for evaluation of environmental conditions in the forest. The complex life cycle of this group usually involves an aquatic stage. The permeability of their skin and eggs make them especially sensitive to habitat change. Among the vertebrate fauna groups, they may serve as one of the parameters for determining environmental changes of a forest. In this context, the status and distribution amphibian fauna need to be re-evaluated in the Pasoh FR and the associated surrounding ecotypes to obtain a better understanding of the environmental factors in the area. For the objective to be realized, a more systematically approach is needed, so that statistically, valid conclusions can be drawn.

ACKNOWLEDGEMENTS

The authors wish to thank the field team from the Medical Ecology Unit of the Institute of Medical Research, Kuala Lumpur for their assistance in the field-trapping programme. The assistance of the four orang asli (aboriginal people) from the Ulu Gombak Forest Reserve in the field activities is also acknowledged. The zoological team from the Forest Research Institute Malaysia (FRIM), Kepong are also thanked for the assistance in the recent fieldwork. Finally, we thanked Messrs. Kishokumar Jeyaraj and Tunku Nazim Tunku Yaakob, Malaysia Environmental Consultants (MEC), for their critical comments on the manuscript.

REFERENCES

Berry, P. Y. (1975) The Amphibian Fauna of Peninsular Malaysia. Tropical Press, Kuala Lumpur, 130pp.

Boulenger, G. A. (1912) A vertebrate fauna of the Malay Peninsula from the Isthmus of Kra to Singapore including the adjacent islands. Taylor and Francis, London, 294pp.

Cranbrook, Earl. (1988) Key Environments. Malaysia. Pergamonn Press.

Francis, C. (1989) Notes on fruit bats (Chiroptera, Pteropodidae) from Malaysia and Brunei, with description of a new subspecies of *Megaerops wetmorei* Taylor, 1934. Can. J. Zool. 67: 2878-2882.

Francis, C. (1990) Trophic structure of bat communities in the understorey of lowland dipterocarp rainforest in Malaysia. J. Trop. Ecol. 6: 421-431.

Harrison, J. L. & Lim, B. L. (1957) Monitor lizards of Malaya. Malay. Nat. J. 12: 1-10.

Kemper, C. (1998) The mammal of Pasoh Forest Reserve, Peninsular Malaysia. Malay. Nat. J. 42: 1-19.

Kemper, C. & Bell, D. T. (1985) Small mammals and habitat structure in lowland rain forest of Peninsular Malaysia. J. Trop. Ecol. 1: 15-22.

Kiew, B. H. (1978) List of amphibians of the Pasoh Forest Reserve. Prepared by IBP, Pasoh Newsletter.

Kiew, B. H. (1984) A new species of sticky frog (*Kalophyrnus palmatissiumus* n. sp.) from Peninsular Malaysia. Malay. Nat. J. 37: 145-152.

Kiew, B. H., Lim, B. L. & Lambert, M. R. K. (1996) To determine the effects of logging and timber extraction, and conservation of primary forest to tree crop plantations, on herpetofaunal diversity in Peninsular Malaysia. Br. Herpeteol. Soc. Bull. 57: 2-20.

Lim, B. L. (1955) Snakes collected near Kuala Lumpur. Malay. Nat. J. 9: 122-125.

Lim, B. L. (1974) Snakes as a natural predators of rats in oil palm estates. Malay. Nat. J. 27: 114-117.

Lim, B. L. (1991) Poisonous snakes of Peninsular Malaysia (3rd ed.). Malayan Nature Society.

Lim, B. L., Ratnam, L. & Hussein, Nor Azman (1995) Snakes examined from the Sungai Singgor area of Temenggor Forest Reserve, Hulu Perak, Malaysia. Malay. Nat. J. 48: 357-364.

Norsham, Y. (2000) In Search of a Prince. FRIM in Focus. August 2000, pp.4-5.

Pan, K A. (1990) Malayan turtles. J. Wildl. Parks 9: 20-31.

Smith, M. A. (1931) The reptilia and amphibia of the Malay Peninsula. Bull. Raffles Mus. 3: 1-129.

Tweedie, M. W. F. (1983) The snakes of Malaya (1st ed.). Government Printing Press, Singapore.

Wells, D. R. (1978) Habitat and biomass of insectivorous birds in the understorey of rainforest at Pasoh forest. Malay. Nat. J. 30: 353-362.

Wong, N. (1985) Understorey birds as indicators of regeneration in a patch of selectively logged West Malaysian rainforest. ICBP Technical Publ. 4: 249-263.

Wong, N. (1986) Trophic organizaton of understory birds in a Malaysian dipterocarp forest. Auk 103: 100-116.

Yasuda, M. (1998) Community ecology of small mammals in a tropical rainforest of Malaysia with special reference to habitat preference, frugivory and population dynamics. Ph.D. diss., Univ. Tokyo.

29 Small Mammals Diversity in Pasoh Forest Reserve, Negeri Sembilan, Peninsular Malaysia

Lim Boo Liat[1], Louis Ratnam[2] & Nor Azman Hussein[2]

Abstract: This paper presents results of historical data (1968–1974) of small mammals trapping, mist netting of bats, and direct observations in the Pasoh Forest Reserve (Pasoh FR). Eighty-five mammalian species belonging to 8 orders from 24 families has been catalogued. Of these, 8 species of fruit bats and 21 species of insectivorous bats were netted. Among the insectivorous bats, four species are of significant findings. These are *Myotis ridleyi*, *Pipistrellus stenopterus*, *Murina cylotes* and *Phoniscus atrox*, which are new localities records for Pasoh FR. The remaining 56 species comprised small mammals (insectivores, tree shrews, rodents, primate, carnivores and herbivores). Among the mammals, two species, *Pithecheir parvus* and *Ptilocercus lowii* are also new locality records for Pasoh FR. An additional of 27 species were recorded in later studies by Kemper (1988) and Francis (1990) that were not recorded in our study made up a total of 112 species altogether, thus far recorded for Pasoh FR. This represents 56% of the total mammalian fauna in Peninsular Malaysia. There is a marked difference in the species diversity between primary forest (PF) and regenerating forest (RF). Of the 85 species recorded in Pasoh, 46 species were found in RF compared to 78 in PF. The higher number of species diversity reported in PF was reflected by more efforts made in PF as opposed to RF. This study is far from conclusive, nevertheless results revealed sign of mammalian fauna recovering in the regenerating forest if left alone without further disturbance.

Key words: bat diversity, new locality records, primary forest, regenerating forest.

1. INTRODUCTION

The documentation on the biodiversity of Pasoh Forest Reserve (Pasoh FR) has started since its establishment as a research station in 1971 (Lee 1995). Studies of the mammal fauna of this area are still somewhat incomplete. While the bat fauna of the area was fairly well catalogued (Francis 1989, 1990), information on the other taxa are based on relatively short-term studies (Chivers 1980; Kemper 1988; Kemper &Bell 1985; Marsh & Wilson 1981; Wells 1978; Wong 1985, 1986) and also from observations made by naturalists who have visited the reserve. A long-term study on the terrestrial small mammal communities was documented by Yasuda (1998). These studies thus far showed that the area did contain fairly good vertebrate species diversity.

This paper presents the compilation of mammalian fauna based on the work carried out by the first author in 1968, 1970 and 1974. The senior author was then working for the Institute of Medical Research (IMR), Kuala Lumpur. The main

[1] Department of Wildlife and National Parks (DWNP), 50664 Kuala Lumpur, Malaysia.
E-mail : limbooliat@yahoo.co.uk
[2] Forest Research Institute Malaysia (FRIM), Malaysia.

objective of his work on the trapping and mist-netting small mammals was for biomedical study. The record on the number and species composition of the small mammal in the study area nevertheless provide a baseline data on abundance of species, although the methods used do not lend themselves to any faunal statistical analysis.

2. METHODS AND MATERIALS

The study on the mammalian diversity in the study area was carried out the regenerating forest, as well as in the primary forest at Pasoh FR. The latter site is near to the present 50-hectare plot. The three work periods were from the 10th–24th January 1968, 13th–25th May 1969 and 30th November to 14th December 1974. The data gathered from this study were based on (1) small mammal trapping, (2) mist netting and (3) direct observation.

2.1 Small mammal trapping

Two hundred wire mesh live-traps, measuring 30 cm × 15 cm × 15 cm were used in each of the regenerating forest (RF) and primary forest (PF). The traps were set randomly at ground level. The choice of placement of the traps was based on the habitat most likely to occupy an animal or where sign of animals were observed. The traps were set up at different location after two days. The main purpose of changing the location of the traps was to obtain as many animals as possible. The number of trap nights for each period was 14, and in total 400 traps were deployed. Thus, for each site, 8,400 trap nights were generated giving a total of 16,800 trap nights for the entire exercise.

Bananas were used as bait for this study. The traps were checked twice daily (8:00–12:00 and 17:00–19:00). Each individuals of animals trapped was anaesthetized with ether before being weighed, measured and their reproductive condition noted. All the animals were released except for murids, which were scarified for the endoparasite studies for the IMR, Kuala Lumpur.

2.2 Bat netting

Bats were captured with 12 m × 2 m mist nets with 36 mm mesh. The nets were set up about 2 m above ground along forest paths randomly. The distance of each mist nest ranges from 40 m to 70 m. The nets were manned just before dusk and checked at every two hours from 18:00 to 22:00. Six nets were deployed for eight nights for each trapping period at both the primary and regenerating forest sites. This results in a total of 432 mist net hours for each site.

Captured bats were weighed, measured and their reproductive condition recorded and then released. A reference collection of each of the species was preserved and deposited in the collection of the IMR, Kuala Lumpur.

2.3 Direct Observation

A record was maintained for any mammals encountered during the day when the teams were working in the field. Any mammals seen at night during frogging were also noted. The team also included four orang asli (aborigines), who set string traps of small mammals on an opportunistic basis. Any small mammals caught particularly the larger species such as the wild cats, civets and porcupines were identified and released while rodents and insectivore were sacrificed as part of the biomedical investigations. Other mode of detections includes animal's scats, faeces, footprints, nest, skull and vocalisation.

Table 1 Summary of mammals caught using small mammal traps (Values are the number of individuals trapped). RF: regenerating forest, PF: primary forest.

Taxa/Family/Species	Common Names	RF	PF	Total
INSECTIVORA				
Erinaceidae				
Enchinosorex gymnurus	Moonrat		6	6
SCANDENTIA				
Tupaidae				
Tupaia glis	Common treeshrew	16	15	31
T. minor	Lesser treeshrew		4	4
Ptilocercus lowii	Pen-tailed treeshrew		2	2
PRIMATES				
Lorisidae				
Nycticebus coucang	Slow loris	1	1	2
RODENTIA				
Sciuridae				
Callosciurus notatus	Plantain squirrel	9	10	19
C. caniceps	Grey-bellied squirrel	1	5	6
C. nigrovittatus	Black-banded squirrel		4	4
C. prevostii	Prevost's squirrel	3	4	7
Sundasciurus hippurus	Horse-tailed squirrel		2	2
S. lowii	Low's squirrel	2	5	7
S. tenuis	Slender squirrel	2	7	9
Lariscus insignis	Three-striped ground	10	18	28
Rhinosciurus laticaudatus	Shrew-faced ground squirrel	1	8	9
Rhizomidae				
Rhizomys sumatrensis	Large bamboo rat		1	1
Muridae				
Sundamys muelleri	Mueller's rat		8	8
Leopoldamys sabanus	Long-tailed giant rat	19	34	53
Maxomys surifer	Red spiny rat	8	25	33
Maxomys rajah	Rajah's rat	15	28	43
Maxomys whiteheadii	Whitehead's rat	2	1	3
Niviventer cremoriventer	Dark-tailed tree rat	2	12	14
Pitcheir parvus	Monkey footed rat		2	2
Lenothrix malaisae	Grey tree rat		2	2
Rattus tiomanicus	Malaysian wood rat	4		4
Hystricidae				
Hystrix brachyurus	Large porcupine	1	1	2
Atherurus macrourus	Brush-tailed porcupine	1	2	3
Trichys lipura	Long-tailed porcupine		1	1
CARNIVORA				
Mustelidae				
Mustela nudipes	Malay weasel		1	1
Viveridae				
Prionodon linsang	Banded linsang		1	1
Paradoxurus hermaphroditus	Common palm civet	1	2	3
Arctogalidia trivirgata	Small-tooth palm civet		1	1
Hespestidae				
Hespetes brachyura	Short-tailed mongoose		1	1
Number of individuals		98	214	312
Trap success		1.16	2.55	1.86
Number of species		18	30	32

Table 2 List of mammals recorded in the study area. RF: regenerating forest, PF: primary forest.

Taxa/Family/Species	Common Names	RF	PF	Remarks
INSECTIVORA				
Soricidae				
Crocidura fuliginosa	Southeast Asian white-toothed shrew		C	On forest floor
DERMOPTERA				
Cynocephalidae				
Cynocephalus variegatus	Flying lemur	S	S	
PRIMATES				
Cercopithecidae				
Macaca fascicularis	Long-tailed macaque	S	S	
Presbytis obscurus	Dusky leaf monkey	S	S	
Hylobatidae				
Hylobates lar	White-handed gibbon	V	V	
PHOLIDOTA				
Manidae				
Manis javanicus	Pangolin	C		On forest floor
RODENTIA				
Sciuridae				
Ratufa bicolour	Black giant squirrel	S	S	
R. affinis	Cream giant squirrel	S	S	
Petauristinae				
Aeromys tephromelas	Black giant flying squirrel	S	S	
Hylopetes lepidus	Red-cheeked flying		C	In tree holes
Petinomys setosus	Temminck's flying squirrel		C	In tree holes
Iomys horsfieldii	Horsefieldii flying squirrel	C		In tree holes
Petaurista petaurista	Giant red flying squirrel	S	S	
CARNIVORA				
Ursidae				
Helartos malayanus	Malayan sun bear		T	Claw marks on bark
Mustelidae				
Amblonyx cinerea	Small-clawed otter		T	Scat and footprints
Martes flavigula	Yellow-throated marten		A	
Viverridae				
Arctitis binturong	Bear cat		S	
Viverra tangalunga	Malay civet		A	
Paguma lavarta	Masked palm civet		S	
Felidae				
Felis bengalensis	Leopard cat		A	
F. planiceps	Flat-headed cat		A	
ARTIODACTYLA				
Suidae				
Sus scrofa	Common wild pig	T	T	Footprints/feeding sign
Tragulidae				
Tragulus javanicus	Lesser mouse deer	A	A	
Cervidae				
Cervus unicolor	Sambar deer		T	

Keys: A=Caught by Orang Asli, C=Hand Caught, T=Tracks, S=Sighted, V=Vocalisation

3. RESULTS

The overall results of the mammalian fauna inventory obtained by various methods are summarised in Tables 1 to 3. A total of 85 mammal species are catalogued here. In these tables a single numbering system runs through from Table 1 to Table 3 so that in each table only new species are given a number under the heading "No".

3.1 Mammals identified through standard trapping methods

Trapping yielded 312 individual small mammals comprising 32 species from both the PF and RF. Of these animals, 68.58% of the total number was trapped in the PF while the remaining 31.41% was trapped in the RF. This is reflected in the trap success of PF and RF being 2.55% and 1.16% respectively. Within the PF, 93.8% of the species were recorded while within the RF the figure is 56.3% (Table 1). Following is the account of species trapped according to their families.

Insectivora

Only a single insectivore species, the Moon rat (*Enchinosorex gymnurus*) was trapped in the PF.

Scandentia

Of the three species of tree shrews, *Tupaia glis* was found to be equally abundant in both habitats, while *T. minor* and *Ptilocercus lowii* appear to be restricted to the PF. The latter species is not common.

Primates

Only a single nocturnal primate was caught. One individual of the Slow loris (*Nycticebus coucang*) was trapped in each of the RF and PF.

Rodentia

Among the 32 species, rodents (squirrels, murids, rhizomyids, porcupines) constituted 68.8% of species diversity as compared to 31.2% of the remaining mammals, which comprised the insectivores (tree shrews), primates (slow loris) and civet. Of the 22 rodent species, those species with fairly large numbers caught were confined to six species only (Table 1). Five of them (*Leopoldamys sabanus*, *Maxomys rajah*, *M. surifer*, *Lariscus insignis* and *Niviventer cremoriventer*) were more abundant in the PF while *Callosciurus notatus* is equally abundant in both areas. The numbers of the remaining 16 rodent species trapped were relatively low.

Carnivora

Five species of carnivores belong to three familes were trapped: *Mustela nudipes*, *Prionodon linsang*, *Paradoxurus hermaphroditus*, *Arctogalidia trivirgata* and *Herpestes brachyura*.

3.2 Other mammals recorded through direct observations

In addition to the trapping results, 24 species of mammals were identified through other methods of collection (Table 2). Of these, 12 species were found only in the RF as compared to 22 in the PF. This does not necessarily indicate a higher diversity in the PF because most of the identification activities were carried out more in the PF. There are ten species found in both of the PF and RF. Some species were hand caught by orang asli. These include the tiny *Crocidura fuliginosa*, *Manis javanica*

Table 3 List of bat species recorded in the Pasoh FR, Negeri Sembilan. RF: regenerating forest, PF: primary forest.

Taxa/Family/Species	Common Names	RF	PF	Total
CHIROPTERA FRUIT				
Pteropodidae				
Cynopetrus brachyotis	Dog-faced fruit bat	38	7	45
C. horsefieldii	Horsefield's fruit bat	22	3	25
Balionycteris maculatus	Spotted winged fruit bat	4	18	22
Chironax melanocephalus	Black-caped fruit bat	2	14	16
Penthetor lucasi	Dusky fruit bat	2	13	15
Macroglossus minimus	Common long-tounged fruit bat	4	17	21
Rousettus amplexicaudatus	Geoffroys' rousette		4	4
Megaerops ecaudatus	Tailess fruit bat		3	3
INSECT BATS				
Emballonuridae				
Taphozous melanopogon	Black-bearded tomb bat	4		4
Emballonura monticola	Lesser sheath-tailed bat	3	15	18
Megadermatidae				
Megaderma spasma	Malaysian false vampire		3	3
Nycteridae				
Nycteris javanica	Hollow-faced bat		2	2
Rhinolophidae				
Rhinolophus affinis	Intermediate horseshoe bat	8	17	25
R. lepidus	Glossy horseshoe bat	2	16	18
R. trifoliatus	Trefoil horseshoe bat	7	3	10
R. luctus	Great eastern horseshoe bat		2	2
Hipposideros galeritus	Cantor's roundleaf horseshoe bat		8	8
H. armiger	Great roundleaf horseshoe bat	1	5	6
H. diadema	Diadem roundleaf horseshoe bat	1	7	8
Vespertilionidae				
Myotis muricola	Whiskered bat	5	4	9
M. ridleyi	Ridley's bat		2	2
Scotophilus kuhli	House bat	4		4
Pipistrellus stenopterus	Malaysian noctule		2	2
Murina cyclotis	Round-eared tube nosed bat		4	4
Kerivoula papilosa	Papilose bat		2	2
K. hardwickii	Hardwicke's forest bat		4	4
Phoniscus atrox	Grooved toothed bat		2	2
Molossidae				
Tadaria mops	Free tailed bat	3		3
Cheiromeles torquatus	Hairless bat		1	1
Number of species		16	26	29
Total number of individuals		110	178	288
Number of bats/nets hour		0.25	0.41	0.33

Table 4 Additional mammals recorded in Pasoh FR by Kemper (1988) and Francis (1990).

TAXA / Family / Species	Common Names
BATS	
Pteropodidae	
Pteropus vampyrus	Flying fox
Dyacopterus spadiceus	Dyak fruit bat
Megaerops wetmorei	White-collared fruit bat
Rhinolophidae	
Rhinolophus borneensis	Bornean horseshoe bat
R. sedulus	Lesser woolly horseshoe bat
R. robinsoni	Robinson's horseshoe bat
Hipposideros cinereus	Least roundleaf horseshoe bat
H. larvatus	Large roundleaf horseshoe bat
H. bicolor	Bicolour roundleaf horseshoe bat
H. sabanus	Lawas roundleaf horseshoe bat
H. ridleyi	Singapore roundleaf horseshoe bat
Vespertilionidae	
Kerivoula pellucida	Clear winged bat
K. intermedia	Small woolly bat
Murina rozendaali	Gilded tube-nosed bat
Harpiocephala harpia	Hairy-winged bat
Pipistrellus cuprosus	Coppery pipistrelle
OTHER MAMMALS	
Cercopithecidae	
Macaca nemestrina	Pig-tailed macaque
Hylobatidae	
Symphangulus syndactylus *	Siamang
Sciuridae	
Petaurista elegans *	Spotted giant flying squirrel
Muridae	
Rattus exulans	Polynesian rat
Viverridae	
Hemigalus derbyanus	Banded palm civet
Felidae	
Panthera pardus *	Panther/Leopard
Elephantidae	
Elephas maximus *	Asiatic elephant
Tapiridae	
Tapirus indicus *	Tapir
Suidae	
Sus barbatus *	Bearded pig
Tragulidae	
Tragulus napu	Large mouse deer
Cervidae	
Muntiacus muntjak *	Barking deer

* species that are not permanent resident within the area

(pangolin) and three species of flying squirrels (*Hylopetes lepidus*, *Petinomys setosus* and *Iomys horsefieldi*). The larger wild species such as the wild boar (*Sus scrofa*), bear (*Helarctos malayanus*) and deer (*Cervus unicolor*) were also identified.

Among the three primate species recorded in the study area, the Long-tailed macaque (*Macaca fascicularis*) and the Dusky leaf monkey (*Presbytis obscurus*) are more frequently encountered than the White-handed gibbon (*Hylobates lar*). The presence of the gibbon was based on vocalisation and no individual was sighted during the study period.

3.3 Bat netting

A total of 29 species of bat were netted, which comprised 8 species of frugivorous bats and 21 species of insectivorous bats (Table 3). Of the 288 individuals caught, 38.2% (16 species) were netted in the RF while 61.8 (26 species) were netted in the PF. There is a marked difference in the number of bats netted in RF compared to PF. Of the 29 species of bat netted 55.2% was netted in the RF while 89% was netted in the PF. All the frugivorous bats are common and abundant species with two exceptions (*Roussettus amplexicaudatus* and *Megaerops ecaudatus*). These two species were also netted in the RF. All species of fruit bat were more abundant in the PF except for the Dog faced frugivorous bat species, i.e. *Cynopterus brachyitis* and *C. horsefieldii* which were netted in much higher numbers in the RF. Of the 21 species of insectivorous bats, 18 species were netted in PF compared to 10 in RF. Among these species, only three were caught in high numbers and were represented by more than 15 individuals (Table 3). These are *Emballonura monticola, Rhinolophus affinis* and *R. lepidus*. Numbers of the other species vary from 10 to 2 individuals. Three species of insectivorous bats were netted in the RF only.

4. DISCUSSION AND CONCLUSION

The present study recorded a total of 85 species of mammal present in the research area of the Pasoh FR of which 29 species were bats. In comparison, Kemper (1988) using small mammal traps and mist nets over a six-month period and night observations recorded 65 species, which included 12 species of bat. Francis (1990) published a list of 37 species of bat caught in 1981, 1986 and 1987 over a total of 27 nights. He deployed harp traps in addition to mist nets. Table 4 listed the mammal species recorded in Pasoh FR by Kemper (1988) and Francis (1990), which were not recorded in this study. Thus far, a total of 112 species of mammalian fauna was catalogued in Pasoh FR. This represents 52% of 215 mammalian fauna in Peninsular Malaysia.

Amongst the mammal species recorded which are of significant importance are the bats. Four species (*Myotis ridleyi*, *Pipistrellus stenopterus*, *Murina cyclotis* and *Phoniscus atrox*) are significant findings as they are not very common in Peninsular Malaysia (Medway 1983).

The results from this study showed that the number and composition of the mammal's species in the two forest types (PF and RF) are different. However, this work is far from conclusive and two obvious steps need to be carried out: (1) systematic studies on the community of mammals in Pasoh FR. (2) Other sites should be identified where the same kind of comparison can be made. Both these studies should be designed from the outset of statistical analysis and attention must be paid to investigate the factors that regulate fluctuation of populations and mammal species diversity in a logged-over forest. This information is needed critically, because it will allow for the development of management systems for the production of lowland dipterocarp forest as well as other forest types that will enable such forests to play an optimal role in the conservation of animal diversity of the country.

ACKNOWLEDGEMENTS

The authors wish to thank the field team from the Medical Ecology Unit of the Institute of Medical Research, Kuala Lumpur for their assistance in the field-trapping programme. The assistance of the four orang asli (aboriginal people) from the Ulu Gombak Forest Reserve in the field activities is also acknowledged. The zoological team from the Forest Research Institute Malaysia (FRIM), Kepong are also thanked for the assistance in the recent fieldwork. Finally, we thank Messrs. Kishokumar Jeyaraj and Tunku Nazim Tunku Yaakob, Malaysia Environmental Consultants (MEC), for their critical comments on the manuscript.

REFERENCES

Chivers, D. J. (1980) Malayan Forest primates. Ten years' study in tropical rainforest. Chivers, Plneum Press, New York.

Kemper, C. (1988) The mammals of Pasoh Forest Reserve, Peninsular Malaysia. Malay. Nat. J. 42: 1-19.

Kemper, C & Bell, D. T. (1985) Small mammals and habitat structure in lowland rain forest of Peninsular Malaysia. J. Trop. Ecol. 1: 15-22.

Francis, C. (1989) Notes on fruit bats (Chiroptera, Pteropodidae) from Malaysia and Brunei, with the description of a new subsepecies of Megaerops wetmorei, Taylor, 1934. Can. J. Zool. 67: 2878-2882.

Francis, C. (1990) Tropic structure of bat communities in the understory of lowland dipterocarp rainforest in Malaysia. J. Trop. Ecol. 6: 421-431.

Lee, S. S. (1995) A guidebook to Pasoh, FRIM Technical Information Handbook No. 3. Forest Research Institute Malaysia, Kuala Lumpur, 73pp.

Marsh, C. W. & Wilson, W. L. (1981) A survey of primates in Peninsular Malaysia forests. Malayan Primate Research Programme. Report 1.

Medway, Lord. (1983) The wild mammals of Malaya (Peninsular Malaysia) and Singapore. Oxford University Press, Kuala Lumpur, 131pp.

Wells, D. R. (1978) Habitat and biomass of insectivorous birds in the understorey of rainforest at Pasoh forest. Malay. Nat. J. 30: 353-362.

Wong, M. (1985) Understorey birds as indicators of regeneration in a patch of selectively logged crest Malayan rain forest. ICBP Technical Publ. 4: 249-263.

Wong, M. (1986) Trophic organization of understory birds in a Malaysian dipterocarp forest. Auk 103: 100-116.

Yasuda, M. (1998) Community ecology of small mammals in a tropical rainforest of Malaysia with special reference to habitat preference, frugivory and population dynamics. Ph.D. diss., Univ. Tokyo.

Part V

Plant-Animal Interactions

30 Insect Herbivores on Tropical Dipterocarp Seedlings

Naoya Osawa[1] **& Toshinori Okuda**[2]

Abstract: We conducted a field experiment in the Pasoh Forest Reserve (Pasoh FR), Peninsular Malaysia, from 1995 to 1996 to investigate seasonal variation in the numbers of herbivorous insects and their insect predators; to clarify the difference in insect density in forest canopy gaps relative to closed canopies; and to determine the effect of artificial defoliation (0, 50 and 100% defoliation of seedlings) on the predator-prey insect community associated with tree seedlings. We noted a trend in herbivore numbers, which peaked in August 1995, then decreased toward March 1996. Predator numbers lagged behind prey numbers; they increased from November 1995 and peaked in January 1996. The densities of both the herbivorous insects and their predators were greater on seedlings in canopy gaps than in closed forest. Herbivores differed in plant species preference, and also showed a preference for seedlings that had not been defoliated. The results are discussed in the context of tropical rain forest regeneration.

Key words: canopy gap, clipping treatment, herbivore, insect predator, regeneration.

1. INTRODUCTION

Current knowledge of the patterns and causes of insect population fluctuation is based almost exclusively on studies conducted in temperate zones (e.g. Wolda 1978). Insect populations in seasonal tropical climates are no more stable than in wet temperate climates; in both regions, stability tends to be low when rainfall is low and unpredictable. In non-seasonal tropical climates, however, insect populations should be more stable (Wolda 1978). Kato et al. (1995) used ultraviolet light traps to show that in tropical lowland dipterocarp forests in the weakly seasonal climate of Sarawak (Malaysia), insect populations did fluctuate and followed a recognizably seasonal trend that was influenced by fluctuating levels of rainfall. However, predator-prey interactions are also predicted to be important factors influencing the stability of insect communities (Begon et al. 1986; Kuno 1987). Theoretical and empirical research suggests that predation can be both a regulating and a disturbing factor in insect populations (Kuno 1987). Obviously, environmental and biological factors interact to determine seasonal variations in the size of insect populations, including the numbers of herbivores and the resulting level of herbivores on forest plants. However, the role of insect predator-prey interactions in population size variability in tropical regions has been less well documented than the corresponding role of climate.

Coley & Barone (1996) reviewed the ecological and evolutionary consequences of plant-herbivore interactions in tropical forests. In primary tropical forests, new tree growth is enhanced by gaps in the canopy. Levels of herbivores

[1] Japan International Cooperation Agency (JICA) Present address: Kyoto University, Kyoto 606-8502, Japan. E-mail: osawa@kais.kyoto-u.ac.jp
[2] National Institute for Environmental Studies (NIES), Japan.

may also be affected by the presence or absence of a canopy gap; if so, then the success of new seedlings is determined in part by both direct and indirect effects of canopy gaps. We know of no information on the variability of herbivores in open and closed tropical forest canopies. Sabelis & Van de Bann (1983) showed that under laboratory conditions the degree of leaf damage by herbivores was correlated with an increase in subsequent herbivore attacks. This has not, to our knowledge, been demonstrated to occur in tropical forests under field conditions.

We conducted field experiments on artificially transplanted seedlings of eight tree species in a lowland dipterocarp forest in Peninsular Malaysia. The aims of this study were: 1) to clarify the seasonal variations of herbivore and predator population sizes, 2) to determine whether canopy gaps influence insect density, and 3) to discover what effect artificial defoliation (mimicking herbivory) has on the insect community.

2. MATERIALS AND METHODS

2.1 Study site and tree species

This research was carried out at the Pasoh Forest Reserve (Pasoh FR) in Negeri Sembilan state, in Peninsular Malaysia (2°59' N, 102°18' S). The mean annual rainfall in this area, measured at the nearest meteorological station (Pasoh II oil plantation, 4 km to the south of the reserve) and averaged over 1974 and 1992, was 1842 mm, with two distinct peaks, one in April/May and the other in November/December. The overall vegetation type in the reserve was lowland dipterocarp forest, characterized by a high proportion of members of the Dipterocarpaceae (see Symington 1943; Wyatt-Smith 1961). The soils in the reserve consisted mainly of a parent material of shale, granite, and fluviatile granitic alluvium (Allbrook 1973).

The seeds of eight species, including six dipterocarps (*Neobalanocarpus heimii, Shorea acuminata, S. parvifolia, S. macrophylla, S. macroptera* and *S. maxima*) and two non-dipterocarp species (*Santiria tomentosa* and *Sapium baccatum*) were collected in July and August 1994, from several sites within Negeri Sembilan and Selangor States. *Neobalanocarpus heimii* is regarded as a shade-tolerant, slow-growing species, while the other dipterocarp species are medium to fast-growing trees. *Sapium baccatum* is usually found in open habitats and it is regarded as a pioneer species. The native distribution of *S. macrophylla* is Borneo (Lee 1998; Soerianegara & Lemmens 1994), but the other seven species are common in the Pasoh FR. The seeds were established at the FRIM (Forest Research Institute Malaysia) nursery, at Kepong, and then transplanted to the Pasoh Forest Reserve.

2.2 Experimental design and plot description

Two 6 × 6-m transplantation plots were situated under a canopy gap, while a third (12 × 12 m in size) was under closed canopy (Fig. 1). The seedlings were planted in rows, leaving 30 to 40 cm between plants. All plots were enclosed by zinc sheets to prevent damage by wild boar. The relative light intensity was 6.4 and 35.8% under the closed canopy and canopy gap, respectively. The forest in the study plots was secondary growth forest, having been logged once in the early 1950s (Chap.2). Most of the canopy trees were 30 to 40 m high, with a DBH (diameter at breast height) of 30 to 40 cm. The density of emergent trees in the experimental plots was much less than in the core area of the reserve.

Some of the seedlings transplanted to the plots were artificially defoliated by cutting their leaves off with scissors. To simulate 50% defoliation, all leaves,

Fig. 1 The artificial transplanted seedlings at the research site in the Pasoh FR.

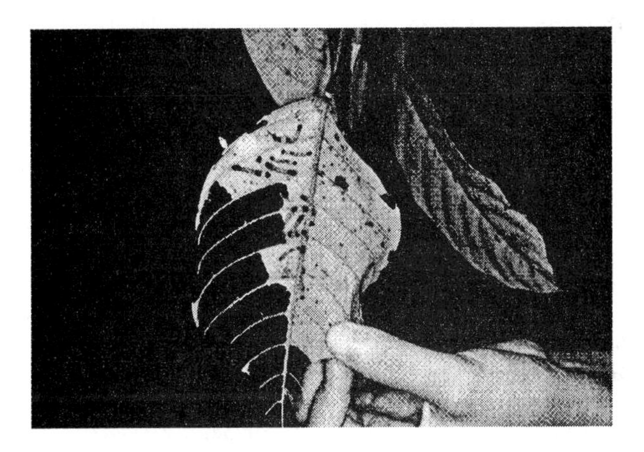

Fig. 2 Group feeding of herbivores on *Shorea macrophylla*.

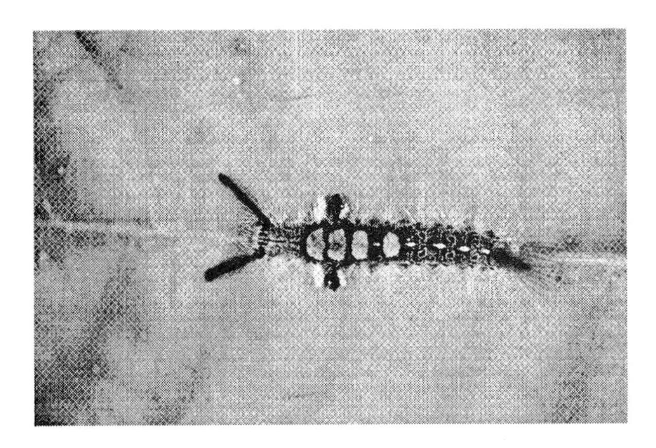

Fig. 3 A typical herbivore on *Shorea macrophylla*.

except newly emerged folded leaves, were clipped perpendicular to the midrib at the broadest point; 100% defoliation was simulated by clipping the leaves at their base, leaving petioles and a small portion of the lamina (< 5% of total area) intact.

To clarify the population dynamics and community structure of insects on the seedlings, all the insects (except ants) on individually marked seedlings were sampled monthly from May 1995 to June 1996 in each plot (Figs. 2 and 3). Date and seedling number were recorded during all observations, and the time for searching for insects was set at one minute per plant. Insects were divided into three categories: herbivore, predator, or visitor, based on direct observations of their behavior during the study period.

2.3 Meteorological data

Rainfall, humidity, and air temperature were monitored at the top of a 52-m observation tower situated about 300 m from the research site. Temperature was measured every minute, and 30-minute averages were automatically recorded by a recorder (CR10, Campbell); the humidity was measured using a humidity meter (HMP 35C, Vaisala). The amount of rainfall was measured every 30 minutes using a rain gauge (B-011-00, Yokogawa Weathac Corporation) and also recorded.

3. RESULTS

In total, 143 insects were collected during the study period (Table 1). Of the insects collected, 87.4% were herbivores, 5.6% were predators and 7.0% were classified as visitors. Among the herbivores, 44.8% were leaf eaters and 38.4% were sap feeders (Table 1).

Fig. 4 shows the observed seasonal changes of insect numbers on the seedlings, together with rainfall and average temperature. The number of herbivores peaked in August 1995 and decreased toward March 1996, while the number of predators increased from November 1995 to January 1996, at which time the herbivore population had already peaked, though their density was low. No significant relationship was found between the total number of herbivores and the monthly rainfall ($df = 1, F = 0.48, P = 0.4986$), or between herbivore numbers and total rainfall in the previous month ($df = 1, F = 0.0051, P = 0.944$).

The number of herbivores collected from the open canopy plots was significantly greater than that from the closed canopy plot (χ^2-test: $df = 1$, $\chi^2 = 9.194$, $P < 0.005$; see Fig. 5); the same pattern was observed in the numbers of

Table 1 The list of insects on the seedlings.

	Order	Food type	No. of insects
Herbivore	Coleoptera	Borer	3
		Leaf eater	2
	Hemiptera	Sap feeder	48
	Lepidoptera	Leaf eater	56
	Orthoptera	Leaf eater	14
	Phasmida	Leaf eater	2
Predator	Coleoptera		2
	Hemiptera		6
Visitor	Blattaria		7
	Coleoptera		1
	Diptera		1
	Hymenoptera		1
			Total 143

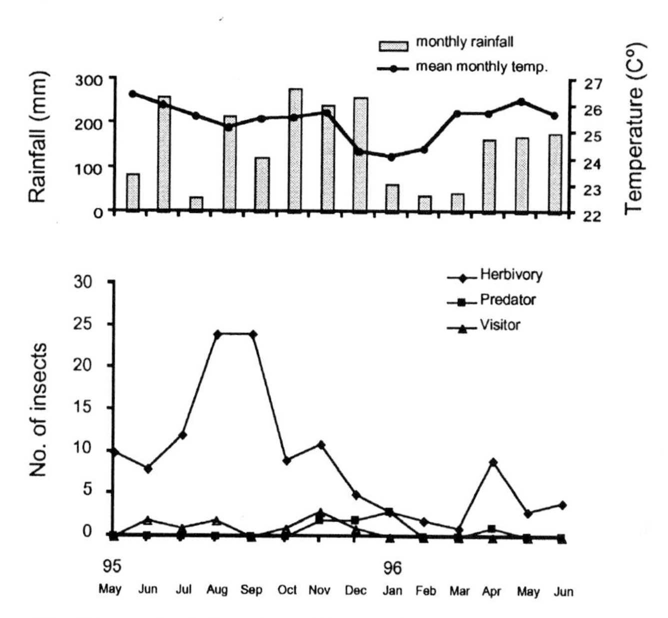

Fig. 4 Seasonal variations of monthly rainfall (total) and temperature (mean) (upper), and the number of insects on the planted seedlings (total) (lower) at the Pasoh FR.

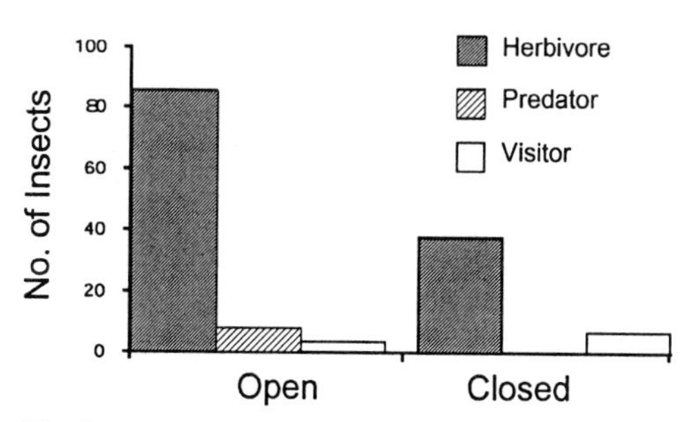

Fig. 5 The difference in number of insects at the tropical seedlings between the open and the closed canopy plot.

predators (Fisher's exact test: $df = 1, P < 0.05$). However, visitor numbers did not differ significantly among the plots (Fisher's exact test: $df = 1, P = 0.4099$).

Herbivores significantly favored unclipped seedlings over defoliated seedlings (χ^2-test: 50% and 100% defoliation data combined, $df = 1, \chi^2 = 13.086, P < 0.0005$; see Table 2). However, predator and visitor numbers were not affected by seedling defoliation (χ^2-test: 50% and 100% defoliation data combined, $df = 1, \chi^2 = 0.16818, P > 0.9$).

Defoliation of seedlings significantly affected the number of herbivores on the plants (two-way ANOVA: $df = 1, F = 43.41, P = 0.0223$; see Table 3). The effects of light intensity and its interaction with defoliation were only marginally significant (two-way ANOVA: $df = 1, F = 12.82, P = 0.0699$; $df = 1, F = 16.08, P = 0.0569$, respectively). Defoliation was also a significant factor affecting the numbers of predators and visitors (two-way ANOVA: $df = 1, F = 5.60, P = 0.0455$). However, light intensity and the interaction between defoliation and light intensity were not significant (two-way ANOVA: $df = 1, F = 0.62, P = 0.4530, df = 1, F = 0.32, P = 0.5886$, respectively).

Finally, significantly more herbivores were found on *Shorea macrophylla* and *Sapium baccatum* seedlings, while fewer herbivores were found on *S. macroptera*, *S. maxima*, *S. parvifolia*, *S. acuminata*, *Neobalanocarpus heimii* and *Santiria tomentosa* seedlings (χ^2-test: $df = 7, \chi^2 = 141.055, P < 0.0001$; see Fig. 6).

Table 2 The difference in number of insects on the seedlings of each clipping treatments.

Clipping treatment (%)	0	50	100
Herbivore	103	22	0
Predator	6	2	0
Visitor	7	2	1

Table 3 Two-way ANOVA of total insect number on the seedlings *versus* light intensity and clipping treatment.

Herbivore

Source	*df*	*SS*	*F*	*P*
Clipping treatment*(C)	1	2821.33	43.41	0.0223
Light intensity**(L)	1	833.33	12.82	0.0699
C*L	1	1045.33	16.08	0.0569

Predator and visitor combined

Source	*df*	*SS*	*F*	*P*
Clipping treatment*(C)	1	18.38	5.6	0.0455
Light intensity**(L)	1	2.04	0.62	0.453
C*L	1	1.04	0.32	0.5886

*clipping treatment was divided into two categories: clipping and unclipping.
**light intensity was divided into two categories: strong (open area) and weak (closed area).

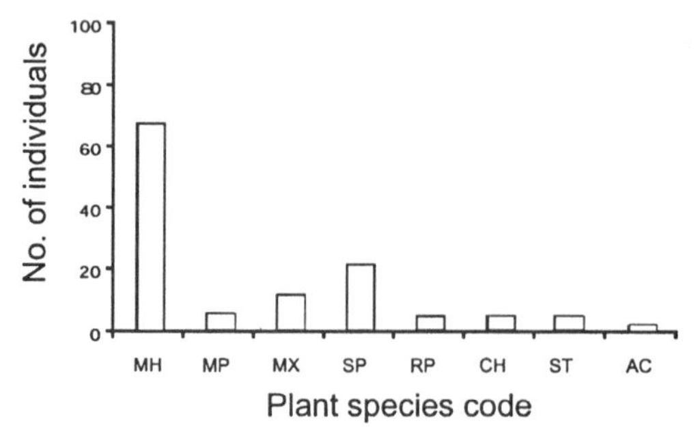

Fig. 6 Species-difference in number of visiting insects during the study period. Tree species codes are as follows: MH, *Shorea macrophylla* (N = 178); MP, *S. macroptera* (N =150); MX, *S. maxima* (N =144); SP, *Sapium baccatum* (N =88); RP, *S. parvifolia* (N =174); CH, *Neobalanocarpus heimii* (N =112); ST, *Sanitria tomentosa* (N =117); AC, *S. acuminata* (N =176).

4. DISCUSSION

This study showed that the insect community associated with tree seedlings in the Pasoh FR experiences seasonal fluctuations in population size; the number of herbivores peaked in August 1995 and decreased toward March 1996. Predator numbers matched the increase, but lagged by several months. However, no relationship was found between the number of herbivores and recent rainfall, which contradicts the results of Kato et al. (1995). However, Kato et al. (1995) used ultraviolet light traps to sample insects, which may have attracted insects from a wider area than the local study plot. In our study, all the insects were sampled directly from the seedlings. Perhaps rainfall is not an important factor in the seasonal variation of tropical insect numbers in small, localized areas, because the quality and quantity of resources, predator-prey interactions, and variations in microhabitats affect the distribution of insects in these small areas. It was also observed that the density of insects was low throughout the year. In total, only 143 insects were sampled from 1,139 seedlings. This implies that typically the density of insects on dipterocarp seedlings in the Pasoh FR (and possibly in tropical dipterocarp forests in general) is low. This may be due to plant defenses against herbivorers, but we do not yet have enough data to make this conclusion.

This study suggests that numbers of herbivores and predators are greater under canopy gaps than under closed canopies. It also suggests that undamaged seedlings are favored by herbivores over defoliated seedlings. This suggests that undamaged seedlings in canopy gaps may be comparatively favored by herbivores and predators over defoliated seedlings under a closed forest canopy, implying that the faster-growing seedlings in open areas are under stronger foraging pressure, and that herbivores on them are simultaneously under stronger predatory pressure than in closed forest in the same system. This study showed that the clipping treatment was a significant factor affecting the total number of insects (herbivores, predators, and visitors) on the seedlings. However, light intensity and the interaction between defoliation and light intensity were marginally significant for herbivores, while not significant for predator and visitor frequency. This suggests that environmental factors and the quantity of resources influence the density of herbivores more strongly than the density of predators and visitors. More data are required to state this conclusively, especially because the seasonal growth variation and species difference of the seedlings were not taken into consideration in this study.

Seedlings under canopy gaps have been observed to produce more leaves than those under closed canopies (Okuda et al. unpubl.). This suggests that in open areas seedlings invest more resources in growth than do those under closed forest. Generally, herbivores on seedlings is thought to be important in maintaining high tree species diversity in the tropics (Connell 1971; Janzen 1970). The results of this study are consistent with earlier studies, in that they suggest a trade-off between allocation of resources to seedling growth and to chemical defenses against herbivores (e.g. Dudt & Shure 1994; Shure & Wilson 1993).

In this study, 54.40% ($N = 68$) of the herbivores were observed on *Shorea macrophylla* seedlings. Moreover, *S. macrophylla* seedlings grew the most rapidly of all the seedlings in the study area (Okuda et al. unpubl.). This also suggests a trade-off between growth and defense in *S. macrophylla* seedlings. Narrow-host-range insect herbivores are responsible for most leaf damage in tropical forests, and likely were and still are the source of selection pressure, resulting in the evolution of plant chemical, developmental and phenological defenses (Coley &

Barone 1996). However, *S. macrophylla* is native to the island of Borneo, not to Peninsular Malaysia (Lee 1998; Soerianegara & Lemmens 1994), implying that many of the herbivores on *S. macrophylla* may be generalists, and that generalist herbivores may be abundant in tropical forests. Therefore, defenses against herbivores in plants in tropical communities may be the result of a close and evolutionarily long interaction between plants and insects.

ACKNOWLEDGEMENTS

We are very indebted to Drs. Laurence G. Kirton, Jurie Intachat and N. Manokaran, Forest Research Institute Malaysia (FRIM) to conduct this study. Thanks are also due to Drs. Makoto Tani, Forest and Forest Product Research Institute (FFPRI) and A. Rahim Nik, FRIM, for their permission to use the meterological data. We are also thanking Messrs. Affendy bin Mohamud and Sharihzal bin Shamudin for helping the field research.

REFERENCES

Allbrook, R. F. (1973) The soils of Pasoh Forest Reserve, Negeri Sembilan. Malay. For. 36: 22-33.

Begon, M., Harper, J. L. & Townsend, C. (1986) Ecology: Individuals, Populations and Community. Blackwell Scientific Publications, Oxford.

Coley, P. D. & Barone, J. A. (1996) Herbivory and plant defenses in tropical forests. Annu. Rev. Ecol. Syst. 27: 305-335.

Connell, J. H. (1971) On the role of natural enemies in preventing competitive exclusion in some marine animals and in rain forest. In Den Boer P. J. & Gradwell, G. R. (eds). Dynamics of Populations, Proceedings of the Advanced Study Institute on Dynamics of Numbers in Population, Oosterbeek. Wageningen, The Netherlands: Center for Agricultural Publishing and Documentation, pp.298-312.

Dudt, J. F. & Shure, D. J. (1994) The influence of light and nutrients on foliar phenolics and insect herbivory. Ecology 75: 86-98.

Janzen, D. H. (1970) Herbivore and the number of tree species in tropical forests. Am. Nat. 105: 97-112.

Kato, M., Inoue, T., Hamid, A. A., Nagamitsu, T., Merdek, M. B., Nona, A. R., Itino, T., Yamane, S. & Yumoto, T. (1995) Seasonarity and vertical structure of light-attracted insect communities in a Dipterocarp forest in Sarawak. Res. Popul. Ecol. 37: 59-79.

Kuno, E. (1987) Principals of predator-prey interaction in theoretical, experimental, and natural population systems. Adv. Ecol. Res. 16: 249-337.

Lee, S. S. (1998) Root symbiosis and nutrition. In Appanah, S. & Turnbull, J. M. (eds). A Review of Dipterocarps Taxonomy, Ecology and Silviculture. Center for International Forest Research, Bogor, Indonesia, pp.99-114.

Sabelis, M. W. & Van de Baan, H. E. (1983) Location of distant spider mite colonies by phytoseiid predators: demonstration of specific kairomones emitted by *Tetranychus urticae* and *Panonychus ulmi.* Entomol. Exp. Appl. 33: 303-314.

Soerianegara, I. & Lemmens, R. H. M. J. (1994) Plant Resources of South-East Asia. No. 5(1). Prosea Foundation, Bogor.

Shure, D. J. & Wilson, L. A. (1993) Patch-size effects on plant phenolics in successional openings of the Southern Appalachians. Ecology 74: 55-67.

Symington, C. F. (1943) Forester's Manual of Dipterocarps. Malayan Forest Records No. 16. Penerbit Universiti Malaya, Kuala Lumpur, 244pp.

Wolda, H. (1978) Fluctuations in abundance of tropical insects. Am. Nat. 112: 1017-1045.

Wyatt-Smith, J. (1961) A note on the fresh-water swamp, lowland and hill forest types of Malaya. Malay. For. 24: 110-121.

31 Spatial Distribution of Flower Visiting Beetles in Pasoh Forest Reserve and Its Study Technique

Kenji Fukuyama[1], Kaoru Maeto[2] & Ahmad S. Sajap[3]

Abstract: To recognize how to maintain sustainable biodiversity in tropical forests, the spatial distributions of flower-visiting beetles were investigated using flower fragrance traps in the Pasoh Forest Reserve (Pasoh FR). 1) The vertical distribution of flower-visiting beetles was investigated using the balloon system and tree-towers. Most flower-visiting beetles (Scarabaeidae) were trapped from 10 m to 25 m above ground, indicating that they are active in the continuous middle layer of the canopy, where most trees and climbers bear flowers. *Mecinonota regia sumatrana* were, however, distributed in the relatively low layer, while *Dasyvalgus dohlni* was distributed in the upper part of the canopy. 2) The horizontal distribution of flower-visiting beetles was investigated using flower fragrance traps in the Pasoh FR to analyze fragmentation and edge effects. Since this forest is surrounded by oil palm plantations and rubber plantations, it remains as an island of forest within them. The number of *Dasyvalgus*, one of the main flower-visiting groups, was significantly greater in the primary forest area than in the regenerating area whilst *Mecinonota* showed no difference. The community structure of scarabaeid beetles was different between the primary forest area and the regenerating forest area. However, difference of beetle fauna between the primary forest area and the regenerating forest area might be more affected by selective logging ca. 40 years ago than the fragmentation or edge effects. This suggests that the effects of selective logging on the beetle fauna still remain even after more than 40 years. Thus, both the observations on vertical and horizontal distributions of flower-visiting beetles imply that management of tropical rain forests should aim to conserve the complex spatial structure of primary forests. 3) Comparison among the trapping systems for canopy fauna in tropical forests were discussed. Balloon trap system and tree-tower with fragrance traps of linalool, eugenol and methyl benzoate were effective for collection of flower-visiting insects on tropical rain forest canopy. The effectiveness of the balloon trap system and tower systems were not different.

Key words: balloon system, canopy fauna, edge effect, flower fragrance trap, Scarabaeidae.

1. INTRODUCTION

Recently, human impacts such as logging or changing to cultivation have quickly reduced insect diversity in tropical forests probably due to decreased complexity of spatial forest structure (Whitmore 1990). A large number of insect species live in tropical rain forests and they play important roles in tropical forest ecosystems (Whitmore 1984). Forest canopies harbor the highest species diversity and

[1] Forestry and Forest Products Research Institute (FFPRI), Tsukuba 305-8687 Japan. E-mail: fukuchan@affrc.go.jp
[2] Shikoku Research Center, FFPRI, Japan.
[3] Universiti Putra Malaysia (UPM), Malaysia.

ecological function of insects (Basset 1990; Paoletti et al. 1991). The vertical heterogeneity of forest structure is one of the guarantees to maintain the diversity of the invertebrate community (Koike et al. 1998; Sutton & Hudson 1980; Toda 1987, 1992). Also, there is positive correlation between species numbers and area size (MacArthur & Wilson 1966), indicating that isolation and partial fragmentation causes a decrease of species diversity. Recent fragmentation of natural forests threatens the persistence of biodiversity and rudimental ecological processes, but empirical data are too limited to understand and assess the impacts of logging and other forestry treatments on the organic communities of forests, especially tropical rain forests. We can appraise such impacts in fragmented primary or managed regenerating forests, however, we need to understand the pattern of spatial structure of arthropod communities in primary rain forest for development of forest management system in a tropical forest (Sutton 1989). Though many studies have been carried out concerning insect distribution in the forest canopy (Kato et al. 1995; Koike et al. 1998; Sutton & Hudson 1980; Sutton et al. 1983), there have been few studies conducted on the relationship between insect communities and human impacts.

The canopy insect fauna consists of several insect groups; i.e. herbivores feeding on leaves and other living tissues, decomposers of dead branches, and temporary visitors at flowers. Social bees as pollinators in Asian tropical rain forest have also been investigated (Nagamitsu & Inoue 1997, 1998; Nagamitsu et al. 1999), however, we do not have enough information about flower-visiting beetles in Asian tropical rain forests. Scarabaeid beetles include many typical flower-visitors. Their body sizes are relatively large (e.g. 5–20 mm) and hard among the flower-visiting beetles and systematic studies are well advanced in tropical areas. Thus, identification of this group is easier than for many other insect groups. Their larvae mainly feed on decaying wood on the ground and the adults feed on tree flowers in the forest canopy. Therefore, flower-visiting beetles can be a good indicator of human impacts on tropical forests because of their important role in pollination (Bawa 1990; Bawa et al. 1985; Kress & Beach 1994; Momose et al. 1998) and decomposition which are biotic interactions affected by changes in forest structure (Aoki 1979). Another benefit of flower-visiting beetles as a biological indicator is their hard body. This is because specimens caught by traps are hard to break and these beetles are easy to collect using attractant collision traps (Ikeda et al. 1993; Iwata & Makihara 1994; Makihara et al. 1989; Sakakibara et al. 1997, 1998).

In this report, we describe the vertical distribution of flower-visiting beetles (Scarabaeidae) associated with forest canopy structure and the horizontal distributions concerned with changes caused by human impacts in the Pasoh Forest Reserve (Pasoh FR), based on surveys with chemical attractant traps.

We therefore developed an inexpensive and movable balloon system in combination with attractant traps and examined the effectiveness of the system in a lowland tropical rain forest (Fukuyama & Maeto 1994) and compared with tower system. We also compared trapping effectiveness for flower-visiting beetles between several flower fragrance chemicals at a UPM (Universiti Putra Malaysia) forest (Maeto et al. 1995).

2. MATERIALS AND METHODS

2.1 Vertical distribution of flower-visiting beetles in Pasoh FR

Vertical surveys were carried out on a tower system built in Pasoh FR. The tower

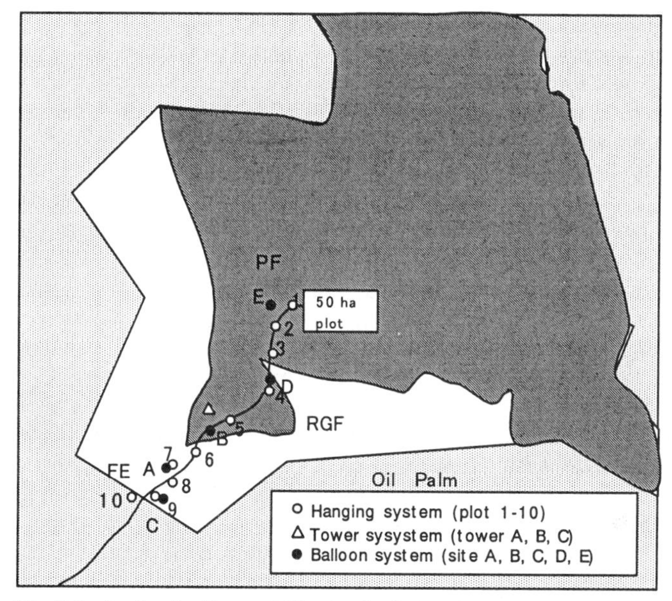

Fig. 1 Study sites for flower visiting beetles in Pasoh FR. Numbers in the figure represent the plot for hanging system. PF, primary forest; RGF, regenerating forest; FE, forest edge.

consists of two towers (towers A and B) of 32 m height and one tower (tower C) of 52 m height (Fig. 1, Color plate 3).

1) Four chemical attractant traps for scarabaeid beetles were set on each tower at heights of 1 m, 10 m, 18 m and 26 m above the ground in 1993. Trapping was conducted for two weeks from November to December, 1993.
2) In August 1995, we also set traps on tower A at heights of 10 m, 18 m and 30 m and on tower C at heights of 10 m, 18 m and 30 m. We set traps on 1 August and collected from them on 20 August 1995. The catches of beetles were collected every day.
3) We also set attractant traps on tower A at heights of 1.5 m, 10 m, 20 m and 30 m, and on tower C at heights of 1.5 m, 10 m, 20 m, 30 m, 40 m and 50 m, in December 1995. We set traps on 2 December 1995 and collected from them on 12 December 1995. The trap used was an improved collision trap that attached to a transparent umbrella to protect it from rain .

Linalool was attached to the trap as an attractant for flower-visiting beetles in 1993 and in August 1995. The collision plate of the trap was surrounded with sticky paper to catch the beetles attracted by the chemical in 1993 and in August 1995. Sticky paper was also put in the collecting vessel. The traps used were collision type with a water vessel and the attractant chemicals were linalool and eugenol in December 1995.

2.2 Horizontal distribution of flower-visiting beetles from edge to primary forest area

We established 10 study plots from the edge to the primary forest area along a path in Pasoh FR. We set a pair of flower fragrance traps at each study plot. Five study plots were established in regenerating forest (RGF), including forest edge (FE) and other five were established in primary forest (PF) (Fig. 1). Every two traps were hung from branches of emergent trees in the forest canopy at different heights; the upper trap was 10 m higher than lower one. The height of upper traps were varied from 14 m to 25 m. Trap type was a white, plastic collision trap. Attractant chemicals were linalool and eugenol. Traps were set on 3 December 1995 and catches were collected 12 December 1995.

2.3 Comparison of the balloon system with other trapping systems

A comparison of trap systems to catch flower-visiting beetles was carried out at Pasoh FR in 1995. Five sets of the balloon trap system (Balloon system) (sites A–E) and 9 sets of a hanging trap system (Hanging system) (plots 1–9) were established from the edge to the core area along a trail in the Pasoh FR (Fig. 1). Hanging system was where a trap was hung by rope from a branch of emergent tree. Three traps were established each on towers A and C at height of 10 m, 18 m and 30 m (Tower system). Balloon systems were set on 31 July, Hanging systems were set on 4 August and Tower systems were set on 1 August 1995. Trap heights of Hanging system varied from 11 m to 25 m and those of Balloon system were from 12 m to 22 m. The attractant trap used in all the trap systems was a collision type with plastic umbrella above the collision plate for protection from rain and with sticky paper instead of a water vessel. The attractant chemical was linalool. The beetles caught by traps were collected every day until 19 August 1995. The balloons were re-filled with helium gas after 10 days. The balloon was a sphere 2.5 m in diameter and made of white colored PVC film 0.12 mm in thickness. The weight of the balloon including ropes was about 3 kg. Pure lift of the balloon was about 4 kg when each balloon was filled to capacity with 7 cubic meters of helium. The balloon was moored with three pieces of thin rope in three different directions to prevent the traps from swaying (Fig. 2).

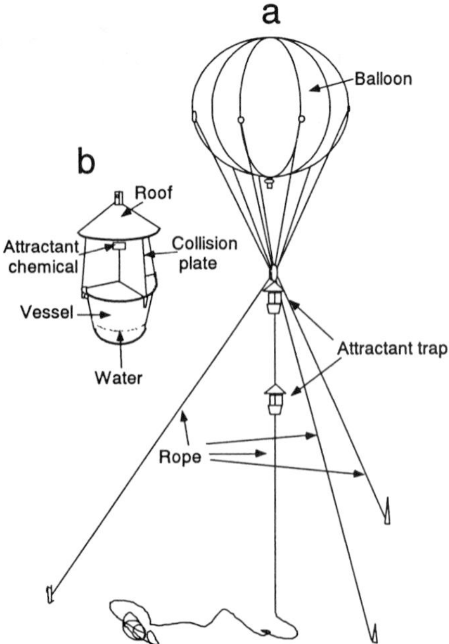

Fig. 2 Schematic representation of (a) the balloon-suspended trap system and (b) an individual attractant trap (Fukuyama & Maeto 1994).

3. RESULTS

3.1 Vertical distribution of flower-visiting beetles in Pasoh FR

The abundance of main flower-visiting genus or family groups caught by the traps in December 1995 were different among trap heights (Fig. 3). Spearman's correlation of flower visiting beetles communities between different height traps were significantly positive correlation among between same or neighbored height and negative correlation between understory and canopy layer (Table 1). Scarabaeidae were caught in the canopy and sub-canopy, Melyridae were caught in the canopy and *Monolepta* spp. (Chrysomelidae) were caught above the canopy (Fig. 3).

Eight species of Scarabaeidae including two genera, *Dasyvalgus* and *Mecinonota*, were caught by the traps in 1993 and 1995. Two dominant species, *Dasyvalgus vethii* and *D. niger*, were mainly collected at a height of 18–20 m which was in the continuous layer of the canopy (Fig. 4). *D. dohlni* was most abundant at

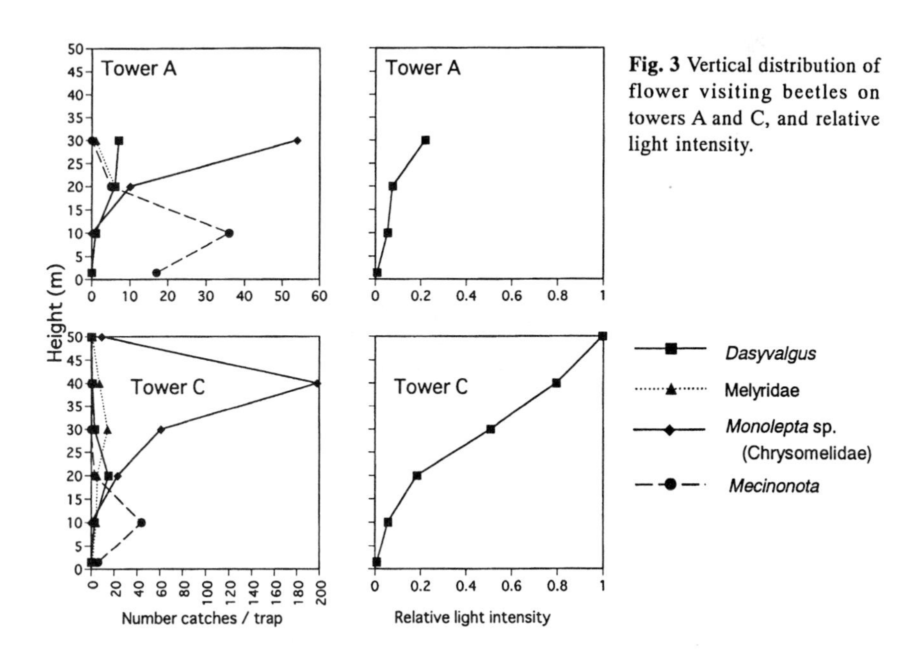

Fig. 3 Vertical distribution of flower visiting beetles on towers A and C, and relative light intensity.

Table 1 Spearman's correlation of flower visiting beetles communities among traps on towers A and C in 1995.

	A10m	A20m	A30m	C1.5	C10	C20	C30	C40	C50
A1.5m	0.68	0.2	−0.5	0.76 *	0.64	−0.2	−0.4	−0.4	−0.3
A10m		0.42	−0.5	0.82 *	0.93 *	−0.1	0	−0.1	−0.1
A20m			0.27	0.45	0.26	0.55	0.63	0.58	0.8 *
A30m				−0.6	−0.8 *	0.77 *	0.28	0.23	0.41
C1.5m					0.83 *	−0.2	−0.1	−0.1	0.11
C10m						−0.3	0.04	0.02	−0.1
C20m							0.49	0.43	0.54
C30m								0.99 *	0.83 *
C40m									0.84 *

*$P < 0.05$, Spearman's correlation

a height of 26–30 m, just above the continuous canopy layer. *Mecinonota regia sumatrana*, the third dominant species, was mainly collected at a height of 10 m (Fig. 4). There were few catches in traps set at 1 m. The results indicate that the flower-visiting Scarabaeidae mainly act in the canopy and their active layers were different among species.

The numbers of Scarabaeidae varied among towers much more at the 26 m height than at 10 m or 18 m height (Fig. 5). The relative light intensity at the 26 m height were different between tower A and C, compared with those at lower points (Fig. 6). It seems to be that the activity of the Scarabaeidae is affected by the light condition of the canopy.

3.2 Horizontal distribution of flower-visiting beetles from edge to primary forest area

The abundance of main flower-visiting groups caught by traps were different among the study plots (Fig. 7). Numbers of *Monolepta* spp. (Chrysomelidae) were higher at the primary forest area (plot 1) and quickly decreased in number in the regenerating area and relatively increased again at the forest edge. Numbers of the Curculionidae were relatively more large at the primary forest area (plots 1–5) and less at the regenerating area (plots 6–10) while those of Melyridae were small in number at the edge area. Mordellidae numbers were larger at the intermediate area (plots 3–7) and the Scarabaeidae were large at all areas, except for plot 1. Relative abundance of scarabaeid beetles was lowest at the primary forest and gradually increased from the primary forest to the forest edge. There were no significant differences between primary forest (plots 1–5) and regenerating forest (plots 6–10) except for Curculionidae (t-test, $N = 5, P = 0.038$).

When species composition in the Scarabaeidae was examined, 9 species of scarabaeid beetles were collected in this study, including 8 species of *Dasyvalgus* and 1 species of *Mecinonota*. The proportion of *Mecinonota* was lower at the primary forest area and higher at the regenerating and edge areas (Fig. 8). The number of *Dasyvalgus* spp., one of the main flower-visiting groups in family Scarabaeidae, was significantly greater in the primary forest area than in the

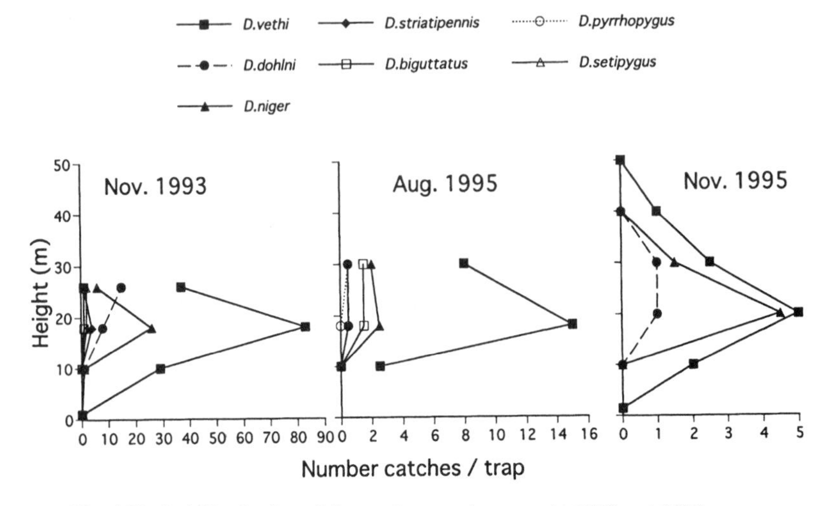

Fig. 4 Vertical distribution of *Dasyvalgus* on the tower in 1993 and 1995.

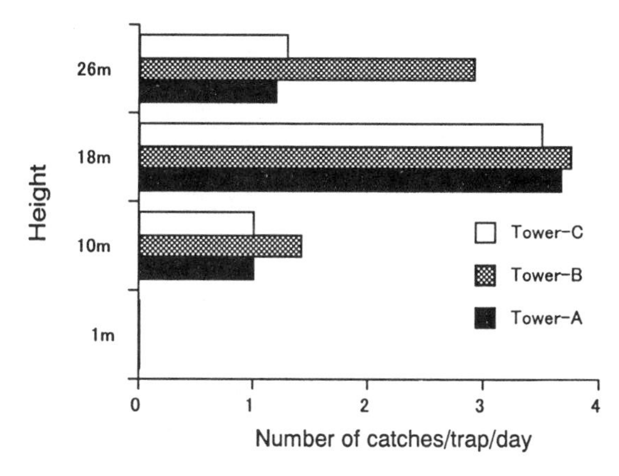

Fig. 5 Fluctuation in the number of Scarabaeidae among three individual towers in 1993.

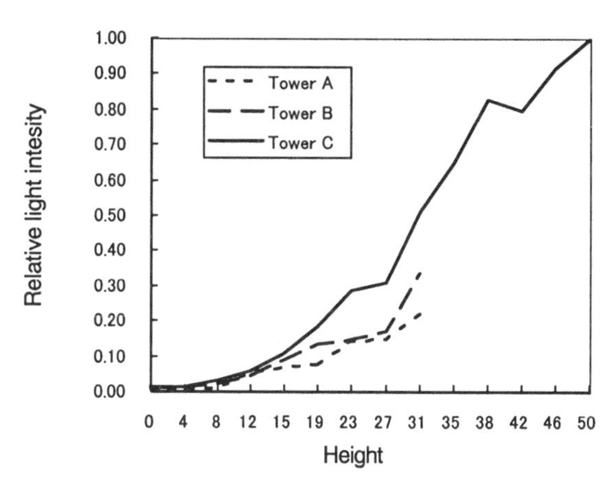

Fig. 6 Relative light intensities on the three towers in 1995.

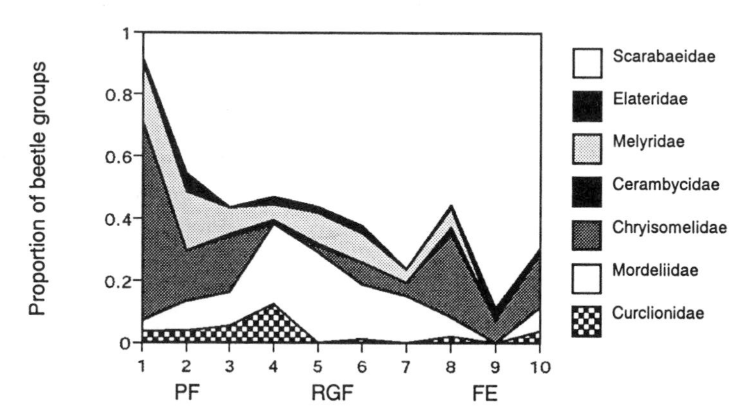

Fig. 7 Proportion of flower visiting beetles from PF to FE (Fig. 1) in Dec. 1995.

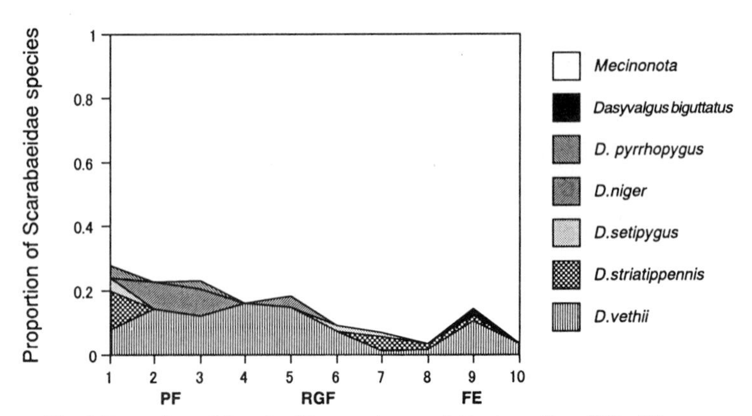

Fig. 8 Proportion of Scarabaeidae species caught by traps from PF to FE (Fig.1) in Dec. 1995.

Table 2 Flower-visiting beetles caught by attractant traps hanging on tree canopy in PF (primary forest) and RGF (regenerating forest).

Species	Number of catches	
	Core (PF)	Buffer (RGF)
Dasyvalgus vethii	34	13
D. niger	12	0
D. setipygus	2	0
D. sp.1	1	0
D. sellatus	4	4
D. biguttatus	0	1
D. sumatranus	0	1
D. striatippennis	0	2
Mecinonota regia sumatrana	196	260

regenerating forest area (t-test, $N = 5$, $P = 0.015$), whilst *Mecinonota,* another dominant flower visitor, showed no difference (t-test, $N = 5$, $P = 0.134$). *Dasyvalgus* species structures were different between the primary and the regenerating area (Table 2).

Difference of beetles fauna between the primary forest area and the regenerating forest area might be more affected by selective logging 40 years ago than the fragmentation or edge effects. This suggests that the effects of selective logging still remain even after more than 40 years.

3.3 Comparison of the balloon system with other trapping systems

Two balloons were accidentally broken on 10 and 13 August. We changed two broken Balloon systems to Hanging system after the accident occurred. However, only one scarabaeid beetle was caught by the two broken systems after changing to the hanging system. Therefore, we could still compare the trapping effects among the three systems.

Nine species of Scarabaeidae were caught by the three trap systems during the study period (Table 3). Though the total numbers of individuals per trap of scarabaeid beetles caught by traps were smallest for Hanging system, there was no difference between Balloon system and Tower system. The species composition of the Scarabaeidae caught by Balloon system was quite similar to Tower system but

Table 3 Species structure of scarabaeid beetles caught by attractant traps in the three systems.

	Balloon	Hanging	Tower
Dasyvalgus vethii	8.60	2.44	8.50
D. niger	0.40	0.78	1.50
D. dohlni	0.20	0.00	0.33
D. striatipennis	0.20	0.22	0.33
D. sp.	0.00	0.11	0.00
D. biguttatus	0.00	0.00	1.00
D. sumatranus	0.00	0.11	0.00
D. pyrrhopygus	0.20	0.00	0.17
Mecinonota reiga sumatrana	4.20	0.11	1.33
Total	13.80	3.78	13.17
Species number	5	6	7
Sample number (Number of traps)	5	9	6

relatively different from Hanging system. The number of species caught for Hanging system was not so large despite 9 traps being set at various places whilst only five traps were set for Balloon system. *D. dohlni* and *D. pyrrhopygus* were not caught with Hanging system. The vertical distribution study using the tower system showed that *D. dohlni* and *D. pyrrhopygus* were relatively distributed in the higher part of the forest canopy (Fig. 4). This suggests that Hanging system is not so effective in catching beetles which act in the upper part of the canopy and that Balloon system is useful in collecting these beetles because we can move the trap up above the canopy layer.

One balloon was broken after 11 days and another was broken after 14 days. The other 3 balloons were continuously moored for 20 days.

The results of the comparative study among the three trap systems showed that Balloon system was as useful as Tower system and it could be continued for more than 10 days in a tropical rain forest.

4. DISCUSSION

4.1 The trap systems for canopy fauna

One of the reasons for the few reports on flower-visiting insects is the great canopy height in tropical forests. Methods such as large towers or cranes (Mitchell 1982; Parker et al. 1992), chemical fogging (Basset 1990; Hijii 1983; Ochi et al. 1968), and climbing techniques (Lowman 1992; Nadkarni & Matelson 1991; Perry 1978) have been used to study canopy insects. Hallè (1990) developed a giant balloon air ship to bring researchers on a flexible raft up to the forest canopy in a South American tropical rain forest. However, each method has disadvantages (Table 4; Fukuyama et al. 1994). Building a tower or crane in a tropical forest is extremely costly and disturbs the forest ecosystem. Tree climbing methods are dangerous and require climbing skills. Chemical fogging is inappropriate for determining the vertical distribution of insects and may disturb the natural environment. Hallè's air ship system is expensive and requires a large number of staff. Many studies on vertical distribution of insects in tropical forests using light traps or sticky traps have been carried out (Kato et al. 1995; Koike et al. 1998; Sutton 1989; Sutton & Hudson 1980; Sutton et al. 1983). Kato et al. (1995) researched the vertical distribution of forest insects in Sarawak using light traps. The light trap mainly caught night moving insects and not flower-visiting insects such as Scarabaeidae. Although sticky traps could catch general flying insects in the forest canopy, collection effectiveness for flower-visiting beetles was not so high. Attractant traps are useful for collecting certain groups of insects (i.e. flower visitors) and some pollinating beetles are attracted to floral volatile compounds (Ikeda et al. 1993; Lajis et al. 1985).

Field test of the balloon trap system (Fukuyama et al. 1994) showed the traps yielded beetles from thirteen families but the dominant flower-visiting genera attracted to benzyl acetate were *Dasyvalgus, Mecinonota* (Scarabaeidae) and *Endaeus* (Curculionidae). Table 5 shows the total number of individuals for each species from these dominant genera attracted to benzyl acetate in traps placed over consecutive days at three different heights. A much larger number of individuals were trapped at 15 m than at 1.5 m over nine parallel trapping days, particularly of *Dasyvalgus* sp.1, whilst an intermediate number were obtained in a trap placed at a height of 7 m for two days during this period.

Comparison of floral fragrance chemicals as attractants for flower-visiting beetles (Maeto et al. 1995) showed that Scarabaeidae were caught in the traps with eugenol, methyl benzoate or linalool and much less frequently in those with benzyl acetate (Fig. 9). Mordellidae were caught in the traps with benzyl acetate, methyl benzoate or linalool, though a majority of them (70%) were caught in linalool traps. Almost the same number of Curculionidae were caught with benzyl acetate, methyl benzoate and linalool. Only a total 37 individuals of Cerambycidae were collected. A few other Coleoptera such as Nitidulidae, Staphilinoidea and Oedemeridae were collected, but they were excluded from the analysis.

The number of individuals caught in the traps varied significantly with the chemicals in species with sufficient sample size, except for *Paragymnopleurus manurus* (Scarabaeidae) (Maeto et al. 1995). This indicates that the species were

Table 4 Comparison among research methods on the canopy insect fauna (Fukuyama 1995).

	Cost	Maintenace	Technic	Staffs	Movablility	Direct observation	Impact on environmen	Risk	Collection species	Duration	Stability
Tower crane	high	high	high	median	limit	well	large	low	large	long	high
Air ship with laft	high	high	high	large	well	well	small	high	large	short	median
Fogging	low	low	low	small	well	no	large	median	large	short	low
Climb	low	low	high	small	well	well	small	high	small	short	low
Hanging	low	low	median	small	possible	no	small	low	median	long	high
Tower	high	low	low	small	no	well	large	median	small	long	high
Walk way on canopy	high	low	low	small	limit	well	median	high	large	long	high
Fell down	median	low	median	small	possible	possible	large	high	small	short	low
Cut down	low	low	high	small	well	possible	small	low	small	short	low
Balloon trap	low	low	low	small	well	no	small	low	median	short	median

Table 5 The number of flower visiting beetles collected using benzyl acetate as an attractant in traps placed simultaneously at different heights or at different sites over consecutive days (Fukuyama et al. 1994).

Species	Traps at different heights in site A					Traps in different sites at 15 m	
	over a period of 9 days		over a period of 2 days			over a period of 2 days	
	1.5m	15m	1.5m	7m	15m	Site A	Site B
Mecinonota sp.	1	6	1	1	2	1	2
Dasyvalgus sp.1	0	29	0	1	13	7	5
Dasyvalgus sp.2	0	1	0	1	1	0	0
Dasyvalgus sp.3	0	1	0	1	0	0	0
Endaeus sp.1	0	2	0	0	0	1	8
Endaeus sp.2	0	2	0	0	1	1	0
Total No. of individuals	1	41	1	4	17	10	15
Total No. of species	1	6	1	4	4	4	3

selectively attracted to certain chemicals. Figure 10 shows the changes in the proportion of individuals collected in each month for major species with more than 50 specimens in total. Most species occurred throughout the year, but 90% of *Endaenidius* sp. and 65% of *Glipostenoda* sp.1 were caught in May. Nearly 50% of *Dasyvalgus vethii* were collected in November. Also, more than 70% of *Glipostenoda* spp. and *Amorphoides* sp. were caught in May and August. For all species, catches in February were less than 20% of the total catch.

Scarabaeids, mordellids, cerambycids and curculionids were collected during the trial period. Eugenol was a weak attractant, but two scarabaeids, *Mecinonota regia sumatrana* and *Dasyvalgus vethii,* were frequently caught in the traps with eugenol. Most mordellid species and all the dominant curculionid species were caught in the traps with benzyl acetate. *Callistethus maculatus* and *Dasyvalgus*

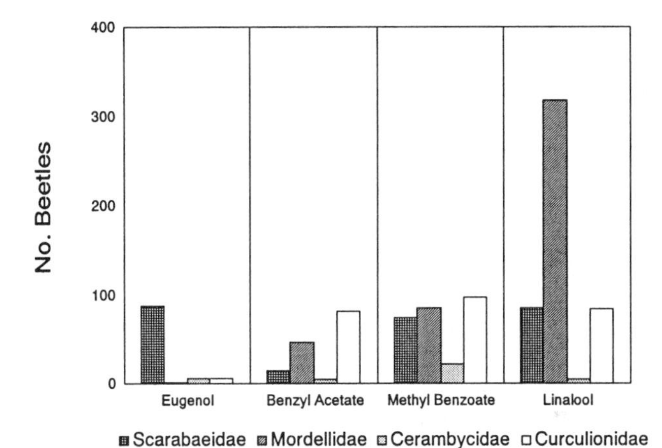

Fig. 9 Total number of Scarabaeidae, Mordellidae, Cerambycidae and Curculionidae caught in the traps with various chemicals (Maeto et al. 1995).

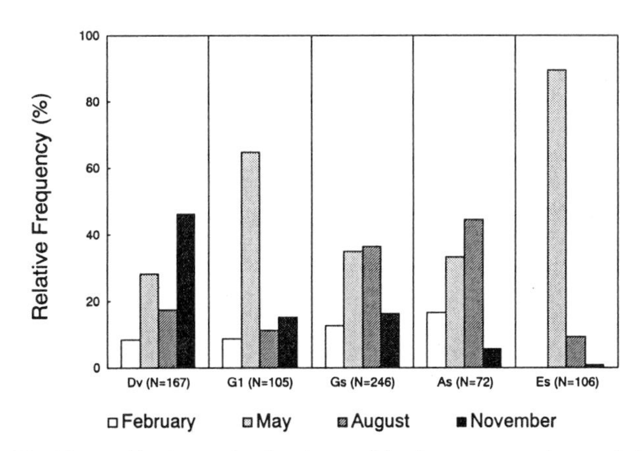

Fig. 10 Monthly changes in abundance of the five major species caught in all the traps. Dv: *Dasyvalgus vethii* (Scarabaeidae), Gl: *Glipostenoda* sp.1 (Mordellidae), Gs: *Glipostenoda* spp. (Mordellidae), As: *Amorphoides* sp. (Curculionidae), Es: *Endaenidius* sp. (Curculionidae) (Maeto et al. 1995).

spp. (Scarabaeidae), *Glipostenoda* spp. (Mordellidae), *Longipalpus apicalis* (Cerambycidae) and *Amorphoides* sp. (Curculionidae) were frequently collected with methyl benzoate. A large proportion of *Dasyvalgus vethii* (Scarabaeidae), *Glipostenoda* spp. (Mordellidae) and *Endaenidius* sp. (Curculionidae) were caught in linalool traps.

The species composition of beetle caught by balloon traps was similar to the flower-visiting beetles observed in the forest canopy of a lowland rain forest in Sarawak, Malaysia (Momose et al. 1998; T. Nagamitsu pers. comm.). Our data demonstrate that the balloon trap system can be used to study the vertical distribution of insects on a spatial and temporal scale. This system proved to be inexpensive, portable and required fewer operators in comparison to cranes, towers or canopy fogging methods. It therefore enables many replications in carrying out ecological studies on insects inhabiting forest canopies. Another advantage of the system is that it can be set up even in very small canopy gaps. The method can also be adapted for a variety of traps, such as sticky traps, light traps, suction traps and other attractant traps, as long as they are within the weight limit that the balloon system can support them. It can also be used in mark-recapture methods for the study of populations, dispersal and migration.

This study has shown that fragrance chemicals were useful for sampling flower-visiting beetles in tropical rain forests. Sajap et al. (1997) reported that the fragrance chemical trap was useful to study lacewings (Neuroptera: Chrysopidae) in lowland tropical forests. The collected beetles included pollen or nectar feeders (Scarabaeidae, Mordellidae, Cerambycidae) and certain groups of Curculionidae which probably lay eggs in flower-buds or ovaries (Hayashi et al. 1984; Morimoto pers. comm.). All are expected to play some role in pollinating flowers. Trapping techniques using floral fragrance chemicals will surely enhance ecological research on pollinating beetles in the canopy.

The number of trapped individuals of *Dasyvalgus* spp. on the tower decreased in the latter half of the study period in 1993 (Fig. 11). One of the reasons for this may be that most of the Scarabaeidae at the study site had been caught by the traps. The adults caught by traps frequently had soil attached to the body surface. This suggests that the adults of *Dasyvalgus* were quickly trapped just after emergence and hints that the trap has high attraction to *Dasyvalgus*.

Momose et al. (1998) showed that the main flower visitors of Dipterocarpaceae, especially *Shorea*, were beetles. The fauna caught by the

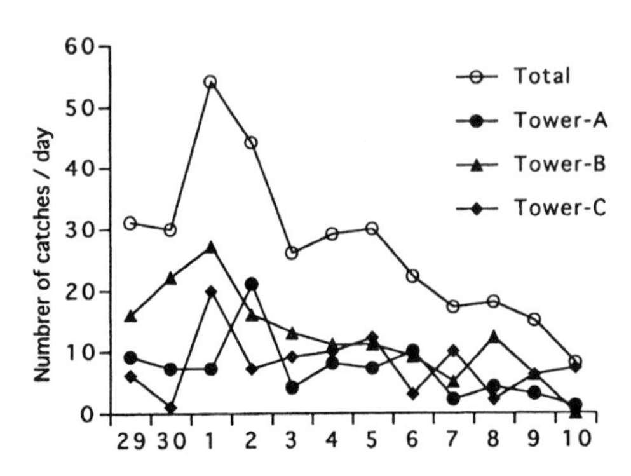

Fig. 11 Daily changes in the number of Scarabaeidae during the study period in 1993.

attractant traps in this study were similar to the list of flower-visiting beetles observed in Sarawak (Momose et al. 1998). Thus, we are able to study flower-visiting beetles with this trapping method using attractant chemicals. Our results for vertical distribution suggests that flower-visiting beetles usually act on certain canopy layers and this differs among the species and groups. The relationship between relative light intensity and abundance of beetles suggests that their active layers are probably decided by relative light intensity in the canopy. The difference in catches among trap heights means that the effective distance of the attractant traps in which to lure the beetles is less than 10 m. This also shows that we should set the traps near the canopy layer where target beetles inhabit.

4.2 Spatial distribution of flower visiting beetles

Unfortunately, we could obtain only few specimens of *Dasyvalgus* using traps. The data are thus inadequate to clearly analyze the vertical structure of *Dasyvalgus* species. But, we could not trap *D. dohlni* using the hanging system while we could catch them at upper canopy on the tower system and the balloon system. This indicates that *D. dohlni* is usually inhabit in the upper canopy layer, since we could only set the hanging trap under the canopy.

Scarabaeid beetles caught by traps in this study are typical flower visitors (Bawa 1990; Bawa et al. 1985; Momose et al. 1998). Though we do not have enough information about *Dasyvalgus* and *Mecinonota* (Scarabaeidae) in tropical forests, the habitat of their larvae is probably decayed wood. *Valgus* and *Nipponovalgus* belonging to the Valginae, as does *Dasyvalgus*, were found from termite nests (Iwata & Naomi 1998; Ritcher 1958). This may suggests that *Dasyvalgus* may also live in termite nests. Abundance and species diversity of this group are probably limited by the amount of decayed wood on the forest floor. The frequency of large tree falls in regenerating forests as they are selectively logged might be low because of the lack of old big trees. This will cause a lower abundance of fallen trees on the forest floor and a lower density of scarabaeid beetles. The results of our comparative study between primary and regenerating forests support this hypotheses. Tree falls also enrich the complex canopy structure in primary forest and this might partly explain why the diversity of flower-visiting beetles was higher in primary forest than in regenerating forest. The assemblage of these beetles will be a useful as a biological indicator of biodiversity in tropical rain forests.

Dominant flower-visiting beetles other than the Scarabaeidae caught by traps in our study were *Monolepta* spp. (Chrysomelidae), Endaeini (Curculionidae) and Mordellidae. *Monolepta* spp. was caught over the canopy layer and its relative abundance was higher than other beetles. *Monolepta* spp. are the main pollinator of *Shorea parvifolia* in Sarawak, Malaysia (Sakai et al. 1999). Endaeini were also caught within or above the canopy, and are the main flower visitor on Annonaceae in Sarawak (Momose et al. 1998). These facts suggest that the flower-visiting beetles attracted by floral fragrance chemicals are important pollinators in tropical rain forests. These species probably inhabit fresh leaves, shoots and flowers (Sakai et al. 1999). Momose et al. (1998) reported that chrysomelid beetles fed on dipterocarp leaves in flowerless seasons and suggested that the group pollinating some dipterocarps in Sarawak fed on the leaves of dipterocarps in the non-general flowering period and shifted resources to floral tissues in the general flowering period. However, there is little information concerned with the life history and biomass of these beetles in tropical forests. Further studies on this are therefore much needed.

ACKNOWLEDGEMENT
We greatly thank Dr. Laurence G. Kirton (FRIM) for his assistance in our study in the Pasoh FR. We thank Drs. Y. Miyake, A. Nobuchi, K. Morimoto and Mr. H. Kojima (Kyushu University), Drs. S. Shiyake (Osaka Museum of Natural History), H. Makihara and M. Isono (FFPRI) for the identification of the Scarabaeidae, Scolytidae, Curculionidae, Mordellidae, Cerambycidae and Chrysomelidae respectively. We are grateful to Messrs. T. Matsumura and T. Hattori for their help in field work and Dr. Y. Maruyama for his assistance in carrying out this study. We thank Drs. S. Lawson (Queensland Forestry Research Institute), K. Ozaki and T. Nagamitsu for their critical checking our manuscript. This study was a part of a joint research project between the NIES, FRIM and UPM under the Global Environment Research Programme funded by the Japan Environment Agency (Grant No. E-2).

REFERENCES
Aoki, J. (1979) Difference in sensitivities of Oribatid families to environmental change by human impacts. Rev. Ecol. Biol. Sol. 16: 415-422.

Basset, Y. (1990) The arboreal fauna of the rainforest tree *Argyrodendron actinophyllum* as sampled with restricted canopy fogging: composition of the fauna. Entomologist 109: 173-183.

Bawa, K. S. (1990) Plant-pollinator interactions in tropical rain forests. Annu. Rev. Ecol. Syst. 21: 399-422.

Bawa, K. S., Bullock, S. H., Perry, D. R., Coville, R. E. & Grayum, M. H. (1985) Reproductive biology of tropical lowland rain forest trees. II. Pollination systems. Am. J. Bot. 72: 346-356.

Fukuyama, K. (1995) How to study the canopy beetles in a tropical rain forest. Research trial in the Pasoh Forest Reserve, Peninsular Malaysia. Tropics 4: 317-326 (in Japanese).

Fukuyama, K., Maeto, K. & Kirton, L. G. (1994) Field tests of a balloon-suspended trap system for studying insects in the canopy of tropical rainforests. Ecol. Res. 9: 357-360.

Hallè , F. (1990) A raft atop the rain forest. Natl. Geogr. 178: 128-138.

Hayashi, M., Morimoto, K. & Kimoto, S. (eds). (1984) The Coleoptera of Japan in color. vol. IV. Hoikusha publishing, Osaka 438pp. (in Japanese).

Hijii, N. (1983) Arboreal arthropod fauna in a forest. I. Preliminary observations on seasonal fluctuations in density, biomass, and faunal composition in a *Chamaecyparis obtusa* plantation. Jpn. J. Ecol. 33: 435-444.

Ikeda, T., Ohya, E., Makihara, H., Nakashima, T., Saito, T., Tate, T. & Kojima, K. (1993) Olfactory responses of *Anaglyptus subfasciatus* Pic and *Demonax transilis* Bates (Coleoptera: Cerambycidae) to flower scents. J. Jpn. For. Soc. 75: 108-112.

Iwata, R. & Makihara, H. (1994) How to collect insects with the insect traps and attractants, marketed for the control and census of forest pest insects. Gekkan Mushi 281: 18-23 (in Japanese).

Iwata, R. & Naomi, S.-I. (1998) Coleopterous fauna of the Japanese termites' nests. Jpn. J. Entomol. 1: 69-82.

Kato, M., Inoue, T., Hamid, A. A., Nagamitsu, T., Merded, M. B., Nona, A. R., Itioka, T., Yamane, S. & Yumoto, T. (1995) Seasonality and vertical structure of light-attracted insect communities in a dipterocarp forest in Sarawak. Res. Popul. Ecol. 37: 59-79.

Koike, F., Riswan, S., Partomihardjo, T., Suzuki, E. & Hotta, M. (1998) Canopy structure and insect community distribution in a tropical rain forest of West Kalimantan. Selbyana 19: 147-154.

Kress, W. J. & Beach, J. H. (1994) Flowering plant reproductive systems. In McDade, L. A., Bawa, K. S., Hespenheide, H. A. & Hartshorn, G. S. (eds). La Selva, ecology and natural history of a neotropical rain forest. University of Chicago Press, Chicago, IL. pp.161-182.

Lajis, N. H., Hussein, M. Y. & Toia, R. F. (1985) Extraction and identification of the main compound present in *Elaeis Guineensis* flower volatiles. Pertanika 8: 105-108.

Lowman, M. D. (1992) Leaf growth dynamics and herbivory in five species of Australian rainforest canopy trees. J. Ecol. 80: 433-447.

MacArthur, R. H. & Wilson, E. O. (1967) The Theory of island biogeography. Princeton University Press, Princeton, 203pp.

Maeto, K., Fukuyama, K., Sajap, A. S. & Wahab, Y. A. (1995) Selective attraction of flower-visiting beetles (Coleoptera) to floral fragrance chemicals in a tropical rain forest. Jpn. J. Entomol. 63: 851-859.

Makihara, H., Kamata, N., Fukuyama, K., Goto, T., Tabata, K., Ito, K. & Hosoda, R. (1989) Comparison of cerambycid-fauna captured at various stands in suburbs. Trans. 100th Anun. Meet. Jpn. For. Soc. pp.599-600 (in Japanese).

Mitchell, A. W. (1982) Reaching the rain forest roof / A Handbook on Techniques of Access and Study in the Canopy. UNEP.

Momose, K., Yumoto, T., Nagamitsu, T., Kato, M., Nagamasu, H., Sakai, S., Harrison, R. D., Itioka, T., Hamid, A. A. & Inoue, T. (1998) Pollination biology in a lowland dipterocarp forest in Sarawak, Malaysia. I. Characteristics of the plant-pollinator community in a lowland dipterocarp forest. Am. J. Bot. 85: 1477-1501.

Nadkarni, N. & Matelson, T. (1991) Fine litter dynamics within the tree canopy of a tropical cloud forest. Ecology 72: 2071-2082.

Nagamitsu, T. & Inoue, T. (1997) Aggressive foraging of social bees as a mechanism of floral resource partitioning in an Asian tropical rainforest. Oecologia 110: 432-439.

Nagamitsu, T. & Inoue, T. (1998) Interspecific morphological variation in stingless bees (Hymenoptera: Apidae, Meliponinae) associated with floral shape and location in an Asian tropical rainforest. Entomol. Sci. 1: 189-194.

Nagamitsu, T., Momose, K., Inoue, T. & Roubik, D. W. (1999) Preference in flower visits and partitioning in pollen diets of stingless bees in an Asian tropical rain forest. Res. Popul. Ecol. 41: 195-202.

Ochi, K., Katagiri, K. & Kojima, K. (1968) Studies on the composition of the fauna of invertebrate animals in the canopy strata of pine forests. I. On the method of investigation and simple description of the fauna. Bull. Gover. For. Exp. Stat. 217: 167-170.

Paoletti, M. G., Taylor, R. A. J., Stinner, B. R., Stinner, D. H. & Benzing, D. H. (1991) Diversity of soil fauna in the canopy and forest floor of a Venezuelan cloud forest. J. Trop. Ecol. 7: 373-383.

Parker, G. G., Smith, A. P. & Hogan, K. P. (1992) Access to the upper forest canopy with a large tower crane sampling the tree-tops in three dimensions. Bioscience 42: 664-670.

Perry, D. R. (1978) A method of access into the crowns of emergent and canopy trees. Biotropica 10: 155-157.

Ritcher, P. O. (1958) Biology of Scarabaeidae. Annu. Rev. Entomol. 3: 311-334.

Sajap, A. S., Maeto, K., Fukuyama, K., Ahmad, F. B. & Wahab, Y. A. (1997) Chrysopidae attraction to floral fragrance chemicals and its vertical distribution in a Malaysian lowland tropical forest. Malay. Appl. Biol. 26: 75-80.

Sakai, S., Momose, K., Yumoto, T., Kato, M. & Inoue, T. (1999) Beetle pollination of *Shorea parvifolia* (Section Mutica, Dipterocarpaceae) in a general flowering period in Sarawak, Malaysia. Am. J. Bot. 86: 62-69.

Sakakibara, Y., Yamae, A., Iwata, R. & Yamada, F. (1997) Evaluation of beetle capture in traps as compared with manual capture on flowers in a long-term investigation in a beech forest. J. For. Res. 2: 233-236.

Sakakibara, Y., Kikuma, A., Iwata, R. & Yamae, A. (1998) Performances of four chemicals with floral scents as attractants for longicorn beetles (Coleoptera: Cerambycidae) in a broad-leaved forest. J. For. Res. 3: 221-224.

Sutton, S. L. (1989) The spatial distribution of flying insects In Leith, H. & Werger, M. J. A. (eds). Tropical Rain Forest Ecosystems. Elsevier Science Publishers, Amsterdam, pp.427-436.

Sutton, S. L. & Hudson, P. J. (1980) The vertical distribution of small flying insects in the lowland rain forest of Zaire. Zool. J. Linn. Soc. 68: 111-123.

Sutton, S. L., Ash, C. P. & Grundy, A. (1983) The vertical distribution of flying insects in the lowland rain forest of Panama, Papua New Guinea and Brunei. Zool. J. Linn. Soc. 78: 287-297.

Toda, M. (1987) Vertical microdistribution of Drosophilidae (Diptera) within various forests in Hokkaido. III. The Tomakomai experiment forest, Hokkaido University. Research Bulletins of the College Experiment Forests, Faculty of Agriculture, Hokkaido University 44: 611-632.

Toda, M. (1992) Three-dimensional dispersion of drosophilid flies in a cool temperate forest of northern Japan. Ecol. Res. 7: 283-295.

Whitmore, T. C. (1984) Tropical rain forests of the far east (2nd ed.). Clarendon Press, Oxford. 352pp.

Whitmore, T. C. (1990) An introduction to tropical rainforests. Oxford University Press, Oxford.

32 Ant Fauna of the Lower Vegetation Stratum in Pasoh Forest Reserve with Special Reference to the Diversity of Plants with Extrafloral Nectaries and Associated Ants

Brigitte Fiala[1] & Saw Leng Guan[2]

Abstract: Associations of ants with plants can be regarded as one reason for the high abundance and diversity of ants in the tropics. A full spectrum of ant-plant partnerships is realised in the tropical forests, ranging from loose, opportunistically changing associations to obligate symbioses. In the simplest case the plant offers nutrient-rich nectar from special glands (extrafloral nectaries) which attracts ants. These associations are characterised by very different intensities of use of the plants resources and, therefore, also varying mutual dependence of the partners. From Southeast Asia almost no information on plants with extrafloral nectaries (EFN) and visiting ants existed, this is especially true for primary forests. Therefore the species richness and frequency of woody angiosperm plants with EFN were studied in the lowland dipterocarp forest of Pasoh Forest Reserve (Pasoh FR) which has a well-known tree-flora. EFN were present on 12.3% of the 741 species surveyed and comprised 19.7% of all tree individuals of the Pasoh Forest 50 ha plot. Ninety-one plant species belonging to 47 genera and 16 families were found to have EFN. Euphorbiaceae, Dipterocarpaceae, Rosaceae, Leguminosae and Ebenaceae were the families most frequently bearing EFN. Most common were flattened glands associated with the leaf blade. We found an increase in the number of species as well as in the percentage of individuals from the understorey to the canopy emergents. EFN were found more often among the abundant species (species with $N > 500$ trees). The interactions between ants and EFN-bearing plants appeared to be rather facultative and non-specific. In Pasoh FR 38 ant species from 18 genera of 5 subfamilies were collected at EFN. The majority of the EFN-associated ants belonged to the subfamily Formicinae while Ponerines were rare. Most species rich were members of the genera *Polyrhachis*, *Camponotus* and *Crematogaster*. The overlap in ant fauna composition between the different sites was low, suggesting a rather mosaic-like pattern of ant colonisation. In comparison to disturbed habitats less ants were found on EFN in primary forests. Data of diversity and abundance of EFN and associated ant fauna are also presented from 3 other forest areas in Malaysia. In addition, ant fauna on the lower vegetation in general was studied with transect surveys and baiting experiments. In general, ant species richness and activity was lower in the vegetation than on the ground. Baits were detected faster and to a higher percentage in more open areas in Ulu Gombak than in Pasoh. Details on ant diversity, abundance and species interactions at baits are reported. All ants found at EFN were also collected on honey as well as protein baits suggesting a rather generalised use of food resources. Long-term functional studies are needed to begin to reveal the role of non-specific ant-plant interactions in undisturbed compared with secondary habitats.

[1]University of Würzburg, D-97074 Würzburg, Germany.
E-mail: fiala@biozentrum.uni-wuerzburg.de
[2]Forest Research Institute Malaysia (FRIM), Malaysia.

Key words: ant-plant interactions, Malaysia, mutualism, primary forest, SE Asia, tropical forest.

1. INTRODUCTION

Ants are among the most numerous arthropods on the ground as well as in the canopy of tropical forests. Estimates indicate that ants can comprise up to 56% of all arthropod individuals collected by fogging techniques (e.g. Floren & Linsenmair 1997). Due to their astounding numbers and biomass and the various niches they occupy, ants have many effects on other species and therefore play an important role in structuring communities in tropical areas.

The remarkable abundance of ants in tropical forest canopies led Tobin (1991, 1994) to question how ants, which have been presumed to be mainly predators, can be more common than their prey. Tobin hypothesised that ants feed as herbivores rather than predators. However, food preferences of most tropical ants are not yet known. Compared to the Neotropics we are still rather ignorant about the composition of the ant fauna in SE Asia despite an increasing number of studies during the last years (e.g. Agosti et al. 1994; Brühl et al. 1998; Chung & Maryati 1996; Floren & Linsenmair 1997; Maryati et al. 1996; Yamane & Nona 1994; Yamane et al. 1996). However, the results are often hardly comparable because of the different sampling methods and the size and degree of habitat disturbance of the study sites. Most collections provided only little information about ecology and biology (e.g. food preferences) of the ants.

Davidson (1997) and Davidson & Patrell-Kim (1996) in a literature survey found that most of the numerically dominant ant species in canopies indeed seem to strongly rely on plant and homopteran exudates, and provided further evidence which would support Tobin's view in a study on nitrogen origin of ant food. What plant resources occur in tropical forests that can be used by ants? Ants exploit plant products either indirectly (through symbiosis with homopterans or lepidopterans; overview Hölldobler & Wilson 1990) or directly. Symbiotic associations between ants and plants have been reported for many different genera and families. They include so-called myrmecophiles, myrmecophytes and myrmecochores (overview e.g. Beattie 1985). Most of these plants are myrmecophiles with unspecific relationships to ants. They offer food resources in the form of sugar secreting glands called extrafloral nectaries (EFN) or in form of food bodies and thereby attract ants from the vicinity. True myrmecophytic species are more closely adapted to the association and continuously house specialised ant-colonies in nesting space (domatia) offered by the plants (overview on ant-plants interactions e.g. Davidson & McKey 1993).

A great diversity of plant species in a wide variety of habitats possess EFN. These structures of diverse morphology and anatomy (Elias 1983) are located outside the flower and are not involved in pollination. Their function has been discussed for years but many studies have now proved that EFN can have an important role in the plant's defence against herbivores (overview see e.g. Koptur 1992). Ants and parasitoids (certain wasps and flies) are the most frequent visitors of extrafloral nectaries and forage preferably on plants where they attack insect herbivores.

Despite the ecological significance of these glands only little has been known about the taxonomic distribution and abundance of plants with EFN in different geographic regions and floras. EFN can be found on plants of at least 93 families in both the tropics and the temperate zone, although they are reported to be more common in tropical areas (Koptur 1992; Morellato & Oliveira 1991) and can be very abundant in some areas. Almost all studies on EFN in tropical habitats to date were conducted at New world sites (overview e.g. Pemberton 1998). This geographic specialisation reflects our poor understanding of ant-plant relationships in the Oriental tropics. Little information exists on distribution of plants with EFN in the Paleotropics (see Pemberton 1998 for temperate-subtropical East Asia).

From SE Asia we know at present only little work (Nielsen 1992; Rickson & Rickson 1998; Schellerich-Kaaden & Maschwitz 1998; Wong & Puff 1995) about EFN-plants and the associated ant-fauna except our own studies (e.g. Fiala & Maschwitz 1991; Fiala et al. 1994ab, 1996; Fiala & Linsenmair 1995; Fiala & Maschwitz 1995; Heil et al. 2000, 2001; Maschwitz et al. 1994; Merbach et al. 1999; Rabenstein et al. 1999) although Ichino & Yamane (1994) also mentioned the importance of EFN. We are especially ignorant about the diversity of these structures in primary rainforests. Therefore we carried out an extensive survey on the abundance of EFN-plants in primary lowland forest of Pasoh (Fiala & Linsenmair 1995). We will here give an overview about the most important findings and also present for the first time information about the associated ant fauna. In addition to Pasoh Forest Reserve (Pasoh FR), data were also collected for comparison in other forest areas in Malaysia to find out more about abundance and diversity of these structures and also about their ant visitors and the ant fauna in the lower vegetation in general.

2. MATERIALS AND METHODS

2.1 Study site

Detailed field studies were conducted in Pasoh FR (Fig. 1). The location, climate, well-studied tree-flora, etc. of Pasoh FR are described extensively in this book (Chap.

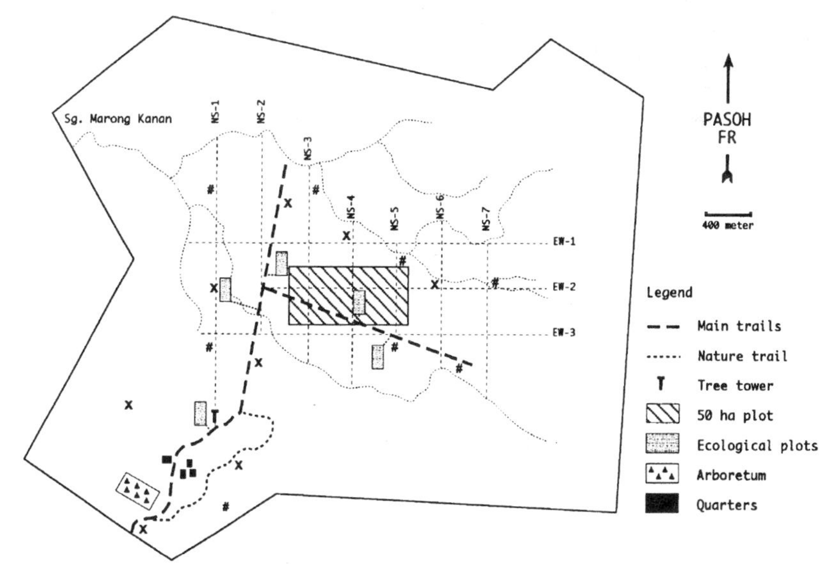

Fig. 1 Distribution of study sites in Pasoh FR. A complete evalution of occurrence and abundance of species with extrafloral nectaries (EFN) took place in the 50 ha plot (see text). Further EFN surveys and ant baitings were conducted along transects within the 50 ha plot (4 sites, not indicated) and outside in different areas of the forest reserve. Sites where EFN surveys as well as ant baitings took place are indicated with 'x', sites with additional EFN surveys with '#'.

1 and 2) and will not be repeated here. The study concentrated on woody plants (except lianas due to the poor taxonomic treatment of this group). Following Kochummen et al. (1990), plant species were assigned to 5 growth categories. The total number of species in Pasoh FR is unknown: even within the 50 ha plot, over 820 species names were listed (Kochummen et al. 1990), but later revised to 814 species (Manokaran et al. 1992).

Some comparative data were collected in the primary forests of Belum, Perak, and Lambir, Sarawak. Further studies were carried out in secondary forest areas in the Ulu Gombak Valley and on Forest Research Institute Malaysia (FRIM) grounds, Kepong. Detailed surveys were also conducted in the Kinabalu National Park, Sabah, however, since this area covers only higher altitudes from 560 m onwards, the results will not be used here for comparison with lowland forest sites.

2.2 Survey of plants

Pasoh FR: Over a period of several years we could examine 741 species (trees and shrubs with a basal trunk diameter of at least 0.5 cm). Our survey was based on a published species list of the 50 ha plot (Manokaran et al. 1992). It concentrated on primary forest although a few species of regenerating forest as well as from adjacent heavily disturbed areas were included when they occurred as saplings in the plot. We tried to check as many species as possible in the field for EFN and for nectar secretion but this was not possible for some rare species which were present only as large trees. We gained additional information by a literature search, and studies of herbarium material (481 species could be checked in the field (comprising all species found to have EFN), for 260 species we had to rely on herbarium specimens). However, often EFN occur only on young leaves and are then normally not mentioned in identification keys. These problems could result in a certain under-estimation of EFN-bearing plants.

We also compared the abundance and diversity of EFN-species along 20 transects. In 10 different locations of the forest reserve we surveyed plants < 2 m height along two transects of about 100 m length (on average 91 plants were inspected on each transect). Nectary secretions of all gland secretions were tested with Dextrostix (Merck) for glucose. Percentage of sugar in the secretions was estimated with a hand-refractometer. Only plant species whose gland secretions gave positive reaction in at least 10 specimens were included in the category of EFN-plants.

Other sites: Time spent in other areas usually was much shorter than in Pasoh, a detailed survey of the whole flora was therefore not possible but had to be limited to a number of transects where the number of plant species and individuals with EFN was recorded.

2.3 Survey of ants

All EFN-plants encountered were checked for ants. Thirty of the most common EFN-species in Pasoh were inspected for their ant visitation twice a day at varying times (at least 10 plants each).

To find out more about ant activity in the lower vegetation, a general survey and baiting experiments were carried out (which ant species forage on the vegetation? How representative are ants species found on EFN for the entire ant fauna using the lower vegetation stratum?). Baiting experiments should also provide information about ant abundances and possible dietary preferences of the ants and especially the use of carbohydrate-rich food sources.

For the survey we walked a number of transects and checked each plant 2 m left and right from the transect for presence of ants (each transect about 500 plants total). Numbers and species diversity were recorded.

Baiting experiments were carried out February–April 1993 also along arbitrarily selected but representative trails that sampled different vegetation from deep primary forest in the 50 ha plot to secondary forest sites. In total baiting was carried out at 12 sites (in the forest and near the entrance road). We used 5% honey water and as protein baits pieces of processed cheese of about 2 cm^3 (as done by Hashimoto et al. 1997 and Gossner 1999 with good results). Droplets of honey water and pieces of cheese were placed on leaves of the understorey vegetation (plants without EFN who had at that time no ants) in about 1–1.5 m height, in each series of at least 10 plants with honey water (or alternating honeywater and cheese) at about 2 m distance. Baiting was carried out mainly from 9 a.m. to 2 p.m. but also took place in the late afternoon up to dusk. Some series were repeated after 14 days at the same site. Plants were monitored during 3 h. In the first hour baits were checked for presence

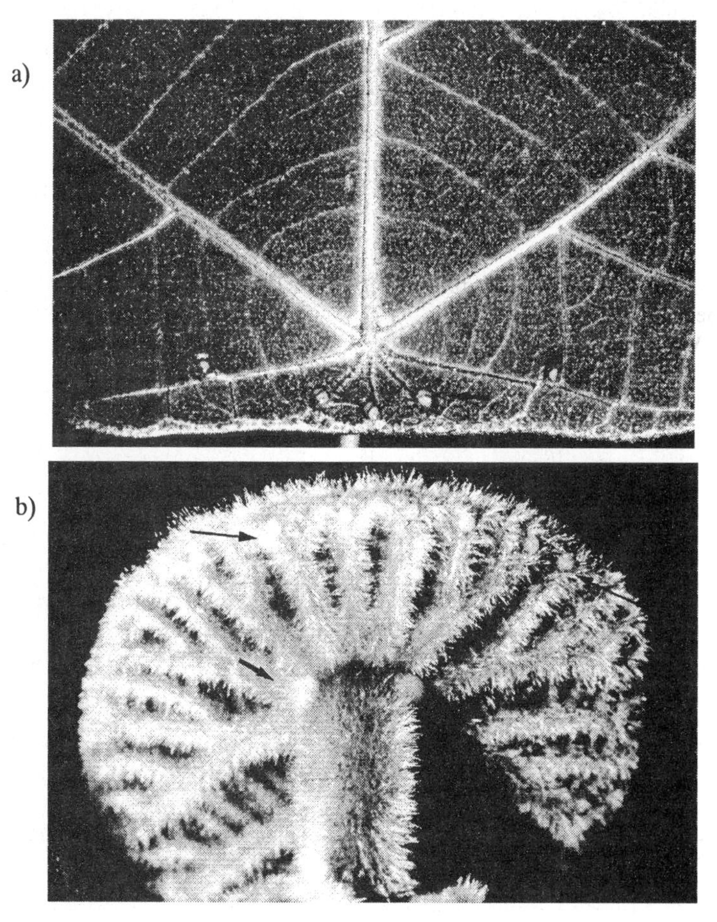

Fig. 2 Different types of extrafloral nectaries Almost structureless but with nectar secretion a) *Macaranga conifera*, b) very conspicuous elevated structures: *Endospermum malaccense.*

of ants every 10 minutes, later every 30 minutes. Time until arrival of the first ant, diversity of ant species, and peak number of ants attracted were recorded at each bait. For comparison baiting experiments were also carried out in secondary forest habitats in Ulu Gombak.

One or a few individuals of all species encountered were collected for identification. A great problem is the insufficient taxonomic treatment of SE Asian ants which only rarely allows species names. Therefore ant specimens were sorted to numbered morphospecies. Ant and also plant voucher specimens are in the collection of the author and at FRIM, Kepong, University Malaya, Kuala Lumpur, and Kinabalu Headquarters of Sabah Parks.

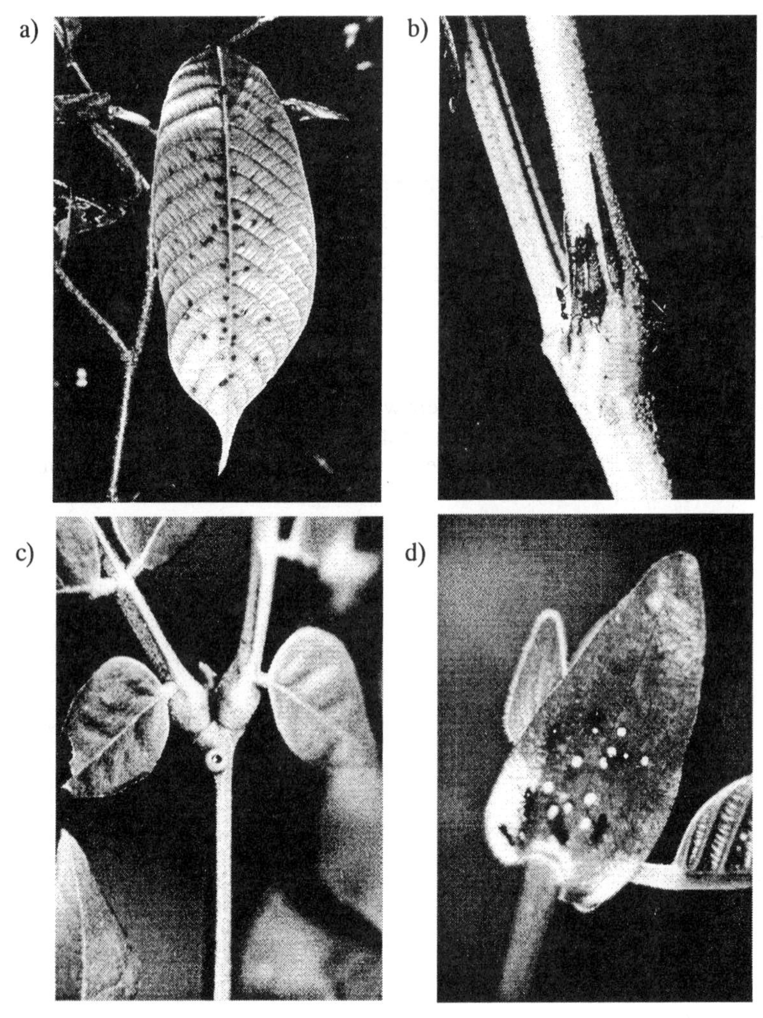

Fig. 3 Different locations of EFN on the plant: a) Flattened nectaries, leaf blade: *Shorea ovalis*, b) elongated nectaries, stem: *Leea indica,* c) elevated nectaries, petiole: *Archidendron contortum*, d) stipules: *Shorea macrophylla* (FRIM).

3. RESULTS

3.1 Survey of plants

We here give an overview on the most important results from our EFN survey (for more details see Fiala & Linsenmair 1995). Structure of EFN in Pasoh FR was rather divers in different taxa. Most commonly glands were located on the leaf blade (70.3%; of those 14.2% were on leaf blade base and 3.3% on the leaf margin). They were also found on the leaf-petiole junction (14.2%), rachis (10.9%), petioles (3.3%) and stem (1.1%) (Fig. 2 and Fig. 3). Some glands were inconspicuous and could often be detected only by a different colour contrasting with the background (Fig. 3a) or by secretion of fluids (Fig. 2a). Those dot glands were especially abundant in the families Dipterocarpaceae, Ebenaceae and Euphorbiaceae. EFN did not only occur on young plants but also on young leaves of mature trees (e.g. *Elateriospermum tapos* Bl., *Macaranga lowii* King ex Hook. f., Euphorbiaceae; *Neobalanocarpus heimii* P.S. Ashton, *Shorea* spp., Dipterocarpaceae). The sugar concentrations of the secreted fluids were rather low with an average of 5–10%.

Of the 741 species surveyed in the whole Forest Reserve, at least 91 species from 47 genera and 16 families were found to have EFN (12.3%) (Table 1). For the species restricted to the 50 ha plot the abundance was 11.4%. That means that EFN were present in 50% of the orders, 22.6% of the families and 17.5% of the genera investigated. The species studied comprised all six subclasses of the Magnoliopsida (Table 2 summarises the data on distribution of EFN among the different subclasses). Species with EFN were concentrated in several families: Euphorbiaceae, Fabaceae, Rosaceae, but were also present in families where they are in general recorded as being rather rare (Elias 1983): Dipterocarpaceae, Ebenaceae, Flacourtiaceae and Sapindaceae. Within one family, genera with and without EFN were found and even within a genus species with and without glands occurred together (e.g. *Shorea*, *Diospyros*, *Aporusa*). Interesting is the abundance of EFN in saplings of the important SE Asian timber wood family Dipterocarpaceae, especially in the genera *Shorea* (10/14 species) and *Vatica* (2/3 species with EFN) (see Table 1).

Whereas only 11.4% of the species surveyed in the 50 ha plot possessed EFN, theoretically 19.7% of the 335,240 tree individuals in the plot belonged to EFN-bearing species (number of trees after Manokaran et al. 1992). This calculation is based on the presumption that also the young leaves of mature trees possess EFN. However, sometimes saplings < 1 cm DBH seemed to be more abundant than the mature trees and therefore the frequency of EFN-plants could even be higher. About half of the EFN-plants belonged to the species with high abundance in the 50 ha plot (which was defined by FRIM botanists as species with more than 300 individuals). We found an increase in the number of species as well as in the percentage of EFN-species from the understorey species to the emergents (Table 3 in Fiala & Linsenmair 1995, based on tree numbers in Manokaran et al. 1992).

Between the 20 different transects no differences in species richness occurred (average 5 ± 2.6 species with EFN among on average 81 plants). Frequency of EFN-plants, however, varied from 7.6% to 22% per transect within the primary forest. The highest frequency was found in light-rich and/or swampy sites e.g. along the north-south trail.

Table 1 Woody plant species with extrafloral nectaries in Pasoh FR (Nomenclature follows Ng (1978, 1989) and Whitmore (1972, 1973); * = not as mature trees in the 50 ha plot; # =the name is actually a misidentification, the species is yet undescribed).

Plant species	Main site of EFN	Growth stature
SUBCLASS HAMAMELIDAE		
Order Urticales		
Moraceae		
Ficus obscura	leaf blade	treelet
SUBCLASS DILLENIIDAE		
Order Theales		
Dipterocarpaceae		
Hopea mengawaran	leaf blade	canopy
Neobalanocarpus heimii	leaf blade	emergent
Shorea acuminata	leaf blade	emergent
Shorea bracteolata	leaf blade	emergent
Shorea dasyphylla	leaf blade	emergent
Shorea guiso	leaf blade	emergent
Shorea lepidota	leaf blade	emergent
Shorea leprosula	leaf blade	emergent
Shorea macroptera	leaf blade, bracts	emergent
Shorea ochrophloia	leaf blade	emergent
Shorea ovalis	leaf blade	emergent
Shorea pauciflora	leaf blade	emergent
Vatica bella	leaf blade	canopy
Vatica pauciflora	leaf blade	canopy
Order Malvales		
Sterculiaceae		
Byttneria sp. *	leaf blade	liana
Order Violales		
Flacourtiaceae		
Homalium caryophyllaceum	junction leaf-petiole	treelet
Homalium dictyoneurum	junction leaf-petiole	canopy
Homalium longifolium	junction leaf-petiole	canopy
Paropsia vareciformis	junction leaf-petiole	understorey
Scolopia macrophylla *	junction leaf-petiole	understorey
Scolopia spinosa	junction leaf-petiole	understorey
Order Ebenales		
Ebenaceae		
Diospyros adenophora	leaf blade	understorey
Diospyros andamanica	leaf blade	treelet
Diospyros apiculata	leaf blade	understorey
Diospyros argentea	leaf blade	treelet
Diospyros diepenhorstii	leaf blade	understorey
Diospyros nutans	leaf blade	treelet
Diospyros "pendula" #	leaf blade	canopy
Diospyros pyrrhocarpa	leaf blade	canopy
Diospyros scortechinii	leaf blade	treelet
Diospyros wallachii	leaf blade base	understorey
SUBCLASS ROSIDAE		
Order Rosales		
Rosaceae		
Licania splendens	leaf blade base	canopy
Maranthes corymbosa	junction leaf-petiole	canopy
Parastemon urophyllus	leaf blade base	canopy
Parinari costata	leaf blade base	canopy
Parinari elmeri	petiole	understorey
Prunus arborea	leaf blade	understorey
Prunus grisea	leaf blade	understorey
Prunus "stipulaceum"	leaf blade base	understorey?
Staphyleaceae		
Turpinia ovalifolia	rachis	understorey
Order Fabales		
Leguminosae (Caesalpiniaceae, Mimosaceae)		
Albizia pedicellata	rachis	canopy
Archidendron bubalinum	rachis	understorey

Table 1 (continued)

Plant species	Main site of EFN	Growth stature
Archidendron clypearia	rachis	understorey
Archidendron contortum	rachis	treelet
Archidendron globosum	rachis	understorey
Archidendron microcarpum	rachis	understorey
Parkia speciosa	rachis	canopy
Pithecellobium clypearia *	rachis	canopy
Pithecellobium splendens	rachis	canopy
Saraca declinata	leaf blade margin	understorey
Saraca thaipingensis	leaf blade margin	understorey
Order Myrtales		
Combretaceae		
Terminalia citrina	junction leaf-petiole	understorey
Terminalia phellocarpa	junction leaf-petiole	canopy
Order Euphorbiales		
Euphorbiaceae		
Alchornea rugosa	leaf blade	shrub
Aporusa aurea	leaf blade base	understorey
Aporusa sp. 1	leaf blade margin	understorey
Baccaurea sumatrana	junction leaf-petiole	understorey
Blumeodendron calophyllum	leaf blade base	canopy
Croton argyratus	leaf blade base	understorey
Croton laevifolius	leaf blade base	understorey
Elateriospermum tapos	petiole, leaf blade	basecanopy
Endospermum malaccense	leaf blade, junction leaf petiole	canopy
Epiprinus malaya nus	junction leaf-petiole	understorey
Fahrenheitia pendula	petiole	canopy
Koilodepas longifolium	leaf blade	understorey
Macaranga conifera	leaf blade	canopy
Macaranga lowii	leaf blade	understorey
Macaranga hosei	leaf blade	understorey
Macaranga recurvata	leaf blade	understorey
Mallotus leucodermis	leaf blade	canopy
Mallotus penangensis	leaf blade	understorey
Mallotus tiliifolius *	leaf blade	treelet
Neoscortechinii nicobarica	leaf blade	understorey
Neoscortechinii sumatrensis	leaf blade	understorey
Pimelodendron griffithianum	leaf blade	canopy
Ptychopyxis costata	leaf blade	understorey
Sapium baccatum	leaf blade base	canopy
Sapium discolor	junction leaf-petiole	understorey
Sapium indicum *	junction leaf-petiole	treelet
Trigonostemon sumatranus *	leaf blade base	treelet
Order Rhamnales		
Vitaceae		
Leea indica	stem	shrub
Order Linales		
Ixonanthaceae		
Ixonanthes icosandra	leaf blade	canopy
Order Polygalales		
Polygalaceae		
Xanthophyllum affine	leaf blade	canopy
Order Sapindales		
Sapindaceae		
Pometia pinnata	leaf blade base, stipules	canopy
Xerospermum laevigatum	leaf blade	canopy
Xerospermum noronhianum	leaf blade	canopy
SUBCLASS ASTERIDAE		
Order Lamiales		
Bignoniaceae		
Deplanchea bancana	leaf blade base	emergent
Radermachera pinnata	leaf blade base	understorey
Verbenaceae		
Clerodendrum sp.*	leaf blade	shrub

Table 2 Proportion of species with EFN among the different subclasses of the Magnoliopsida (*N* = number of species surveyed in this subclass and their percentage of the flora in Pasoh FR; n.p. = not present).

Subclass	*N*	Species with EFN
Magnoliidae	62 (8.3%)	-
Hamamelidae	35 (4.7%)	1.2 %
Caryophyllidae	n.p.	-
Dilleniidae	228 (30.8%)	35.4 %
Rosidae	362 (48.8%)	61.0 %
Asteridae	54 (7.2%)	2.4 %

Table 3 Comparison of ant species found at extrafloral nectaries in 4 study sites in Peninsular Malaysia (Pasoh FR, Negri Sembilan; Belum, Perak; Ulu Gombak, Selangor, FRIM = grounds of the Forest Research Institute, Kepong, Selangor).

	Primary forest		Secondary habitats	
	Pasoh FR	Belum	Ulu Gombak	FRIM
Pseudomyrmecinae				
Tetraponera sp.	+	+	+	-
Ponerinae				
Diacamma 'rugosum'	+	+	+	+
Gnamptogenys sp.	+	+	+	+
Odontomachus rixosus	+	-	-	+
Myrmicinae				
Cataulacus hispidulus	+	-	-	-
Crematogaster inflata	+	-	-	-
Crematogaster	4 (+)spp.	4 (+)spp.	5 (+)spp.	5 (+)spp
Dilobocondyla borneensis	+	-	-	-
Meranoplus mucronatus	+	+	+	+
Myrmicaria cf. *birmana*	+	-	-	+
Pheidole sp.	-	-	+	2 spp.
Tetramorium sp.	+	-	-	+
unident. species	1	2	3	3
Dolichoderinae				
Dolichoderus	2 spp.	2 spp.	2spp.	2 spp.
Philidris sp.	-	+	-	-
Technomyrmex	3 spp.	+	2 spp.	+
unident. species	1 spp.	1 spp.	2 spp.	3 spp.
Formicinae				
Anaplolepis longipes	-	-	+	+
Camponotus	5 spp.	3 spp.	5 spp.	3 spp.
Echinopla sp.	+	-	+	-
Oecophylla smaragdina	-	+	+	+
Paratrechina sp.	+	-	-	-
Polyrhachis	9 spp.	3 spp.	6 spp.	3 spp.

3.2 Habitats with different degrees of disturbance and comparison with other forest areas in Malaysia

We did not find much difference between the core area of primary forest in Pasoh FR and the buffer zone around it. The average percentage of plant individuals with EFN per investigated area were: primary forest 11.7% (± 5.8%), regenerating forest 13.1 (± 5.7%) (10 areas each with total 824 and 794 plants, respectively). The bufferzone had only been selectively logged once already 40 years ago and the number of emergents is still reduced (Chap. 2; Okuda et al. in press). The rather disturbed areas, however, as for instance along the road going into Pasoh FR had an average of 6.5 ± 1.4 species per study plot and 25.2% cover of individual plants with EFN (5 study plots with a total of 321 plants).

Table 4 Percentage of plants with EFN along transects in habitats of different degrees of disturbance (- : habitat not present).

	Study sites in Malaysia				
	Pasoh FR	Belum	Lambir	Gombak	FRIM
No. of plants investigated	1939	330	296	403	499
Habitat:					
Closed primary forest	12.8 ± 5.8	7.4 ± 4.3	8.5 ± 4.5	-	-
Gaps in primary. forest	16.0 ± 6.3	11.6 ± 3.8	19.5 ± 6.6	-	-
Rather disturbed areas	25.2 ± 15.8	17.8 ± 6.3	-	22.0 ± 11.8	23.5±1.3

Table 5 Ant species at extrafloral nectaries in Pasoh FR.

Family	Species
Pseudomyrmecinae	*Tetraponera attenuata*
Ponerinae	*Diacamma cf. 'rugosum'*
	Gnamptogenys cf. *menadensis*
	Odontomachus rixosus
Myrmicinae	*Cataulacus hispidulus*
	Crematogaster inflata
	Crematogaster 4(+) spp.
	Dilobocondyla borneensis
	Meranoplus mucronatus
	Myrmicaria cf. *birmana*
	Tetramorium sp. 1
	unid. species 1(+)
Dolichoderinae	*Dolichoderus beccarii*
	Dolichoderus thoracicus
	Philidris sp. 1
	Technomyrmex 3 spp.
	unid. species 1(+)
Formicinae	*Camponotus* 5(+) spp.
	Echinopla sp. 1
	Paratrechina sp. 8
	Polyrhachis bicolor
	P. (Hemioptica) boltoni
	P. carbonaria
	P. calypso
	P. furcata
	P. mülleri
	P. striata
	Polyrhachis 2 unid. spp.

Abundance of EFN in the primary forest parts seems to be lower than in disturbed areas. Our surveys in understorey of other primary lowland forest (altitude below 300 m) in Belum, Perak, and Lambir, Sarawak yielded even lower values of about 7% and 8.5% of the plants investigated bearing EFN (A detailed comparison was not possible since we did spend only a few days in Belum. However, in that short time we could already find 20% of the plant species with EFN which also occurred in Pasoh). In more open patches in the forest or at the forest edge, however, percentage of plants with EFN did increase (Table 4).

3.3 Ant fauna

3.3.1 Diversity of ants on extrafloral nectarines

At least 38 ant species from 18 genera were found visiting EFN (Table 5). They comprised the most important subfamilies occurring in Malaysia (except of the ground-living predatory subfamilies Cerapachyinae, Dorylinae, Aenictinae, Leptanillinae), but make only small fraction among the 489 ant species which have to date been recorded from Pasoh (Chap. 26). However, only about 160 species of that investigation were sampled on the lower vegetation. The percentage of ant species found at EFN ($N = 37$) from those species collected in the lower vegetation were: 14.3% for Pseudomyrmecinae, 2.7% for Ponerinae, 19.38% for Myrmicinae, 38.8% for Dolichoderinae and 19.37% for Formicinae (data and stratum definition after Rosciszewski 1995; Chap.26). Our data did not allow a statistical treatment of ant composition between different sites (some species were not found visiting EFN frequently enough). However, some species were found only in the primary forest (*Dolichoderus* spp.) and the regenarating forest bufferzone (*Echinopla* spp.),

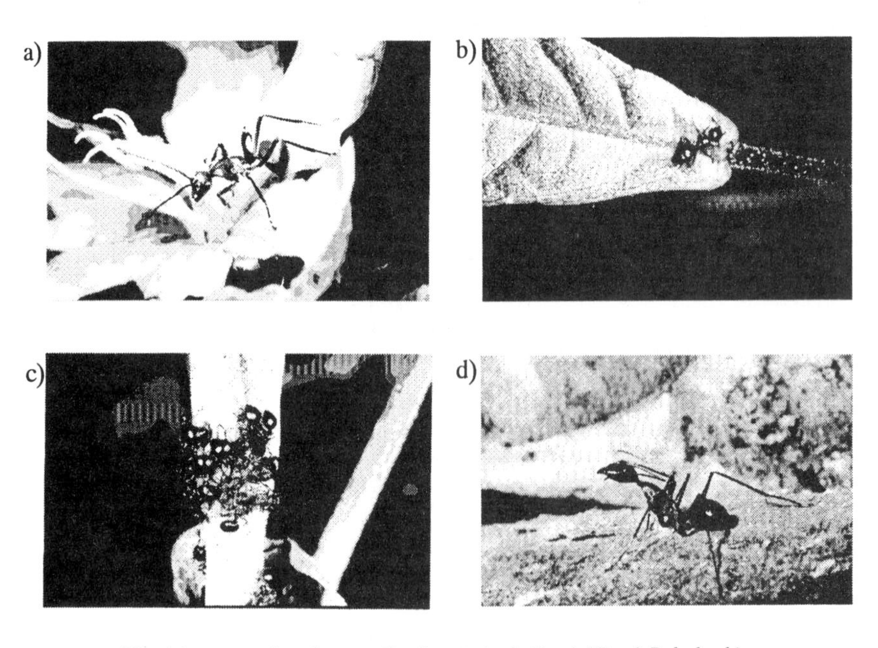

Fig. 4 Ant genera found at extrafloral nectaries in Pasoh FR : a) *Polyrhachis* sp. forages also for flower nectar, b) *Technomyrmex* sp. on nectaries of *Macaranga lowii*, c) an assembly of *Meranoplus mucronatus* on stem nectaries of *Leea indica,* d) *Camponotus* sp. (photo K. E. Linsenmair).

respectively, but this might change with increasing sampling effort.

In general most abundant were *Gnamptogenys menadensis* (Ponerinae, 30.3% of all sightings) and several *Crematogaster* spp. (Myrmicinae, 23% of all sightings), the last being also dominant in number of individuals. Ants were found on 29.8% of the 1179 individuals of the more thoroughly surveyed 30 EFN-plant species (at least 10 specimens of each plant species). Each EFN species was visited by on average 3.7 ant species (range 1–10). There were, however, specific differences: some plant species were always visited by ants whereas on others ants were only rarely present. *Macaranga lowii* (Fig. 4b) was the species with EFN which we controlled very intensively (670 plants of up to 1.60 m height) at 5 sites of different degree of disturbance (forest in the 50 ha plot to entrance road in Pasoh FR). Percentage of individuals visited by ants ranged from 20.2% along the main trail of the 50 ha plot to 49.1 of the plants along the entrance road. Number of workers present at the EFN of the 5 youngest leaves of a sapling varied from 5 to 69. The number of ants seen at a control visit on the other regularly controlled species (per species 2–94 individuals) was rather low, on average 5.1 workers per plant (range 1–8). Exceptions were the gap plants *Endospermum malaccense* M.A. (Fig. 3b) and *Leea indica* (Burm.f.) Merr. (Fig. 3b) with 15.5 and 17.5 ant workers found visiting the EFN at a time. Also quite frequented were *Pometia pinnata* Forst. and *Epiprinus malayanus* Griff., and some of the *Diospyros* species such as *D. andamanica* (Kurz) Bakh. and *D. nutans* K. & G. with more than 50% of the 80 plants checked found with ants visiting the EFN (see also Wong & Puff 1995 for report on the myrmecophilous character of *D. andamanica*). Presence of ants at EFN-plants (of the same species) was locally very different–probably a consequence of patchy nesting of the ants.

There was no indication for specific interactions. The most thoroughly studied EFN-bearing species *Macaranga lowii* for instance was visited by 18 different ant species.

Comparisons between Pasoh FR and our two other study areas in secondary forest in Gombak and Kepong revealed rather similar taxonomic composition of ant genera of EFN visitors, with addition of typical indicators for disturbed areas such *Anaplolepis longipes* Jerdon (Table 3).

We always found a clear numerical dominance of *Crematogaster* spp., *Camponotus* spp. (Fig. 4d), *Polyrhachis* spp. (Fig. 4a) and a *Gnamptogenys* sp. (Table 6). Conspicuous was the abundance of *Meranoplus mucronatus* Smith (Fig. 4c) in Belum which had a much lower abundance in the other areas (Belum 10% versus Pasoh 2%, Gombak 3%, FRIM 2%).

Table 6 Relative abundance of ant taxa at EFN (N = numbers of sightings of at least one worker).

Species	Relative abundance (%)		
	Pasoh FR	Gombak	FRIM
	N = 398	N = 297	N = 261
Gnamptogenys sp.	30.0		
Crematogaster spp.	23.0	46.3	> 61.0
Polyrhachis spp.	11.0	8.8	
Camponotus spp.	8.0	10.2	11.1
Diacamma rugosum		11.7	
Anaplolepis longipes			12.6
All others (% each sp.)	< 8.0	< 8.0	< 5.0

Table 7 Ant activity on the lower vegetation in habitats of different degrees of disturbance (percentage of plants investigated found with ants, mean ± *SD*; - : habitat not present).

	Study sites in Malaysia			
	Pasoh FR	Belum	Gombak	FRIM
Total No. plants	1249	265	389	274
Habitat:				
Closed primary forest	6.6 ± 3.5	5.9 ± 2.9	-	-
Gaps in primary. forest	12.2 ± 8.6	9.1 ± 4.3	-	-
Rather disturbed areas	16.6 ± 14.3	20 ± 16.7	17.8 ± 15.8	14.2 ± 12.4

3.3.2 Ants in the lower vegetation stratum

In Pasoh FR ants were found on every 15th plant on average (= 6.6% of 1,249 plants). The other primary forest also had similar values: Belum, forest interior: on average ants were present on every 17th plant (5.9% of N = 1,250). Ant activity in secondary habitats as the regenerating forest buffer zone was higher (Table 7).

The rather low ant activity on the vegetation of the primary forest was supported by the results of the baiting experiments. In Pasoh on average 4.8 ± 1.9 ant species were found per baiting (1–8 species), while in Gombak 2.8 ± 1.3 (1–6 species) were found at baits. At the 12 baiting sites in Pasoh FR ants were found with different probability of occurrence. Most widely distributed was *Gnamptogenys* sp. which occurred at 58% of the sites. *Philidris* for instance and *Echinopla* however, occurred at only 1 site. No species did show any significant correlation to a special habitat or area. The baiting experiment supported the picture of a rather mosaic pattern of ant colonisation. The overlap was rather low, we found a changing composition from site to site. Repeats of baiting experiments at the same sites after 14 days showed that the ant species were mostly the same as at the first time.

Twenty-two ant species were found at the honey baits in Pasoh FR and 18 at the cheese baits, resulting in total 25 species collected in 2 weeks. It usually took quite some time until the first baits were found (on average 36–64 minutes depending on the site) and the percentage of baits found at all during the duration of the observation (3 h) was about only 35%. The activity was in general higher in Gombak where the average value for detection of the baits was 24.7 minutes, with a total 44% of the baits found. At 6 sites the baits were already found after 5 minutes. At two very open and light rich sites in Gombak (logging road) all baits were found and used within a few minutes. Here also the number of ant species involved was highest (6 species on 50 m × 50 m).

There were no significant differences between honey and protein baits, at least not among those which could be found several times so that a statistical evaluation is possible. Differences in visitation of baits obviously were not caused by preference for a special type of bait but seemed to be determined by different composition of the ant community at the sites.

Protein baits were found more often than honey baits in Pasoh FR but not in Gombak (Pasoh: 30% ± 22.4 of honeybaits found, 42% ± 26.4 of protein baits, Gombak: honey 39.6% ± 28.5, protein 38.7% ± 24). Ants were recruited stronger to protein baits than to honey (Pasoh FR: honey max. 20.12 ± 8 ants per bait , protein max. 51 ± 25). In Gombak recruitment took place twice as often to protein than to honey, the strongest recruitment with most workers, however, took place at a honey bait (max. average 19.8 ants at protein, 37.5 at honey).

3.3.3 Interactions of the ants

Usually only one species occupied the bait. Only rarely several ant species occurred together at one food resource. An important factor was the fast finding of the resource. we could only once observe that a bait was taken over by another species. After Hölldobler & Wilson (1990) we can differentiate several forms of resource use. Species of *Polyrhachis* and *Diacamma* are usually single foragers. An exception was *Polyrhachis* (*Hemioptica*) *boltoni* Dorow & Kohout in Pasoh FR, which recruited very quickly and also defended the food sources (Foraging and recruitment of *Polyrhachis* have been described as rather variable by Liefke et al. 1998).

Most other species also did recruit nestmates, they can be regarded as so-called 'opportunists', which find food sources very quickly and recruit fast. They do not defend the food against other more aggressive ant species (as could be observed e.g. for *Technomyrmex*). Some species seem to be very good food finders, they therewith obtain some time advantage. One example from Pasoh FR was the interaction of *Tetraponera* with *Polyrhachis*, the latter could finally drive away *Tetraponera* from the bait but since this species had been there much earlier it had already exploited a large part of the food.

A defence against food competitors could also be observed e.g. in the recruiting species *Philidris* sp. and *Gnamptogenys menadensis*. Very small ants such as different tiny myrmicines could 'insinuate', they occurred in small numbers only and could participate at the food source without provoking ant attack of the larger species. Thus, the ant assembly at food sources is also determined by the ability of 'timid' ant species to coexist with dominant, aggressive species. Different *Crematogaster*-species were always present in large numbers and could totally occupy the food source. They were not very aggressive but could effectively keep away other ants species with their gland secretions. We could often observe much larger ants (such as *Camponotus* spp.) to shrink back when getting near a *Crematogaster* worker. So there exist different strategies which obviously allow some overlap in food resource exploitation.

4. DISCUSSION

4.1 Diversity, abundance, and taxonomic distribution of plants with EFN

The data from Pasoh FR documented a lower abundance and cover of EFN than in most other tropical areas studied (overview e.g. Pemberton 1998). The results presented here about the cover of EFN plants in the plot are, however, a very conservative estimation. The field survey may have missed glands which produced nectar only occasionally or in small quantities. Some species were only encountered as mature trees which might have stopped EFN production. Therefore we tried to find as many saplings to be able to check for EFN on young and accessible leaves but their number could not be included in the estimation of frequency of individuals since only data for trees > 1 cm exist for the 50 ha plot. However, at present, there exit few studies which can actually be used for comparison: Only rarely studies were conducted in primary tropical forests, most data come from secondary habitats and most studies regard cover and not proportion of species in the local flora. Values of cover of EFN-plants were also higher than their proportion of species in Pasoh FR: about 20% of the trees in the 50 ha plot consist of species with EFN. Comparable to this investigation are data from the cover of EFN-plants in Terra firme forest in Brazil (Morellato & Oliveira 1991) which revealed a similar percentage as in Pasoh FR.

Koptur et al. (1977) report a rather low cover of only 8% EFN-bearing plants from lowland primary forest in Costa Rica. Marquis & Braker (1994) provide a more detailed estimation: Primary understorey forest in Costa Rica: 0.9 to also 8% EFN-plant cover, understorey trees in successional plots: 2.2 to 29.2% (canopy and subcanopy trees and lianas had not been sampled). Preliminary findings from the forest in Cameroon, however, show that EFN can be rather frequent also in primary forests. Although they were most abundant among plants of secondary vegetation, they also occurred on about 30% of the canopy tree samples, but appeared to be less frequent in the understorey (McKey 1992). We could only speculate why EFN-bearing plants were for instance more common among abundant species in Pasoh.

To date, the most extensive study on EFN-plants in tropical forests was conducted by Schupp & Feener (1991) in Barro Colorado Forest in Panama, which is half-secondary and half mature. Their survey had a strong bias to species occurring in gaps, only 60% were plants of closed forest. The results of most surveys so far indicate that plants with EFN should be more common in open areas than in closed habitats. This is supported by our results from Pasoh and other forests in Malaysia where abundance of EFN-plants was always higher in more light-rich habitats.

Species with EFN were not independently distributed among taxa, a pattern which as been found in other surveys as well (e.g. Elias 1983; Schupp & Feener 1991). Species with EFN were most abundant in the three advanced subclasses of the Magnoliopsida. Our data support this picture and add 29 genera to the list of Koptur (1992). However, even if the occurrence of EFN seems to be taxonomically limited, Schupp & Feener (1991) provide arguments that taxonomic distribution of EFN is not simply an outcome of phylogeny but also of convergent selection under similar selective forces. The abundance of plants with EFN increases with decreasing latitude (overview Pemberton 1998). Euphorbiaceae and Malvaceae for instance in which EFN are abundant have also their greatest development in warm areas.

4.2 Ant fauna at extrafloral nectaries

The ant fauna visiting EFN was rather diverse with 38 species from 18 genera. They comprised almost one third of the ant species recorded on the lower vegetation in Pasoh FR. Most species at EFN in Pasoh FR belonged to the subfamily Formicinae (16 species), followed by Myrmicinae (11) and Dolichoderinae (7). A comparison reveals that most of the EFN-visiting ant species in the Neotropics as well as in the Paleotropics belonged to the subfamily Myrmicinae, ponerines were rare and dolichoderines and formicines were involved in about equal proportion. Interactions of ants with EFN-plants were rather unspecific in both tropical regions (overview on New World studies in Oliveira & Brandao 1991).

In regard to total species richness of ants in Pasoh FR (with emphasis on ground living ants, Malsch et al. this volume chap.26) myrmicines were at first position, concerning the total number of species found in the vegetation in their study, however, again formicines had the first rank. This is in accordance with other results of general surveys in the vegetation where formicines always had the greatest percentage of species (Brühl et al. 1998 for Kinabalu Park, Sabah, and Gossner (1999) for lowland forests in Sepilok, Deramakot and Danum, all Sabah). Most species found at EFN were from the genera *Polyrhachis*, *Camponotus* and *Crematogaster*. These also belong to the 4 most species rich genera in the whole ant fauna in Pasoh FR (Chap. 26) but were also found to be among the most species rich in the studies of Brühl et al. (1998 in Poring), Yamane & Nona (1994 in Lambir) and Chung & Maryati (1996 in Danum). Actually they belong to the 4 most speciose genera in the

Indo-Australian region (Bolton 1995).

Our study demonstrates that EFN was a resource used by ants also in primary forests. In comparison to secondary habitats, however, remarkably less ants were visiting EFN. We have rarely found mass assemblies on EFN-plants in primary forest. In gaps much larger aggregations of ants on EFN occurred, and also in trophobiotic associations with homopteran insects ants were more numerous. This is in accordance with our results of greater abundance of EFN plants in open habitats. Some EFN-species in the primary forest were visited only rarely. The reasons for this are still unknown, perhaps different nectar composition and production rate are relevant. Cost for producing nectar could be too high for plants growing in the shady closed forests where light availability limits photosynthesis. Bentley (1976) also found ant abundance and plants with EFN to be more common in clearing and forest edges than in closed forest. Synecological studies have shown a positive correlation between ant abundance and EFN (Bentley 1976; Keeler 1979a,b). Different results, however, were recently published by Feener & Schupp (1998). Although 25% of the plants in gap plots belonged to ant-defended plant species (with EFN and food bodies) whereas only 5% in closed forest plots did so, the authors could not show a difference in ant abundance in gaps compared to closed forest. They therefore concluded that the prevalence of plants with ant-attracting food resources in gap habitats is largely independent of spatial variation in local ant communities. Regardless of ant abundance the use of ants by plants should be more affordable and effective in high-light habitats. It would be interesting to see whether ants are more abundant and active in the upper canopy of closed forests where photosynthesis by plants is highly promoted under high-light conditions. However, the effects of ants on plant demography in primary forests (e.g. by effecting survival of saplings with and without EFN) are to date totally unknown. More long-term studies are needed to begin to reveal the role of those plant trails for plant communities in undisturbed compared to secondary habitats.

4.3 General ant activity on the lower vegetation

Only few studies on ant fauna with emphasis on the vegetation exist from SE Asia most studies concentrated on the ground ant flora or did not differentiate between collections from ground and vegetation (e.g. Chung & Maryati 1996; Maryati et al. 1996; Yamane & Nona 1994; Yamane et al. 1996) so that their results are not comparable to this study. One study which concentrated on different height of vegetation in Lambir (Itino & Yamane 1994) revealed results more similar to our study: at sugar baits in the understorey and subcanopy also mainly *Camponotus*, *Crematogaster* and *Polyrhachis* were found.

Compared with all the collections made on the ground the present study also revealed a rather low general ant activity on the lower vegetation. In the understorey vegetation of the primary forest in Pasoh FR on average only every 15th plant was found with ants. In gaps and more disturbed areas (like trails in the regenerating buffer zone), however, ant abundance was higher (on average every 6th plant). Also baiting experiments on the vegetation supported this picture. Baits on the ground are usually found faster. Obviously there is a higher density of ants on the ground and a greater accessibility of the baits relative to vegetation (e.g. Feener & Schupp 1998 from BCI; and own observations). In general also less species have been recorded from the (lower) vegetation than from the ground, however, this may also be due to greater sampling effort. Some preliminary data from different forest in Malaysia (however, not directly comparable due to different sampling designs and study periods): Pasoh FR (Chap. 26: 244 species on the ground (different methods, 120 species with restriction to Winkler method), 130 species on the lower vegetation (hand collection); Sabah , Borneo: primary lowland forest in Danum Valley: leaf litter (Winkler method and baits) 176 species, lower vegetation 87 species (hand collection); Deramakot Forest reserve (secondary forest) 123 species on the ground (Winkler method and baits), 61 species on lower vegetation (hand collection) (Gossner 1999 and C. Brühl, pers. comm.).

The attractivness of the various types of baits in the vegetation did not differ much, a result which was also supported by other studies (e.g. Gossner 1999; Hashimoto et al. 1997; Maryati et al. 1996; Yamane et al. 1996). There was even no significant difference in the number of species collected at honey and protein baits (also in other studies e.g. Rosciszewski 1995; Chap. 26) (but see also Feener & Schupp for Panama, where traps with lipid-rich baits attracted 3 times as many ground living ants as those baited with honey, however, this fits in the picture that ground-living ants may be more predatory than those which use the vegetation strata.)

We know of only one study from SE Asia so far which tried to approach Tobin's hypothesis (Tobin 1991, 1994) as canopy ants being mainly herbivores (Hashimoto et al. 1997). They found in a study on food preferences of 19 ant species that honey and as well as protein was accepted by all freely foraging ants. This supports our results from ants on the lower vegetation. In general it appears that most ants tended to accept and utilise a variety of food items and therefore have at least not yet completely abandoned carnivory. An important finding was that all ants that we found at EFN also occurred at protein baits and can therefore to be regarded as generalists. Obviously we have to deal with an opportunistic kind of resource use, however, different food resources could gain different importance at various seasons so we need surveys at different time of the year. It has also to be mentioned that one key resource of ants influencing their abundance on vegetation has not been studied in Pasoh: honeydew secretion from homopterans. These may also influence the ant assembly at least on the lower vegetation (see e.g. Blüthgen et al. 2000).

5. CONCLUSION

That mutualistic interactions of plants with ants do play an important role in tropical forest has now been proved by many studies. The abundance of these interactions can also be quite high (see e.g. Fonseca & Ganade 1996; Rico-Gray 1993). A number of studies also gave evidence that these facultative interactions can vary considerably even over short distances among habitats in the number, diversity and seasonal distribution of ant-plant interactions (Rico-Gray et al. 1998). Therefore the benefits to the plants vary both geographically and temporally and depend largely on the protective abilities of the visiting ant species. Evidence for ants visiting extrafloral nectaries and providing protection against herbivores is now abundant (overview e.g. De la Fuente & Marquis 1999; Koptur 1992). We could so far demonstrate positive effects of unspecific ants visiting EFN in several cases in our own investigations in Malaysia (Fiala et al. 1994a, 1996; Heil et al. 2000, 2001; Merbach et al. unpubl. results). In the frame of the studies in Pasoh FR we have started first preliminary studies on protective function of ants on *Macaranga lowii* (a primary forest species of the section *Pseudorottlera*) and members of the genus *Shorea* which can not be reported here in detail. Effects become often visible only after months, depending on seasonal variation in herbivore pressure and density of ant colonies. Factors contributing to the variability in ant efficacy, quality of nectar, nutritional requirements and availability of alternative food resources need to be examined in order to understand patterns of ant-plant mutualisms. The scope of the study reported here was to survey the existing structures as a basis for further investigations which we have already started on several species. It was not possible to cover in addition the ecology of plant and ants species involved. Therefore inter- and intraspecific variability will have to be explored in further studies. Pasoh FR with its high species richness of plants and ants will hopefully remain one of the most suitable forest areas in Malaysia for such investigations. Tropical forests are becoming fragmented into small relict patches or simplified in structure and composition. The complex ecosystems in which the mutualistic systems evolved are at present being diminished. The rapid conversion of primary dipterocarp forests can be expected to alter these associations in future with still totally unknown consequences for the further development and function.

ACKNOWLEDGEMENTS

We are very grateful for the excellent cooperation with FRIM. The former director Dr. Salleh Mohd Nor gave generous permission to use FRIM facilities and to work at Pasoh FR. We thank the staff of the FRIM herbarium for invaluable help with identification of plant specimens and all other members of the FRIM staff which cannot be listed here for their support. Special thanks are due to the late Dr. K. M. Kochummen. Dr. S. Appanah, Dr. N. Manokaran, and Mr. E. S. Quah gave information on very helpful data of the FRIM 50 ha plot at Pasoh FR. Dr. Wulf Killmann and the GTZ-team at FRIM kindly provided support in many ways which greatly facilitated this study. We thank Dr. James LaFrankie for his manifold logistic and scientific support and Dr. Sean Thomas for plant identification and other valuable information. Dr. Krzysztof Rosciszewski is thanked for support in ant identification. Carsten Brühl contributed unpublished data, Drs. T. Okuda and J. Moog gave valuable comments on the manuscript. The Economic Planning Unit, Prime Minister's Department granted permission to conduct research in Malaysia. Financial support from the German Research Foundation is gratefully acknowledged.

We are grateful for opportunity to use data from the 50 ha plot. The large scale forest plot (i.e. the 50 ha plot) at the Pasoh FR is an ongoing project of the Malaysian Government, and was initiated by the Forest Research Institute Malaysia through its former Director-General, Dato' Dr. Salleh Mohd. Nor, and under the leadership of Dr. N. Manokaran, Dr. Peter S. Ashton and Dr. Stephen P. Hubbell. Supplementary funding was provided by the National Science Foundation (USA); the Conservation, Food, and Health Foundation, Inc. (USA); the United Nations, through its Man and the Biosphere (MAB) program; UNESCO-MAB grants; UNESCO-ROSTSEA; and the continuing support of the Smithsonian Tropical Research Institute (Barro Colorado Island, Panama) and the Center for Global Environmental Research (CGER) at the National Institute for Environmental Studies (NIES), Japan.

REFERENCES

Agosti, D., Maryati, M. & Chung, A. Y. (1994) Has the diversity of tropical ant fauna been underestimated? An indication from leaf litter studies in a West Malaysian lowland rain forest. Trop. Biodiversity 2: 270-275.

Beattie, A. J. (1985) The evolutionary ecology of ant-plant mutualisms. Cambridge University Press, Cambridge, UK.

Bentley, B. L. (1976) Plants bearing extrafloral nectaries and the associated ant community: interhabitat differences in the reduction of herbivore damage. Ecology 57: 815-820.

Blüthgen, N., Verhaagh, M. , Goitia, W., Jaffé, K., Morawetz, W. & Barthlott, W. (2000) How plants shape the ant community in the Amazonian rainforest canopy: the key role of extrafloral nectaries and homopteran honeydew. Oecologia 125: 229-240.

Bolton, B (1995) A taxonomic and zoogeographical census of the extant ant taxa (Hymenoptera: Formicidae). J. Nat. Hist. 29: 1037-1056.

Brühl, C. A., Linsenmair, K. E. & Gunik, G. (1998) Stratification of ants (Hymenoptera, Formicidae) in a primary rain forest in Sabah, Borneo. J. Trop. Ecol. 14: 285-297.

Chung, A. Y. & Maryati, M. (1996) A comparative study of the ant fauna in a primary and secondary forest in Sabah, Malaysia. In Edwards, D. S., Booth, W. E. & Choy, S. C. (eds). Tropical Rainforest Research - Current Issues. Kluwer Academic Publishers, Netherlands, pp.357-366.

Davidson, D. W. (1997) The role of resource imbalances in the evolutionary ecology of tropical arboreal ants. Biol. J. Linn. Soc. 61: 153-181.

Davidson, D. W. & McKey, D. (1993) The evolutionary ecology of symbiotic ant-plant relationships. J. Hymenoptera Res. 2: 13-83.

Davidson, D. W. & Patrell-Kim, L. (1996) Tropical arboreal ants: Why so abundant? In Gibson, A. C. (ed). Neotropical biodiversity and conservation. Mildred E. Mathias Botanical Garden, University of California, Los Angeles, California, USA, pp.127-140.

De la Fuente, M. A. S. & Marquis, R. J. (1999) The role of ant-tended extrafloral nectaries in the protection and benefit of a Neotropical rainforest tree. Oecologia 118: 192-202.

Elias, T. S. (1983) Extrafloral nectaries: their structure and distribution. In Bentley, B. L. & Elias, T. S. (eds). The biology of nectaries. Columbia University Press, New York, USA, pp.174-203.

Feener, D. H. & Schupp, E. W. (1998) Effect of treefall gaps on the patchiness and species richness of Neotropical ant assemblages. Oecologia 116: 191-201.

Fiala, B. & Linsenmair, K. E. (1995) Distribution and abundance of plants with extrafloral nectaries in the woody flora of a lowland primary forest in Malaysia. Biodiversity Conserv. 4: 165-182.

Fiala, B. & Maschwitz, U. (1991) Extrafloral nectaries in the genus *Macaranga* (Euphorbiaceae) in Malaysia: comparative studies of their possible significance as predispositions for myrmecophytism. Biol. J. Linn. Soc. 44: 287-305.

Fiala, B. & Maschwitz, U. (1995) Mutualistic associations between ants and plants in SE Asian rainforests. Wallaceana 75: 1-5.

Fiala, B., Grunsky, H., Maschwitz, U. & Linsenmair, K. E. (1994a) Diversity of ant-plant interactions: protective efficacy in *Macaranga* species with different degrees of ant-association. Oecologia 97: 186-192.

Fiala, B., Rabenstein, R. & Maschwitz, U. (1994b) Ant-attracting plant-structures: food bodies of SE Asian Vitaceae. In Lenoir, A., Arnold, G. & Lepage, M. (eds). Les insectes sociaux. Proc. 12th IUSSI-Congress Paris, France, p.174.

Fiala, B., Krebs, S. A., Barlow, H. S. & Maschwitz, U. (1996) Interactions between *Thunbergia grandiflora, Dolichoderus* sp. and *Xylopa latipes*: a mutualistic association. Malay. Nat. J. 50: 1-14.

Floren, A. & Linsenmair, K. E. (1997). Diversity and recolonisation dynamics of arthropod communities with special reference to the Formicidae-fauna on different tree species in a lowland rain forest in Sabah, Malaysia. In Stork, N. E., Adis, J. & Didham, R. K. (eds). Canopy arthropods. Chapman & Hall, London. pp.344-381.

Fonseca, C. R. & Ganade, G. (1996) Asymmetries, compartments and null interactions in an Amazonian ant-plant community. J. Anim. Ecol. 65: 339-347.

Gossner, M. (1999) Vergleich von Diversität und Artenzusammensetzung der Ameisenzönosen der unteren Vegetation zwischen Primär- und Sekundärwaldflächen im Tieflandregenwald von Sabah, Borneo. M. thesis, Univ. Würzburg (unpubl.).

Hashimoto, Y., Yamane, S. & Ichioka, T. (1997) A preliminary study on dietary habits of ants in a Bornean rain forest. Jpn. J. Entomol. 65: 688-695.

Heil, M., Fiala, B., Baumann, B. & Linsenmair, K. E. (2000) Temporal, spatial and biotic variations in extrafloral nectar secretion by *Macaranga tanarius*. Funct. Ecol. 14: 749-757.

Heil, M., Koch, T., Hilpert, A., Fiala, B., Boland, W. & Linsenmair, K. E. (2001) Extrafloral nectar production of the ant-associated plant, *Macaranga tanarius*, is an induced indirect defensive response elicited by jasmonic acid. Proc. Natl. Acad. Sci. USA 98:1083-1088.

Hölldobler, B. & Wilson, E. O. (1990) The Ants. Belknap Press, Cambridge, Massachusetts, 732pp.

Ichino, T. & Yamane, S. (1994) Vertical distribution of ants in the canopy of a lowland mixed dipterocarp forest of Sarawak. In Inoue, T. & Abang Abdul Hamid (eds). Plant reproductive systems and animal seasonal dynamics. Canopy Biology Program in Sarawak. Series I. Kyoto University, pp.227-230.

Keeler, K. H. (1979a) Distribution of plants with extrafloral nectaries in a temperate flora (Nebraska). Prairie Naturalist 11: 33-37.

Keeler, K. H. (1979b) Distribution of extrafloral nectaries and ants at two elevations in Jamaica. Biotropica 11: 152-154.

Kochummen, K. M., LaFrankie, J. V. & Manokaran, N. (1990) Floristic composition of Pasoh Forest Reserve, a lowland rain forest in Peninsular Malaysia. J. Trop. For. Sci. 3: 1-13.

Koptur, S. (1992) Extrafloral nectary-mediated interactions between insects and plants. In Bernays, E. (ed). Insect-plant interactions. vol. IV. CRC Press, Boca Raton, Florida, USA, pp.82-129

Koptur, S., Dillon, P; Foster C. (1977) A comparison of ant activity and extrafloral nectaries at various sites in Costa Rica. O.T.S. Tropical Biology Coursebook 77-3: 299.

Liefke, C., Dorow, W. H. O., Hölldobler, B. & Maschwitz, U. (1998) Nesting and food resources of syntopic species of the ant genus *Polyrhachis* (Hymenoptera, Formicidae) in West Malaysia. Insectes Sociaux 45: 411-425.

Manokaran, N., LaFrankie, J. V., Kochummen, K. M., Quah, E. S., Klahn, J. E., Ashton, P. S. & Hubbell, S. P. (1992) Stand table and distribution of species in the 50-ha Research Plot at Pasoh Forest Reserve. FRIM Research Data No. 1. Kepong, Malaysia

Marquis, R. J. & Braker, H. E. (1994) Plant-herbivore interactions: diversity, specificity, and impact. In McDade, L. A., Bawa, K. S., Hespenheide, H. A. & Hartshorn, G. S. (eds). La Selva: Ecology and Natural History of a Neotropical Rainforest., University of Chicago Press, Chicago, pp.261-281.

Maryati, M., Azizah, H. & Arbain, K. (1996) Terrestrial ants (Hymenoptera: Formicidae) of Poring, Kinabalu Park, Sabah. In Edwards, D. S., Booth, W. E. & Choy, S. C. (eds). Tropical Rainforest Research - Current Issues. Kluwer Academic Publishers, Netherlands, pp.117-123

Maschwitz, U., Fiala, B. & Linsenmair, K. E. (1994) *Clerodendrum fistulosum* (Verbenaceae) un unspecific myrmecophyte from Borneo. Blumea 39: 143-150.

McKey, D. (1992) Interactions between ants and plants: comparison of canopy, understorey and clearing environments. In Hallé, F. & Pascal, O. (eds). Biologie d'une canopée de foret équatoriale II. Fondation elf, Paris, France, pp.66-73

Merbach, M. A., Zizka, G., Fiala, B., Merbach, D. & Maschwitz, U. (1999) Giant nectaries in the peristome thorns of the pitcher plant *Nepenthes bicalcarata* Hook F. (Nepenthaceae): anatomy and functional aspects. Ecotropica 5: 45-50.

Morellato, L. P. C. & Oliveira, P. S. (1991) Distribution of extrafloral nectaries in different vegetation types of Amazonian Brazil. Flora 185: 33-38.

Ng, F. S. P. (ed). (1978, 1989) Tree Flora of Malaya. vol. 3 & 4.

Nielsen, I. C. (1992) Mimosaceae (Leguminosae-Mimosoideae). Flora Malesiana, Series 1, 11; 1-226.

Okuda, T., Suzuki, M., Adachi, N., Quah, E. S., Hussein, N. A. & Manokaran, N. (2003) Effect of selective logging on canopy and stand structure in a lowland dipterocarp forest in Peninsular Malaysia. For. Ecol. Manage. 175: 297-320.

Oliveira, P. S. & Brandao, R. F. 1991. The ant community associated with extrafloral nectaries in the Brazilian cerrados. In Huxley, C. R. & Cutler, D. F. (eds). Ant-plant interactions.

Oxford University Press, Oxford, UK, pp.198-212.
Pemberton, R. W. (1998) The occurrence and abundance of plants with extrafloral nectaries, the basis for antiherbivore defensive mutualisms, along a latitudinal gradient in east Asia. J. Biogeogr. 25: 661-668.
Rabenstein, R., Liefke, C., Maschwitz, U., Fiala, B. & Rosli bin Hashim (1999) Ants like it sweet - Plant based food resources of the ant *Polyrhachis olybria*. Malay. Nat. J. 52: 5-9.
Rickson, F. R. & Rickson, M. M. (1998). The cashew nut, *Anacardium occidentale* (Anacardiaceae), and its perennial asociation with ants: extrafloral nectary location and the potential for ant defense. Am. J. Bot. 85: 835-849.
Rico-Gray, V. (1993) Use of plant-derived food resources by ants in the dry tropical lowlands of coastal Veracruz, Mexico. Biotropica 25: 301-315.
Rico-Gray, V., García-Franco, J. G., Palacios-Rios, M., Díaz-Castelazo, C., Parra-Tabla, V. & Navarro, J. A. (1998) Geographical and seasonal variation in the richness of ant-plant interactions in México. Biotropica 30: 190-200.
Rosciszewski, K. (1995) Die Ameisenfauna eines tropischen Tieflandregenwaldes in Südostasien: Eine faunistisch-ökologische Bestandsaufnahme. Ph.D. diss., Univ. Frankfurt.
Schellerich-Kaaden, A. & Maschwitz, U. (1998) Extrafloral nectaries on culm sheath auricles: Observations on four Southeast Asian giant bamboo species (Poaceae-Bambusoideae). Sandakania 11: 61-68.
Schupp, E. W. & Feener, D. H. (1991) Phylogeny, lifeform, and habitat dependence of ant-defended plants in a Panamanian forest. In Huxley, C. R. & Cutler, D. F. (eds). Ant-plant interactions. Oxford University Press, Oxford, UK, pp.175-197.
Tobin, J. E. (1991) A neotropical rainforest canopy ant community: some ecological considerations. In Huxley, C. R. & Cutler, D. F. (eds). Ant-plant interactions. Oxford University Press, Oxford, UK, pp.536-538.
Tobin, J. E. (1994) Ants as primary consumers: diet and abundance in the Formicidae. In Hunt, J. H. & Nalepa, C. A. (eds). Nourishment and evolution in insect societies. Westview, Boulder, pp.279-307.
Whitmore, T. (ed). (1972, 1973) Tree flora of Malaya. vol. 1,2. Longman, Kuala Lumpur.
Wong, K. M. & Puff, C. (1995) Notes on a heterophyllous *Diospyros* (Ebenaceae). Sandakania 6: 55-62.
Yamane, S. & Nona, A. R. (1994) Ants from Lambir Park, Sarawak. In Inoue, T. & Abang Abdul Hamid (eds). Plant reproductive systems and animal seasonal dynamics. Canopy Biology Program in Sarawak. Series I. Kyoto University, pp.222-226.
Yamane, S., Ichino, T. & Nona, A. R. (1996) Ground ant fauna in a Bornean dipterocarp forest. Raffles Bull. Zool. 44: 253-262.

33 Ant-Plant Diversity in Peninsular Malaysia, with Special Reference to the Pasoh Forest Reserve

Joachim Moog[1], Brigitte Fiala[2], Michael Werner[2], Andreas Weissflog[2], Saw Leng Guan[3] & Ulrich Maschwitz[2]

Abstract: The overview of ant-plants in Peninsular Malaysia presented here covers: (i) diversity of participating ant and plant taxa; (ii) specificity of associations; (iii) morphological diversity of plant structures used for ant-housing; (iv) plant growth habit; and (v) food type of ants directly or indirectly derived from host plants. Included are unpublished observations both from field and herbarium studies and records of ant-plants obtained from a literature survey. It is the first comprehensive account of Malayan ant-plants on species level since many decades. At least 45 'true' myrmecophytic species belonging to 20 genera and 14 families are recognized. Twenty-two additional species (from 17 genera) are looked upon as potential myrmecophytes but evidence is still anecdotal and incomplete. The review presented here allows a conservative estimate of 'true' ant-plants in Pasoh Forest Reserve (Pasoh FR). The reserve is home to at least one third of the known myrmecophytes occurring on the peninsula.

Key words: biodiversity, *Camponotus*, *Cladomyrma*, *Crematogaster*, domatia myrmecophytes.

1. INTRODUCTION

Mutualistic interactions between plants and ants have become a major focus of tropical ecology in recent years and they have been suggested as a factor driving diversification and specialization in associated plants and ants (Fiala et al. 1999). This kind of relationship reaches a high degree of sophistication in the so-called myrmecophytes (ant-plants), which offer housing and food for their ant partners. In return, the ants are known or presumed to protect the plants from herbivory. Additional benefits to the plant may include protection against pathogenic fungi, removal of vines, nutrient addition through breakdown of collected debris in domiciles and absorption of ant-respired carbon dioxide (e.g. Beattie 1985; Heil et al. 2000; Jolivet 1996; Letourneau 1998; Sagers et al. 2000; Treseder et al. 1995).

However, existing evidence to assess fitness consequences of particular associations is often too meagre. Therefore, the term 'myrmecophyte' is used here sensu Davidson & McKey (1993), i.e. it only describes plants regularly inhabited by ants, without implying that host plants actually benefit from the ant tenants. Another definition of the term 'myrmecophyte' is based on a morphological trait, i.e., it describes plant species that have evolved specialized structures (domatia) that house ants. However, the term is used here (sensu Longino & Hanson 1995) also for plants that, in the absence of known *specialized* structures, are nevertheless

---

[1] Department of Zoology, Johann Wolfgang Goethe-University, 60054 Frankfurt, Germany. E-mail: J.Moog@zoology.uni-frankfurt.de
[2] University of Würzburg, Germany.
[3] Forest Research Institute Malaysia (FRIM), Malaysia.

regularly inhabited by one or more plant-ant species. Even these two definitions cannot cover all the associations we find between plants and inhabiting ants since a continuum of specialization towards myrmecophytism exists. Variations in the degree of specialization are well illustrated e.g. in the bamboo/ant or plant/*Cladomyrma* associations (see below).

Another difficulty of the definition 'myrmecophyte' lies in the condition of 'regular' ant-inhabitation. Generally, ant occupancy rates of south-east Asian ant-plants are, with exception of myrmecophytic *Macaranga*, poorly studied. In the literature we find ant colonization of particular plants often vaguely expressed as 'usually', 'frequently', 'often' or 'sometimes'. Especially the latter term is prone to misinterpretations. It may either just state occasional colonizations by opportunistic ants or it is a reflection of the temporal and spatial variation of ant occupancy found in several ant-associated plants, e.g. colonization frequencies depend on the developmental stage of the plant (Fiala & Maschwitz 1992a; Gay 1993; Moog et al. 1998) or is influenced by habitat and/or by proximity to foundress sources (Yu & Davidson 1997).

Many plant species only establish comparatively weak, unspecific relationships with ants by providing food (e.g. extrafloral nectar) for ant visitors but no nesting space. These so-called 'myrmecophiles' are treated by Fiala & Saw (Chap. 32). Excluded from the overview presented here are also the 'ant nest-garden epiphytes' (Benzing 1990). While ant-garden epiphytes possess root systems that are usually colonized by ants, ant-house epiphytes not only are frequently and often exclusively associated with ants but also exhibit unique vegetative structures promoting ant-housing. However, there is not always a clear distinction between ant-garden and ant-house epiphytes since (i) the definitions are not entirely exclusive and (ii) evidence for special ant-housing adaptation is often indirect and incomplete (Davidson & Epstein 1989). This is reflected in conflicting views concerning e.g. some Malayan *Dischidia* species on their being a myrmecophyte (Kerr 1912; Kiew & Anthonysamy 1995; Ridley 1910; Weir & Kiew 1986).

Apart from the difficulties of categorizing ant-associated plants, our knowledge on the diversity of ant-plant mutualisms in the Oriental region has long been surprisingly poor compared to the Neotropics, but recent proliferation of work has helped to improve our understanding of ant-plant relationships in SE Asia (e.g. Blattner et al. 2001; Brouat & McKey 2000; Federle et al. 1997, 1998a,b, 1999, 2000, 2001; Feldhaar et al. 2000; Fiala et al. 1991, 1994, 1996, 1999; Heckroth et al. 1998; Heil et al. 1997, 1998, 1999, 2001a,b; Itioka et al. 2000; Janka et al. 2000; Leo et al. 1999; Maschwitz & Fiala 1995; Maschwitz et al. 1992, 1994a,b, 1996a,b; Mattes et al. 1998; Merbach et al. 1999; Moog et al. 1998; Nomura et al. 2000; Werner et al. 1996; Wong & Puff 1995).

Here we provide the first overview at species level of the ant-plant associations in the Peninsular Malaysia and Pasoh Forest Reserve (Pasoh FR), focusing principally on non-epiphytic myrmecophytes. Ant-plant diversity is described in two ways, (i) by presenting the taxonomic variety (species richness) of myrmecophytes and their ant inhabitants and (ii) by pointing out the morphological and structural differences of myrmecophytic traits (e.g. domatia types, facultative versus specific colonization). In addition, we present a tabular survey (Table 1) of the known records (including our own hitherto unpublished work) of Malayan ant-plants and inhabiting ants.

We aimed at summarizing and classifying the often fragmentary and scattered records of (supposedly) ant-plants of Peninsular Malaysia. This is a

Table 1 List of recorded ant-plants and inhabiting ants of Peninsula Malaysia. Missing data are indicated by question marks. Columns: ***Host Plant Taxa***: Plant species which we consider 'true' myrmecophytes are given in bold letters. Asterisks * denote plant species whose specialization resp. regular ant inhabitation is uncertain. Square brackets [ ] indicate plant species which do not come within our definition of myrmecophytes (see text); ***Growth Form*** (including tree strata group according to Manokaran et al. 1992): B = bamboo, E = epiphyte, H = hemi-epiphyte, L = liana or climber, R = rattan, S = shrub or small treelet, T = tree, c = canopy, e = emergent, u = understory (incl. treelets); ***Ant Housing Structures***: Ap = pair of auricles, erect narrow extensions of the leaf sheath, closely adpressed to the stem, Bi = insect borings, Cl = carton structures build around clusters of cup-shaped leaves, Dh = swollen, naturally hollow stem domatia due to a (nearly) complete pith degeneration, Dp = swollen, pithy stem domatia, hollowed by ants (the pith is usually soft), (Dp) = swollen domatia present in saplings, in later stages not (or rarely) conspicuous, Es = self-opening, slit-like entrance holes, Fc = convex basal fronds, Gs = galleries enclosed by interlocking combs of spines, forming collars on leaf sheaths, Hi = inflated hypocotyl, forming chambered tuber, Lb = buds covered by large scale-like leaves, Lc = leaf cluster, forming an ovoid multichambered cavity, Lf = flat or lens-shaped leaves, Lp = pitcher-like leaves, Lr = lowermost leaflets reflexed back across the stem thus forming a more or less secluded shelter, Ls = convex, shell-like leaves closely appressed to surface of host tree, Lt = tuft of very close-set leaves, with leaf bases forming a dense cluster, Nt = thickened nodes, Od = elongated ocrea (proximal extension of leaf sheath beyond petiole), diverging from the stem, Oi = inflated ocrea, Pb = pseudobulbs swollen at base with galleries, Pro = prostomata (thin zone in stem wall facilitating entry of ants into domatium), Ps = pseudostipules (stipule-like basal pair of leaflets) reflexed backward and clasping the twig, Pw = petiole broadly winged on each side, with the wings folded towards the adaxial side of the petiole, Rh = inflated, hollow rhizomes, Sh = naturally hollow stems (at least some of the plants total internodes are hollow), not distinctly swollen, Sp = pithy stems, not distinctly swollen, hollowed by ants, St = (semi-) persistent stipules; ***Food Source***: (directly or indirectly derived from host plant): fb = food bodies, fn = floral nectar, efn = extrafloral nectar, (efn) = only present in saplings, h = honeydew, i.e. exudates of trophobiotic coccoids or aphids; ***Ant Taxa***: bold letters indicate obligate plant-ant species (for *Crematogaster (Decacrema)* species inhabiting *Macaranga* plants only main ant occupants are listed). Asterisks * denote ant species whose specialization is uncertain; ***Pasoh***: plant species occurring in Pasoh FR y = yes, n = no; ***References***: numbered references refer to the list given below. Included are only main references.

Host Plant Taxa	Growth Form	Ant Housing Structures	Food Source	Trophobiont	Ant Taxa	Pasoh	References
ANACARDIACEAE							
[*Lannea coromandelica* (Houttuyn) Merrill]	T, u	Sh or Sp	?	?	undetermined	?	27, 60
ANNONACEAE							
[*Goniothalamus ridleyi* King]	T, u	see text	fn	?	undet.	?	46
ASCLEPIADACEAE							
Dischidia albiflora Griffith (= *D. collyris* Wallich)	E	Ls	fn	?	undet.	?	29, 47
Dischidia astephana Scort. ex King & Gamble	E	Ls	fn	?	**Crematogaster treubi* Emery, *C*. sp.	n	19, 29, 47, 50
Dischidia cochleata Blume (= *D. coccinea* Griffith)	E	Ls	fn	?	undet.	?	29, 47

Table 1 (continued 1)

Host Plant Taxa	Growth Form	Ant Housing Structures	Food Source	Trophobiont	Ant Taxa	Pasoh	References
Dischidia complex Griffith (= *D. shelfordii* Pearson)	E	Lp	fn	?	undet.	?	4, 46, 47, but see 29
Dischidia imbricata (Bl.) Steud. (= *D. depressa* Clarke ex King & Gamble)	E	Ls	fn	?	undet.	?	47, but see 29
Dischidia longepedunculata Ridley	E	Ls	fn	?	undet.	n	29, 47, 50
[*Dischidia nummularia* R. Br.] (= *D. gaudichaudii* Decne.)	E	Lf	fn	?	*Crematogaster* sp., *Pheidole* sp., **Philidris* sp.	?	26, 28, 29, 47, 62
Dischidia major (Vahl) Merrill (= *D. rafflesiana* Wallich)	E	Lp	fn, h	coccoids (rare)	**Philidris cordatus* (F. Smith)	?	26, 28, 29, 46, 47
[*Dischidia parvifolia* Ridley]	E	Lf	fn	?	**Crematogaster* sp.	n	29, 50
Hoya mitrata Kerr	E	Lc	h	coccids	various species	?	51
CRYPTERONIACEAE							
Crypteronia griffithii C.B. Clarke	T, u, (c)	Sp, Nt	h	pseudococcids coccids	*Cladomyrma maschwitzi* Agosti	y	32, 38
DIPTEROCARPACEAE							
**Shorea acuminata* Dyer	T, (c), e	St	efn	?	**Technomyrmex* sp.	y	33, 57, 58, 59
EBENACEAE							
Diospyros andamanica (Kurz) Bakh.	T, u	Cl	efn, h	coccoids	various species	y	54
EUPHORBIACEAE							
[*Agrostistachys sessilifolia* (Kurz) Pax & Hoffm.]	S, u	Lt	?	?	undet.	n	7
Drypetes longifolia (Bl.) Pax & K. Hoffm.	T, u	Sp, (Dp)	h	pseudococcids	*Cladomyrma petalae* Agosti, *C. nudidorsalis* Agosti, Moog &	y	2, 39, 60
[*Drypetes pendula* Ridley]	T, u	Sp, Bi	h	pseudococcids, coccids	various species (rarely occupied)	y	33, 58, 59, 60
Macaranga caladiifolia Beccari	S, u	Dh, Es	efn, fb	-	various species	n	17
Macaranga constricta Whitmore & Airy Shaw	T, u	Dh	fb, h	coccids	*Crematogaster* (*Decacrema*) sp.5 +	n	16, 18
Macaranga hosei King ex Hook. f.	T, u	Dp	(efn), fb, h	coccids	*C.* (*D.*) sp.2	y	15, 16, 18
Macaranga hullettii King ex Hook. f.	T, u	Dh, Pro	fb, h	coccids	*Crematogaster* (*D.*) sp.3 + sp.4	n	14, 15, 16, 18
Macaranga hypoleuca (Reichb. f. & Zoll) Muell. Arg.	T, u	Dh	fb, h	coccids	*Crematogaster* (*D.*) sp.1 + sp.6	y	15, 16, 18
Macaranga kingii Hook. f. var. *kingii*	T, u	Dh, Pro	fb, h	coccids	*Crematogaster* (*D.*) sp.4	n	14, 16, 18, 53
Macaranga motleyana Muell. Arg. subsp. *griffithiana*	T, u	Dh, (Pro)	fb, h	coccids	*Crematogaster* (*D.*) sp.1 + sp.5	y	14, 16, 18

Table 1 (continued 2)

Host Plant Taxa	Growth Form	Ant Housing Structures	Food Source	Trophobiont	Ant Taxa	Pasoh	References
Macaranga pruinosa (Miq.) Muell. Arg.	T, u	Dp, St	efn, fb, h	coccids	for Dp: *Crematogaster* (*D.*) sp.1 + sp.2 for St: *Technomyrmex* sp.A and various species	n	13, 18
Macaranga puncticulata Gage	T, u	Dh	efn, fb	-	*Camponotus* (*Colobopsis*) sp.1, various species	n	13, 18
Macaranga triloba (Bl.) Muell. Arg.	T, u	Dh, Pro	fb, h	coccids	*Crematogaster* (*D.*) sp.3 + sp.4	n	14, 15, 18, 49
FABACEAE (Leguminosae)							
[*Archidendron clypearia* (Jack) Nielsen]	T, u	Bi	efn	?	various species	y	12, 60
[*Archidendron ellipticum* (Bl.) Nielsen]	T, u	?Sp, ?Bi	efn	?	'red tailor-ants' (Sabah)	n	41
Saraca thaipingensis Cantley ex Prain	T, u	Sp, Nt	h	pseudococcids	*Cladomyrma petalae* ,	y	32, 36, 40, 60
Spatholobus bracteolatus Prain ex King	L	Sp, (Dp)	h	pseudococcids	*Cladomyrma petalae*	n	36, 37, 60
FLACOURTIACEAE							
Ryparosa fasciculata King	T, u	Sp, Nt	h	pseudococcids	*Cladomyrma petalae* , *C.*	y	2, 37, 60
LAURACEAE							
**Actinodaphne sesquipedalis* Hk. f. & Thoms. ex Meisn.	T, u	Lb	?	?	undet.	y	7
LOGANIACEAE							
**Strychnos vanprukii* Craib	L	Sp, Sh	h	pseudococcids	*Cladomyrma petalae*	?	2, 36
MELIACEAE							
**Chisocheton tomentosus* (Roxb.) Mabb.	T, u	Sp, ?Bi	?	?	'small black ants'	y	42, 60
MORACEAE							
Ficus obscura Blume var. *borneensis* (Miq.) Corner	H	Sh, Es	efn, (h)	pseudococcids (rare)	various species	y	34
MYRTACEAE							
**Leptospermum flavescens* Smith	T	Sp, Bi	?	?	**Crematogaster* sp.	n	27, 50
ORCHIDACEAE							
**Grammatophyllum speciosum* Blume	E	Pb	?	?	undet.	?	5, 24
PALMAE							
**Calamus javensis* Blume	R	Lr	h	aphids	various species	y	10, 52
**Calamus laevigatus* Mart.	R	Lr	h	aphids	various species	n	10, 52
Calamus polystachys Beccari	R	Gs	h	aphids	*C.* (*Myrmoplatys*) *beccarii* Emery	(y)	10, 52, 61
Daemonorops macrophylla Beccari	R	Gs	h	aphids	*C.* (*M.*) sp. mw0286 near *beccarii*	y	10, 45, 52, 61
Daemonorops oligophylla Beccari	R	Gs	h	aphids	*C.* (*M.*) sp. mw0286 near *beccarii*	n	10, 52, 61

Table 1 (continued 3)

Host Plant Taxa	Growth Form	Ant Housing Structures	Food Source	Trophobiont	Ant Taxa	Pasoh	References
Daemonorops sabut Beccari	R	Gs	h	aphids	*C.* (*M.*) sp. mw0286 near *beccarii*	y	10, 52, 61
Daemonorops verticillaris (Griff.) Mart.	R	Gs	h	aphids	*C.* (*M.*) sp. mw0286 near *beccarii*	y	10, 45, 52, 61
Korthalsia echinometra Beccari	R	Oi	h	aphids	*C.* (*M.*) *contractus* Mayr, *C.* (*M.*) sp. kortL, *C.* (*M.*) sp. mw0989hos, *C.* sp. haC (subgenus unknown)	n	10, 52, 61
Korthalsia hispida Beccari	R	Od	h	aphids	*C.* (*M.*) sp. A, *C.* (*M.*) sp.B	n	9, 61
Korthalsia rostrata Blume	R	Oi	h	aphids	*C.* (*M.*) sp. hosD + hosH, *C.* (*M.*)	(y)	10, 52, 61
Korthalsia scortechinii Beccari	R	Oi	h	aphids	*C.* (*M.*) *contractus*, *C.* (*M.*) sp. kortL + kr49kort, *C.* (*M.*) sp. hosD + mw0542hos, *C.* sp. haC + haN	y	10, 31, 52, 61
**Pogonotium ursinum* (Becc.) J. Dransf.	R	Ap	?	?	undet.	n	11
POACEAE							
**Gigantochloa ligulata* Gamble	B	Sh, Bi	h	pseudococcids	*Tetraponera binghami* Forel	?	6
**Gigantochloa scortechinii* Gamble]	B	Sh, Bi	(efn), h	pseudococcids, aphids	*Tetraponera binghami*, *Polyrhachis arachne* Emery, *P. hodgsoni* Forel, *Cataulacus muticus* Emery	?	6, 8, 30, 59
**Gigantochloa thoii* Wong	B	Sh, Bi	(efn), h	pseudococcids	*Polyrhachis schellerichae* Dorow	?	48
**Schizostachyum* Nees 1 sp. (undet.)	B	Sh, Bi	(efn), h	pseudococcids, aphids	*Polyrhachis arachne*, *P. hodgsoni*	?	8
POLYPODIACEAE							
Lecanopteris crustacea Copeland	E	Rh	h	coccoids (on adjacent	*Crematogaster* sp.,	?	19, 21
Lecanopteris sinuosa (Wall. ex Hook.) Copel. (= *Phymatodes sinuosa* (Wall. ex Hook.) J. Sm.)	E	Rh	h	coccoids (on adjacent plants)	**Philidris cordatus*, (**Crematogaster treubi*),	?	19, 21, 46, 55
Lecanopteris pumila Blume	E	Rh	h	coccoids (on adjacent plants)	**Crematogaster treubi*, *Crematogaster biformis* Andre,	n	19, 20, 21, 55
**Platycerium coronarium* (Koenig) Desv.	E	Fc	efn	?	*Pheidole* sp., various species	y	43, 60
**Platycerium ridleyi* H. Christ	E	Fc	efn	?	undet.	?	21
RUBIACEAE							
Hydnophytum formicarum Jack	E	Hi	h	coccoids (mostly on	**Philidris cordatus*	?	4, 25, 26, 44
Myrmecodia tuberosa Jack	E	Hi	h	coccoids (mostly on	**Philidris cordatus*	?	4, 25, 26
**Uncaria* Schreb. 2 spp. (undet.)	L	Sh, Sp, Bi	h	coccids	*Crematogaster* sp., *Camponotus* (*Colobopsis*) sp.	y	23, 56, 59

Table 1 (continued 4)

Host Plant Taxa	Growth Form	Ant Housing Structures	Food Source	Trophobiont	Ant Taxa	Pasoh	References
RUTACEAE							
Luvunga Buch.-Ham. 1 sp. (undet.)	L	Sp, (Dp)	h	pseudococcids	*Cladomyrma petalae*	?	2, 39, 60
Zanthoxylum myriacanthum Wall. ex Hook. f.	T, c	Sh, Es	efn, h	coccoids (rare)	various species	n	22, 33
Zanthoxylum limonella (Dennst.) Alston (= *Z. rhetsa* (Roxb.) DC.)	T, c	Sh, ?Es	?	?	?	n	22, 33
SAPINDACEAE							
**Lepisanthes alata* (Bl.) Leenhouts	T, u	Ps	efn	?	undet.	n	1
**Lepisanthes amoena* (Hassk.) Leenhouts	T, u	Ps	efn	?	undet.	n	1
Pometia pinnata Forst. forma *glabra* (Bl.) Jacobs	T, c	Ps	efn, h	pseudococcids, coccids	various species	y	1, 59, 60
SCROPHULARIACEAE							
**Wightia borneensis* Hook. f.	H	Sh or Sp	?	?	undet.	?	24, 27
URTICACEAE							
[*Poikilospermum suaveolens* (Bl.) Merr.]	H	St	efn, ?fb, h	pseudococcids, coccids, membracids	various species	y	27, 56, 60
[*Poikilospermum microstachys* (Barg.-Petr.) Merr.]	H	St	fb	?	various species	?	12
VERBENACEAE							
**Clerodendrum breviflorum* Ridley	S, u	Sh, ?Bi	?	?	?	n	5, 27, 46
Clerodendrum deflexum Wallich	S, u	Sh, Bi	efn, h	coccids (on plant surface)	various species	y	35
Clerodendrum phyllomega Steud. var. *myrmecophilum* (Ridl.) (= *C. myrmecophilum* Ridley)	S, u	Sh, Bi	efn, h	coccids (on plant surface)	various species	n	35, 46
**Teijsmanniodendron pteropodum* (Miq.) Bakh. (= *Vitex peralata* King)	T, u	Pw	?	?	undet.	n	7
VIOLACEAE							
[*Rinorea javanica* (Bl.) Kuntze]	T, u	?	?	?	?	n	3, 59

Note added in proof: The myrmecophytic *Macaranga triloba* (Blume) Muell. Arg. has been re-named as *M. bancana* (Miq.) Muell. Arg., see Davies 2001: Systematics of *Macaranga* sects. *Pachystemon* and *Pruinosae* (Euphorbiaceae). Harvard Papers Bot. 6: 371-448.

Source (references Table 1) *Publications:* 1.Adema et al. 1994, 2.Agosti et al. 1999, 3.Balgooy 1997, 4.Beccari 1884-1886, 5.Bequaert 1922, 6.Buschinger et al. 1994, 7.Corner 1998, 8.Dorow & Maschwitz 1990, 9.Dransfield 1973, 10.Dransfield 1979, 11.Dransfield 1980, 12.Federle 1998, 13.Federle et al. 1998a, 14.Federle et al. 2001, 15.Fiala et al. 1989, 16.Fiala & Maschwitz 1992a, 17.Fiala et al. 1996, 18.Fiala et al. 1999, 19.Gay & Hensen 1992, 20.Gay 1993, 21.Gay et al. 1993, 22.Hartley 1966, 23.Heckroth et al. 1998, 24.Hoelldobler & Wilson 1990, 25.Huxley 1980, 26.Janzen 1974, 27.Jolivet 1996, 28.Kerr 1912, 29.Kiew & Anthonysamy 1995, 30.Klein et al. 1992, 31.Lehmann 1998, 32.Maschwitz et al. 1991, 33.Maschwitz et al. 1992, 34.Maschwitz et al. 1994b, 35.Maschwitz et al. 1994a, 36.Moog & Maschwitz 1994, 37.Moog et al. 1997, 38.Moog et al. 1998, 39.Moog & Maschwitz 1999, 40.Moog & Maschwitz 2000, 41.Nielsen 1992, 42.Pannell 1992, 43.Paterson 1982, 44.Rickson 1979, 45.Rickson & Rickson 1986, 46.Ridley 1910, 47.Rintz 1980, 48.Schellerich-Kaad. et al. 1997b, 49.Smith 1903, 50.Weir & Kiew 1986, 51.Weissflog et al. 1999, 52.Werner 1993, 53.Whitmore 1973, 54.Wong & Puff 1995, 55.Yapp 1902. *Pers. comm.*: 56.H.-P. Heckroth, 57.Tho Y.P. & F.R. Rickson. *Pers. obs. of the authors*: 58. B. Fiala, 59. U. Maschwitz, 60.J. Moog, 61. M. Werner, 62. A. Weissflog

prerequisite for a reliable estimate of the number and proportion of ant-plants occuring in Pasoh FR. Our review thus provides the basis for future research on ant/plant co-existence systems in Pasoh FR.

2. MATERIALS AND METHODS

The study was in part carried out in Pasoh FR (Negeri Sembilan), but comparative data on ant-plant diversity were also collected in other parts of Peninsular Malaysia (Fig. 1), e.g. in Ulu Gombak (Selangor), Genting Highlands, Fraser's Hill (Pahang), Bukit Larut (= Maxwell Hills), Belum (Perak), and Endau-Rompin (Johore). In Pasoh FR the survey took place (February–March 1993, December 1995, January 1996, June 1999) both in the 50 ha plot and the surrounding area (for details see Fig. 1). Additional information was obtained from herbarium and literature studies.

Various aspects of the ant-plant associations were studied in the field, including: myrmecophytic characteristics (e.g. occurrence of extrafloral nectaries, food bodies or domatia), the specificity of the relationships between the plants and their ant inhabitants, and the occurrence of trophobiotic homopterans (scale insects, aphids). In cases where plants could not be identified in the field, dried specimens were compared with herbarium material at Forest Research Institute Malaysia (FRIM), Kepong, Malaysia, Rijksherbarium Leiden, The Netherlands, and Royal Botanic Gardens, Kew, UK. For each host plant, voucher specimens from each ant colony were collected for identification. The ant fauna of the region is poorly described, so several ant species were identified to generic level and are presented as morphotypes (e.g. *Crematogaster* sp.1). Taxa identified only by a code number (e.g. *Camponotus* sp. mw0286) may represent undescribed species whose descriptions and new names will be found in future publications (Werner, in prep.).

In the present contribution the term 'myrmecophyte' is used to describe plants *regularly* inhabited by *nesting* ants. This definition does not automatically imply (i) that host plants actually benefit from their ant inhabitants and (ii) that plants are known to have evolved specialized structures for ant housing (see Introduction). The terms 'inhabitation', 'colonization' and 'occupation' are generally used here to signify that ants *nest* on the plant, i.e. that ant brood has been found.

3. RESULTS AND DISCUSSION

The overview of Malayan myrmecophytes is presented in alphabetical order of plant families. Our comments on ant-plants recorded in the literature are included therein. At the end of each paragraph the occurrence of plant species in the 50 ha plot in Pasoh FR is reported according to Manokaran et al. (1992).

At first, however, we will deviate from this order of treatment by giving priority to the plant-ant genus *Cladomyrma* Wheeler, which inhabits many different plant species of several plant families. That way we avoid too much redundancy and emphasize the importance of 'host plant choice' by *Cladomyrma* ants.

3.1 *Cladomyrma* host plants

All *Cladomyrma*/plant associations are three-partner-systems, consisting of a plant, an ant and a phloem-sucking trophobiotic scale insect. In total 19 plant species from 9 different genera and 7 families are known to be inhabited by *Cladomyrma* ants (Agosti et al. 1999; Maschwitz et al. 1991; Moog & Maschwitz 1994, 1999, unpubl. results). The caulinary lodgings utilized by *Cladomyrma* vary among plant species from internodes with thickened nodes (*Saraca*) over 'intermediate' forms with swollen internodes (e.g. *Drypetes*, Fig. 2G) to markedly swollen domatia

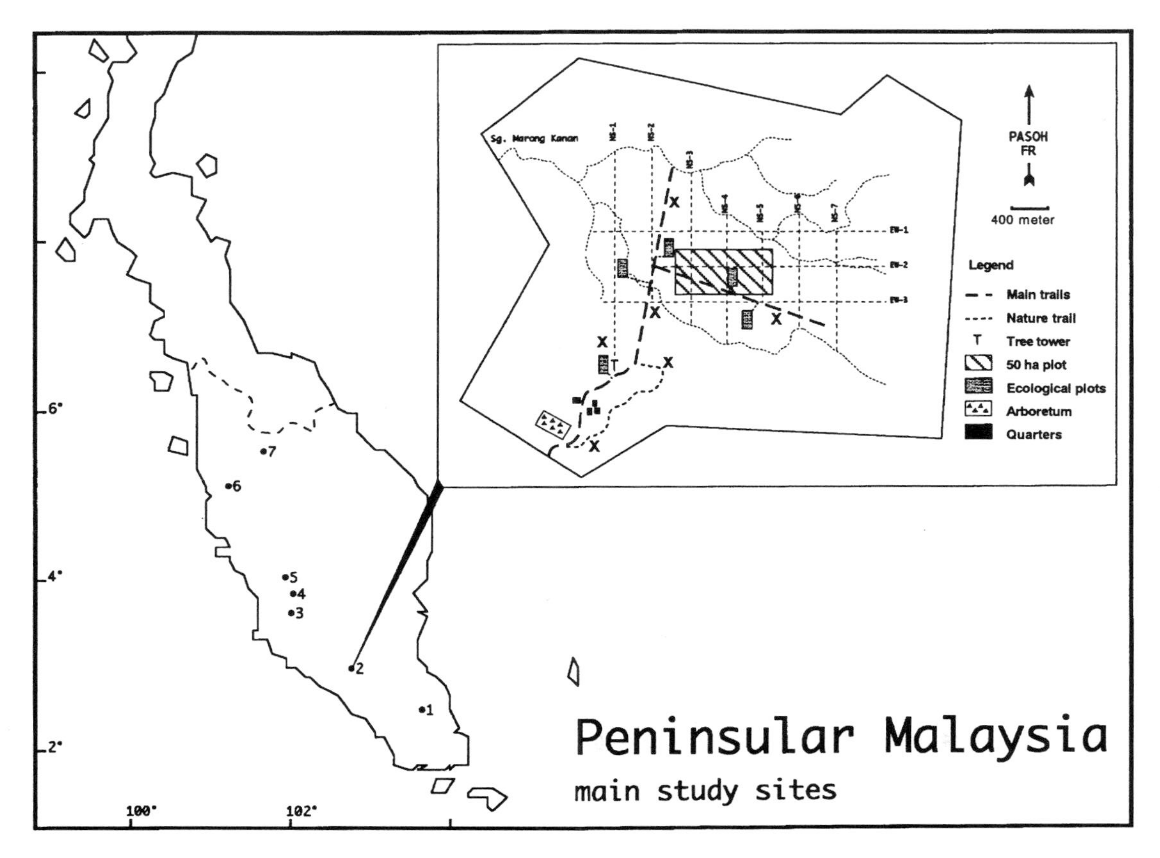

Fig. 1 Main study sites on Peninsular Malaysia: 1. Endau-Rompin NP (Johore); 2. Pasoh FR (Negeri Sembilan); 3. Ulu Gombak Valley (Selangor); 4. Genting Highlands (Pahang); 5. Bukit Fraser (Pahang); 6. Bukit Larut (Perak); 7. Belum (Perak). Inset: X indicate study sites outside the 50 ha plot in Pasoh FR (for clarity study sites within the 50 ha plot are omitted). Map of Pasoh FR modified from Lee 1995.

Fig. 2 Ant-domatia of Malayan plants (examples). Drawn by J. Moog (A = modified from Weissflog et al. 1999; E = modified from Huxley 1982). Scale bar = 1 cm. A) *Hoya mitrata* type: leaf cluster forming an ovoid multichambered cavity. B) *Zanthoxylum myriacanthum* type: naturally hollow stem with self-opening, slit-like entrance holes. The diagram shows one newly produced slit and an older one which is partly closed by callus growth except for a small roundish hole kept open by the ant inhabitants. C) *Pometia pinnata* type: lowermost pair of leaflets (pseudostipules) reflexed backward and clasping the stem thus forming a chamber with open margins which are later usually closed with ant carton. D) *Macaranga triloba* type: swollen stem domatia in a sapling c. 15 cm in height. Ant queens enter these inflated, naturally hollow internodes to found a colony. E) *Myrmecodia tuberosa* type: swollen hypocotyl forming a chambered tuber. The chambers may have smooth or warted surfaces. The ants usually keep their brood in the smooth chambers and place debris in the warted chambers. F) *Korthalsia hispida* type: the strongly elongated ocrea (= extension of the leaf sheath) diverges from the stem and forms a tube-like chamber with inrolled margins. Ant entrances are located at the base of the ocrea, just beyond the point of origin of the leaf petiole, and at the tip of the ocrea. These openings are later sealed by the ant inhabitants. G) *Drypetes longifolia* type: swollen twig with soft pith. In young plants each (slightly) swollen area stretches out over about two internodes. Ants hollow out these parts to construct nest cavities. The diameter of the intervening twig parts is still too small for ant colonization but in older plants the twigs are hollowed by ants throughout their length. H) *Spatholobus bracteolatus* type: the strongly swollen, pithy stems develop already in very young liana saplings whose natural primary stem diameter is far too small to allow ant colonization. In mature lianas the ants colonize the entire stem. I) *Korthalsia scortechinii* type: the ocrea is inflated and appressed to the plant surface. Ants gnaw holes to gain access into the interior.

(Bornean *Neonauclea*), indicating different levels of myrmecophytism. According to a new revision, 11 *Cladomyrma* species are recognized (Agosti et al. 1999). As a rule, colonization patterns are not strictly species-specific. Most host-plants have a small set of ant partners and vice versa.

A characteristic feature of all known species of *Cladomyrma* is the utilization of live pithy stems as nest sites: Colony founding queens gnaw entrance holes in suitable soft young internodes and excavate a chamber in which they rear their first brood in isolation. The initial founding chamber will be later expanded by the emerging workers (Maschwitz et al. 1991; Moog et al. 1998). Multiple colonizations of an individual plant sapling in different internodes are the rule, but eventually a single colony monopolizes the entire host plant. The monogynous colonies protect young foliage of their host plants against herbivores and prune young plant tips of encroaching vegetation (Moog et al. 1994, 1998, unpubl. data). All *Cladomyrma* species always tend coccoids inside the nest hollows and feed on honeydew excreted by their trophobionts (unpubl. results). The majority of the coccoid partners are Pseudococcidae (rarely Coccidae) and the involved taxa belong to a large variety of species e.g. from the genera *Paraputo*, *Planococcus*, *Pseudococcus* and *Maconellicoccus* (D. Williams, pers. comm.).

Cladomyrma is restricted in its distribution to the ever-wet part of the West Malesian floristic region, comprising the Malay Peninsula, Borneo and Sumatra. Only *C. maschwitzi* Agosti is known to occur in all three geographical areas. The centre of *Cladomyrma* evolution seems to be Borneo with 9 species whereas from Peninsular Malaysia only 3 species, inhabiting 7 host plant species, are recorded.

a) *Crypteronia griffithii* Clarke in Hook. f. (Crypteroniaceae)

Crypteronia griffithii is a Bornean, Malayan and Sumatran tree growing near tree fall gaps, between bamboo stands and along logging roads in primary and advanced secondary lowland forest, reaching a height of up to 40 m. The twigs are swollen at the nodes. Nearly 40% of saplings < 1 m were already found to be inhabited by queens of *Cladomyrma maschwitzi* (Moog et al. 1998). Trees of various size were checked for inhabitation by *C. maschwitzi* at different sites on the Malay Peninsula. Overall, colonization rate was 82.5% (99/120) and increased to 100% for plants ranging in height from ≧1 to 8 m (N = 86). In adult flowering trees (> 20 m) the association with *C. maschwitzi* seems not to be developed any more. A detailed account of the association is given by Moog et al. (1998).

Pasoh FR: *C. griffithii* is a rare tree in the 50 ha plot, with only 32 individuals (≧ 1 cm diameter) reported. However, in the buffer zone of regenerating forest surrounding the plot *C. griffithii* appears to be rather common. Along a transect of about 200m (following a path) 21 trees ranging from 1.1 to 8 m (median 2.6 m) in height were found. All plants harboured a colony of *C. maschwitzi*. It is the most common *Cladomyrma*/plant association in Pasoh FR.

b) *Drypetes longifolia* (Bl.) Pax & Hoffm. (Euphorbiaceae)

This small tree of primary and advanced secondary forest is regularly inhabited by *Cladomyrma* both in Borneo and in the Peninsular Malaysia (Moog & Maschwitz 1999). Stem swellings are obvious (Fig. 2G), especially in young saplings, and the pith tissue inside the domatia structures consists of white and soft parenchymatic cells compared to the rather hard and compact ones found in other stem parts. Ant workers of growing colonies enlarge their initial nest chambers by excavating also the less soft pith of non-dilated stem parts. In Peninsular Malaysia the two

Cladomyrma species, *C. petalae* and *C. nudidorsalis* Agosti, Moog & Maschwitz, are known to occupy *D. longifolia* at very high frequencies (92%, unpubl. results). However, we could not confirm regular ant inhabitation of *D. pendula* Ridl. (Maschwitz et al. 1992), which has also been recorded to be an ant-tree (Corner 1988). The description given by Corner (vol. 1, p.285) matches the colonization of *D. longifolia* by *Cladomyrma*, indicating that Corner probably did not distinguish between sterile young plants of *D. pendula* and *D. longifolia*. Examination of herbarium specimens of *D. pendula* corroborate the findings obtained by us from different localities on the peninsula that this species is not a myrmecophyte (unpubl. results).

Pasoh FR: *D. longifolia* is a not uncommon tree in the 50 ha plot, with 406 plants ($\geqq$ 1 cm diameter) reported. Most plants we came across were too high to be checked for ant inhabitation, however, in each of 2 small saplings we found the remains of a *Cladomyrma* queen in a colony-founding chamber, probably belonging to *C. nudidorsalis*. Again, we have no indications that *C. petalae*, usually an inhabitant of *D. longifolia* in other parts of its distribution, does occur in Pasoh FR.

c) *Saraca thaipingensis* Cantley ex Prain (Fabaceae-Caesalpinioideae)

In Peninsular Malaysia *Saraca thaipingensis* is a small and common tree in stream valleys forming a typical element of the Malayan flora. The twigs of *S. thaipingensis* do not provide swollen parts or 'weak spots' (so-called prostomata) to promote ant inhabitation and access, yet they have internodes gradually but distinctly increasing in diameter towards the nodes which support the big-sized pinnate leaves. Usually, the primary stem diameter of *Saraca* saplings (from 1 m onwards) is sufficient to allow colonization by queens of *Cladomyrma petalae* Agosti. The ant associate *C. petalae* reduces herbivore damage levels of young expanding leaves significantly (Moog & Maschwitz 1994, unpubl. results). The association has been found in all parts of the peninsula from southern Thailand to Singapore. Occupancy rates vary probably due to tree age, habitat and interspecific competition (e.g. Ulu Gombak: total 55%, and 88% in trees 1–10 m in height, N = 280). In small, restricted local habitats *C. petalae* may be replaced mainly by a particular *Crematogaster* species, but colony-founding queens of the latter lack the ability to tunnel nest chambers into their host, thus they rely on preformed shelters such as stipules or hollows made by stem-boring insects (Moog & Maschwitz 2000). Also, *Crematogaster* is not specific to a particular host plant but uses many riverside plant species in an opportunistic way. In contrast, *Cladomyrma* is always the primary colonizer of *S. thaipingensis* (unpubl. results).

Pasoh FR: 241 trees are recorded from the 50 ha plot. In the few trees checked by us in the field we could not establish the occurrence of *C. petalae* in Pasoh FR. However, existing evidence is yet too scarce to confirm a general absence of *C. petalae* in Pasoh FR. *C. petalae* has been observed in *Saraca* trees found in the Pasoh area outside the forest reserve.

d) *Spatholobus bracteolatus* Prain ex King (Fabaceae-Papilionoideae)

Two *Spatholobus* species are hosts of *Cladomyrma* ants. One is restricted to Borneo (*S. oblongifolius* Merr.) the other to Peninsular Malaysia; *S. bracteolatus* is a big woody climber of primary and old secondary forest, reaching a length over 30 m. The liana is colonized by *Cladomyrma petalae* at an early developmental stage (from 0.2 – 1.6 m in length) although the primary stem diameter is still far too small to harbour ant queens. However, young plants provide some stem internodes with distinctly enlarged diameter (domatia, Fig. 2H) to permit colonization by colony-founding queens (one plant, 2.4 m in length, already had produced 12 domatia). Later, in larger plants, the diameter of the stem is sufficient to allow the ants to excavate nest chambers. The swollen domatia are not seen in herbarium material since collectors typically ignore young plants and collect only flowering branches of mature plants. Occupancy rate of *S. bracteolatus* by *C. petalae* was found to be 98% (40/41); only one plant of c. 1 m length lacked domatia and hence could not be colonized by colony-founding queens.

Pasoh FR: *S. bracteolatus* appears to be restricted to elevations of c. 800 to 1300 m a.s.l. (pers. obs.) and thus it is most probably absent from the forest reserve.

e) *Ryparosa fasciculata* King (Flacourtiaceae)

Ryparosa fasciculata, endemic to the Malay Peninsula, is an uncommon and small tree, occasionally up to 20 m, with fairly long leaves and swollen nodes. The regular colonization by *Cladomyrma petalae* or *C. nudidorsalis* was both seen in herbaria and field specimens. Colony-founding *Cladomyrma* queens were already observed in the enlarged internodial apex of saplings < 0.5 m. Percentage of ant occupation was 90% (36/40), and 100% for plants > 0.62 m in height.

Pasoh FR: *R. fasciculata* is a rare tree with only 81 individuals in the plot. A survey of the plants in the plot has not been carried out.

f) *Strychnos vanprukii* Craib (Loganiaceae)

Strychnos vanprukii is a common woody climber in the Gombak area northeast of Kuala Lumpur, growing sympatrically with the other host plants *Saraca thaipingensis*, *Luvunga* sp. and *Ryparosa fasciculata*. This climber species takes up a special position among *Cladomyrma* hosts. Colony foundations and small colonies (up to several hundred workers) of *C. petalae* were observed in only about 36% of all plants (31/86, 0.65 – 6.6 m in length) and mature colonies with sexuals have not been found. Data from localities other than Gombak are lacking, nonetheless, the percentage of ant occupation excludes the possibility that colonization by *Cladomyrma* is an isolated and fortuitous event. The factors leading to such an interesting colonization pattern are not yet understood. Each of two *Strychnos* plants which climbed up *Saraca* trees were found to be inhabited by the same colony which nested in the respective *Saraca*. In both cases the queen was located in the *Saraca*, indicating that the ability to select among potential host plants is shown both by *Cladomyrma* queens and workers. Needless to say, other climber species that may occur on a host tree are not colonized by *Cladomyrma*. Although it is premature to account for the low frequency of *Strychnos* occupation by *Cladomyrma*, we hypothesize that the regularly observed 'erroneous' inhabitation expresses the present state of a host expansion process.

Pasoh FR: data on distribution not available. Not seen by us in the forest reserve.

g) *Luvunga* sp. Buch.-Ham. (Rutaceae)

Luvunga species, which are all very similar in general appearance, are woody vines that cling to forest trees by means of strong, recurved spines in the axils of trifoliolate leaves. *Luvunga* is a well-characterized, readily recognizable genus, but it is very difficult to identify sterile collections. We were not able to collect flowering material of the rare host plant of *Cladomyrma petalae* and, therefore, could not identify the plant to species level with confidence. Herbarium specimens which had typical *Cladomyrma* entrance holes in their stems (and apparently belonging to the same species) were identified and labeled as 3 different *Luvunga* spp. A critical taxonomic study of the genus is badly needed. Young plants with swollen internodes (domatia) are often colonized by *Cladomyrma* queens already at a height of only 0.2 to 0.5 m. The climbing habit of this *Luvunga* species is visible only after the plant reaches a height of about 2 meters. In addition, the leaves are unifoliolate, not trifoliolate prior to the same height. Occupancy rate of *Luvunga* by *C. petalae* was found to be high, with 78% overall (21/27, 0.2 to 8 m in length) and 100% for plants > 0.75 m.

Pasoh FR: no data available. We did not find the climber in the reserve.

3.2 Anacardiaceae

Lannea coromandelica (Houtthuyn) Merrill: According to the literature (Hoelldobler & Wilson 1990; Jolivet 1996) plants found in Vietnam have hollow stems inhabited by undetermined ants. It is unclear if these caulinary cavities are true domatia or just galls occupied anew by ants. *L. coromandelica* has been introduced to Peninsular Malaysia and is more or less naturalized (Hou 1978). We were able to check 6 herbarium specimens collected in Peninsular Malaysia ($N = 5$) and Borneo ($N = 1$) for traits of ant inhabitation. None of these specimens showed any signs of ant colonization.

Pasoh FR: not reported.

3.3 Annonaceae

Goniothalamus ridleyi King: Ridley (1910) reported that the flower masses at the base of the trunk are almost invariably covered by a nest of small black ants, which accumulate soil particles all over the inflorescences (usually before buds open) so that they are often quite concealed. The flowers are thus inaccessible to other pollinators like bees and butterflies, yet the plants fruit regularly and heavily. The ants are attracted by the floral nectar, and Ridley believes that the ants take part in the pollination of the flowers. We have not seen the plant in the field but evidently *G. ridleyi* is not a myrmecophyte.

Pasoh FR: not reported.

3.4 Dipterocarpaceae

Shorea acuminata Dyer produces pairs of large semi-persistent stipules at the tips of young shoots. They are often utilized as nesting sites, mostly by *Technomyrmex* sp. (Maschwitz et al. 1992; Tho, Y. P. & Rickson, F. R., pers. comm.). The ants build carton structures around the stipules and thus utilize them as a frame for their nest construction. It appears that the ability of ants to build carton structures is a prerequisite for establishing nesting sites under the stipules of *S. acuminata*. The ant *Technomyrmex* sp. is most probably specialized to utilize such supporting plant structures rather than being specialized on *S. acuminata* itself, since we have observed this ant species colonizing the foliage of several other tree species. A similar use of preformed, leaf-derived shelters by various ant species is commonly observed, e.g. in *Diospyros* and *Pometia* (see Table 1).

Pasoh FR: All observations on the relationship of *S. acuminata* with ants were made on a few trees ($N = 14$) from one locality only, i.e. in Pasoh FR. Therefore, we have not enough grounds to conclude that *S. acuminata* is an ant-plant or not. In the 50 ha plot the tree is very common and widely distributed (> 2,000 individuals). The plot is perfectly suited to study the ecological nature of the *Shorea*-ant relationship in necessary detail.

3.5 Euphorbiaceae

Agrostistachys sessilifolia (Kurz) Pax & Hoffm. is a shrub or treelet with leaves set in a dense tuft at the apex of the stem, with only very short sections of the stem visible between. According to Corner (1988) this species "is, in a sense, an ant-plant, not that its twigs are hollow and tenanted by ants ..., but that small black ants habitually make their nests on the outside of the stem among the close-set leaf-bases and fiercely resent interference with their abode." The Tree Flora of Malaya (Whitmore 1973) reduces *A. sessilifolia* to *A. longifolia* (Wight) Benth., but both are recognized by Airy Shaw (1975). In the experience of Corner the two species are readily distinguished and "this confusion may account for the disagreement over the ants which so often infest *A. sessilifolia*." However, a footnote in Corner (1988) casts doubt on his estimation of this plant species being an ant-plant: In one location (Sg. Menyala Forest Reserve) ants were found by Ng, F. S. P. only once, and on a second trip to the same site "there were no ants at all despite an intensive search on many plants of this species." We think that ants inhabiting the dense leaf-base clusters of *A. sessilifolia* only take advantage of a structure generally suitable for nesting, thus leading to a both temporal and spatial highly variable colonization pattern by opportunistic ants.

Pasoh FR: not reported.

Macaranga is the world's largest genus of pioneers (Whitmore 1984) and also the plant genus in the Oriental and Australian region with the greatest radiation of myrmecophytes (with 23(+) myrmecophytic species, only rivalled by the rubiaceous genus *Neonauclea* with about 17 myrmecophytic species in the Malay Archipelago). *Macaranga* is probably also the best-known myrmecophytic system of SE Asia. For more than ten years, our research teams have been studying a variety of aspects of these ant-plant symbioses (reviewed in Fiala 1996 and Fiala et al. 1999, see references therein). Most occupants are ants of the genus *Crematogaster* (subfamily Myrmicinae). What appeared at the beginning of studies on a few *Macaranga* species to be a single ant species, named *Crematogaster borneensis* André, has since turned out to be a large number of similar species. With one exception they belong to the subgenus *Decacrema*. The ants are at present being described but, since this is still incomplete, we will use morphospecies numbers here. Besides the associations with the dominant *Crematogaster*, only few *Macaranga* species are known to be inhabited by specialized ants from other taxa (all from the subfamily Formicinae, e.g. Federle et al. 1998a,b; Maschwitz et al. 1996b). Our recent investigations show that this myrmecophytic system is much more complex concerning life types and species diversity than previously supposed (review of different life types in Fiala 1996, for ant colonization Fiala et al. 1999). Here we can only present a very brief summary of the basic features of the *Macaranga-Crematogaster* associations:

Macaranga provides nesting space in the form of hollow or easily excavatable stems (Fig. 2D) and nutritious food bodies, small epidermal emergences containing lipids, carbohydrates and amino acids (Fiala & Maschwitz 1992b). The ants protect their *Macaranga* host against herbivores and plant competition. They are so narrowly specialized to living on *Macaranga* that they perish soon if no such plant is available (Fiala & Maschwitz 1990). There is an additional, third group of partners in *Crematogaster-Macaranga* symbiosis: more than 20 species of honeydew-producing scale insects, found nowhere else, are cultivated by the ants inside the *Macaranga* stems (Heckroth et al. 1998). The centre of evolution of myrmecophytic species seems to be Borneo whereas from Peninsular Malaysia only 9 ant-inhabited species are recorded (see Table 1, the remaining *Macaranga* species are not colonised by ants [non-myrmecophytes] but they often attract ants by extrafloral nectaries and food bodies).

In one species, *M. caladiifolia* Beccari, the domatia open by themselves (in contrast to all other myrmecophytic *Macaranga* species) and provide access to a great number of unspecific, opportunistic arboreal ants (Fiala et al. 1996). These observations were obtained from Bornean plants and it is unclear if they hold for Malayan plants as well. Whitmore (1973, 1975) reported *M. caladiifolia* to occur on the peninsula but also regarded this species as synonymous with Malayan *M. puncticulata* Gage, though the two species differ markedly in their myrmecophytic characters (Federle et al. 1998b, see Table 1). Hence it remains unclear if Whitmore's Malayan *M. caladiifolia* is actually *M. puncticulata*. We did not yet find *M. caladiifolia* on the peninsula and thus the occurrrence of *M. caladiifolia* there appears to be questionable. As a first step towards elucidating the co-evolution of ant-plant interactions in the *Macaranga-Crematogaster* system, we have initiated a molecular investigation of the plant partners phylogeny (Blattner et al. 2001).

Pasoh FR: In Pasoh FR and surroundings we found 7 *Macaranga* species, of which 3 were myrmecophytic. Most abundant in the 50 ha plot was the myrmecophilic (but non-myrmecophytic) species *M. lowii* King Ex Hook. f. (N = 2108) which has, as the following non-myrmecophytic species, extrafloral nectaries that attract unspecific ants (for details see Fiala & Saw, Chap. 32). As could be expected for a primary forest, all other species occurring in the plot were rather rare: *M. recurvata* Gage (N = 79), *M. conifera* (Zoll.) Muell. Arg. (N = 40) and *M. gigantea* (Reichb. f. & Zoll.) Muell. Arg. (N = 5). The latter species, however, was very abundant outside the forest, especially small saplings were found in a high number throughout the buffer zone. The large stipules especially of young *M. gigantea* were occasionally used by unspecific ants as nesting site.

The three myrmecophytic *Macaranga* species were not found in the closed forest itself but in (mostly) larger gaps. *M. motleyana* Muell. Arg. subsp. *griffithiana* (Muell. Arg.) Whitmore, which can become very gregarious outside the forest, was only found with 3 saplings in the regenerating forest at buffer zone and never in the 50 ha plot. Also *M. hosei* King ex Hook. f. was rather rare occurring only in larger gaps (N = 8, one of the nine mentioned in Manokaran et al. 1992 was actually a *M. hypoleuca*). Despite thorough search we have found only three saplings of *M. hosei* in the plot, whereas in the regenerating buffer zone in total 41 saplings could be found. The only really abundant myrmecophytic species in the primary forest was *M. hypoleuca* (Reichb. f. & Zoll.) Muell. Arg. (N = 131 in the plot). This species was also mostly found in gaps but is distributed throughout the whole 50 ha plot. Small saplings were rather common.

The consequences of isolation of gaps for the colonization ability of *Crematogaster (Decacrema)* ants were difficult to evaluate. The density of gaps in Pasoh FR at the time of our study was surprisingly high and therefore the distance between the gaps was not very large (max. 300 m). We could find and check 43 gaps for presence of *Macaranga* saplings. 17 small gaps were without any *Macaranga* plants. From 91 *M. hypoleuca* plants monitored (0.1–4 m) only 7 were not inhabited by ants (all smaller than 50 cm). Also very small saplings in dense undergrowth in gaps were already inhabited and no plant > 80 cm was found without ants. *M. hypoleuca* was 89% inhabited by its specific partner *Crematogaster*

(Decacrema) msp. 6 and rarely (mainly in large trees) by msp. 1.

In contrast, only 30% (13/41) of the *M. hosei* plants (0.2 – 4 m) were inhabited (always by queens of *C. (Decacrema)* msp. 2 which is the typical colonizer of this species). This is, however, probably caused by intrinsic characteristics of this species where a stem that is too thin prevents earlier colonization, rather than due to the inability of ants to find the small plants (Fiala & Maschwitz 1992a). The 3 persistent large trees in the plot and the 4 saplings found at the edge of the plot were all inhabited despite their very isolated location. Only 5 saplings of *M. motleyana* ssp. *griffithiana* could be found in the forest. Of these only one was occupied by a queen of *C. (Decacrema)* msp. 2 which is not the main colonizer of this species (msp. 1).

It was astonishing that even at rather isolated sites small saplings of *M. hypoleuca* and *M. hosei* growing side by side were indeed colonized by their specific ant partners. Further studies at other sites where more than 5 species occurred closely together (such as in Lambir, Sarawak) have supported these results (Fiala et al. 1999): specificity of colonization in myrmecophytic *Macaranga* is also maintained at rather isolated sites, such as forest gaps, despite rather distant and hidden occurrence of individuals. This requires differentiated and very effective mechanisms of host plant location and points to use of chemical cues. The 50 ha plot in Pasoh FR with its grid of trails allows easy orientation and would offer good opportunities to study experimentally influence of distances and isolation of plants for host plant finding.

3.6 Fabaceae (Mimosoideae)

The genus *Archidendron* contains several New Guinean myrmecophytic species, but only two Malaysian species, *A. clypearia* (Jack) Nielsen and *A. ellipticum* (Bl.) Nielsen, have been reported to have branches that are sometimes hollow and inhabited by ants (Federle 1998; Gossner 1999; Nielsen 1992). Herbarium specimens of *A. ellipticum* examined from both Malay Peninsular ($N = 30$) and Borneo ($N = 26$) showed no regular signs of ant colonization. However, 14% of herbarium specimens of Malayan *A. clypearia* (Jack) Nielsen possessed internodes with short hollows and ant entrance holes. In one chamber the remains of a *Crematogaster* species were found. Field observations in a peat swamp forest (Sg. Buloh) revealed that a few branches of several trees are regularly attacked by lepidopteran larvae which produce short stem borings (c. 2 cm long) which were later occupied by *Crematogaster*, *Tapinoma*, and an unidentified myrmicine species. We suspect that ant workers (of some species) are able to enlarge the preformed cavities since the pith of the twigs is of a rather soft texture. The ants, probably attracted by the extrafloral nectaries at which they feed, used these chambers only as shelters; we never found brood inside ($N > 20$ chambers). If the specimens we examined are representative, this plant is myrmecophilic but not myrmecophytic. Trees from other localities should be studied to verifiy these findings.

Pasoh FR: Only 75 individuals of *A. clypearia* were reported from the 50 ha plot. A few plants of a very rare third species, *A. contortum* (Mart.) Nielsen, were found in Pasoh FR. Again, some branches had hollow cavities containing *Crematogaster* workers, but this appeared not to be a regular phenomenon.

3.7 Lauraceae

The ‘ant laurel’ *Actinodaphne sesquipedalis* Hk. f. & Thoms. ex Meisn. has buds which are covered by a bunch of large green scale-leaves resembling under-sized foliage leaves. Corner (1988) reported that these large leaf-like scales "are usually inhabited by ants which find them convenient places to build their nests." No further details are provided on the identity of ants or on colonization rate. This poorly known fact needs to be verified anew.

Pasoh FR: 118 individuals in the 50 ha plot.

3.8 Meliaceae

In his monograph of *Aglaia* (Meliaceae), Pannell (1992) mentioned that all the trees of *Chisocheton tomentosus* (Roxb.) Mabberley which he examined at Kuala Lompat, Krau Game Reserve, had hollowed twigs and petioles in which there were colonies of 'black ants' with larvae. In New Guinea a number of *Chisocheton* species are regularly ant-inhabited (Stevens 1978) but ants had not been recorded living inside the twigs of *Chisocheton* species in Malaysia. We have no personal data on ant-occupation of *C. tomentosus* but our experience with *C. ceramicus* (Miq.) DC. both in the field and from herbarium specimens indicates that association with ants of *C. tomentosus* might be loose and fortuitous (as in *C. ceramicus*), mediated through occasional hollows made by insect larvae (see also *Archidendron* above).

Pasoh FR: *C. tomentosus* is not an uncommon small tree in the 50 ha plot (nearly 200 individuals), hence we recommend a survey there to clarify the relationship of this species with ants.

3.9 Palmae

Spiny climbing palms are a characteristic element of Malaysian rain forests. These scandent species of the Old World palm tribe Calamae are known as rattan palms. The rattan habit differs considerably from the growth form we usually associate with palms–an upright stem with a terminal leaf crown. In high-climbing rattans the sequentially arranged tube-like leaf sheaths cover both a strongly elongated stem and, partially, the following sheaths, so that the construction creates the impression of a flexible, extended antenna stretching through the vegetation.

Occurrence of ants on rattans is a very common phenomenon, though often casual, and several palm structures appear to be directly related to associations with ants (Beccari 1884 – 1886; Dransfield 1979). External ant domatia have evolved from leaf sheath structures in the three largest and most widespread rattan genera, *Calamus*, *Daemonorops*, and *Korthalsia*. Three types of domatia can be distinguished: (i) interlocking spine combs forming horizontal galleries around the leaf sheath (*Calamus*, *Daemonorops*), (ii) inflated extensions (swollen ocreas) of the leath sheath (*Korthalsia*, Fig. 2I), and (iii) elongated, tube-like ocreas diverging from the stem (*Korthalsia*, Fig. 2F).

In Peninsular Malaysia a total of 9 so-called ant-rattans are known (4 *Korthalsia* species, 1 *Calamus* species, 4 *Daemonorops* species, see Table 1). At least eight more species are recorded mainly from Sundaland (Beccari 1884–1886; Uhl & Dransfield 1987). Present knowledge indicates that all 17 of these ant-rattan species–even at the geographical borders of their distribution–are colonized at very high frequency and intensity by about 16 species of specialized plant-ants. Fourteen of these belong to the genus *Camponotus* Mayr subgenus *Myrmoplatys* Forel; the subgenus of the remaining 2 species is still uncertain (M. Werner, unpubl. results). The systematic status of *Myrmoplatys* is currently under investigation by one of the authors (MW), and it is unclear whether this taxon is monophyletic. The species of *Camponotus* (*Myrmoplatys*) can be grouped in three lineages which are strictly linked to domatia type (spine galleries, inflated and diverging ocreas) provided by the palms (M. Werner, unpubl. results). In all ant-rattan associations the ants tend within the domatia trophobiotic aphids (Hormaphididae, incl. *Cerataphis* spp.) whose honeydew serves as the principal food source since the rattan palms do not provide extrafloral nectar or food bodies for their ant partners.

These 'truly' myrmecophytic associations account for just a fraction of the relationships rattans form with ants. As mentioned above, ants are frequently observed occupying rattan species. Several unspecialized or only slightly modified structures generally favor their utilization by ants. (i) The inflorescences of some *Daemonorops* and *Ceratolobus* species are (partially) enclosed by persistent bracts and this space is well suited to be utilized as a nest site (Ridley 1910, unpubl. results). (ii) Small, narrow extensions of the leaf sheath, called auricles, have been recorded from the genus *Pogonotium*. These slightly convex structures, closely appressed to the stem, form tunnels which can only be inhabited by minute ants (Dransfield 1979, 1980). (iii) Sessile leaflets may be reflexed to form a chamber in which ants either nest and/or tend aphids (*Calamus laevigatus* Martius, *C. javensis* Blume). In the latter two species a large variety of ant species readily use these shelters. However, none of the ants found belongs to species known to be specialized on myrmecophytic rattans (Werner 1993).

Pasoh FR: The forest reserve is home to a considerable portion (56%, 5/9) of the ant-rattans of Peninsular Malaysia. About half of the known *Myrmoplatys* species (6/14) has been collected from Pasoh FR and its vicinity. Apparently the most common ant-rattan species is *Daemonorops verticillaris* (Griff.) Mart., followed by *Korthalsia scortechinii* Becc. Of the remaining species, *Calamus polystachys* Becc. and *Daemonorops macrophylla* Becc., were each found once, whereas the occurrence of *D. sabut* Becc. in Pasoh FR is known to us only from a herbarium specimen.

3.10 Poaceae

Giant bamboos, e.g. *Gigantochloa scortechinii* Gamble, *G. ligulata* Gamble, *G. thoii* Wong and a species of *Schizostachyum* Nees, have been reported to be inhabited by ant species specialized on bamboo (Buschinger et al. 1994; Dorow & Maschwitz 1990; Klein et al. 1992; Kovac 1994; Schellerich-Kaaden et al. 1997b). Entrances into the remarkably hard and spacious hollow culm internodes are produced by the activities of wood-boring beetles, woodpeckers or mammals (Kovac & Streit 1996). These cavities and other bamboo structures (shelters formed by branch and leaf insertions) offer a readily accepted microhabitat for a large variety of animals, especially ants (e.g. Kovac & Streit 1996; Schellerich-Kaaden et al. 1997a). In Ulu Gombak (Selangor) more than 80 different ant species were observed nesting on *Gigantochloa scortechinii* or feeding at extrafloral nectaries of young shoots (A. Schellerich-Kaaden, unpubl. results).

Some of these bamboo-dwelling ants have never been found away from giant bamboo and, therefore, are assumed to be bamboo specialists. They include *Tetraponera binghami* Forel (= *T.* sp. PSW-80 in earlier publications: Buschinger et al. 1994; Klein et al. 1992; Kovac 1994), *Polyrhachis arachne* Emery, *P. hodgsoni* Forel (Dorow & Maschwitz 1990), *P. schellerichae* (Dorow 1996; Schellerich-Kaaden et al. 1997b) and *Cataulacus muticus* Emery (Maschwitz & Moog 2000; Maschwitz et al. 2000). The ants keep mealybugs (Pseudococcidae) inside the internodes (*Tetraponera binghami, Polyrhachis schellerichae, P. arachne,* in part *Cataulacus muticus*) or cultivate pseudococcids and aphids in silk pavilions constructed on bamboo leaves (*Polyrhachis arachne, P. hodgsoni*). Colony-foundresses of the culm-inhabiting species are not able to force an entry into the internodes, hence they rely on preformed holes to gain access into the hard-walled bamboo stems. Only the workers of *Tetraponera binghami* have the ability to chew holes into the culms of young bamboo shoots; and this laborious work takes up to 3 to 5 days (Kovac 1994).

No evidence indicates that these 'bamboo-ants' benefit their host e.g. by deterring or killing herbivores. In particular, *Tetraponera binghami, Polyrhachis schellerichae* and *Cataulacus muticus* show extremely low levels of foraging or patrolling activity on the plant surface and, while they possess morphological and behavioural traits connected to living in bamboo, are looked upon as being more likely parasitic than mutualistic associates of giant bamboo (Buschinger et al. 1994; Maschwitz et al. 2000; Schellerich-Kaaden et al. 1997b). On the plant side, no mutualistic adaptations such as the providing of specialized nesting space (domatia) could be recognized, except for the extrafloral nectaries on culm sheath auricles (Schellerich-Kaaden & Maschwitz 1998). Apparently the association between these 5 ant species and giant bamboo is a case of one-sided specialization.

Pasoh FR: not reported. The authors have not made observations on bamboos in Pasoh FR.

3.11 Rubiaceae

During our field studies in Peninsular Malaysia we repeatedly came across 1 or 2 different climber species of the genus *Uncaria* Schreb., which we did not identify to species level. They both appeared to be regularly attacked by beetle larvae of the genus *Anadastus* sp. (Languriidae; det.: Richard Leschen). Various ant species, including *Crematogaster* spp., established nest sites in the stem hollows produced by the shootborer. Only one ant species, *Camponotus* (*Colobopsis*) sp., regularly occupied *Uncaria* lianas without being dependent on the languriid beetle for stem access. The workers were able to penetrate into young internodes and to construct continuous nest cavities by perforating the nodal septa. Young *Uncaria* climbers were rarely inhabited by ants whereas older, mature lianas regularly harboured ants in at least some sections of their stem. However, these casual observations are not sufficient to evaluate how tight the *Uncaria*/ant relationship is. A thorough investigation of this phenomenon has still to come.

Pasoh FR: We have observed ant-occupied *Uncaria* climbers in the regenerating forest buffer surrounding the core area. Data on *Uncaria* are lacking for the 50ha plot (Manokaran et al. 1992 list tree species only).

3.12 Sapindaceae

This plant family is known to include several genera with myrmecophytic species, e.g. *Alectryon*, *Guioa*, *Harpullia* and *Sarcopteryx* (Adema et al. 1994), with all ant-plant species occurring in New Guinea. The branches are usually swollen below the nodes and ant nest openings can be found in the swellings. Malaysian members of the family which (sometimes) exhibit associations with ants, *Lepisanthes* and *Pometia*, possess ant shelters different from those observed in New Guinea: the lowermost pair of leaflets is reduced in size and reflexed back to the twig, thus forming a shelter covering twig and petiole base (Fig. 2C). While only occasional observations on the ant association with *Lepisanthes* exist, we can present preliminary results of a study on the relationship between *Pometia pinnata* Forst. forma *glabra* (Bl.) Jacobs and ants. Overall, 63% of all trees studied (60/95, tree height 0.3 – 20 m) in Ulu Gombak, Krau Game Reserve and Pasoh FR harboured ants of a large variety of species in their pseudostipules. The ants partly used these shelters not as nesting sites but for feeding at extrafloral nectaries or tending trophobiotic scale insects. In older plants (> 7 m in height) ant occupancy rate increased: virtually every tree was occupied (42/43), and proportion of trees harbouring nesting ants reached about 70% (30/43). The reason for increased ant frequencies on older plants is the rather 'late' development of ant-sheltering structures. Only from a height of about 4 m onwards did young trees regularly begin to produce pseudostipules large enough to house ants. Nevertheless, young plants of less than 0.5 m in height already attract ant visitors, due to the extrafloral nectaries active on developing leaves. First analyses of data indicates that herbivore damage can be significantly reduced if ants are present on both saplings and older trees of *P. pinnata*. Degree of protection, however, clearly varied greatly among alternative ant species.

Pasoh FR: *P. pinnata* f. *alnifolia (*Bl.) Jacobs is common and widely distributed in the 50 ha plot, but this form of *P. pinnata* lacks the stipule-like leaflets. Since we have not searched for this form in the plot itself we do not know if they harbour ants. However, trees of forma *glabra* (all > 5 m in height) examined by us in the vicinity of the field station had a mean ant occupancy rate of 80% (8/10), i.e. the pseudostipules were utilized both by nesting and coccoid-tending ants.

3.13 Scrophulariaceae

Wightia borneensis Hook. f.: In Peninsular Malaysia the plant species has been found in hill forests of Perak, Pahang and Melaka. Jolivet (1996) reported that the upper swollen internodes are hollow and colonized by undetermined ants. The cigar-shaped cavities are possibly excavated by the ants. The status of these plant structures as true domatia is not established (Hoelldobler & Wilson 1990). Herbarium specimens not seen.

Pasoh FR: not reported.

3.14 Urticaceae

Jolivet (1996) reports that the stipules and capitulate inflorescences of *Poikilospermum suaveolens* (Bl.) Merr. usually harbour ants, e.g. *Dolichoderus* sp. (*thoracicus* group). In his estimation the plant has a myrmecophilic disposition, but no details are given on the nature of the association with ants. *P. suaveolens* has nectaries at the base of the inflorescences and spacious (semi-)persistent stipules (Chew 1963, own obs.) suitable for inhabitation by carton-building ants which are able to close the remaining gaps between stipule margins and plant surface. And indeed, nearly all plants (6/7) we came across in the forest had at least a few of the stipules occupied by ants. They belonged to the genera *Crematogaster*, *Dolichoderus* and *Technomyrmex*. However, in five plants we found no ant brood inside the stipules but scale insects tended by ants. The ants thus used the stipules merely as shelters and not as nesting sites. At present the relationship of *P. suaveolens* with ants reminds us of some sort of 'self-service restaurant'. In Malaya a second species, *P. microstachys* (Barg.-Petr.) Merr., with persistent stipules has been reported by Federle (1998) to be inhabited by various ant species. Of the shoot apices examined 28% (14/50) harboured ants. The author also observed food bodies on the leaf lamina of *P. microstachys*. Further studies are needed to clarify the nature of the *Poikilospermum*/ant association.

Pasoh FR: Though *P. suaveolens* is present in Pasoh FR (pers. obs.), especially in regenerating forest, we cannot provide data on its abundance and the regularity of ant colonisation in the 50 ha plot.

3.15 Verbenaceae

Two Malayan *Clerodendrum* species, *C. breviflorum* Ridley and *C. phyllomega* Steud. var. *myrmecophilum* (Ridley) Moldenke, were reported as myrmecophytes (Ridley 1910). A third species, *C. deflexum* Wallich, was suspected by Maschwitz et al. (1994a) to display myrmecophytic characters as well. Since then we have been able to check the plant in the field (Ulu Gombak and Labis Forest Reserves). It is a small, little-branched or unbranched shrub, rarely exceeding 1.5 m in height, whose internodes are mostly hollow due to a nearly complete pith disintegration, leaving only thin, horizontal lamellate discs. The internodes often exhibit a slight increase (of c. 1 mm) in outer diameter towards the node. Usually the internode is rounded at the base and gradually becomes somewhat quadrangular towards the apex. Between the hollow stem sections sometimes short, solid internodes, with a slight decrease in diameter (0.5–1 mm), are inserted. In contrast to Bornean *C. fistulosum* Becc., *C. deflexum* does not develop slit-like self-opening holes. However, in total we found 46 plants of which 20 (43%) had entrance holes leading into the hollow internodes. 30% ($N = 14$) harboured ants inside, principally *Crematogaster* (3 spp.) and 2 unidentified myrmicine species. Additional examination of herbarium specimens of *C. deflexum* (Rijksherbarium Leiden and FRIM) corroborate the findings obtained in the field. Typical ant entrance holes were detected in 59% (10/17) of the specimens.

Ants are not the only tenants of *C. deflexum*: a beetle larva was observed feeding on the tissue inside a very young internode whose pith had not yet collapsed. In other *Clerodendrum* species, *C. phyllomega* var. *myrmecophilum* and an unidentified species from Borneo, we found very similar larvae, some pupae and freshly hatched adults. These shootborers were identified as members of the beetle family Languriidae (det.: M. Schoeller, R. Leschen) which are also known to hollow *Uncaria* stems (see above). Of the Bornean *Clerodendrum* plants found (undet. species), 88% (15/17) were inhabited by ants and 35% (6/17) had languriid larvae or adults in the internodes. Present evidence therefore favors the hypothesis that beetle borings serve to promote ant inhabitation in several *Clerodendrum* species by providing holes, hence access to long-termed nesting sites for ants.

Pasoh FR: *C. deflexum* is reported to occur in considerable numbers in the 50 ha plot (170 individuals). Thus future field studies might usefully be carried out in Pasoh FR. Herbarium specimens of *C. brevifolium* ($N = 4$), reported to be usually inhabited by ants (Ridley 1910), showed hollow internodes but no signs of former ant inhabitation. At present, the status of *C. brevifolium* as ant-plant remains uncertain.

Clerodendrum is not the only genus in the family that has been reported to include ant-plants in southeast Asia. The moderate-sized tree *Teijsmanniodendron pteropodum* (Miq.) Bakh. (= *Vitex peralata* King) has leaf-stalks which develop long wing-like appendages on each side, the free margins of which curl up to either side of the petiole and form a channel-like chamber with a more or less open 'roof'. The wings or flanges of the petiole are so large that leaves seem to be joined across the twig, and Corner (1988) observed that "in the sheathing base, thus formed, small biting ants, apparently identical with those which inhabit the twigs of the Mahang trees (*Macaranga*), make their nests so that this *Vitex* (= *Teijsmanniodendron*) must be regarded as one of our ant trees". Certainly, the ant species mentioned by Corner (1988) is not a typical *Crematogaster* (*Decacrema*) species of the group that inhabits exclusively several *Macaranga* species (see above). However, Corner(1988) may be correct in assuming that the species observed belonged to the genus *Crematogaster*. Many species of these arboricolous ants opportunistically nest in such plant structures. Data on occupation rate were not provided by Corner (1988) but the 'nature' of this plant structure used as an ant shelter points to a rather 'loose' association with facultative ants.

Pasoh FR: not recorded.

3.16 Violaceae

Rinorea javanica (Bl.) Kuntze is listed as a Malesian ant-plant by Balgooy (1997), but no details on plant structures used for ant housing are given. On all 12 herbarium specimens checked (11 from Peninsular Malaysia, 1 from Borneo) we were unable to detect any myrmecophytic traits.

Pasoh FR: not recorded.

3.17 Ant-Epiphytes

'Ant-garden' epiphytes, which will not be treated here in detail, are far more diverse and common in Peninsular Malaysia than previously thought. They include several species of *Dischidia* (Asclepiadaceae), e.g. *D. albida* Griffith, *D. hirsuta* (Bl.) Decaisne, *D. nummularia* R. Br., *D. parvifolia* Ridley, probably 5 species of *Hoya* (also Asclepiadaceae), *H. endauensis* R. Kiew, *H. elliptica* Hook. f., *H. lacunosa* Bl., *H. obtusifolia* Wight, *H. pusilla* Rintz; several species of *Aeschynanthus* (Gesneriaceae) and *Pachycentria* (Melastomataceae), a few orchid species, e.g. of the genera *Dendrobium* and *Liparis*, some ferns, and an ericaceous species, *Vaccinium bancanum* Miq. (Clausing 1998; Kaufmann et al. 2001; Kiew & Anthonysamy 1987, 1995; Kiew 1989; Rintz 1978).

Ant-epiphyte associations may arise when ants colonize the root system of epiphytes or retrieve attractive epiphyte seeds and carry them into their nest. This latter character is usually regarded as the condition for the use of the term 'ant-garden' epiphyte, i.e. it comprises ant-dispersed epiphytic angiosperms growing from arboreal carton ant nests (Davidson & Epstein 1989). Seed dispersal mechanisms of Malayan ant-epiphytes are still poorly documented, but forthcoming studies will greatly expand our knowledge of Malayan ant-associated epiphytes (Kaufmann et al. 2001; Weissflog & Maschwitz in prep.). Ant occupancy of epiphytes sometimes is encouraged by ant attraction to nectaries or other plant-derived food sources, such as nutritious appendages of seeds. According to the observations of E. Kaufmann and A. Weissflog (pers. comm.), the main ant colonizers of Malayan ant-gardens belong to the ant genera *Camponotus*, *Crematogaster*, *Pheidole* and *Philidris*.

The so-called 'ant-house' epiphytes form another substantial element of the ant-associated epiphyte community. In contrast to ant-garden epiphytes, however, these plants provide ant nest cavities which are derived from a diversity of plant structures such as fleshy tubers, rhizomes and leaves (see Table 1). In Peninsular Malaysia they include the well-known rubiaceous species *Myrmecodia tuberosa* Jack, *Hydnophytum formicarum* Jack, a few *Lecanopteris* fern species and several members of the family Asclepiadaceae from the genera *Hoya* and *Dischidia* (see Table 1). In 'ant-house' *Dischidia* species (*D. albiflora* Griffith, *D. astephana* Scort. ex King & Gamble, *D. cochleata* Blume, *D. complex* Griffith, *D. imbricata* (Bl.) Steud. and *D. longepedunculata* Ridley) the ants sometimes may

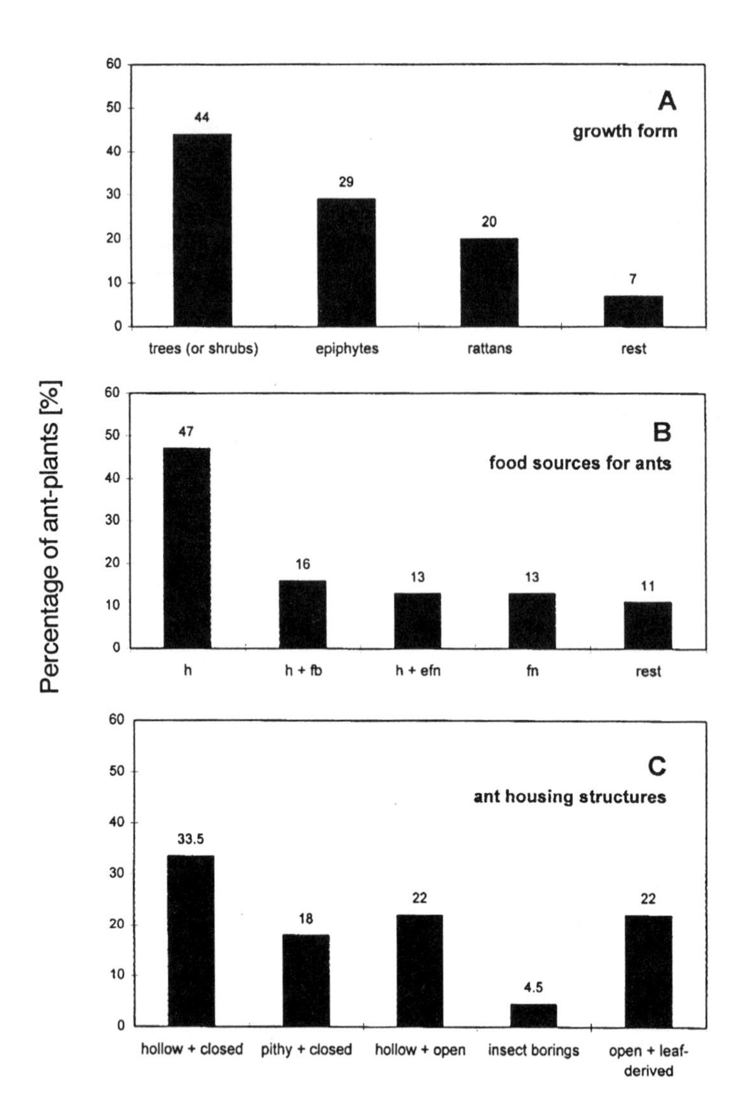

Fig. 3 Percentage of 'true' ant-plant species (N = 45) from Peninsular Malaysia according to (A) growth form, (B) food sources for ants (single or in combinations), and (C) ant housing structures. A: rest includes the growth forms 'lianas' (2x) and 'hemiepiphytes' (1x). B: h = honeydew, efn = extrafloral nectar, fb = food bodies, fn = floral nectar, rest includes rare combinations of food sources such as efn + fb (2x), efn + fb + h (1x), fn + h (1x), or unknown food sources (1x). C: ant housing structures are grouped into 5 categories according to 'ant accessibility' (see text): hollow + closed stem domatia, pithy + closed stem domatia, hollow + open stem domatia, cavities and entrance holes created by twig-borers other than ants (insect borings), and open + leaf-derived domatia. For details see Table 1.

nest not only in or under the modified leaves but also in the epiphytes root system or in bark crevices or twig hollows of the host tree.

Pasoh FR: We are not aware of studies on epiphyte diversity in Pasoh FR. With the exception of *Platycerium coronarium* (Koenig) Desv., which is a common fern on Pasoh FR trees, we cannot provide valid data on ant-epiphyte distribution from this location. However, a few ant-associated species are definitely absent from Pasoh FR due to their restricted altitudinal range. They include the montane species *D. astephana*, *D. longepedunculata* and *Lecanopteris pumila* Blume.

4. CONCLUSIONS

4.1 Diversity of plants

The evidence available at this time only allows a rough estimate of the true number of regular ant-plant associations in the area. Our overview shows that at least 45 'true' myrmecophytic plant species (belonging to 20 genera and 14 plant families) occur in Peninsular Malaysia, i.e. these plant species are regularly inhabited by ants and mostly provide modified structures to house ants (see Table 1, plant taxa given in bold letters). Of these 45 true myrmecophytes, 44% (20) are of 'tree' or 'shrub' growth habits whereas 29% (13) are epiphytes (Fig. 3A). The third major type of growth habit among ant-plants is represented by rattans (climbing palms) with 20% (9) of all species. The ant-plants directly or indirectly provide food for ants in several forms and combinations. The principal food sources are food bodies, extrafloral nectar, and honeydew excretions of trophobiotic homopterans. Myrmecophytic plants providing honeydew alone account for the major portion (21/45 or 47%), followed by plants supplying food bodies + honeydew (7/45 or 16%), extrafloral nectar + honeydew (6/45 or 13%), or floral nectar (13%). For more details see Fig. 3B and Table 1.

In addition to the 45 'true' myrmecophytes, a considerable number of plant species (22) cannot yet be classified as 'ant-plants' since our knowledge of the regularity of ant colonization is poor and based solely on occasional observations recorded in the literature (Table 1, marked with an asterisk). They belong to 17 genera of which only 3 genera are already represented among the known true myrmecophytic plants. Thus the number of known myrmecophytes from Peninsular Malaysia might clearly be increased at both species and genus level by results of work in progress.

For 9 plant species reported to be associated with ants, we could either reject the hypothesis that they are regularly colonized by ants or show–as in the case of the typical 'ant-garden' plants such as *Dischidia nummularia* and *D. parvifolia*–that they do not fall within our definition of true myrmecophytes (see Introduction), although the root system of some individuals within any site is inhabited by ants (Kiew & Anthonysamy 1995; E. Kaufmann pers. comm.; A. Weissflog pers. obs.).

Four 'groups' of plants are recognized as the major components of the myrmecophytic associations found on the peninsula: (i) the *Macaranga* ant-plants (at least 9 species), (ii) the myrmecophytic rattans of the genera *Calamus*, *Korthalsia* and *Daemonorops* (9 species), (iii) the asclepiad epiphytes *Dischidia* and *Hoya* that house ants in or under modified leaves (c. 6 + 1 species), and (iv) the host plants of *Cladomyrma* ants (7 species). The last group, however, includes plant species belonging to a diverse range of unrelated genera and families, thus they are 'grouped' not according to their phylogenetic affinity but rather to the 'host choice' exerted by *Cladomyrma* ants.

Several plant genera, containing many ant-plant species in other parts of the Malay Archipelago, have no or only few myrmecophytic representatives in Peninsular Malaysia, e.g. the epiphytes *Lecanopteris*, *Hydnophytum*, *Myrmecodia*, *Anthorrhiza*, and the trees *Neonauclea, Myristica* and *Chisocheton*.

Pasoh FR is home to at least one-third of the known myrmecophytes on Peninsular Malaysia. It is a very conservative estimate because for a major part of ant-plant species, especially those with a non-tree growth habit, their occurrence in Pasoh FR is not known. Despite its small area, the flora of Pasoh FR includes roughly 25% of all Malayan tree species (Kochummen 1997). Regarding myrmecophytes with a tree growth habit (because their occurrence or absence from Pasoh FR is comparatively good documented) we estimate that 50% (10/20) of the myrmecophytic tree (or shrub) species of Peninsular Malaysia are known to occur in Pasoh FR. Overall, the proportion of 'true' myrmecophytic tree species to the total number of tree species found in the Pasoh FR 50 ha plot is 1.2% (10/814).

4.2 Diversity of ants

The diversity of obligate Malayan 'plant-ants' regularly inhabiting one or a small set of host plant species cannot be assessed accurately because many arboricolous ants opportunistically take advantage of any suitable structure for nesting. Thus a distinction between generalists and specialized plant-ants is dependent on a thorough study of several characteristics of their life history, and on knowledge which is too often lacking. Characteristics of specialists include: (i) colony foundation by ant queens in a particular host plant, (ii) a more or less strongly developed host specificity, (iii) host fidelity, defined here as the tendency of the ant colony to stay on the plant throughout the life of the colony, (iv) high occupancy rate, and (v) intraspecific competition for host plants. Taken together, these traits have been shown in Peninsular Malaysia for 3 major groups of plant-ants, namely 7 *Crematogaster* (*Decacrema*) (morpho)species inhabiting eight species of *Macaranga*, about 14 species of *Camponotus* (mostly *Myrmoplatys*) colonizing nine rattan species, and 3 species of *Cladomyrma* living inside the stems of seven host plants from diverse taxa. One *Cladomyrma* species, *C. petalae,* is notably catholic in its host plant choice; it inhabits 6 different plant species from 6 genera.

Another 'type' of plant-ant is exemplified by some species of *Polyrhachis*, *Tetraponera* and *Cataulacus*, all of which are known to exclusively colonize giant bamboo (see above). Apparently these associations are a case of 'one-sided' specialization. The ant queens lack the ability to penetrate the hard-walled, hollow culm internodes by themselves, and must rely on holes made by other insects to enter the hollow bamboo stems. Also, the ants usually do not monopolize the entire host plant, while priority of access and subsequent monopolization of the host is regularly observed in plant-ants inhabiting 'true' myrmecophytes.

Species of *Philidris* (formerly part of *Iridomyrmex*) are regularly found inhabiting several ant-house epiphytes, and apparently exhibit a general tendency to colonize these epiphytes. However, they are probably not confined to living in ant-plants (Gay & Hensen 1992; Huxley 1978). These ants typically place large quantities of debris and insect remains into the plants which absorb the decomposition products (Treseder et al. 1995). However, species within the genus *Philidris* have received little attention, either biologically or taxonomically (Shattuck 1992). Thus they may include both specialized plant-ants and opportunists so that a distinction cannot be made at the present time. In Malaysia ants of the species (group) *Philidris cordatus* (F. Smith) are recorded as the dominant inhabitants of the ant-epiphytes *Lecanopteris sinuosa* (Wall. ex Hook.) Copel., *Myrmecodia tuberosa* and *Hydnophytum formicarum* (Gay & Hensen 1992; Gay et al. 1993; Huxley 1980; Janzen 1974; Maeyama et al. 1997). The life history of another ant species, *Crematogaster treubi* Emery, recorded as dominating the montane fern *Lecanopteris pumila* in Peninsular Malaysia, is not known (Gay & Hensen 1992).

Future studies may show that species-specific colonization patterns of ant-epiphytes are determined mainly by habitat effects (open sunny habitats, closed humid forests, altitude) rather than by extreme host specificity of the ants; in other words habitat dependencies may lead to ecological 'species sorting' (Davidson et al. 1991). In New Guinea, species-pairings observed between *Philidris* and epiphytes have been interpreted as an indication of co-ecological preferences rather than as a result of co-evolution (Jebb 1991).

4.3 Diversity of domatia

Structures for housing ants are extremely variable ranging for example from fistulose stems over shelter-forming leaves and persistent stipules to inflated hypocotyls (chambered tubers) or spine galleries, with each type represented by morphologically diverse examples (see Table 1 and Fig. 2). These domatia may be grouped according to their morphological origin and structure, their similar appearance or by their functional affinity. Here, we present a simplified classification based on a combination of these features, with special emphasis on the ants' point of view (an important functional aspect):

a) hollow, closed domatia: Ant queens and workers must gnaw an entrance hole to get inside a preformed chamber located in naturally hollow plant structures. These latter structures may consist of (i) swollen stems with degenerated pith, (ii) inflated ocreas (proximal extensions of the leaf sheath), and (iii) galleries enclosed by interlocking combs of spines, forming collars on leaf sheaths, as found in some rattan species. This combination of features is usually seen in highly specialized ant-plant systems such as the *Macaranga-Crematogaster* or the *Camponotus-Korthalsia* resp. *Camponotus-Daemonorops/Calamus* associations. The spine galleries of the ant-rattans *Daemonorops* and *Calamus* are at first sometimes not completely closed, but through the ants' activity the remaining gaps are sealed with cut off spines and fluffy indument harvested at the apex of the rattan. Plant species with hollow, closed domatia account for the major portion of the 'true' ant-plant species on Peninsular Malaysia (15/45 or 33.5%, see Fig. 3C).

Some ant-plants facilitate hole-boring of ants by providing especially thin zones in the domatium wall, so-called prostomata, where ants preferably chew their entrance holes. Well-known examples are neotropical *Cecropia* and *Triplaris*, and African *Leonardoxa* (McKey 1991; Schimper 1888; Wheeler 1942). In Peninsular Malaysia prostomata have been recently described from 3 (4) *Macaranga* ant-plants, *M. kingii* Hook. f. var. *kingii*, *M. hullettii* King ex Hook. f., *M. triloba* (Bl.) Muell. Arg. and *M. motleyana* Muell. Arg. subsp. *griffithiana* (Muell. Arg.) Whitmore (Federle et al. 2001). The last species apparently represents an intermediate stage with weakly developed prostomata. In *Macaranga* the prostoma has the form of a straight line running all over the internode. This prostoma line is characterized by a reduced stem wall thickness, the absence of vascular bundles and latex vessels, and by a delayed cambium formation. Federle et al. (2001) suggest that prostomata not only facilitate entry of ant associates, but that they act as ecological filter by favoring specialized ants before opportunistically nesting ants.

b) pithy domatia: Not all *Macaranga* species provide self-hollowing domatia. In two Malayan species, *M. hosei* King ex Hook f. and *M. pruinosa* (Miq.) Muell.Arg., the stem's interior remains solid. The pith, however, is soft and dry, and the *Crematogaster* ants can remove it. In contrast, pithy domatia are a trait common to all host plants of *Cladomyrma*. Ant queens and workers force their entry into the stems and remove the pith to construct a nest chamber. Softening of pith is observed in several host plants of *Cladomyrma* (e.g. in swollen stem parts of *Spatholobus* and *Drypetes*, Fig. 2G,H). However, the pith never breaks down but consists of white and soft parenchymatic cells. This pattern of stem utilization by *Cladomyrma* is correlated with a diverse range of host plant taxa, indicating that the ability of the ant queen to hollow out live stems for colony foundation facilitates the acquisition of new hosts. Nevertheless, host specificity is generally high with species pairings ranging from 1:1 for *Cladomyrma maschwitzi* (*Crypteronia*), 1:2 for *C. nudidorsalis* (*Drypetes, Ryparosa*) to 1:6 for *C. petalae* (*Saraca*, *Spatholobus*, *Drypetes*, *Ryparosa, Luvunga and Strychnos*).

c) hollow, open domatia: This type of domatium is seen in various modifications in Malayan plants offering (i) chambered tubers or rhizomes (Fig. 2E, *Myrmecodia, Hydnophytum, Lecanopteris*), (ii) elongated, tube-like ocreas with inrolled edges (Fig. 2F, *Korthalsia hispida*), and (iii) stems that possess naturally occurring slits which allow access to the cavity (*Ficus obscura* Blume var. *borneensis* (Miq.) Corner, *Macaranga caladiifolia* Beccari and *Zanthoxylum myriacanthum* Wall. ex Hook. f., Fig. 2B). Among the 45 'true' myrmecophytes recognized for Peninsular Malaysia 22% of these species possess ant housing structures grouped as hollow, open domatia (Fig. 3C).

Usually, the 'openness' of the domatia is correlated with high ant species richness colonizing these plants because access to the nesting space is not restricted. Thus stochastic colonization events in the unpredictable mosaic of potentially available ant tenants play a more important role in structuring the ant community inhabiting ant-plants with 'open' domatia. Such a relationship between freedom of domatia access and ant occupant richness has been shown for species of Malayan *Ficus*, *Macaranga* and *Zanthoxylum,* as well as for Bornean *Capparis, Myrmeconauclea*, *Clerodendrum* and Costa Rican *Tetrathylacium* (Fiala et al. 1996; Maschwitz et al. 1989, 1992, 1994a,b, 1996a; Tennant 1989).

However, things are not simple. Specificity of ant species is often high within a population or a given habitat, especially for occupants of ant-house epiphytes (see above). Outside the ranges of regular inhabitants other ant occupants are recorded quite frequently, though not on a regular basis (Gay et al. 1993; Huxley 1982; Jolivet 1973). Similarly, high rates of occupancy of plants with self-opening stem cavities by only one or few ant species have been observed in Sri Lankan *Humboldtia laurifolia* Vahl (Krombein et al. 1999) and, in some localities, for Bornean *Myrmeconauclea strigosa* Merr. (unpubl. results).

These examples make clear that domatia access is a major but not the only factor affecting the composition of ant occupants in a particular host plant. Other diverse factors, such as habitat dependencies, colonization behaviour, or ant dominance, may influence the outcome of ant/plant occupation patterns.

d) cavities and entrance holes created by insects: Stem-boring insects other than ants commonly pave the way for subsequent colonization of live stem cavities by opportunistic ants. Such nesting opportunities for ants are, however, usually not predictable and not sufficiently large to support the reception of larger colonies. Although they may serve as founding chambers for ant queens the nesting space does not grow with the developing ant colony, unless the workers are able to enlarge the nest into solid plant parts or to maintain nest cavities distributed over a large area. However, in some plants insect borings appear to occur consistently, leading to rather high rates of ant occupancy. These plants include Malayan *Uncaria* climbers and *Clerodendrum* shrubs, as well as giant bamboo (see above). Other examples of plants in which insect borings provide nesting space for ants in live plant parts are recorded from Africa (*Cuviera*; Bequaert 1922), the neotropics (e.g. *Acacia, Avicennia, Clibadium, Remijia, Tachigali, Tecoma, Witheringia*; Bailey 1923; Benson 1985; Huxley 1986; Nesom & Stuessy 1982; Ward 1991; Wheeler 1942), SE Asia (*Acacia, Archidendron, Chisocheton, Leptospermum*; this publ.; Pijl 1955; Weir & Kiew 1986), and Australia and New Guinea (*Avicennia, Eucalyptus, Syzygium, Timonius*; Gullan et al. 1993; Sands & House 1990; Ward 1991). Although the regularity of such borings in particular plants is rarely documented, a hypothesis has been proposed by Ward (1991) that such occasional inhabitations of live, insect-bored cavities may be crucial to the inception of more specialized interactions with plants. In Peninsular Malaysia only two 'true' ant-plant species (Fig. 3C), *Clerodendrum phyllomega* var. *myrmecophilum* and *C. deflexum*, appear to be consistently entered by coleopteran stem-borers, thus apparently promoting ant inhabitation by providing entrance holes (see above).

e) open, leaf-derived domatia: Ant-shelters formed by leaf-derived structures are extremely diverse (see Table 1). Examples are shell- or pitcher-like leaves (*Dischidia*), leaf clusters (Fig. 2A, *Hoya*), pseudostipules or leaflets reflexed back across the stem (Fig. 2C, *Pometia, Lepisanthes, Calamus*), broadly winged petioles (*Teijsmanniodendron*), convex basal fronds (*Platycerium*) or semi-persistent stipules (*Poikilospermum*). However, many arboricolous ants take advantage of any suitable plant structure for nesting. Thus structures devoid of any myrmecophytic disposition nonetheless are often inhabited. An impressive example has been recorded by Way & Bolton (1997), who found 85 different ant species nesting in the leaf axils and spadices of coconut palms. The natural tendency of ants to occupy vacant spaces makes it often difficult to determine the status of a plant and its relationship to ants, especially where details of natural history are sparse. However, for one tree species in Peninsular Malaysia with open, leaf-derived domatia we can confirm a close relationship with ants. The shelters formed by reflexed pseudostipules of *Pometia pinnata* f. *glabra* trees (> 7 m) were regularly colonized by ants. Preliminary results indicate that resident ants may reduce leaf damage by herbivores, but level of biotic defense varied greatly between ant species present. Details of relationships with ants in other plant taxa with this type of structure (e.g. *Shorea, Actinodaphne, Platycerium, Lepisanthes* and *Teijsmanniodendron*; see Table 1) cannot be clarified at this stage. Nevertheless, existing evidence supports the conception that leaf-derived shelters can be consistently occupied by ants.

We are now beginning to accumulate the information necessary to generalize about ant-plant relationships in SE Asia and to reveal their role in the ecology of tropical forests. More extensive investigation in the area will continue to broaden the perspective on plants and their ants. We make a strong plea here for studies on ant-plant associations in Pasoh FR. Available information from other studies conducted in Pasoh FR (this volume) will help to determine the relative impact of ants on vegetation structure and dynamics.

ACKNOWLEDGEMENTS

We are very grateful for the good cooperation with FRIM. The former director Dr. Salleh Mohd Nor gave generous permission to use FRIM facilities and to work at Pasoh FR. We thank the staff of the FRIM herbarium for invaluable help with identification of plant specimens. Special thanks are due to Dr. K. Mat-Salleh and the late Dr. K. M. Kochummen. The manuscript was improved by discussions with Dr. Doyle McKey and by comments of three anonymous reviewers. The Economic Planning Unit, Prime Minister's Department granted permission to conduct research in Malaysia. Financial support from the German Research Foundation is gratefully acknowledged.

REFERENCES

Adema, F., Leenhouts, P. W. & Welzen, van, P. C. (1994) Sapindaceae. Flora Malesiana, Series 1, 11: 419-768.

Agosti, D., Moog, J. & Maschwitz, U. (1999) Revision of the Oriental plant-ant genus *Cladomyrma*. Am. Mus. Novit. 3283: 1-24.

Airy Shaw, H. K. (1975) The Euphorbiaceae of Borneo. Kew Bulletin Additional Series 4: 1-245.

Bailey, I. W. (1923) Notes on neotropical ant-plants. II. *Tachigalia paniculata* Aubl. Bot. Gaz. 75: 27-41.

Balgooy, M. M. J., van (1997) Malesian seed plants. vol. 1-spot characters (an aid for identification of families and genera). Rijksherbarium-Hortus Botanicus, Leiden, The Netherlands, 154pp.

Beattie, A. J. (1985) The evolutionary ecology of ant-plant mutualism. Cambridge University Press, Cambridge, UK, 175pp.

Beccari, O. (1884-1886) Piante ospitatrici, ossia piante formicarie della Malesia e della Papuasia. Malesia, Vol. 2. Raccolta di osservazini botaniche intorno alla Piante dell'Archipelago Indo-malese e Papuano. R. Instituto di studi superiori pratici e di perfezinomento (Firenze), Genova, pp.1-340.

Benson, W. W. (1985) Amazon ant-plants. In Prance, G. & Lovejoy, T. (eds). Amazonia, Pergamon Press, New York, pp.239-266.

Benzing, D. H. (1990) Vascular epiphytes. General biology and related biota. Cambridge tropical biology series, vol. 1. Cambridge University Press, Cambridge, 354pp.

Bequaert, J. (1922) Ants in their diverse relations to the plant world. Bull. Am. Mus. Nat. Hist. 45: 333-584.

Blattner, F. R., Weising, K., Bänfer, G., Maschwitz, U. & Fiala, B. (2001) Molecular analysis of phylogenetic relationships among myrmecophytic *Macaranga* species (Euphorbiaceae). Mol. Phylogenet. Evol. 19: 331-344.

Brouat, C. & McKey, D. (2000) Origin of caulinary ant domatia and timing of their onset in plant ontogeny: evolution of a key trait in horizontally transmitted ant-plant symbioses. Biol. J. Linn. Soc. 71: 801-819.

Buschinger, A., Klein, R. W. & Maschwitz, U. (1994) Colony structure of a bamboo-dwelling *Tetraponera* sp. (Hymenoptera: Formicidae: Pseudomyrmecinae) from Malaysia. Insectes Sociaux 41: 29-41.

Chew, W.-L. (1963) A revision of the genus *Poikilospermum* (Urticaceae). Garden's Bulletin Singapore 20: 1-103.

Clausing, G. (1998) Observations on ant-plant interactions in *Pachycentria* and other genera of the Dissochaeteae (Melastomataceae) in Sabah and Sarawak. Flora Jena 193: 361-368.

Corner, E. J. H. (1988) Wayside trees of Malaya (3rd ed.), 2 volumes. Malayan Nature Society, Kuala Lumpur, 861pp.
Davidson, D. W. & Epstein, W. W. (1989) Epiphytic associations with ants. In Lüttge, U. (ed). Vascular plants as epiphytes, Springer Verlag, Berlin, pp.200-233.
Davidson, D. W. & McKey, D. (1993) The evolutionary ecology of symbiontic ant-plant relationships. J. Hymenoptera Res. 2: 13-83.
Davidson, D. W., Foster, R. B., Snelling, R. R. & Lozada, P. W. (1991) Variable composition of some tropical ant-plant symbioses. In Price, P. W., Lewinsohn, T. M., Fernandes, G. W. & Benson, W. W. (eds). Plant-animal interactions: evolutionary ecology in tropical and temperate regions, John Wiley & Sons, New York, pp.145-162.
Dorow, W. H. O (1996) *Polyrhachis* (*Myrmhopla*) *schellerichae* n. sp., a new ant of the hector-group from the Malay Peninsula. Senckenbergiana biologica 76: 121-127.
Dorow, W. H. O. & Maschwitz, U. (1990) The arachne-group of *Polyrhachis* (Formicidae, Formicinae): weaver ants cultivating homoptera on bamboo. Insectes Sociaux 37: 73-89.
Dransfield, J. (1973) *Korthalsia hispida* Becc. in Malaya. Garden's Bulletin Singapore 26: 239-244.
Dransfield, J. (1979) A manual of rattans of the Malay Peninsula. Malayan Forest Records No. 29. Forest Department, Kuala Lumpur, 270pp.
Dransfield, J. (1980) *Pogonotium* (Palmae: Lepidocaryoideae), a new genus related to *Daemonorops*. Kew Bulletin 34: 761-768.
Federle, W. (1998) Strukturmechanismen der Lebensgemeinschaft auf einer Ameisenpflanze. Dissertation, Bayerische Julius-Maximilians-University, Wuerzburg, Germany, 169pp.
Federle, W., Maschwitz, U., Fiala, B., Riederer, M. & Hoelldobler, B. (1997) Slippery ant-plants and skilful climbers: selection and protection of specific ant partners by epicuticular wax blooms in *Macaranga* (Euphorbiaceae). Oecologia 112: 217-224.
Federle, W., Fiala, B. & Maschwitz, U. (1998a) *Camponotus* (Colobopsis) (Mayr 1861) and *Macaranga* (Thouars 1806): a specific two-partner ant-plant system from Malaysia. Trop. Zool. 11: 83-94.
Federle, W., Maschwitz, U. & Fiala, B. (1998b) The two-partner ant-plant system of *Camponotus* (Colobopsis) sp.1 and *Macaranga puncticulata* (Euphorbiaceae): Natural history of the exceptional ant partner. Insectes Sociaux 45: 1-16.
Federle, W., Leo, A., Moog, J., Azarae, H. I. & Maschwitz, U. (1999) Myrmecophagy undermines ant-plant mutualisms: ant-eating *Callosciurus* spp. squirrels (Rodentia: Sciuridae) damage ant-plants in SE Asia. Ecotropica 5: 35-43.
Federle, W., Rohrseitz, K. & Hölldobler, B. (2000) Attachment forces of ants measured with a centrifuge: better 'wax-runners' have a poorer attachment to a smooth surface. J. Exp. Biol. 203: 505-512.
Federle, W., Fiala, B., Zizka, G & Maschwitz, U. (2001) Incident daylights as orientation cue for hole-boring ants: prostomata in *Macaranga* ant-plants. Insectes Sociaux 48: 165-177.
Feldhaar, H., Fiala, B., Rosli, H. & Maschwitz, U. (2000) Maintaining an ant-plant symbiosis: secondary polygyny in the *Macaranga triloba* - *Crematogaster* sp. association. Naturwissenschaften 87: 408–411.
Fiala, B. (1996) Ants benefit pioneer trees: the genus *Macaranga* as an example of ant-plant associations in dipterocarp forest ecosystems. In Schulte, A. & Schoene, D. (eds). Dipterocarp Forest Ecosystems: Towards sustainable management, World Scientific, Singapore, pp.102-123.
Fiala, B. & Maschwitz, U. (1990) Studies on the South East Asian ant-plant association *Crematogaster borneensis*/*Macaranga*: adaptations of the ant partner. Insectes Sociaux 37: 212-231.
Fiala, B. & Maschwitz, U. (1992a) Domatia as most important adaptations in the evolution of myrmecophytes in the paleotropical tree genus *Macaranga* (Euphorbiaceae). Plant Syst. Evol. 180: 53-64.
Fiala, B. & Maschwitz, U. (1992b) Food bodies and their significance for obligate ant-association

in the tree genus *Macaranga* (Euphorbiaceae). Bot. J. Linn. Soc. 10: 61-75.
Fiala, B., Maschwitz, U., Pong, T. Y. & Helbig, A. J. (1989) Studies of a South East Asian ant-plant association: protection of *Macaranga* trees by *Crematogaster borneensis*. Oecologia 79: 463-470.
Fiala, B., Maschwitz, U. & Tho, Y. P. (1991) The association between *Macaranga* trees and ants in South East Asia. In Huxley, C. R. & Cutler, D. F. (eds). Ant-plant interactions, Oxford University Press, Oxford, pp.263-270.
Fiala, B., Grunsky, H., Maschwitz, U. & Linsenmair, K. E. (1994) Diversity of ant-plant interactions: protective efficacy in *Macaranga* species with different degrees of ant association. Oecologia 97: 186-192.
Fiala, B., Maschwitz, U. & Linsenmair, K. E. (1996) *Macaranga caladiifolia*, a new type of ant-plant among Southeast Asian myrmecophytic *Macaranga* species. Biotropica 28: 408-412.
Fiala, B., Jakob, A., Maschwitz, U. & Linsenmair, K. E. (1999) Diversity, evolutionary specialization and geographic distribution of a mutualistic ant-plant complex: *Macaranga* and *Crematogaster* in South East Asia. Biol. J. Linn. Soc. 66: 305-331.
Gay, H. (1993) Animal-fed plants: an investigation into the uptake of ant-derived nutrients by the far-eastern epiphytic fern *Lecanopteris* Reinw. (Polypodiaceae). Biol. J. Linn. Soc. 50: 221-233.
Gay, H. & Hensen, R. (1992) Ant specificity and behaviour in mutualisms with epiphytes: the case of *Lecanopteris* (Polypodiaceae). Biol. J. Linn. Soc. 47: 261-284.
Gay, H., Hennipman, E., Huxley, C. R. & Parrott, F. J. E. (1993) The taxonomy, distribution and ecology of the epiphytic Malesian ant-fern *Lecanopteris* Reinw. (Polypodiaceae). Garden's Bulletin Singapore 45: 293-335.
Gossner, M. (1999) Vergleich von Diversitaet und Artenzusammensetzung der Ameisenzoenosen der unteren Vegetation zwischen Primaer- und Sekundaerwaldflaechen im Tieflandregenwald von Sabah, Borneo. M. thesis, Univ. Wuerzburg (unpubl.).
Gullan, P. J., Buckley, R. C. & Ward, P. S. (1993) Ant-tended scale insects (Hemiptera: Coccidae: Myzolecanium) within lowland rain forest trees in Papua New Guinea. J. Trop. Ecol. 9: 81-91.
Hartley, T. G (1966) A revision of the Malesian species of *Zanthoxylum* (Rutaceae). J. Arnold Arboretum 47: 171-221.
Heckroth, H.-P., Fiala, B., Gullan, P. J., Azarae, H. I. & Maschwitz, U. (1998) The soft scale (Coccidae) associates of Malaysian ant-plants. J. Trop. Ecol. 14: 427-443.
Heil, M., Fiala, B., Linsemair, K. E., Zotz, G., Menke, P. & Maschwitz, U. (1997) Food body production in *Macaranga triloba* (Euphorbiaceae): A plant investment in anti-herbivore defence via symbiotic ant partners. J. Ecol. 85: 847-861.
Heil, M., Fiala, B., Kaiser, W. & Linsenmair, K. E. (1998) Chemical contents of *Macaranga* food bodies: adaptations to their role in ant attraction and nutrition. Funct. Ecol. 12: 118-122.
Heil, M., Fiala, B. & Linsenmair, K. E. (1999) Reduced chitinase activities in ant plants of the genus *Macaranga*. Naturwissenschaften 86: 146-149.
Heil, M., Staehelin, C. & McKey, D. (2000) Low chitinase activity in *Acacia* myrmecophytes: a potential trade-off between biotic and chemical defences? Naturwissenschaften 87: 555-558.
Heil, M., Fiala, B., Maschwitz, U. & Linsenmair, K. E. (2001a) On benefits of indirect defence: short- and long-term studies of antiherbivore protection via mutualistic ants. Oecologia 126: 395-403.
Heil, M., Hilpert, A., Fiala, B. & Linsenmair, K. E. (2001b) Nutrient availability and indirect (biotic) defence in a Malaysian ant-plant. Oecologia 126: 404-408.
Hölldobler, B. & Wilson, E. O. (1990) The Ants. Belknap Press, Cambridge, Massachusetts, 732pp.
Hou, D. (1978) Anacardiaceae. Flora Malesiana, Series 1, 8: 395-548.
Huxley, C. R. (1978) The ant-plants *Myrmecodia* and *Hydnophytum* (Rubiaceae), and the relationships between their morphology, ant occupants, physiology and ecology. New

Phytol. 80: 231-268.
Huxley, C. R. (1980) Symbiosis between ants and epiphytes. Biol. Rev. Camb. Philos. Soc. 55: 321-340.
Huxley, C. R. (1982) Ant-epiphytes of Australia. In Buckley, R. C. (ed). Ant-plant interactions in Australia, vol. 4, Geobotany. Dr. W. Junk Publishers, The Hague, Boston, London, pp.63-73.
Huxley, C. R. (1986) Evolution of benevolent ant-plant relationships. In Juniper, B. & Southwood, T. R. E. (eds). Insects and the plant surface, Edward Arnold, Publ., London, pp.257-282.
Itioka, T., Nomura, M., Inui, Y., Itino. T., & Inoue, T. (2000) Difference in intensity of ant defense among three species of *Macaranga* myrmecophytes in a Southeast Asian dipterocarp forest. Biotropica 32: 318-326.
Janka, H. I., Zizka, G., Moog, J. & Maschwitz, U. (2000) *Callicarpa saccata*, eine Ameisenpflanze aus Borneo mit Blattdomatien und extrafloralen Nektarien - und das Verbreitungsraetsel von Blatttaschen-Ameisenpflanzen. Der Palmengarten 64: 38-47.
Janzen, D. H. (1974) Epiphytic myrmecophytes in Sarawak: mutualism through the feeding of plants by ants. Biotropica 6: 237-259.
Jebb, J. (1991) Cavity structure and function in the tuberous Rubiaceae. In Huxley, C. R. & Cutler, D. F. (eds). Ant-plant interactions, Oxford University Press, Oxford, New York, Tokyo, pp.374-389.
Jolivet, P. (1973) Les Plantes myrmécophiles du Sud-Est Asiatique. Cahiers du Pacifique 17: 41-69.
Jolivet, P. (1996) Ants and plants. An example of coevolution, enlarged edition (transl. from the French), Backhuys Publ., Leiden, 303pp.
Kaufmann, E., Weissflog, A., Hashim, R. & Maschwitz, U. (2001) Ant-gardens on the giant bamboo *Gigantochloa scortechinii* (Poaceae) in West-Malaysia. Insectes Sociaux 48: 125-133.
Kerr, A. F. G. (1912) Notes on *Dischidia rafflesiana* Wall., and *Dischidia nummularia* Br. Sci. Proc. R. Dublin Soc. 13: 293-308.
Kiew, R. (1989) *Hoya endauensis* (Asclepiadaceae) and *Licuala dransfieldii* (Palmae), two new species from the Ulu Endau, Peninsular Malaysia. Malay. Nat. J. 42: 261-265.
Kiew, R. & Anthonysamy, S. (1987) A comparative study of vascular epiphytes in three epiphyte-rich habitats at Ulu Endau, Johore, Malaysia. Malay. Nat. J. 41: 303-315.
Kiew, R. & Anthonysamy, S. (1995) Ant-garden and ant-tree associations involving *Dischidia* species (Asclepiadaceae) in Peninsular Malaysia. In Kiew, R. (ed). The taxonomy and phytochemistry of the Asclepiadaceae in tropical Asia, Universiti Pertanian Malaysia, pp. 95-109.
Klein, R. W., Kovac, D., Schellerich, A. & Maschwitz, U. (1992) Mealybug-carrying by swarming queens of a Southeast Asian bamboo-inhabiting ant. Naturwissenschaften 79: 422-423.
Kochummen, K. M. (1997) Tree flora of Pasoh forest. Malaysian Forest Records No. 44. Forest Research Institute Malaysia, Kepong, Kuala Lumpur, 461pp.
Kovac, D. (1994) Die Tierwelt des Bambus: Ein Modell für komplexe tropische Lebensgemeinschaften. Natur und Museum 124: 119-136.
Kovac, D. & Streit, B. (1996) The arthropod community of bamboo internodes in Peninsular Malaysia: microzonation and trophic structure. In Edwards, D. S., Booth, W. E. & Choy, S. C. (eds). Tropical Rainforest Research - Current Issues, Kluwer Academic Publishers, Dordrecht, Boston, London, pp.85-99.
Krombein, K. V., Norden, B. B., Rickson, M. M. & Rickson, F. R. (1999) Biodiversity of domatia occupants (ants, wasps, bees, and others) of the Sri Lankan myrmecophyte *Humboldtia laurifolia* Vahl (Fabaceae). Smithson. Contrib. Zool. 603: 1-34.
Lee, S. S. (1995) A guidebook to Pasoh. FRIM Technical Information Handbook, No. 3. Forest Research Institute Malaysia, Kepong, 73pp.
Lehmann, M. (1998) Interaktionen zwischen Rattanpalmen und Ameisen unter besonderer Beruecksichtigung der Assoziation zwischen *Korthalsia scortechinii* und ihren *Camponotus*-Partnern. M. thesis, Johann-Wolfgang-Goethe-Univ., 161pp. (unpubl.).

Leo, A., Federle, W. & Maschwitz, U. (1999) Damage to *Macaranga* ant-plants by a myrmecophagous squirrel (*Callosciurus notatus*, Rodentia, Sciuridae) in West Malaysia. 12. Jahrestagung der Deutschen Gesellschaft für Tropenoekologie (17.-19.ii.1999).

Letourneau, D. K. (1998) Ants, stem-borers, and fungal pathogens: Experimental tests of a fitness advantage in *Piper* ant-plants. Ecology 79: 593-603.

Longino, J. T. & Hanson, P. E. (1995) The ants (Formicidae). In Hanson, P. E. & Gauld, I. D. (eds). The Hymenoptera of Costa Rica, The Natural History Museum, London, Oxford University Press, New York, Oxford, Tokyo, pp.588-620.

Maeyama, T., Terayama, M. & Matsumoto, T. (1997) Comparative studies of various symbiotic relationships between rubiacerous epiphytic myrmecophytes and their inhabitant ant species (Hymenoptera: Formicidae). Sociobiology 30: 169-174.

Manokaran, N., LaFrankie, J. V., Kochummen, K. M., Quah, E. S., Klahn, J. E., Ashton, P. S. & Hubbell, S. P. (1992) Stand table and distribution of species in the 50-ha research plot at Pasoh Forest Reserve, vol. 1. FRIM Research Data. Forest Research Institute Malaysia, Kepong, Malaysia, 454pp.

Maschwitz, U. & Fiala, B. (1995) Investigations on ant-plant associations in the South-East-Asian genus *Neonauclea* Merr. (Rubiaceae). Acta Oecologica 16: 3-18.

Maschwitz, U. & Moog, J. (2000) Communal peeing: a new mode of flood control in ants. Naturwissenschaften 87: 563-565.

Maschwitz, U., Fiala, B., Lee, Y. F., Chey, V. K. & Tan, F. L. (1989) New and little-known myrmecophytic associations from Bornean rain forests. Malay. Nat. J. 43: 106-115.

Maschwitz, U., Fiala, B., Moog, J. & Saw, L. G. (1991) Two new myrmecophytic associations from the Malay Peninsula: ants of the genus *Cladomyrma* (Formicidae, Camponotinae) as partners of *Saraca thaipingensis* (Caesalpiniaceae) and *Crypteronia griffithii* (Crypteroniaceae). I. Colony foundation and acquisition of trophobionts. Insectes Sociaux 38: 27-35.

Maschwitz, U., Fiala, B. & Linsenmair, K. E. (1992) A new ant-tree from SE Asia: *Zanthoxylum myriacanthum* (Rutaceae), the thorny Ivy-Rue. Malay. Nat. J. 46: 101-109.

Maschwitz, U., Fiala, B. & Linsenmair, K. E. (1994a) *Clerodendrum fistulosum* (Verbenaceae), an unspecific myrmecophyte from Borneo with spontaneously opening domatia. Blumea 39: 143-150.

Maschwitz, U., Fiala, B., Saw, L. G., Norma-Rashid, Y. & Azarae, H. I. (1994b) *Ficus obscura* var. *borneensis* (Moraceae), a new non-specific ant-plant from Malesia. Malay. Nat. J. 47: 409-416.

Maschwitz, U., Dumpert, K., Moog, J., LaFrankie, J. V. & Azarae, H. I. (1996a) *Capparis buwaldae* Jacobs (Capparaceae), a new myrmecophyte from Borneo. Blumea 41: 223-230.

Maschwitz, U., Fiala, B., Davies, S. J. & Linsenmair, K. E. (1996b) A South-east Asian myrmecophyte with two alternative inhabitants: *Camponotus* or *Crematogaster* as partners of *Macaranga lamellata*. Ecotropica 2: 29-40.

Maschwitz, U., Dorow, W. H. O, Schellerich-Kaaden, A. L., Buschinger, A. & Azarae, H. I. (2000) *Cataulacus muticus* Emery 1889 a new case of a Southeast Asian arboreal ant, non mutualistically specialized on giant bamboo (Insecta, Hymenoptera, Formicidae, Myrmicinae). Senckenbergiana biologica 80: 165-173.

Mattes, M., Moog, J., Werner, M., Fiala, B., Nais, J. & Maschwitz, U. (1998) The rattan palm *Korthalsia robusta* Bl. and its ant and aphid partners: studies of a myrmecophytic association in the Kinabalu Park. Sabah Parks Nat. J. 1: 47-60.

McKey, D. (1991) Phylogenetic analysis of the evolution of a mutualism: *Leonardoxa* (Caesalpiniaceae) and its associated ants. In Huxley, C. R. & Cutler, D. F. (eds). Ant-plant interactions, Oxford University Press, Oxford, New York, Tokyo, pp.310-334.

Merbach, M. A., Zizka, G., Fiala, B., Merbach, D. & Maschwitz, U. (1999) Giant nectaries in the peristome thorns of the pitcher plant *Nepenthes bicalcarata* Hooker f. (Nepenthaceae): anatomy and functional aspects. Ecotropica 5: 45-50.

Moog, J. & Maschwitz, U. (1994) Associations of *Cladomyrma* (Hym., Formicidae, Formicinae) with plants in SE Asia. In Lenoir, A., Arnold, G. & Lepage, M. (eds). Les Insectes

Sociaux, Abstract volume, 12th congress of the IUSSI, Université Paris Nord, Paris, Sorbonne, 21-27 August, 173.
Moog, J. & Maschwitz, U. (1999) The Oriental plant-ant genus *Cladomyrma* (Hym., Formicinae) and its host plants. Poster, Seminar Zoology & Ecology (Zoologi & Ekologi), Faculty of Science (Fakulti Sains), Universiti Malaya (20.iii.1999).
Moog, J. & Maschwitz, U. (2000) The secret of the Saraca tree. Malay. Nat. 53: 18-23.
Moog, J., Drude, T., Agosti, D. & Maschwitz, U. (1997) Flood control by ants: water-bailing behaviour in the south-east Asian plant-ant genus *Cladomyrma* Wheeler (Formicidae, Formicinae). Naturwissenschaften 84: 242-245.
Moog, J., Drude, T. & Maschwitz, U. (1998) Protective function of the plant-ant *Cladomyrma maschwitzi* to its host, *Crypteronia griffithii*, and the dissolution of the mutualism (Hymenoptera: Formicidae). Sociobiology 31: 105-129.
Nesom, G. L. & Stuessy, T. F. (1982) Nesting of beetles and ants in *Clibadium microcephalum* S. F. Blake (Compositae, Heliantheae). Rhodora 84/837: 117-124.
Nielsen, I. C. (1992) Mimosaceae (Leguminosae-Mimosoideae). Flora Malesiana, Series 1, 11: 1-226.
Nomura, M., Itioka, T. & Itino, T. (2000) Variations in abiotic defense within myrmecophytic and non-myrmecophytic species of *Macaranga* in a Bornean dipterocarp forest. Ecol. Res. 15: 1-11.
Pannell, C. M. (1992) A taxonomic monograph of the genus *Aglaia* Lour. (Meliaceae), Kew Bulletin Additional Series 16, Royal Botanic Gardens, Kew, London, pp.1-379.
Paterson, S. (1982) Observations on ant associations with rainforest ferns in Borneo. Fern Gaz. 12: 243-245.
Pijl, L. van der (1955) Some remarks on myrmecophytes. Phytomorphology 5: 190-200.
Rickson, F. R. (1979) Absorption of animal tissue breakdown products into a plant stem - the feeding of a plant by ants. Am. J. Bot. 66: 87-90.
Rickson, F. R. & Rickson, M. M. (1986) Nutrient acquisition facilitated by litter collection and ant colonies on two Malaysian palms. Biotropica 18: 337-343.
Ridley, H. N. (1910) Symbiosis of ants and plants. Ann. Bot. 24/64: 457-483.
Rintz, R. E. (1978) The Peninsular Malaysian species of *Hoya* (Asclepiadaceae). Malay. Nat. J. 30: 467-522.
Rintz, R. E. (1980) The Peninsular Malayan species of *Dischidia* (Asclepiadaceae). Blumea 26: 81-126.
Sagers, C. L., Ginger, S. M. & Evans, R. D. (2000) Carbon and nitrogen isotopes trace nutrient exchange in an ant-plant mutualism. Oecologia 123: 582-586.
Sands, D. & House, S. (1990) Plant/insect interactions. Food webs and breeding systems. In Webb, L. J. & Kikkawa, J. (eds). Australian Tropical Rainforests, CSIRO, Melbourne, pp. 88-97.
Schellerich-Kaaden, A. & Maschwitz, U. (1998) Extrafloral nectaries on culm sheath auricles: Observations on four Southeast Asian giant bamboo species (Poaceae: Bambusoideae). Sandakania 11: 61-68.
Schellerich-Kaaden, A., Stein, S. & Maschwitz, U. (1997a) Die Tierwelt des Bambus in West-Malaysia. In Jenny, M. (ed). In der Welt des Bambus, Sonderheft 25, Palmengarten, Frankfurt am Main, Germany, pp.53-57.
Schellerich-Kaaden, A. L., Dorow, W. H. O., Liefke, C., Klein, R. W. & Maschwitz, U. (1997b) *Polyrhachis schellerichae* (Hymenoptera: Formicidae), a specialized bamboo dwelling ant species from the Malay Peninsula. Senckenbergiana biologica 77: 77-87.
Schimper, A. F. W. (1888) Die Wechselbeziehungen zwischen Pflanzen und Ameisen im tropischen Amerika. Botanische Mitteilungen aus den Tropen 1: 1-95.
Shattuck, S. O. (1992) Generic revision of the ant subfamily Dolichoderinae (Hymenoptera: Formicidae). Sociobiology 21: 1-181.
Smith, W. (1903) *Macaranga triloba*: a new myrmecophilous plant. New Phytol. 2: 79-82.
Stevens, P. F. (1978) Meliaceae. In Womersley, J. S. (ed). Handbooks of the flora of Papua New Guinea, vol. 1. Melbourne University Press (reprint 1995), Carleton, Victoria, pp.135-174.

Tennant, L. E. (1989) A new ant-plant, *Tetrathylacium costaricense*. Symposium: Interactions between ants and plants, Oxford, p.27.
Treseder, K. K., Davidson, D. W. & Ehleringer, J. R. (1995) Absorption of ant-provided carbon dioxide and nitrogen by a tropical epiphyte. Nature 375: 137-139.
Uhl, N. W. & Dransfield, J. (1987) Genera Palmarum: a classification of palms based on the work of H. E. Moore Jr., L. H. Bailey, Hortorium & the International Palm Society. Lawrence, Kansas, USA, 610pp.
Ward, P. S. (1991) Phylogenetic analysis of pseudomyrmecine ants associatied with domatia-bearing plants. In Huxley, C. R. & Cutler, D. F. (eds). Ant-plant interactions, Oxford University Press, Oxford, pp.335-352.
Way, M. J. & Bolton, B. (1997) Competition between ants for coconut palm nesting sites. Nat. Hist. 31: 439-455.
Weir, J. S. & Kiew, R. (1986) A reassessment of the relations in Malaysia between ants (*Crematogaster*) on trees (*Leptospermum* and *Dacrydium*) and epiphytes of the genus *Dischidia* (Asclepiadaceae) including 'ant-plants'. Biol. J. Linn. Soc. 27: 113-132.
Weissflog, A., Moog, J., Federle, W., Werner, M., Rosli, H. & Maschwitz, U. (1999) *Hoya mitrata* Kerr (Asclepiadaceae): a new myrmecotrophic epiphyte from south-east Asia with a unique multileaved domatium. Ecotropica 5: 221-225.
Werner, M. (1993) Suedostasiatische Palmen und ihre Ameisenfauna. M. thesis, Univ. Kaiserslautern, 73pp. (unpubl.).
Werner, M., Dumpert, K. & Maschwitz, U. (1996) Specificity and diversity within South East Asien ant-rattan associations. Abstract volume 'Global Biodiversity Research in Europe', International Senckenberg Conference, Dec. 9-13, 1996 (supported by The Linnean Society, London, Societé Francaise de Systematique), Frankfurt am Main, Germany, p.80.
Wheeler, W. M. (1942) Studies of Neotropical ant-plants and their ants. Bulletin of the Museum of Comparative Zoology at Harvard University 90: 1-262.
Whitmore, T. C. (1973) Tree flora of Malaya vol. 2, Malayan Forest Records No. 26. Longmans, Sdh. Bhad., Kuala Lumpur.
Whitmore, T. C. (1975) *Macaranga* Thou. In Airy Shaw, H. K. (ed). The Euphorbiaceae of Borneo, Kew Bulletin Additional Series 4, pp.140-159.
Whitmore, T. C. (1984) Tropical forests of the Far East (2nd ed.), Clarendon Press, Oxford, 352pp.
Wong, K. M. & Puff, C. (1995) Notes on a myrmecophytic heterophyllous *Diospyros* (Ebenaceae). Sandakania 6: 55-62.
Yapp, R. H. (1902) Two Malayan 'myrmecophilous' ferns, *Polypodium* (*Lecanopteris*) *carnosum* (Blume), and *Polypodium sinuosum* Wall. Ann. Bot. 16: 185-231.
Yu, D. W. & Davidson, D. W. (1997) Experimental studies of species-specificity in *Cecropia*-ant relationships. Ecol. Monogr. 67: 273-294.

34 Leaf Herbivory and Defenses of Dipterocarp Seedlings in the Pasoh Forest Reserve

Shinya Numata[1], Naoki Kachi[2], Toshinori Okuda[3] & N. Manokaran[4]

Abstract: Leaf herbivores can potentially reduce the growth and survival of individual tree seedlings, thereby influencing the composition and species diversity of plant communities in forests. Plants have developed various defense mechanisms against herbivory. Since the habitat of tree seedlings is highly heterogeneous in tropical rain forests, changes in the environment would affect both plant performance and herbivory. In this study, we demonstrate the relationships between herbivory and the defenses of dipterocarp seedlings under different light regimes in the Pasoh Forest Reserve (Pasoh FR) of Malaysia. In all four species studied, two putative carbon-based defenses (total phenol concentrations in leaves and leaf toughness) were generally higher in plants growing in microsites beneath gaps in the forest canopy. The lack of significant differences in rates of leaf area loss between plants growing beneath canopy gaps and a closed canopy suggests that the potential pressure of leaf herbivory is greater in canopy gaps than beneath a closed canopy. Even though dipterocarp trees are generally regarded as climax shade-tolerant species, the species with the longest leaf lifespan tended to have the greatest concentration of phenols in their leaves and greater leaf toughness than species with the shortest leaf lifespan, irrespective of the light regime. This indicates that the species-specific patterns of leaf defensive traits may reflect evolutionary responses to seedling light requirements.

Key words: chemical defense, dipterocarp species, herbivory, leaf damage, light environment, phenolic compounds, Southeast Asia.

1. INTRODUCTION

Seedling establishment is the most important stage in the life history of a tree for determining regeneration success. During the early stages of regeneration under natural conditions, tropical tree seedlings suffer from environmental resource limitations and from biotic stresses such as leaf herbivory (Coley & Barone 1996). Seedling leaves show large differences in both the amounts and the types of defenses against leaf herbivores (Coley & Barone 1996; Coley & Kursar 1996; Rhoades 1979). Leaves accumulate various chemical substances that act as chemical defenses and lower the palatability of leaf tissues. For example, leaf phenolic compounds are effective defensive compounds against diverse herbivores because they form tannins that decrease the availability of nitrogen to the herbivores (Baldwin & Schultz 1988; Coley 1986; Feeny 1970, 1976; Fleck & Tomback 1996; Martin &

[1] Tokyo Metropolitan University. Present address: National Institute for Environmental Studies (NIES), Tsukuba 305-8506, Japan. E-mail: numata.shinya@nies.go.jp
[2] Tokyo Metropolitan University, Japan.
[3] NIES, Japan.
[4] Forest Research Institute Malaysia (FRIM), Malaysia

Martin 1982, 1983; Rhoades & Cates 1976). As well, leaf toughness mechanically impedes chewing by insect herbivores (Beck 1965; Choong et al. 1992; Schowalter et al. 1986). Leaf nutritional values may also be partly shaped by selection in response to herbivores and pathogens (Moran & Hamilton 1980), since low leaf nitrogen levels have been repeatedly associated with reduced preference of insects for these leaves and reduced performance of insects that have consumed these leaves (Coley & Barone 1996; Hartley & Jones 1997; Mattson 1980).

Strong inter- and intraspecific patterns in the defensive traits of seedling leaves have been observed under natural conditions, and the resource availability theory has been proposed as an explanation of interspecific variation in leaf defensive traits. This theory states that resource availability in the environment is the major determinant of both the magnitude and the nature of a plant's defenses (Coley et al. 1985). This theory suggests that plants with inherently slow growth would have a competitive advantage over those with fast growth when resources are limited. Because slow-growing plants are more at risk from herbivory than faster-growing plants, understory seedlings of slow-growing species tend to have more leaf defenses and lower rates of herbivory than fast-growing species (Coley 1983). In contrast, responses to variable light and nutrient conditions are more important in explaining intraspecific variations in leaf defensive traits (Bryant et al. 1983). Because canopy openings (gaps) enhance light availability to seedlings, they play an important role in the regeneration of many tropical rain forest species (Bazzaz & Pickett 1980; Denslow 1987; Whitmore 1984) and in intraspecific variation in leaf defensive traits (Bryant et al. 1983). Since the investment in defensive mechanisms can increase in resource-rich environments, where some resources can be invested in leaf defenses, carbon-based defenses such as lignins or phenolic compounds may increase for plants of a species exposed to high light levels (Dudt & Shure 1994; Shure & Wilson 1993).

For these reasons, interactions between plants and herbivores could change within and among species in response to environmental changes. Nevertheless, little is known about the variation in leaf defensive traits in dipterocarp tree seedlings. The present study examined species-specific responses of leaf traits and herbivory to canopy gaps. For four *Shorea* species (Dipterocarpaceae), we compared growth, putative leaf defensive traits, and the rate of leaf area loss under regimes with high and low light levels. Knowledge of the responses of seedling defensive traits to canopy gaps should improve our understanding of coexistence mechanisms in tropical tree species.

2. MATERIALS AND METHODS

2.1 Study species

Dipterocarp trees are widely distributed throughout tropical region in Asia, and are common in forests as canopy and emergent trees (Appanah & Turnbull 1998). *Shorea* is a common genus in the dipterocarps (Symington 1943), and 14 *Shorea* species have been recorded in a 50-ha plot in the Pasoh Forest Reserve (Pasoh FR). Seedlings of four *Shorea* species (*Shorea leprosula* Miq., *S. macroptera* Dyer, *S. maxwelliana* King and *S. pauciflora* King) were studied in this experiment. These species are mainly distributed in lowland forests throughout Peninsular Malaysia. *S. leprosula*, *S. macroptera* and *S. pauciflora* produce timber classified as light hardwood (Appanah & Weinland 1993; Symington 1943; Turner 1989; Vincent

1961). *S. maxwelliana* is a member of the heavy hardwoods (Appanah & Weinland 1993; Symington 1943), and is generally described as a more slow-growing species than the other three species (Symington 1943). These four species are known to contain phenolic compounds, condensed tannins, or both in their leaves (Bate-Smith & Whitmore 1959; Becker 1981; Numata et al. 2000). Alkaloids were not detected in the leaves of several *Shorea* seedlings in a prior study (Becker 1981).

2.2 Experimental design

Sound fruits of the four species were collected at the time of dispersal during a mast-fruiting event in 1996. We germinated the fruits in four plastic trays (80 cm × 40 cm × 10 cm deep) in September 1996. Sixty 4-weeks-old seedlings of each species were transplanted into plastic pots (10 cm diameter × 15 cm deep).

We chose two different light regimes (canopy gap and closed canopy) in a selectively logged forest along nature trails within the Pasoh FR. We placed 30 potted seedlings per species in microsites under each light regime. The relative photon flux density (RPFD) in the canopy-gap regime was approximately 35%, versus 6.5% in the closed-canopy regime (Okuda et al. unpubl. data). The two groups of plants were surrounded by zinc fences to prevent disturbance by mammals. Seedling performance was monitored in November 1996, and in March, May and September 1997.

2.3 Leaf characteristics, demographics and area loss

All seedlings were harvested in September 1997 so we could measure leaf toughness, total phenolic content, nitrogen concentration, and rates of leaf damage. Leaf toughness was measured for fresh leaves with a digital force gauge (FGC-0.5 SHIMPO, Japan). This device measures the force (in Newtons) required to punch a 0.5-mm-diameter rod through the leaf. The sampled seedlings were dried at 80°C for one day in an electric oven, then weighed.

Dried leaves were powdered to permit the measurement of total phenol concentrations and carbon to nitrogen ratios. Phenolic compounds were extracted with 70% acetone, then the samples were centrifuged at 0.1 g to prevent contamination of the supernatant liquid with solid materials. Extraction and centrifugation were repeated four times, and all extracts were combined for each leaf sample. Total phenol concentrations in each extract were determined by means of a Folin-Denis assay (Waterman & Mole 1994). Nitrogen and carbon concentrations in the leaf material were determined with a C/N Analyzer (Sumigraph NC-90, SUMITOMO Chem. Ind., Japan). Since the masses of phenolic compounds and of nitrogen are directly proportional to the amount of leaf tissue present, they are strongly affected by herbivory (Feeny 1970), and we calculated the total phenol and nitrogen concentrations on the basis of leaf dry mass rather than leaf area. These measurements were repeated three times in principle. Estimates were done either once or twice because the size of some samples was insufficient to permit a full series of these chemical analyses.

All leaves of the harvested seedlings were photocopied before the chemical analyses were performed. The images were then digitized (Color Pixel Duo scanner, CANON, Japan and NIH Image software, National Institute of Health, USA) to determine the total leaf area and the percentage of leaf area lost to insect herbivores, including strip feeders and pit feeders. To examine rates of leaf herbivory, the values of four leaf demographic variables were calculated:

1. Leaf increase rate (LIR) equaled the number of leaves added per month for each seedling.
2. Leaf production rate (LPR) equaled the number of leaves remaining on the seedling plus the number of leaf scars.
3. Estimated leaf lifetime (LLT) was calculated as 1/(LPR - LIR).
4. Mean leaf area loss rate (LAL, per month) was estimated as (mean LAL/LLT).

2.4 Statistical analysis

All statistical comparisons were performed by using StatView J-4.5 (ABACUS Concepts, Inc., Berkeley, CA, 1992). Comparisons of seedling survival rates were analyzed by the chi-squared test. For comparisons of leaf demographic traits, leaf traits, and LAL, we used the Kruskal-Wallis test and Mann-Whitney's U-test.

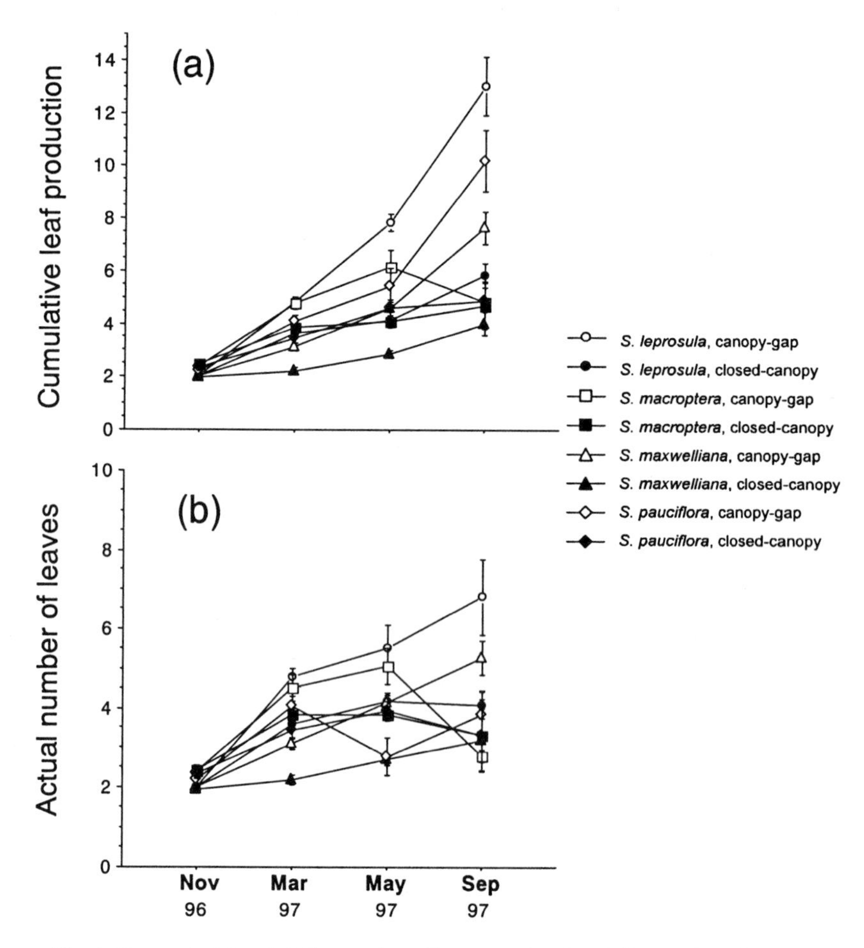

Fig. 1 Changes in (a) cumulative number of leaves produced and (b) actual numbers of leaves on seedlings of four *Shorea* species under the canopy-gap and closed-canopy light regimes. Symbols and bars indicate mean values and *SE*. Open symbols represent the canopy-gap regime and closed symbols represent the closed-canopy regime.

3. RESULTS

3.1 Seedling performance under closed canopy and in canopy gaps

Leaf demographic traits for the seedlings differed significantly between plants and species in the canopy-gap and closed-canopy regimes (Fig. 1). LPR was significantly higher under the canopy-gap regime than under the closed-canopy regime for all species ($Z = -5.7$, $P < 0.0001$), but the responses to canopy gaps differed among species. Only three of the species (*S. leprosula*, *S. maxwelliana* and *S. pauciflora*) showed significant differences between the canopy-gap and closed-canopy regimes (Table 1). Although LIR was generally higher under the canopy-gap regime than under the closed-canopy regime ($Z = -2.8$, $P = 0.005$), only *S. maxwelliana* showed significant differences in LIR between these regimes (Table 1). *Shorea macroptera* had a lower LIR under the canopy-gap regime than under the closed-canopy regime.

Only three species (*S. leprosula*, *S. maxwelliana* and *S. pauciflora*) showed significant differences in the LLT between the canopy-gap and closed-canopy regimes. Mean leaf lifetime ranged from 1.48 (*S. pauciflora*) to 4.49 months (*S. macroptera*) under the canopy-gap regime, versus 5.66 (*S. pauciflora*) to 8.47 months (S. *macroptera*) under the closed-canopy regime (Table 1).

3.2 Putative leaf defensive traits and LAL

Table 2 summarizes the differences in levels of putative leaf defensive traits (phenol concentration, toughness, and C/N ratio) between the canopy-gap and closed-canopy regimes for each species (Table 2). Of the four species, seedlings of *S. macroptera* had the highest concentrations of leaf phenolic compounds, irrespective of light environment. Although these concentrations were generally higher under the canopy-gap regime than under the closed-canopy regime, only *S. pauciflora* showed significant differences between the two regimes. Leaf toughness was lower for *S. leprosula* than for the other species under both regimes. The leaf toughness increased significantly under the canopy-gap regime for all species except *S.*

Table 1 Estimated values of three leaf demographic variables (LPR: leaf production rate, LLR: leaf loss rate, and LLT: leaf lifetime) for the seedlings of four *Shorea* species. Values shown are means ± standard deviations. Asterisks indicate statistically significant differences between the values of a variable under the canopy-gap (gap) and closed-canopy (closed) regimes for a given species (Mann-Whitney's U-test: ns: $P \geqq 0.05$; * $P < 0.05$; ** $P < 0.01$; *** $P < 0.001$; **** $P < 0.0001$).

Species	Regime	Leaf production rate (mo^{-1})	Leaf increasing rate (mo^{-1})	Leaf life time (mo)
S. leprosula	Gap	1.25 ± 0.38***	0.50 ± 0.39 ^{ns}	1.77 ± 0.89****
	Closed	0.42 ± 0.16	0.23 ± 0.13	6.67 ± 3.43
S. macroptera	Gap	0.32 ± 0.27^{ns}	0.04 ± 0.13 ^{ns}	4.49 ± 3.90 ^{ns}
	Closed	0.28 ± 0.12	0.14 ± 0.08	8.47 ± 2.93
S. maxwelliana	Gap	0.67 ± 0.27***	0.39 ± 0.21****	4.15 ± 2.54*
	Closed	0.24 ± 0.18	0.15 ± 0.09	6.95 ± 2.95
S. pauciflora	Gap	0.96 ± 0.36*	0.12 ± 0.26 ^{ns}	1.48 ± 0.84**
	Closed	0.29 ± 0.19	0.09 ± 0.18	5.66 ± 2.73

Table 2 Leaf area, total concentration of phenolic compounds, leaf toughness and C/N ratio for the sampled seedlings of four *Shorea* species. Leaf toughness was measured for all leaves of surviving seedlings in September 1997. Values shown are means ± standard deviations. Asterisks indicate statistically significant differences between the values of a variable under the canopy-gap (gap) and closed-canopy (closed) regimes for a given species (Mann-Whitney's U-test: ns: $P \geqq 0.05$; * $P < 0.05$; ** $P < 0.01$; **** $P < 0.0001$).

Species	Regime	Leaf size (cm^2)	Total concentration of phenolic compounds (mg/g DW) (N = 4)	Toughness (Newtons)	C/N ratio (N = 6)
S. leprosula	Canopy gap	28.1 ± 4.1 ****	77.1 ± 4.1 ns	0.081 ± 0.016**	20.7 ± 1.7 ns
	Closed-canopy	13.5 ± 5.4	72.7 ± 5.8	0.071 ± 0.015	20.8 ± 2.0
S. macroptera	Canopy gap	21.8 ± 6.9 ns	100.8 ± 3.5 ns	0.176 ± 0.034*	32.0 ± 5.5 ns
	Closed-canopy	18.7 ± 12.5	90.0 ± 12.6	0.143 ± 0.039	27.1 ± 2.9
S. maxwelliana	Canopy gap	9.3 ± 7.4 ****	72.7 ± 5.3 ns	0.121 ± 0.022*	21.1 ± 2.6 ns
	Closed-canopy	4.2 ± 1.8	64.8 ± 15.2	0.106 ± 0.020	22.5 ± 1.4
S. pauciflora	Canopy gap	24.7 ± 11.0 **	68.6 ± 2.6*	0.148 ± 0.026 ns	27.0 ± 5.9 ns
	Closed-canopy	14.4 ± 6.1	55.6 ± 9.2	0.133 ± 0.022	23.8 ± 2.7

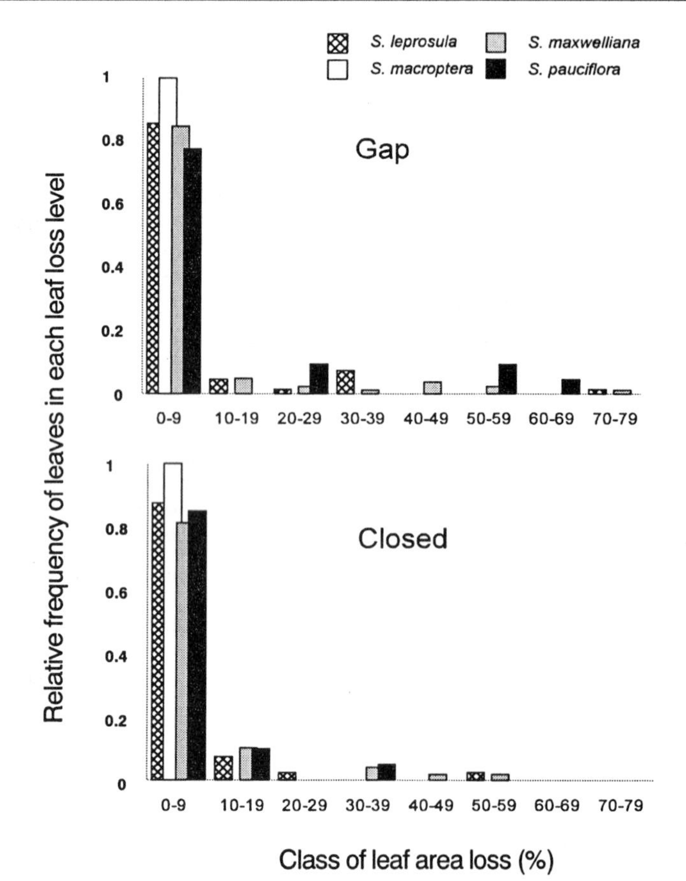

Fig. 2 Relative frequencies on leaf area loss of each leaves in four *Shorea* species under the canopy-gap and closed-canopy regimes.

pauciflora. The leaf C/N ratio was lower for *S. leprosula* and *S. maxwelliana* than for the other species under both regimes. We found no significant differences in leaf C/N ratios between regimes for any species.

Individual leaf size increased significantly under the canopy-gap regime for all species except *S. macroptera* (Table 2). Estimated LAL varied greatly, though most LAL values (expressed as a percentage) were less than 10% (Fig. 2). *S. macroptera* showed a much lower level of LAL than the other species under both light regimes. The mean LAL ranged from 0.07 (*S. macroptera*) to 11.2% (*S. pauciflora*) under the canopy-gap regime, versus 0.3 (*S. macroptera*) to 1.9% (*S. leprosula*) under the closed-canopy regime. Leaf loss rates did not differ significantly between the canopy-gap and closed-canopy regimes.

Fig. 3 summarizes seedling responses to canopy gaps in terms of the relationships between LLT and LAL, concentration of phenols, leaf toughness, and C/N ratio. In comparison with changes in LLT, changes in LAL and the other three leaf traits were small. LAL generally decreased with increasing LLT for all the species. The values for the three putative leaf defensive traits generally increased with increasing LLT within each light regime, but there was no apparent correlation between these leaf traits and LLT across all species.

4. DISCUSSION

The values for two putative carbon-based defenses (total concentrations of phenolic compounds in leaves and leaf toughness) were generally higher under the canopy-gap regime in all species. This result was consistent with the carbon and nutrient balance theory, which proposes that increasing light availability lets plants increase resource allocation to defenses against herbivory (Aide & Zimmerman 1990; Dudt & Shure 1994; Folgarait & Davidson 1994; Shure & Wilson 1993; Waring et al. 1985). The accumulation of leaf phenolic compounds at high levels of light availability could be enhanced by the development of a carbon surplus in the leaves (Mole et al. 1988). Increased leaf toughness under canopy gaps can also be understood in terms of allocation of photosynthate to structural rather than metabolic biomass to high levels of light availability (e.g. Turner 2001).

Although LAL for the seedlings was mostly less than 10%, the estimation of LAL by harvesting discrete samples of leaves may underestimate real losses to herbivory, because this approach does not account for leaves that were totally eaten (Lowman 1984). However, most leaf damage occurs during the expansion of leaves (Coley & Kursar 1996; Howlett & Davidson 2001). For example, leaf loss by herbivores was often observed in young expanding leaves on dipterocarp seedlings, whereas damage in mature leaves was often caused by necrosis due to pathogens rather than by insect herbivores (Numata unpubl. data). Thus, although the present study may have underestimated the real amount of leaf herbivory, the magnitude of this difference is likely to be relatively small.

Regardless of any possible underestimation, the lack of significant differences in LAL between the canopy-gap and closed-canopy regimes, combined with the larger leaf areas under the canopy-gap regime, suggests that potential herbivory pressure is greater in the canopy-gap regime. Herbivore population density and behavior could be important factors in determining whether herbivory rates differ between the canopy-gap and closed-canopy regimes (Howlett & Davidson 2001; Lincoln & Mooney 1984). Since more new leaves were available as food resources under the canopy-gap regime, the high light availability would let insect herbivores feed more rapidly or more efficiently (Benson 1978; Lincoln

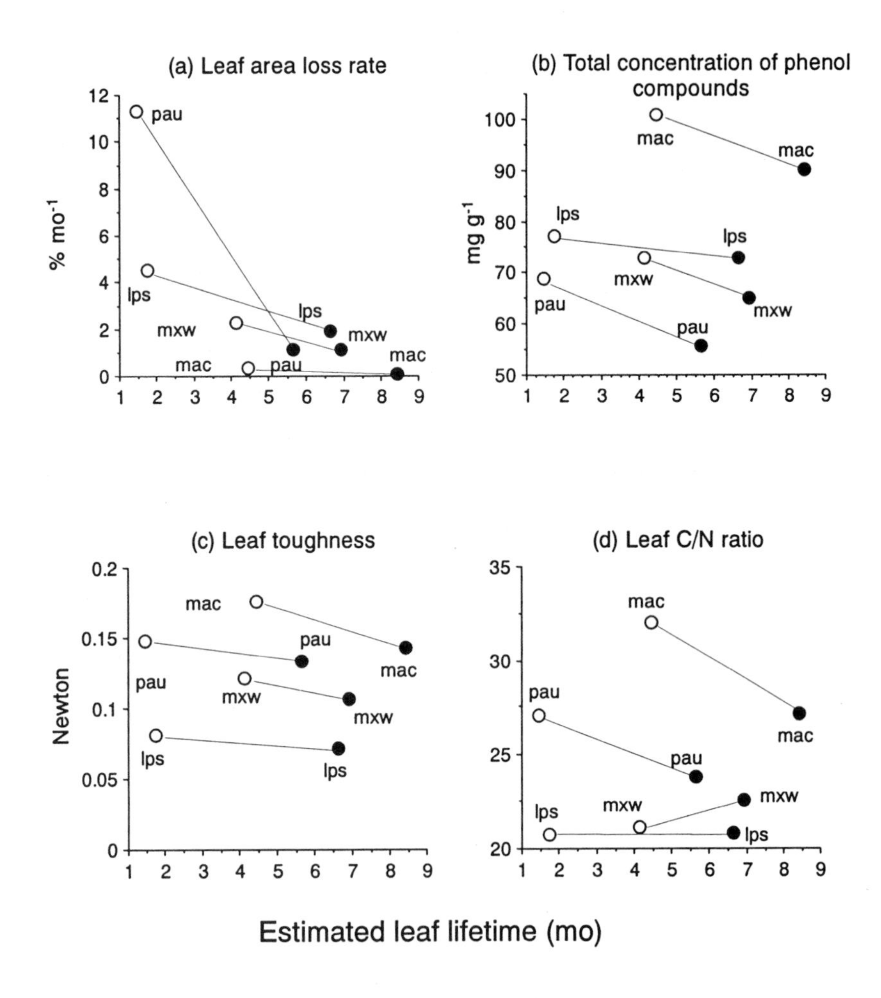

Fig. 3 Relationships between estimated leaf lifetime and (a) mean leaf area loss rate per month, (b) total concentration of leaf phenolic compounds, (c) leaf toughness, and (d) leaf C/N ratio for the four *Shorea* species. Open circles represent seedlings under the canopy-gap regime, and closed circles represent seedlings under the closed-canopy regime. Abbreviations for the species names are as follows: lps (*S. leprosula*), mac (*S. macroptera*), mxw (*S. maxwelliana*), pau (*S. pauciflora*).

& Mooney 1984). Risks of predation on insect herbivores may also differ between sunny and shaded regimes (Maiorana 1981). Therefore, comprehensive studies of the predators, parasitoids, and pathogens of herbivores will be important in understanding general patterns of variation in herbivory across different environmental gradients.

Species-specific patterns in leaf defensive traits may reflect evolutionary responses to seedling light requirements (e.g. Coley & Barone 1996). This relationship may arise from the effects of leaf damage by herbivores, which are greater for long-lived leaves than for short-lived leaves from the viewpoint of total carbon gain during the plant's lifetime (Chabot & Hicks 1982). Even though dipterocarp trees are shade-tolerant climax canopy species, the species in the present study with the longest leaf lifetime (*S. macroptera*) tended to have higher concentrations of leaf phenolic compounds and greater leaf toughness than the species with the shortest leaf lifetime (*S. pauciflora*), irrespective of light regime. These results are consistent with the resource availability theory.

The modest observed rates of leaf herbivory suggest that herbivory may have little effect on the survival rates of dipterocarp seedlings. For example, Becker (1983) showed that one-time artificial defoliation (25% removal) did not decrease mortality of *S. leprosula* seedlings. This means that only heavy leaf damage directly affects the survival of established dipterocarp seedlings. However, the survival of dipterocarp seedlings under natural conditions depends on many other factors, including intrinsic defenses, the abundance of herbivory, and growth responses to the light environment. Since larger seedlings of some dipterocarp species have a survival advantage over smaller seedlings under shaded conditions (S. Numata unpubl. data), reduced growth as a result of herbivory would indirectly affect the regeneration process of these species. Thus, we expect that leaf herbivory will have greater impacts on seedling growth and survival under the shaded conditions in the understory than under the higher light levels beneath canopy gaps.

ACKNOWLEDGEMENTS

We thank Drs. S. Thomas and K. Niiyama for valuable comments and suggestions on an earlier version of this manuscript. We also thank Dr. S. Ohara for advice on chemical analysis. This study was financially supported by the Joint Research Project between the NIES, FRIM and UPM (Global Environmental Research Program, supported by Japan Environmental Agency, Grant No. E-2-3).

REFERENCES

Aide, T. M. & Zimmerman, J. K. (1990) Patterns of insect herbivory, growth, and survivorship in juveniles of a neotropical liana. Ecology 71: 1412-1421.

Appanah, S. & Turnbull, J. M. (1998) A review of dipterocarps. Taxonomy, ecology and silviculture. Center for International Forestry Research, Bogor.

Appanah, S. & Weinland, G. (1993) Planting quality timber trees in Peninsular Malaysia-a review. Forest Research Institute Malaysia, Kuala Lumpur.

Baldwin, I. T. & Schultz, J. C. (1988) Phylogeny and the patterns of leaf phenolics in gap- and forest-adapted *Piper* and *Miconia* understory shrubs. Oecologia 75: 105-109.

Bate-Smith, E. C. & Whitmore, T. C. (1959) Chemistry and taxonomy in the Dipterocarpaceae. Nature 184: 795-796.

Bazzaz, F. A. & Pickett, S. T. A. (1980) Physiological ecology of tropical succession: a comparative review. Annu. Rev. Ecol. Syst. 10: 287-310.

Beck, S. D. (1965) Resistance of plants to insects. Annu. Rev. Entomol. 10: 207-232.

Becker, P. (1981) Potential physical and chemical defenses of *Shorea* seedling leaves against insects. Malay. For. 44: 346-356.

Becker, P. (1983) Effect of insect herbivory and artificial defoliation on survival of *Shorea* seedlings. In Sutton, S. L., Whitmore, T. C. & Chadwick, A. C. (eds). Tropical rain forest: ecological and management. Special publication Number 2 of the British Ecological Society, Blackwell Scientific Publications, Oxford, pp.241-252.

Benson, W. W. (1978) Resource partitioning in passion vine butterflies. Evolution 32: 493-518.

Bryant, J. P., Chapin, F. S. & Klein, D. R. (1983) Carbon/nutrient balance of boreal plants in relation to vertebrate herbivory. Oikos 40: 357-368.

Chabot, B. F. & Hicks, D. J. (1982) The ecology of leaf life spans. Annu. Rev. Ecol. Syst. 13: 229-259.

Choong, M. F., Lucas, P. W., Ong, J. S. Y., Pereira, B., Tan, H. T. W. & Turner, I. M. (1992) Leaf

fracture toughness and sclerophylly—their correlations and ecological implications. New Phytol. 121: 597-610.
Coley, P. D. (1983) Herbivory and defensive characteristics of tree species in a lowland tropical forest. Ecol. Monogr. 53: 209-233.
Coley, P. D. (1986) Costs and benefits of defense by tannins in a neotropical tree. Oecologia 70: 238-241.
Coley, P. D. & Barone, J. A. (1996) Herbivory and plant defenses in tropical forests. Annu. Rev. Ecol. Syst. 27: 305-335.
Coley, P. D. & Kursar, T. A. (1996) Anti-herbivore defenses of young tropical leaves: Physiological constraints and ecological trade-offs colonization. In Mulkey, S. S., Chazdon, R. L. & Smith, A. P. (eds). Tropical Forest Plant Ecophysiology. Chapman & Hall, New York, pp.305-336.
Coley, P. D., Bryant, J. P. & Chapin, F. S. (1985) Resource availability and plant antiherbivore defense. Science 230: 895-899.
Denslow, J. S. (1987) Tropical rainforest gaps and tree species diversity. Annu. Rev. Ecol. Syst. 18: 431-451.
Dudt, J. F. & Shure, D. J. (1994) The influence of light and nutrients on foliar phenolics and insect herbivory. Ecology 75: 86-98.
Feeny, P. (1970) Seasonal changes in oak leaf tannins and nutrients as a cause of spring feeding by winter moth caterpillars. Ecology 51: 565-581.
Feeny, P. (1976) Plant apparency and chemical defense. In Wallace, J. & Mansell, R. L. (eds.) Biochemical interactions between plants and insects. Recent Advances in Phytochemistry. Plenum Press, New York, pp.1-40.
Fleck, D. C. & Tomback, D. F. (1996) Tannin and protein in the diet of a food-hoarding granivore, the Western Scrub-Jay. Condor 98: 474-482.
Folgarait, P. J. & Davidson, D. W. (1994) Antiherbivore defenses of myrmecophytic *Cecropia* under different light regimes. Oikos 71: 305-320.
Hartley, S. E. & Jones, C. G (1997) Plant chemistry and herbivory, or why the world is green. In Crawley M. J. (ed.) Plant Ecology, Blackwell Scientific Publications, Oxford, pp.284-324.
Howlett, B. E. & Davidson, D. W. (2001) Herbivory and planted dipterocarp seedlings in secondary logged forests and primary forests of Sabah, Malaysia. J. Trop. Ecol. 17: 285-302.
Lincoln, D. E. & Mooney, H. A. (1984) Herbivory on *Diplacus aurantiacus* shrubs in sun and shade. Oecologia 64: 173-176.
Lowman, M. D. (1984) An assessment of techniques for measuring herbivory: is rainforest defoliation more intense than we thought? Biotropica 16: 264-268.
Maiorana, V. C. (1981) Herbivory in sun and shade. Biol. J. Linn. Soc. 15: 151-156.
Martin, J. S. & Martin, M. M. (1982) Tannin assays in ecological studies: lack of correlation between phenolics, proanthocyanidins and protein-precipitating constituents in mature foliage of six oak species. Oecologia 54: 205-211.
Martin, J. S. & Martin, M. M. (1983) Tannin assays in ecological studies, preparation of ribulose-1.5-bisphosphate carboxylase/oxygenase by tannic acid, quebracho, and oak foliage extracts. J. Chem. Ecol. 9: 285-294.
Mattson, W. J. J. (1980) Herbivory in relation to plant nitrogen content. Annu. Rev. Ecol. Syst. 11: 119-161.
Mole, S., Ross, J. A. M. & Waterman, P. G (1988) Light-induced variation in phenolic levels on foliage of rain-forest plants. 1. Chemical changes. J. Chem. Ecol. 14: 1-21.
Moran, N. & Hamilton, W. D. (1980) Low nutritive quality as defense against herbivores. J. Theor. Biol. 86: 247-254.
Numata, S., Kachi, N., Okuda, T. & Manokaran, N. (2000) Leaf damage and traits of dipterocarp seedlings in a lowland rain forest in Peninsular Malaysia. Tropics 9: 237-243.
Rhoades, D. F. & Cates, R. G (1976) Towards a general theory of plant anti-herbivore chemistry. In Wallace, J. & Mansell, R. L. (eds). Biochemical Interactions between Plants and Insects. Recent Advances in Phytochemistry. Plenum Press, New York, pp.168-213.

Rhoades, D. F. (1979) Evolution of plant chemical defense against herbivores. In Rosenthal, G. A. & Janzen, D. H. (eds). Herbivores- their Interaction with Secondary Plant Metabolites. Academic Press, New York, pp.3-54.
Schowalter, T. D., Hargrove, W. W. & Crossley, D. A. (1986) Herbivory in forested ecosystems. Annu. Rev. Entomol. 31: 177-196.
Shure, D. J. & Wilson, L. A. (1993) Patch-size effects on plant phenolics in successional openings of the Southern Appalachians. Ecology 74: 55-67.
Symington, C. F. (1943) Forester's Manual of Dipterocarps. Malayan Forest Records No. 16. Penerbit Universiti Malaya, Kuala Lumpur, 244pp.
Turner, I. M. (1989) A shading experiment on some tropical rain forest tree seedlings. J. Trop. For. Sci. 1: 383-389.
Turner, I. M. (2001) The Ecology of Trees in the Tropical Rain Forest. Cambridge University Press, Cambridge.
Vincent, A. J. (1961) A note on the growth of *Shorea macroptera* Dyer (*Meranti melantai*). Malay. For. 24: 190-209.
Waring, R. H., McDonald, A. J. S., Larsson, S., Ericsson, T., Wiren, A. & Arwidsson, E. (1985) Differences in chemical composition of plants grown at constant relative growth rates with stable mineral nutrition. Oecologia 66: 157-160.
Waterman, P. G & Mole, S. (1994) Analysis of phenolic plant metabolites. Blackwell Scientific Publications, Inc., Oxford.
Whitmore, T. C. (1984) Tropical rain forest of the far east Asia. Oxford University Press, New York, 238pp.

35 Native, Wild Pigs (*Sus scrofa*) at Pasoh and Their Impacts on the Plant Community

Kalan Ickes[1] & Sean C. Thomas[2]

Abstract: Although many large-bodied terrestrial mammals are presently extinct or exceedingly rare at Pasoh Forest Reserve (Pasoh FR), the native, wild pigs (*Sus scrofa*) are thriving. Line transect surveys conducted in 1996 and 1998 yielded density estimates of 47.0 and 27.0 pigs/km^2, respectively. These are among the highest density estimates ever recorded for this species. Important factors contributing to the maintenance of such high pig density at Pasoh FR are likely the absence of large carnivores and an abundant year-round food supply in the oil palm plantations that virtually surround the reserve. Several studies have recently addressed some of the effects that such a high density of pigs may have on the understory plant community at Pasoh FR. To quantify the effects of soil rooting and seed predation by pigs on woody saplings, pig exclosures were constructed in the primary forest at the center of the reserve. After two years the number of woody plant recruits, total stem density, species richness, and height growth were greater inside enclosed areas than paired control plots to which pigs had access. Another study examined the prevalence of nest building by pigs. When ready to deliver young, pregnant females snap off or uproot tree and liana saplings 40–350 cm in height. These stems are meticulously piled into dome-shaped structures under which the female gives birth. Annual surveys of pig nests were conducted in the western 25-ha of the permanent 50-ha tree plot. More than 600 nests were located over a four-year period, for an average density of 6.0 pig nests constructed/ha/year. Based on examinations of 10 nests and damage to the surrounding areas, each nest contained on average 145 uprooted stems and an additional 122 stems that were snapped off, leaving behind stumps. Pigs gathered these stems from an average area of 244 m^2 surrounding each nest, damaging or killing 53% of woody understory vegetation > 70 cm tall and < 2.0 cm diameter at breast height. We estimated that pigs caused 29% of the total mortality of trees in the 1–2 cm Diameter at Breast Height (DBH) size class. Pigs also preferentially used saplings of the Dipterocarpaceae as nest construction material. Nest construction created > 85,000 stumps/km^2 at Pasoh FR, suggesting that understory regeneration and future species composition may be influenced considerably by the resprouting abilities of damaged plants. More than 1,800 stumps were examined for 36 months to investigate resprouting. There were large differences in resprouting success among species with different life history characteristics and taxonomic associations. Stumps of the Dipterocarpaceae had by a wide margin the lowest survivorship at 36 months of the 19 most common families in the study. Overall, the data suggest that if pigs continue to be hyper-abundant in the reserve there could be a shift away from the economically and ecologically paramount Dipterocarpaceae.

Key words: edge effect, exclosure, mammal density, nest building, pigs, resprouting, rooting.

[1] Department of Biological Sciences, Louisiana State University, Baton Rouge, LA 70803, USA.
Present address: 1274 Overlook Drive, Washington, PA 15301, USA.
E-mail: kalan42@hotmail. com
[2] University of Toronto, Canada.

1. THE LARGER MAMMALS OF PASOH FOREST RESERVE

The mammal community at Pasoh Forest Reserve (Pasoh FR) has undergone significant changes in recent decades, as has occurred in other relatively small, isolated forest reserves around the world (Woodroffe & Ginsberg 1998). The landscape within and around Pasoh FR has been altered from continuous primary forest to a mosaic of primary and selectively logged forest, extensive tree plantations, agricultural land planted with annual crops, and human habitation. These changes have decreased the extent of forested landscape such that mammals with large home ranges no longer occur at Pasoh FR, or are exceedingly rare. In addition, the proximity of human habitation and agricultural areas to the reserve requires that dangerous mammals be shot or relocated elsewhere when they wander outside the forest boundary. Table 1 lists the non-flying, medium to large bodied mammals formerly known or presumed to have occurred at Pasoh FR. There are a number of factors that have probably contributed to the decrease in abundance or occurrence of many of these mammal species in this reserve. These factors include the small size of the reserve, relatively homogenous habitat within the protected area, poaching, habitat fragmentation in the surrounding area, relative isolation,

Table 1 Status of large mammals found historically at (or near) Pasoh FR. Information presented here comes from our own experience at Pasoh FR (from 1989–1999), Kemper (1985), Kochummen (1997), discussions with other visiting scientists, and the researchers and staff of the Forestry Research Institute Malaysia (FRIM).

Species	Status	Comments
Potential predators of *Sus scrofa*		
Tiger (*Panthera tigris*)	Absent	Apparently resident at Pasoh until the 1960s. There have been unconfirmed reports of vagrant individuals near the reserve in the 1990s.
Clouded leopard (*Neofelis nebulosa*)	Absent	Present in nearby hill forest areas.
Panther (*Panthera pardus*)	Extremely rare	
Malayan sun bear (*Helarctos malayanus*)	Absent	
Red dog (*Cuon alpinus*)	Absent	
Potential competitors of *Sus scrofa*		
Malayan tapir (*Tapirus indicus*)	Extremely rare	Only one individual known to be present as of 1998.
Sambar (*Cervus unicolor*)	Extremely rare	
Indian Elephant (*Elephas maximus*)	Absent	The last elephant was removed by the Wildlife Department in 1989.
Javan rhinoceros (*Rhinoceros sondaicus*)	Absent	Historical status at Pasoh FR unknown.
Sumatran rhinoceros (*Dicerorhinus sumatrensis*)	Absent	Historical status at Pasoh FR unknown.
Gaur (*Bos gaurus*)	Absent	Historical status at Pasoh FR unknown.
Barking deer (*Muntiacus muntjak*)	Extremely rare if still present	
Greater Mouse-deer (*Tragulus napu*)	Absent	
Lesser Mouse-deer (*Tragulus javanicus*)	Common	
Malayan porcupine (*Hystrix brachyura*)	Present but status unknown	
Brush-tailed porcupine (*Atherurus macrourus*)	Present but status unknown	
Pangolin (*Manis javanica*)	Common	
Bearded pig (*Sus barbatus*)	Extremely rare visitor	A group of 18 pigs (6 adult, 12 young) was seen once at Pasoh during a masting event in 1996.
Wild pig (*Sus scrofa*)	Abundant	
White-handed Gibbon (*Hylobates lar*)	Common	
Banded Leaf Monkey (*Presbytis melalophos*)	Common	
Dusky Leaf Monkey (*Presbytis obscura*)	Common	
Pig-tailed Macaque (*Macaca nemestrina*)	Common	At least two large troops occupy the western portion of the reserve, frequently seen feeding in the Palm oil tree plantations and carrying fruits back into the forest.
Long-tailed Macaque (*Macaca fascicularis*)	Uncommon	
Siamang (*Hylobates syndactylus*)	Present, but status unknown	This species can be heard vocalizing from the hilly portion of the reserve to the northeast, but we are aware of no data regarding their status.

and removal of potentially dangerous animals from the surrounding oil palm plantations. However, not all larger mammal species have decreased in population size or disappeared from the reserve. One notable exception is the native pig, *Sus scrofa*, which is thought to have increased dramatically in density at Pasoh FR over the last few decades.

2. THE UBIQUITOUS PIGS OF PASOH FR

Sus scrofa naturally occurs throughout Europe and Asia, extending as far south as the Malay Peninsula, Sumatra and Java. Though *S. scrofa* commonly inhabits primary forests, the species also thrives in human-altered landscapes throughout its native and introduced range, often becoming an agricultural pest (Boitani et al. 1994; Cargnelutti et al. 1992). Ickes (2001a) conducted line transect surveys at Pasoh FR from May–October 1996 and during the same period in 1998 to estimate pig density. The same 13 transects were used in both survey years and the total length of surveyed transects was approximately 80 km in each year. In 1996, 44 sightings of *S. scrofa* consisting of 166 individuals were recorded. In 1998, 39

Table 2 Density estimates of *Sus scrofa* within its native range. Locations are organized by country in alphabetical order. Data from Pasoh FR are shown in bold face.

Country	Location	Vegetation type	Density N / km^2	Source
Tropical and subtropical sites:				
India	Various locations	Riparian forest, grassland, marsh	1.2 - 2.9	Spillett 1967a,b,c
India	Nagarahole National Park	Tropical moist deciduous and dry deciduous forest	4.2	Karanth & Sunquist 1992
India	Gir Forest, Gujarat	Tropical dry deciduous forest	0.1	Berwick & Jordan 1971
India	Gir Forest, Gujarat	Tropical dry deciduous forest	< 1.0	Khan et al. 1996
India	Kanha National Park	Tropical dry deciduous forest, Sal forest, grassland	1.2	Schaller 1967
Indonesia	Peucang Island, Ujung Kulong Nat Park	Coastal, disturbed, and dipterocarp forest; swamp	27 - 32	Pauwels 1980
Malaysia	Pasoh Forest Reserve	Lowland dipterocarp rain forest	27 and 47	Ickes 2001a
Malaysia	Dindings district, Perak	Dipterocarp and coastal forest, agriculture	< 1.0	Diong 1973
Nepal	Royal Chitwan National Park, Java	Riverine forest, tall grassland	5.8	Seidensticker 1976
Nepal	Royal Karnali-Bardia Wildlife Reserve	Sal, riverine, and hardwood forests; grassland	4.2	Dinerstein 1980
Nepal	Royal Karnali-Bardia Wildlife Reserve	Savanna, grassland	3.8	Dinerstein 1980
Pakistan	Chiniot, Sargodha District	Sugarcane fields	32 and 72	Shafi & Khokhar 1985
Pakistan	Changa Manga Forest	Irrigated forest plantation	10.4	Inayatullah 1973
Pakistan	Thatta District,	Riparian forest, agricultural fields, swamp	3.7	Smiet et al. 1979
Sri Lanka	Ruhuna National Park	Lowland monsoon forest, savanna woodland, grassland	0.7	Santiapillai & Chambers 1980
Sri Lanka	Wilpattu National Park	Lowland monsoon and monsoon scrub forest	0.3 - 1.2	Eisenberg & Lockhart 1972
Sri Lanka	Gal Oya National Park	Lowland monsoon forest, savannah woodland, grassland	0.6	McKay 1973
Thailand	Huai Kha Khaeng Wildlife Sanctuary	Several dry forest habitats	< 0.5	Srikosamatara 1993
Temperate sites:				
Italy	Maremma Natural Park	Mediterranean macchia	5.1 - 6.1	Massei & Tonini 1992
Poland	Various locations	Deciduous forest / agriculture	1.2 - 1.8	Mackin 1970
Poland	Various locations	Coniferous forest	0.2 - 3.5	Pucek et al. 1975
Poland	Various locations	Deciduous forest	0.4 - 3.1	Pucek et al. 1975

sightings were made of 129 individuals. Population density estimates were based on analysis of the distribution of perpendicular distances of pigs from transect lines, using the half-normal/cosine detection function of Buckland et al. (1993), as implemented by the Distance program of Thomas et al. (1998). Estimated densities for the two survey periods were 47.0 pigs/km^2 in 1996 (with a 95% confidence interval of 28.2–78.6 pigs/km^2), and 27.0 pigs/km^2 in 1998 (95% C.I. of 16.2–44.7 pigs/km^2). Similar numbers of adult pigs were recorded in 1996 and 1998 (93 and 99 individuals, respectively), but 2.5 times as many young pigs were seen in 1996 (73 vs. 30 young). 1996 was a mast year at Pasoh FR, and more *S. scrofa* individuals reproduced during that period compared with 1998, as evidenced by higher numbers of recently constructed reproductive nests built by females immediately prior to giving birth.

Density estimates of *S. scrofa* from a diverse range of habitats and locations across its native range are given in Table 2. Clearly, the estimates of *S. scrofa* density recorded at Pasoh FR are considerably higher than at other forest locations in mainland Europe and Asia. Furthermore, evidence from two other studies in Peninsular Malaysia suggests that the *S. scrofa* density in lowland dipterocarp rain forests of Peninsular Malaysia is not generally this high. Working in Northwest Peninsular Malaysia, Diong (1973) estimated the entire pig density of the Dindings district to be 800 pigs, or less than 1.0/km^2. Laidlaw (1994) conducted mammal surveys in seven Virgin Jungle Reserves within Peninsular Malaysia (Pasoh FR not among them) and reported the average number of pig sightings/km of transect walked. Pigs were observed seven times more frequently at Pasoh FR when compared to these seven Virgin Jungle Reserves (Ickes 2001a). Therefore, it appears that changes in and around Pasoh FR have contributed to a dramatic increase in *S. scrofa* density within this forest reserve compared to other areas within the native range of this species, including other lowland dipterocarp forests within Peninsular Malaysia.

One factor that has likely contributed to increased pig densities at Pasoh FR is the local extinction of the natural predators of *S. scrofa* (Table 1). Tigers (*Panthera tigris*) have been essentially absent from Pasoh FR for many years, and there is no evidence that panthers (*Panthera pardus*) or clouded leopards (*Neofelis nebulosa*) are still present. Pythons (*Python reticulatus*) are extant in the reserve and are known to predate *S. scrofa* (Shine et al. 1998) but seem unlikely to have a large impact. However, even if there is currently little or no natural predatory pressure on pigs, population density could still be constrained by food supply. The local extinction of several potential competitors (Table 1) may thus also contribute to higher pig density. Also, fallen fruits of the African oil palm trees (*Elaeis guineensis* Jacq.) in the extensive plantations that border the forest reserve provide a tremendous year-round food supply. Pigs can often be observed consuming oil palm fruits near the reserve edge (K. Ickes & S. Thomas, pers. obs.). It seems likely that this combination of an absence of natural predators and a considerable year-round food supply has allowed *S. scrofa* to proliferate at Pasoh FR, attaining some of the highest densities in the world for this species.

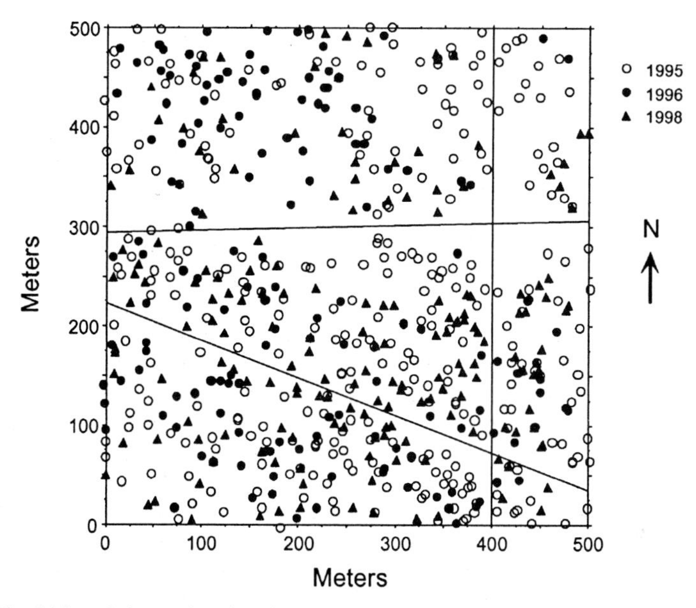

Fig. 1 Map of pig nest locations in the western half of the 50-ha plot at Pasoh FR. Solid lines are the three trails through the study area.

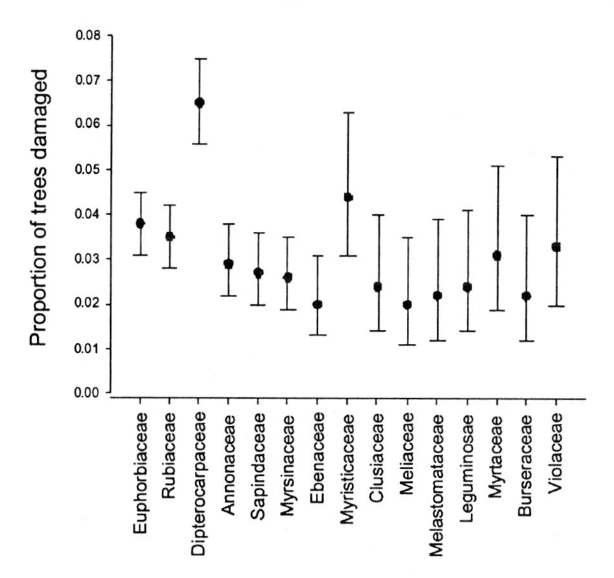

Fig. 2 Proportion of stems within a 15-m radius of each pig nest that were used in nest construction. All nests from the 1996 survey were used in calculating the proportion taken. Families are arranged from left to right on the x-axis in decreasing order of the number of stems from that family available to pigs for nest construction: 2,970 saplings 1–2 cm in Diamter at Breast Height (DBH) were available from the Euphorbiaceae while 515 were available from the Violaceae. Only families with > 500 stems available are shown. Error bars represent the 95% confidence interval of the proportion taken, based on the normal approximation to the binomial distribution with continuity correction.

3. NEST BUILDING BY PIGS AND IMPACTS ON UNDERSTORY PLANTS

Throughout the range of *S. scrofa*, females seek out or construct nesting sites shortly before giving birth. The amount and type of material gathered for nest building depends on environmental conditions and availability (Jensen 1989), but in most habitats the young are birthed under dense vegetation such as shrubs, or herbaceous plants are gathered into a pile. In both cases the impact on the surrounding vegetation is minimal. However, there are no dense shrub species in lowland rain forest understory and herbaceous ground cover is rare at Pasoh FR. Consequently, female pigs predominately use woody vegetation when building nests. She grasps a 40–350 cm tall sapling in her jaws and twists her head to the side, either snapping the stem and leaving behind a stump or uprooting the sapling entirely. Stems are then arrayed in a radial fashion over a shallow, crater-like depression in the soil, with the foliage in the center and the snapped or uprooted ends facing outward (Medway 1983). More than one individual may work to construct a nest, but on the few occasions when this was observed the second pig was much smaller, possibly a daughter of the larger pig from a previous birth. The pigs utilized different sizes of woody plants: the larger pig snapped off or uprooted taller, larger diameter plants while the smaller pig gathered smaller stems. Plants that are uprooted entirely will die, but those that are snapped off may resprout.

Ickes et al. (2001b) made use of the > 330,000 trees tagged in the Pasoh FR 50-ha forest dynamics plot to investigate the impacts of nest building by *S. scrofa* on the plant community. The 25-ha area comprising the western half of the plot was surveyed for pig nests in May–October 1995, 1996 and 1998, and each nest located was mapped on the 50-ha plot coordinate system. When a pig uses a tree of > 1.0 cm Diamter at Breast Height (DBH) for nest construction from within the 50-ha plot, the tag is either still attached to the tree by it's ribbon or on the ground around the stump that is left behind when a tree snaps. Therefore, each nest and the stumps in the immediate vicinity were searched for tree tags and the numbers were recorded. In cases where tags were still attached to stumps, it was noted whether or not the stump had successfully resprouted or was dead at the time of the pig-nest survey.

A total of 643 nests were located and mapped in the three surveys. An average of 6.0 nests ha^{-1} y^{-1} were constructed at Pasoh FR. The area of understory vegetation affected by construction of a single nest was 244 m^2. Each nest was composed of an average of 145 snapped saplings and 122 uprooted saplings. Overall, pigs damaged or killed 53% of all woody saplings < 70 cm tall and < 2.0 cm in DBH within the 244 m^2 affected area around each nest (Ickes et al. 2001b).

Nests were clumped spatially within years, but no spatial clumping of nest location data pooled among years was detected. These results indicate that pigs constructed nests throughout the area, but that there were local hotspots of nesting activity each year. Pigs also clearly avoided construction of nests near the three main trails used by researchers in the plot (Fig. 1).

More than 3,400 tags from 50-ha plot trees were recovered from the 643 pig nests or from the nearby stumps. From these tags it was estimated that pigs caused 0.53% annual mortality of trees 1–2 cm in DBH, or 29% of the total mortality for trees in this size class. However, saplings were not gathered randomly by pigs. Saplings from the Dipterocarpaceae were used significantly more often than saplings from other families (Fig. 2 ; Ickes et al. 2001b).

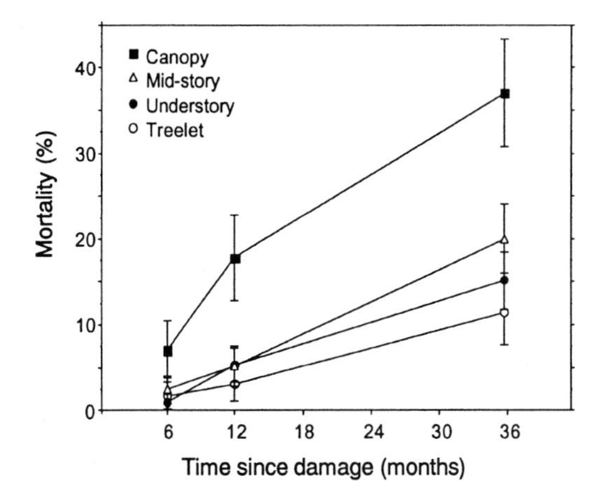

Fig. 3 Mortality as a function of time since damage for trees grouped by stature at reproductive maturity (modified from Ickes et al. 2003). Error bars represent 95% confidence limits based on a normal approximation. "Treelet" were defined as species that rarely exceed 2.0 cm in DBH, "Understory" species were those that regularly reach maximum DBH 2–10 cm, "Mid-story" 10–30 cm, and "Canopy" < 30 cm.

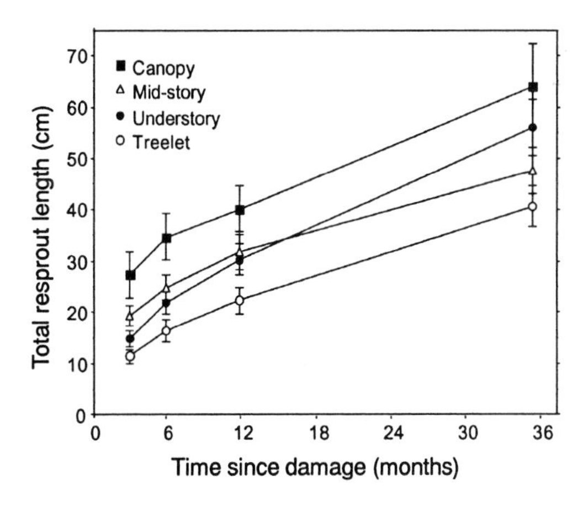

Fig. 4 Total resprout length (the sum of the lengths of all resprouted branches) as a function of time since damage for trees grouped by stature at reproductive maturity (modified from Ickes et al. 2003). Error bars represent 95% confidence limits. See Fig. 3 for size class descriptions.

4. STUMP RESPROUTING FOLLOWING PIG DISTURBANCE

With pig density and the resulting number of new nests constructed each year at Pasoh FR so high, we estimate that there are more than 85,000 saplings/km^2 being converted to stumps annually at Pasoh FR. In areas of the highest pig nest density, the understory consists largely of a dense carpet of thousands of these resprouting stumps, and visibility in the normally dense understory may be 75 meters or more. As a consequence of this large number of stumps created annually, differences in resprouting ability among the more than 1,200 tree and liana species that occur at Pasoh FR are likely to play an important role in determining the future composition of the forest understory.

Ickes et al. (2003) investigated resprouting at Pasoh FR by tagging and recensusing more than 1,450 stumps formed during pig nest construction and an additional 350 experimentally created stumps. Stumps were surveyed at 3, 6, 12 and 36 months post damage with all resprouting branches measured and leaves counted. More than 90% of stumps were alive 3 months post-damage, but new shoot production was slow: 15% of live plants had not yet produced any new shoots, while 42% did not have any fully expanded leaves. Despite the high initial rate of resprouting for all stems, by 36 months mortality was considerably higher for stumps compared to adjacent undamaged saplings (33% and 9%, respectively). Stump basal diameter was strongly, positively correlated with survival, number of new shoots produced, and total new shoot length.

Interspecific patterns of growth and survivorship following damage suggest that the tree community at Pasoh FR may change predictably in response to pig-related damage. Tree species that grow to canopy size trees had significantly lower survivorship following damage than species that attain smaller sizes at maturity (Fig. 3). Interestingly, individuals of canopy species that did survive stem snap had significantly greater new shoot production (Fig. 4). There was also considerable variation in resprouting ability among the most common species and families at Pasoh FR. Of particular note was the low resprouting ability of species in the families Dipterocarpaceae and Euphorbiaceae.

Clark & Clark (1991; see also Guariguata 1998), suggested that one adaptational strategy to stem damage is to maximize resource allocation to growth, since the probability of mortality following damage is generally inversely related to tree size. Although originally applied to damage resulting from falling debris, this hypothesis may also pertain to damage from pigs during nest construction in that pigs seldom take saplings > 2.5 cm in DBH. At Pasoh FR, the faster-growing species possibly favored by this mechanism might be expected to include larger-statured canopy trees, which tend to show more rapid growth than do understory species (Thomas 1996). This "escape" factor may thus partially offset declines among faster-growing canopy species (such as the red meranti group of *Shorea*) at Pasoh FR.

5. SOIL ROOTING AND SEED PREDATION BY PIGS

Although no studies on the diet of *S. scrofa* have been conducted in the tropical extent of it's range, studies from temperate areas suggest that more than 75% of its diet is subterranean in origin. This seems to be the case in the forest at Pasoh FR as well. Pigs can be heard and often seen rooting through the soil, and large patches of ground in or along the trails are rooted on a daily basis. Soil is routinely churned to a depth of 15 cm, and often as deep as 30 cm. In these repeatedly rooted locations there is no surface leaf litter, since the rooting process buries freshly fallen leaves.

Kemper & Bell (1985) found that small terrestrial mammals were rare or absent at Pasoh FR in heavily rooted-areas, and suggested that the leaf-litter invertebrates and vertebrates upon which they feed could not survive without an intact litter layer. This lack of a litter layer also has repercussions for the physical environment experienced by germinating seeds and seedlings, as well as soil nutrient dynamics. The physical action of digging has a direct impact on the seedling community in that smaller plants may be knocked over and trampled. These rooted areas are characterized by bare, churned soil and a scarcity or lack of small seedlings.

Ickes et al. (2001) quantified the impacts of soil rooting on understory plant density and diversity by experimentally excluding pigs using fences. Eight exclosures, 7 m × 7 m × 1.5 m, were constructed at 40 m intervals immediately south of the 50-ha plot, in the primary forest near the center of the reserve. Exclosures were of an open top design, and consisted of 4 cm × 4 cm mesh chain link fencing. Therefore, only terrestrial mammals larger than the mesh size could not enter the exclosures. Centered within each exclosure were 5 m × 5 m plots, paired with two control plots outside each fenced area. Within plots all free-standing woody plants were tagged and surveyed in 1996, and again in 1998.

After two years, no differences were observed in sapling mortality, but three times as many plants had recruited into exclosures compared with the control plots (Table 3). Stem density was highly correlated with species richness, which also increased significantly inside exclosures. Height growth of plants was 53% greater in the exclosures for trees between 101–700 cm tall. Trees less than 30–100 cm tall, however, exhibited no growth differences between treatments (Table 3). This exclosure study continues, and recent observations indicate persistent treatment effects on stem density, growth, and species diversity.

Table 3 Results of ANOVAs of the effects of pig exclusions on plant mortality, recruitment, recruitment height, number of seedlings, stem density, species richness, Fisher's α, liana proportion of stems, and growth. Average values are LS Means calculated in SAS using Proc Mixed. All variables had a covariate in the ANOVA except height of recruits, number of seedlings, and proportion of recruits made up of lianas. Modified from Ickes et al. (2001).

	LS Means ± S.E.		
Variable	Exclosure	Control	*P*
Stand Structure			
Mortality (%)	8.7 ± 2.37	10.0 ± 1.67	ns
Recruits (> 30 cm tall)	48.8 ± 10.6	14.9 ± 8.4	0.006
Height of recruits (cm)	36.1 ± 0.72	34.5 ± 0.59	0.05
Seedlings (< 30 cm tall) 1998	118 ± 21.0	75.5 ± 17.8	0.04
Stem density 1998	142 ± 11.3	107.1 ± 8.4	0.009
Species Diversity			
Species richness 1998	55.7 ± 1.05	50.6 ± 1.06	0.02
Fisher's α 1998	39.7 ± 2.81	45.8 ± 2.31	0.04
Height Growth (cm)			
Trees initially 101 - 700 cm tall	19.6 ± 3.04	12.88 ± 2.41	0.04
Trees initially 30 - 100 cm tall	10.1 ± 1.39	8.23 ± 1.11	ns

6. POTENTIAL LONG-TERM CONSEQUENCES OF PIGS ON FOREST DYNAMICS AT PASOH FR

Elevated pig densities clearly alter patterns of tree recruitment, and over the long term this may be expected to result in changes in forest structure, species composition, and diversity. That pig activities result in reduced density of vegetation in the understory has been confirmed experimentally. We speculate that pig activities are likely to result in: (1) greater light penetration in the understory, favoring more light-demanding tree species; (2) a shift in species composition toward trees that can, as saplings, successfully resprout and grow following removal of leaf and stem tissues by pigs. These two kinds of effects may be in conflict, if light-demanding species tend not to show a high capacity for resprouting. Also, species that show rapid growth in the understory may also potentially escape pig impacts. To the degree that some species are favored over others under high pig densities, a reduction in the diversity of tree species seems likely. The ecologically and economically dominant Dipterocarpaceae seems to be at particular risk from elevated pig activity at Pasoh FR. Not only were dipterocarp saplings used twice as often as stems from other families for nest construction, but dipterocarps had by far the lowest survivorship of snapped stems. If elevated pig densities continue there could be a long-term shift away from the currently dominant Dipterocarpaceae.

7. SYNTHESIS AND CONSERVATION IMPLICATIONS

That large-bodied mammal species can have a substantial role in structuring plant communities has now been well documented in many forest systems (e.g. Leopold 1936; McNaughton & Sabuni 1988; McInnes et al. 1992; Schreiner et al. 1996). In tropical forests particular attention has been given to the role of very large species such as Elephants, whose activities may commonly retard forest regeneration, maintaining forests in an early-successional state (Struhsaker et al. 1996). The impacts of wild pigs on forest vegetation have mainly been investigated where the species has been introduced, such as in the forests of Hawaii (Stone & Loope 1987). The work summarized in this chapter constitutes the first set of studies on *S. scrofa* impacts on vegetation within the species' native range.

Both observational and experimental studies suggest a considerable impact of pigs on forest vegetation at Pasoh FR. One interesting aspect of these pig-tree interactions is that no direct trophic relationship is involved. Pigs do not use trees as a food resource, but rather impact trees as an incidental side effect of foraging and nesting behavior. It also seems unlikely that trees have specifically evolved defense mechanisms that might discourage pig impacts. It is actually difficult to imagine what kind of defense would be effective: for example, even species with extremely spiny stems, such as *Scolopia spinosa*, were frequently taken by nest-building females, as were species that produce copious latex or resin when the stem is cut. There is, however, large interspecific variation in resprouting capacity of saplings. Given the high current pig densities at Pasoh FR, increased dominance of species with higher allocation to roots, and possibly other physiological attributes that enable effective resprouting, seems very likely.

There has been considerable recent debate concerning the existence and nature of large-scale edge effects on ecological processes in tropical forests (e.g. Curran 1999; Woodroffe & Ginsberg 1998), and increased pig abundance at Pasoh FR has been cited as an example of such a large-scale edge effect (Laurance 2000a,b). Whether or not an edge framework is appropriate depends critically on what mechanism(s) account for increased pig abundance (Ickes & Williamson 2000).

Specifically, if access to oil palm fruits is the main cause of increased abundance and pig home-range sizes are sufficiently small, an edge framework may be appropriate. However, if predators or competition with other large mammals is of greater importance, then such a framework may not be entirely appropriate. Wide-ranging species such as tigers likely respond mainly to overall changes in forest cover at the regional scale, and not to fragmentation per se (cf. Fahrig 1997). Simply put, tigers are increasingly less likely to set foot in Pasoh FR as a consequence of forest impacts throughout the Main Range of the peninsula and their interactions with humans in the "matrix" habitat, rather than as a result of any kind of change in habitat within or adjacent to Pasoh FR itself.

Although natural predators of wild pigs are scarce, there is currently hunting of pigs and other animals along the oil palm roads near the edge of Pasoh FR. In light of the negative impacts of pigs within the reserve, hunting activities may be ecologically beneficial by reducing pig population to some extent. Moreover, hunting at the forest edge specifically targets those animals relying on oil palms as a food resource, and generally does not involve entry of hunters beyond the periphery of the reserve (K. Ickes & S. Thomas, pers. obs.). It is even possible that the ease of hunting near the reserve edge may act to discourage illegal poaching within the reserve. It seems likely that the current default policy of "benign neglect" (i.e. turning a blind eye to pig hunting along the reserve margin) may be having some impact, but has obviously not prevented pig populations from reaching levels that are seriously impacting the forest community as a whole. It would clearly be of interest to have quantitative details on pig hunting in and around Pasoh FR, and also on the level and nature of large mammal impacts on vegetation within core areas of very large reserves.

The future of pigs at Pasoh FR promises to be of considerable interest. The oil palm plantations that surround the reserve, planted in the early 1970s, were scheduled for removal and replanting in 2002. If high pig densities in the reserve are largely a result of diet supplementation by oil-palm fruits, then we might expect pig densities to rapidly decrease after this time. On the other hand, there may also be an influx of pigs and other animals resident in the oil palms into the reserve. Regardless of future circumstances, further studies and monitoring of Pasoh FR pigs will be necessary to understand current patterns and temporal changes of a wide range of ecological processes within the reserve.

ACKNOWLEDGEMENTS

We thank Roy and Judi Ickes, Kate Walker, Tristram Seidler, and Alison Li for help during vegetation surveys and mapping in 1996, and Sadali bin Sahat for help with fence construction. We also express gratitude to Dr. Abdul Razak, Director General of the Forestry Research Institute Malaysia (FRIM), for permission to conduct research at Pasoh FR, and to Samad bin Latif for logistical support in and around Pasoh FR. Eric Gardette and various botanists in the FRIM herbarium helped with botanical identifications. We also thank Barry Moser for statistical advice, and G. Bruce Williamson, who wrote the SAS program to calculate Fisher's α and made helpful comments on manuscripts. This research was funded by the Center for Tropical Forest Studies, under the Smithsonian Tropical Research Institute, by a Grant-in-Aid of Research to KI from the National Academy of Sciences, through Sigma Xi, and by an NSF post-doctoral fellowship to SCT. The 50-ha forest plot in the Pasoh FR is an ongoing project initiated by the Forest Research Institute Malaysia. Supplementary funding was provided by the National Science Foundation

(USA); the Conservation, Food, and Health Foundation, Inc. (USA); the United Nations, through its Man and the Biosphere (MAB) program; UNESCO-MAB grants; UNESCO-ROSTSEA; and the continuing support of the Smithsonian Tropical Research Institute (Barro Colorado Island, Panama) and the Center for Global Environmental Research (CGER) at Japan's National Institute for Environmental Studies (NIES).

REFERENCES

Berwick, S. H. & Jordan, P. S. (1971) First report of the Yale-Bombay Natural History society studies of wild ungulates at the Gir Forest, Gujarat, India. J. Bombay Nat. Hist. Soc. 68: 412-23.

Boitani, L., Mattei, L., Nonis, D. & Corsi, F. (1994) Spatial and activity patterns of wild boars in Tuscany, Italy. J. Mammal. 75: 600-612.

Buckland, S. T., Anderson, D. R., Burnham, K. P. & Laake, J. L. (1993) Distance Sampling: Estimating abundance of biological populations. Chapman & Hall, London.

Cargnelutti, B., Spitz, F. & Valet, G. (1992) Analysis of the dispersion of Wild Boar (*Sus scrofa*) in Southern France. In Spitz, F., Janeau, G., Gonzalez, G. & Aulagnier, S. (eds). Ongulés/ ungulates 91. Institute Techerche Grand Mammiferes, Paris-Toulouse, France, pp.419-421.

Clark, D. B. & Clark, D. A. (1991) The impact of physical damage on canopy tree regeneration in tropical rain forest. J. Ecol. 79: 447-457.

Curran, L. M., Caniago, I., Paoli, G. D., Astianti, D., Kusneti, M., Leighton, M., Nirarita, C. E. & Haeruman, H. (1999) Impact of El Niño and logging on canopy tree recruitment in Borneo. Science 286: 2184-2188.

Dinerstein, E. (1980) An ecological survey of Royal Karnali-Bardia wildlife reserve, Nepal, Part III: ungulate populations. Biol. Conserv. 18: 5-38.

Diong, C. H. (1973) Studies of the Malayan Wild Pig in Perak and Johor. Malay. Nat. J. 26: 120-151.

Eisenberg, J. F. & Lockhart, M. (1972) An ecological reconnaissance of Wilpattu National Park, Ceylon. Smithson. Contrib. Zool. 101: 1-118.

Fahrig, L. (1997) Relative effects of habitat loss and fragmentation on population extinction. J. Wildl. Manage. 61: 603-610.

Genov, P. (1981) Significance of natural biocenoses and agrocenoses as the source of food for wild boar. Ekologia Polska 29: 117-136.

Guariguata, M. R. (1998) Response of forest tree saplings to experimental mechanical damage in lowland Panama. For. Ecol. Manage. 102: 103-111.

Ickes, K. (2001a) Density of Wild pigs (*Sus scrofa*) in a lowland dipterocarp forest of Peninsular Malaysia. Biotropica 33: 682-690.

Ickes, K. (2001b) The effects of wild pigs(Sus scrofa) on woody understory vegetation in a lowland rain forest of Malaysia. Ph.D. diss., Louisiana State University.

Ickes, K. & Williamson, G. B. (2001) Edge effects and ecological processes: are they on the same scale? Trends Ecol. Evol. 15: 373.

Ickes, K., DeWalt, S. J. & Appanah, S. (2001) Effects of native pigs (*Sus scrofa*) on the understorey vegetation in a Malaysian lowland rain forest: an exclosure study. J. Trop. Ecol. 17: 191-206.

Ickes, K., DeWalt, S. J. & Thomas, S. C. (2003) Resprouting of woody saplings following stem snap by pigs in a Malaysian rain forest. J. Ecol. (in press).

Inayatullah, C. (1973) Wild boar in West Pakistan. Bulletin No. 1. Pakistan Forest Institute, Peshawar.

Karanth, K. U. & Sunquist, M. E. (1992) Population structure, density and biomass of large herbivores in the tropical forests of Nagarahole, India. J. Trop. Ecol. 8: 21-35.

Kemper, C. (1988) The mammals of Pasoh Forest Reserve, Peninsular Malaysia. Malay. Nat. J. 42: 1-19.

Kemper, C. M. & Bell, D. T. (1985) Small mammals and habitat structure in lowland rain forest

of Peninsular Malaysia. J. Trop. Ecol. 1: 5-22.
Khan, J. A., Chellam, R., Rodgers, W. A. & Johnsingh, A. J. T. (1996) Ungulate densities and biomass in the tropical dry deciduous forests of Gir, Gujarat, India. J. Trop. Ecol. 12: 149-162.
Kochummen, K. M. (1997) Tree flora of Pasoh forest. Forest Research Institute Malaysia, Kuala Lumpur.
Laidlaw, R. K. (1994) The Virgin Jungle Reserves of Peninsular Malaysia: the ecology and dynamics of small protected areas in managed forest. Ph.D. diss., Cambridge Univ.
Laurance, W. F. (2000a) Do edge effects occur over large spatial scales? Trends Ecol. Evol. 15: 134-135.
Laurance, W. F. (2000b) Edge effects and ecological processes: are they on the same scale? Reply. Trends Ecol. Evol. 15: 373.
Leopold, A. (1936) Deer and Dauerwald in Germany, 2. Ecology and policy. J. For. 34: 460-466.
Mackin, R. (1970) Dynamics of damage caused by wild boar to different agriculture crops. Acta Theriologica 15: 447-458.
Massei, G. & Tonini, L. (1992) The management of Wild Boar in the Maremma Natural Park. In Spitz, F., Janeau, G., Gonzalez, G. & Aulagnier, S. (eds). Ongulés/ungulates 91. Institute Techerche Grand Mammiferes, Paris-Toulouse, France, pp.443-445.
McInnes, P. F., Naiman, R. J., Pastor, J. & Cohen, Y. (1992) Effects of Moose browsing on vegetation and litter of the Boreal Forest, Isle Royale, Michigan, USA. Ecology 73: 2059-2075.
McKay, G. M. (1973) Behavior and ecology of the Asiatic elephant in Southeastern Ceylon. Smithson. Contrib. Zool. 125: 1-113.
McNaughton, S. J. & Sabuni, G. A. (1988) Large African mammals as regulators of vegetation structure. In Werger, M. J. A., Van der Aart, P. J. M., During, H. J. & Verhoeven, J. T. A. (eds). Plant form and vegetation structure. Academic Publishing, The Hague, pp.339-354.
Medway, L. (1983) The wild mammals of Malaya (Peninsular Malaysia) and Singapore. Oxford University Press, Kuala Lumpur.
Pauwels, W. (1980) Study of *Sus scrofa vittatus*, its ecology and behavior in Ujung Kulon Nature Reserve, Java, Indonesia. Ph.D. diss., Univ. Basel.
Pucek, Z., Bobek, B., Labudzki, L., Mitkowski, L., Murow, K. & Tomek, A. (1975) Estimates of density and number of ungulates. Polish Ecol. Studies 1: 121-135.
Santiapillai, C. & Chambers, M. R. (1980) Aspects of the population dynamics of the wild pig (*Sus scrofa* Linnaeus, 1758) in the Ruhuna National Park, Sri Lanka. Spixiana 3: 239-250.
Schaller, G. B. (1967) The deer and the tiger: a study of wildlife in India University of Chicago Press, Chicago.
Schreiner, E. G., Krueger, K. A., Happe, P. J. & Houston, D. B. (1996) Understory patch dynamics and ungulate herbivory in old-growth forests of Olympic national park, Washington. Can. J. For. Res. 26: 255-265.
Seidensticker, J. (1976) Ungulate populations in Chitwan Valley, Nepal. Biol. Conserv. 10: 183-210.
Shafi, M. M. & Khokhar, A. R. (1985) Some observations on wild boar (*Sus scrofa*) and its control in sugarcane areas of Punjab, Pakistan. J. Bombay Nat. Hist. Soc. 83: 63-67.
Shine, R., Harlow, P. S., Keogh, J. S. & Boeadi (1998) The influence of sex and body size on food habits of a giant tropical snake, *Python reticulatus*. Funct. Ecol. 12: 248-258.
Singer, F. J. (1981) Wild pig populations in national parks. Environ. Manage. 5: 263-270.
Smiet, A. C., Fulk, G. W. & Lathiya, S. B. (1979) Wild boar ecology in Thatta district: a preliminary study. Pakistan J. Zool. 11: 295-302.
Spillett, J. J. (1967a) A report on wild life surveys in North India and Southern Nepal: The Jaldapara Wild Life Sanctuary, West Bengal. J. Bombay Nat. Hist. Soc. 63: 534-556.
Spillett, J. J. (1967b) A report on wild life surveys in North India and Southern Nepal: The Kaziranga Wild Life Sanctuary, Assam. J. Bombay Nat. Hist. Soc. 63: 494-528.
Spillett, J. J. (1967c) A report on wild life surveys in North India and Southern Nepal: The large

mammals of the Keoladeo Ghana Sanctuary, Rajasthan. J. Bombay Nat. Hist. Soc. 63: 602-607.
Srikosamatara, S. (1993) Density and biomass of large herbivores and other mammals in a dry tropical forest, western Thailand. J. Trop. Ecol. 13: 33-43.
Stone, C. P. & Loope, L. L. (1987) Reducing negative effects of introduced animals on native biotas in Hawaii: what is being done, what needs doing, and the role of national parks. Environ. Conserv. 14: 245-258.
Struhsaker, T. T., Lwanga, J. S. & Kasenene, J. M. (1996) Elephants, selective logging and forest regeneration in the Kibale Forest, Uganda. J. Trop. Ecol. 12: 45-64.
Thomas, S. C. (1996) Asymptotic height as a predictor of growth and allometric characteristics of Malaysian rain forest trees. Am. J. Bot. 83: 556-566.
Thomas, L., Laake, J. L., Derry, J. F., Buckland, S. T., Borchers, D. L., Anderson, D. R., Burnham, K. P., Strindberg, S., Hedley, S. L., Burt, M. L., Marques, F., Pollard, J. H. & Fewster, R. M. (1998) Distance 3.5. Research Unit for Wildlife Population Assessment, University of St. Andrews, UK.
Woodroffe, R. & Ginsberg, J. R. (1998) Edge effects and the extinction of populations inside protected areas. Science 280: 2126-2128.

Part VI

Anthropogenic Impacts and Forest Management

36 Is the Termite Community Disturbed by Logging?

Kenzi Takamura[1]

Abstract: Censuses of wood-litter-feeding termites and lichen-feeding *Hospitalitermes* were made in primary and regenerating forests at Pasoh. The taxonomic composition of wood-litter-feeding termites was more diverse in primary forests than in regenerating forests. *Macrotermes malaccensis* (Haviland) was the most dominant in both forests. Termites of the Macrotermitinae were more dominant in regenerating forest. Termites of the Nasutitermitinae occurred more frequently in primary forest. Two lichen-feeding termites, *H. umbrinus* (Haviland) and *H. medioflavus* (Holmgren), were recorded from both primary and regenerating forests. For both species, the total number of colonies found was higher, although not significantly, in primary forest than in regenerating forest. The number of colonies engaged in foraging differed significantly for neither species between both forests. The size distribution of trees used for feeding and nesting by the two species indicated a preference for larger trees, which could be inhibited by loss of those trees by logging.

Key words: lichen, mound, regenerating forest, termite, wood decay.

1. INTRODUCTION

Tropical lowland rain forerst is a rich source of natural products. Timber is always the first to exploit. Thus, every forest is threatened with being deprived of the large mature trees that are the source of regeneration. Vast areas of primary forest on continents and islands have been selectively logged or entirely converted to other land uses. Even in the case of selective logging, the forest environment is greatly altered owing not only to loss of giant trees, but also to disturbance through road construction, log-landing, skid trailing and so on (Nussbaum et al. 1995).

Termites are among the most important organisms in the ecosystem of lowland tropical rain forersts such as at Pasoh FR, especially for decomposition (Abe 1978, 1979, 1980; Collins 1983; Matsumoto & Abe 1979; Yamashita & Takeda 1998). The absence of termites greatly slows decomposition of plant litter (Takamura 2001; Takamura & Kirton 1999; Yamashita & Takeda 1998). As shown in surveys of termite communities in an area of lightly to highly disturbed semi-deciduous forest in Cameroon (Eggleton et al. 1995, 1996; Lawton et al. 1998) and of primary to selectively logged lowland rain forest in East Malaysia (Burghouts et al. 1992), logging disturbance may affect their occurrence. I envisaged two direct effects of logging on termites. One is that large trees that are used as feeding and nesting sites are logged. The other is that most parts of logged trees are brought outside the forest and are not used by termites as food resources. Therefore, I censused two groups of termites - wood-litter-feeding termites and lichen-feeding termites of the genus *Hospitalitermes*- to evaluate the effects of logging disturbance on them.

[1] National Institute for Environmental Studies (NIES), Tsukuba 305-8506, Japan.
E-mail: takaken@nies.go.jp

2. METHODS

2.1 Wood-litter-feeding termites

A plot of primary forest (PF) and another of regenerating forest (RF) were set in the Pasoh Forest Reserve (Pasoh FR). Both plots were square measuring 100 m east-west by 100 m north-south. PF covered the southern half of Plot 1 for permanent vegetation-census. RF, 400 m north of PF, was in an area selectively logged in 1958–1959.

Every dead trunk and limb measuring ≧10 cm in diameter of and ≧ 2 m in length on the forest floor were mapped and marked by Drs. Hattori, Fukuyama & Maeto. The abundance of fallen trees was lower in RF than in PF (Chap. 12). I collected termites from each log (68 logs in PF and 59 in RF) by cutting, digging, scraping and turning over, but left each log largely intact for further census. On most occasions I found worker termites first and continued collecting until I found soldiers. I collected once in 1996 and once in 1997. I also collected termites from twigs < 10 cm in diameter on the forest floor along trails (36 twigs in PF and 41 in RF), dissecting the twigs thoroughly to get the specimens.

Specimens were preserved in 99.5% ethanol and dissected under a stereomicroscope to inspect morphological characteristics, especially mandible structure of soldiers. Specimens were identified according to Tho (1992) and Ahmad (1958, 1965) to species level as much as possible.

Air temperature at 10–20 cm above the ground was monitored under the closed canopies at two points in each forest plot with a water-resistant temperature data recorder (MDS-T, Alec Electronics Co., Kobe)

2.2 Lichen-feeding termites of *Hospitalitermes*

The same plots as in 2.1 were used for censusing *Hospitalitermes*. I searched for a procession between 08:00 A.M. and noon by walking around in a zigzag pattern in a 20 m × 20 m subplot sectioned from the plot of 100 m × 100 m. When I found a procession of *Hospitalitermes*, I followed it to both the nest and to the feeding site and recorded the locations on a map. In many cases, feeding places could not be seen because the termites climbed up to the canopy. In these cases, I identified the trees they were climbing as feeding sites unless diverged routes from them through veins and branches were found.

Six censuses were performed at intervals of three to six months. One to three mornings were spent on each census in a plot. On the first morning, I searched thoroughly. On later mornings I checked previously found colonies and recorded any newly found one.

Although *Hospitalitermes* is most active not in the morning, but at midnight (Collins 1979; Miura & Matsumoto 1997a), processions cannot be efficiently censused in the dark. And because rain disturbs processions (Collins 1979; Miura & Matsumoto 1997a), I did not census on a morning following the preceding night rainfall.

From every procession, I collected several soldiers and workers and preserved them in 99.5% ethanol. Specimens were identified on the basis of body coloration and head length and width (Tho 1992).

Trees with a diameter at breast height (DBH) of ≧ 5 cm in PF had already been censused by the researchers of the Forest and Forestry Products Research Institute of Japan (FFPRI) and the Forest Research Institute Malaysia (FRIM). (Chap. 39) I censused the trees in RF. As *Hospitalitermes* termites are rarely seen to use trees with a DBH of < 5 cm, I set a minimum DBH of 2 cm. I censused trees with a DBH of ≧ 10 cm all over the plot, but counted smaller trees in a central area of 10 m × 10 m within each 20 m × 20 m subplot.

3. RESULTS

3.1 Wood-litter-feeding termites

The daily maximum, average, and minimum air temperatures just over the ground varied seasonally from 20.6 to 33.6 °C (Fig. 1). The peak of the variation lay in April and May. The valley lay in December and January. The daily maximum fluctuated more widely than the average and minimum did. The daily maximum was a bit higher in RF than in PF on nearly all days: the difference of daily maximum temperature was 3.95 °C at most and 0.59 °C on average. All the maximum, average and minimum temperatures were significantly higher in RF than in PF (Wilcoxon's signed-rank test, Z = -16.31, -6.10 and -18.11, respectively; $P < 0.0001$)

Termites themselves or signs of their activity were found in 51 logs (75.0%) in PF and 50 logs (84.7%) in RF (Fig. 2). In both forests, *M. malaccensis* was found most frequently. The second most abundant species was *Odontotermes sarawakensis* Holmgren. Twelve (PF) and seven (RF) other species or taxa were found at low frequencies. Consequently, species diversity indices were higher in PF (MacArthur's H' [diversity] =2.896, E [equitability] =0.761) than in RF (H' =2.036, E=0.642). Macrotermitinae were predominant in both forests, including *Macrotermes*, *Odontotermes*, *Ancistrotermes* and *Hypotermes* species. The primary forest was characterized by the richness of Nasutitermitinae including *Nasutitermes*, *Hirtitermes* and *Havilanditermes* species. Two species of Rhinotermitidae (*Coptotermes curvignathus*, *Schedorhinotermes medioobscurus*) were found in PF and two of Termitinae (*Dicuspiditermes nemorosus*, *Pericapritermes*

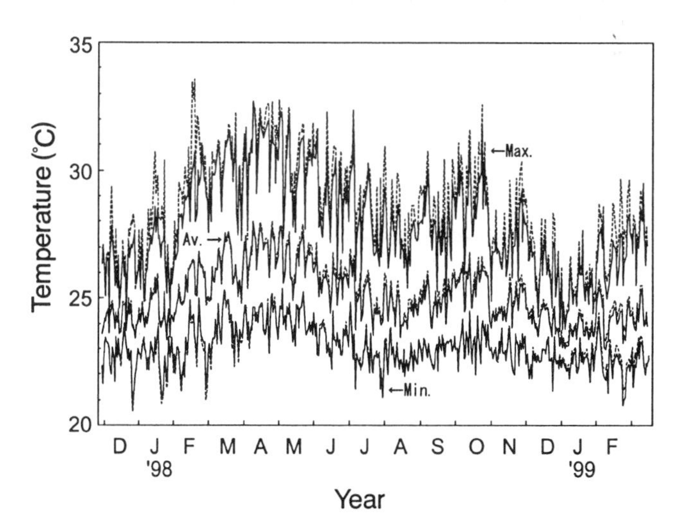

Fig. 1 Air temperature of the forest floor in the PF and RF of Pasoh FR during December 1997–February 1999. The maximum, average and minimum temperatures on the daily basis are shown from above to below. Solid line, PF; dotted line, RF.

dolichocephalus) were found in RF.

Termites collected from twigs are listed in Fig. 3. Here again, the dominance of Macrotermitinae was clear, especially in RF. The occurrence of *Ancistrotermes pakistanicus* (Ahmad) was much higher than in logs.

In spite of these differences in species composition, the overall occurrences of wood-litter-feeding termites did not differ significantly between PF and RF in either logs or twigs (Wilcoxon's signed-rank test, $Z = -0.53$ and -0.14, $P = 0.596$ and

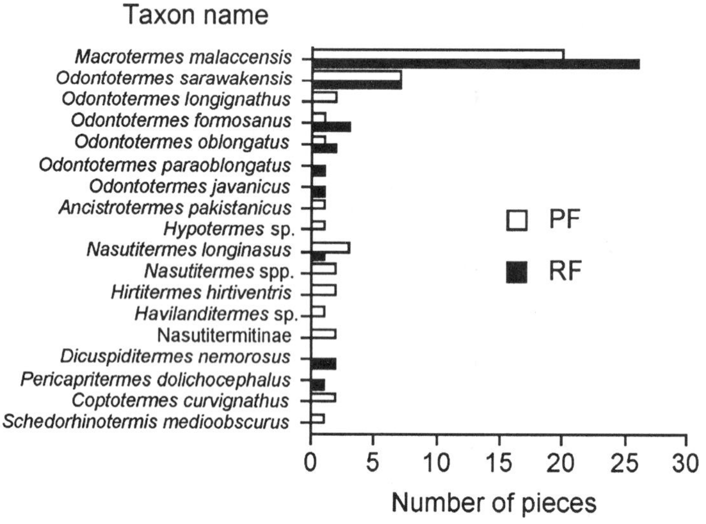

Fig. 2 Occurrence frequency of termite taxa in wood litter with the diameter ≧ 10 cm in the PF and RF.

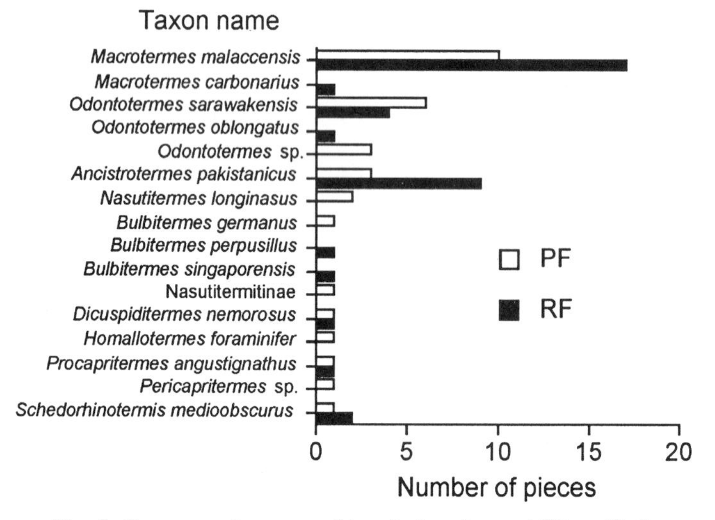

Fig. 3 Occurrence frequency of termite taxa in wood litter with the diameter < 10 cm in the PF and RF.

0.886). There were, however, differences in occurrence at higher taxonomic levels. The dominance of Macrotermitinae was significantly higher in RF than in PF in logs (*G*-test, $G = 4.069$, $P = 0.044$). Most Nasutitermitinae found in logs were confined to PF (Fig. 2) whereas those found in twigs did not clearly differed between the plots (Fig. 3). The occurrence of Nasutitermitinae in logs was significantly greater in PF than in RF (*G*-test, $G = 6.819$, $P = 0.009$). Termitinae were collected only in RF from logs, but in both plots from twigs. Occurrence of these minor termites, however, did not differ clearly between PF and RF (Figs. 2 , 3).

3.2 Lichen-feeding termites of *Hospitalitermes*

Two species of *Hospitalitermes* were found. The dominant one was *H. umbrinus* (Haviland). This species was identified based on its brown color of a whole body except paler abdominal sternites. The other was *H. medioflavus* (Holmgren), which differed from *H. umbrinus* in having fairly darker color of a whole body except for yellowish-white abdominal tergites and sternites.

The number of *H. umbrinus* colonies engaged in foraging was higher in PF than in RF in April, September and December 1997 (Fig. 4). It was the same in both plots in April 1998, but, after then, it was higher in RF than in PF. The difference was not significant between plots (Wilcoxon's signed-rank test, $Z_6 = -0.542$, $P = 0.58$). The total numbers of colonies found were 15 in PF and 12 in RF (Fig. 4).

The number of *H. medioflavus* colonies engaged in foraging was higher in

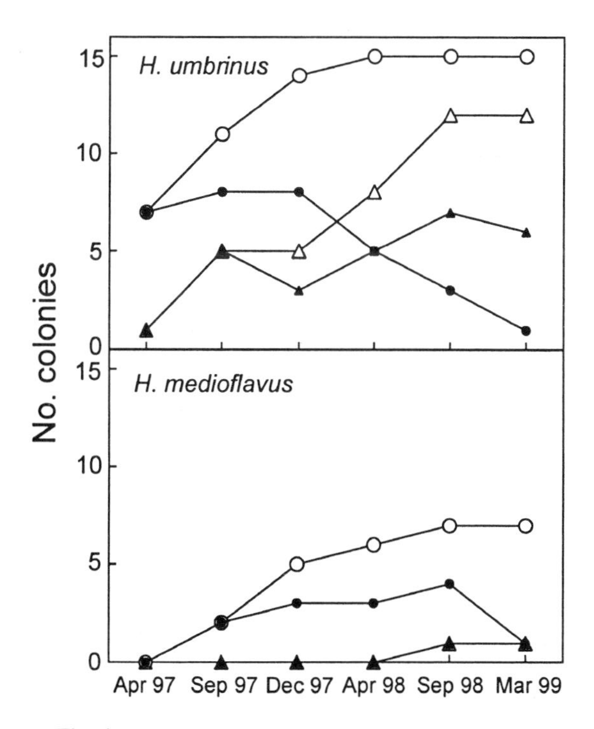

Fig. 4 Number of colonies of *Hospitalitermes umbrinus* and *H. medioflavus* found in the PF and RF. Circles, PF; triangles, RF; open symbols, total number of colonies found; closed symbols, number of colonies observed to be engaged in foraging.

PF than in RF in September and December 1997 and in April and September 1998 (Fig. 4). On these occasions, two to four colonies performed open foraging in PF, but only one colony did so in RF. In total, I found seven colonies in PF and one colony in RF (Fig. 4). The difference in the numbers of active colonies was slightly insignificant (Wilcoxon's signed-rank test, $Z_6 = -1.89$, $P = 0.059$).

In PF, *H. umbrinus* was recorded feeding on 47 trees and *H. medioflavus* on seven trees. *H. umbrinus* preferred trees with a DBH of 10–20 cm, and *H. medioflavus* preferred trees with a DBH of ≧ 100 cm (Fig. 5). The DBH of trees used for feeding was significantly larger than the mean of all trees in the plot (Kolmogorov-Smirnov test, $D_{511,47} = 0.384$, $P < 0.0001$ for *H. umbrinus*; $D_{511,7} = 0.744$, $P = 0.001$ for *H. medioflavus*). I found *H. umbrinus* in 11 nest trees and *H. medioflavus* in five. The peak was in the 20–30 cm DBH class (Fig. 5). The diameters of trees used for nesting were significantly larger than the mean of all trees in the plot for *H. umbrinus* (Kolmogorov-Smirnov test, $D_{511,11} = 0.608$, $P = 0.0007$ for *H. umbrinus*; $D_{511,5} = 0.553$, $P = 0.096$ for *H. medioflavus*).

In RF, *H. umbrinus* was recorded feeding on 39 trees and *H. medioflavus* on four. *H. umbrinus* preferred trees with a DBH of 10–20 cm (Fig. 6). *H. medioflavus* showed no preference. The diameters of trees used for feeding were significantly

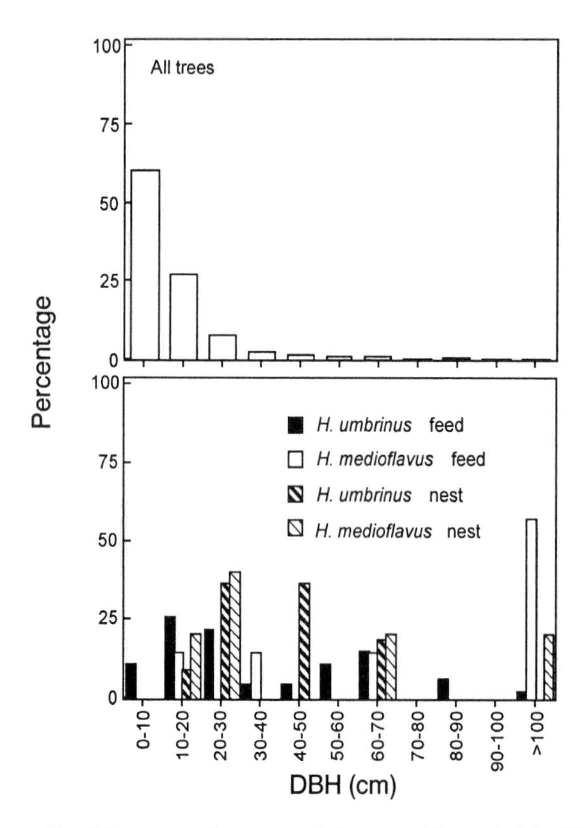

Fig. 5 Percentage frequency of diameter at breast height (DBH) of trees used for feeding and nesting by two *Hospitalitermes* species in the primary forest plot. The percentage frequency of all trees with the DBH ≧ 5 cm is shown above.

larger than the mean of all trees in the plot (Kolmogorov-Smirnov test, $D_{726,39}$=0.221, P = 0.0388 for *H. umbrinus*; $D_{726,4}$ = 0.792, P = 0.0136 for *H. medioflavus*). I found the nest of *H. umbrinus* in six trees and *H. medioflavus* in one. Trees with a DBH of 70–90 cm were intensively used (Fig. 6). The diameters of trees used for nesting were significantly larger than the mean of all trees in the plot (Kolmogorov-Smirnov test, $D_{726,6}$ = 0.708, P = 0.0051 for *H. umbrinus*). In both forests, *Hospitalitermes* almost preferred larger trees for foraging and nesting.

Hospitalitermes foraged mostly on live trees, rarely on dead trees (one case each of *H. umbrinus* in RF and *H. medioflavus* in PF). However, nests were more frequently located in dead trees (two of *H. umbrinus* in PF and five in RF; four nests of *H. medioflavus* in PF). One nest each of *H. medioflavus* in PF and *H. umbrinus* in RF was bare and solitary.

Interspecific comparison revealed a very marginally non-significant difference in feeding tree size in PF (Kolmogorov-Smirnov test, $D_{47,7}$ = 0.550, P = 0.0500). *H. medioflavus* probably used larger trees for feeding more frequently than *H. umbrinus* did (Fig. 6). Size frequency distributions of live trees in PF and RF resembled each other. The frequency distributions of DBH ≧ 10 cm were not

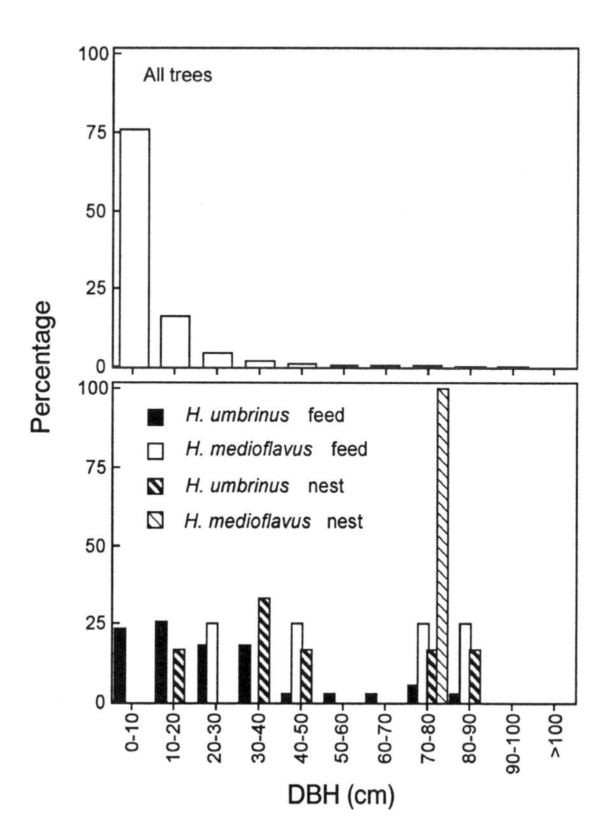

Fig. 6 Percentage frequency of diameter at breast height (DBH) of trees used for feeding or nesting by two *Hospitalitermes* species in the regenerating forest plot. The percentage frequency of all trees with the DBH ≧ 2 cm is shown above.

significantly different between the plots (Kolmogorov-Smirnov test, $D_{511,726} = 0.039$, $P = 0.8014$). Although RF was selectively logged 40 years ago, no discernible difference in tree size distribution was present between the plots.

PF contained 366 tree species. *Hospitalitermes* used 35 for feeding and 15 for nesting. The most abundant species in PF was *Mallotus griffithianus*. However, only one nest of *H. medioflavus* was found in 60 trees of this species. Feeding of *H. umbrinus* was observed four times and two of *H. umbrinus* nests and one of *H. medioflavus* nest was found from trees of *Dipterocarpus sublamellatus*. This tree species was the 12th most abundant in PF. *H. umbrinus* foraged most frequently on *Elateriospermum tapos* represented by only three trees in PF. Although several species were used for feeding and nesting, the frequency of tree use by *Hospitalitermes* termites was not proportional to the occurrence of tree species.

4. DISCUSSION

4.1 Wood-litter-feeding termites

The Macrotermitinae termites were predominant in wood litter of both primary and regenerating forests. The dominance was more remarkable and was reflected in a lower species diversity index in regenerating forest. Thus, their dominance indicates disturbance by selective logging.

An abundance of Nasutitermitinae can indicate low disturbance. Most Nasutitermitinae feed on rotten wood and soil except for *Hospitalitermes* species (Abe 1979; Collins 1984; Jones & Brendell 1998). They prefer well-decayed plant litter that is often hard to discriminate from soil. The warmer environment of the regenerating forest may reduce moisture availability, which reduces the decay rate of wood litter into soil-like material.

Termitinae also depend on well-decayed plant material. Those I collected are all soil feeders except for *Homallotermes foraminifer* (Haviland) which feeds on soil-like decayed wood (soil/wood feeder) (Abe 1979; Jones & Brendell 1998). They are also likely to prefer primary forest to regenerating forest, as the Nasutitermitinae do. Although their occurrence was not as high as that of the Nasutitermitinae, their abundance in Pasoh is not negligible (Abe & Matsumoto 1979; Matsumoto 1976). Abe & Matsumoto (1979) recorded 131–210 nests/ha of *Dicuspiditermes nemorosus* (Haviland) from a census of primary forest in Pasoh FR. I found 16 mounds of *D. nemorosus* from a regenerating forest area of 1/3 ha (unpublished data). That equals about 50 nests/ha. The discrepancy between these two records might signify the preference of this group of termites for primary forest.

The dominance of the Macrotermitinae is high in arid deciduous forest in S. E. Asia (Davies 1997) and in savanna in West Africa (Collins 1981). Warmer temperatures resulting from logging disturbance may make the forest environment drier, favoring the Macrotermitinae, as my results suggest. The Macrotermitinae have the advantage in being able to use fresh wood litter (Collins 1981). Their earlier exploition of fresh wood litter may reduce the preferred food resources of rotten wood and soil feeders.

4.2 Lichen-feeding termites of *Hospitalitermes*

Hospitalitermes feed on live photosynthetic organisms such as lichens, mosses, and algae so they are regarded as primary consumers in the ecosystem of the S. E. Asian lowland rain forerst. This unique ecology has been the focus of attention by termite researchers (Abe 1979; Collins 1979). This group of termites collects organic matter produced photosynthetically on trees, transports it to the ground, and consumes it there. Their nest material is relatively rich in nutrients and is likely to be foraged by *Termes* (Miura & Matsumoto 1997b). *Hospitalitermes* themselves may be preyed upon during open foraging or in their nests by other animals. In this way, the termites may generate a rapid transfer of energy and nutrients from primary producers to predators and decomposers. This transfer may be as fast as that achieved by herbivorous insects and mammals in Pasoh FR, and can be even faster because these termites actively bring their food down to the ground. The high nitrogen content of food balls that *Hospitalitermes* workers carry (Miura & Matsumoto 1997b) indicates that the transfer of nutrients is more concentrated than that achieved through decomposition of tree litter. Moreover, the organisms they take encrust tree trunks and leaves, and thus harvesting these competitors might benefit the trees.

Hospitalitermes may directly suffer from logging, although my results did not reveal any clearly significant difference in the colony density between primary and regenerating forests. Tall large trees may be suitable for foraging because their sun-lit canopies are good places for photosynthesis by lichens, mosses, and algae. Further, nests of *Hospitalitermes* are often built at the foot of large trees. These large trees are the primary targets of logging. There was a tendency for *Hospitalitermes* to prefer larger trees for foraging and nesting. So, loss of large trees by logging may severely deteriorate the habitat conditions of the termites.

There was no significant difference between PF and RF in the tree diameter. The tree heights of the two plots are not known, but they would be a better indicator of the habitat conditions of *Hospitalitermes*. A much more thorough census of the trees of Pasoh FR. (Chap.2) reported that trees were larger in both height and diameter in the primary area than in the regenerating area. Therefore, logging may have negatively affected *Hospitalitermes* in ways that are detectable on a broader scale.

Colony densities of *Hospitalitermes* have been censused by several researchers (Abe 1979; Abe & Matsumoto 1979; Collins 1979; Jones & Gathorne-Hardy 1995). The total numbers of *Hospitalitermes* colonies in my plots are much higher than those reported: 0–5 colonies/ha in the Pasoh FR (Abe 1979), 3 colonies /ha in Mulu forest, Sarawak (Collins 1979) and 3 colonies/ha in Batu Apoi forest, Sabah (Jones & Gathorne-Hardy 1997 *H. hospitalis* only). These values are much closer to the numbers of colonies engaged in foraging in my results. The difference in colony densities between my study and the others may have two causes. First, different time schedules of censuses may have produced part of the difference. In my study, censuses were repeated at long intervals over two years. In the most intensive case of the other studies, censuses were taken during three months (Jones & Gathorne-Hardy 1997). Birth and death of colonies are more likely to have occurred during the longer time span of my study. Such dynamics of colonies may increase the cumulative count of colonies. Second, I observed the same colonies engaged in foraging after 7–16 months in the course of my study. This indicates that more colonies may have existed at the times of censuses. Censuses repeated over several years may find colonies not found in shorter censuses packed within a few months.

So, are the colony densities obtained in my study higher than those obtained in other study areas and previously in Pasoh FR? Jones & Gathorne-Hardy (1995) and Collins (1979) focused on processional activities, not on colony census. Abe

(1979) focused his efforts on colony census, but he counted colonies by finding nests on the ground. In my study, I found colonies by following termites to their nest entrances, some of which were without conspicuous nest structures at the foot of trees or on the ground. This approach is likely to find colonies of *Hospitalitermes* efficiently.

My higher counts of *Hospitalitermes* colonies may stem in part from the census method used. Even if *Hospitalitermes* colonies were more frequent in the Pasoh FR than in other forests and previously at Pasoh, I have no way to explain the higher density. All these forests are lowland mixed dipterocarp forests, so I cannot expect any remarkable vegetation difference to affect the colony density. Repeated censuses conducted by finding processions may reveal variations in colony densities among forests showing variations in habitat environment.

ACKNOWLEDGEMENTS

This article is based on work performed in the NIES, FRIM and UPM Joint Research Project financed by the Environmental Agency of Japan (Global Environment Research Fund E-3.3 in 1990–92, E-3.2 in 1993-95 and E-3.1 in 1996–98). The Director General of FRIM gave me permission to use the Institute's facilities and other FRIM and UPM staff gave me various kinds of logistic support. Drs. Abdul Rahim Nick, Chan Hung Tuck, N. Manokaran and Muhamad Awang supervised that support. Dr. Laurence G. Kirton, my counterpart in FRIM, gave me sincere advice and support. Dr. Jurie Intachat also stimulated me through discussion. Dr. Yap Son Kheong gave me warm support in the early stage of my work. Japanese members of the joint project, especially Drs. Akio Furukawa, Yoshitaka Tsubaki, Naoki Kachi, Yutaka Maruyama, Toshinori Okuda, Naoya Osawa, Sanei Ichikawa, Hisashi Nagata and Tang Yan Hong gave me suggestions and supported my work. Drs. Tsutomu Hattori, Kenji Fukuyama and Kaoru Maeto allowed me to use the map of fallen trees in the forest plots. Drs. Abdul Rahman Kassim, Azizi Ripin, Simmathiri Appanah, Kaoru Niiyama, Shigeo Iida, and Katsuhiko Kimura allowed me to use the tree census data of Plot 1. Dr. Masatoshi Yasuda allowed me to use the tree census data of the regenerating area. Drs. Hiroshi Takeda, Tamon Yamashita, the late Drs. Takuya Abe, Tadao Matsumoto, and Toru Miura gave me discussion and information on studies of litter decomposition and termite ecology. I am grateful to all of these persons.

REFERENCES

Abe, T. (1978) Studies on the distribution and ecological role of termites in a lowland rain forest of West Malaysia (1) Faunal composition, size, colouration and nest of termites in Pasoh Forest Reserve. Kontyû 46: 273-290.

Abe, T. (1979) Studies on the distribution and ecological role of termites in a lowland rain forest of West Malaysia (2) Food and feeding habits of termites in Pasoh Forest Reserve. Jpn. J. Ecol. 29: 121-135.

Abe, T. (1980) Studies on the distribution and ecological role of termites in a lowland rain forest of West Malaysia (4) The role of termites in the process of wood decomposition in Pasoh Forest Reserve. Revue d'Ecologie et de Biologie du Sol 17: 23-40.

Abe, T. & Matsumoto, T. (1979) Studies on the distribution and ecological role of termites in a lowland rain forest of West Malaysia (3) Distribution and abundance of termites in Pasoh Forest Reserve. Jpn. J. Ecol. 29: 337-351.

Ahmad, M. (1958) Key to the Indomalayan termites. Biologia 4: 33-198.

Ahmad, M. (1965) Termites (Isoptera) of Thailand. Bull. Am. Mus. Nat. Hist. 131: 1-113.

Burghouts, T., Ernsting, G., Korthals, G. & De Vries, T. (1992) Litterfall, leaf litter decomposition and litter invertebrates in primary and selectively logged dipterocarp forest in Sabah, Malaysia. Philos. Trans. R. Soc. London Ser. B 335: 407-416.

Collins, N. M. (1979) Observations on the foraging activity of *Hospitalitermes umbrinus* (Haviland), (Isoptera: Termitidae) in the Gunung Mulu National Park, Sarawak. Ecol. Entomol. 4: 231-238.

Collins, N. M. (1981) The role of termites in the decomposition of wood and leaf litter in the Southern Guinea savanna of Nigeria. Oecologia 51: 389-399.

Collins, N. M. (1983) Termite populations and their role in litter removal in Malaysian rain forests. In Sutton, S. L., Whitmore, T. C. & Chadwick, A. C. (eds). Tropical Rain Forest: Ecology and Management, Blackwell Scientific Publications, pp.311-325.

Collins, N. M. (1984) The termites (Isoptera) of the Gunung Mulu National Park with a key to the genera known from Sarawak. In Jermy, A. C. & Kavanagh, K. P. (eds). Gunung Mulu National Park, Sarawak Museum Journal, 30 (51), Special Issue No. 2, pp.65-87.

Davies, R. G. (1997) Termite species richness in fire-prone and fire-protected dry deciduous dipterocarp forest in Doi Suthep-Pui National Park, northern Thailand. J. Trop. Ecol. 13: 153-160.

Eggleton, P., Bignell, D. E., Sands, W. A., Waite, B., Wood, T. G. & Lawton, J. H. (1995) The species richness of termites (Isoptera) under differing levels of forest disturbance in the Mbalmayo Forest Reserve, southern Cameroon. J. Trop. Ecol. 11: 85-98

Eggleton, P, Bignell, D. E., Sands, W. A., Mawdsley, N. A., Lawton, J. H., Wood, T. G. & Bingell, N. C. (1996) The diversity, abundance and biomass of termites under differing levels of disturbance in the Mbalmayo Forest Reserve, southern Cameroon. Philos. Trans. R. Soc. London Ser. B 351: 51-68

Jones, D. T. & Brendell, M. J. D. (1998) The termite (Insecta: Isoptera) fauna of Pasoh Forest Reserve, Malaysia. Raffles Bull. Zool. 46: 79-91.

Jones, D. T. & Gathorme-Hardy, F. (1997) Foraging activity of the processional termite *Hospitalitermes hospitalis* (Termitidae: Nasutitermitinae) in the rain forest of Brunei, north-west Borneo. Insects Sociaux 42: 359-369.

Lawton, J. H., Bignell, D. E., Bolton, B., Bloemers, G. F., Eggleton, P., Mammond, P. M., Hodda, M., Holt, R. D., Larsen, T. B., Mawdsley, N. A., Stork, N. E., Srivastava, D. S. & Watt, A. D. (1998) Biodiversity inventories, indicator taxa and effects of habitat modification in tropical forest. Nature 391: 72-76.

Matsumoto, T. (1976) The role of termites in an equatorial rain forest ecosystem of West Malaysia I. Population density, biomass, carbon, nitrogen and calorific content and respiration rate. Oecologia 22: 153-178.

Matsumoto, T. & Abe, T. (1979) The role of termites in an equatorial rain forest ecosystem of West Malaysia II. Leaf litter consumption on the forest floor. Oecologia 38: 261-274.

Miura, T. & Matsumoto, T. (1997a) Foraging organization of the open-air processional lichen-feeding termite *Hospitalitermes* (Isoptera, Termitidae) in Borneo. Insects Sociaux 44: 1-16.

Miura, T. & Matsumoto, T. (1997b) Diet and nest material of the processional termite *Hospitalitermes*, and cohabitation of *Termes* (Isoptera, Termitidae) on Borneo Island. Insects Sociaux 44: 267-275.

Nussbaum, R., Anderson, J. & Spencer, T. (1995) Effects of selective logging on soil characteristics and growth of planted dipterocarp seedlings in Sabah. In Primack, R. B. & Lovejoy, T. E. (eds). Ecology, conservation, and management of Southeast Asian rainforersts, Yale University Press, New York, pp.105-115.

Takamura, K. (2001) Effects of termite exclusion on decay of heavy and light hardwood in a tropical rainforest of Peninsular Malaysia. J. Trop. Ecol. 17: 541-548.

Takamura, K. & Kirton, L. G. (1999) Effects of termite exclusion on decay of a high-density wood in tropical rain forests of Peninsular Malaysia. Pedobiologia 43: 289-296.

Tho, Y. P. (1992) Termites of Peninsular Malaysia. In Kirton, L. G. (ed). Malayan Forest Records, No. 36. Forest Research Institute Malaysia, Kuala Lumpur, 224pp.

Yamashita, T. & Takeda, H. (1998) Decomposition and nutrient dynamics of leaf litter in litter bags of two mesh sizes set in two dipterocarp forest sites in Peninsular Malaysia. Pedobiologia 42: 11-21.

37 Small Mammal Community: Habitat Preference and Effects after Selective Logging

Masatoshi Yasuda[1], Nobuo Ishii[2], Toshinori Okuda[3] & Nor Azman Hussein[4]

Abstract: The present study of small mammals in a Malaysian tropical forest attempts to gain some insight into the correlation between their habitat preferences. A 10 ha study plot including primary, old regenerating, and riparian forests were established in the Pasoh Forest Reserve (Pasoh FR) and monthly live trapping employing a mark-and-recapture method was conducted over a period of four years. Thirteen out of the 17 species of the small mammals predominating at the Pasoh FR, including moonrat, treeshrew, squirrels, rats and porcupines, were subjected to statistical analysis to determine their habitat preferences. Based on their abundance in the three habitats, they were divided into four categories: 1) primary forest species, 2) regenerating forest species, 3) riparian species and 4) ubiquitous species, with all the squirrel species categorised as primary forest species. Several environmental factors may explain the observed habitat preferences of small mammals. The present results indicated that the primary forest habitat is rich in food resources and spatial heterogeneity, and should thus provide a higher carrying capacity than the other two habitats. Although 40 years have elapsed since selective logging was carried out in the 1950s, there are still distinct differences between disturbed and undisturbed forests with respect to the small mammal community. Food resources and spatial heterogeneity may account for the differences in mammalian community among the habitats.

Key words: community structure, long-term population census, small mammals, species composition.

1. INTRODUCTION

Most of the studies on mammalian communities carried out in Malaysian forests have been fairly limited in their scope of community ecological concepts (Duff et al. 1984; Harrison 1957; Kemper 1988; Kemper & Bell 1985; Langham 1983; Lim 1970; Lim et al. 1995; Medway 1966, 1972, 1978; Payne et al. 1985; Stevens 1968; Stuebing & Gasis 1989; Zubaid & M. Khairul 1997; Zubaid & Rizal 1995). Some of these studies were based on qualitative (presence/absence) information of species but not on quantitative aspects in terms of population density, while others dealt with a few dominant species only. There were also studies that focussed on the comparison of mammalian fauna in different habitats, but were carried out in study sites far away from each other. Furthermore, in most cases, environmental conditions within the sites were insufficiently described as to infer the factors regulating animal populations in the habitat.

---

[1] Forestry and Forest Products Research Institute (FFPRI), Tsukuba 305-8687, Japan. E-mail: myasuda@ffpri.affrc.go.jp
[2] Japan Wildlife Research Center (JWRC), Japan.
[3] National Institute for Environmental Studies (NIES), Japan.
[4] Forest Research Institute Malaysia (FRIM), Malaysia.

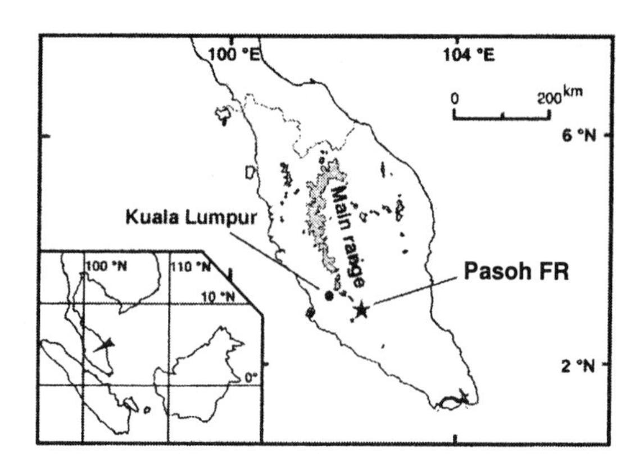

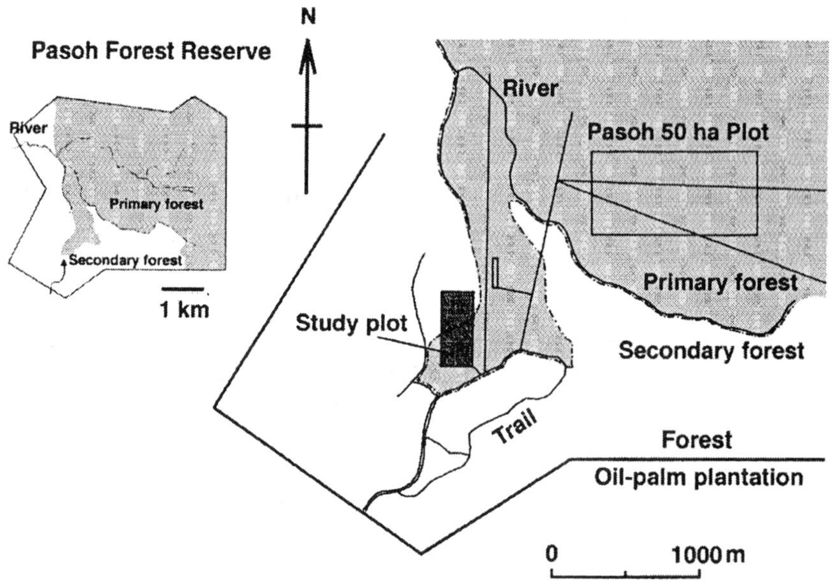

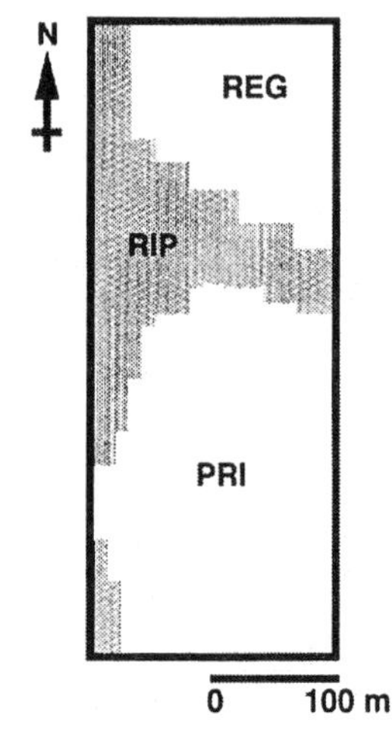

Fig. 1 Location of study plot and distribution of vegetation in the plot. PRI, REG and RIP represent primary forest, old regenerating forest, and riparian forest habitats. Shaded area (RIP) indicates where the forest floor covered with water except during severe droughts.

The present study provides more detailed information on species abundance of small mammals in certain habitats in relation to their habitat preference. In this study, surveys of small mammals were carried out in three different forest types, i.e. primary, old regenerating and riparian forests in a Malaysian lowland forest.

2. MATERIALS AND METHODS

2.1 Study site

A 10 ha (500 m × 200 m) study plot was established in the boundary area between primary and regenerating forests in the Pasoh Forest Reserve (Pasoh FR), Negeri Sembilan, Malaysia (Fig. 1). The study plot consisted of three forest types: primary forest (hereafter PRI), old regenerating forest (REG) and riparian forest (RIP). The PRI habitat is an intact primary forest. The REG habitat is a 40-year regenerating forest that was selectively logged in 1958–1959 under the Malayan Uniform System and has been regenerating naturally (Appanah & Mohd. Rasol 1990; Caldecott 1986; Manokaran & Swaine 1994). The RIP habitat, in this context, is established in an area with small streams running between PRI and REG. It has a water-logged forest floor that occasionally dries out during severe drought. The border areas within 10 m of the water surface were considered RIP habitat, irrespective of forest types. Thus RIP contains the edges of both PRI and REG, though the logging history of RIP is unknown. PRI, REG and RIP in the 10 ha plot accounted for 43.2%, 22.4% and 34.4% of the total area, respectively. The three habitats are adjacent to each other and there were no apparent obstacles blocking the free movement of small mammals within the plot.

2.2 Trapping of small mammals

Trapping was carried out monthly from June 1992 to May 1996, during which no mast fruiting event occurred. Two hundred and fifty live-traps were set systematically in a grid design on the forest floor in the 10 ha study plot. The distance between adjacent traps was 20 m. The number of traps in PRI, REG and RIP was 108, 56 and 86, respectively. The size of each trap was 17 cm × 17 cm × 44 cm, for width, height and depth, respectively (made by Environmental Supplies and Services, Selangor, Malaysia). Oil palm fruit (*Elaeis guineensis* Jacq.) was used as bait. One trapping session consisted of four consecutive days. Traps were checked once a day in the morning.

Animals captured were anesthetized with ether and toe-clipped individually for permanent identification. Each animal was identified to species according to Corbet & Hill (1992). At the initial capture of an individual, the following measurements were taken: length of head and body (HB), tail (T), hind foot (HF), ear (E) and body mass (W).

2.3 Analyses

A data set was compiled to analyze the habitat use of small mammals. If an individual was captured repeatedly in different habitats in a trapping session, the presence of that individual was allocated among the habitats according to the frequency of captures. For instance, if an individual was captured twice in PRI and once in REG and RIP, the presence probabilities of the individual for that species were allocated as 0.5, 0.25 and 0.25 to the respective habitats. The summation of the presence probabilities for all individuals of the species in a habitat gives the number of individuals captured (hereafter *NI*) of that species in a given habitat in a given trapping session.

Since disturbance of traps by large animals, e.g. monkeys, was sometimes severe and the intensity was different among habitats, we employed an empirical saturating regression to eliminate the effect of disturbance on the trapping data derived from a ten-day trapping session in the same plot (Yasuda 1998), as

$$NI_{adj(i,j)} = \frac{1.37T_j}{T_j + 891} NI_{(i,j)}$$

where $NI_{(i,j)}$ and $NI_{adj(i,j)}$ are the monthly NI for the ith species in the jth habitat respectively, and T_j is the total number of effective traps in the jth habitat in the trapping session. The coefficient of determination (r^2) for the regression was 0.992 ($P < 0.01$).

The density of a species in a habitat was determined simply by dividing NI_{adj} by the area of a given habitat (4.32, 2.24 and 3.44 ha for PRI, REG and RIP, respectively). Biomass in a given habitat was calculated by multiplying the above density by the average body mass of adults of the species.

Based on the four-year average of species density in each habitat (see Table 2), percentage similarity (Whittaker 1952) in terms of the relative abundance of species among the three habitats were calculated between species, and a similarity matrix was obtained. Then the 13 species of small mammals were plotted using the similarity matrix and an ordination method, multidimensional scaling (Kruskal & Wish 1978). SYSTAT for the Macintosh version 5.2.1 (Systat, Inc., Evanston, IL, USA) was used for the analysis.

3. RESULTS

3.1 Mammalian fauna

In total, 22 species of mammals belonging to eight families of six orders were recorded in the consecutive four-year censuses. The most dominant order was Rodentia (17 spp.), consisting of 8 spp. of squirrels, 5 spp. of rats, 2 spp. of flying squirrels and 2 spp. of porcupines. Table 1 shows the body dimensions of the most dominant 17 species of small mammals with their locomotion and daily activity cycle. Some arboreal species were frequently captured on the ground. The number of species recorded in PRI, REG and RIP was 17, 15 and 15, respectively. Two species were recorded only in PRI: the Prevost's squirrel, *Callosciurus prevostii* (Desmarest), and the red-cheeked flying squirrel, *Hylopetes spadiceus* (Blyth). Except for some rarely captured species (NI_{adj} < 0.5 individuals month^{-1} on average), 13 species of relatively common small mammals were subjected to the analysis below. They were *Echinosorex gymnurus* (Raffles) (Echinaceidae, Insectivora), *Tupaia glis* (Diard) (Tupaiidae, Scandentia), *Callosciurus notatus* (Boddaert), *Callosciurus nigrovittatus* (Horsfield), *Sundasciurus lowii* (Thomas), *Lariscus insignis* (Cuvier), *Rhinosciurus laticaudatus* (Müller) (Sciuridae, Rodentia), *Rattus tiomanicus* (Miller), *Leopoldamys sabanus* (Thomas), *Maxomys rajah* (Thomas), *Maxomys surifer* (Miller), *Maxomys whiteheadi* (Thomas) (Muridae, Rodentia) and *Trichys fasciculata* (Shaw) (Hystricidae, Rodentia).

3.2 Species composition among habitats

The density and biomass of the 13 species in the three habitats are shown in Tables 2 and 3. There were four remarkable differences in community structure among the habitats. First, total density and biomass decreased in the order PRI > REG > RIP;

Table 1 Body dimensions of small mammals in the Pasoh FR.

Locomotion / Activity cycle / Species	Family	Sex	Body measurements (mm)								Body mass (g)		
			HB	*SD*	T	*SD*	HF	*SD*	E	*SD*	W	*SD*	*N*
Arboreal													
Diurnal													
Callosciurus caniceps	Sciuridae	female	210	na	190	na	46	na	15	na	225	na	1
		male	198	6	203	8	47	2	15	1	232	13	3
Callosciurus prevostii	Sciuridae	female	221	na	240	na	54	na	11	na	324	na	1
		male	255	na	263	na	57	na	18	na	438	na	2
Callosciurus notatus	Sciuridae	female	199	8	186	18	44	5	15	1	250	18	8
		male	197	10	189	14	45	4	15	2	250	23	14
Callosciurus nigrovittatus	Sciuridae	female	192	3	169	15	41	6	14	3	222	30	3
		male	206	3	192	21	45	4	16	4	235	16	4
Sundasciurus lowii	Sciuridae	female	136	11	108	18	33	5	12	2	85	14	12
		male	141	8	110	9	33	6	12	2	88	12	7
Nocturnal													
Ptilocercus lowii	Tupaiidae	female	138	6	183	10	26	2	19	3	55	7	3
		male	na	na	na	na	na	na	na	na	na	na	0
Hylopetes spadiceus	Sciuridae	female	150	na	145	na	26	na	15	na	95	na	1
		male	na	na	na	na	na	na	na	na	na	na	0
Terrestrial													
Diurnal													
Tupaia glis	Tupaiidae	female	175	22	162	10	42	3	12	2	134	20	36
		male	181	12	161	16	42	3	14	2	150	20	22
Lariscus insignis	Sciuridae	female	185	9	97	31	45	4	13	2	210	16	17
		male	191	10	110	19	46	2	14	3	211	26	33
Rhinosciurus laticaudatus	Sciuridae	female	209	12	125	24	43	3	16	3	232	9	6
		male	217	7	114	24	44	2	17	3	248	24	12
Nocturnal													
Echinosorex gymnurus	Echinaceidae	female	342	40	254	18	60	4	26	2	837	75	3
		male	362	15	249	15	62	2	26	2	867	101	3
Rattus tiomanicus	Muridae	female	141	13	143	15	30	2	16	2	74	15	12
		male	148	18	143	18	30	1	16	1	83	25	10
Leopoldamys sabanus	Muridae	female	226	10	358	23	45	2	25	2	331	52	21
		male	239	19	366	32	47	2	24	2	353	67	50
Maxomys rajah	Muridae	female	178	na	177	na	37	na	20	na	145	na	2
		male	190	23	184	14	39	3	20	2	158	17	6
Maxomys surifer	Muridae	female	167	16	169	17	38	2	19	2	141	50	7
		male	189	19	187	15	41	1	21	3	159	35	13
Maxomys whiteheadi	Muridae	female	114	14	96	7	25	1	14	2	44	9	7
		male	119	10	101	10	26	3	14	2	45	11	25
Trichys fasciculata	Hystricidae	female	380	na	190	na	62	na	27	na	1680	na	2
		male	na	na	na	na	na	na	na	na	na	na	0

HB: Head and body, T: Tail, HF: Hind foot, E: Ear, W: Weight, na: not available.

four-year averages of 7.31 ± 1.89 *SD*, 5.43 ± 1.77 *SD* and 4.75 ± 1.36 *SD* individuals ha^{-1} and 1,819, 1,525 and 1,369 g ha^{-1}, respectively. Second, the total density of diurnal species, including all the squirrels and the common treeshrew (*T. glis*), were lowest in the REG habitat, in the order PRI (3.13 individuals ha^{-1}) > RIP (2.43) > REG (1.18). The proportion of the diurnal community was only 21.7% of the whole community in REG, versus 42.8% and 51.2% in PRI and RIP, respectively. In contrast, the proportion of the nocturnal community was highest in REG, in the order REG (78.3%) > PRI (57.2%) > RIP (48.9%). This tendency resulted from the relatively high density of *Leopoldamys sabanus* (the long-tailed giant rat) in the REG habitat. Finally, the densities of *Echinosorex gymnurus* (the moonrat) and *Trichys fasciculata* (the long-tailed porcupine) were highest in RIP: 0.31 and 0.10 individuals ha^{-1}, respectively. They accounted for 6.5% and 2.0% of the whole community of RIP in terms of density and 19.2% and 11.7% in terms of biomass.

As shown in Table 4, ten species of small mammals showed significant differences in population density among habitats (Friedman test, $P < 0.05$ for all ten species, $N = 48$) and *Callosciurus nigrovittatus* (the black-striped squirrel) and *Trichys fasciculata* showed marginal significance ($P = 0.052$ and 0.090, respectively). Wilcoxon signed rank test was used as a post hoc test to examine the preference of species between two habitats. The significance level was set 0.05. All six diurnal species and two nocturnal species of rats, *Maxomys rajah* and *Maxomys whiteheadi*, showed preference for primary forest habitat over regenerating forest habitat (PRI > REG). In contrast, two species of rats, *Leopoldamys sabanus* and *Maxomys surifer*, showed preference for both primary forest habitat and regenerating forest habitat, and avoided the riparian forest habitat (REG > PRI > RIP and REG = PRI > RIP, respectively). *Echinosorex gymnurus* and *Trichys*

Table 2 Densities and biomass of common small mammals in the three habitats. (Species are sorted in the order of population density in the primary forest habitat).

		Animal density (individuals ha^{-1})						Biomass (g ha^{-1})		
		PRI		REG		RIP		PRI	REG	RIP
Species	W (g)	mean	*SD*	mean	*SD*	mean	*SD*			
Diurnal community										
Tupaia glis	140	1.07	0.62	0.72	0.62	1.11	0.57	149	101	155
Lariscus insignis	211	0.79	0.46	0.20	0.28	0.66	0.42	167	42	140
Callosciurus notatus	250	0.62	0.29	0.16	0.25	0.33	0.26	155	40	82
Rhinosciurus laticaudatus	243	0.36	0.31	0.06	0.18	0.15	0.19	87	14	36
Sundasciurus lowii	86	0.19	0.27	0.04	0.13	0.10	0.17	17	3	9
Callosciurus nigrovittatus	229	0.10	0.14	0.01	0.06	0.08	0.16	22	2	19
Subtotal		3.13	1.21	1.18	0.74	2.43	0.89	597	201	441
Nocturnal community										
Leopoldamys sabanus	347	2.13	0.72	2.62	0.94	1.30	0.62	739	909	450
Maxomys surifer	153	0.83	0.32	0.77	0.65	0.16	0.22	127	118	25
Maxomys rajah	155	0.44	0.32	0.23	0.26	0.05	0.11	69	36	8
Maxomys whiteheadi	44	0.42	0.28	0.26	0.36	0.29	0.27	18	11	13
Echinosorex gymnurus	852	0.21	0.21	0.26	0.35	0.31	0.27	177	219	263
Rattus tiomanicus	78	0.10	0.16	0.10	0.24	0.12	0.21	8	8	9
Trichys fasciculata	1680	0.05	0.10	0.01	0.07	0.10	0.15	83	24	161
Subtotal		4.18	1.04	4.25	1.41	2.32	0.96	1221	1324	928
Total		7.31	1.89	5.43	1.77	4.75	1.36	1819	1525	1369

Table 3 Species composition of small mammal community in the three habitats.

Species	PRI density (%)	PRI biomass (%)	REG density (%)	REG biomass (%)	RIP density (%)	RIP biomass (%)
Diurnal community						
Tupaia glis	14.6	8.2	13.2	6.6	23.3	11.3
Lariscus insignis	10.8	9.2	3.6	2.7	13.9	10.2
Callosciurus notatus	8.5	8.5	2.9	2.6	6.9	6.0
Rhinosciurus laticaudatus	4.9	4.8	1.0	0.9	3.1	2.6
Sundasciurus lowii	2.6	0.9	0.7	0.2	2.1	0.6
Callosciurus nigrovittatus	1.3	1.2	0.2	0.1	1.8	1.4
Subtotal	42.8	32.8	21.7	13.2	51.1	32.2
Nocturnal community						
Leopoldamys sabanus	29.2	40.7	48.2	59.6	27.3	32.9
Maxomys surifer	11.3	7.0	14.2	7.7	3.4	1.8
Maxomys rajah	6.1	3.8	4.3	2.4	1.1	0.6
Maxomys whiteheadi	5.7	1.0	4.7	0.7	6.1	0.9
Echinosorex gymnurus	2.8	9.7	4.7	14.4	6.5	19.2
Rattus tiomanicus	1.4	0.4	1.9	0.5	2.5	0.7
Trichys fasciculata	0.7	4.6	0.3	1.5	2.0	11.7
Subtotal	57.2	67.2	78.3	86.8	48.9	67.8
Total	100.0	100.0	100.0	100.0	100.0	100.0

Table 4 Habitat preference of small mammals. Habitat types shown in table represent the more preferred habitat over another.

Species	Family	Habitat preference: PRI vs. REG	PRI vs. RIP	REG vs. RIP	*P*
Diurnal					
Tupaia glis	Tupaiidae	PRI	n.s.	RIP	< 0.001
Callosciurus nigrovittatus	Sciuridae	PRI	n.s.	RIP	0.052
Callosciurus notatus	Sciuridae	PRI	PRI	RIP	< 0.001
Lariscus insignis	Sciuridae	PRI	n.s.	RIP	< 0.001
Rhinosciurus laticaudatus	Sciuridae	PRI	PRI	RIP	< 0.001
Sundasciurus lowii	Sciuridae	PRI	n.s.	n.s.	0.008
Nocturnal					
Leopoldamys sabanus	Muridae	REG	PRI	REG	< 0.001
Maxomys rajah	Muridae	PRI	PRI	REG	< 0.001
Maxomys surifer	Muridae	n.s.	PRI	REG	< 0.001
Maxomys whiteheadi	Muridae	PRI	n.s.	n.s.	0.016
Rattus tiomanicus	Muridae	-	-	-	0.825
Trichys fasciculata	Hystricidae	n.s.	RIP	RIP	0.090
Echinosorex gymnurus	Echinaceidae	n.s.	RIP	n.s.	0.023

n.s.: not significant ($P > 0.05$)

fasciculata showed preference for the riparian habitat over the other two habitats (RIP > PRI = REG). One species, *Rattus tiomanicus*, showed no preference for any of the three habitats (PRI = REG = RIP).

3.3 Ordination of species distribution among habitats

The multidimensional scaling performed well (Fig. 2), as the badness-of-fit criterion was 0.020. Relative abundance of species in a certain habitat was shown in the figure. The first dimension related to the relative abundance of species in the RIP habitat, while the second dimension was associated with that in the PRI habitat.

4. DISCUSSION

4.1 Grouping of small mammals according to habitat preferences

There is no apparent obstacle preventing small mammals from moving freely among the three habitats and no mast fruiting, which may significantly affect small mammal populations, occurred during the study period. Coefficients of variation (CV) on a population density basis were small, being 30.4%, 24.9% and 22.0%, for diurnal,

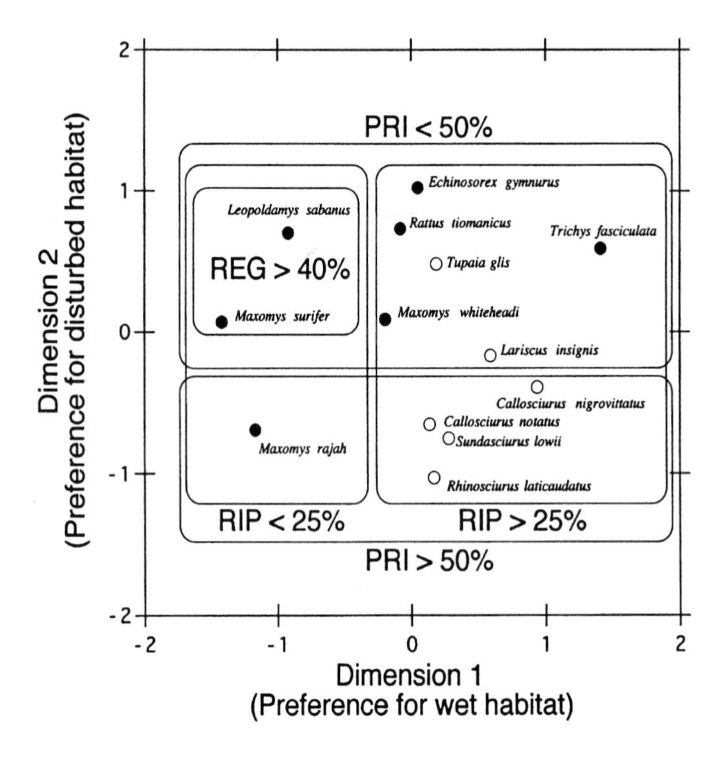

Fig. 2 Habitat preference of small mammals. Multidimensional scaling was applied to a percentage similarity index based on relative abundance of species in three habitats. PRI, REG and RIP represent primary forest, old regenerating forest and riparian forest habitats. Percentages shown in figure represent relative abundance of species in the habitat. Open circle: diurnal species, Closed circle: nocturnal species.

Table 5 Summary of the habitat preference of small mammals and some other related factors.

Species group / Species	Family	Locomotion*	Activity cycle*	Density	Habitat preference	Correlation with environmental factors** Fruit avairability	Fallen logs	Canopy gaps	Food habit*
(1) Primary forest species									
Tupaia glis	Tupaiidae	Terrestrial	Diurnal	high	PRI = RIP > REG	Positive			insect, fruits
Callosciurus nigrovittatus	Sciuridae	Arboreal	Diurnal	low	PRI = RIP > REG				insect, fruits
Callosciurus notatus	Sciuridae	Arboreal	Diurnal	medium	PRI > RIP > REG				insect, fruits
Lariscus insignis	Sciuridae	Terrestrial	Diurnal	high	PRI = RIP > REG	Positive	Positive		insect, fruits
Rhinosciurus laticaudatus	Sciuridae	Terrestrial	Diurnal	low	PRI > RIP > REG		Positive		insect
Sundasciurus lowii	Sciuridae	Arboreal	Diurnal	low	PRI > REG		Positive		insect, fruits
Maxomys rajah	Muridae	Terrestrial	Nocturnal	medium	PRI > REG > RIP				insect, fruits
Maxomys whiteheadi	Muridae	Terrestrial	Nocturnal	medium	PRI > REG		Positive		insect, fruits
(2) Secondary forest species									
Leopoldamys sabanus	Muridae	Terrestrial	Nocturnal	high	REG > PRI > RIP	Positive		Negative	insect, fruits
Maxomys surifer	Muridae	Terrestrial	Nocturnal	high	REG = PRI > RIP				insect, fruits
(3) Riparian species									
Echinosorex gymnurus	Echinaceidae	Terrestrial	Nocturnal	medium	RIP > PRI = REG				insect, fruits, aquatic animals
Trichys fasciculata	Hystricidae	Terrestrial	Nocturnal	low	RIP > PRI = REG				insect, fruits
(4) Ubiquitous species									
Rattus tiomanicus	Muridae	Terrestrial	Nocturnal	low	PRI = REG = RIP				insect, fruits

* cited from Medway (1978); and ** from Yasuda (1998), $P < 0.05$.

nocturnal and whole communities, respectively (Yasuda 1998). This implies that the population density of small mammals in a habitat is rather stable. These data allow us to examine a species' habitat preference based on the differences in population density among habitats.

Eight out of the 13 species (61.5%) showed significant preference for primary forest habitat over regenerating forest habitat, while two species of rats (15.4%) tended to favor the regenerating forest habitat (Table 4). The 13 species of small mammals can be categorized into four species groups; i.e. primary forest species, regenerating forest species, riparian species and ubiquitous species (Table 5). This conclusion was well coincident with that of multidimensional scaling, as two dimensions of habitat preference, one is for wet habitat and another is for disturbed habitat, were important to determine the species composition of small mammals in a certain habitat (Fig. 2). The details of each grouping are as follows:

4.1.1 Primary forest species

This group consists of eight species: *Tupaia glis*, *Callosciurus nigrovittatus*, *C. notatus*, *Lariscus insignis*, *Rhinosciurus laticaudatus*, *Sundasciurus lowii*, *Maxomys rajah* and *M. whiteheadi*. All the diurnal species, including a treeshrew and all squirrels, belong to this group. Their population densities were significantly higher in PRI than in REG (Tables 2, 4), indicating a preference for primary forest habitat.

4.1.2 Regenerating forest species

This group consists of two species of rats, *Leopoldamys sabanus* and *Maxomys surifer*. Their population densities were high in REG as well as in PRI, and low in RIP (Tables 2, 4). Both of these species prefer the primary and regenerating forest habitats to the riparian habitat, which concurs with the findings of Harrison (1957) and Lim (1970).

4.1.3 Riparian species

This group is represented by the moonrat and a porcupine, *Echinosorex gymnurus* and *Trichys fasciculata*, respectively. They showed a higher density in RIP than in the other two habitats, though they were captured in all the habitat types (Table 2). The population density of *E. gymnurus* was significantly higher in RIP than in PRI (Tables 2, 4). This agrees with the findings of Lim (1967) and Medway (1978) in which *E. gymnurus* was usually trapped near streams and preferred swampy forest. The population density of *T. fasciculata* tended to be higher in RIP than in the other two habitats (Table 2), though this pattern was only marginally significant ($P = 0.09$, Table 4).

4.1.4 Ubiquitous species

A medium-sized rat, *Rattus tiomanicus*, belongs to this group. The species was captured equally in all the forest habitats in the study plot, with a density of 0.10, 0.10 and 0.12 individuals ha^{-1} in PRI, REG and RIP, respectively (Table 2). This species seems to be unaffected by either logging disturbance or the water conditions of the forest floor.

Stevens (1968) compiled the available information of the mammalian fauna in various habitats in Peninsular Malaysia and observed that 29 out of 51 species (56.9%) of small mammals were confined to primary forests. The results of the present study, based on statistical analysis, showed 61.5% (8 out of 13) of small mammal species are primary forest dwellers confirmed his observation. Previous studies by other researchers also showed that most of the Malaysian small mammals prefer primary forest habitat (Harrison 1969; Lim 1970; Zubaid & M. Khairul 1997; Zubaid & Rizal 1995). Thus, the present study further confirmed the observations of previous researchers.

4.2 Potential factors affecting the distribution of small mammals
4.2.1 Primary forest species

Yasuda (1998) found that the availability and diversity of fruits was higher in PRI than in REG in the Pasoh FR. Lavelle & Pashanasi (1989) showed that the biomass of ground invertebrates is larger in a primary forest than in an old regenerating forest in Peruvian Amazonia. As most terrestrial mammals are omnivorous (Harrison 1961; Lim 1970; Medway 1978, summarized in Table 5), the primary forest habitat should be rich in food.

Secondly, the density of large fallen logs on the forest floor was higher in PRI than in REG (Yasuda 1998). As most terrestrial small mammals nest in fallen trees or hollows under dead trees (Medway 1978), the abundance of large fallen logs in PRI thus provide more foraging, hiding, resting and nesting sites for them. Furthermore PRI habitat was characterized by well-developed canopy (vertical) structure of the forest (Yasuda 1998). It can be concluded that the primary forest habitat is rich in food resources and spatial heterogeneity, and therefore can provide a higher carrying capacity for small mammals, both arboreal and terrestrial.

4.2.2 Regenerating forest species

Leopoldamys sabanus and *Maxomys surifer* were common in both disturbed and undisturbed forest habitats (Table 4). Apparently, these species are not so dependent on fallen logs as a resource for nests because they usually nest underground (Lim 1970). This implies that these species can survive in disturbed forest habitats with fewer fallen logs, but may not inhabit areas covered with water. These two species seem to be pre-adapted to disturbed environments.

4.2.3 Riparian species

Echinosorex gymnurus is known to prefer swampy forests or areas nearby streams as habitat (Lim 1967; Medway 1978). A telemetric study in the Pasoh FR revealed that *E. gymnurus* nests underground in dry places both in PRI and REG, but not in RIP (Yasuda, unpubl. data). They have two or more nests and move among them frequently at intervals of several days. This species is said to depend on aquatic animals for food (Banks 1931; Davis 1962; Gould 1978; Lim 1966, 1967; Medway 1978). Davis (1962) reported that the species feeds mainly on terrestrial invertebrates, such as earthworms, termites, beetles and other arthropods. Wild boars (*Sus scrofa*) intensively dig up the forest floor in wet areas near streams to search for food, probably earthworms (Kemper & Bell 1985; Payne et al. 1985), implying that invertebrate food resources are potentially abundant in the surface soil in this humid habitat. To maintain its large body, *E. gymnurus* may need to search for food over a large area, and the wet area near streams might be an important feeding site for the species.

A telemetric study revealed that *Trichys fasciculata* nests underground in large fallen logs in dry places only (Yasuda, unpubl. data). Using an automatic camera system (Miura et al. 1997), we found that *T. fasciculata* frequently comes to feed on fallen fruits on the ground in RIP as well as in PRI and REG (Yasuda, unpublished data). Since the species had also been caught in RIP, it is quite probable that *T. fasciculata* utilizes such wet areas as a part of its feeding range.

In addition, the large body size of *E. gymnurus* and *T. fasciculata* (Table 1) may allow them to forage for food in wet areas, perhaps because it insulates them against heat loss when they are wet.

4.2.4 Ubiquitous species

In contrast with the other rats, *Rattus tiomanicus* was captured in both wet areas and in dry forests, and its population density did not differ among habitats (Tables 2, 4). According to previous works, *R. tiomanicus* is a species usually found in cleared land, disturbed or fringing forest (Harrison 1958), but also occasionally in virgin forests (Medway 1972). Harrison (1958) and Sanderson & Sanderson (1964) carried out their studies in grasslands and estimated the size of the home ranges of the species as 73–102 m and 120 m long, respectively. Although Harrison (1958) argued that *R. tiomanicus* has territoriality, our results did not concur. We had 56 catches of 44 individuals of *R. tiomanicus* in total during the four-year census, but 37 individuals (84%) were caught only once. This suggests that, at least in the forest habitat, few individuals of the species stay in their home ranges for more than a few months and that they wander in the forest extensively. Such behavior in forest habitat may account for the ubiquitous distribution of the species found in the present study.

4.3 Effects of selective logging in small mammal community and the forest

As discussed above, there was a large difference in small mammal communities between the primary forest and the old regenerating forest, even though small mammals generally have high reproductive capacity and mobility. With the exception of a few species, the regenerating forest appeared to have a lower carrying capacity for small mammals than the adjacent primary forest. It is obvious that the inferior habitat of the regenerating forest was resulted from the selective logging carried out about 40 years ago. The effects of the logging still remains in the small mammal community there, even though the forest has been untouched since then. It can be concluded that 40 years is still insufficient for small mammal communities of tropical rain forests to recover from the disturbance of selective logging.

Small mammals have the potential to contribute to the seed dispersal of plants through their foraging behavior, i.e. food hoarding (Howe & Westley 1988; Vander Wall 1990). In Malaysian forests, Becker et al. (1985) and MacKinnon (1978) reported that arboreal squirrels hoard fruits at the canopy level, and recently Yasuda et al. (2000) found that some terrestrial rodents in Malaysia also have food hoarding behavior. They are important seed dispersers in tropical rain forests. The present study pointed out that squirrels are among the most vulnerable to logging disturbance in the small mammal community, because they are members of the primary forest species. The loss of dispersal agents as a consequence of logging could reduce the reproductive success of trees in the forest.

4.4 Concluding remarks

It is our firm belief that management plans for tropical rain forests should consider their animal inhabitants as well as trees. Frugivores may contribute little of the ecosystem's energy flow but are essential parts of the ecosystem as seed dispersers.

Logging and forest fragmentation have strong impacts on the frugivore community, even after several decades. In order to restore and maintain the overall vitality of forests, the most necessary and reasonable measure would be to artificially enhance degraded forests to promote better habitats for frugivores. This could be achieved through the construction of corridors to promote animal movement among isolated forest fragments, enrichment of the forest with indigenous fruit-bearing trees, and artificial thinning to improve forest structure.

ACKNOWLEDGEMENTS

The success of this study owes many things to many people, but in particular to Dato' Dr. Salleh Mohd. Nor, Dato' Dr. Abdul Razak Mohd Ali, Dr. N. Manokaran and Mr. Louis C. Ratnam in Forest Research Insititute Malaysia (FRIM). We also thank Drs. Lim Boo Liat, Akio Furukawa, Yoshitaka Tsubaki, Shingo Miura and Kimito Furuta who supported and supervised our work. Special thanks are due to Messrs. Zulhamli Bin Jamaluddin, Ahmad Bin Awang and Adnan Bin Awang for their field assistance in the Pasoh FR.

REFERENCES

Appanah, S. & Mohd. Rasol Abd. Manaf (1990) Small trees can fruit in logged dipterocarp forests. J. Trop. For. Sci. 3: 80-87.

Banks, E. (1931) A popular account of the mammals of Borneo. J. Malay. Br. R. Asiatic Soc. 9(2): 1-139.

Becker, P., Leighton, M. & Payne, J. B. (1985) Why tropical squirrels carry seeds out of source crowns? J. Trop. Ecol. 1: 183-186.

Caldecott, J. O. (1986) An ecological and behavioural study of the pig-tailed macaque. Contrib. Primatol. 21:1-259.

Corbet, G B. & Hill, J. E. (1992) The mammals of the Indomalayan region: a systematic review. Oxford University Press, New York.

Davis D. D. (1962) Mammals of the lowland rain-forest of North Borneo. Bull. Nat. Mus. (Singapore) 31: 1-129.

Duff, A. B., Hall, R. A. & Marsh C. W. (1984) A survey of wildlife in and around a commercial tree plantation in Sabah. Malay. For. 47: 197-213.

Gould, E. (1978) The behavior of the moonrat, *Echinosorex gymnurus* (Erinaceidae) and the pentail shrew, *Ptilocercus lowi* (Tupaiidae) with comments on the behavior of other Insectivora. Z. Tierpsychol. 48: 1-27.

Harrison, J. L. (1957) Habitat studies of some Malayan rats. Proc. Zool. Soc. London 128: 1-21.

Harrison, J. L. (1958) Range of movement of some Malayan rats. J. Mamm. 39: 190-206.

Harrison, J. L. (1961) The natural food of some Malayan mammals. Bull. Nat. Mus. (Singapore) 30: 5-18.

Harrison, J. L. (1969) The abundance and population density of mammals in Malaysian lowland forests. Malay. Nat. J. 22: 174-178.

Howe, H. F. & Westley, L. C. (1988) Ecological relationships of plants and animals. Oxford University Press, New York.

Kemper, C. (1988) The mammals of Pasoh Forest Reserve, Peninsular Malaysia. Malay. Nat. J. 42: 1-19.

Kemper, C. & Bell, D. T. (1985) Small mammals and habitat structure in lowland rain forest of Peninsular Malaysia. J. Trop. Ecol. 1: 5-22.

Kruskal, J. B. & Wish, M. (1978) Multidimensional scaling. Sage publications, Beverly Hills, London.

Langham, N. P. E. (1983) Distribution and ecology of small mammals in three rain forest localities of Peninsular Malaysia with particular references to Kedah Peak. Biotropica 15: 199-206.

Lavelle, P. & Pashanasi, B. (1989) Soil macrofauna and land management in Peruvian Amazonia

(Yurimaguas, Loreto). Pedobiologia 33: 283-291.
Lim, B. L. (1966) Land molluscs as food of Malayan rodents and insectivores. J. Zool. London 148: 554-560.
Lim, B. L. (1967) Note on the food habits of *Ptilocercus lowii* Gray (Pentail tree-shrew) and *Echinosorex gymnurus* (Raffles) (Moonrat) in a Malaya with remarks on "ecological labeling" by parasite patterns. J. Zool. London 152: 375-379.
Lim, B. L. (1970) Distribution, relative abundance, food habits, and parasite patterns of giant rats (Rattus) in west Malaysia. J. Mamm. 51: 730-740.
Lim, B. L., Ratnam, L. & Francis, C. (1995) Vertebrate fauna. Herpetofauna, birds and mammals. In Lee, S. S. (ed). A guide book to Pasoh, FRIM Technical Information Handbook No. 3. Forest Research Institute Malaysia, Kuala Lumpur.
MacKinnon, K. S. (1978) Stratification and feeding differences among Malayan squirrels. Malay. Nat. J. 30: 593-608.
Manokaran, N. & Swaine, M. D. (1994) Population dynamics of trees in dipterocarp forests of Peninsular Malaysia. Malayan Forest Records No. 40. Forest Research Institute Malaysia, Kuala Lumpur.
Medway, L. (1966) Fauna of Pulau Tioman: The mammals. Bull. Nat. Mus. (Singapore) 34: 9-32.
Medway, L. (1972) The distribution and altitudinal zonation of birds and mammals on Gunong Benom. Bull. Br. Mus. Nat. Hist. Zool. 23: 105-154.
Medway, L. (1978) The wild mammals of Malaya (Peninsular Malaysia) and Singapore (2nd ed., reprinted with corrections in 1983). Oxford University Press, Kuala Lumpur.
Miura, S., Yasuda, M. & Ratnam, L. (1997) Who steals the fruit? Monitoring frugivory of mammals in a tropical rain forest. Malay. Nat. J. 50: 183-193.
Payne, J., Francis, C. M. & Phillipps, K. (1985) A field guide to the mammals of Borneo. The Sabah Society, Kota Kinabalu.
Sanderson, G. C. & Sanderson, B. C. (1964) Radio-tracking rats in Malaya–a preliminary study. J. Wildl. Manage. 28: 752-768.
Stevens, W. E. (1968) Habitat requirements of Malayan mammals. Malay. Nat. J. 22: 3-9.
Stuebing, R. B. & Gasis, J. (1989) A survey of small mammals within a Sabah tree plantation in Malaysia. J. Trop. Ecol. 5: 203-214.
Vander Wall, S. B. (1990) Food hoarding in animals. University of Chicago Press, Chicago.
Whittaker, R. H. (1952) A study of summer foliage insect communities in the Great Smoky Mountains. Ecol. Monogr. 22: 1-44.
Yasuda, M. (1998) Community ecology of small mammals in a tropical rain forest of Malaysia, with special reference to habitat preference, frugivory and population dynamics. Ph. D. diss., Univ. Tokyo.
Yasuda, M., Miura, S. & Nor Azman, H. (2000) Evidence for food hoarding behaviour in terrestrial rodents in Pasoh Forest Reserve, a Malaysian lowland rain forest. J. Trop. For. Sci. 20: 164-173.
Zubaid A. & M. Khairul Effendi Ariffin (1997) A comparison of small mammal abundance between a primary and disturbed lowland rain forest in Peninsular Malaysia. Malay. Nat. J. 50: 201-206.
Zubaid, A. & Rizal, M. (1995) The relative abundance of rats between two forest types in Peninsular Malaysia. Malay. Nat. J. 49: 139-142.

38 Woodpeckers (Picidae) at Pasoh: Foraging Ecology, Flocking and the Impacts of Logging on Abundance and Diversity

Alison R. Styring[1] & Kalan Ickes[2]

Abstract: Woodpeckers are extremely diverse in the lowland rain forests of Southeast Asia. In these forests, up to 16 species can be found, representing the highest syntopic diversity for woodpeckers in the world. Understanding how foraging resources are partitioned in this species-rich guild is of both ecological and management concern. We investigated the foraging ecology of 13 species of woodpecker at Pasoh Forest Reserve (Pasoh FR) from May–July 1998. We obtained interpretable sample sizes for four species (*Meiglyptes tristis*, *M. tukki*, *Picus mentalis* and *Reinwardtipicus validus*). These four species exhibited distinct foraging behaviors and partitioned resources by using different substrates or attack maneuvers, or by foraging at different heights. Although the overall abundance of woodpeckers was similar between logged and unlogged forest, there were significant differences in the relative abundance of individual species. The most likely reason for these differences is the lack of heterogeneity in the mature logged forest compared to the unlogged forest. The relatively low abundance of snags, treefalls, and treefall gaps in the logged forest probably limit the number of woodpeckers that use these resources. While conducting this study, we also documented flocking interactions among several woodpecker species, and between woodpeckers and the Greater Racket-tailed Drongo (*Dicrurus paradiseus*). Understanding the complex foraging associations of this diverse guild warrants further study.

Key words: *Meiglyptes tristis*, mixed-species flock, *M. tukki*, *Picus mentalis*, *Reinwardtipicus validus,* selective logging.

1. INTRODUCTION

Lowland dipterocarp rain forests of the Greater Sundas are unique in that up to 16 species of woodpecker (Picidae) may occur sympatrically (Short 1978; Styring & Ickes 2001a; Wells 1999). Woodpeckers comprise the majority of the bark-foraging guild in this region and also comprise a larger percentage of the total avifauna than in other tropical forests (Table 1). Consequently, studies on the biology and foraging ecology of woodpeckers in aseasonal dipterocarp forests can address questions ranging from species packing and niche differentiation to conservation biology and the impacts of logging on insectivorous birds.

How is that 16 ecologically similar species can coexist within one forest type? How are the feeding niches differentiated? Short (1978) qualitatively described the foraging and nesting behavior of 13 woodpecker species at Pasoh Forest Reserve (Pasoh FR) and several other sites in Peninsular Malaysia. Though quantitative data on foraging was lacking, his observations suggested that a

[1] Department of Biological Science, Museum of Natural Science, Louisiana State University, Louisiana 70803, USA. E-mail: astyring@aol.com
[2] Louisiana State University, USA.

Table 1 Comparison of woodpecker diversity in relation to overall bird diversity from selected tropical locations.

	Location	Total no. of woodpecker species	Total no. of bird species	Percent of avifauna comprised	Source	Forest Type
NEOTROPICAL SITES	Barro Colorado Island, Panama	5	197	2.5	Karr et al. 1990	Seasonal lowland rainforest
	La Selva, Costa Rica	5	170	2.9	Karr et al. 1990	Aseasonal lowland rainforest
	Manaus, Brazil	12	293	4.1	Cohn-Haft et al. 1997	Seasonal lowland rainforest
	Manu, Peru	10	311	3.2	Karr et al. 1990	Seasonal lowland rainforest
AFROTROPICAL SITES	Basse Cassamance NP, Senegal	2	64	3.1	Thiollay 1985	Semi-deciduous wet (lowland)
	Makokou-Belinga area, Gabon	7	250	2.8	Thiollay 1985	Lowland rainforest
	Tai NP, Ivory Coast	5	233	2.1	Thiollay 1985	Lowland
	Budongo Forest Reserve, Uganda	2	73	2.7	Owiunjii & Plumptre 1998	Lowland rainforest
SOUTHEAST-ASIAN SITES	Danum Valley Cons. Area, Sabah	16	254	6.2	Lambert 1990	Aseasonal lowland rain
	Pasoh Forest Reserve, Malaysia	15	186	8.1	Ickes unpubl. data	Aseasonal lowland rain
	Sungai Tekam, Malaysia	11	193	5.7	Johns 1986	Aseasonal lowland rainforest
	Similajau, Sarawak	10	193	5.2	Duckworth et al. 1996	Aseasonal lowland rainforest

diversity of foraging strategies were employed by woodpeckers in Malaysia and that this may allow for niche differentiation among this uniform group. We tested this hypothesis of niche differentiation within the foraging strategies of woodpeckers by making detailed, quantified observations on the foraging ecology of this speciose group of birds at Pasoh FR (Chap. 27).

Another issue we addressed is how this group of species responds to logging. Many woodpeckers are known to forage on dead limbs and standing dead trees known as snags, often preferentially on large snags. In addition, most woodpecker species nest in tree cavities, a resource often found in older trees and snags. Not surprisingly, both snags and tree cavities are thought to be less common in logged tropical forests (Gibbs et al. 1993; Pattanavibool & Edge 1996). Given the importance of large trees and snags to the feeding and nesting biology of woodpeckers, this family of birds may be relatively more susceptible to habitat degradation following logging than other groups. Several studies have investigated the effects of selective logging on bird communities in both Peninsular Malaysia and Borneo and have provided evidence that woodpeckers may indeed be more sensitive to logging than other feeding guilds (e.g. smaller frugivores: Johns 1989; Lambert 1992; Wong 1985). However, these studies lacked sufficient data to identify the most affected woodpecker species. We tried to determine which species at PFR, if any were more susceptible to logging impacts by comparing the abundance and diversity of woodpeckers in the unlogged and selectively logged areas of the reserve.

Table 2 Description of attack variables used in this study (taken from Remsen & Robinson 1990).

Maneuver	Purpose
Glean	To pick food from a nearby substrate. Can be reached without full extension of legs or neck without the involvement of acrobatic movements.
Hang-down	To hang, head down in order to reach food not obtainable by any other perched position.
Reach-out	To reach laterally by extending legs and neck. Used to pick prey from nearby leaves.
Probe	To insert bill into cracks, holes, or soft substrates to capture hidden food.
Flake	To brush aside loose substrate with a sideways, sweeping motion.
Peck	To drive the bill against a substrate to remove some of the exterior of the substrate.
Chisel	Like "peck" but the bill is aimed obliquely rather than perpendicularly at the substrate.
Hammer	To deliver a series of pecks. Used to excavate deep holes to reach bark or wood-dwelling insects or sap.

While studying woodpeckers at Pasoh FR, two undocumented associations became apparent. Woodpeckers were frequently seen associating in mixed-species flocks with other woodpecker species, including congeners, and the Greater Racket-tailed Drongo (*Dicrurus paradiseus*) was often seen following woodpeckers. We quantified the strength of these foraging associations.

Pasoh FR is an ideal research site for studying the foraging ecology of woodpeckers because it is home to a large permanent tree plot (Chap. 2). Within the unlogged, 600 ha core of the reserve is a 50-ha tree plot, with all stems ≥ 1.0 cm diameter at breast height (DBH) tagged, measured, mapped and identified to species. With over 300,000 trees in the data set mapped over a 50-ha area (Chaps. 1, 2), such large-scale tree plots are advantageous for studies of plant-animal interactions for several reasons: (1) By noting which trees a vertebrate study species is foraging in and recording the tree tag number in the field, it is possible to later obtain the size, location, and species determination of that tree from the data set. (2) It is possible to follow study focal organisms off trails throughout the plot without getting lost. (3) The plots greatly facilitate the determination of habitat usage because of the large area covered.

Pasoh FR is also a good location for investigating the impacts of logging on vertebrate species because the 600 ha core of the reserve is adjoined by 700 ha of forest that was selectively logged from 1955–1959 (Chap. 2). Thus, a comparison of species composition between forest types can be made (as in Wong 1985, 1986). Previous studies of the effects of logging on bird communities in Malaysia have focused on stands logged less than 25 years ago. Investigating older managed stands such as Pasoh FR is important, however, as many physical characteristics of logged forests change over time. For example, at least one study in managed forest in a temperate region has shown that snags are decreasingly abundant in managed stands through time (Moorman et al. 1999), and others have shown that some woodpecker species may be negatively affected (Flemming et al. 1999; Virkkala et al. 1993). We combined the data on foraging ecology with comparative census data from the primary and selected logged forest to explain the observed species-specific responses to logging.

2. METHODS

2.1 Foraging observations

Foraging data were collected in the 50-ha plot at Pasoh FR. Because the plot is clearly marked in 20 m × 20 m quadrants, we were able to move in an approximately straight lines through the plot length of 1,000 m without having to cut transects. Four such lines, 100 m from the edge of the plot and 100 m apart, were followed. These lines were walked slowly (about 0.5 km/hour), and the following observations were recorded vocally with a microcassette recorder on foraging birds encountered opportunistically: food item (if determined), estimated height above ground, attack method and substrate type. Food items were classified as ant, termite, larva, or other (and identified as specifically as possible). "Attack method" followed the classification scheme of Remsen & Robinson (1990) (Table 2). Birds were followed for as long as possible, but only the initial observation was used during statistical analysis to avoid problems with non-independent data.

Data were analyzed using correspondence analysis using SYSTAT 8.0 (Chicago, Illinois, USA). Correspondence analysis is a multivariate technique that reduces data belonging to multiple categories into a few, more easily interpreted, dimensions. The continuous variable "foraging height" was grouped into three categories for analysis: understory (0–10 m), midstory (11–25 m) and canopy (26 m and above). Correspondence analysis was performed on the parameters "attack method", "substrate" and "foraging height." Several variables within these two parameters were grouped for the analysis. Under the parameter "Attack method", the variable "excavate" included data classified at "chisel", "hammer", "peck" and "flake". The variable "glean" included "hang upside down", "side reach" and "glean". For the parameter "Substrate type", the variables "dead liana", "dead trunk" and "dead branch" were combined into the single variable "dead wood." Likewise, the variables "live liana", "live trunk" and "live branch" were pooled together in the variable "live wood".

2.2 Woodpecker surveys in logged and unlogged forest

We designated already established trails totaling 4 km in length in each forest type (logged and unlogged) as census transects. Surveys were conducted at dawn and dusk. Each transect was walked six times: three times in the morning and three times in the afternoon, for a total of 48 surveys. Transects were surveyed slowly, at a rate of about 0.6 km/hour. Attempts were made to locate visually every woodpecker detected. Birds that could not be located visually were included in the data, but recorded as "heard only". It is important to point out that, although four transects were used in each forest type, only two continuous and adjacent stands of forest were studied. To avoid repetition in counting woodpeckers, all transects were placed at least 100 m apart and only birds that fell within 50 m of a transect were counted. However, woodpeckers, like most birds, are relatively mobile, and can certainly move the distance of transect both within and between forest types.

A likelihood-ratio Chi-square test was computed for the six most commonly recorded species to detect distributional differences of species between logged and primary forest. To determine overall difference in woodpecker abundance as well as species-specific differences, binomial tests were conducted for overall totals and on species encountered at least five times in a forest type (for more information, see Styring & Ickes 2001a).

Table 3 Parameters and variables used in correspondence analysis. Species codes: RV = *Reinwardtipicus validus*, PM = *Picus mentalis*, MTU = *Meiglyptes tukki*, MT = *Meiglyptes tristis*.

Attack method	Species	Glean	Hang upside down	Side reach	Probe	Flake	Peck	Chisel	Hammer	Mean ht. (m)	*N*
	RV	0	0	0	8.3	4.2	37.5	4.2	45.8	12.8 ± 8.2	24
	PM	5.3	0	0	63.2	0	10.5	0	21.1	13.6 ± 3.3	19
	MTU	14.3	0	0	7.1	7.1	42.9	7.1	21.4	7.7 ± 5.3	14
	MT	9.6	26.9	5.8	11.5	21.2	5.8	1.9	17.3	9.5 ± 5.0	52

Substrate type		Trunk	Dead trunk	Branch	Dead branch	Liana	Dead liana	Ant nest	Leaf	*N*
	RV	29.2	37.1	8.3	12.5	8.3	4.2	0	0	24
	PM	36.8	21.1	10.6	10.5	10.5	5.3	0	5.3	19
	MTU	35.7	21.4	0	7.1	0	0	28.6	7.1	14
	MT	5.8	1.9	38.5	0	5.8	13.5	9.6	25	52

2.3 Mixed-species flocking associations

While collecting foraging and census data, interspecific interactions among woodpecker species, and between woodpeckers and the Greater Racket-tailed Drongo (*Dicrurus paradiseus*) were recorded. When woodpeckers were observed in foraging flocks, we noted the presence of other woodpecker species as well as the presence of drongos. For purposes of this study, a flock is defined as a group of two or more individuals of two or more species in close proximity (approximately < 15 m) to one another, moving through the habitat and foraging together. Agonistic interactions and other interspecific-dependent behaviors were also noted. All drongos seen or heard were also followed and the presence / absence of woodpeckers and other species was noted (Styring & Ickes 2001b,c).

Binomial probabilities were calculated for all species observed at least 10 times during the 2-month study period. The expected probability for seeing a woodpecker or drongo in association with another species was set at 0.1. This is taken from McClure (1967) who considered birds flock attendants (regular, usual or habitual) if they were observed in flocks in at least 10% of all sitings. However, the expected probability of seeing a woodpecker with other woodpecker species was set at 0.47. This value was calculated by dividing the total number of woodpeckers seen at Pasoh FR ($N = 15$) by the number of flock attendants observed by McClure (1967). Only birds listed in the appendix of McClure (1967) that also occurred at Pasoh FR, and only those found in 12% or more of all flocks ($N = 32$) were included to make the estimate conservative. These same probabilities were used in calculating the binomial probability for drongos both in association with flocks in general (0.1) and with woodpeckers in particular (0.47).

3. RESULTS

3.1 Foraging observations

With the exception of *Meiglyptes tristis*, the number of foraging observations for individual species was rather low. We have presented data for only the four most commonly observed species, which were also among the six most commonly recorded species on transects.

There was considerable variation among substrates used and attack methods (Table 3) although all species analyzed preferred some type of wood as a substrate to leaves or ant nests. Preference for dead wood varied considerably, but overall frequency of use was near 50% across species. None of the species analyzed seemed to show taxonomic preferences for trees. For example, there were 34

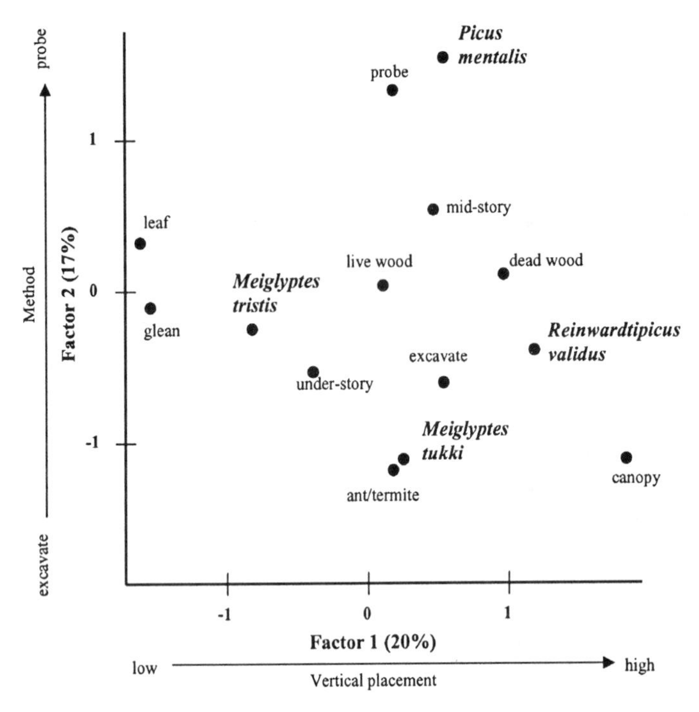

Fig. 1 Correspondence plot of woodpecker species with foraging height (under-story = 0–10 m, mid-story = 11–25 m, canopy = 26 m and higher), substrate (ant/termite = external ant and termite nests and tunnels, leaf = live and dead leaves, live wood = live trunks, branches and lianas, dead wood = dead trunks, branches and lianas), and foraging method (glean, probe and excavate). Boldface words indicate species.

Table 4 Relative abundance of fifteen woodpecker species present at Pasoh FR (from Styring & Ickes 2001a with permission). Numbers indicate the number of individuals seen in each forest type. Birds seen in a given forest type, but not counted during transects are denoted with a *. *P*-values are the binomial probability calculated for species encountered at least five times in a single forest type. Weight data taken from Short (1978, 1982).

species	English name	mean weight (g)	logged	unlogged	*P*
Sasia abnormis	Rufous Piculet	9	0	1	-
Hemicircus concretus	Grey-and-Buff Woodpecker	32	1	*	-
Meiglyptes tristis	Buff-rumped Woodpecker	46	4	14	0
Meiglyptes tukki	Buff-necked Woodpecker	53	4	5	0.3
Picoides canicapillus	Grey-capped Woodpecker	27	0	*	-
Celeus brachyurus	Rufous Woodpecker	66	2	4	-
Picus puniceus	Crimson-winged Woodpecker	84	12	7	0.1
Picus mentalis	Checker-throated Woodpecker	98	20	1	0
Picus miniaceus	Banded Woodpecker	89	4	2	-
Blythipicus rubiginosus	Maroon Woodpecker	83	1	4	-
Dinopium javanense	Common Goldenback	79	4	*	-
Dinopium rafflesii	Olive-backed Woodpecker	96	0	2	-
Reinwardtipicus validus	Orange-backed Woodpecker	155	6	15	0
Dryocopus javensis	White-bellied Woodpecker	225	1	6	0.1
Mulleripicus pulverulentus	Great Slaty Woodpecker	430	0	1	-
TOTAL			59	62	0.1

Table 5 Woodpecker participation in mixed species flocks, and with other woodpecker species at Pasoh FR.

Species	*M. tristis*	*M. tukki*	*P. puniceus*	*P. mentalis*	*P. miniaceus*	*R. validus*	*D. javensis*
Total sitings	40	23	22	48	22	30	14
Sitings in Flocks	24	11	16	38	19	18	6
Binomial Probability (for flocking)	< 0.001	< 0.001	< 0.001	< 0.001	< 0.001	< 0.001	0.001
Sitings with other Woodpecker species	17	8	10	29	12	12	4
Binomial Probability (for flocking with other woodpecker species)	0.01	0.06	0.09	< 0.001	0.07	0.09	0.21

independent foraging observations of *M. tristis*, of which 31 one were from different tree species.

Correspondence analysis of species and individual parameters (Fig. 1) revealed that variation in the data set is relatively evenly divided among parameters (with species, attack method, substrate and foraging height accounting for about 25% of the variation each). Not surprisingly, the substrate "leaf" and the attack method "glean" were more closely linked than other substrates and attack methods. Another tight association was seen between the attack method "probe" and the species *Picus mentalis*. Probing was not a preferred attack method for any other species.

3.2 Woodpecker surveys in logged and unlogged forest

Results of transect data analyses (Table 4) revealed that overall woodpecker abundance differed only slightly between forest types. However, distribution of species in the two forest types differed significantly ($G = 36.3$, $df = 5$, $P < 0.0001$) with *Meiglyptes tristis*, *Reinwardtipicus validus* and *Dryocopus javensis* significantly more abundant in unlogged forest (with *Picus puniceus* more abundant but not significantly so), and *Picus mentalis* significantly more abundant in logged forest.

3.3 Mixed-species flocking associations

More than 50% of sightings of all three *Picus* species, *Meiglyptes tristis* and *Reinwardtipicus validus* were of individuals in mixed-species flocks (Table 5). The six species for which there were > 10 sightings in mixed-species feeding flocks were associated with at least one heterospecific woodpecker 62%–76% of the time (Styring & Ickes 2001c).

Dicrurus paradiseus showed a strong preference to associate with flocks in general and woodpeckers in particular. Of the 79 times this species was observed, 76 were in flocks ($P < 0.0001$), and 70 of these flocks included woodpeckers ($P < 0.0001$). While in association with woodpeckers, drongos often perched within 2 meters of a woodpecker, sallying out to capture prey flushed from the bark surface as the woodpecker moved along the substrate. On one occasion, a drongo sallied very close to a *Picus mentalis* and hovered over the bird as it caught an invertebrate in flight. It was in such close proximity that its wings repeatedly made contact with the head of *P. mentalis*, which remained still but cowered low to avoid the contact (Styring & Ickes 2001b).

4. DISCUSSION

The number of foraging observations transect sightings in logged vs. unlogged forest, and sightings in mixed-species foraging flocks were all greatest for *M. tristis*,

M. tukki, *P. mentalis* and *R. validus*. An examination of the habits of these four species may provide clues as to how so many closely related species can occur sympatrically.

Reinwardtipicus validus shows a strong preference for excavating food items from the trunks of dead trees (Table 3). Other species foraged on dead wood with similar frequency, but *R. validus* spent a greater portion of its foraging time on this substrate (82.1% compared to 12.9%, 12.2% and 51.0% for *M. tristis*, *M. tukki* and *P. mentalis* respectively). It is, therefore, not unexpected that *R. validus* was less abundant in logged forest. Lower occurrence of standing dead and dying trees in logged forest may cause this species to persist at lower population densities than those found in primary forest. Evidence from the literature and our own observations for *D. javensis* suggests that this species forages in a mode similar to *R. validus*, foraging primarily on standing dead wood (Short 1978; Wells 1999). It may therefore also be expected to occur in logged areas at lower population density than in primary forest. Although our sample size for habitat preference was small for this species, it did appear that *D. javensis* was in fact less common in the selectively logged forest at Pasoh FR (Table 4).

Meiglyptes tristis is not a traditional "woodpecking" species. Rather than excavating wood to obtain food items, it gleans bark and leaf surfaces. Particularly unusual is the frequency that it leaf-gleans (23% of all observations), a behavior virtually undocumented in woodpeckers. Although *M. tristis* spends a lot of time on branches and leaves, its foraging height is relatively low (Table 3), making it an understory species. This combination of factors suggests that this species prefers areas with relatively dense understory vegetation, which is found most often in forest gaps. Given this foraging strategy, we would predict that this species would do well in recently disturbed forests, which are often characterized by higher light levels and lower vegetation stature. The managed forest at Pasoh FR was logged more than 40 years ago, however, and over time the stand architecture must have changed considerably. Stand structure in the selectively logged forest at Pasoh FR during the time of this study was typical of older, managed forests in which lianas and rattans have also been removed: uniformly-aged with relatively few snags and climbers and thus few large gaps. *M. tristis* occurs frequently in primary forest (Styring & Ickes 2001a) with associated large gaps and young secondary forests (Styring unpubl. data) with higher light levels and dense vegetation closer to the ground. Conversely, *M. tristis* is much less common in older logged forests (Styring & Ickes 2001a) and tree plantations (Mitra & Sheldon 1993) where the vegetation is more uniform and gaps are less common.

Meiglyptes tukki showed no difference between forest types in terms of relative abundance, but the sample size was small. This bird appears to be restricted to the forest understory, and it forages frequently on ant nests found on tree trunks and lianas. It is possible that this resource is not severely affected by logging, but more data is needed to confirm this suggestion.

Only one species, *Picus mentalis*, was significantly more abundant in logged forest, although *P. puniceus* may also be more common in this forest type. *P. mentalis* did not use different substrates preferentially, but the preferred attack method was probing with gleaning and probing, accounting for nearly 70% of all feeding observations. We can offer no concrete explanation for how this combination of foraging characteristics can explain the highly significant difference in abundance recorded between logged and unlogged forest at Pasoh FR. Perhaps the lower abundance of other species has beneficial effects for *P. mentalis*.

An unexpected, but ecologically interesting outcome of this study was the frequency that woodpeckers were observed flocking together. While woodpeckers are known to participate in mixed species foraging flocks, the presence of flocks containing multiple woodpecker species is virtually undocumented (Styring & Ickes 2001c). The only documented case is a description of a single flock observed in Sumatra that consisted of 5 species: *Meiglyptes tukki*, *Picus mentalis*, *Picus miniaceus*, *Dinopium rafflesii* and *Reinwardtipicus validus* (Jepson & Brinkman 1993). With the exception of *D. rafflesii*, the same species were shown to preferentially associate with other woodpeckers at Pasoh FR. Species that participate in mixed species insectivorous flocks are believed to have minimal niche overlap, which allows them to avoid competition while foraging in close proximity (Buskirk 1976; Powell 1985). Woodpeckers are a rather homogenous family of birds with similar specialized morphology. It might be expected, therefore, that closely related species that flock together demonstrate differences in foraging behavior, thereby minimizing competition for food. Correspondence analysis supported this expectation. Behavioral differences were found among several of the flocking species in foraging methods, substrate choice, and foraging height.

Another finding made while investigating the foraging habits of woodpeckers at Pasoh FR was the strong association between the Greater Racket-tailed *Drongo (Dicrurus paradiseus*) and woodpeckers. Drongos (Dicruridae) are a small family (22 species) of sub-tropical- tropical birds that range from Africa to Asia, Australia and the Solomon Islands. At least five species of drongo were previously known to follow large mammals and bird flocks in order to capture flushed prey (Herremans & Herremans-Tonnoeyr 1997; Hino 1998; Robson 2000; Simpson & Day 1993; Veena & Lokesha 1993). *Dicrurus paradiseus* is now the sixth drongo documented to associate with other vertebrates. Laman (1992) suggested that an association between drongos and *Picus* woodpeckers exists, but the study lacked conclusive data. At Pasoh FR, *D. paradiseus* showed a very strong association with woodpeckers: 70 out of 79 observations were of this species following at least one woodpecker. Although *D. paradiseus* often associated with *Picus* woodpeckers, we were unable to demonstrate that they followed these species for frequently than several other species (Styring & Ickes 2001b). Further studies of *D. paradiseus* -woodpecker interactions in other forests should give some indication of the strength of this association and preferences for certain species.

5. CONCLUSIONS AND SUGGESTIONS FOR FURTHER RESEARCH

Due to the high diversity of woodpeckers at Pasoh FR and the difficulty in obtaining foraging and transect observations on the less common species, species-specific sample sizes in these studies were low in some cases. Nonetheless, the data do illustrate some of the variation in foraging behavior displayed by of the woodpeckers of lowland Malaysia, and may, in part, explain why so many can persist in a single habitat. *Reinwardtipicus validus* and *Meiglyptes tristis* may represent ends of a broad spectrum of substrate preference and foraging maneuvers that allow so many species to persist in sympatry.

Increasing the number of observations of all species will give us a more solid understanding of the microhabitat preferences of woodpeckers in these forests. Such data, combined with relative abundance or density estimates of these species in logged forests, should allow us to identify key factors with regard habitat that are important to this guild when managing logged stands.

In temperate forests, logging impacts on woodpeckers are intensely studied, and in some areas, management guidelines are almost exclusively aimed at maintaining woodpecker (particularly large bodied woodpeckers) and secondary cavity nester populations. Much of the focus has been on management of nesting habitat for large woodpeckers (Flemming et al. 1999). However, other studies indicate that understanding the foraging ecology of woodpeckers is at least equally as important to forest management (Conner et al. 1994; Welsh & Capen 1992). The preliminary results of this study suggest that many of the findings for temperate woodpeckers (maintaining mature and dead trees in logged stands) also apply in the tropics. Numerous secondary cavity nesters such as hornbills, owls, and several mammal species stand to benefit from such management practices.

ACKNOWLEDGEMENTS

We are grateful to Dr. Abdul RazakMohd. Ali, Director General of the Forestry Research Institute Malaysia (FRIM), for permission to conduct research at Pasoh FR. J. V. Remsen, Jr. made constructive comments on the manuscript. ARS was funded in part by: the Frank M. Chapman Memorial Fund from the American Museum of Natural History, the Charles M. Fugler Fellowship in Tropical Vertebrate Biology from the Louisiana State University Museum of Natural Science, the LSU Museum of Natural Science Tropical Bird Research Fund, and the LSU Museum of Natural Science Research Award. KI was funded in part by a Grant-in-Aid of Research from the National Academy of Sciences, through Sigma Xi.

REFERENCES

Buskirk, W. H. (1976) Social systems in a tropical forest avifauna. Am. Nat. 110: 293-305.

Cohn-Haft, M., Whittaker, A. & Stouffer, P. C. (1997) A new look at the "species-poor" central Amazon: the avifauna north of Manaus, Brazil. Ornithological Monogr. 48: 205-235.

Conner, R. N., Jones, S. D. & Jones, G. D. (1994) Snag condition and woodpecker foraging ecology in a bottomland hardwood forest. Wilson Bull. 106: 242-257.

Duckworth, J. W., Wilkinson, R. J., Tizard, R. J., Kelsh, R. N., Irvin, S. A., Evans, M. I. & Orrell, T. D. (1996) Bird records from Similajau National Park, Sarawak. Forktail 12: 159-196.

Flemming, S. P., Holloway, G. L., Watts, E. J. & Lawrance, P. S. (1999) Characteristics of foraging trees selected by pileated woodpeckers in New Brunswick. J. Wildl. Menage. 63: 461-469.

Gibbs, J. P., Hunter, M. L. Jr. & Melvin, S. M. (1993) Snag availability and communities of cavity nesting birds in tropical versus temperate forests. Biotropica 25: 236-241.

Herremans, M. & Herremans-Tonnoeyr, D. (1997) Social foraging in the Forktailed Drongo *Dicruris adsimilis*: beater effect or kleptoparasitism? Bird Behav. 12: 41-45.

Hino, T. (1998) Mutualistic and commensal organization of avian mixed-species foraging flocks in a forest of western Madagascar. J. Avian Biol. 29: 17-24.

Jepson, P. & Brinkman, J. J. (1993) Notes on a mixed woodpecker flock at Bukit Barisan Selatan National Park, Lampung, Sumatra. Kukila 6: 135.

Johns, A. D. (1986) Effects of selective logging on the ecological organization of a Peninsular Malaysian rainforest avifauna. Forktail 1: 65-79.

Johns, A. D. (1989) Recovery of a Peninsular Malaysian rainforest avifauna following selective timber logging: the first twelve years. Forktail 4: 89-105.

Karr, J. R., Robinson, S. K., Blake, J. G. & Bierregaard Jr., R. O. (1990) Birds of four Neotropical forests. In Gentry, A. H. (ed). Four Neotropical Rainforests, Yale University.

Laman, T. G. (1992) Composition of Mixed-species Foraging Flocks in a Bornean Rainforest. Malay. Nat. J. 46: 131-144.

Lambert, F. R. (1990) Avifaunal changes following selective logging of a North Bornean rain

forest. Royal Society South-east Asian Rain Forest Research Committee (Final Report), Not published, 59pp.
Lambert, F. R. (1992) The consequences of selective logging for Bornean lowland forest birds. Philos. Trans. R. Soc. London Ser. B 335: 443-457.
McClure, H. E. (1967) The composition of mixed species flocks in lowland and sub-montane forests of Malay. Wilson Bull. 79: 131-154.
Mitra, S. S. & Sheldon, F. H. (1993) Use of an exotic tree plantation by Bornean lowland forest birds. Auk 110: 529-540.
Moorman, C. E., Russell, K. R., Sabin, G. R. & Guynn Jr., D. C. (1999) Snag dynamics and cavity occurrence in the South Carolina Piedmont. For. Ecol. Manage. 118: 37-48.
Owiunjii, I. & Plumptre, A. J. (1998) Bird communities in logged and unlogged compartments in Budongo Forest, Uganda. For. Ecol. Manage. 108: 115-126.
Pattanavibool, A. & Edge, W. D. (1996) Single-tree selection silviculture affects cavity resources in mixed deciduous forests in Thailand. J. Wildl. Manage. 60: 67-73.
Powell, G. V. N. (1985) Sociobiology and adaptive significance of interspecific foraging flocks in the Neotropics. Ornithological Monogr. 36: 713-732.
Remsen Jr., J. V. & Robinson, S. K. (1990) A classification scheme for foraging behavior of birds in terrestrial habitats. Studies in Avian Biology 13: 144-160.
Robson, C. (2000) A Guide to Birds of Southeast Asia: Thailand, Peninsular Malaysia, Singapore, Myanmar, Laos, Vietnam, Cambodia, Princeton University Press, Princeton, 504pp.
Short, L. L. (1978) Sympatry in woodpeckers of lowland Malayan forest. Biotropica 10: 122-133.
Short, L. L. (1982) Woodpeckers of the World. Delaware Museum of Natural History, Greenville.
Simpson, K. D. & Day, N. (1993) Field Guide to the Birds of Australia. Penguin Books, Ringwood, 392pp.
Styring, A. R. & Ickes, K. (2001a) Woodpecker abundance in a logged (40 years ago) vs. unlogged lowland dipterocarp forest in Peninsular Malaysia. J. Trop. Ecol. 17: 261-268.
Styring, A. R.& Ickes, K. (2001b) Interactions between the Greater Racket-tailed Drongo *Dicrurus paradiseus* and woodpeckers in a lowland Malaysian rainforest. Forktail 17: 119-120.
Styring, A. R. & Ickes, K. (2001c) Woodpecker participation in mixed species flocks in Peninsular Malaysia. Wilson Bull. 113:342-345.
Thiollay, J. M. (1985) The West African forest avifauna: a review. In Diamond, A. W. & Lovejoy, T. E. (eds). Conservation of Tropical Forest Birds. ICBP, Cambrigde, 318pp.
Veena, T. & Lokesha, R. (1993) Association of drongos with myna flocks: are drongos benefited? J. Biosci. 18: 111-119.
Virkkala, R., Alanko, T., Laine, T. & Tiainen, J. (1993) Population contraction of the White-backed Woodpecker *Dendrocopos leucotos* in Finland as a consequence of habitat alteration. Biol. Conserv. 66: 47-53.
Wells, D. R. (1999) The Birds of the Thai-Malay Peninsula, Academic Press, London, 648pp.
Welsh, C. J. E. & Capen, E. (1992) Availability of nesting sites as a limit to woodpecker populations. For. Ecol. Manage. 48: 31-41.
Wong, M. (1985) Understory birds as indicators of regeneration in a patch of selectively logged West Malaysian rainforest. In Diamond, A. W. & Lovejoy, T. E. (eds), Conservation of Tropical Forest Birds, ICBP, Cambridge, 318pp.
Wong, M. (1986) Trophic organization of understory birds in a Malaysian dipterocarp forest. Auk 103: 100-116.

39 Regeneration of a Clear-Cut Plot in a Lowland Dipterocarp Forest in Pasoh Forest Reserve, Peninsular Malaysia

Kaoru Niiyama[1], Abdul Rahman Kassim[2], Shigeo Iida[3], Katsuhiko Kimura[4], Azizi Ripin[2] & Simmithiri Appanah[2]

Abstract: Regeneration after clear cutting was investigated using a data set of tree censuses in 1971 and 1996 at plot 1 in Pasoh Forest Reserve (Pasoh FR), Peninsular Malaysia. Species composition and tree size larger than 10 cm in DBH (Diameter at Breast Height) were recorded in a plot (20 m × 100 m) cleared in 1971. Above ground parts of this plot were clearly cut and weighed for biomass estimation under the International Biological Program (IBP) in 1973. In 1994, we established a 6-ha plot including the above cleared plot. All trees ≧ 5 cm in DBH were tagged, measured, and identified in 1994, 1996, and 1998. We compared the species composition and spatial distribution of trees in 1996 with those in 1971. The species composition of the regenerating trees in 1996 was largely altered from the original composition before cutting. Primary forest tree species having dormant seeds, e.g. *Melicope glabra* and *Pternandra echinata*, and pioneers, *Endospermum diadenum, Macaranga conifera,* and *M. hosei*, dominated in the cleared plot. These pioneers were rare in the surrounding intact part of the 6-ha plot. Original canopy species before cutting, e.g. *Koompassia malaccensis* and *Dipterocarpus cornutus*, were very rare in the cleared plot. Nevertheless, fast-growing *Shorea* species such as *S. leprosula* and *S. parvifolia* were common in the plot. After clear cutting, 59 of 82 original tree species failed to return in this cleared site. These results reveal that the species composition of a primary lowland dipterocarp forest is drastically altered after a clear cutting, even though its area was small.

Key words: clear cutting, IBP project, pioneer species, spatial distribution, species composition.

1. INTRODUCTION

Logging in Peninsular Malaysia has disturbed a large area of primary tropical rain forests. Most parts of these disturbed forests are regenerating into secondary forests, and many climax species have been regenerated in these forests. To monitor and improve these regenerating forests is crucial for the sustainable forest management. However, precise prediction of the regeneration process is very difficult. Disturbances and kinds of impact on the forest ecosystem differ in size and magnitude. Nutrient conditions and abundance of advanced regeneration in a forest differ from site to site. Furthermore, the amount of seeds in the soil and dispersing from the surrounding vegetation also differs among sites. Such diversity of impacts, habitats, and seed sources prevents generalization concerning secondary succession

[1] Forestry and Forest Products Research Institute (FFPRI), Tsukuba 305-8687, Japan. E-mail: niiya@ffpri.affrc.go.jp
[2] Forest Research Institute Malaysia (FRIM), Malaysia.
[3] Hokkaido Research Center, FFPRI, Japan.
[4] Fukushima University, Japan.

in the tropical forests. Many case studies, therefore, are needed to clarify the secondary succession of tropical forests.

Clear cutting is one of the most severe disturbances on the primary forest. However, if a cleared area does not suffer soil erosion the site has the legacy of a soil seed bank (Liew 1973; Kanzaki et al. 1997; Metcalfe & Turner 1998; Putz & Appanah 1987). If such a cleared site is small, fresh seeds are supplied from the surrounding intact primary forest. In this aspect, a small sized clearing is similar to a canopy gap. In 1973, above ground parts of all trees in a small plot (20 m × 100 m) in Pasoh Forest Reserve (Pasoh FR) was clearly cut and weighed for biomass estimation under the International Biological Program (IBP) project of Malaysia, Japan and UK (Kato et al. 1978; Chap. 1). Many studies on secondary succession lack the data of original species composition in the study site (Kochummen 1966; Kochummen & Ng 1977; Symington 1933; Wyatt-Smith 1949, 1954, 1955). This clearing for biomass estimation provided a rare chance to start the study of secondary succession from the original species composition. An incomplete species list and some photographs in the cleared plot were already published (Ashton 1978; Richards 1996). In this chapter, first, the species composition of regenerated trees in 1996 is compared with that of the original trees in 1971. Second, the spatial distribution of successfully regenerated tree species in the plot will be analysed.

2. MATERIALS AND METHODS

The regeneration after clear cutting was investigated using the data sets of tree censuses in 1971 and 1996 at plot 1 (Fig. 1) in Pasoh Forest Reserve (Pasoh FR), Peninsular Malaysia. The species composition of trees larger than 10 cm in DBH (Diameter at Breast Height) had been recorded in the cleared plot (20 m × 100 m) in 1971 (Ashton unpublished data; Yamakura personal communication). For biomass estimation under the IBP project in March–April 1973, different sampling areas were adopted for six tree size classes in the plot (Kato et al. 1978). For example, large trees (≧ 40 cm DBH) were cut in the whole area, 20 m × 100 m. Medium trees (20 cm ≦ DBH < 40 cm) were cut in a part of the plot, 10 m × 100 m. Other size trees were cut in smaller areas in the plot. After biomass measurement, all residual trees were cut down near the ground in the plot (Yoneda personal communication). This cleared plot (20 m × 100 m) was located within the plot 1 (100 m × 200 m), which was established by Wong & Whitmore (1970) and expanded by Ashton (1976). We further expanded

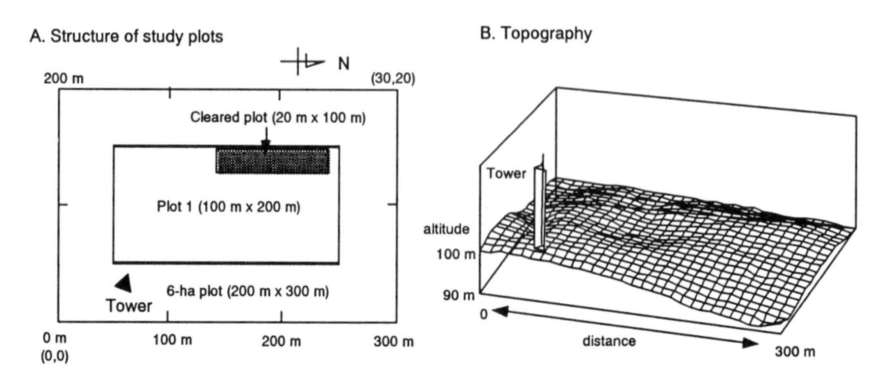

Fig. 1 Study site. A: structure of three study plots; cleared plot, plot 1 and 6-ha plot. B: Topography and altitude of the 6-ha plot.

the plot 1 to a 6-ha long-term ecological research plot in 1994 (Fig. 1). All trees ≧ 5 cm in DBH in the 6-ha plot were tagged, measured, and identified in 1994, 1996 and 1998. We use the data set of 1996 in this chapter.

Degree of aggregation of regenerated trees in the cleared plot was tested by the binomial distribution test (Sokal & Rohlf 1995). The distribution of eight most abundant tree species (> seven individuals) in the cleared plot, and the largest species, *Koompassia malaccensis*, and three dipteocarps, *Dipterocarpus cornutus, Shorea macroptera* and *Shorea multiflora*, were tested by the binomial distribution test.

Tree species were categorized to five ecological species groups: emergent, main canopy, understorey, pioneer and late-seral (Manokaran & Swaine 1994). This classification is partly depend on descriptions in Tree Flora of Malaya (Ng 1978, 1989; Whitmore 1972a,b) and Tree Flora of Pasoh Forest (Kochummen 1997). Nomenclature follows Turner (1995).

3. RESULTS

The species composition of the regenerating trees in 1996 (Table 2) was largely altered from the pre-logging species composition (Table 1). The compositions of the 20 abundant species in basal area were completely different between 1971 and 1996. Dominant families changed from Leguminosae and Dipterocarpaceae in 1971 to Rutaceae, Euphorbiaceae, Annonaceae, Elaeocarpaceae and Melastomataceae

Table 1 Original composition of dominant tree species (DBH ≧ 10 cm) in 1971 in the cleared plot (20 m × 100 m) which were cleared for IBP project in 1973. Tree species are listed in order of basal area, BA (cm^2).

No.	Species	Family	Species group*	*N*	Max. DBH (cm)	BA
1	*Koompassia malaccensis* Maing. ex Benth.	Leguminosae	E	1	100.9	7988
2	*Dipterocarpus cornutus* Dyer	Dipterocarpaceae	E	5	77.6	7122
3	*Ganua* sp.A	Sapotaceae		8	32.8	3581
4	*Mesua* sp.	Guttiferae		1	65.6	3380
5	*Nephelium ramboutan-ake* (Labill.) Leenh.	Sapindaceae	MC	2	53.2	3212
6	*Pentaspadon motleyi* Hook. f.	Anacardiaceae	MC	2	57.9	2980
7	*Sindora* sp.	Leguminosae		1	59.8	2810
8	*Neoscortechinia kingii* (Hook. f.) Pax & K. Hoffm.	Euphorbiaceae	MC	2	52.5	2806
9	*Syzygium hoseanum* (King) Merr. & L.M. Perry	Myrtaceae	U	2	42.0	1699
10	*Knema* sp.	Myristicaceae		1	45.7	1640
11	*Sarcotheca griffithii* (Planch. ex Hook. f.) Hallier f.	Oxalidaceae	MC	1	44.8	1578
12	*Lansium domesticum* Correa	Meliaceae	U	4	34.1	1372
13	*Shorea macroptera* Dyer	Dipterocarpaceae	E	1	40.7	1300
14	*Shorea dasyphylla* Foxw.	Dipterocarpaceae	E	1	38.3	1155
15	*Ryparosa wallichii* Ridl.	Flacourtiaceae	U	1	38.1	1141
16	*Shorea pauciflora* King	Dipterocarpaceae	E	1	37.9	1129
17	*Gymnacranthera farquhariana* (Hook. f. & Thomson) Warb. var. *eugeniifolia* (A. DC.) R.T.A.Schout	Myristicaceae	MC	1	37.2	1086
18	*Pimelodendron griffithianum* (Mull Arg.) Benth.	Euphorbiaceae	MC	1	37.0	1075
19	*Shorea multiflora* (Burck) Symington	Dipterocarpaceae	E	2	32.2	1075
20	*Dialium procerum* (Steenis) Steyaert	Leguminosae	MC	3	33.0	1073
	Others			81	35.8	17846
	Total			122	100.9	67047

* Species group: E = emergent, MC = main canopy, U = understorey, P = pioneer, LS = late-seral (Manokaran & Swaine 1994).

in 1996. These families contained pioneer, late-seral, main canopy and understorey species (Table 2). Pioneers or late-seral species, e.g. *Endospermum diadenum*, *Macaranga hosei* and *M. conifera* were abundant in 1996. However, main canopy species, *Melicope glabra* was the most abundant tree species in 1996. *Pternandra echinata* is one of the putative primary understorey species with small seed (Metcalfe & Turner 1998) yet was very abundant in the cleared plot. Nevertheless, the maximum DBH of this species was the smallest among the abundant twenty species in 1996 (Table 2). The cleared plot contained *Shorea leprosula* and *S. parvifolia* that are categorized into light-demanding, fast growing emergent species. *Koompassia malaccensis* , *Dipterocarpus cornutus* and *Shorea macroptera*, which are common emergent tree species in Pasoh FR (Manokaran & Kochummen 1987; Manokaran & LaFrankie 1990; Manokaran et al. 1992a,b) , were not common in 1996.

Under the same DBH threshold of 10 cm, only four of the original species were observed in 1996 (Fig. 2A). However, under different diameter threshold, 5 cm in 1996 and 10 cm in 1971, 23 species were observed in both the censuses (Table 3, Fig. 2B). Eight species showed significant concentration in the cleared plot (Figs. 3, 4). In particular, *Melicope glabra, Macaranga conifera* and *Macaranga hosei* were remarkably restricted to the cleared plot. Another four species, *Pternandra echinata*, *Xylopia ferruginea*, *Porterandia anisophyllea* and *Shorea leprosula*, also showed significant aggregation in the cleared plot. Their mother trees were distributed in the surrounding intact part of forest (Fig. 4). Abundant canopy species before cutting,

Table 2 Species composition of regenerated trees (DBH ≧ 5cm) dominated in 1996, 23 years after clearing in the cleared plot (20 m × 100 m). Tree species are listed in order of basal area, BA (cm^2).

No.	Species	Family	Species group*	*N*	Max.DBH (cm)	BA (cm^2)
1	*Melicope glabra* (Blume) T.G. Hartley	Rutaceae	MC	26	36.0	5786
2	*Endospermum diadenum* (Miq.) Airy Shaw	Euphorbiaceae	LS	19	38.6	4587
3	*Xylopia ferruginea* (Hook. f. & Thomson) Hook. f. & Thomson var. *ferruginea*	Annonaceae	MC	19	18.6	2015
4	*Macaranga hosei* King ex Hook. f.	Euphorbiaceae	P	9	18.1	1577
5	*Elaeocarpus stipularis* Blume var. *stipularis*	Elaeocarpaceae	MC	3	28.6	1389
6	*Pternandra echinata* Jack	Melastomataceae	U	25	10.4	1005
7	*Macaranga conifera* (Zoll.) M.Arg.	Euphorbiaceae	LS	12	12.1	705
8	*Porterandia anisophyllea* (Jack ex Roxb.) Ridl.	Rubiaceae	U	11	11.5	455
9	*Shorea leprosula* Miq.	Dipterocarpaceae	E	8	11.5	451
10	*Shorea parvifolia* Dyer ssp. *parvifolia*	Dipterocarpaceae	E	2	17.1	402
11	*Croton argyratus* Blume	Euphorbiaceae	U	6	13.9	338
12	*Xylopia ferruginea* (Hook. f. & Thomson) Hook. f. & Thomson var. *oxyantha* (Hook. f. & Thomson) J. Sinclair	Annonaceae	MC	6	15.3	327
13	*Melicope lunu-ankenda* (Gaertn.) T.G. Hartley	Rutaceae	MC	4	15.1	325
14	*Litsea castanea* Hook. f.	Lauraceae	MC	4	14.8	308
15	*Lithocarpus curtisii* (King ex Hook. f.) A. Camus	Fagaceae	MC	1	18.9	281
16	*Litsea costalis* (Nees) Kosterm.	Lauraceae	MC	2	13.5	279
17	*Artocarpus scortechinii* King	Moraceae	MC	4	11.2	279
18	*Nephelium cuspidatum* Blume var. *eriopetalum* (Miq.) Leenh.	Sapindaceae	MC	4	12.9	275
19	*Elaeocarpus nitidus* Jack var. *nitidus*	Elaeocarpaceae	MC	1	17.5	241
20	*Mezzettia parviflora* Becc.	Annonaceae	MC	2	14.8	239
	Others			138	17.4	7097
	Total			306	38.6	28362

*Species group: E = emergent, MC = main canopy, U = understorey, P = pioneer, LS = late-seral (Manokaran & Swaine 1994).

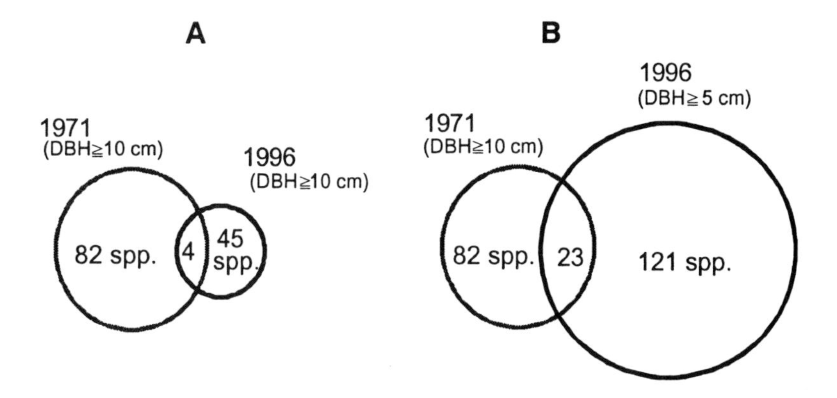

Fig. 2 Number of tree species being common to both data sets of tree census of 1971 and 1996 in the cleared plot (20 m × 100 m). A: Number of species co-occurred in both the data set (DBH ≧ 10 cm) of 1971 and that (DBH ≧ 10 cm) of 1996. B: Number of species co-occurred in both the data set (DBH ≧ 10 cm) of 1971 and that (DBH ≧ 5 cm) of 1996.

Table 3 Co-occurred 23 tree species both in 1971 before clearing and in 1996, 23 years after clearing in the cleared plot (20 m × 100 m) . Tree species are listed in the order of the alphabet.

No.	Species	Family	Max. DBH (cm)		Number of trees	
			1971	1996	1971	1996
1	*Aporusa bracteosa* Pax & K. Hoffm.	Euphorbiaceae	10.5	5.3	1	2
2	*Artocarpus dadah* Miq.	Moraceae	14.2	7.5	1	1
3	*Dacryodes puberula* (Benn.) H.J. Lam	Burseraceae	32.0	5.3	1	1
4	*Dacryodes rostrata* (Blume) H.J. Lam	Burseraceae	15.2	5.8	2	1
5	*Dacryodes rugosa* (Blume) H.J. Lam	Burseraceae	24.2	10.2	3	1
6	*Dipterocarpus cornutus* Dyer	Dipterocarpaceae	77.6	7.6	5	1
7	*Garcinia parvifolia* (Miq.) Miq.	Guttiferae	11.9	7.9	1	1
8	*Gymnacranthera farquhariana* (Hook. f. & Thomson) Warb. *var. eugeniifolia* (A. DC.) R.T.A. Schout	Myristicaceae	37.2	4.8	1	1
9	*Horsfieldia polyspherula* (Hook. f.) J. Sinclair var. *polyspherula*	Myristicaceae	12.9	10.5	1	2
10	*Ixonanthes icosandra* Jack	Linaceae	35.8	4.9	1	1
11	*Knema furfuracea* (Hook. f. & Thomson) Warb.	Myristicaceae	13.8	5.0	1	1
12	*Lansium domesticum* Correa	Meliaceae	34.1	5.1	4	1
13	*Madhuca malaccensis* (C.B. Clarke) H.J. Lam	Sapotaceae	13.0	5.6	1	1
14	*Mallotus griffithianus* Hook. f.	Euphorbiaceae	13.6	8.0	5	1
15	*Mesua ferrea* L.	Guttiferae	17.4	5.9	1	1
16	*Monocarpia marginalis* (Scheff.) J. Sinclair	Annoaceae	16.3	11.3	1	2
17	*Ochanostachys amentacea* Mast.	Olacaceae	21.1	6.6	2	1
18	*Payena lucida* A. DC.	Sapotaceae	10.6	6.9	1	2
19	*Polyalthia jenkensii* (Hook. f. & Thomson) Hook. f. & Thomson	Annoaceae	20.0	6.4	2	1
20	*Scaphium macropodum* (Miq.) Beumee ex Heyne	Sterculiaceae	10.8	10.8	1	2
21	*Shorea macroptera* Dyer	Dipterocarpaceae	40.7	9.5	1	4
22	*Shorea multiflora* (Burck) Symington	Dipterocarpaceae	32.2	8.4	2	1
23	*Shorea pauciflora* King	Dipterocarpaceae	37.9	6.8	1	3

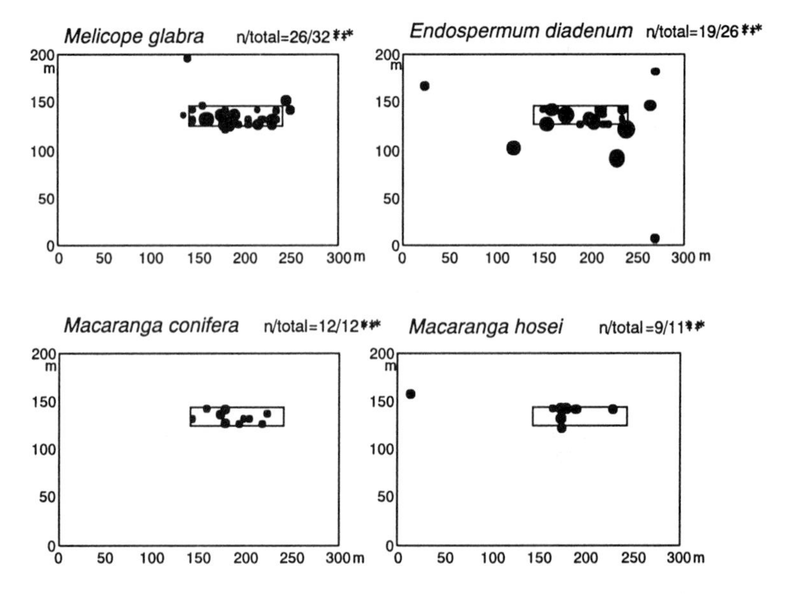

Fig. 3 Spatial distribution of four tree species which strongly concentrated on the cleared plot (**$P < 0.01$). The large rectangle shows the 6-ha plot, 200 m × 300 m. A small rectangular within the 6-ha plot indicates the cleared plot, 20 m × 100 m (see Fig. 1). A number of trees in the cleared plot and total number of trees occurring in the 6-ha plot are described in each figure, e.g. n/total=26/32, n/total=19/26, etc. Diameter of closed circle represents tree diameter at breast height.

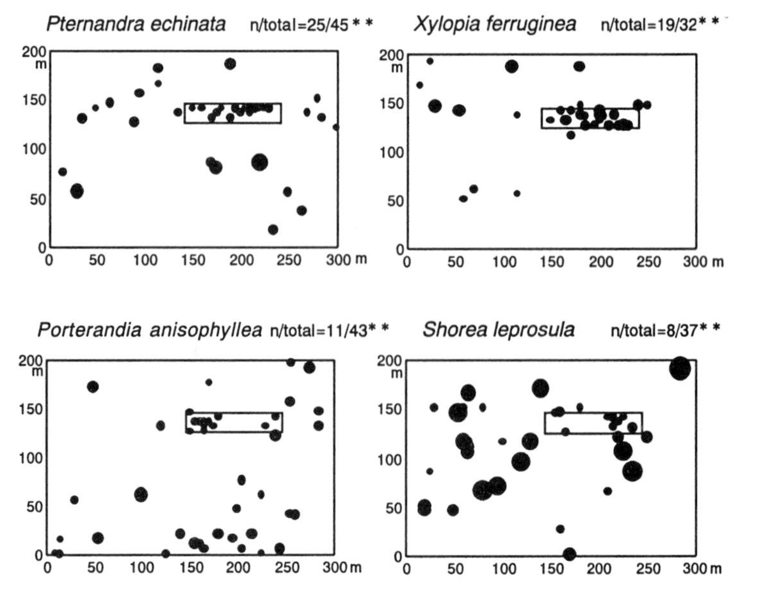

Fig. 4 Spatial distribution of other four tree species which concentrated on the cleared plot. Legends are the same as Fig. 3 (**$P < 0.01$).

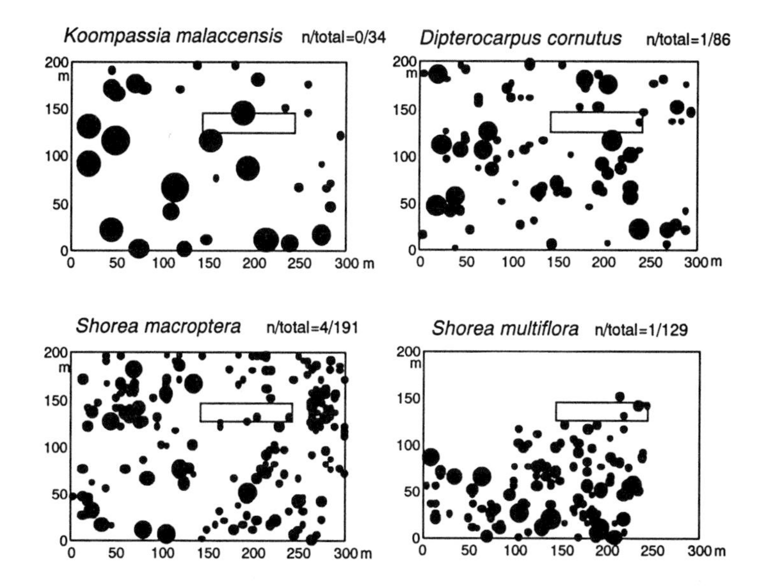

Fig. 5 Spatial distribution of four tree species which did not relate to the cleared plot. Plots and circle size are explained in Figs. 1 and 3.

e.g. *Koompassia malaccensis, Dipterocarpus cornutus*, *Shorea macroptera* and *S. multiflora* did not regenerate well in the cleared plot (Fig. 5). They did not show distinct concentration into the cleared plot.

4. DISCUSSION

The artificial clearing in the IBP provides a good opportunity for regeneration of pioneer tree species in Pasoh FR. Furthermore, the slender shape of the cleared plot well simulates the domino effect of multi-trees fall, and provides opportunities for effective seed dispersal from adult trees located on surrounding intact parts of the forest. Although the plot was cleared for biomass estimation, it became a case study for secondary succession in a lowland dipterocarp forest in Peninsular Malaysia.

The size of cleared plot, 2,000 m^2 (20 m × 100 m), in this study exceeds most of canopy gaps formed by emergent trees in lowland dipterocarp forests; single falling canopy tree creates a gap of about 400 m^2 (Whitmore 1984). The size of the cleared plot corresponds to a canopy gap created by five or more canopy trees, which are rare in Pasoh FR (Putz & Appanah 1987). However, we have information on a rare huge storm created large gaps on 13 July 1982 (N. Manokaran in pers. comm.).

First tropical tree species were divided into two ecological species groups, pioneer and climax (Swaine & Whitmore 1988). These groups differ in regeneration mode. Pioneers only germinate in canopy gaps open to the sky, and cannot survive in shade below canopy. Climax can germinate below canopy and survive there to some extent. Pioneer subdivided into short-lived pioneer and long-lived pioneers (late-seral). "Large pioneer" and "late-seral" indicate the same ecological group (Manokaran & Swaine 1994). *Macaranga hosei* is categorized into small pioneer, and *Endospermum diadenum* and *Macaranga conifera* into large pioneers (Manokaran & Swaine 1994).

The cleared plot contained many "secondary forest species" that has been used empirically by several researchers (Kochummen 1966; Wyatt-Smith 1966). For example, *Melicope glabra, Endospermum diadenum, Elaeocarpus stipularis* and *Artocarpus scortechinii* have been called secondary forest species. However, it is an ill-defined term (Swaine & Whitmore 1988). Secondary species contains both small and large pioneers and light-demanding climax species that grow fast and have a low wood density. *Melicope glabra* and *Endospermum diadenum* have seed banks as does *Macaranga hosei*, (Putz & Appanah 1987), though *Melicope glabra* is categorized into main canopy in the primary and *Endospermum diadenum* into late-seral (Manokaran & Swaine 1994). *Pternandra echinata* and *Xylopia ferruginea* have been defined as secondary forest species (Wyatt-Smith 1966), but they are re-defined as climax species (Manokaran & Swaine 1994). These results suggest that several climax species well regenerate after cutting and dominate in the regenerating forest as same as pioneers.

What ecological traits do encourage the dominance of climax species after cutting? A putative primary species, *Pternandra echinata,* having small seed, regenerated very well in the cleared plot (Table 2). It has been reported that *Pternandra echinata* has dormant seed banks and regenerate from seed banks (Kanzaki et al. 1997; Metcalfe & Turner 1998). Such small-seeded shade-tolerant species germinates most frequently on steep and disturbed soils in primary forests (Metcalfe & Grubb 1997). Metcalfe & Turner (1998) called this type of species litter-gap demanders. The cleared plot is flat but the clear cutting and measuring for biomass probably disturbed the litter and soil surface. This may be one of the reasons why shade-tolerant primary species *Pternandra echinata* could regenerate well in the cleared plot. *Pternandra echinata* and *Porterandia anisophyllea* were common in another regenerating plot after selective logging under the Malayan Uniform System (MUS) in Pasoh FR (Manokaran 1998; Chap. 2).

Typical light demanding pioneer species such as *Macaranga gigantea* and *Melastoma malabathricum* have been emphasized to dominate in the early stage of secondary succession in Malaysia (see Richards 1996; Whitmore 1984). These species were absent in the cleared plot in 1996. However, *Macaranga gigantea* had been growing in 1977, when 4-years-old secondary forest was observed in the plot (Ashton 1978). This fact shows that *Macaranga gigantea* had been lost during 20 years between 1977 to 1996. Such decreasing pioneer populations and alternation of dominant trees among pioneers were also reported in a succession after farming in Kepong in Peninsular Malaysia (Kochummen 1977).

The cleared plot contained *Shorea leprosula* and *S.parvifolia* that are categorized into light-demanding and fast growing light hardwood dipterocarp (Manokaran & Swaine 1994). These dipterocarp species are reported as early invaders in other secondary succession after farming in Peninsular Malaysia (Kochummen 1966; Kochummen & Ng 1977; Wyatt-Smith 1954, 1955). These species showed higher mortality than other dipterocarps in the 50 ha plot in Pasoh FR (Manokaran et al. 1992a). Among dipterocarps there is a large difference in utilization of open sites after cutting.

These results suggest that the cleared plot is in a stste of succession from a small pioneer stage to the mixture of large pioneers (late-seral) and light-demanding, fast growing climax species. Slow growing climax species have probably been regenerating below the present tree canopy, but they did not reach the measuring limit of 5 cm DBH. Light-demanding dipterocarps, *Shorea leprosula* and *Shorea parvifolia* growing under the present canopy might dominate the next stage of succession. However, it needs one hundred or more years to observe the alternation of dominant species in the cleared plot from large pioneers to slow growing climax species through the dominance of light demanding dipterocarp species.

ACKNOWLEDGEMENTS

We thank the Director General of Forest Research Institute Malaysia (FRIM) and the staff of FRIM and Forest Department, Negeri Sembilan and Malacca for permission to use the Pasoh FR. We deeply thank Drs. P. S. Ashton, T. Yamakura, T. Kira, H. Ogawa and other IBP members of Malaysia, UK and Japan teams for permission to use the cleared tree data in IBP. We deeply thank the late Dr. K. Yoda for his encouragement to our studies in Pasoh. We are grateful to Drs. T. Okuda, N. Manokaran and P. S. Ashton who provided critical comments on the manuscript. This study was funded by the Global Environmental Research Program (Grant No. E-1) by the Japan Environmental Agency.

REFERENCES

Ashton, P. S. (1976) Mixed dipterocarp forest and its variation with habitat in the Malayan lowlands: a re-evaluation at Pasoh. Malay. For. 39: 56-72.

Ashton, P. S. (1978) Crown characteristics of tropical trees. In Tomlinson, P. B. & Zimmermann, M. H. (eds). Tropical Trees as a living systems, Cambridge University Press, Cambridge, London, pp.591-615.

Kanzaki, M., Yap, S. K., Kimura, K., Okauchi, U. & Yamakura, T. (1997) Survival and germination of buried seeds of non-dipterocarp species in a tropical rain forest at Pasoh, West Malaysia. Tropics 7: 9-20.

Kato, R., Tadaki, Y. & Ogawa, H. (1978) Plant biomass and growth increment studies in Pasoh forest. Malay. Nat. J. 30: 211-224.

Kochummen, K. M. (1966) Natural plant succession after farming in SG. Kroh. Malay. For. 29: 170-181.

Kochummen, K. M. (1997) Tree Flora of Pasoh Forest. Malayan Forest Records No. 44. Forest Research Institute Malaysia, Kepong, 461pp.

Kochummen, K. M. & Ng, F. S. P. (1977) Natural succession after farming in Kepong. Malay. For. 40: 61-78.

Liew, T. C. (1973) Occurrence of seeds in virgin forest top soil with particular reference to secondary species in Sabah. Malay. For. 36: 185-193.

Manokaran, N. (1998) Effect, 34 years later, of selective logging in the lowland dipterocarp forest at Pasoh, Peninsular Malaysia, and implications on present-day logging in the hill forests. In Lee, S. S., Dan, Y. M., Gauld, I. D. & Bishop, J. (eds). Conservation, management and development of forest resources: proceedings of the Malaysia-United Kingdom Programme workshop. Forest research Institute Malaysia. Kuala Lumpur, Malaysia. pp.41-60.

Manokaran, N. & Kochummen, K. M. (1987) Recruitment, growth and mortality of tree species in a lowland dipterocarp forest in Peninsular Malaysia. J. Trop. Ecol. 3: 315-330.

Manokaran, N. & LaFrankie, J. V. (1990). Stand structure of Pasoh Forest Reserve, a lowland rain forest in Peninsular Malaysia. J. Trop. For. Sci. 3: 14-24.

Manokaran, N. & Swaine, M. D. (1994) Population dynamics of trees in dipterocarp forests of Peninsular Malaysia. Malayan Forest Records No. 40. Forest Research Institute Malaysia, Kuala Lumpur, 173pp.

Manokaran, N., Abd. Rahman Kassim, Azman Hassan, Quah, E. S. & Chong, P. F. (1992a) Short-term population dynamics of dipterocarp trees in a lowland rain forest in Peninsular Malaysia. J. Trop. For. Sci. 5: 97-112.

Manokaran, N., LaFrankie, J. V., Kochummen, K. M., Quah, E. S., Klahn, J. E., Ashton, P. S. & Hubbell, S. P. (1992b) Stand table and distribution of species in the 50-ha research plot at Pasoh Forest Reserve. FRIM Research Data No. 1. Forest Research Institute Malaysia,

Kepong.
Metcalfe, D. J. & Grubb, P. J. (1997) The responses to shade of seedlings of very small-seeded tree and shrub species from tropical rain forest in Singapore. Funct. Ecol. 11: 215-221.
Metcalfe, D. J. & Turner, I. M. (1998) Soil seed bank from lowland rain forest in Singapore: canopy-gap and litter-gap demanders. J. Trop. Ecol. 14: 103-108.
Ng, F. S. P. (1978) Tree Flora of Malaya. vol. 3. Malayan Forest Records No. 26. Longmans Malaysia Sdn. Bhd., Kuala Lumpur, 339pp.
Ng, F. S. P. (1989) Tree Flora of Malaya. vol. 4. Malayan Forest Records No. 26. Longmans Malaysia Sdn. Bhd., Kuala Lumpur, 549pp.
Putz, F. E. & Appanah, S. (1987) Buried seeds, newly dispersed seeds, and dynamics of a lowland forest in Malaysia. Biotropica 19: 326-333.
Richards, P. W. (1996) The tropical rain forest (2nd ed.). Cambridge University Press, Cambridge, 575pp.
Sokal, R. R. & Rohlf, F. J. (1995) Biometry (3rd ed.). W. H. Freeman and Company, New York, 850pp.
Swaine, M. D. & Whitmore, T. C. (1988) On the definition of ecological species groups in tropical rain forests. Vegetatio 75: 81-86.
Symington, C. E. (1933) The study of secondary growth on rain forest sites. Malay. For. 2: 107-117.
Turner, I. M. (1995) A catalogue of the vascular plants of Malaya. The Garden's Bulletin Singapore 47 (Part 1 & 2).
Whitmore, T. C. (1972a) Tree Flora of Malaya vol. 1. Malayan Forest Records No. 26. Longmans Malaysia Sdn. Bhd., Kuala Lumpur, 473pp.
Whitmore, T. C. (1972b) Tree Flora of Malaya vol. 2. Malayan Forest Records No. 26. Longmans Malaysia Sdn. Bhd., Kuala Lumpur, 444pp.
Whitmore, T. C. (1984) Tropical rainforest of the Far East (2nd ed.). Clarendon Press, Oxford.
Wong, Y. K. & Whitmore, T. C. (1970) On the influence of soil properties on species distribution in a Malayan lowland dipterocarp forest. Malay. For. 33: 42-54.
Wyatt-Smith, J. (1949) Natural plant succession. Malay. For. 12: 148-152.
Wyatt-Smith, J. (1954) Storm forest in Kelantan. Malay. For. 17: 5-11.
Wyatt-Smith, J. (1955) Changes in composition in early natural plant succession. Malay. For. 18: 44-49.
Wyatt-Smith, J. (1966) Ecological studies on Malayan Forests. I Composition of and dynamic studies in Lowland Evergreen Rain-Forest in two 5-acre Plots in Bukit Lagong and Sungai Menyala Forest Reserve and in two Half-acre Plots in Sungai Menyala Forest Reserve, 1947-59. Research Pamphlet No. 101. Forest Research Institute Malaysia, Kepong.

40 Prospects and Priorities for Research Towards Forest Policy Reform and the Sustainable Management of Biodiverse Tropical Forests

Toshinori Okuda[1] & Peter S. Ashton[2]

Abstract: To facilitate the sustainable management of biodiverse tropical forests, we propose that the ecological services and goods of these forests be assessed. The assessment should start with a review of existing ecological data, to be supplemented with new data on ecological service values in order to improve the flexibility of evaluation. In this highly complicated ecosystem, conflicting values have precluded finding a solution to deforestation and depletion. We believe that ecological service assessment will play an important role in connecting different fields of ecological studies and linking forest managers and field ecologists. We propose setting up a model site at Pasoh Forest region that includes both pristine forest and other land-use types (e.g. production forest, plantations). By integrating and mapping ecological service data at the landscape level, we will be able to prepare a risk management program, a landscape zoning plan, and other planning instruments to optimize the ecological service values of the model site. We also propose other prospective research, such as long-term monitoring of the reproduction and genetic diversity of tree species and the use of remote sensing to improve ecosystem monitoring, which we believe will facilitate integrated ecosystem assessment.

Key words: eological services, integrated ecosystem assessment, landscape.

1. INTRODUCTION

1.1 Current situation and problems

Unlike most global environmental research, which studies the atmosphere and climate or the oceans and in which the issues and conclusions are predominantly global in scope, research that studies the fate of terrestrial ecosystems must be based on site-specific systems. The characteristics of many such systems are highly complex and unique to the site or of limited application; identification of generalizable conclusions requires rigorous research design and analysis. For these same reasons there is a great diversity of appropriate research fields. The research issue of sustaining tropical forests provides an extreme example of the challenge. Furthermore, although there are a variety of social factors behind the dwindling of the tropical rain forests, the direct causes of this decrease and degradation are mainly logging (including illegal logging) and a shift of land use from forest to agriculture (e.g. plantations, shifting cultivation). Research on the conditions that lead to land degradation and deforestation, which are as much socioeconomic as

[1] National Institute for Environmental Studies (NIES), Tsukuba 305-8506, Japan.
E-mail: okuda@nies.go.jp
[2] Harvard University, USA

ecological, is not the sort that many researchers, whose training is overwhelmingly theoretical in orientation, are interested in pursuing. Thus, if one were to dedicate oneself to basic research, dealing with the loss of forests would necessarily become an indirect concern.

If we are to address environmental problems in tropical forest regions, research might focus on two points: 1) how to comprehensively assess the factors that encourage deforestation and those that encourage sustainable use of forest resources; and 2) how to diagnose those characteristics of rain forest ecosystems without which they cannot function and survive, and how to restore that essential structure and function of the original ecosystem most quickly and optimally after it has been depleted. Unless research findings are offered in a form that can be incorporated into policy for tropical forest conservation, they are of no avail, and it might be impossible to regenerate the forests, let alone stop their loss.

So far, most research projects have proceeded with the idea of first understanding the functioning of tropical forest ecosystems. For more than a century now, researchers have studied diverse specialized areas such as meteorology, hydrology, and various fields of ecology, including animal and vegetation ecology and coevolution, in order to better understand how the biological diversity of tropical forests is sustained in nature. But much remains to be known about these forests, including their processes of regeneration and maintaining diversity, which will require much time and effort. At the same time, the more that basic research progresses in each specialized field, the more difficult it is to synthesize a general understanding. Such being the case, we are unfortunately still seldom able to offer specific proposals on the sustainable management of forests. Bearing in mind these problems, this chapter presents a vision for research to help achieve the conservation and management of tropical forests.

1.2 Why tropical forests?

Why, of all the forest ecosystems on the planet, must tropical forests be the focus of global environmental research at this time? Until sponsors and supporters are as aware of the reasons as researchers, project integration will be impossible. It is therefore possible that we will fail to gain a general understanding of forest function, and instead accumulate a mere encyclopedia of disconnected facts, while continuing to document the spread of deforestation. Tropical forests must be the object of research because they embody the problems involved in worldwide deforestation in their most extreme form on account of their extreme biological complexity and habitat fragility. Through research on them, we can best understand the history and present state of the global forest environment. By extension, through international collaboration in research aimed at achieving sustainable management of biodiverse tropical forest, such shared experience will lead to the development of mutual relationships of trust among nations. But first to some hard facts:

1)Tropical forests are disappearing faster than the forests in any other climatic zone. Natural forest cover in developing countries decreased at the rate of 15.4 million ha per year (9% of total) during the decade 1980 to 1990 and at 13.7 million ha per year during 1990–1995, which shows that deforestation continues apace (FAO 1999). By contrast, the loss of forests in the developed world has finally stopped, and recently began to reverse (a nearly 2% increase in forest cover). Further, tropical forests are not being restored after being destroyed. While some afforestation has been carried out with high-cash-value tree species, trade in dipterocarps and many other non-timber products (e.g. rattan) is still dependent even now on harvesting from natural forests.

2)Supplying wood from natural forests currently involves considerable ecosystem damage. This damage can largely be avoided by careful and thorough management, but concessionaires are not persuaded that the extra cost involved will gain them extra profits or other benefits. Their chances of retaining licenses over more than a single felling cycle are uncertain, and the high discount rates in developing economies discourage investment in long-lived crops such as natural forest stands by owners and concessionaires alike. No economically reliable system has been established for planting trees to secure wood resources, so wood is harvested by selective logging of natural forests. Even this selective logging causes serious impacts on forest environments owing to the large size of trees and because logging roads for timber removal crisscross the forests without careful planning. In the process, soil runoff upsets river ecosystems, the soil surface of logging roads becomes crusty, and the regeneration of vegetation cover is delayed. Furthermore, the desirable dipterocarp and other canopy tree species take many years to bear seeds, which cannot be stored owing to lack of dormancy. Their removal leads to scarcity of seed sources for regeneration. Instead, dense growth of pioneer woody species of no economic value can take over, shading out the regeneration of economic old-growth species for many years even if it survives harvesting. These are thus other factors that prevent cost reduction in afforestation operations, in some cases making it impossible to stably supply the major commercial timber species on a dependable logging rotation cycle (Kurpick et al. 1977). Unlike forest ecosystems in other climatic zones, there is no guarantee that forests will return to their original state by merely allowing vegetational succession to run its course.

3)The depletion of tropical forests is not a technical silvi-ecological problem of the forests but rather a manifestation of the social, economic, and political climate. One of the main causes of deforestation in Southeast Asia is the conversion of forest lands to other uses. This includes housing for the rapidly increasing numbers of poor, and the policy of transmigration to the forests which so often results in slash-and-burn farming by inexperienced newcomers who lack the capital to invest in permanent crops. It was a problem waiting to arise. Tropical forests are also converted by commercial interests to commodity plantations, which are still favored by the perceived opportunity costs even when most are in danger of overproduction. Certainly, such land use can provide greater employment than natural forest, however managed. In Indonesia for instance, colonists under the government's transmigration policy first practice slash-and-burn farming, which eventually fails, then the forests are cut for purposes such as large-scale corporate plantations. Burning for purposes of slash-and-burn farming and clearing forests spreads to natural forests and wetlands, destroying large areas to no purpose and causing smoke damage in nearby countries. Another factor that makes forest destruction more serious is that countries with tropical forests are in the process of economic development and are therefore prone to selling off their existing natural resources as a means of earning foreign currency. The fundamental and ubiquitous reason for these forest conversions is a lack of appreciation of the true, total value of tropical forests by their owners-generally governments-further hampered by the difficulty of capturing their full revenue potential, especially of their service values, the beneficiaries of which are largely free-riders in the industrial North.

4)There is a greater woody aboveground biomass in forests than in any other ecosystem; whereas tropical rainforests cover only about 12% of Earth's land area, they account for 45% of its above-ground biomass, the highest proportion of biomass. Accordingly, they sequester the largest amount of CO_2 per unit area in above-ground organic form. (Boreal forests have the biggest land area and sequester the most carbon overall, but 60% to 70% of their stock is underground; IPCC 2000) The source of the current increase in atmospheric carbon is mostly the industrialized North, though developing economies such as China are playing an increasing part. So far, there are few signs that this service value can be captured as rental by the forest owners.

5)Tropical forests are rich in genetic resources. According to a review of biodiversity in the tropics (Rice et al. 1997), species living in tropical forests are thought to account for as much as 90% of all Earth's species. For example, some of Peru's moist lowland forests have 300 woody plant species per hectare, approaching half the number of tree species recorded for all of temperate North America. In a patch of Peruvian moist forest 5 km on a side, there were 1,300 butterfly species and over 600 bird species (by contrast, there are 400 recorded butterfly species and 700 bird species for the entire USA). In other regions of tropical rainforests, over 150 tree species per hectare have been found (Gentry 1988; Richards 1996; Whitmore 1984). Therefore, tropical forests are not only resources for timber, they also have diverse genetic resources that represent an irreplaceable information data-base for genetic engineers.The theoretical research objective-to understand how the rich biota coexists in tropical rainforests-is therefore at the same time basic to the sustainable management of this genetic diversity and also nothing less than elucidating how living things have co-evolved on our planet.

In some areas, in Southeast Asia in particular, there are places where it appears at first glance as though all loggable forests have been cut, and so the destruction of forests in conjunction with economic development has halted; but as in the case of Indonesia's forest fires, there are also places where tropical forests continue to be destroyed and lost through intentional burning, which often gets out of control. Because selectively logged secondary forests are also categorized as indigenous forest, they are often not accounted for in forest area lost; but as their stands differ considerably in composition from natural forests (Chap. 2), the decline of forests in that sense continues even now. In view of these circumstances, current discourse on forest decline focuses primarily on tropical deforestation. In consequence of the underlying causes, the most serious socioeconomic problems associated with forestry are concentrated in the tropics. It is therefore necessary to elucidate the ecological characteristics of tropical rainforests, gain the scientific knowledge and develop the techniques needed to regenerate them, and, from the perspective of the global environment, quickly take steps to stop their loss and to expedite their regeneration.

2. WHAT IS SUSTAINABLE MANAGEMENT?

Sustainable management is a form of forest utilization–and methods for its maintenance and management–which production is maintained by means of the natural recovery capacity while making the best effort to maintain the forest in its present natural state. Beginning in 1985, $400 million was invested per year as financial assistance for this sustainable management under the Tropical Forestry Action Program (Plan) by UN organizations (FAO, UNDP, World Bank and World Resource Center) ; the amount was raised to $1.3 billion in 1990 (Oksanen et al.

1993). But the plan was not successful and has not reduced the deforestation rates (Oksanen et al. 1993; Ross & Donovan 1986). In the late 1980s, just 0.1% of the remaining natural forest was under active sustainable management (Holmberg et al. 1991). Even today, forests certified by eco-labeling are merely 0.5% of the world's forest, most of them in the temperate region.

Is sustainable development of forest resources really possible in the tropics (say, for timber production)? Technically, yes. However, all timber species are members of the natural forest, and thus their regeneration is influenced by the interactive relationships among species and the forest environment. We compared the flora of a natural forest in Malaysia's Pasoh FR and a regenerating forest that had been selectively logged in the 1950s, and discovered differences between canopy height, structure, and surface area in the reserve (Chap. 2; Okuda et al. in press). It was also found large differences in characteristics such as their insects, birds and mammals between the regenerating and primary forests (Chaps. 27, 28, 29, 31, 36, 37, 38). Similar examples reported in recent years are leading to a reassessment of logging standards (Manokaran 1998). According to a report from Uganda by Chapman & Chapman (1997), even 25 years after logging there were discernible differences in characteristics such as number of trees and number of mortalities. In selectively logged secondary forests, Chapman observed a high rate of canopy gap formation due to post-harvest mortality associated with logging damage and other factors, which apparently delayed regeneration. Once the forest canopy structure is broken, it revives only with great difficulty even in places with few artificial disturbances. According to a simulation model of regeneration after logging in Malaysian dipterocarp forests (Kurpick et al. 1997), the optimum logging cycle appears to be 100 years if the goal is to maximize timber production; however, this assumes that there is no damage to the forest floor during logging. Furthermore, it takes no account of the economically optimal cycle length. To restore the species composition to its unlogged state, a regeneration period of 200 years after logging would be needed (Kurpick et al. 1997). If one only seeks sustainable management in terms of timber production, the logging cycle may be shorter than in the case of ecologically sustainable management. However, as timber production by selective logging is totally dependent on the natural generation of the forest, logging impacts on ecological aspects such as forest structure and tree species composition, tree distribution, phenology of tree species, and seed dispersal composition (Chaps. 2, 39) could reduce the growth and reproductive potential of commercial timber species (Chaps. 15, 21). Reduced genetic heterogeneity due to the removal of mature trees definitely reduces regeneration potential (Chap. 21). Unfortunately, we have very little scientific data to demonstrate how such a chain of interaction would influence sustainable timber production. Instead of our current forest economy, which is based on the profligate use of natural resources, it is therefore essential that we practice very conservative resource use and management that hold waste down to the absolute minimum and are based on the close calculation of resource stock, turnover rate, and consumption rate. In this light, it seems that the term "sustainable management of forest resources" has been used as a smoke screen meant to cover the development of forests by just cutting them down, thus implying that such popular catchphrases have been used arbitrarily without any scientific backing. Considering how fast tropical forests are disappearing, we have hardly any time to reform. What we can and must do now is to value forest accurately and draw up a management plan with a scientific basis.

A major reason for the dire state of the rain forest economy is the exclusive concentration on harvesting for timber (e.g. Vincent 1992). There is serious doubt as to whether sustainable management for timber alone is economically viable, bearing in mind the necessary long cycle and the high discount rates general in developing countries (Howard et al. 1996; Panayotou & Ashton 1992; Rice et al. 1997). On timber alone, opportunity costs of land under tropical rain forest will rarely support retention of the natural forest. But rain forests provide a wealth of other goods and services. If these too are valued and incorporated into sustainable management protocols, it appears likely that biodiverse rain forest may often be favored economically over other uses of the land. In traditional and impoverished rural economies, the diverse non-timber products are of value in the village economy and may be more widely traded. In more developed economies, service values — water for downstream industrial and domestic uses, and for recreation, for instance — become important. And in all economies, the increasing value of forest services of importance to the world community, especially carbon sequestration, climate amelioration, and biodiversity conservation, may tip the balance in favor of sustainable management. The challenge to the forest owner is to recognize within-country values that are of indirect benefit to development but do not result directly in financial profit, and to find a means to capture values that accrue to beneficiaries overseas who have heretofore taken them for granted — as free riders.

3. ASSESSMENT OF ECOLOGICAL SERVICES OF TROPICAL FORESTS

Before setting up a forest management plan, one must therefore correctly understand the extent of forest functions. In recent years, some economic assessments have been made of the ecosystem service values provided by forests (e.g. Costanza et al. 1997; Myers 1996). The background for these assessments was that the ecological service value of forests (besides commercial timbers) has usually been neglected, while commercial crop value on farmland has been accurately calculated and has overwhelmed the existing and optional values provided by forest. In addition, the value of natural forests has not been objectively assessed. This is a leading reason why it is difficult to explain the necessity and urgency of protecting natural resources and diversity; it instead leads decision-makers to jump to the question of why biodiversity is necessary. One reason for this is the insufficiency of scientific knowledge on the functional roles of forests. Another is that our techniques for economic analysis cannot yet grasp the potential capabilities of forests, let alone comprehensively assess them. As a result, assessments of the ecosystem service value of forests remain unclear and without justification. It is comparatively easy to calculate the asset value of forests as wood — calculations of the costs and benefits of afforestation and logging operations have been done for years; but with the mixture of diverse tree species, size classes, and growth rates in tropical forests it is difficult to assess what potential economic value each tree has, let alone the indirect benefits gained from the forests' existence. Researchers have performed direct and indirect economic assessments of the services provided by forests (e.g. Brown & Pearce 1994; Myers 1994; Panayotou & Ashton 1992), but forests are continually changing value with the change in markets and the process of economic development, and there is as yet little research on fluctuations in asset value that takes in the dynamism of forests.

Ecological services of forests can be summarized as listed below, though the classification and categories of services vary with authors, targeted ecosystem, and regional background (e.g. Costanza et al. 1997; Ehrlich 1989; Ehrlich & Mooney

1983; Folke 1991; Myers 1996). We consider that the following categories and functional aspects of ecological services can be applied to tropical forests.

1) Protection of watershed ecosystems (water supply, regulating river flow, preventing floods, ensuring flow of good-quality water to communities and industry downstream, protecting wildlife habitat; maintaining the biodiversity of freshwater, brackish and seawater ecosystems).
2) Providing biological resources (timber and non-timber resources, wildlife and fishery resources).
3) Supplying fuel wood, non-timber products (e.g. rattan), pharmaceuticals, and other chemical products.
4) Providing genetic resources and libraries (discovery of woody commercial plants, gene transfer, developing pharmaceuticals).
5) Climate regulation (regulation of global and local temperature, precipitation, and other biologically mediated process), especially as a carbon sink.
6) Nutrient cycling (storage, cycling, and acquisition of nutrients).
7) Gas regulation in atmospheric chemistry (e.g. CO_2, O_3, SO_x).
8) Soil formation and protection (preventing soil erosion and nutrient loss).
9) Recreation and cultural resources (providing opportunities for ecotourism, outdoor activities, education, scientific research, artistic activities, and other amenity values).
10) Social and educational heritage (environmental education, spiritual (religious) value).

According to Costanza et al. (1997), the total value of the world's ecological (ecosystem) services was estimated to be in the range of US$16–54 trillion ($10^{12}$) per year, with an average of US$33 trillion per year. Of this, the global value of tropical forest was US$3.813 trillion per year (US$ 2,007 yr^{-1} ha^{-1}). However, several problems and criticisms of this economic valuing of the services have been pointed out:

1) As the social and economic values of forests are continually changing, owing to the factors mentioned earlier and in a broader context to society's surplus economic capacity, assessments made with one-time criteria of economic value make little sense.
2) When deforestation occurs because of poverty, illegal logging, slash-and-burn, and other factors deriving from local communities, economic assessments are powerless unless they take reality into account. Which must come first, then, the economic research or the policy reform?
3) Once an economic assessment has been performed, it is also highly possible that the figures will be used by forest managers, developers, and policy makers without further scientific review of the values.

However, economic assessments are not able to evaluate all potential service values and goods of an ecosystem; they only offer a basis for extracting some variables that influence the costs and benefits through proposed landscape changes and provide some alternative options for land use and management systems. In this regard, of course, one time assessment is insufficient and revision of service value assessment methodology and reassessment should be undertaken repeatedly, highlighting changes in volume, conditions and market prices of natural resources and conservation policy. Such well revised total economic assessments immediately attract public interest, as they have social impact, and therefore facilitate media coverage and encourage best-practice landscape management.

Current selective logging methods are destructive, and no quantitative overall evaluations of them have been performed. Both simulations and follow-up assessments are needed. Yet logged sites are usually left without any further reassessment or reclamation, and vast areas of forest have been converted to crop land, both illegally and legally without forecasts of environmental risks and future economics that may offset all the efforts spent on land-use change and development. For example, in countries with tropical forests, standards of living are rising in conjunction with economic development, and as the unit cost of labor rises, there will come a time in the near future when it will no longer pay economically to grow oil palm or to maintain rubber plantations. In other words, there will necessarily come a point in time when continuing to run plantations leads to the accumulation of economic losses, and the balance will continue to tip in favor of the service values accrued by maintaining present natural and secondary forests.

Considering this history and background and the high rate of deforestation and depletion in the tropics, we are unlikely to convince policy-makers of the realities of how much risk underlies proposed land-use development and why conservation of target areas is necessary by providing only a biological assessment of the target ecosystem. Instead, we have to be able to answer the following questions: How do the service values of forests alter opportunity costs of land in their favor? Might this lead to criteria to decide how much forest to leave untouched? And how much must be reforested?

We do believe that the one dimensional analyses on ecological service values will not solve the deforestation or ecosystem degradation problems. In fact, there are apparently trade-off relationships among some criteria of economic goods and ecological services; for example, timber production vs. protection of watershed and biodiversity. However, with improved recognition of functional aspects and

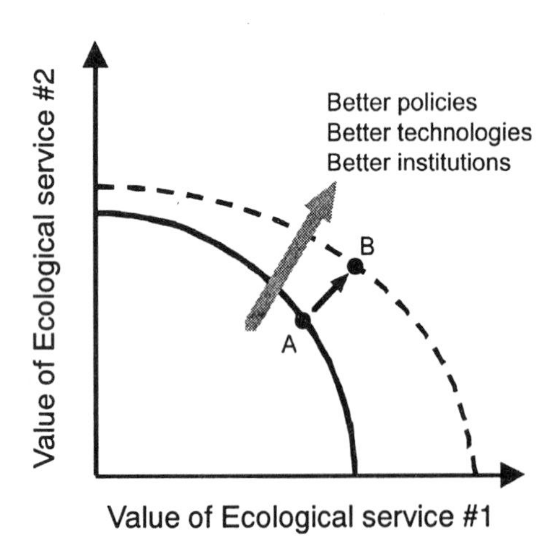

Fig. 1 The trade-off relationship between two conflicting ecological services. Each line represents combined value of differing proportions of two ecological services. Given the existing policies and technologies, every point on the line maximizes the combined value of differing proportions of two ecological services. The value could increase from "A" to "B", if new technologies (Irrigation system, reduced impact logging system) or new policies (Wood price controls, Forest management certification) are introduced (Adopted and partly modified from Millennium Ecosystem Assessment Committee (1998)).

ecological services, as well as the introduction of new technology and policies, the total ecosystem service value determined by the balance between conflicting service criteria may be increased (Fig. 1). When such trade-off relationships among services were unimportant, a sectoral approach made some sense, but today, nations expect that ecosystem management must meet some conflicting goals (Millennium Ecosystem Assessment Committee 1998). Therefore, assessments should be improved by incorporating new criteria or guidelines for ecological services. As well, ecologists should continue to search for new criteria, guidelines, and technologies that may improve the trade-off relationships among conflicting services. At the same time, ecologists and economists should cooperate to find tools that can make a linkage between dynamic changes in ecological services and socioeconomy at local or global scale.

There are expectations that such economic assessments of ecological services could serve as a bridge between forest management and ecological studies. They could also play an important role in integrating research on the various tropical forest components.

4. STRATEGY FOR INTEGRATED ECOSYSTEM MANAGEMENT

4.1 Feasibility of ecological service assessment in Pasoh

Let us now consider the feasibility and applicability of such an assessment at Pasoh Forest Region (see Color plate 1, Fig. 2 of Chap. 2,). The changes in forest value observed in this region are paradigmatic of all landscape in undergoing economic development in Peninsular Malaysia. At the beginning of the twentieth century, Pasoh Forest Reserve (Pasoh FR) was part of a vast forest in a region sparsely settled by several minorities, who valued the forest for a wealth of products, some of which, notably dammar (a tree resin used for varnish), they traded for salt and other essential commodities. From the late 1960s to the early 1970s, the first stage of development began as the forest surrounding the research forest was selectively logged (Chaps. 1, 2). Although this forest was managed to favor natural regeneration, the owner – the government – soon saw more value in conversion to smallholder commodity (oil palm) plantations. This and other policies brought increasing prosperity to the Pasoh region. By the last decade, the once diverse non-timber forest economy was reduced to a handful of products such as buah petai (*Parkia*) fruits, gaharu (diseased wood of *Aquilaria*), and rattan, but timber exploitation remained a major financial asset. Meanwhile, the service values have become increasingly apparent, but there has as yet been no attempt to either recognize or capture them. It is generally recognized for instance – though no monitoring has taken place – that the water table in the forest has become lower since conversion of the surrounding forests, which implies that the remaining forest may still have an ameliorating hydrological influence on the surrounding plantations. Or again, Pasoh FR is a refuge for wild boar, which plague the oil palm plantations, eating the fallen fruit; they also may severely affect natural forest regeneration (Chap. 35). This could be turned to advantage by licensing parties to hunt, thereby bringing a scourge under control at a profit; but this has yet to be done.

These changes in forest values, from predominance of products to predominance of services, have been observed during the course of successful economic development (Panayotou & Ashton 1992), but they have yet to be

Table 1 Sample structure of an ecological services database to be prepared for an integrated ecosystem assessment.

Service criteria	Controlling factor	Data set
Production of biological resources	Topography, soils, logging history, disturbance	Land-use classification, market price of the crop
Carbon sequestration potential	Vegetation type, logging history, tree composition, soil respiration	Biomass (canopy height, canopy structure) Soil carbon balance (respiration, solution into aquifer)
Conservation of biological diversity	Habitat heterogeneity, food resources	Forest structure, species composition, gap dynamics, topography
Protection of watershed ecosystems	Soil erosion, nutrient leaching	USLE,* vegetation type, slope aspects, soil type, precipitation, soil nutrients
Recreation	Community size, population, employment, racial background, gender	Option value, existing value

*USLE: Universal Soil Loss Equation (Wischmeier & Smith 1965, 1978).

documented in detail. This must be a top research priority, for the fate of the rain forest depends on the recognition, and where possible capture, of service values, and the policy reforms necessary to accomplish this.

4.2 Database of ecological services and values

To manage an ecosystem, we need to begin by establishing a database of ecological services. Table 1 shows a sample structure of a database comprising some service criteria and relevant ecological data. The criteria listed include conflicting goods and services to be traded-off and exclude overlapping service values (double counting) (Costanza et al. 1997). We can start the evaluation of ecological services with an existing database that was acquired through an ecological survey of the study site at Pasoh region (Chap. 2). The existing data need to be reviewed and checked first to see whether they could be incorporated into the database. Pasoh FR makes an appropriate model site because of its long history of ecological study. As a part of a NIES-FRIM-UPM project which started in 1991 (Chap. 1), we have focused on diagnosing the degradation of forest functions and value owing to transformation from primary to regenerationg forests after selective logging finished in the late 1950s. Many intriguing ecological findings were obtained through comparative studies here (Chaps. 2, 12, 21, 31, 36, 37). These findings can lay the foundation for developing general criteria and guidelines for ecological service assessment. We believe that such knowledge can be extrapolated to other study sites in the Southeast Asian tropics. Meanwhile, other service criteria should be reviewed and included in the database. It is also necessary to constantly update the database with data from ecological field studies, such as those of biological interactions among species, dynamic changes in the forest structure and composition, and other functional aspects of forest ecology.

4.3 Mapping of ecological services for integrated ecosystem management

Considering that we can identify many ecological services whose values vary according to local community, race and other socioeconomic backgrounds, we can see that there is no point focusing only on the protection of a fragmented pristine forest. Indeed, it would be very difficult to protect the forest from illegal logging or hunting, even if the forest were protected by fences or guards, unless we were to provide a management plan that could optimize the costs and benefits derived from different social and economic demands. We should recognize that the problems lying behind the deforestation, depletion of the ecosystem and land-use changes no longer relate to forestry alone (Myers 1996). Sustainable management will never be achieved unless local communities are given incentives to better use natural resources and improve the environment. Thus, to study the ecological service value of a forest, we propose to broaden the target area from patches of pristine forest (e.g. Pasoh FR) to the landscape level, in which primary forest, production forest, agricultural crop land and other types of land use are all included (Fig. 2, Chap. 2).

As a first step, a land-use zoning map should be created, incorporating the ecological service values of each land-use type. This should be done with the cooperation of the local administrative agency, and could be prepared through the use of computer simulation, the database of ecological services and a GIS (Geographical Information System) (e.g. Jessen 1992; de Meijee et al. 1988; Nagasawa 1994). Clearing forests for commodity plantations in remote areas, on hillsides, or at higher elevations creates costs associated with soil erosion, road construction and transportation of crops. If we count the service values of carbon sequestration and nutrient cycling in the forest, the costs may then outweigh the benefits. Development does not compensate for the loss of ecological service value of natural forest. In light of these conflicts, a land-use zoning map can be used to analyze how to optimize the land-use value by balancing the ecological service value and economic benefits of development. Thus, this information can be used as a decision-making tool for landscape management. With the assistance of a GIS, we can then design the best spatial pattern of land use under a given profile of ecological services and local economic conditions and provide for the protection of biodiversity by defining protected areas, buffer zones and wildlife corridors. Ideally but not essentially, these models should then be implemented through administrative operations. The important point is that through the use of mapping and zoning by ecological service assessment, policy-makers and land-use managers can always know how the present land use differs from the optimum land use, and what environmental risks underlie the present land use.

We can make best use of the ecological service database in a risk management program that shows the present status of the targeted ecosystem and its service value and what risks could be expected in a proposed land-use development. The potential risks and benefits of the development can be altered by altering the form of development and subsequent management by reclamation or rehabilitation. Kumari (1995) showed that the total economic value of forests at Sungai Karan and in the Raja Forest Reserve in Peninsular Malaysia increased by 18% with the introduction of benign methods of logging that use tramline or cable yarding, instead of conventional logging methods by bulldozer. Such alternatives can be incorporated into the risk management program. The program should also show the total economic value of targeted ecosystems over time after logging or conversion to farmland, depending on the local and global economies and other social factors.

5. ADDITIONAL APPROACHES TO INTEGRATED ECOSYSTEM MANAGEMENT

In addition to the ecological service assessment described above, the approaches listed below could be carried out in the target area. Sections 5.1 through 5.4 present items that we recommend be incorporated into integrated ecosystem management as tools. Implementing them in conjunction with ecological services makes possible an overall overview of the changes in ecological service values and goods in a study area, and allows a number of environmental management options to be presented in more concrete forms. Logistics and environmental education, discussed in section 5.5, transcend the category of a single research project, but they are important for ensuring that local communities benefit through research achievements, and for the continuity of long-term research. Needless to say, inter-regional networking of a project (section 5.6) is indispensable for global ecosystem assessment. The result may be powerful enough to prevent an environmental crisis, to society's benefit.

5.1 Follow-up studies on the process of post-logging restoration

Post-logging follow-up studies are needed to determine the extent to which the sustainable use of tropical forests is possible, and whether the current selective cutting cycle and cutting limits are appropriate(Chaps. 2, 39). Forest managers have been shortening the selective logging cycle and reducing the minimum diameter of trees to be selectively logged, which is leaving forests in worse condition. The situation is analogous to the marine fisheries, where common access and the desire to maximize short-term profits has led to population crashes in species that take a long time to reach reproductive maturity. As long as the circumstances forcing people to sell off their forests persist, borders have to be drawn somewhere for buffer zones between natural forests and logging forests, and studies are needed to determine the extent to which forests recover in what period of time, how wide buffer zones should be, and how they should be arranged. It should be possible to take advantage of existing findings of basic research comparing the structure, composition, and regeneration of natural and secondary forests, and research on canopy gap dynamics. It will also be necessary to establish plots in stands about to be selectively logged, and to compare forest functions before and after selective logging. Central to the success of such research will be the collaboration of silviculturists with economists and basic ecologists. The scientific evidence obtained through follow-up assessment will be indispensable in promoting and strengthening eco-labeling campaigns (e.g. certification by the Forest Stewardship Council). New standards can be proposed on the basis of the pre- and post-logging assessments to allow the implementation of ecologically sustainable management that complements eco-labelling.

5.2 Impacts of fragmentation on wildlife

A look at forest vegetation in tropical regions shows that contiguous expanses of old-growth forest are rarely left. Usually they are divided by large plantations, and remaining patches are isolated. Around the Pasoh FR, for example, lowland forests remain only in patches. How does wildlife move and breed among the few divided stepping-stone-like patches of forest? Studies on this would reveal the relationship between animal migration routes and stand structure, and serve as a guideline for environmental management on, for example, what level of forest stands to leave, and in what arrangement. Forest fragmentation and secondary vegetation may severely affect many specialists and wide-ranging species, but provide preferred habitat to many generalist vertebrates; some of the latter may become serious crop pests or browse forest regeneration. Again, some of these herbivores are candidates for recreational hunting; their control may be achieved while simultaneously providing revenue with hunting licenses. The issues are not simple but are relatively circumscribed, and therefore provide excellent cases for targeted collaborative interdisciplinary research.

5.3 Long-term monitoring on reproductive ecology and genetic diversity in relation to disturbance and climate change

When planning tropical forest conservation, and especially when there are particular species to be conserved, we must assess how large an area (or how many individual trees) is needed to maintain the population size. Because dipterocarps and many other species that are commercially logged irregularly bear seeds that have no dormancy, replacements are only erratically available. When a species has been repeatedly logged, the shortage of replacements is exacerbated by a reduction in the number of reproductive individuals, and there is likely to be a reduction in genetic diversity. Consequent inbreeding can result in low resistance of offspring to disease, pests and environmental stress, so that the reduced density of mature reproductive trees caused by selective logging and fragmentation will trigger the genetic degradation of the population (Chap. 21). Studying this process of genetic degradation will provide guidelines on cutting limits, logging cycle, post-logging reclamation, buffer zones and landscape zoning (see section 4.3 of this chapter).

5.4 Remote sensing and GIS studies

Research based on selected areas and sites requires a means to spatially extrapolate results. Remote sensing technology, indispensable for obtaining a general view of vegetation, land-use distribution, and other attributes of study areas, has this potential. Unfortunately, it is at present not possible to classify the highly diverse and dense forest types by species composition only from satellite images, but it is to some extent possible to estimate canopy height and structure and to analyze canopy layer dynamics by using aerial photography (Chap. 2; Okuda et al. in press) and synthetic aperture radar. If geomorphological features recognizable by GIS are correlated with floristic compositional variation, then they can be mapped as proxies for forest types. As ultra-high-resolution and ultra-multiband data become available, it will be soon possible to measure physiological parameters of trees, identify the canopy tree species, and detect the micro-scale environment of the forest floor.

5.5 Support for environmental education, ecotourism and other community outreach

While meriting long-term research, in reality the issue of whether tropical forests can continue to exist depends on whether the communities who own the forests, especially those who live nearby, perceive that highly diverse tropical forests are justified by the benefits they yield. At the least, representative forests would require museum facilities to acquaint visitors with forest ecosystems and functions. Long-term considerations include the training of parataxonomists, apprentice curator and the establishment of biodiversity centers, as in Costa Rica. These can be arranged in connection with official development associates in developed countries. It is important to recognize that a research project will function soundly only when research, administration and NGOs mesh well.

5.6 Networking tropical research

An extraordinary amount of forest research is under way throughout the tropics. Diversity in the biotic environments and communities of each place makes sampling replication difficult, but replication is essential to allow comparisons and information exchange with study plots in other regions. For that purpose researchers are working on common procedures for study sampling. Specific examples are the International Long-Term Ecological Research Network, an initiative sponsored by the US National Science Foundation to link long-term observation plots throughout the world; and the network being facilitated through the Center of Tropical Forest Science (CTFS) at the Smithsonian Tropical Research Institute, in which large plots have been established in Central and South America, Africa, India, Southeast Asia and other places for long-term forest dynamics research using the same observation methods. The Center is actively promoting international collaboration for a global network of long-term control plots large enough to represent whole forest communities and for tree demographic research, in the major types of biodiverse lowland evergreen forests. The 50-ha Pasoh FR plot (Chaps. 1, 2, 3, 7, 13, 27, 35, 38), which was the first in Asia, can be seen as the flagship control plot for the mixed evergreen rain forests of Southeast Asia, and is already attracting researchers from a wide range of disciplines and an increasing number of students. Such networking is indispensable in order to promote the integration of science in tropical forests with knowledge and techniques for the sustainable management of the tropical forest ecosystem.

6. CONCLUDING REMARKS

Up to this point we have stressed the ideals of a long-term strategy for the development of policy-relevant research for tropical forests. Returning for a moment to the broader significance of tropical forest research, one recognizes many questions and the moral responsibility we have as researchers. Despite the mountains of data that have resulted from tropical forest research and the accumulation of scholarly articles, the loss of the forests has not slowed a bit. Yet decision-makers still give researchers and technical experts the continuing and still only partly resolved mission of elucidating the mystery of how species diversity is maintained in tropical forests, which is certainly a prerequisite of long-term conservation. But considering the short time left for the survival of tropical forests at current rates of conversion, and the fact that the survival of even research sites is in jeopardy, there is considerable doubt that the usual cataloguing of compartmentalized research themes and the running of projects according to each researcher's individual agenda can contribute to "conservation" as described in research proposals. That is to say, connecting the voluminous data accumulated so far on forest ecosystems with ecosystem management, including that of tropical forests, is a very real problem. As a way of addressing this we propose the assessment of forests' ecological services and goods as a means to use and integrate basic and applied research into formulating management plans and policies. But it makes no sense if each of these activities is conducted independently, in different places, and on differing scales. In view of the diverse social factors in tropical regions and the special nature of the environments where tropical forests exist, it also makes no sense to do nothing but present general and theoretical solutions.

We suggest the importance of choosing model sites, generating ideas from the academic research perspective for sustainable management, using these to suggest how ecosystem functions can be assessed, and conducting research aimed at landscape management, including the improvement of logging operations and the reclamation of degraded areas.

Integrated ecosystem management plans should include landscape management (zoning), watershed management, and biodiversity management. All of these can be covered and arranged in a risk management program (see section 4.3 of this Chapter). The program should be used to predict concretely and visibly, when the data are integrated and a resource management option is adopted, what the results will be and what kind of options should be chosen to provide long-term, stable supplies of natural resources and agricultural produce in an area, and to increase asset value. For that purpose it is important, as shown in the first half of this chapter, to assess and review ecological services of forest ecosystems to the best extent from the knowledge already available, and then to predict in advance the asset value of the forest.

Forests also have an important role in providing a basis for environmental education and recreation, which become increasingly important as countries and regions develop economically. For that purpose, building venues to practice environmental education in model areas provides a good opportunity to extricate ourselves from the research colonialism still too often practiced by well-meaning researchers from developed countries. Under these circumstances, we believe that proposing integrated ecosystem management in a model area is indeed timely joint international research.

REFERENCES

Brown, K. & Pearce, D. W. (eds). (1994) The Causes of Tropical Deforestation: The Economic and Statistical Analysis of Factors Giving Rise to the Loss of the Tropical Forest. London: University College London Press, 338pp.

Chapman, C. A. & Chapman, L. J. (1997) Forest regeneration in logged and unlogged forests of Kibale National Park, Uganda. Biotropica 29: 396-412.

Costanza, R., d'Rage, R., de Grout, R., Farber, S., Grass, M., Hannon, B., Limburger, K., Name, S., O'Neill, R. V., Paulo, J., Ruskin, R. G., Sutton, P. & van den Belt, M. (1997) The value of the world's ecosystem services and natural capital. Nature 387: 253-260.

de Meijee, J. C., Mardanus, B. & van de Kasteele, A. M. (1988) Land use modeling of the upper Komering watershed. ITC Journal 7: 91-95.

Ehrlich, P. R. (1989) The limits to substitution: metaresource depletion and a new economic-ecologic paradigm. Ecol. Econ. 1: 9-16.

Ehrlich, P. R. & Mooney, H. A. (1983) Extinction, substitution and ecosystem services. Bioscience 33: 248-254.

FAO (1999) State of the World's Forests. Rome: Food and Agriculture Organization.

Folke, C. (1991) Socioeconomic dependence on the life-supporting environment. In Folke, C. & Kaberger, T. (eds). Linking the Natural Environment and the Economy: Essay form the Eco-Eco group. Dordrecht: Kluwer Academic Publishers, pp.77-94.

Gentry, A. H. (1988) Tree species richness of upper Amazonian forest. Proc. Natl. Acad. Sci. USA 85: 156-159.

Holmberg, J., Bass, S. & Timberlake, L. (1991) Defending the Future: A Guide to Sustainable Development. London: IIED/Earthscan.

Howard, A. F., Rice, R. E. & Gullison, R. E. (1996) Simulated financial returns and selected environmental impacts from four alternative silvicultural prescriptions applied in the Neotropics. For. Ecol. Manage. 89: 43-57.

IPCC (2000) A special report of the Intergovernmental Panel on Climate Change. In Watson, R. T., Noble, I. R., Bolin, B., Ravindranath, N. H., Verardo, D. J. & Dokken, D. J. (eds). Land Use, Land-Use Change, and Forestry, Cambridge: Cambridge University Press, pp.1-20.

Jessen, M. R. (1992) Land resource survey of the Pijiharjo Sub-subwatershed, upper Solo watershed, Central Java. DSIR Land Resources Scientific Report 35, Ministry of Forestry, Indonesia.

Kumari, K. (1995) An Environmental and Economic Assessment of Forest Management Options: A Case Study in Malaysia, 49pp.

Kurpick, P., Kurpick, U. & Huth, A. (1997) The influence of logging on Malaysian dipterocarp rain forest: a study using a forest gap model. J. Theor. Biol. 185: 47-54.

Manokaran, N. (1998) Effects, 34 years later, of selective logging in the lowland dipterocarp forest at Pasoh, Peninsular Malaysia, and implications on present day logging in the hill forests. In Lee, S. S., Dan, Y. M., Gauld, I. D. & Bishop, J. (eds). Conservation, Management and Development of Forest Resources. Forest Research Institute Malaysia, Kepong, pp.41-60.

Millennium Ecosystem Assessment Committee (1998) Millennium ecosystem assessment: strengthening capacity to manage ecosystems for human development. http://www.ma-secretariat.org.

Myers, N. (1994) Past-debate statement. In Myers, N. & Simon, J. (eds). Scarcity or Abundance: A Debate on the Environment. New York: W.W. Norton, 254pp.

Myers, N. (1996) The world's forest: problems and potentials. Environ. Conserv. 23: 156-168.

Nagasawa, R. (1994). Monitoring and evaluation system development on Central Sumatra Forest Rehabilitation Project. DGRLR Report 7. Ministry of Forestry, Indonesia.

Oksanen, T., Heering, M., Cabarle, B. & Sargent, C. (1993) A Study on Coordination. In Sustainable Forestry Development, Report to the Tropical Forestry Action Program Forestry Advisers' Group.

Okuda, T., Suzuki, M., Adachi, N., Quah, E. S., Nor Azman, H., Manokaran, N. (2003) Effects of selective logging on canopy and stand structure and tree species composition in a lowland dipterocarp forest in Peninsular Malaysia. For. Ecol. Manage. 175: 297-320.

Panayotou, T. & Ashton, P. (1992) Not by timber alone–economics and ecology for sustaining tropical forests. Washington, DC: Island Press, 282pp.

Rice, R. E., Gullison, R. E. & Reid, J. W. (1997) Can sustainable management save tropical forests? Scientific American, April, pp.44-49.

Richards, P. W. (1996) The tropical rain forest (2nd ed.). Cambridge: Cambridge University Press, 575pp.

Ross, M. S. & Donovan, D. G. (1986) The world tropical forestry action plan: Can it save the tropical forests? J. World For. Resource Manage. 2: 119-136.

Vincent, J. (1992) The tropical timber trade and sustainable development. Science 256: 1651-1655.

Whitmore, T. C. (1984) Tropical rain forests of the far east. Oxford: Oxford University Press, 352pp.

Wischmeier, W. H. & Smith, D. D. (1965) Predicting rainfall-erosion losses from cropland east of the Rocky Mountains—guide for selection of practices for soil and water conservation. Agriculture Handbook No. 282. Washington, DC: US Department of Agriculture.

Wischmeier, W. H. & Smith, D. D. (1978) Predicting rainfall erosion losses—a guide to conservation planning. Agriculture Handbook No. 537. Washington, DC: US Department of Agriculture.

References: Ecological Studies in Pasoh Forest Reserve (1964—)

Abdul Rahim, N., Baharuddin, K., Noguchi, S. & Tani, M. (2001) Roles of tropical rain forest in hydrological processes and climatic events in forested catchments, Peninsular Malaysia. In Proceedings of the International Symposium Canopy Processes and Ecological Roles of Tropical Rain Forest. 12-13 March 2001, Miri, Sarawak, Malaysia, pp.64-71.

Abe, T. (1978) Studies on the distribution and ecological role of termites in a lowland rain forest of West Malaysia. (1) Faunal composition, size, colouration and nest of termites in Pasoh Forest Reserve. Kontyû 46: 273-290.

Abe, T. (1978) The role of termites in the breakdown of dead wood in the forest floor of Pasoh Forest Reserve. Malay. Nat. J. 30: 391-404.

Abe, T. (1979) Studies on the distribution and ecological role of termites in a lowland rain forest of West Malaysia. (2) Food and feeding habits of termites in Pasoh Forest Reserve. Jpn. J. Ecol. 29: 121-135.

Abe, T. (1980) Studies on the distribution and ecological role of termites in a lowland rain forest of West Malaysia. (4) The role of termites in the process of wood decomposition in Pasoh Forest Reserve. Revue d'Ecologie et de Biologie du Sol 17: 23-40.

Abe, T. & Matsumoto, T. (1978) Distribution of termites in Pasoh Forest Reserve. Malay. Nat. J. 30: 325-334.

Abe, T. & Matsumoto, T. (1979) Studies on the distribution and ecological roles of termites in a lowland rain forest of West Malaysia. (3) Distribution and abundance of termites in Pasoh Forest Reserve. Jpn. J. Ecol. 29: 337-351.

Abe, Y., Hattori, T., Maziah, Z. & Lee, S. S. (1999) Wood-decay fungi of lowland forests in Malaysia. List of collected cultures and a small database for their cultural characters. In Proceedings of the International Conference on Asian Network on Microbial Research, 29 November-1 December 1999, Chiang Mai, Thailand, pp.879-886.

Allbrook, R. F. (1973) The soils of Pasoh Forest Reserve, Negeri Sembilan. Malay. For. 36: 22-33.

Aminah, H., Mohd. Afendi Hussin & Mohd. Jaffar Sharri (1995) A note on the germination of *Dryobalanops aromatica* and *Shorea macroptera* in different sowing media. J. Trop. For. Sci. 7: 507-510.

Amir Husni, M. S. & Miller, H. G. (1990) *Shorea leprosula* as an indicator species for site fertility evaluation in dipterocarp forests of Peninsular Malaysia. J. Trop. For. Sci. 3: 101-110.

Amir Husni, M. S., Miller, H. G. & Appanah, S. (1991) Soil fertility and tree species diversity in two Malaysian forests. J. Trop. For. Sci. 3: 318-331.

Ang, L. H. & Maruyama, Y. (1995) Survival and early growth of *Shorea platyclados, Shorea macroptera, Shorea assamica* and *Hopea nervosa* in open planting. J. Trop. For. Sci. 7: 541-557.

Aoki, M., Yabuki, K. & Koyama, H. (1975) Micrometeorology and assessment of primary production of a tropical rain forest in West Malaysia. J. Agric. Meteorol. 31: 115-124.
Aoki, M., Yabuki, K. & Koyama, H. (1978) Micrometeorology of Pasoh forest. Malay. Nat. J. 30: 149-160.
Appanah, S. (1979) The Ecology of insect pollination of some tropical rain forest trees. Ph.D. diss., Univ. Malaya.
Appanah, S. (1980) Pollination in Malaysian primary forests. In Furtado, J. I., Morgan, W. B., Pfafflin, J. R. & Ruddle, K. (eds). Tropical ecology and development: Proceedings of the 5th International Symposium of Tropical Ecology, 16-21 April 1979, Kuala Lumpur, Malaysia, pp.177-182.
Appanah, S. (1981) Pollination in Malaysian primary forest. Malay. For. 44: 37-42.
Appanah, S. (1982) Pollination of androdioecious *Xerospermum intermedium* Radlk. (Sapindaceae) in a rain forest. Biol. J. Linn. Soc. 18: 11-34.
Appanah, S. (1985) General flowering in the climax rain forests of South-east Asia. J. Trop. Ecol. 1: 225-240.
Appanah, S. (1987) Insect pollination and diversity of dipterocarps. In Kosterman, A. J. G. H. (ed). Proceedings of the 3rd International Roundtable Conference on Dipterocarps, UNESCO, Jakarta, pp.277-29
Appanah, S. & Chan, H. T. (1981) Thrips: The pollinators of some dipterocarps. Malay. For. 44: 234-252.
Appanah, S. & Mohd. Rasol, A. M. (1990) Smaller trees can fruit in logged dipterocarp forests. J. Trop. For. Sci. 3: 80-87.
Appanah, S. & Putz, F. E. (1984) Climber abundance in virgin dipterocarp forest and the effect of pre-felling climber cutting on logging damage. Malay. For. 47: 335-342.
Appanah, S. & Weinland, G. (1993) A preliminary analysis of the 50-hectare Pasoh demography plot: I. Dipterocarpaceae. FRIM Research Pamphlet No. 112. Kepong, Malayshia, 183pp.
Appanah, S., Willemstein, S. C. & Marshall, A. G. (1986) Pollen foraging by two *Trigona* colonies in a Malaysian rain forest. Malay. Nat. J. 39: 177-191.
Appanah, S., Gentry, A. H. & LaFrankie, J. V. (1993) Liana diversity and species richness of Malaysian rain forests. J. Trop. For. Sci. 6: 116-123.
Ashton, P. S. (1971) Pasoh Forest Reserve Vegetation Survey International Biological Program, Malayan Project, 2nd Report. (Mimeographed) Limited Circulation. 51pp.
Ashton, P. S. (1976) An approach to the study of breeding systems, populations structure and taxonomy of tropical trees. In Burley, J. & Styles, B. T. (eds). Tropical Trees: Variation, Breeding and Conservation. Linn. Soc. Symp. Series No. 2, Academic Press, London, pp.35-42.
Ashton, P. S. (1976) Factors affecting the development and conservation of tree genetic resources in Southeast Asia. In Burley, J. & Styles, B. T. (eds). Tropical Trees: Variation, Breeding and Conservation. Linn. Soc. Symp. Series No. 2. Academic Press, London, pp.189-198.
Ashton, P. S. (1976) Mixed dipterocarp forest and its variation with habitat in the Malayan lowlands: A re-evaluation at Pasoh. Malay. For. 39: 56-72.
Ashton, P. S. (1977) A contribution of rain forest research to evolutionary theory. Ann. Mo. Bot. Gard. 64: 694-705.
Ashton, P. S. (1978)Vegetation and soil association in tropical forests. Malay. Nat. J. 30: 225-228.
Ashton, P. S. (1984) Biosystematics of tropical forest plants: A problem of rare species. In Grant, W. F. (ed). Plant Biosystematics. Academic Press, Toronto, pp.497-518.
Ashton, P. S. (1988) Systematics and ecology of rain forest trees. Taxon 37: 622-629.
Ashton, P. S. (1988) Dipterocarp biology as a window to the understanding of tropical forest structure. Annu. Rev. Ecol. Syst. 19: 347-370.
Ashton, P. S. (1989) Dipterocarp reproductive biology. In Lieth, H. & Werger, M. J. A. (eds). Tropical Rain Forest Ecosystems - Biogeographical and Ecological Studies. Ecosystems of the World Series, 14B. Elsevier, Amsterdam, pp.219-240.

Ashton, P. S. (2001) Patterns of species variation in West and East Malaysia. Malay. Nat. J. 55: 181-16.
Ashton, P. S. & LaFrankie, J. V. (1998) Patterns of tree species richness among tropical rain forests. Proceedings of the Symposium for the International Prize in Biology. Hayan, Japan.
Ashton, P. S., Soepadmo, E. & Yap, S. K. (1977) Current research into the breeding systems of rainforest tree and its implications. In Brunig, E. F. (ed). Transaction International MAB-IUFRO Workshop Tropical Rainforest Ecology Resources, Hamburg-Reinbek, pp.187-192.
Ashton, P. S., Gan, Y. Y. & Robertson, F. W. (1984) Electrophoretic and morphological comparisons in ten rain forest species of *Shorea* (Dipterocarpaceae). Bot. J. Linn. Soc. 89: 293-304.
Ashton, P. S., Givnish, T. J. & Appanah, S. (1988) Staggered flowering in the Dipterocarpaceae: New insights into floral induction and the evolution of mast fruiting in the aseasonal tropics. Am. Nat. 132: 44-66.
Becker, P. (1981) Potential physical and chemical defenses of *Shorea* seedling leaves against insects. Malay. For. 44: 346-356.
Becker, P. (1983) Effects of insect herbivory and artificial defoliation on survival of *Shorea* seedlings. In Sutton, S. L., Whitmore, T. C. & Chadwick A. C. (eds). Tropical Rain Forest: Ecology and Management, Blackwell Scientific Publications, Oxford, pp.241-252.
Becker, P. (1985) Catastrophic mortality of *Shorea leprosula* juveniles in a small gap. Malay. For. 48: 263-265.
Becker, P. & Martin, J. S. (1982) Protein-precipitating capacity of tannins in *Shorea* (Dipterocarpaceae) seedling leaves. J. Chem. Ecol. 8: 1353-1367.
Becker, P. & Wong, M. (1985) Seed dispersal, seed predation, and juvenile mortality of *Aglaia* sp. (Meliaceae) in lowland dipterocarp rainforest. Biotropica 17: 230-237.
Boscolo, M. (1999) Strategies for multiple use management of tropical forests: An assessment of alternative options. Center for International Development, Harvard University, Working Paper.
Boscolo, M. (in press) Non-convexities and multiple use management in tropical forests. In Losos, E. & Leigh, E. G. Jr. (eds). Forest diversity and dynamism: Findings from a network of large-scale tropical forest plots. Chicago University Press, Chicago.
Boscolo, M. & J. Buongiorno. (1997) Managing a tropical rainforest for timber, carbon storage and tree diversity. Commonwealth For. Rev. 76: 246-254.
Boscolo, M. & Buongiorno, J. (1998) Carbon storage, income, and habitat diversity in managed tropical forests. 7th Symposium on Systems Analysis in Forest Resources, May 28-31, 1997, Bellaire, MI, USA.
Boscolo, M., Buongiorno, J. & Panayotou, T. (1997) Simulating options for carbon sequestration through improved management of a lowland tropical rainforest. Environ. Dev. Econ. 2: 241-263.
Boscolo, M. & Vincent, J. R. (1998) Promoting better logging practices in tropical forests. The World Bank Development Research Group, Washington, DC.
Boscolo, M., Vincent, J. R. & Panayotou, T. (1998) Discounting costs and benefits in carbon sequestration projects. Harvard Institute for International Development, Cambridge, Massachusetts, USA.
Bullock, J. A. (1971/1972) Pasoh: Forest ecosystem study. Malay. Sci. 6: 78-85.
Bullock, J. A. (1972) IBP-Malaysian Project: Annual Report for 1971.
Bullock, J. A. (1978) A contribution to the estimation of litter production and tree loss in Pasoh Forest Reserve. Malay. Nat. J. 30: 363-365.
Burgess, N. A. (1978) The conservation of Pasoh Forest Reserve. Malay. Nat. J. 30: 445-447.
Burslem, D. F. R. P., Garwood, N. C. & Thomas, S. C. (2001) Tropical forest diversity—the plot thickens. Science 291: 606-607.
Chadwick, A. R. (ed). (1983) Tropical Rain Forest: Ecology and Management. Blackwell Scientific Publications, Oxford, 498pp.
Chan, H. T. (1977) Reproductive biology of some Malaysian dipterocarps. Ph.D. diss., Univ.

Aberdeen.

Chan, H. T. (1980) Reproductive biology of some Malaysian dipterocarps. II. Fruiting biology and seedling studies. Malay. For. 43: 438-451.

Chan, H. T. (1980) Reproductive biology of some Malaysian dipterocarps. In Morgan, W. B., Pfafflin, J. R. & Ruddle, K. (eds). Tropical ecology and development: proceedings of the 5th International Symposium of Tropical Ecology, 16-21 April 1979, Kuala Lumpur, Malaysia, pp.169-175.

Chan, H. T. (1981) Reproductive biology of some Malaysian dipterocarps. III. Breeding systems. Malay. For. 44: 28-36.

Chan, H. T. (1982) Reproductive biology of some Malaysian dipterocarps. IV. An assessment of gene flow within a natural population of *Shorea leprosula* using leaf morphological characters. Malay. For. 45: 354-360.

Chan, H. T. & Appanah, S. (1980) Reproductive biology of some Malaysian dipterocarps. I. Flowering biology. Malay. For. 43: 132-143.

Chen, S. H. (1972) A preliminary ecological study of soil nematodes in lowland dipterocarp forest. B. thesis, Univ. Malaya.

Chiba, S. (1978) Numbers, biomass and metabolism of soil animals in Pasoh Forest Reserve. Malay. Nat. J. 30: 313-324.

Chiba, S., Abe, T., Aoki, J., Imadate, G., Ishikawa, K., Kondoh, M., Shiba, M. & Watanabe, H. (1975) Studies on the productivity of soil animals in Pasoh Forest Reserve, West Malaysia. I. Seasonal change in the density of soil mesofauna: *Acari collembola* and others. Science Report of Hirosaki University 22: 87-124.

Chua, L. S. L., Hawthrone, W., Saw, L. G. & Quah, E. S. (1998) Biodiversity database and assessment of logging impacts. In Lee, S. S., Dan, Y. M., Gauld, I. D. & Bishop, J. (eds). Conservation, Management and Development of Forest Resources. Proceedings of the Malaysia-United Kingdom Programme Workshop, 21-24 October 1996, Kuala Lumpur, Malaysia, pp.30-40.

Condit, R. (1995) Research in large, long-term tropical forest plots. Tresnds Ecol. Evol. 10: 18-22.

Condit, R. (1996) Defining and mapping vegetation types in mega-diverse tropical forests. Trends. Ecol. Evol. 11: 4-5.

Condit, R. (1998) Tropical Forest Census Plots: Methods and results from Barro Colorado Island, Panama and a comparison with other plots. Springer-Verlag, Berlin.

Condit, R., Hubbell, S. P., LaFrankie, J. V., Sukumar, R., Manokaran, N., Foster, R. B. & Ashton, P. S. (1996) Species-area and Species-individual relationships for tropical trees: a comparison of three 50-ha plots. J. Ecol. 84: 549-562.

Condit, R., Ashton, P. S., Manokaran, N., LaFrankie, J. V., Hubbell, S. P. & Foster, R. (1999) Dynamics of the forest communities at Pasoh and Barro Colorado: comparing two 50-ha plots. Philos. Trans. R. Soc. London B. 354: 1739-1748.

Condit, R., Ashton, P. S., Baker P., Bunyavejchewin, S., Gunatilleke, S., Gunatilleke, N., Hubbell, S. P., Foster, R. B., Itoh, A., LaFrankie, J. V., Lee, H. S., Losos, E., Manokaran N., Sukumar, R. & Yamakura, T. (2000) Spatial patterns in the distribution of tropical tree species. Science 288: 1414-1418.

Condit, R., Hubbell, S. P., Foster, R. B., Manokaran, N., Ashton, P. S. & LaFrankie, J. V. (in press) A direct test for density dependent population change in two rainforest plots. In Losos, E. & Leigh, E. G. Jr. (eds). Forest diversity and dynamism: Findings from a network of large-scale tropical forest plots. University of Chicago Press, Chicago.

Debski, I., Burslem, D. F. R. P., Palmiotto, P. A., LaFrankie, J. V., Lee, H. S. & Manokaran, N. (in press) The spatial distribution and habitat preferences of *Aporosa* in two Malaysian rain forests: implications for abundance and co-existenceology. Ecology.

Fiala, B. & Linsenmair, K. E. (1995) Distribution and abundance of plants with extrafloral nectaries in the woody flora of a lowland primary forest in Malaysia. Biodiversity Conserv. 4: 165-182.

Fisher, B. L., Malsch, A. K. F., Gadagkar, R., Delabie, J. H. C., Vasconcelos, H. L. & Majer, J. D. (2000) Applying the ALL protocol. Selected case studies. In Agosti, D., Majer, J. D.,

Alonso, L. E. & Schultz, T. R. (eds). Ants. Standard methods for measuring and monitoring biodiversity. Smithsonian Institution Press, Washington, London, pp.207-214.

Fong, O. M. (1973) A preliminary survey of the Formicidae at Pasoh Forest Reserve. B. thesis, Univ. Malaya.

Fukuyama, K. (1993) Twenty days in Malaysia (1) To Pasoh Forest Reserve. Hoppo-ringyo (Northern forestry) 45: 65-68 (in Japanese).

Fukuyama, K. (1993) Twenty days in Malaysia (2) Balloon in a tropical forest. Hoppo-ringyo (Northern forestry) 45: 107-110 (in Japanese).

Fukuyama, K. (1995) How to research actives of beetles in a forest canopy in a tropical forest. Tropics 4: 317-326 (in Japanese).

Fukuyama, K. (1997) Biodiversity of insects in canopy. Shinrin Kagaku 20: 24-26 (in Japanese).

Fukuyama, K., Maeto, K. & Kirton, L. G. (1994) Field tests of a balloon-suspended trap system for studying insects in the canopy of tropical rainforests. Ecol. Res. 9: 357-360.

Furukawa, A., Toma, T., Maruyama, Y., Matsumoto, Y., Uemura, A., Abdulah, A. M. & Awang, M. (2001) Photosynthetic rates of four tree species in the upper canopy of a tropical rain forest at the Pasoh Forest Reserve in Peninsular Malaysia. Tropics 10: 519-527.

Gan, Y. Y. (1976) Population and phylogenetic studies on species of Malaysian rain forest tress. Ph.D. diss., Univ. Aberdeen.

Gan, Y. Y. (1980) A novel approach to the study of genetic variation in forest trees. In Furtado, J. I., Morgan, W. B., Pfafflin, J. R. & Ruddle, K. (eds). Tropical ecology and development: Proceedings of the 5th International Symposium of Tropical Ecology, 16-21 April 1979, Kuala Lumpur, Malaysia, pp.203-206.

Gan, Y. Y., Robertson, F. W., Ashton, P. S., Soepadmo, E. & Lee, D. W. (1977) Genetic variation in wild population of rainforest trees. Nature 269: 323-325.

Gan, Y. Y., Robertson, F. W. & Soepadmo, E. (1981) Isozyme variation in some rainforest trees. Biotropica 13: 20-28.

Gardette, E. (1998) The effect of selective timber logging on the diversity of woody climbers at Pasoh. In Lee, S. S., Dan, Y. M., Gauld, I. D. & Bishop, J. (eds). Conservation, Management and Development of Forest Resources. Proceedings of the Malaysia-United Kingdom Programme Workshop, 21-24 October 1996, Kuala Lumpur, Malaysia, pp.115-125.

Gobbett, D. J. (1972) Geological map of the Malay Peninsular. Geol. Soc. Malaysia, Kuala Lumpur.

Gong, W. K. (1972) Studies on the rates of fall, decomposition and nutrient element release of leaf litter of representative species in lowland dipterocarp forest. B. thesis, Univ. Malaya.

Ha, C. O. (1978) Embryological and cytological aspects of the reproductive biology of some understorey rain forest trees. Ph.D. diss., Univ. Malaya.

Ha, C. O. (1980) Some aspects of the reproductive biology of understorey tree in the tropical rain forest. In Furtado, J. I., Morgan, W. B., Pfafflin, J. R. & Ruddle, K. (eds). Tropical ecology and development: Proceedings of the 5th International Symposium of Tropical Ecology, 16-21 April 1979, Kuala Lumpur, Malaysia, pp.199-202.

Ha, C. O., Sands, V. E., Soepadmo, E. & Jong, K. (1988) Reproductive patterns of selected understorey trees in the Malaysian rainforest: Sexual species. Bot. J. Linn. Soc. 97: 295-316.

Ha, C. O., Sands, V. E., Soepadmo, E. & Jong, K. (1988) Reproductive patterns of selected understorey trees in Malaysian rain forest: The apomictic species. Bot. J. Linn. Soc. 97: 317-331.

Hall, P., Ashton, P. S., Condit, R., Manokaran, N. & Hubbell, S. P. (1998) Signal and noise in sampling tropical forest structure and dynamics. In Dallmeier, F. & Comisky, J. A. (eds). Forest Biodiversity Research, Monitoring and Modeling: Conceptual Background and Old World Case Studies. Parthenon/UNESCO, Paris, France.

Hattori, T. & Lee, S. S. (1999) Two new species of *Perenniporia* described from a lowland rainforest of Malaysia. Mycologia 91: 525-531.

Hattori, T. & Lee, S. S. (1999) Host specificity of wood decay basidiomycetes in a tropical rainforest in Malaysia. In Proceedings of the International Joint Workshop for Studies on Biodiversity Species 2000. Value of Information for 21st Century. CGER/NIES Japan, 14-16 July, Tsukuba, Japan, 54pp.

He, F-L. & Legendre, P. (1996) On species-area relations. Am. Nat. 148: 719-737.

He, F. & LaFrankie, J. V. (in press) Scale dependence of tree abudance and tree species richness in a tropical rain forest. In Losos, E. & Leigh, E. G. Jr. (eds). Forest diversity and dynamism: Findings from a network of large-scale tropical forest plots. University of Chicago Press, Chicago.

He, F-L., Legendre, P., Bellehumeur, C. & LaFrankie, J. V. (1994) Diversity pattern and spatial scale: A study of a tropical rain forest in Malaysia. Environ. Ecol. Sci. 1: 265-286.

He, F. L., Legendre, P. & LaFrankie, J. V. (1996) Spatial pattern of diversity in a tropical rain forest in Malaysia. J. Biogeogr. 23: 57-74.

He, F., Legendre, P. & LaFrankie, J. V. (1997) Distribution patterns of tree species in a Malaysian tropical rain forest. J. Vegetation Sci. 8: 105-114.

He, F-L., P. Legendre & LaFrankie, J. V. (in press) Stand structure analysis of a tropical rain forest with respect to abundance and diameter size. J. Trop. Ecol.

Hill-Rowley, R. L., Kirton, L., Ratnam, L. & Appanah, S. (1996) The use of a grid-based geographic information system to examine ecological relationships within the Pasoh 50 ha research plot. J. Trop. For. Sci. 8: 570-572.

Ibrahim J., Abdul Rashid Ahmad & Abu Said Ahmad (1995) The musilagenous extract of Cinnomomum iners. J. Trop. For. Prod. 1: 189-193.

Ickes, K. (in press) Hyper-abundance of native wild pigs (*Sus scrofa*) in a lowland Dipterocarp rain forest of Peninsular Malaysia. Biotropica.

Ickes, K., Dewalt, S. J. & Appanah, S. (2001) Effects of native pigs (*Sus scrofa*) on woody understorey vegetation in a Malaysian lowland rain forest. J. Trop. Ecol. 17: 191-206.

Ikeda, E. (1997) A new genus and a new species of Tetrastichinae (Hymenoptera, Eulophidae) from Malaysia and Japan. Jpn. J. Entomol. 65: 721-727.

IIwata, H., Konuma, A. & Tsumura, Y. (2000) Development of microsatellite markers in the tropical tree *Neobalanocarpus heimii* (Dipterocarpaceae). Mol. Ecol. 9: 1684-1686.

Jong, K. (1976) Cytology of the Dipterocarpaceae. In Burley, J. & Styles, B. T. (eds). Tropical trees: variation, breeding and conservation. Linn. Soc. Symp. Series No. 2. Academic Press, London, pp.79-84.

Jong, K. (1980) A cytoembryological approach to the study of variation and evolution in rainforest tree species. In Furtado, J. I., Morgan, W. B., Pfafflin, J. R. & Ruddle, K. (eds). Tropical ecology and development: Proceedings of the 5th International Symposium of Tropical Ecology, 16-21 April 1979, Kuala Lumpur, Malaysia, pp.213-218.

Jong, K. & Kaur, A. (1979) A cytotaxonomic view of the Dipterocarpaceae with some comments on polyploidy and apomixis. Mem. Mus. Nation. d'hist. Nat., N. S., B26: 41-49; 157-158.

Jong, K., Stone, B. C. & Soepadmo, E. (1973) Malaysian tropical forest: An underexploited genetic reservoir of edible fruit tree species. In Soepadmo, E. & Singh, K. G. (eds). Proceeding of symposium of biological resources and natural development, Malayan Nature Society, Kuala Lumpur, pp.113-121.

Kachi, N. (1992) On-going project on tropical forests in Peninsular Malaysia. Trop. Ecol. Lett. 5: 1-4 (in Japanese).

Kachi, N. (1993) Seedling establishment as affected by predators. Bull. FRIM 2: 5.

Kachi, N. (1993) Present of Pasoh Forest Rreserve. Trop. Ecol. Lett. 10: 9-13 (in Japanese).

Kachi, N. (1994) Degradation of tropical forests and biodiversity. Kurashinoki 3: 24-25 (in Japanese).

Kachi, N. (1996) Diversity of dipterocarps and mast flowering. In Global Environment Division, Japan Environment Agency (ed). The future of the global environments: Degradation of tropical forests, Chuohoki-Shuppan, Tokyo, pp.74-85 (in Japanese).

Kachi, N. (1998) A mechanisms for coexistence of canopy tree species in a tropical rain forest. Global Environ. Res. 3: 29-36 (in Japanese).

Kachi, N., Okuda, T. & Yap, S. K. (1993) Seedling establishment of a canopy tree species in Malaysian tropical rain forests. Plant Species Biol. 8: 167-174.
Kachi, N., Okuda, T. & Yap, S. K. (1995) Effect of herbivory on seedling establishment of *Dryobalanops aromatica* (Dipterocarpaceae) under plantation forest in Peninsular Malaysia. J. Trop. For. Sci. 8: 59-70.
Kachi, N., Okuda, T., Yap, S. K. & Manokaran, N. (1995) Biodiversity and regeneration of canopy tree species in a tropical rain forest in Southeast Asia. J. Environ. Sci. 9: 17-36.
Kanzaki, M., Yap, S. K., Kimura, K., Okauchi, Y. & Yamakura, T. (1997) Survival and germination of buried seeds of non-dipterocarp species in a tropical rain forest at Pasoh, West Malaysia. Tropics 7: 9-20.
Kato, R., Tadaki, Y. & Ogawa, H. (1978) Plant biomass and growth increment studies in Pasoh Forest. Malay. Nat. J. 30: 211-224.
Kaur, A. (1978) Embryological and cytological studies in some of the Dipterocarpaceae. Ph.D. diss., Univ. Aberdeen.
Kaur, A. (1980) Cytological and embryological studies on some Dipterocarpaceae. In Furtado, J. I., Morgan, W. B., Pfafflin, J. R. & Ruddle, K. (eds). Tropical ecology and development: Proceedings of the 5th International Symposium of Tropical Ecology, 16-21 April 1979, Kuala Lumpur, Malaysia, pp.207-212.
Kaur, A., Ha, C. O., Jong, K., Sands, V. E., Chan, H. T., Soepadmo, E. & Ashton, P. S. (1978) Apomixis may be widespread among trees of the climax rain forest. Nature 271: 440-442.
Kaur, A., Jong, K., Sands, V. E. & Soepadmo, E. (1986) Cytoembryology of some Malaysian dipterocarps with some evidence of apomixis. Bot. J. Linn. Soc. 92: 75-88.
Kemper, C. M. (1985) Small mammals and habitat structure in lowland rainforest of Peninsular Malaysia. J. Trop. Ecol. 1: 5-22.
Kemper, C. M. (1988) The mammals of Pasoh Forest Reserve, Peninsular Malaysia. Malay. Nat. J. 42: 1-19.
Khozirah, S. & Waterman, P. G (1995) Chemical constituents on some Malaysian Flacourtiaceae species. In Conference on Forestry and Forest Products Research 1, pp.104-109.
Kira, T. (1975) Primary production of forests. In Cooper, J. P. (ed). Photosynthesis and Production in Different Environments, Cambridge University Press, Cambridge, vol. 3. pp.5-40.
Kira, T. (1976) Pasoh Forest Reserve in Negeri Sembilan, West Malaysia. Background information for IBP soil fauna studies. Nature & Life in Southeast Asia 7: 1-8.
Kira, T. (1978) Primary productivity of Pasoh forest - A synthesis. Malay. Nat. J. 30: 291-297.
Kira, T. (1978) Community architecture and organic matter dynamics in tropical lowland rain forests of Southeast Asia with special reference to Pasoh Forest, West Malaysia. In Tomlinson, P. B. & Zimmermann, M. H. (eds). Tropical Trees as Living Systems. Cambridge University Press, pp.561-590.
Kira, T. (1978) Studies on biological production in forest and freshwater ecosystem in West Malaysia. In Tamiya, H. (ed). Summary report on the contribution of the Japanese national committee for IBP, 1964-1974, JIBP Synthesis 20. University of Tokyo Press, Tokyo, pp.225-234.
Kira, T. (1987) Primary production and carbon cycling in a primeval lowland rainforest of Peninsular Malaysia. In Sethuraj, M. R. & Raghavendra, A. S. (eds). Tree Crop Physiology. Elsevier Science Publishers, Amsterdam, pp.99-119.
Kira, T. & Yoda, K. (1989) Vertical stratification in microclimate. In Lieth, H. & Werger, M. J. A. (eds). Tropical Rainforest Ecosystems-Biogeographical and Ecological Studies. Ecosystems of the world 14B, Elsevier Science Publishers, Amsterdam, pp.55-71.
Kochummen, K. M. (1997) Tree Flora of Pasoh Forest. Malayan Forest Records No. 44. Forest Reseaarch Institue Malaysia, Kepong, 461pp.
Kochummen, K. M., LaFrankie, J. V. & Manokaran, N. (1990) Floristic composition of Pasoh Forest Reserve, a lowland rain forest in Peninsular Malaysia. J. Trop. For. Sci. 3: 1-13.
Kochummen, K. M., LaFrankie, J. V. & Manokaran, N. (1992) Diversity of trees and shrubs in Malaya at regional and local levels. Malay. Nat. J. 45: 545-554.

Kochummen, K. M., LaFrankie, J. V. & Manokaran, N. (1992) Representation of Malayan trees and shrubs at Pasoh Forest Reserve. In Yap, S. K. & Lee, S. W. (eds). Harmony with Nature: Proceedings of an International Symposium on the Conservation of Biodiversity. Forest Research Institute of Malaysia, Kuala Lampur, Malaysia.

Kondoh, M., Watanabe, H., Chiba, S., Abe, T., Shiba, M. & Saito, S. (1980) Studies on the productivity of soil animals in Pasoh Forest Reserve, West Malaysia. V. Seasonal change in the density and biomass of soil macrofauna: *Oligochaeta, Hirundinea,* and *Arthropoda,* Mem. Shiraume Gakuen Coll. No. 16., pp.1-26.

Konuma, A., Tsumura, Y., Lee, C. T., Lee, S. L. & Okuda, T. (2000) Estimation of gene flow in the tropical-rainforest tree *Neobalanocarpus heimii* (Dipterocarpaceae), inferred from paternity analysis. Mol. Ecol. 9: 1843-1852.

Koyama, H. (1978) Photosynthesis studies in Pasoh forest. Malay. Nat. J. 30: 253-258.

Koyama, H. (1981) Photosynthesis rates in lowland rainforest trees of Peninsular Malaysia. Jpn. J. Ecol. 31: 361-369.

Kubo, T., Kohyama, T., Potts, M. D. & Ashton, P. S. (2000) Mortality rate estimation when inter-census intervals vary. J. Trop. Ecol. 16: 753-756.

LaFrankie, J. V. (1994) Population dynamics of some tropical trees that yield non-timber forest products. Econ. Bot. 48: 301-309.

LaFrankie, J. V. (1996) Distribution and abundance of Malayan trees: Significance of family characteristics for conservation. Garden's Bulletin Singapore 48: 75-87.

LaFrankie, J. V. (1996) The contribution of large-scale forest dynamic plots to theoretical community ecology. In Turner, I. M., Diong, C. H., Lim, S. S. L. & Ng, P. K. L. (eds). Biodiversity and the Dynamics of Ecosystems. DIWPA Series vol. 1. Singapore.

LaFrankie, J. V. Jr. & Chan, H. T. (1991) Confirmation of sequential flowering in *Shorea* (Dipterocarpaceae). Biotropica 23: 200-203.

Lee, S. L., Ng, K. K. S., Saw, L. G, Norwati, A., Salwana, M. H. S., Lee, C. T. & Norwati, M. (2002) Population genetics of *Intsia palembanica* (Leguminosae) and genetic conservation of Virgin Jungle Reserves in Peninsular Malaysia. Am. J. Bot. 89: 447-459.

Lee, S. S. (1995) Fungi mushrooms. In Lee, S. S. (ed). A Guide Book to Pasoh. FRIM Technical Information Handbook No. 3, pp.66-72.

Lee, S. S. (ed). (1995) A Guide Book to Pasoh. FRIM Technical Information Handbook No. 3, 73pp.

Lee, S. S., Watling, R. & Turnbull, E. (1995) Ectomycorrhizal fungi as possible bio-indicators in forest management. In Conference on Forestry and Forest Products Research 1, pp.63-68.

Lee, S. S., Dan, Y. M., Gauld, I. D. & Bishop, J. (eds). (1998) Conservation, Management and Development of Forest Resources. Proceedings of the Malaysia-United Kingdom Programme Workshop, 21-24 October 1996, Kuala Lumpur, Malaysia, 392pp.

Lee, S. S., Watling, R. & Noraini Sikin Yahya. (2002) Ectomycorrhizal basidiomata fruiting in lowland rain forests of Peninsular Malaysia. Bois et Forêts des Tropiques 274: 33-43.

Leigh, C. (1978) Slope hydrology and denudation in the Pasoh Forest Reserve. I. Surfacewash: Experimental techniques and some preliminary results. Malay. Nat. J. 30: 179-197.

Leigh, C. (1978) Slope hydrology and denudation in the Pasoh Forest Reserve. II. Throughflow: Experimental techniques and some preliminary results. Malay. Nat. J. 30: 199-210.

Leigh, C. (1982) Sediment transport by surfacewash and throughflow at the Pasoh Forest Reserve, Negeri Sembilan, Peninsular Malaysia. Geogr. Ann. 64A: 171-180.

Leow, K. C. I. (1973) A study of the variation in acarine populations at different sampling sites at Pasoh Forest Reserve, Malaysia (with emphasis on the *Cryptostigmata).* B. thesis, Univ. Malaya.

Leow, K. C. I. (1975) Meiofauna of decomposing leaves of some rainforest trees. M. thesis, Univ. Malaya.

Leow, I. (1978) Population studies of soil meiofauna (particularly *Acari*) in Pasoh Forest. Malay. Nat. J. 30: 307-312.

Liang, N., Tang, Y. & Okuda, T. (2001) Is elevation of carbon dioxide concentration beneficial to seedling photosynthesis in the understory of tropical rain forests? Tree Physiol. 21:

1047-1055.

Lim, H. F. & Ismail, J. (1994) The uses of non-timber forests products in Pasoh Forest Reserve, Malaysia. vol. 113. Forest Research Institute of Malaysia Research Pamphlets. Forest Research Institute of Malaysia, Kuala Lampur, Malaysia.

Lim, M. T. (1972) Studies on some aspects of the cycle of nutrient element in *Licuala* sp. B. thesis, Univ. Malaya.

Lim, M. T. (1978) Litterfall and mineral content of litter in Pasoh Forest. Malay. Nat. J. 30: 375-380.

Linatoc, A. (1998) Life history studies of *Mangifera* (Anacardiaceae. species in the Pasoh 50 ha plot. Ph.D. diss., National Univ. Malaysia.

Liu, J. & Ashton, P. S. (1995) Individual-based simulation models for forest succession and management. For. Ecol. Manage. 73: 157-175.

Liu, J. & Ashton, P. S. (1998) FORMOSAIC: an individual-based spatially explicit model for simulating forest dynamics in landscape mosaics. Ecol. Model. 106: 177-200.

Liu, J. & Ashton, P. S. (1999) Simulating effects of landscape context and timber harvest on tree species diversity. Ecol. Appl. 9: 186-201.

Liu, J., Ickes, K., Ashton, P. S., LaFrankie, J. V. & Manokaran, N. (1999) Spatial and temporal impacts of adjacent areas on the dynamics of species diversity in a primary forest. In Mladenoff, D. & Baker, W. (eds). Spatial Modeling of Forest Landscape Change: Approaches and Applications. Cambridge University Press, Cambridge, UK.

Losos, E. & Leigh, E. G. Jr. (eds). (in press) Forest diversity and dynamism: Findings from a network of large-scale tropical forest plots. University of Chicago Press, Chicago.

Low, K. S. (1974) Interception loss of precipitation in Lowland Dipterocarp Fprest. International Biological Programme Synthesis Meeting, Kuala Lumpur.

Maeto, K., Fukuyama, K., Sajap, A. S. & Wahab, Y. A. (1995) Selective attraction of flower-visiting beetles (Coleoptera) to floral fragrance chemicals in a tropical rain forest. Jpn. J. Entomol. 63: 851-859.

Maeto, K., Fukuyama, K. & Kirton, L. G. (1999) Edge effects on ambrosia beetle assemblages in lowland rain forest bordering oil palm plantations in Peninsular Malaysia. J. Trop. For. Sci. 11: 537-547.

Maeto, K., Osawa, N. & Fukuyama, K. (2001) Impact of mast fruiting on abundance of vespid wasps (Hymenoptera) in a lowland tropical forest of Peninsular Malay. Entomol. Sci. 4: 247-250.

Malsch, A. (2000) Investigations of the diversity of leaf-litter inhabiting ants in Pasoh, Malaysia. In Agosti, D., Majer, J. D., Alonso, L. & Schultz, T. (eds). Sampling ground-dwelling ants: case studies from the worlds' rainforests. Curtin University, Perth, Australia, pp.31-40.

Manokaran, N. (1977) Nutrients in various phases of water movement in a lowland tropical rainforest in Peninsular Malaysia. M. thesis, Univ. Malaya.

Manokaran, N. (1978) Nutrient concentration in precipitation, throughfall and stemflow in a lowland tropical rain forest in Peninsular Malaysia. Malay. Nat. J. 30: 423-432.

Manokaran, N. (1979) Stemflow, throughfall and rainfall interception in a lowland tropical rain forest in Peninsular Malaysia. Malay. For. 42: 174-201.

Manokaran, N. (1980) The nutrient contents of precipitation, throughfall and stemflow in a lowland tropical rain forest in Peninsular Malaysia. Malay. For. 43: 266-289.

Manokaran, N. (1986) Clustering in *Calamus tumidus* (Rotan manau tikus). RIC Bull. 5: 1-3.

Manokaran, N. (1988) Population dynamics of tropical forest trees. Ph.D. diss., Univ. Aberdeen.

Manokaran, N. (1998) Effect, 34 years later of selective logging in the lowland dipterocarp forest at Pasoh, Peninsular Malaysia, and implications on present-day logging in the hill forests. In Lee, S. S., Dan, Y. M., Gauld, I. D. & Bishop, J. (eds). Conservation, Management and Development of Forest Resources. Proceedings of the Malaysia-United Kingdom Programme Workshop, 21-24 October 1996, Kuala Lumpur, Malaysia, pp.41-60.

Manokaran, N. & Kochummen, K. M. (1987) Recruitment, growth and mortality of tree species in a lowland dipterocarp forest in Peninsula Malaysia. J. Trop. Ecol. **3**: 315-

330.
Manokaran, N. & Kochummen, K. M. (1990) A re-examination of data on structure and floristic composition of hill and lowland dipterocarp forest in Peninsular Malaysia. Malay. Nat. J. 44: 61-75.
Manokaran, N. & Kochummen, K. M. (1994) Tree growth in primary lowland and hill dipterocarp forests. J. Trop. For. Sci. 6: 332-345.
Manokaran, N. & LaFrankie, J. V. (1990) Stand structure of Pasoh Forest Reserve, a lowland rainforest in Peninsular Malaysia. J. Trop. For. Sci. 3: 14-24.
Manokaran, N. & Swaine, M. D. (1994) Population Dynamics of Trees in Dipterocarp Forests of Peninsular Malaysia. Malayan Forest Records No. 40. Forest Research Institute Malaysia, Kepong.
Manokaran, N., LaFrankie, J. V., Kochummen, K. M., Quah, E. S., Klahn, J., Ashton, P. S. & Hubbell, S. P. (1990) Methodology for 50 ha research plot at Pasoh Forest Reserve. FRIM Research Pamphlet No. 104, Kepong, Malaysia, 69pp.
Manokaran, N., LaFrankie, J. V. & Ismail, R. (1991) Structure and composition of the Dipterocarpaceae in a lowland rain forest in Peninsular Malaysia. Biotropica 41: 317-331.
Manokaran, N., LaFrankie, J. V. & Ismail, R. (1991) Structure and composition of Dipterocarpaceae in a lowland rain forest in peninsular Malaysia. In Tjitrisoma, S. S., Umay, R. C. & Umboh, I. M. (eds). Proceedings of the Fourth Round-Table Conference on Dipterocarps. Biotrop Special Publications 41. SEAMEO-BIOTROP, Bogor, Indonesia.
Manokaran, N., Abd. Rahman Kassim, Azman Hassan, Quah, E. S. & Chong, P. F. (1992) Short-term population dynamics of dipterocarp trees in a lowland rain forest in Peninsular Malaysia. J. Trop. For. Sci. 5: 97-112.
Manokaran, N., LaFrankie, J. V., Kochummen, K. M., Quah, E. S., Klahn, J., Ashton, P. S . & Hubbell, S. P. (1992) Stand table and distribution of species in the fifty hectare research plot at Pasoh Forest Reserve. FRIM Research Data No.1, Forest Research Institute Malaysia, Kepong, Malaysia, 454pp.
Maruyama, Y. (1997) Ecophysiological study on adaptability of dipterocarps to environmental stress. Trop. For. 40: 45-54 (in Japanese).
Maruyama, Y., Uemura, A., Shigenaga, H., Ang, L. H. & Matsumoto, Y. (1998) Photosynthesis, transpiration, stomatal conductance and leaf water potential of several tropical tree species. In Mansur Fatawi, H. S. (ed). Proceedings of the 2nd International Symposium on Asian Tropical Forest Management, PUSREHUT-UNMUL & JICA, pp.263-275.
Maschwitz, U. & Dorow, W. H. O. (1993) Nesttarnung bei tropischen Ameisen. Naturwissenschaftliche Rundschau 46: 237-239.
Maschwitz, U., Fiala, B. & Linsenmair, K. E. (1992) A new ant-tree from SE Asia: *Zanthoxylum myriacanthum* (Rutaceae), the thorny Ivy-Rue. Malay. Nat. J. 46: 101-109.
Matsumoto, T. (1976) The role of termites in an equatorial rainforest ecosystem of West Malaysia. I. Population density, biomass, carbon, nitrogen and calorific content and respiration rate. Oecologia 22: 153-178.
Matsumoto, T. (1978) Population density, biomass, nitrogen and carbon content, energy value and respiration rate of four species of termites in Pasoh Forest Reserve. Malay. Nat. J. 30: 335-351.
Matsumoto, T. (1978) The role of termites in the decomposition of leaf litter on the forest floor of Pasoh study area. Malay. Nat. J. 30: 405-413.
Matsumoto, T. & Abe, T. (1979) The role of termites in an equatorial rain forest ecosystem of West Malaysia. II. Leaf litter consumption on the forest floor. Oecologia 38: 261-274.
Matsumoto, Y., Maruyama, Y. & Ang, L. H. (2000) Maximum gas exchange rates and osmotic potential in sun leaves of tropical tree species. Tropics 9: 195-209 (in Japanese with English summary).
May, R. M. & Stumpf, M. P. H. (2000) Species-area relations in tropical forests. Science 290: 2084-2086.
Menon, K. D. (1968) International Biological Programme in Malaysia. Malay. Sci. 1: 18-22.
Menon, K. D. (1969) Malaysian National Programme for IBP: A review of highlight. Malay.

For. 32: 395-400.
Miura, S. & Yabe, T. (1998) Conservation of wildlife in the tropical rain forest. Int. Coop. Agri. For. 20: 21-35 (in Japanese).
Miura, S., Yasuda, M. & Ratnam, L. C. (1997) Who steals the fruits?: Monitoring frugivory of mammals in a tropical rain forest. Malay. Nat. J. 50: 183-193.
Miyamoto, K. (1998) Natural regeneration of dipterocarps in tropical lowland forest of Malaysia. M. thesis, Hiroshima Univ.
Moad, A. S. (1992) Dipterocarp juvenile growth and understorey light availability in Malaysian tropical forest. Ph.D. diss., Harvard Univ.
Nagamitsu, T., Ichikawa, S., Ozawa, M., Shimamura, R., Kachi, N., Tsumura, Y. & Muhammad, N. (2001) Microsatellite analysis of the breeding system and seed dispersal in *Shorea leprosula* (Dipterocarpaceae). Int. J. Plant Sci. 162: 155-159.
Nazre, M. (1999) Life history studies of *Garcinia* (Clusiaceae. species in the Pasoh 50 ha plot. Ph.D. diss., National Univ. Malaysia.
Ng, R. (1978) The vertical distribution of aerial insects in Pasoh Forest Reserve. Malay. Nat. J. 30: 299-305.
Ng, R. & Lee, S. S. (1980) Environmental factors affecting the vertical distribution of *Diptera* in a tropical primary lowland dipterocarp forest in Malaysia. In Furtado, J. I., Morgan, W. B., Pfafflin, J. R. & Ruddle, K. (eds). Tropical ecology and development: proceedings of the 5th International Symposium of Tropical Ecology, 16-21 April 1979, Kuala Lumpur, Malaysia, pp.123-129.
Ng, R. & Lee, S. S. (1982) The vertical distribution of insects in a tropical primary lowland dipterocarp forest in Malaysia. Malaysian J. Sci. 7: 37-52.
Ng, S. C. R. (1972) A preliminary survey of the distribution of acari (with the emphasis on *Cryptostigmata*) at Pasoh Forest Reserve, Malaysia. B. thesis, Univ. Malaya.
Noguchi, S., Adachi, N., Tang, Y., Osada, N. & Abudul Rahim, M. Spatial distribution of surface soil moisture in a tropical rain forest, Pasoh Forest Reserve. Proceedings of 4th Asian Science and Technology Congress 2002, 25-27th April 2002, Kuala Lumpur, Malaysia. Malaysian Scientific Association and Federation of Asian Scientific Academies and Societies, Kuala Lumpur (in press).
Numata, S. (1998) Herbivory and defenses at early stages of regeneration of dipterocarps in Malaysia. M. thesis, Tokyo Metropolitan Univ.
Numata, S. (2001) Comparative Ecology of regeneration process of dipterocarp trees in a lowland rain forest, Southeast Asia. Ph.D. diss., Tokyo Metropolitan Univ.
Numata, S., Kachi, N., Okuda, T. & Manokaran, N. (1999) Chemical defences of fruits and mast-fruiting of dipterocarps. J. Trop. Ecol. 15: 695-700.
Numata, S., Kachi, N., Okuda, T. & Manokaran, N. (2000) Leaf damage and traits of dipterocarp seedlings in a lowland rain forest in Peninsular Malaysia. Tropics 9: 237-243.
Nur Supardi, M. N. (1999) The impact of logging on the community of palms (Arecaceae) in the lowland dipterocarp forest of Pasoh, Peninsular Malaysia. Ph.D. diss., Univ. Reading.
Nur Supardi, M. N. (1999) Biodiversity and Economic Assessment of Palms in Pasoh Forest Reserve. Malaysian Science and Technology Congress '99. Kuching. 8-10 November 1999.
Nur Supardi, M. N., Shalihin, S. & Aminuddin, M. (1995) The Rattan Composition and Distribution in the Lowland Forest of Pasoh Forest Reserve. Poster presentation at the Conference on Forestry and Forest Products Research. Kepong, 3-4 October 1995, Forest Research Institute Malaysia.
Nur Supardi, M. N., Shalihin, S. & Aminuddin, M. (1995) Sampling methods for rattan inventory. In Conference on Forestry and Forest Products Research 1, pp.45-55.
Nur Supardi, M. N., Dransfield, J. & Pickersgill, B. (1998) Preliminary observation on the species diversity of Palms (rattan) in Pasoh Forest Reserve. In Lee, S. S., Dan, Y. M., Gauld, I. D. & Bishop, J. (eds). Conservation, Management and Development of Forest Resources. Proceedings of the Malaysia-United Kingdom Programme Workshop, 21-24 October 1996, Kuala Lumpur, Malaysia, pp.105-104
Nur Supardi, M. N., Dransfield, J. & Pickersgill, B. (1999) The Species Diversity of Rattans

and Other Palms in the Unlogged Lowland Forest of Pasoh Forest Reserve, Negeri Sembilan, Malaysia. In Rattan Plantation Seminar. Final Meeting on the Conservation, Genetic Improvement & Silviculture of Rattan Species in Southeast Asia (4th EEC-STD3 Steering Committee Meeting - Rattan Project) CIRAD-Forest /RBG/FRIM/ ICSB/FD (Sabah), Kuala Lumpur. 12-14 May 1998.

Obayashi, K., Tsumura, Y., Ihara-Ujino, T., Niiyama, K., Tanouchi, H., Suyama, Y., Washitani, I., Lee, C. T., Lee, S. L. & Muhammad, N. (2002) Genetic diversity and outcrossing rate between undisturbed and selectively logged forests of *Shorea curtisii* (Dipterocarpaceae) using microsatellite DNA analysis. Int. J. Plant Sci. 163: 151-158.

Ogawa, H. (1978) Litter production and carbon cycling in Pasoh Forest. Malay. Nat. J. 30: 367-373.

Ohtani, Y., Okano, M., Tani, M., Yamanoi, K., Watnabe, T., Yasuda, Y. & Abdul Rahim, N. (1997) Energy and CO_2 fluxes above a tropical rainforest in Peninsular Malaysia - under-estimation of eddy correlation fluxes during low wind speed conditions. J. Agric. Meteorol. 52: 453-456.

Okuda, T. (1993) Role of herbivore in maintaining species diversity in tropical forest. In Washitani, I. & Ohgushi, T. (eds). Plant and animal Interaction. Heibonsha, Tokyo, pp.190-206 (in Japanese).

Okuda, T. (1993) Monitoring on tropical forest. In Tagawa, H., Yasuoka, Y., Tsubaki, Y., Kachi, N. & Okuda, T. (eds). Vanishing Tropical Forest, Chuo-hoki, Tokyo, pp.87-117 (in Japanese).

Okuda, T. & Kachi, N. (1992) Does offspring away from mother tree grow well? Vegetation Sci. 8: 25-36 (in Japanese).

Okuda, T. & Manokaran, N. (1997) Forest community dynamics in lowland dipterocarp forest in Pasoh, Malaysia. J. Popul. Ecol. 54: 41-46 (in Japanese).

Okuda, T., Kachi, N., Yap, S. K. & Manokaran, N. (1994) Spatial pattern of saplings and adult trees of canopy- and sub-canopy- forming species in a lowland rain forest in Peninsular Malaysia. In Yasuno, M. & Watanabe, M. (eds). Biodiversity: Its Complexity and Role, Global Environmental Forum, Tokyo, pp.99-110.

Okuda, T., Kachi, N., Yap, S. K. & Manokaran, N. (1995) Spatial pattern of adult trees and seedling survivorship of *Pentaspadon motleyi* in a lowland rain forest in Peninsular Malaysia. J. Trop. For. Sci. 7: 475-489.

Okuda, T., Kachi, N. & Manokaran, N. (1997) Forest dynamics in lowland dipterocarp forest in Malaysia. J. Popul. Ecol. 54: 41-46 (in Japanese).

Okuda, T., Kachi, N., Yap, S. K. & Manokaran, N. (1997) Tree distribution pattern and fate of juveniles in a lowland tropical rain forest - implications for regeneration and maintenance of species diversity. Plant Ecol. 131: 155-171.

Okuda, T., Kachi, N. & Manokaran, N. (1998) Equilibrium status of species diversity and composition in lowland dipterocarp forest in Peninsular Malaysia. Global Environ. Res. 3: 81-91 (in Japanese).

Okuda, T., Kachi, N., Yap, S. K. & Manokaran, N. (1998) The distribution pattern of juvenile and adult trees in a Malaysian lowland tropical rain forest in Peninsular Malaysia. CTFS News: 12-13.

Okuda, T., Yoshida, K. & Adachi, N. (2002) Studies on ecological services of tropical forest—Laying the groundwork for ecological studies and integrated ecosystem management. Tropics 11: 193-204 (in Japanese with English summary).

Okuda, T., Suzuki, M., Adachi, N., Quah, E. S., Hussein, N. A. & Manokaran, N. (2003) Effect of selective logging on canopy and stand structure in a lowland dipterocarp forest in Peninsular Malaysia. For. Ecol. Manage. 175: 297-320.

Okuda, T., Nor Azman, H., Manokaran, N., Saw, L. Q., Amir, H. M. S. & Ashton, P. S. (in press) Local variation of canopy structure in relation to soils and topography and the implications for species diversity in a rain forest of Peninsular Malaysia. In Losos, E. & Leigh, E. G Jr. (eds). Forest diversity and dynamism: Findings from a network of large-scale tropical forest plots. University of Chicago Press, Chicago.

Osada, N. (2001) Leaf dynamics and maintenance of tree crowns in a Malaysian rain forest.

Ph.D. diss., Kyoto Univ.
Osada, N. & Takeda, H. (in press) Branch architecture, light interception, and crown development in saplings of a plagiotropically branching tropical tree, *Polyalthia jenkinsii* (Annonaceae). Ann. Bot.
Osada, N., Takeda, H., Awang, M. & Furukawa, A. (1998) Leaf dynamics of several tree species in a lowland dipterocarp forest in Peninsular Malaysia. Sci. Int. 10: 177-179.
Osada, N., Takeda, H., Awang, M. & Furukawa, A. (2000) Crown dynamics of several species in a tropical rain forest. In Appanah, S., Yusoff, M., Jasery, A. W. & Choon, K. K. (eds.) Conference on Forestry and Forest Product Research 1997. Forest Reserch Institute Malaysia , Kuala Lumpur, Malaysia, pp.262-270.
Osada, N., Takeda, H., Furukawa, A. & Awang, M. (2001) Leaf dynamics and maintenance of tree crowns in a Malaysian rain forest stand. J. Ecol. 89: 774-782.
Osada, N., Takeda, H., Furukawa, A. & Awang, M. (2001) Fruit dispersal of two dipterocarp species in a Malaysian rain forest. J. Trop. Ecol. 17: 911-917.
Osada, N., Takeda, H., Furukawa, A. & Awang, M. (2002) Ontogenetic changes in leaf phenology of a canopy species, *Elateriospermum tapos* (Euphorbiaceae), in a Malaysian rain forest. J. Trop. Ecol. 18: 91-105.
Osada, N., Takeda, H., Furukawa, A. & Awang, M. (2002) Changes in shoot allometry with increasing tree height in a tropical canopy species, *Elateriospermum tapos*. Tree Physiol. 22: 625-632.
Osada, N., Takeda, H., Kawaguchi, H. Furukawa, A. & Awang, M. (in press) Estimation of crown characters and leaf biomass from leaf litter in a Malaysian canopy species, *Elateriospermum tapos* (Euphorbiaceae). For. Ecol. Manage.
Osawa, N. & Kirton, L. G. (1997) A new host record of the common posy, *Drupadia ravindra moorei* (Lepidoptera, Lycaenidae), in Malaysia. Jpn. J. Entomol. 65: 853-854.
Peh, C. H. (1976) Rates of sediment transport by surface wash in three forested areas of Peninsular Malaysia. M. thesis, Univ. Malaya.
Peh, C. H. (1978) Rates of sediment transport by surfacewash in three forested areas in Peninsular Malaysia. Occasional Paper No. 3. Department of Geography, University of Malaya, Kuala Lumpur.
Plotkin, J., Potts, M., Leslie, N., Manokaran, N., LaFrankie, J. V. & Ashton, P. S. (2000) Species-area curves, spatial aggregation, and habitat specialization in tropical forests. J. Theor. Biol. 207: 81-99.
Plotkin, J. B., Potts, M. D., Yu, D. W., Bunyavejchewin, S., Condit, R., Foster, R., Hubbell, S., LaFrankie, J., Manokaran, N., Lee, H. S., Sukumar, R., Nowak, M. A. & Ashton, P. S. (2000) Predicting species diversity in tropical forests. Proc. Natl. Acad. Sci. USA 97: 10850-10854.
Plotkin, J. B., Chave, J. & Ashton, P. S. (2002) Cluster analysis of spatial patterns in Malaysian tree species. Am. Nat. 160: 629-644.
Potts, M. D., Plotkin, J. B., Lee, H. S., Manokaran, N., Ashton, P. S. & Bossert, W. H. (2001) Sampling biodiversity: effects of plot shape. Malay. For. 64: 29-34.
Putz, F. E. & Appanah, S. (1987) Buried seeds, newly dispersed seeds and the dynamics of a lowland dipterocarp forest in Malaysia. Biotropica 19: 326-333.
Richards, P. W. (1978) Pasoh in perspective. Malay. Nat. J. 30: 145-148.
Rickson, F. R. & Rickson, M. M. (1986) Nutrient acquisition facilitated by litter collection and ant colonies on two Malaysian palms. Biotropica 18: 337-343.
Rogstad, S. H. (1990) The biosystematics and evolution of the *Polyalthia hypoleuca* species complex (Annonaceae) of Malesia. II. Comparative distributional ecology. J. Trop. Ecol. 6: 387-408.
Rosciszewski, K. (1995) Die Ameisenfauna eines tropischen Tieflandregenwaldes in Südostasien: Eine faunistisch-ökologische Bestandsaufnahme. Ph.D. diss., J. W. Goethe Univ., Germany, 184pp.
Rutter, G. & Watling, R. (1997) Taxonomic and floristic notes on some larger Malaysian fungi II. Malay. Nat. J. 50: 229-234.
Saifuddin, S., Abdul Rahim, N. & LaFrankie, J. V. (1994) Pasoh climatic summary (1991-1993).

FRIM Research Data 3: 31.

Saito, S. (1976) Studies on the productivity of soil animals in Pasoh Forest Reserve, West Malaysia. IV. Growth respiration and food consumption of some cockroaches. Jpn. J. Entomol. 26: 37-42.

Sajap, A. S., Maeto, K., Fukuyama, K., Fauan, B. H. A. & Yaacob, A. W. (1997) Chrysopidae attraction to floral fragrance chemicals and its vertical distribution in a Malaysian lowland tropical forest. Malay. Appl. Biol. 26: 75-80.

Salleh, M. N. (1968) Forest resources reconnaissance survey of Malaya-Jelebu district, Negeri Sembilan. FRIM - Report No. 21, Kepong, Malaysia.

Sato, T., Itoh, H., Kudo, G., Yap, S. K. & Furukawa, A. (1996) Species composition and structure of epiphytic fern community on oil palm trunks in Malay Archipelago. Tropics 6: 139-148.

Sato, T., Guan, S. L. & Furukawa, A. (2000) A quantitative comparison of pteridophytes diversity in small scales among different climatic regions in Eastern Asia. Tropics 9: 83-90.

Saw, L. G (1995) Taxonomy and ecology of Licuala in Peninsular Malaysia. Ph.D. diss., Univ. Reading.

Saw, L. G & Chan, H. T. (1995) The diversity of plant life. In Lee, S. S. (ed). A Guide Book to Pasoh. FRIM Technical Information Handbook No. 3, pp.7-33.

Saw, L. G, LaFrankie, J. V., Kochummen, K. M. & Yap, S. K. (1991) Fruit trees in a Malaysian rainforest. Econ. Bot. 45: 120-136.

Sham, S. (1983) Microclimate aspects of a lowland dipterocarp forest in Pasoh, Negeri Sembilan, Malaysia. Report, Malaysian MAB Committee, UNESCO. Jakarta.

Soepadmo, E. (1973) Progress report (1970-1972) on IBP-PT Project at Pasoh, Negeri Sembilan, Malaysia. In Mori, S. & Kira, T. (eds). Proceeding of the East Asian Regional Seminar for the International Biological Programme, 30 January-1 February 1973, Kyoto, Japanese National Committee for IBP, Kyoto, pp.29-39.

Soepadmo, E. (1973) IBP-PT Pasoh Project, Negeri Sembilan, Malaysia. Annual Report for 1972.

Soepadmo, E. (1978) Introduction to the Malaysian IBP Synthesis Meeting. Malay. Nat. J. 30: 119-124.

Soepadmo, E. (1979) Genetic resources of Malaysian fruit trees. Malay. Appl. Biol. 8: 33-42.

Soepadmo, E. (1989) Conbtribution of reproductive biological studies towards the conservation and development of Malaysian plant genetic resources. In Zakri, A. H. (ed). Genetic Resources of Underutilized Plants in Malaysia, MNCPGR Kuala Lumpur, pp.1-41.

Soepadmo, E. & Eow, B. K. (1976) The reproductive biology of *Durio zibethinus* Murr. The Garden's Bulletin of Singapore 29: 25-33.

Soepadmo, E. & Kira, T. (1977) Contribution of the IBP-PT research project to the understanding of Malaysian forest ecology. In Sastry, C. B., Srivastava, P. B. L. & Abd. Manap Ahmad (eds). A New Era in Malaysian Forestry, University Pertanian Malaysia, Serdang, pp.63-94.

Styring, A. R. & Ickes, K. (2001) Interactions among Woodpeckers and the Greater Racket-tailed Drongo in a lowland Malaysian rainforest. Forktail 17: 119-120.

Styring, A. R. & Ickes, K. (2001) Woodpecker abundance in a logged (40 years ago) vs. unlogged lowland dipterocarp forest in Peninsular Malaysia. J. Trop. Ecol. 17: 261-268.

Styring, A. R. & Ickes, K. (in press) Mixed-species woodpecker flocks in Malaysia. Wilson Bulletin.

Takamura, K. (2001) Effects of termite exclusion on decay of heavy and light hardwood in a tropical rainforest of Peninsular Malaysia. J. Trop. Ecol. 17: 541-548.

Takamura, K. & Kirton, L. G (1999) Effects of termite exclusion on decay of a high-density wood in tropical rain forests of Peninsular Malaysia. Pedobiologia 43: 289-296.

Takeda, H. (1996) Templates for the organization of soil animal communities in tropical forests. In Turner, I. M., Diong, C. H., Lim, S. S. L. & Ng, P. K. L. (eds). Biodiversity and the Dynamics of Eecosystems. DIWPA Series vol. 1, pp. 217-226.

Tang, Y. (1999) Heterogeneity of light availability and its effects on simulated carbon gain of tree

leaves in a small gap and the understory in a tropical rain forest. Biotropica 31: 268-278.
Tang, Y. & Kachi, N. (1997) A measuring system for characterizing spatial and temporal variation of photon flux density within plant canopies. For. Ecol. Manage. 97: 174-201.
Tang, Y., Kachi, N., Furukawa, A. & Awang, M. (1996) Light reduction by regional haze and its effect on simulated leaf photosynthesis in a tropical forest of Malaysia. For. Ecol. Manage. 89: 205-211.
Tani, M. & Abdul Rahim, N. (1995) Characteristics of micrometeorology monitored in a tropical rainforest of Peninsular Malaysia. In Proceeding the Second International Study Conference on GEWEX in Asia and GAME, pp.343-346.
Tani, M., Abdul Rahim, N. & Ohtani, Y. (1996) Estimating energy budget above a tropical rain forest in Peninsular Malaysia using Bowen method. In Proceeding of IGBP/BAHC-LUCC Joint Inter-Core Projects Symposium, pp.127-130.
Taylor, C. (1982) Reproductive biology and Ecology of some tropical pioneer trees. Ph.D. diss., Univ. Aberdeen.
Tho, Y. P. (1982) Gap formation by the termite *Microtermes duinus* in lowland forests of Peninsular Malaysia. Malay. For. 45: 184-192.
Tho, Y. P. & Maschwitz, U. (1988) The use of prefabricated plugs for emergency entrance sealing. Naturwissenschaften 75: 527-528.
Thomas, S. C. (1993) Interspecific allometry in Malaysian rain forest trees. Ph.D. diss., Harvard Univ.
Thomas, S. C. (1993) Sex, size and inter year variation in flowering among dioecious trees of the Malayan rain forests. Ecology 74: 1529-1537.
Thomas, S. C. (1995) Ontogenetic changes in leaf size in the Malaysian rain forest trees. Biotropica 27: 427-434.
Thomas, S. C. (1996) Asymptotic height as a predictor of growth and allometric characteristics in Malaysian rain forest trees. Am. J. Bot. 83: 556-566.
Thomas, S. C. (1996) Relative size at the onset of maturity in rain forest trees: a comparative analysis of 37 Malaysian species. Oikos 76: 145-154.
Thomas, S. C. (1996) Reproductive allometry in Malaysian rain forest trees: biomechanics vs. optimal allocation. Evol. Ecol. 10: 517-530.
Thomas, S. C. (1997) Geographic parthenogenesis in a tropical forest tree. Am. J. Bot. 84: 1012-1015.
Thomas, S. C. (1999) Asymptotic height as a predictor of photosynthetic characteristics in Malaysian rain forest trees. Ecology 80: 1607-1622.
Thomas, S. C. (2001) Structure and species diversity of forest fragments near Pasoh Forest Reserve, West Malaysia. Inside CTFS Summer 2001: 9, 16.
Thomas, S. C. (in press) Ecological correlates of tree species persistence in tropical forest fragments. In Losos, E. C. & Leigh, E. G. Jr. (eds). Forest diversity and dynamism: Findings from a network of large-scale tropical forest plots. University of Chicago Press, Chicago.
Thomas, S. C. & LaFrankie, J. V. (1993) Sex, size, and inter-year variation in flowering among dioecious trees of the rain forest understorey. Ecology 74: 1529-1537.
Thomas, S. C. & Appanah, S. (1995) On the statistical analysis of reproductive size thresholds in dipterocarp forest. J. Trop. For. Sci. 7: 412-418.
Thomas, S. C. & Ickes, K. (1995) Ontogenetic changes in leaf size in Malaysian rain forest trees. Biotropica 27: 427-434.
Thomas, S. C. & Bazzaz, F. A. (1999) Asymptotic height as a predictor of photosynthetic characteristics in Malaysian rain forest trees. Ecology 80: 1607-1622.
Thomas, S. C. & Winner, W. E. (in press) Photosynthetic differences between saplings and adult trees: An integration of field results via meta-analysis. Tree Physiol. 22.
Toy, R. (1991) Interspecific flowering patterns in the Dipterocarpaceae in West Malesia: Implications for predator satiation. J. Trop. Ecol. 7: 49-57.
Toy, R. J. & Toy, S. J. (1992) Oviposition preferences and egg survival in *Nanophyes shoreae* (Coleoptera, Apionidae), a weevil fruit-predator in South-east Asian rain forest. J. Trop. Ecol. 8: 195-203.

Toy, R. J., Marshall, A. G. & Tho, Y. P. (1992) Fruiting phenology and the survival of insect fruit predators: a case study from the South-east Asian Dipterocarpaceae. Philos. Trans. R. Soc. London Ser. B 335: 417-423.

Tsubaki, Y. & Intachat, J. 1994. Dung beetle community structure in a Malaysian Rain Forest: Gradient from edge to core area. In Yauno, M. & Watanabe, M. M. (eds). Biodiversity: Its Complexity and Role. Global Environmental Forum, Tokyo. pp.139-145.

Tsumura, Y., Kawahara, T., Wickneswari, R. & Yoshimura, K. (1996) Molecular phylogeny of Dipterocarpaceae in Southeast Asia using PCR-RFLP analysis-of chloroplast genes. Theor. Appl. Genet. 93: 22-29.

Wan Ahmad, W. M. S. (1997) Verification and validation of sampling intensities in the pre-felling inventory of a tropical forest in Peninsular Malaysia. M. thesis, Univ. Putra.

Wan Ahmad, W. M. S., Wan Mohd, W. R. & Muktar. A. (1997) Natural forest dynamics. Homogeneity of species distribution. J. Trop. For. Sci. 10: 1-9.

Wan Mohd, W. R. & Wan Ahmad, W. M. S. (1997) Natural forest dynamics. II. Sampling of tree volume using quadrats in tropical forests of Peninsular Malaysia. J. Trop. For. Sci. 10: 141-154.

Wan Mohd, W. R. & Wan Ahmad, W. M. S. (1999) An evaluation of statistical reliability in SMS's pre-felling inventory: The case for confidence and error-levels. J. Trop. For. Sci. 11: 11-25.

Watling, R. & Lee, S. S. (1995) Ectomycorrhizal fungi associated with members of the Dipterocarpaceae in Peninsular Malaysia- I. J. Trop. For. Sci. 7: 657-669.

Watling, R. & Lee, S. S. (1998) Ectomycorrhizal fungi associated with members of the Dipterocarpaceae in Peninsular Malaysia-II. J. Trop. For. Sci. 10: 421-430.

Watling, R., Lee, S. S. & Turnbull, E. (1996) Putative ectomycorrhizal fungi of Pasoh Forest Reserve, Negeri Sembilan, Malaysia. In Lee, S. S., Dan, Y. M., Gauld, I. D. & Bishop, J. (eds). Conservation, Management and Development of Forest Resources. Proceedings of the Malaysia-United Kingdom Programme Workshop, 21-24 October 1996, Kuala Lumpur, Malaysia, pp.96-104.

Watling, R., Lee, S. S. & Turnbull, E. (2002) The occurrence and distribution of putative ectomycorrhizal basidiomycetes in a regenerating South-east Asian Rainforest. In Watling, R., Frankland, J. C., Ainsworth, A. M., Isaac, S. & Robinson C. H. (eds). Tropical Mycology vol. 1, Macromycetes. CABI Publishing: Wallingford, Oxon, pp.25-43.

Wells, D. R. (1978) Number and biomass of insectivorous birds in the understorey of rainforest at Pasoh Forest. Malay. Nat. J. 30: 353-362.

Wickneswari, R., Lee, S. L., Mahani, M. C., Lim, A. L., Fatimah, M., Harris, S. A., Jong, K. & Zakri, A. H. (1998) Genetics and reproductive biology of dipterocarps. In Lee, S. S., Dan, Y. M., Gauld, I. D. & Bishop, J. (eds). Conservation, Management and Development of Forest Resources. Proceedings of the Malaysia-United Kingdom Programme Workshop, 21-24 October 1996, Kuala Lumpur, Malaysia, pp.160-168.

Wills, C. (in press) Comparable non-random forces act to maintain diversity in both a New and an Old World rainforest plot. In Losos, E. C. & Leigh, E. G. Jr. (eds). Forest diversity and dynamism: Findings from a network of large-scale tropical forest plots. University of Chicago Press, Chicago.

Wills, C. & Condit, R. C. (1999) Similar non-random processes maintain diversity in two tropical rainforests. Proc. R. Soc. London B. 266: 1445-1452.

Wolsely, P., Ellis, L., Harrington A. & Moncrieff, C. (1998) Epyphytic Criptogams at Pasoh Forest Reserve, Negri Sembilan, Malaysia -Quantitative and qualitative sampling of epiphytic cryptograms at Pasoh Forest reserve- In Lee, S. S., Dan, Y. M., Gauld, I. D. & Bishop, J. (eds). Conservation, Management and Development of Forest Resources. Proceedings of the Malaysia-United Kingdom Programme Workshop, 21-24 October 1996, Kuala Lumpur, Malaysia, pp.61-83.

Wong, M. (1981) Impact of dipterocarp seedlings on the vegetative and reproductive characteristics of *Labisia pumila* (Myrsinaceae) in the understorey. Malay. For. 44: 370-376.

Wong, M. (1983) Understory phenology of the virgin and regenerating habitats in Pasoh Forest Reserve, Negeri Sembilan, West Malaysia. Malay. For. 46: 197-223.

Wong, M. (1985) Understory birds as indicators of regeneration in a patch of selectively logged Malaysian rainforest. ICBP Techn. Publ. 4: 249-263.
Wong, M. (1986)Tropic organization of understorey birds in a Malaysian dipterocarp forest. Auk 103: 100-116.
Wong, Y. K. & Whitmore, T. C. (1970) On the influence of soil properties on species distribution in a Malayan lowland dipterocarp rainforest. Malay. For. 33: 42-54.
Wyatt-Smith, J. (1964) A preliminary vegetation map of Malaya with descriptions of the vegetation types. J. Trop. Geogr. 18: 200-213.
Wyatt-Smith, J. (1987) Manual of Malayan silviculture for inland forest, Part 3 - Chapter 8. Red meranti - keruing forest. FRIM Research Pamphlet No. 101, Kepong Malaysia, 89pp.
Yabuki, K. & Aoki, M. (1978) Micrometeorological assessment of primary production rate of Pasoh Forest. Malay. Nat. J. 30: 281-289.
Yamada, T., Okuda, T., Abdullah Makmom, Awang Muhamad & Furukawa, A. (2000) The leaf development process and its significance for reducing self-shading of a tropical pioneer tree species. Oecologia 125: 476-482.
Yamashita, T. & Takeda, H. (1998) Decomposition and nutrient dynamics of leaf litter in litter bags of two mesh sizes set in two dipterocarp forest sites in Peninsular Malaysia. Pedobiologia 42: 11-21.
Yap, S. K. (1976) The reproductive biology of some understorey fruit tree species in the lowland dipterocarp forest of West Malaysia. Ph.D. diss., Univ. Malaya.
Yap, S. K. (1980) Phenological behaviour of some fruit tree species in a lowland dipterocarp forest of West Malaysia. In Furtado, J. I., Morgan, W. B., Pfafflin, J. R. & Ruddle, K. (eds). Tropical ecology and development: Proceedings of the 5th International Symposium of Tropical Ecology, 16-21 April 1979, Kuala Lumpur, Malaysia, pp.161-167.
Yap, S. K. (1981) Collection, germination and storage of dipterocarp seeds. Malay. For. 44: 281-300.
Yap, S. K. (1982) The phenology of some fruit tree species in a lowland dipterocarp forest. Malay. For. 45: 21-35.
Yap, S. K. (1987) Gregarious flowering of dipterocarps: Observation based on fixed tree populations in Selangor and Negeri Sembilan, Malay Peninsula. In Kostermans, A. J. G. H. (ed). Proceeding of the Third International Roundtable Conference on Dipterocarps, UNESCO, Jakarta, pp.305-317.
Yasuda, M. (1998) Community Ecology of small mammals in a tropical rain forest of Malaysia with special reference to habitat preference, frugivory and population dynamics. Ph.D. diss., Tokyo Univ.
Yasuda, M., Matsumoto, J., Osada, N., Ichikawa, S., Kachi, N., Tani, M., Okuda, T., Furukawa, A., Abdul Rahim, K. & Manokaran, N. (1999) The mechanism of general flowering in Dipterocarpaceae in the Malay Peninsula. J. Trop. Ecol. 15: 437-449.
Yasuda, M., Miura, S. & Nor Azman, H. (2001) Evidence for food hoarding behavior in terrestrial rodents in Pasoh Forest Reserve, a Malaysian lowland rain forest. J. Trop. For. Sci. 12: 164-173.
Yasuda, Y., Ohtani, Y., Watanabe, T., Okano, M., Yokota, T., Naishen Liang, Tang, Y., Abudl Rahim, N., Tani, M. & Okuda, T. (2003) Measurement of CO_2 flux above a tropical rain forest at Pasoh in penisular Malaysia. Agric. For. Meteorol. 114: 235-244.
Yoda, K. (1974) Three dimensional distribution of light intensity in tropical rainforest of West Malaysia. Jpn. J. Ecol. 24: 247-254.
Yoda, K. (1978) Three dimensional distribution of light intensity in a Malaysian tropical forest. Malay. Nat. J. 30: 161-177.
Yoda, K. (1978) Organic carbon, nitrogen and mineral nutrients stock in soils of Pasoh Forest. Malay. Nat. J. 30: 229-251.
Yoda, K. (1978) Respiration studies in Pasoh Forest Plants. Malay. Nat. J. 30: 269-279.
Yoda, K. (1983) Community respiration in a lowland rainforest at Pasoh, Peninsular Malaysia. Jpn. J. Ecol. 33: 183-197.
Yoda, K. & Kira, T. (1982) Accumulation of organic matter, carbon, nitrogen and other nutrient

32: 275-291.

Yoneda, T. (1995) Dynamics of aboveground big woody organs in tropical rain forest ecosystems. In Box E. O., Peet, R. K., Masuzawa, T., Yamada, I., Fujiwara, K. & Maycock, P. F. (eds). Vegetation science in Forestry: Global Perspective Based on Forest Ecosystems of East and Southeast Asia: Papers from Four Symposia from the international, Kluwer Academic Publishers, Dordrecht, pp.545-556.

Yoneda, T., Yoda, K. & Kira, T. (1977) Accumulation and decomposition of big wood litter of *Shorea*-dipterocarpus type in Pasoh Forest, West Malaysia. Jpn. J. Ecol. 27: 53-60.

Yoneda, T., Yoda, K. & Kira, T. (1978) Accumulation and decomposition of wood litter in Pasoh Forest. Malay. Nat. J. 30: 381-389.

Contributors

Adachi, Naoki
National Institute for Environmental Studies (NIES)
Present address: Naka-Ochiai 1-6-13-503, Shinjuku-ku, Tokyo 161-0032, Japan

Ang, Khoon Cheng
Forest Research Institute Malaysia (FRIM)
Kepong, 52109 Kuala Lumpur, Malaysia

Ang, Lai Hoe
Forest Research Institute Malaysia (FRIM)
Kepong, 52109 Kuala Lumpur, Malaysia

Appanah, Simmithiri
Forest Research Institute Malaysia (FRIM)
Kepong, 52109 Kuala Lumpur, Malaysia

Ashton, Peter S.
Organismic and Evolutionary Biology
Harvard University
Cambridge, Massachusetts 02138, USA

Awang, Muhamad
Department of Environmental Science
Faculty of Science and Environmental Studies
Universiti Putra Malaysia (UPM)
43400 UPM Serdang, Selanger, Malaysia

Blanc, Lilian
University Claude Bernard
Lyon I, 43 Bd du 11 novembre 1918, 69622 Villeurbanne Cedex, France

Davies, Stuart J.
Center for Tropical Forest Science - Arnold Arborteum Asia Program
Harvard University
22 Divinity Avenue, Cambridge, MA 02138 USA

Elouard, Claire
Laboratoire de Biologie Evolutive et Biometrie
University C. Bernard Lyon I
Villeurbanne, France.

Fiala, Brigitte
Animal Ecology and Tropical Biology (Zoologie III), Biozentrum
University of Würzburg Am Hubland
97074 Wüzburg, Germany

Francis, Charles M.
National Wildlife Research Centre
Canadian Wildlife Service
Ottawa, Ontario, K1A OH3 Canada

Fukuyama, Kenji
Forestry and Forest Products Research Institute (FFPRI)
Matsunosato 1, Tsukuba, Ibaraki 305-8687, Japan

Furukawa, Akio
Kyousei Science Center for Life and Nature
Nara Women's University
Kitauoya-Nishimachi, Nara, 630-8506, Japan

Harayama, Hisanori
Forestry and Forest Products Research Institute (FFPRI)
Matsunosato 1, Tsukuba, Ibaraki 305-8687, Japan

Hashim, Mazlan
Universiti Teknologi Malaysia (UTM)
81310 UTM, Skudai, Johor, Malaysia

Hattori, Tsutomu
Forestry and Forest Products Research Institute (FFPRI)
Matsunosato 1, Tsukuba, Ibaraki 305-8687, Japan

Holloway, Jeremy Daniel
Department of Entomology
The Natural History Museum
Cromwell Road, London SW7 5BD, UK

Hussein, Nor Azman
Forest Research Institute Malaysia (FRIM)
Kepong, 52109 Kuala Lumpur, Malaysia

Ickes, Kalan
Department of Biological Sciences
Louisiana State University
Present address : 1274 overlook Dr. washington, PA 15301, USA

Iida, Shigeo
Forestry and Forest Products Research Institute (FFPRI)
Matsunosato 1, Tsukuba, Ibaraki 305-8687, Japan

Intachat, Jurie
Forest Research Institute Malaysia (FRIM)
Present address : Owley Farm, South Brent, Devon TQ10 9HN, UK

Ishida, Atsushi
Forestry and Forest Products Research Institute (FFPRI)
Matsunosato 1, Tsukuba, Ibaraki 305-8687, Japan

Ishii, Nobuo
Japan Wildlife Research Center (JWRC)
Shitaya 3-10-10, Taito-ku, Tokyo 110-8676, Japan

Kachi, Naoki
Tokyo Metropolitan University
Hachiouji, Tokyo 192-0397, Japan

Kajita, Tadashi
Botanical Gardens
Graduate School of Science, University of Tokyo
Hakusan 3-7-1, Bunkyo, Tokyo 112-0001, Japan

Kanzaki, Mamoru
Graduate School of Agriculture
Kyoto University
Kitashirakawa, Sakyo-Ku, Kyoto 606-8502, Japan

Kassim, Abdul Rahman
Forest Research Institute Malaysia (FRIM)
Kepong, 52109 Kuala Lumpur, Malaysia

Kasran, Baharuddin
Forest Research Institute Malaysia (FRIM)
Kepong, 52109 Kuala Lumpur, Malaysia

Kasuya, Nobuhiko
Faculty of Agriculture
Kyoto Prefectural University
Shimogamo, Sakyo, Kyoto 606-8522, Japan

Kawarasaki, Satoko
Ibaraki University,
Bunkyo 2-1-1, Mito 310-8512, Japan

Kimura, Katsuhiko
Department of Biology, Faculty of Education
Fukushima University
Kanayagawa 1, Fukushima 960-1296, Japan

Konuma, Akihiro
Graduate School of Science and Technology
Niigata University
Niigata 950-2101, Japan

LaFrankie, James V.
Center for Tropical Forest Science-Arnold Arboretum Asia Program
National Institute for Education
Jurong, Singapore.

Lee, Su See
Forest Research Institute Malaysia (FRIM)
Kepong, 52109 Kuala Lumpur, Malaysia

Lim, Boo Liat
Department of Wildlife and National Parks (DWNP)
KM 10, Jalan Cheras, 50664 Kuala Lumpur, Malaysia

Maeto, Kaoru
Faculty of Agriculture
Kobe University
Rokkodai 1, Nada-ku, Kobe, Hyogo 657-8501, Japan

Malsch, Annette K. F.
Faculty of Public Health
University of Bielefeld
Postfach 100131, 33501 Bielefeld, Germany

Md. Noor, Nur Supardi
Forest Research Institute Malaysia (FRIM)
Kepong, 52109 Kuala Lumpur, Malaysia

Manokaran, N.
Forest Research Institute Malaysia (FRIM)
Present address: No.4, Jalan SS 22/29, Damansara Jaya, 47400 Petaling Jaya Selangor, Malaysia

Maruyama, Yutaka
Hokkaido Research Center
Forestry and Forest Products Research Institute (FFPRI)
Hitsujigaoka 7, Toyohira, Sapporo, Hokkaido 062-8516, Japan

Mansor, Marzalina
Forest Research Institute Malaysia (FRIM)
Kepong, 52109 Kuala Lumpur, Malaysia

Maschwitz, Ulrich
Department of Zoology, Ethoecology
Johann Wolfgang Goethe-University of Frankfurt
60054 Frankfurt, Germany

Masumoto, Toshiya
Graduate School of Agriculture and Life Sciences,
University of Tokyo
Present address: Otani-cho D-6, 6, Otsu, Shiga 520-0062, Japan

Matsumoto, Yoosuke
Forestry and Forest Products Research Institute (FFPRI)
Matsunosato 1, Tsukuba, Ibaraki 305-8687, Japan

Md. Sahat, Mohd
Forest Research Institute Malaysia (FRIM)
Kepong, 52109 Kuala Lumpur, Malaysia

Moog, Joachim
Department of Zoology
Johann Wolfgang Goethe-University
Siesmayerstr. 70, 60054 Frankfurt am Main, Germany

Nadarajan, Jayanthi
Forest Research Institute Malaysia (FRIM)
Kepong, 52109 Kuala Lumpur, Malaysia

Nagamitsu, Teruyoshi
Hokkaido Research Center
Forestry and Forest Products Research Institute (FFPRI)
Hitsujigaoka 7, Toyohira, Sapporo, Hokkaido 062-8516, Japan

Nakano, Takashi
Yamanashi Institute of Environmental Science
Fuji-Yoshida, Yamanashi 403-0005, Japan

Niiyama, Kaoru
Forestry and Forest Products Research Institute (FFPRI)
Matsunosato 1, Tsukuba, Ibaraki 305-8687, Japan

Nik, Abdul Rahim
Forest Research Institute Malaysia (FRIM)
Kepong, 52109 Kuala Lumpur, Malaysia

Noguchi, Shoji
Japan International Research Center for Agricultural Sciences (JIRCAS)
Ohwashi 1-1, Tsukuba, Ibaraki 305-8686, Japan

Numata, Shinya
National Institute for Environmental Studies (NIES)
Onogawa 16-2, Tsukuba, Ibaraki 305-8506, Japan

Obayashi, Kyoko
Oficina de JICA
en R.D.A.P.1163, JICA,Av.Sarasota No.20 Torre Empresarial AIRD, 7mo piso,
La Julia Santo Domingo, Republica Dominicana

Ohtani, Yoshikazu
Forestry and Forest Products Research Institute (FFPRI)
Matsunosato 1, Tsukuba, Ibaraki 305-8687, Japan

Okauchi, Yuka
Graduate School of Science
Osaka City University
Sumiyoshi, Osaka 558-8585, Japan

Okuda, Shiro
Shikoku Research Center
Forestry and Forest Products Research Institute (FFPRI)
Asakuranishimachi 2-915, Kochi 780-8077, Japan

Okuda, Toshinori
National Institute for Environmental Studies (NIES)
Onogawa 16-2, Tsukuba, Ibaraki 305-8506, Japan

Osada, Noriyuki
Nikko Botanical Garden
Graduate School of Science, The University of Tokyo
Hanaishi 1842, Nikko, Tochigi 321-1435, Japan

Osawa, Naoya
Graduate School of Agriculture
Kyoto University
Kitashirakawa, Sakyo-Ku, Kyoto 606-8502, Japan

Quah, Eng Seng
Forest Research Institute Malaysia (FRIM)
Kepong, 52109 Kuala Lumpur, Malaysia

Ratnam, Louis
Forest Research Institute Malaysia (FRIM)
Present address: 91 Taman Hijau, Jalan Reko 43000 Kajang, Selangor, Malaysia

Ripin, Azizi
Forest Research Institute Malaysia (FRIM)
Kepong, 52109 Kuala Lumpur, Malaysia

Rosciszewski, Krzysztof
Department of Zoology, Ethoecology
Johann Wolfgang Goethe-University of Frankfurt
60054 Frankfurt, Germany

Sajap, Ahmad S.
Universiti Putra Malaysia (UPM)
Serdang Selangor Darul Ehsan, 43400, Malaysia

Saw, Leng Guan
Forest Research Institute Malaysia (FRIM)
Kepong, 52109 Kuala Lumpur, Malaysia

Shigenaga, Hidetoshi
Forestry and Forest Products Research Institute (FFPRI)
Matsunosato 1, Tsukuba, Ibaraki 305-8687, Japan

Shimizu, Michiru
Graduate School of Agricultural and Life Science
University of Tokyo
Tokyo 113-8657, Japan

Styring, Alison R.
Department of Biological Science, Museum of Natural Science
Louisiana State University
LSU-MNS, 119 Foster Hall, Baton Rouge, LA 70803, USA

Suzuki, Mariko
National Institute for Environmental Studies (NIES)
Onogawa 16-2, Tsukuba, Ibaraki 305-8506, Japan

Takamura, Kenzi
National Institute for Environmental Studies (NIES)
Onogawa 16-2, Tsukuba, Ibaraki 305-8506, Japan

Takanashi, Satoru
Graduate School of Agriculture
Kyoto University
Kitashirakawa, Sakyo-Ku, Kyoto 606-8502, Japan

Takeda, Hiroshi
Graduate School of Agriculture
Kyoto University
Kitashirakawa, Sakyo-Ku, Kyoto 606-8502, Japan

Tang, Yanhong
National Institute for Environmental Studies (NIES)
Onogawa 16-2, Tsukuba, Ibaraki 305-8506, Japan

Tani, Makoto
Graduate School of Agriculture
Kyoto University
Kitashirakawa, Sakyo-Ku, Kyoto 606-8502, Japan

Thomas, Sean C.
Faculty of Forestry
University of Toronto
33 Willcocks St. Toronto, Ontario, M5S 3B3, Canada

Tsubaki, Yoshitaka
National Institute for Environmental Studies (NIES)
Onogawa 16-2, Tsukuba, Ibaraki 305-8506, Japan

Tsumura, Yoshihiko
Forestry and Forest Products Research Institute (FFPRI)
Matsunosato 1, Tsukuba, Ibaraki 305-8687, Japan

Turnbull, Evelyn
Royal Botanic Garden
Edinburgh EH3 5LR, UK

Uemura, Akira
Forestry and Forest Products Research Institute (FFPRI)
Matsunosato 1, Tsukuba, Ibaraki 305-8687, Japan

Ujino-Ihara, Tokuko
Forestry and Forest Products Research Institute (FFPRI)
Matsunosato 1, Tsukuba, Ibaraki 305-8687, Japan

Wan Chik, Suhaimi
Forest Research Institute Malaysia (FRIM)
Kepong, 52109 Kuala Lumpur, Malaysia

Wan Rasidah, Kadir,
Forest Research Institute Malaysia (FRIM)
Kepong, 52109 Kuala Lumpur, Malaysia

Watanabe, Tsutomu
Forestry and Forest Products Research Institute (FFPRI)
Matsunosato 1, Tsukuba, Ibaraki 305-8687, Japan

Watling, Roy
Royal Botanic Garden
Edinburgh EH3 5LR, United Kingdom

Weissflog, Andreas
Department of Zoology
Johann Wolfgang Goethe-University
Siesmayerstr. 70, 60054 Frankfurt am Main, Germany

Wells, David R.
Serendip, Old Farm, Illington, Thetford, Norfolk IP24 1RP, UK

Werner, Michael
Department of Zoology
Johann Wolfgang Goethe-University
Siesmayerstr. 70, 60054 Frankfurt am Main, Germany

Yaakob, Norsham
Forest Research Institute Malaysia (FRIM)
Kepong, 52109 Selangor, Malaysia

Yamada, Toshihiro
Faculty of Environmental and Symbiotic Sciences
Prefectural University of Kumamoto
Tsukide 3-1-100, Kumamoto 862-8502, Japan

Yamakura, Takuo
Graduate School of Science
Osaka City University
Sumiyoshi, Osaka 558-8585, Japan

Yamashita, Naoko
Hokkaido Research Center
Forestry and Forest Products Research Institute (FFPRI)
Hitsujigaoka 7,Toyohira, Sapporo, Hokkaido 062-8516, Japan

Yamashita, Tamon
Education and Research Centre for Biological Resources
Shimane University
Matsue, Shimane 690-8504, Japan

Yap, Song Kheong
Forest Research Institute Malaysia (FRIM)
Present address : Sunway Nursery & Landscaping Sdn. Bhd, Petaling Jaya, Malaysia.

Yasuda, Masatoshi
Forestry and Forest Products Research Institute (FFRRI)
Matsunosato 1, Tsukuba, Ibaraki 305-8687, Japan

Yasuda, Yukio
Forestry and Forest Products Research Institute (FFPRI)
Matsunosato 1, Tsukuba, Ibaraki 305-8687, Japan

Yoshida, Keiichiro
National Institute for Environmental Studies(NIES)
Onogawa 16-2, Tsukuba, Ibaraki 305-8506, Japan

Yusop, Zulkifli
Universiti Teknologi Malaysia (UTM)
81310 UTM, Skudai, Johor, Malaysia

Index